The Human Revolution

To the memory of:

François H. Bordes
Glyn Ll. Isaac
Charles McBurney
Hallam L. Movius Jr.

The Human Revolution

Behavioural and Biological Perspectives on the Origins of Modern Humans

PAUL MELLARS
AND
CHRIS STRINGER
Editors

PRINCETON UNIVERSITY PRESS
PRINCETON, NEW JERSEY

Published by Princeton University Press,
41 William Street, Princeton, New Jersey 08540

Library of Congress Cataloging-in-Publication Data
The Human revolution : behavioural and biological perspectives on the origins of modern humans / Paul Mellars and Christopher Stringer, editors.
812 p. 23.4 cm.
Includes bibliographies and index.
ISBN 0-691-08539-0 : $65.00
1. Man—Origin. 2. Human evolution. I. Mellars, Paul.
II. Stringer, Christopher, 1947-
GN281.H849 1989
573.2—dc20

Printed in Great Britain

Contents

Preface

Few topics in anthropology have generated more interest and debate over the past few years than the biological and behavioural origins of fully 'modern' human populations. The debates have arisen partly from new discoveries and the application of new dating methods, and partly from the use of more sophisticated approaches to the modelling of human evolutionary processes, both in terms of biological evolution, and the associated (and inevitably interrelated) patterns of cultural change. A central factor in much of this rethinking has of course been the recent developments in molecular genetics, which are now opening up an entirely new perspective on the evolutionary origins of modern human populations.

It was this spate of recent research and lively reappraisal of existing ideas which led to the plans for a major conference on the theme of the 'Origins and Dispersal of Modern Humans', as a way of bringing together specialists in the different fields of human evolution, archaeology and molecular genetics, and attempting a concerted, multi-disciplinary survey of current approaches to these problems. The plans for the conference took shape in 1984, and were originally envisaged as a component of the XIth Congress of the International Union of Prehistoric and Protohistoric Science (the 'World Archaeological Congress') to be held in Southampton in September of 1986. It is now a matter of history that this event was sadly disrupted by political pressures to exclude South African participants, followed by the withdrawal of support by the International Executive of the IUPPS. Reluctantly, but with conviction, the decision was taken to withdraw the symposium from the context of the World Archaeological Congress, and to hold this as a separate, academically-unsegregated meeting, during the Spring of 1987.

The meeting that was eventually held in Cambridge from 22–26 March 1987 brought together fifty-five of the leading international workers in the study of the behavioural and biological aspects of modern human origins, in areas ranging from Western Europe and the USA, through various regions of Africa and Western Asia, to Southeast Asia and Australia. Even with this broad range of expertise there were still a number of notable omissions. In particular it was hoped that Professor Wu Xinzhi would be present to discuss some of the important new discoveries from the People's Republic of China, and that other colleagues such as Erik Trinkaus, Karl Butzer, Bill Farrand, Henri Laville, Jean-Philippe Rigaud, Jean-Jacques Hublin, Jíří Svoboda and others would be there to provide broader coverage of the recent developments in both geochronology and some of the current 'bio-cultural' approaches to human development. In the event, these and a number of other invited participants were prevented by other commitments from attending. Nevertheless, the scope achieved at the conference was broad, and seemed to embrace most of the current approaches to the study of both the biological and behavioural emergence of modern human populations over most areas of the occupied world.

Unfortunately, the sheer number and scale of the papers prepared for the Cambridge meeting has made it impossible to include all of these within the bounds of a single volume. Following a good deal of discussion, it was eventually decided to include in the present volume all the specifically 'biological' papers given at the meeting, together with the more general and broad-ranging studies of the associated behavioural and archaeological data. The remainder of the more specialized case studies of the archaeological evidence will be published in a second, companion volume, under the title of *The Emergence of Modern Humans: an Archaeological Perspective*. All of the contributors were given the opportunity to revise their papers for publication, in the light of the discussions at the conference itself, and to take account of some of the important new discoveries and data which emerged during the period following the meeting. Papers were eventually submitted to their press in their final, revised form, in the autumn of 1988.

It is hardly necessary to emphasise that none of the organization of the Cambridge meeting would have been possible without the generous support and backing of a number of institutions and individuals. The meeting was sponsored formally by three institutions: the University of Cambridge (Department of Archaeology); the British Museum (Natural History); and the Royal Anthropological Institute of Great Britain and Northern Ireland. Financial support for the meeting was obtained from a wide range of sources. In particular, the generous support provided by the L. S. B. Leakey Foundation for Anthropological Research, the Boise Fund of Oxford University, the British Academy, the Royal Society and the Association for Cultural Exchange is warmly acknowledged, and made a substantial contribution towards the travel expenses of many of the foreign participants in the meeting. Amongst the many colleagues who gave encouragement and contributed in various ways to the success of the meeting, special thanks are due to Colin Renfrew, Ernest Gellner, Peter Andrews, Bernard Wood, Geoffrey Harrison, J. Desmond Clark, F. Clark Howell, and David Pilbeam. Most of the burden of the secretarial work involved in the organization of the conference was shouldered, expertly and without complaint, by the Secretary of the Cambridge Archaeology Department, José John. Other invaluable help, in a wide range of capacities, was provided by Anny Mellars, David Phillipson, Tim Reynolds, Colin Shell, Karen Stringer and Tony Sinclair. The Master and Fellows of Corpus

Christi College are to be thanked for hosting the domestic arrangements for the Conference, and for contributing to the initial Reception for participants.

Finally, the editors owe a special debt of thanks to Nigel Holman, Corinne Duhig and Doreen Simpson for their help and expertise in collating and editing the conference papers, and to members of the Edinburgh University Press for their guidance in seeing the volume through to the final publication stage. Doubtless there have been numerous errors and oversights in the editorial procedures, but for these the editors themselves must (by definition!) take the blame.

In closing, there are a number of features of the Cambridge meeting which will no doubt be vividly remembered by participants. There was the occasion, for example, when the lights suddenly went out at the conference Reception, sharply on time at the pre-arranged hour of eight o'clock. Participants will also recall the apparently insurmountable difficulties of adequately blacking out the conference room – though whether this served to obscure or clarify some of the visual presentations of the papers seemed to be a matter of some debate. Above all, colleagues will remember the lively and spirited discussions of the conference papers, which frequently continued well into the night in a number of public houses, college rooms and other venues throughout Cambridge. No doubt all of these things contributed in one way or another to the distinctive atmosphere of the Cambridge meeting. Our thanks are due to all of the contributors for making this such a congenial, as well as scientifically stimulating, event.

PAUL MELLARS
CHRIS STRINGER
May 1989

List of Contributors

R. D. ALEXANDER, Museum of Zoology, University of Michigan, Ann Arbor, Michigan 48109, USA.

B. ARENSBURG, Dept. of Anatomy & Anthropology, Sackler School of Medicine, Tel Aviv University, Tel Aviv 69978, Israel.

O. BAR-YOSEF, Dept. of Anthropology, Peabody Museum, Harvard University, Cambridge, Mass. 02138, USA.

G. BRÄUER, Institut für Humanbiologie, Universität Hamburg, Allende-Platz 2, 2000 Hamburg 13, West Germany.

P. G. CHASE, University Museum (Archaeology), University of Pennsylvania, 33rd & Spruce Streets, Philadelphia, Pennsylvania 19104, USA.

G. A. CLARK, Dept. of Anthropology, Arizona State University, Tempe, Arizona 85287, USA.

J. D. CLARK, Dept. of Anthropology, University of California, Berkeley, California 94720, USA.

H. J. DEACON, Dept. of Archaeology, University of Stellenbosch, Stellenbosch 7600, South Africa.

H. L. DIBBLE, Dept. of Anthropology, 325 University Museum, 33rd & Spruce Streets, Philadelphia, Pennsylvania 19104-6398, USA.

R. A. FOLEY, Dept. of Biological Anthropology, University of Cambridge, Downing Street, Cambridge CB2 3DZ.

D. GAMBIER, Laboratoire d'Anthropologie, Faculté des Sciences, Université de Bordeaux I, 33405 Talence Cedex, Gironde, France.

E. GELLNER, Dept. of Social Anthropology, Free School Lane, Cambridge CB2 3RF.

C. P. GROVES, Dept. of Prehistory and Anthropology, Australian National University, P.O. Box 4, Canberra, ACT 2601, Australia.

P. J. HABGOOD, Dept. of Anthropology, University of Sydney, Sydney, Australia.

F. B. HARROLD, Dept. of Sociology and Anthropology, Box 19599, University of Texas (Arlington), Arlington, Texas 76019, USA.

R. JONES, Dept. of Prehistory, Research School of Pacific Studies, Australian National University, Canberra, ACT 2601, Australia.

R. G. KLEIN, Dept. of Anthropology, University of Chicago, 1126 East 59th Street, Chicago, Illinois 60637, USA.

P. LIEBERMAN, Cognitive and Linguistic Sciences, Box 1978, Brown University, Providence, Rhode Island 02912, USA.

G. LUCOTTE, Laboratoire d'Anthropologie Physique, Collège de France, 11 Place Marcelin Berthelot, 75231 Paris Cedex, France.

P. MELLARS, Dept. of Archaeology, University of Cambridge, Downing Street, Cambridge CB2 3DZ.

G. P. RIGHTMIRE, Dept. of Anthropology, State University of New York, Binghamton, New York 13901, USA.

S. ROUHANI, The Galton Laboratory, University College London, Wolfson House, 4 Stephenson Way, London NW1 2HE.

J. F. SIMEK, Dept. of Anthropology, University of Tennessee, South Stadium Hall, Knoxville, Tennessee 37996, USA.

J. J. SHEA, Dept. of Anthropology, Peabody Museum, Harvard University, 11 Divinity Avenue, Cambridge, Massachusetts 02138, USA.

F. H. SMITH, Dept. of Anthropology, University of Tennessee, South Stadium Hall, Knoxville, Tennessee 37996, USA.

O. SOFFER, Dept. of Anthropology, University of Illinois at Urbana-Champaign, 109 Davenport Hall, 607 South Mathews Avenue, Urbana, Illinois 61801, USA.

M. STONEKING, Dept. of Biochemistry, University of California, Berkeley, California 94720, USA.

C. B. STRINGER, Dept. of Palaeontology, British Museum (Natural History), London SW7 5BD.

A.-M. TILLIER, Laboratoire d'Anthropologie, Faculté des Sciences, Université de Bordeaux I, 33405 Talence Cedex, Gironde, France.

B. VANDERMEERSCH, Laboratoire d'Anthropologie, Faculté des Sciences, Université de Bordeaux I, 33405 Talence Cedex, Gironde, France.

J. S. WAINSCOAT, Dept. of Haematology, John Radcliffe Hospital, Headington, Oxford OX3 9DU.

R. WHALLON, Museum of Anthropology, University Museum, University of Michigan, Ann Arbor, Michigan 48109-1079, USA.

R. WHITE, Dept. of Anthropology, New York University, 25 Waverly Place, New York, NY 10003, USA.

M. H. WOLPOFF, Dept. of Anthropology, University of Michigan, Ann Arbor, Michigan 48109, USA.

E. ZUBROW, Dept. of Anthropology, 581-L Spaulding Quadrangle, State University of New York, Buffalo, New York 14261, USA.

1. Introduction

PAUL MELLARS AND CHRIS STRINGER

The changes in biology and behaviour which led to the emergence of 'modern' humans, and the factors that lay behind those changes, have been subjects of research attention for over a century. But only in the last few years has this topic really taken off to become (currently) one of the most central and controversial in the whole field of palaeoanthropology. The reasons why attention has recently become focused on this area are many and varied, but they certainly include the arrival of a wealth of new information about the course of recent human evolution in the shape of new fossil and archaeological discoveries, and the reinterpretation of old ones. They also include significant improvements in established dating techniques, and the development and application of new ones. The existing data have also been subjected to new studies via innovations in areas such as statistical analysis, microscopy, taphonomy and (for the fossils) cladistic analysis.

Several recent conferences have focused on these exciting developments, but they have generally been restricted in scope, either in terms of the topics dealt with, or the particular periods or geographical areas under review. Furthermore, none of these brought together specialists in palaeontology and archaeology with those in the fast-developing field of what might be termed 'palaeogenetics' – historical reconstruction from present-day genetic data. It is this area which has provided the major stimulus to debate about modern human origins in the last couple of years, and it is one which was an essential component of our plans for the Cambridge conference from the very beginning.

If there was a 'human revolution' with the emergence of 'modern' humans (and it should be said that not all of our participants would even agree that this was so), then two related but distinct aspects must be considered. First, there is the transition (however conceived processually) from non-modern ('archaic') to anatomically modern humans. Second, there are the behavioural changes which more or less correlate with the conventional 'Middle-Upper Palaeolithic transition' in many parts of the Old World, and which are often seen as signalling the arrival of behaviour closely comparable to modern hunter-gatherers, in all its essentials. One of the main aims of the conference was to bring together specialists in both of these 'transitions', in the hope of achieving interaction, and perhaps even some degree of integration, between them.

However, as the conference proceeded, it became clear (if it was not recognised before) that there were numerous differences, not only in the data bases, but also in approach between the 'biological' and 'behavioural' participants (and even within each field of study), and thus many workers felt it best to concentrate on their own areas of expertise without encroaching on those of others. This caution was not necessarily a bad thing during the early stages of the conference, but it meant that much of the most constructive discussion was left until the final sessions, when participants seemed more prepared to discuss the wider implications of their work.

One of the primary reasons for an increasing caution in correlating biological and behavioural changes has been the growing realization that the conventional view of a modern skeletal pattern emerging in close synchrony with the emergence of 'Upper Palaeolithic' technology is, at best, a serious over-simplification. It is now recognised that early anatomically modern humans were producing Middle Palaeolithic tools at Skhūl and Qafzeh in Israel (apparently at a surprisingly early date: Valladas *et al.* 1988; Schwarcz *et al.* 1988; Stringer *et al.* 1989), while a partial Neanderthal skeleton recovered from the Châtelperronian level of Saint-Césaire in western France has shown that archaic humans were apparently producing at least one form of early Upper Palaeolithic industry. This gives the clearest possible indication that the anatomical and behavioural 'transitions', at least as they are commonly perceived, were to a large extent separate phenomena. Of course it may be that all the elements of these transitions did not arrive together, and that a more piecemeal perception of the changes would show more overlap between the biological and behavioural events. Alternatively, the underlying links between biological and behavioural change are perhaps not being recognised in studies centred primarily on lithic technology, which may be unable to reflect important developments in such things as social structure, language or symbolic behaviour. However, such special pleading based on an absence of evidence is not particularly helpful at present!

The arrangement of papers in this publication thus differs from the organization of the sessions at the meeting. There, we preferred a geographical arrangement, grouping speakers who worked on the evidence from particular regions, whether biological or behavioural. Here, we have gathered the papers into separate biological and behavioural subsections, although we have encouraged authors to take as wide a view as possible, and gratifyingly some have provided papers which cross the divide between the subsections.

MAJOR ISSUES

No doubt each of our participants would have his or her own view of the fundamental questions which need to be addressed in a volume such as this. However, we will take the liberty of presenting our own list of the five major issues which we feel came out of the discussions at the meeting, and our reading of the submitted papers. Four of these issues are directly concerned with the study of changes, and the other is concerned with mechanisms and interactions accompanying changes.

1. To what extent was the transition from anatomically 'archaic' to 'modern' populations in different regions of the world associated with a process of major population dispersal and (eventually) population replacement? Alternatively, how far can this transition be accounted for, equally if not more parsimoniously, by an essentially gradual process of continuous demographic and biological development within each region?

2. Was the anatomical transition (or transitions) associated with major changes in human behaviour? If so, what particular aspects of behaviour were involved in these changes, and how similar were the overall patterns of change in different regions of the world?

3. If major changes in behaviour and organisation were involved in the transition from archaic to modern populations, what models can be advanced to account for the origins of both the biological and behavioural changes, and for the varying patterns of change documented in the different inhabited regions? To what extent were there direct cause and effect relationships between changes in different spheres of behaviour (e.g. between subsistence patterns, technology, social organisation, demography etc.), or between changes in behavioural patterns and the biological features of the populations?

4. Similarly, if major dispersals of population were involved (to whatever extent) in the transition from archaic to modern humans, what kind of mechanisms were involved in this process of population expansion? Equally, what kind of interactions can be visualised, or documented, between the expanding populations of modern humans and the local populations of archaic hominids within the different regions?

5. Finally, how far did any behavioural changes associated with the archaic-modern human transition involve not only changes in the range and complexity of cultural patterns, but also some fundamental shift in the mental or cognitive abilities of the human groups? In other words, are we simply dealing with a process of gradual, cumulative change in the total range of cultural expression, or with some kind of radical shift in the innate neurological *capacities* of the human brain to accumulate and organise culture?

NEW PERSPECTIVES

With such a wide range of geographical areas, databases and theoretical perspectives under consideration, it is hard to do justice to the range of discussions at the meeting within the scope of a brief review. It was no doubt predictable that much of the critical debate at the conference would centre on the rapidly accumulating evidence for a primary origin for anatomically and genetically modern human populations within Africa. Two aspects of the evidence in particular generated lively discussion: on the one hand the evidence for essentially 'modern' anatomical forms at a number of sites in Southern and Eastern Africa apparently dating in the region of 70-100 000 BP (see papers by Rightmire, Bräuer, Klein, Clark, Deacon);

and on the other hand the spate of new evidence derived from recent studies of DNA and other aspects of molecular genetics, which appear to point to an origin of the modern human genotype specifically within the African continent (see Stoneking and Cann, Wainscoat *et al.*, Lucotte, Rouhani).

Both of these claims were subjected to a good deal of critical scrutiny during the conference discussions. The evidence for dating many of the African hominid localities is of course controversial, and there is clearly room for a good deal of latitude in both the relative and absolute dating of many of the most important sites (e.g. Klasies River Mouth and Omo). In other cases (as at Border Cave) doubts have been expressed as to the reliability of the claimed associations of the human remains with the geological and archaeological successions in the sites (see Klein). However, most of the contributors seemed to accept that the African evidence taken as a whole does seem to indicate the presence of essentially 'modern' human anatomy extending back to at least 70-100 000 BP, and perhaps in some cases (e.g. Omo) even earlier than this. Fortunately, there are strong indications that many of these current problems in chronology will be resolved shortly by the application of new dating techniques to the sites (including, for example, TL and ESR dating, uranium series methods etc.).

The interpretation of the recent molecular genetic evidence was challenged specifically in the paper by Wolpoff, in a wide-ranging and incisive critique of the whole of the current 'Out of Africa' model for modern human origins. The main thrust of Wolpoff's critique was directed at the evidence for calibrating the rates of genetic mutation of mitochondrial DNA. He argued that by using a rather different set of estimates based on slower rates of change (e.g. Moritz *et al.* 1987), one might push back the date of the inferred population dispersal from Africa to around 850 000 BP – which (as he points out) would correspond closely with the generally accepted date for the initial dispersal of *Homo erectus* populations from Africa in the early Pleistocene. He also challenged the assumption of adaptive neutrality of variations in mtDNA, and suggested that much of the evidence for interbreeding between expanding 'modern' populations and local 'archaic' populations throughout Eurasia could have been lost in the course of later evolution, either as a result of random genetic drift, or through the operation of some specific selective mechanisms. In the period since the Cambridge conference there has also been further debate about the methods used in some of the genetic analyses favouring a recent African origin (Saitou and Omoto 1987; Darlu and Tassy 1987; Sanchez-Mazas and Langaney 1988; Pamilo and Nei 1988). This debate has been concerned with the relative importance of genetic distance analyses (which may give different results from tree-building using sequence data) and with the actual methods of tree construction (e.g. whether maximum parsimony should be used, and its reliability). Some of these objections were anticipated and challenged in the papers by Stoneking and Cann, and Rouhani, from a variety of both genetic and demographic perspectives (see also Cann *et al.* 1987; Stringer and Andrews 1988). In addition, the 'Out of Africa' genetic scenario has recently found further support from the first use of a chimpanzee outgroup for the human mtDNA analyses, supporting the rooting proposed by Stoneking and Cann, and from the most comprehensive analysis of genetic polymorphisms yet produced, using genetic distances (Cavalli-Sforza *et al.* 1988). No doubt all of these debates are likely to continue for some time into the future. Here again, however, there are good grounds for thinking that continued research into the detailed patterns of genetic variability in modern populations (based, for example, on larger samples, and on more precisely controlled estimates of genetic mutation rates) will provide a firm answer to many if not all of these issues over the course of the next few years.

The available archaeological evidence from African sites is discussed fully in the papers by Clark, Klein, Deacon and others, drawing on the results of recent research in many parts of the continent. The major debate in this context seems to centre on the significance of the behavioural transition from the Middle to the Later Stone Age. The paper by Deacon in particular argues that most if not all of the essential features of economic and social organization which can be documented amongst the latest Pleistocene and even Holocene communities of southern Africa can be traced back to the archaeological records of the Middle Stone Age, around 70-80 000 BP. Klein, on the other hand, is inclined to see much more significant changes in both economic and social/demographic patterns coinciding with the transition from the Middle to the Later Stone Age. In particular, Klein suggests that MSA groups may have been less effective in hunting some of the larger and more dangerous species of game (e.g. Cape Buffalo and Bushpig) than their successors, and may have exploited some other resources (such as seals or shellfish) in a less sharply focused and less intensive way. This in turn could indicate, perhaps, improved hunting technology, increased population densities, and higher levels of 'logistical organization' in subsistence and settlement strategies over the period of the Middle-Later Stone Age transition. As Klein points out, these shifts might be seen as broadly comparable in several respects with those which are thought to have characterized the Middle-Upper Palaeolithic transition in Eurasia – and indeed as occurring at broadly the same point in time.

To a European prehistorian, one of the most striking features of the African evidence is the remarkably 'advanced' appearance of some of the stone-tool assemblages recorded from the African Middle Stone Age sites. This is reflected most strikingly in the so-called 'Howiesons Poort' industries, which have been found clearly interstratified with typical MSA industries at a range of sites in southern Africa (as for example at Klasies River Mouth and the Boomplaas Cave in Cape Province: see Deacon, this volume). In these assemblages, relatively high frequencies of blade production are combined not only with typical specimens of both end-scrapers and burins, but also with a range of small, carefully shaped geometric forms (triangles, crescents and trapezes), which must inevitably have formed components in composite, hafted tools (Singer and Wymer 1982). In Europe, this kind of technology would be difficult to parallel before the middle or later stages of the Upper Palaeolithic sequence. From some of the African sites there are also rare but potentially significant examples of carefully shaped and notched bone tools – again closely resembling those found in the Eurasian Upper Palaeolithic. The major controversy at present centres on the chronology of the Howiesons Poort industries; Parkington (this conference, volume 2) in particular has argued for a much more critical approach to the evidence for dating the Howiesons Poort levels beyond a minimum date of around 45-50 000 BP. Other workers prefer a much earlier date of around 70-90 000 BP (see Deacon, this volume). Exactly what relevance these industries may have for the early emergence of anatomically modern hominids in southern Africa is, perhaps, one of the most intriguing questions posed by current research into modern human origins (see Ambrose *et al.*, volume 2).

Similar debates over chronology have recently emerged as a central and crucially important issue in the interpretation of the biological and archaeological evidence from southwest Asia – centred principally on Iraq, Israel, Syria and the adjacent regions of the Levant. This region has of course produced a wealth of hominid remains dating from the earlier part of the late Pleistocene, with particularly rich and well-documented remains from the sites of Tabūn, Skhūl, Kebara, Amūd and Qafzeh in Israel, and the Shanidar Cave in Iraq (see papers by Vandermeersch, Arensburg, Tillier, Bar-Yosef, Shea). Ever since the 1930s it has been recognized that these remains display an unusually wide range of anatomical variation, ranging from relatively typical, robust, 'Neanderthal'-like forms from Tabūn, Amūd, Kebara and Shanidar, to others (most notably the remains from Skhūl and Qafzeh) which are in most respects 'anatomically modern' in form. Obviously, any interpretation of the significance of this variation must depend to a large extent on how the different hominids are arranged in a relative and absolute chronological sequence.

Until recently there has been a fairly widespread assumption that the basic pattern of anatomical development within southwest Asia was from the more archaic, Neanderthal-like forms to those of more 'modern', gracile form. This assumption was initially challenged some ten years ago by Tchernov, Haas and others, on the basis of the faunal associations of the different hominids (especially the rodent faunas) which seemed to indicate a substantially *earlier* date for the anatomically modern forms from Qafzeh than for the more 'archaic' forms from Tabūn and Kebara. Following a good deal of controversy in the literature, this dating has now received strong support from the results of both thermoluminescence and electron-spin-resonance dating for the sites of Skhūl and Qafzeh, which point to a median age for these sites of around 90-100 000 years (Valladas *et al.* 1988; Schwarcz *et al.* 1988; Stringer *et al.* 1989). By contrast, TL and ESR dating of the Mousterian levels associted with the recently discovered (and very robust) Neanderthal skeleton from Kebara has yielded a much younger date of only *c.* 50-60 000 BP (Valladas *et al.* 1987; Grün pers. comm.). The dating of other archaic hominids from Tabūn, Amūd and Shanidar remains more controversial, but there are strong indications that at least some of these must be of broadly similar age to the Kebara skeleton, and therefore more recent than the anatomically modern forms from Skhūl and Qafzeh (see

Bar-Yosef, Vandermeersch).

Clearly, this recent dating evidence has a critical bearing on any assessment of the phylogenetic and other relationships of the southwest Asian hominids. As Vandermeersch (this volume) has pointed out, the anatomically modern hominids from Skhūl and Qafzeh might well be seen as broadly comparable in age with some of the earliest anatomically modern forms from Africa. This need not, of course, undermine claims for an African origin for anatomically modern hominids, but it would suggest that any dispersal of these populations from Africa must extend back to at least the earliest stages of the late Pleistocene (Stringer 1988). The implications for phylogenetic development within southwestern Asia are equally significant. Clearly, any hypothesis of a simple, linear evolution from the broadly Neanderthal morphology of Tabūn and Shanidar to the essentially modern morphology of Skhūl and Qafzeh must now be regarded as implausible. Similarly, unless one is to postulate a remarkable degree of internal variability within the Near Eastern populations, the implication would seem to be that populations of essentially anatomically modern form were already established within this region by around 90-100 000 BP, and subsequently maintained some kind of parallel development alongside the 'Neanderthal' populations of northern and western Eurasia over a period of at least 40-50 000 years. Whichever way the situation is envisaged, the Near Eastern evidence as a whole would seem much more in keeping with the idea of a largely independent development of 'archaic' and 'modern' populations throughout the earlier stages of the late Pleistocene, than with the alternative of a gradual, linear pattern of evolution throughout this time range (see Vandermeersch, Stringer, Bar-Yosef).

All of this new evidence poses an intriguing challenge for the interpretation of the archaeological sequence in the Near East. Despite continuing controversy over the relative and absolute chronology of the Middle Palaeolithic sequence (Bar-Yosef, Marks, Clark and Lindly), no one would question that the lithic assemblages associated with the anatomically modern hominids at both Skhūl and Qafzeh are essentially of 'Middle Palaeolithic' form. By comparison with most of the Mousterian industries in Europe, many of the Middle Palaeolithic industries of the Near East do have a relatively 'advanced' appearance (reflected, for example, by relatively high frequencies of blade production, associated with varying frequencies of both burins and end-scrapers) which are reminiscent of those documented in the African Middle Stone Age industries, and which anticipate in some respects those developed more fully in the later, Upper Palaeolithic industries. Nevertheless, the overall technology of the Middle Eastern sites, relying primarily on various 'Levallois' flaking techniques, and including typical side-scrapers, points, denticulates etc., is undoubtedly Middle Palaeolithic in a general technological sense (see Shea, Bar-Yosef, Clark and Lindly). From this it must be concluded that if the postulated dispersal of anatomically modern humans from Africa was associated with new forms of cultural or behavioural expression, this was not reflected in any simple or direct way in the character of the associated lithic industries.

If one is looking for evidence of more 'complex' behaviour in the Near Eastern sites, the most significant observations may be the well-documented grave offerings found in association with the human burials at both Qafzeh and Skhūl (Vandermeersch 1976). At present, these would appear to be the earliest well-documented examples of grave goods so far recorded from Palaeolithic sites in Eurasia (see Chase and Dibble 1987). It may well be significant that both of these occurrences were associated with hominids of anatomically modern (as opposed to Neanderthal) skeletal form.

Any illusion that the situation in the heavily researched and relatively well-documented regions of Western, Central and Eastern Europe was largely cut and dried would no doubt have been rapidly dispelled by the papers and discussions at the Cambridge meeting. Here, the long-standing divergence between the 'population continuity' and 'population replacement' perspectives reemerged with equal if not renewed vigour. Papers by Wolpoff, Smith *et al.*, Simek and Price, and Clark and Lindly argued the case for a relatively slow, *in situ* transition of Neanderthal to fully modern populations, associated with an equally gradual transition from Middle to Upper Palaeolithic technology. Other contributors (including Gambier, Stringer, Harrold, Kozlowski and Allsworth-Jones), were inclined to see a much sharper break, both demographically and culturally, between the latest Neanderthal and earliest anatomically modern populations. On the biological side it was argued that the recent discovery of apparently typical Neanderthal remains dated to *c.* 33-35 000 BP at the site of Saint-Césaire in western France (Lévêque and Vandermeersch 1980) effectively precludes any hypothesis of a gradual evolution from Neanderthal to anatomically modern populations within Western Europe itself, since the earliest occurrences of characteristically anatomically modern forms can hardly be more than 2-3 000 years later (at most) than the Saint-Césaire Neanderthal (see Gambier, Bräuer, Harrold, this volume; Howell 1984). On the archaeological side it was argued that the period between c. 35 000 and 32 000 BP witnessed a whole spectrum of radical cultural innovations, including a shift from predominantly flake to blade technology, the appearance of many new forms of stone tools, the emergence of complex, elaborately shaped bone and antler artifacts, the appearance of personal ornaments and traded objects (especially marine shells), and the emergence of sophisticated representational art. And in at least one of the best documented regions of Western Europe (western France), these developments can be seen to coincide with a sharp increase in site numbers, an increase in maximum site size, and the appearance of highly specialized hunting strategies focused on the exploitation of reindeer (Mellars 1973, 1989). The association of these features with the 'Aurignacian' cultural phenomenon – which is almost certainly intrusive in Western Europe – and with the earliest occurrence of 'Cro-Magnon' skeletal forms, would appear to argue strongly in favour of a major population replacement in Western Europe, coinciding closely with the traditional interface between Neanderthal and anatomically modern populations. According to this scenario, the Châtelperronian assemblages of France and northern Spain (which occur in association with a Neanderthal at Saint-Césaire, and are interstratified with typical Aurignacian industries at several sites) might be seen most economically as some kind of 'acculturation' phenomenon, reflecting both certain continuities with the earlier Mousterian traditions, combined with certain new elements of specifically Upper Paleolithic form (see Harrold, this volume; Farizy, volume 2).

How far the archaeological records from other regions of Europe can be interpreted in similar terms is still a topic of vigorous debate. As a generalization it is probably true to say that most if not all of the earliest occurrences of anatomically modern hominids in the different regions of Europe have been found (as in western France) in association with industries of essentially 'Aurignacian' type. This is true for example of the well-documented remains from Stetten (i.e. Vogelherd) in south Germany, Mladeč (Czechoslovakia) and Velika Pečina (Yugoslavia) and perhaps also Grotte des Enfants (Italy) and Cioclovina (Romania). It can also be argued that in most if not all regions of Europe, the Aurignacian shows the earliest clear occurence of most of the major cultural and technological features which are generally seen as the distinctive hallmarks of fully developed 'Upper Palaeolithic' culture. As noted above, this includes not only the appearance of increased frequencies (and generally improved technology) of blade production, but also the earliest occurrence of extensively shaped bone and antler tools, the earliest well-documented occurrence of perforated animal teeth, marine shells and other forms of 'personal ornaments', the first systematic and wide-ranging distribution networks in flint, marine shells and other raw materials and – in a number of rare but highly significant contexts – the occurrence of complex naturalistic art (see Kozlowski 1988, and this conference, volume 2). If one were looking for an archaeological expression of a relatively rapid spread of a new human population, there is no doubt that the Aurignacian phenomenon would provide the most plausible candidate for this event (see Howell 1984; Kozlowski 1988; Mellars 1989). Significantly, this striking uniformity in technology seems to have been achieved over the different areas of Europe within a relatively short space of time – ranging between c. 43 000 BP in the Balkans, to c. 36-40 000 BP in both Central Europe and (according to the results of recent radiocarbon accelerator dating from the site of Castillo: Cabrera-Valdes, pers. comm.) the Cantabrian coast. Recent work at Ksar Akil (Lebanon) suggests that closely similar technology was established in the northern parts of the Middle East at broadly the same point in time (Mellars and Tixier 1989).

The major debate in Europe at present centres on several other industrial variants which (as in the case of the French Châtelperronian) seem to show a curious mixture of both distinctively 'Upper Palaeolithic' features, and other features which suggest close links with the latest Middle Palaeolithic industries of the same regions. Associations of this kind are well documented in the 'Szeletian' and related leaf-point industries of Central and Northern Europe, in the 'Uluzzian' industries of Italy, and (apparently) in the 'Streletskaya/Kostenki' industries of south Russia (see Soffer, this volume; Allsworth-Jones, Kozlowski, volume 2). The central problem is to know whether these represent genuinely 'transitional' industries, reflecting a gradual, *in situ* development of Upper Palaeolithic technology from the local Mousterian traditions in the same regions, or whether they represent simply various forms of contact and acculturation between the latest Middle Palaeolithic/Neanderthal populations, and intrusive populations of anatomically modern hominids. The papers by Allsworth-Jones, Kozlowski and others have argued explicitly for the latter interpretation, pointing out that in most regions of Europe there is evidence for the presence of typical 'Aurignacian' industries coinciding closely with the emergence of the apparently 'transitional' industries. Other papers by Clark and Lindly (this volume) and Otte (volume 2), by contrast, have opposed this, and argued for an essentially gradual pattern of technological and cultural change over the whole of the Middle-Upper Palaeolithic succession. The major problem at present lies in the absence of substantial and well-documented human skeletal material in association with these technologically transitional assemblages. In the absence of this evidence, arguments over continuities and discontinuities in the local records of both biological and cultural development will, no doubt, continue.

The archaeological and biological records from further eastwards in central, southern and eastern Asia remain at present poorly documented, poorly dated, and in many cases poorly described. However, information continues to accrue about the important Chinese evidence (Pope 1988), including new radiocarbon accelerator dates for the Zhoukoudian Upper Cave material (Hedges *et al.* 1988). This important sample of anatomically modern hominids (lacking many Mongoloid regional characteristics and apparently associated with an industry of 'Upper Palaeolithic' affinities) is now well bracketed between 13 000 and 33 000 radiocarbon years. By far the most fully documented evidence however derives from the spate of recent research in Australia, New Guinea and the adjacent islands (comprising the ancient continent of 'Sahul') which must presumably be seen as representing the extreme southeastern limit of Pleistocene human colonization throughout the eastern Asian zone (see Jones, Habgood, Groves, this volume; Bowdler, volume 2). Two features at least are now firmly established: that relatively large areas of both Australia and New Guinea must have been colonized by at least 30 000 BP, and probably as early as 40-45 000 BP; and that at least some of these early populations were of essentially modern (and remarkably 'gracile') anatomical form. The sheer feat of the colonization of Australia was of course a remarkable achievement, involving a series of long and difficult sea crossings of up to 90-100 km. There is equally clear evidence that by around 25-30 000 BP these populations were practising a variety of human burial and cremation rites (including liberal use of red ochre) and that by around 20 000 BP relatively sophisticated forms of art were being produced in at least certain regions of Australia. The major debate in Australia at present centres on the significance of the much more robust (or even 'archaic') hominid remains recovered from the sites of Kow Swamp, Cohuna and elsewhere. Opinions at present remain sharply divided on whether these should be seen as the result of an entirely separate phase of colonization (perhaps derived from the earlier *Homo erectus* populations of Indonesia) or whether they could reflect simply the effects of later anatomical divergence within the Australian populations (perhaps accentuated by varying degrees of pathology and head binding or other cultural deformations). Recent dating evidence from Australia suggests that the 'robust' and 'gracile' morphologies were both present in the Willandra Lakes area between 25-30 000 years, since the very robust WLH 50 calvaria has now been dated by ESR to this time range (Caddie *et al.* 1987). The significance of this morphological and metrical varition is disputed, with Wolpoff (this volume) arguing for separate population origins, while Brown (1987) considers that the contrasting morphologies could form part of a single population range. One of the main problems in determining the probable ancestry of these early Australians has been the lack of plausible local antecedents in the early Late Pleistocene. New evidence suporting such an antiquity for the Ngandong material from Java has seemingly accentuated difficulties for the regional continuity model (Bartstra *et al.* 1988). A young age for this sample, which several workers refer to *Homo erectus*, would considerably narrow the time range available for any *in situ* transition to modern *H. sapiens*. However, it certainly leaves open the possibility of gene flow from such relict archaic populations during any dispersal of modern *H. sapiens* into Australia. Clearly, the biological records of the human colonization of Australia are extremely complex, and it will no doubt require much more research (backed up hopefully by new hominid discoveries, and improved dating methods) to resolve these issues in any conclusive way. There is also a further challenge to prehistorians in the strikingly simple appearance of the earliest stone-tool industries recovered from the Australian site. Do these represent some kind of technological adaptation to purely local conditions, or do they reflect a relatively long and complex history of population expansion which commenced well before the emergence of characteristically 'Upper Palaeolithic' blade-and-burin industries in the western and central zones of Eurasia? Whatever the eventual explanation, it is clear that this technology was adequate to achieve the relatively rapid colonization of a remarkably diverse and demanding range of new environments.

All of this of course raises the crucial question of exactly how some form of large-scale population dispersals of the kind implied by the recent genetic data could have occurred, and what kinds of interactions may have taken place between the indigenous and intrusive populations in different regions of Eurasia. At present we have so little reliable information on the social and economic organization of the latest Neanderthal populations in Europe that it is difficult to offer any informed speculations on this point. What does seem clear is that in at least some areas of Western Europe (and perhaps many areas of Central and Eastern Europe) the latest Neanderthal and the earliest anatomically modern populations must have survived in relatively close proximity over a period of at least several centuries (as, for example, the interstratification of Aurignacian and Châtelperronian levels at a number of sites in western France and nothern Spain clearly implies: see Harrold, this volume; Leroyer and Leroi-Gourhan 1983). In this context there would seem to be two sharply contrasting ways of viewing the evidence. Either the late Neanderthal and early anatomically modern populations were adapted to such differing ways of life that there was little direct competition between the two groups for ecological space, resources, or whatever. Or alternatively that, despite some form of direct competition in those spheres, the final Neanderthal populations were nevertheless well equipped to cope with this over a period of many generations. The latter interpretation might suggest that the Neanderthal populations were in fact far better equipped in general cultural, technological and behavioural terms than some recent characterizations of them have been inclined to accept.

Many of the discussions at the conference underscored the difficulties of making clear-cut comparisons of the behavioural adaptations of archaic and modern populations. Exactly what criteria can be used to distinguish objectively between the products of hunting and those of scavenging in faunal assemblages (see Chase, this voume)? How reliable are data based on site dimensions, site location, or site numbers for making inferences about past demography, settlement systems, or social organisation? And how far – if at all – can one derive information about mental abilities or linguistic complexity from studies of stone tools, or even art (Dibble, Lieberman, Gellner, Whallon, Mellars)? There are clearly fundamental and far-reaching problems here in our ability to interpret the archaeological record in meaningful behavioural terms. Worst of all is the danger of applying completely circular arguments which equate the *expression* of culture with the *capacity* for culture – arguments which tend to assume that because a particular form of cultural or behavioural expression is not reflected in the archaeological record at a particular period, the mental (or physical) capacities for it were lacking in the population involved. This is perhaps the ultimate dilemma in studying the earlier phases of human cultural development.

In short, we are now in a position of being able to perceive many of the essential patterns in human cultural and biological development, but far less able to offer clear and convincing explanations for the patterns we observe. Exactly what kind of environmental or behavioural stimuli could have led to the evolution of biologically modern human forms in Africa and their eventual dispersal (apparently) over large areas of the world (see Foley, Zubrow, Deacon, Clark)? What kinds of stimuli were involved in the emergence of new behavioural and cultural patterns that we describe under conventional archaeological labels such as 'the Middle-Upper Palaeolithic transition' (see Mellars, Alexander, White, Clark and Lindly)? And above all, what were the relationships between the successive or simultaneous, but inevitably interrelated processes of biological and behavioural change? The Cambridge conference undoubtedly served to bring many of these issues into much sharper focus, and to clarify some of the important patterns which we can now discern in both the biological and archaeological data. But it left most of the participants with the feeling that there are still many fundamental – and very exciting – problems to be resolved.

REFERENCES

Bartstra, G.-J., Soegondho, S. and Van der Wijk, A. 1988. Ngandong Man: age and artifacts. *Journal of Human Evolution* 17: 325-337.

Brown, P. 1987. Pleistocene homogeneity and Holocene size reduction: the Australian human skeletal evidence. *Archaeology in Oceania* 22: 41-67.

Caddie, D. A., Hunter, D. S., Pomery, P. J. and Hall, H. J. 1987. The ageing chemist – can electron spin resonance (ESR) help? In W. R. Ambrose and J. M. J. Mummery (eds) *Archaeometry: Further Australasian Studies*. Canberra: Australian National University: 67-176.

Cann, R., Stoneking, M. and Wilson, A.C. 1987. Disputed African origin of human populations. *Nature* 327: 111-112.

Cavalli-Sforza, L. L, Piazza, A., Menozzi, P. and Mountain, J. 1988. Reconstruction of human evolution: bringing together genetic, archaeological, and linguistic data. *Proceedings of the National Academy of Sciences* (U.S.A.) 85: 6002-6006.

Chase, P. G. and Dibble, H. L. 1987. Middle Pleistocene symbolism: a review of current evidence and interpretations. *Journal of Anthropological Archeology* 6: 263-293.

Darlu, P. and Tassy, P. 1987. Roots (a comment on the evolution of human mitochondrial DNA and the origins of modern humans). *Human Evolution* 2: 407-412.

Hedges, R. E. M., Housley, R. A., Law, I. A., Perry, C. and Hendy E. 1988. Radiocarbon dates from the Oxford AMS system: Archaeometry Datelist 8. *Archaeometry* 30: 291-305.

Howell, F. C. 1984. Introduction. In F. H. Smith and F. Spencer (eds) *The Origins of Modern Humans: a World Survey of the Fossil Evidence*. New York: Alan R. Liss: xiii-xxii.

Kozlowski, J. K. 1988. L'apparition du Paléolithique supérieur. In M. Otte (ed.) *L'Homme de Néandertal*: Vol. 8: *La Mutation*. Liége: Études et Recherches Archéologiques de l'Université de Liége (ERAUL): 11-21.

Leroyer, C. and Leroi-Gourhan, A. 1983. Problémes de chronologie: le Castelperronien et l'Aurignacien. *Bulletin de la Société Préhistorique Française* 80: 41-44.

Lévêque, F. and Vandermeersch, B. 1980. Découverte de restes humains dans un niveau castelperronien à Saint-Césaire (Charente-Maritime). *Comptes Rendus de l'Academie des Sciences de Paris* (Série II) 291: 187-189.

Mellars, P. A. 1973. The character of the Middle-Upper Palaeolithic transition in southwest France. In A. C. Renfrew (ed.) *The Explanation of Culture Change*. London: Duckworth: 255-276.

Mellars, P. A. 1989. Major issues in the emergence of modern humans. *Current Anthropology* 30: 349-385.

Mellars, P. A. and Tixier, J. 1989. Radiocarbon-accelerator dating of Ksar 'Aqil (Lebanon) and the chronology of the Upper Palaeolithic sequence in the Near East. *Antiquity*. In Press.

Moritz, C., Dowling, T. E. and Brown, W. M. 1987. Evolution of animal mitochondrial DNA: relevance for population biology and systematics. *Annual Review of Ecological Systematics* 18: 269-292.

Pamilo, O. and Nei, M. 1988. Relationships between gene trees and species trees. *Molecular Biology and Evolution* 5: 568-583.

Pope, G. 1988. Recent advances in Far Eastern Palaeoanthropology. *Annual Review of Anthropology*. 17: 43-77.

Saitou, N. and Omoto, K. 1987. Time and place of human origins from mtDNA data. *Nature* 327: 288.

Sanchez-Mazas, A. and Langaney, A. 1988. Common genetic pools between human populations. *Human Genetics* 78: 161-166.

Schwarcz, H. P., Grün, R., Vandermeersch, B., Bar-Yosef, O., Valladas, H. and Tchernov, E. 1988. ESR dates for the hominid burial site of Qafzeh in Israel. *Journal of Human Evolution* 17: 733-737.

Singer, R. and Wymer, J. 1982. *The Middle Stone Age at Klasies River Mouth in South Africa*. Chicago: University of Chicago Press.

Stringer, C. B. 1988. The dates of Eden. *Nature*. 331: 565-566.

Stringer, C. B. and Andrews, P. 1988. Genetic and fossil evidence for the origin of modern humans. *Science* 239: 1263-1268.

Stringer, C. B., Grün, R., Schwarcz, H. P. and Goldberg, P. 1989. ESR dates for the hominid burial site of Es Skhūl in Israel. *Nature* 338: 756-8.

Valladas, H., Joron, J. L., Valladas, B., Arensburg, P., Bar-Yosef, O., Belfer-Cohen, A., Goldberg, P., Laville, H., Meignen, L., Rak, Y., Tchernov, E., Tillier, A.-M. and Vandermeersch, B. 1987. Thermoluminescence dates for the Neanderthal burial site at Kebara (Mount Carmel) in Israel. *Nature* 330: 159-160.

Valladas, H., Reyss, J. L., Joron, J. L., Valladas, G., Bar-Yosef, O. and Vandermeersch, B. 1988. Thermoluminescence dating of Mousterian 'Proto-Cro-Magnon' remains from Israel and the origin of modern man. *Nature* 331: 614-616.

Vandermeersch, B. 1976. Les sépultures néandertaliennes. In H. de Lumley (ed.). *La Préhistoire Française*: Vol. 1: *Les Civilisations Paléolithiques et Mésolithiques*. Paris: Centre National de La Récherche Scientifique: 725-727.

SECTION I

Biological Change

2. African Origin of Human Mitochondrial DNA

MARK STONEKING AND REBECCA L. CANN

INTRODUCTION

Mitochondrial DNA (mtDNA), by virtue of its relative ease of analysis, rapid rate of evolution, and strictly maternal and haploid mode of inheritance, is a source of new perspectives concerning the evolutionary history of our species and the genetic relatedness of human populations (Wilson *et al.* 1985). Recently, we proposed that all mtDNAs in modern human populations are descended from a single common ancestor who lived in Africa some 200 000 years ago (Stoneking *et al.* 1986; Cann *et al.* 1987). This article and a companion paper (Wilson *et al.* 1987) discuss in more detail the evidence leading to this conclusion and its implications for the origin and dispersal of modern humans.

A SINGLE COMMON ANCESTOR

Figure 2.1 is a phylogenetic tree relating the 134 types of mtDNA detected in a worldwide survey of 148 humans (Cann *et al.* 1987). Because mtDNA is inherited strictly through the maternal line with no segregation or recombination, this tree is actually a genealogy reflecting the maternal ancestry of these humans. The inevitable consequence of the parsimony method of phylogenetic analysis and evolutionary principles is that a genealogy must ultimately stem from a single ancestor (indicated by the arrow in Figure 2.1), yet this conclusion has been a source of some confusion. For example, it has been suggested that a group of women with identical mtDNA types, rather than a single common ancestor, may be the source of mtDNA diversity in modern populations (e.g. Wainscoat 1987). Yet, this hypothetical group of women must be descended from a single common ancestor in a preceding generation. It is not biologically feasible to have multiple lines of descent without a common ancestor.

AFRICAN ORIGIN

Clearly, the conclusion that all mtDNA types in present-day human populations trace back to a single common ancestor is less interesting than the geographical and temporal location of that ancestor. Our hypothesis that the common ancestor lived in Africa stems from two lines of evidence: (1) the genealogical analysis of mtDNA types; and (2) comparisons of the

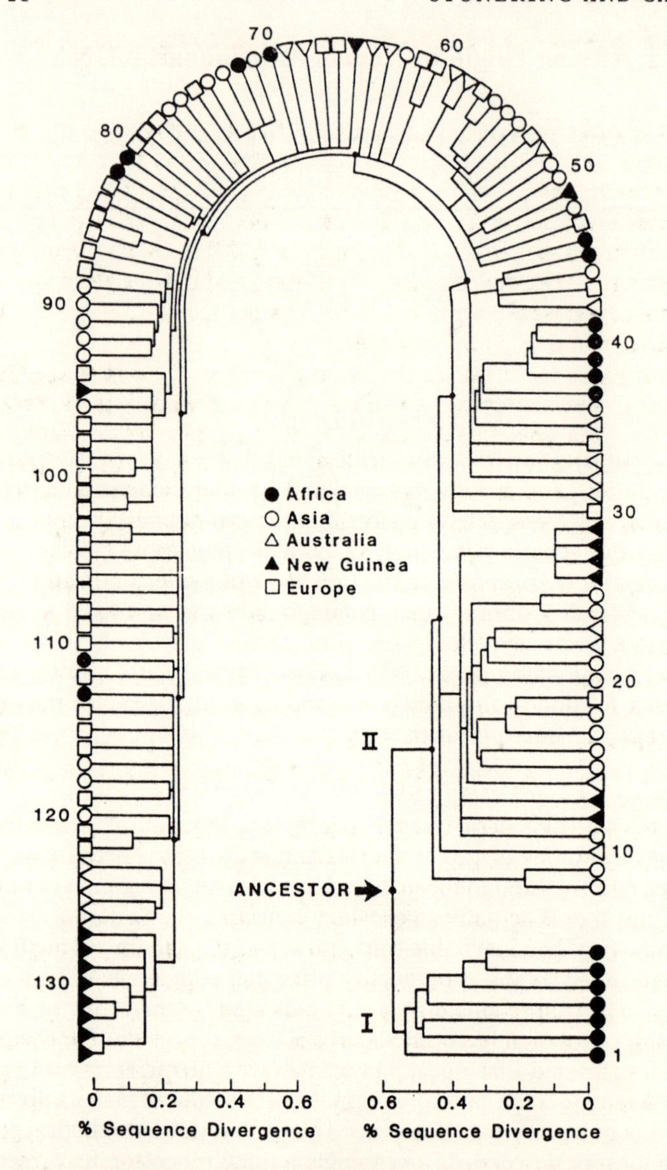

Figure 2.1. Genealogy relating 134 mtDNA types found in a survey of 148 people occupying five different geographic regions. This tree consists of two primary branches (labelled I and II) and was constructed from restriction maps of about 370 cleavage sites per mtDNA, as detailed in Cann *et al.* (1987). The computer program PAUP (Swofford 1985) was used to relate mtDNA types in a branching network that minimized the total number of mutations; the resulting network was converted to a genealogy by placing the ancestor at the midpoint of the longest path connecting any two mtDNA types (i.e. midpoint rooting). Inferred ancestral nodes were then positioned approximately with respect to the scale of sequence divergence by averaging estimates of pairwise divergence (calculated from the restriction maps by the method of Nei and Tajima 1983) for the appropriate descendant mtDNA types.

amount of mtDNA variability in different populations.

Genealogical analysis

The first line of evidence is illustrated by the tree depicted in Figure 2.1, which consists of two main branches, one leading to mtDNA types 1-7 and the other leading to all of the other mtDNA types (types 8-134). Since the seven types comprising the first branch are exclusively of African origin, and since the second branch also contains African types, the simplest interpretation is that the common ancestor was African. Otherwise, additional intercontinental migrations would be needed to explain the geographic distribution of mtDNA types.

This interpretation rests on the assumption that the common ancestor has, indeed, been correctly located on the tree. Positioning the common ancestor requires assigning the direction of change in the characters used to construct the tree. The preferred method of doing this is to use an *outgroup*, i.e. a taxonomic group that is known to be more distantly related to any of the individuals studied than any of the individuals are to each other (Swofford 1985). Unfortunately, the necessary information (i.e. high-resolution mtDNA cleavage maps or sequences) from the most logical outgroups for human mtDNA comparisons (chimpanzees and gorillas) has yet to be obtained.

In the absence of the appropriate outgroup information, we placed the common ancestor at the midpoint of the longest path that connected two mtDNA types (Cann *et al.* 1987). This *midpoint* rooting method introduces the assumption of a constant rate of mtDNA evolution (this assumption will be discussed below).

Two other studies have included mtDNA types of African origin (Greenberg *et al.* 1983; Johnson *et al.* 1983). Greenberg *et al.* (1983) presented sequences of the major noncoding (D loop) region from seven individuals of Caucasian and African (Black American) origin; a genealogy of these seven sequences, rooted by the midpoint method, indicates that the common ancestor was African (see Figure 4 of Cann *et al.* 1987).

Johnson *et al.* (1983) surveyed an average of 39 restriction sites per mtDNA from 200 individuals of African, Asian, Caucasian, and New World (Venezuelan) origin; a midpoint-rooted tree for the resulting 35 mtDNA types appears in Figure 2.2. This tree, like that in Figure 2.1, contains two primary branches, one consisting exclusively of African mtDNA types and the other leading to all of the other mtDNA types (including some African types). Thus, all available genealogical evidence concerning mtDNA types in modern human populations points to an African origin.

One point concerning the tree shown in Figure 2.2 deserves further discussion. Johnson *et al.* (1983) claim that their data support an Asian origin for human mtDNA. They arrive at this conclusion by observing that their mtDNA type 1 is the only type found in all of the populations they surveyed, and hence represents the ancestral mtDNA type. Since type 1 is present in highest frequency in Asian populations, Johnson *et al.* (1983) argue that Asia is the source. However, the midpoint method of rooting

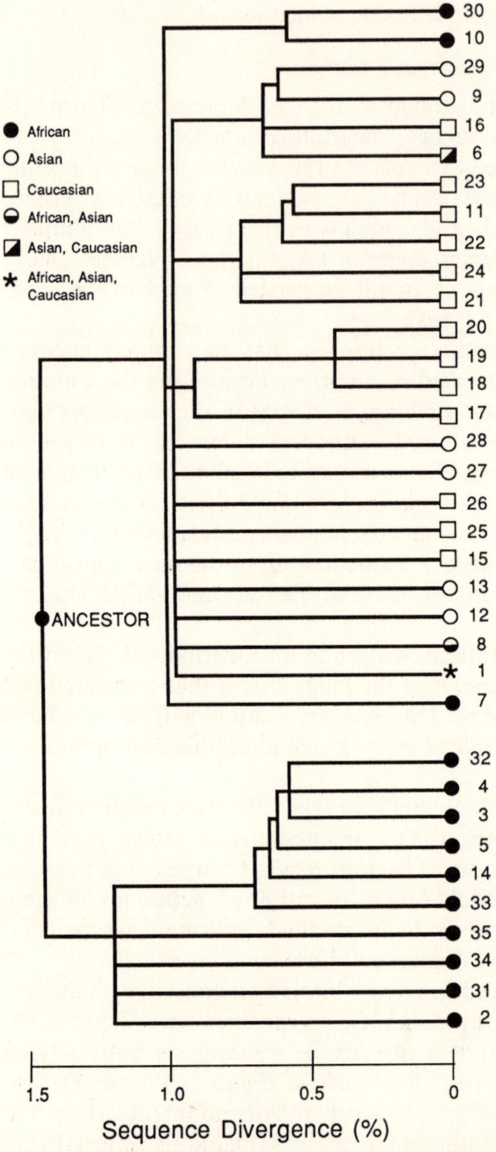

Figure 2.2. Genealogy relating 35 mtDNA types found by Johnson *et al.* (1983) in their survey of 200 mtDNAs from four geographic regions (mtDNAs from Venezuela could not be typed with *Hae*II but were the same as type 1 for the other restriction enzymes). MtDNA types are numbered in accordance with the nomenclature of Johnson *et al.* (1983). This genealogy and associated scale of sequence divergence were constructed in the same fashion as in Figure 2.1, and are one of several equally-parsimonious representations. The tree differs slightly from that presented by Johnson *el al.* (1983) in that from our mapping results (Cann *et al.* 1987) we interpret their *Msp*I morph 2 as representing two separate site losses at positions 8112 and 8150 of the published sequence (Anderson *et al.* 1981). We therefore group mtDNA type 33 with types 3, 4, 5, 14, and 32, whereas Johnson *et al.* (1983) placed type 33 with types 2, 34, and 35. Other equally-parsimonious representations include grouping types 9 and 29 with type 8, type 24 with type 28, and type 31 with types 10 and 30.

The amount of sequence divergence that has accumulated since the common ancestor of these types is 1.47 percent, or about 2.5 times greater than the estimated depth of our tree in Figure 2.1. This discrepancy arises from two factors: (1) stochastic variation due to the much smaller number of restriction sites mapped per individual (an average of 39 per individual for the above tree, over 9 times less than the average number of sites we mapped per individual); and (2) the five restriction enzymes used by Johnson *et al.* (1983) were purposely chosen because they detect significant amounts of variability in human mtDNA, which will lead to a bias, namely overestimating the actual mtDNA variability of human populations. By contrast, the restriction enzymes we used were chosen strictly on the basis of availability and therefore should not be subject to the latter basis. For these two reasons our estimate of 0.57 percent for the amount of sequence divergence since the common mtDNA ancestor is probably more accurate.

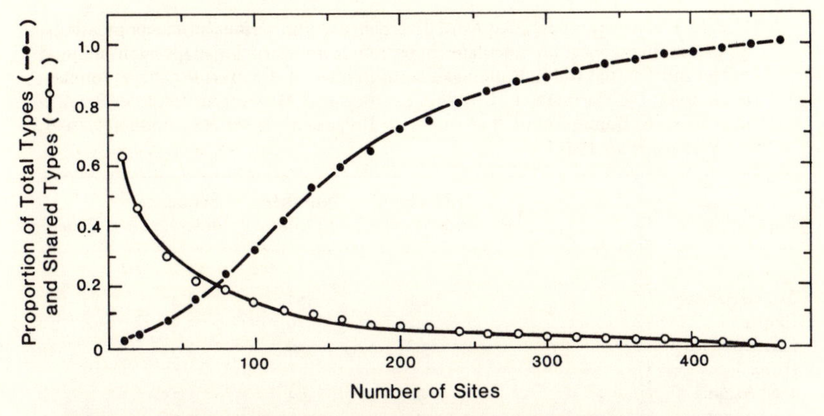

Figure 2.3. Average number of mtDNA types and proportion of types shared between geographic regions, based on random resampling of the restriction site data of Cann *et al.* (1987). Each point represents the average of 100 samples of the indicated number of restriction sites, drawn at random without replacement. The fraction of the total number of 134 mtDNA types detected and the proportion of types shared between geographic regions is plotted for the 148 individuals.

does not indicate that type 1 is the ancestral type (Fig. 2.2); to argue that it is ancestral implies that the rate of mtDNA evolution is faster in African populations than in other human populations. MtDNA evolution proceeds at a fairly uniform rate in vertebrates (Wilson *et a*l. 1985; Higuchi *et al.* 1987; Shields and Wilson 1987) and there is currently no information to support the hypothesis that humans present an exception (Stoneking *et al.* 1986).

Furthermore, we would question the existence of a single geographically widespread mtDNA type (e.g. type 1 of Johnson *et al.* 1983). Our sample of 241 individuals from five geographic regions showed no instances in which the same mtDNA type was found in populations occupying different regions (Stoneking 1986; Cann *et al.* 1987). We surveyed on average over nine times as many restriction sites per individual (363 *versus* 39) than did Johnson *et al.* (1983); we believe our techniques would reveal that their type 1 consists of a multitude of mtDNA types, none of which would occur in more than one geographic region.

Two additional studies support this conclusion. First, Brega *et al.* (1986b) found six subtypes of the Johnson *et al.* (1983) type 1 using just one additional restriction enzyme (approximately 10 additional restriction sites per individual) in their survey of mtDNA variability in Italy and Sardinia. Second, we sampled random subsets of our restriction site data and calculated the total number of mtDNA types and the proportion of identical types found in more than one geographic region (Fig. 2.3). This analysis reveals that, for our data, sampling a total of less than 200-250 restriction sites will result in significant under-representation of mtDNA types and over-representation of the proportion of types shared between geographic regions. Although Johnson *et al.* (1983) used a set of restriction enzymes that tend to detect more variability in human mtDNA than the enzymes we used

Table 2.1. Average mtDNA sequence divergence within various human populations. Sequence divergences are calculated in percent from restriction maps by the method of Nei and Tajima (1983). References are as follows: (1) Cann *et al* 1987; (2) Johnson *et al.* 1983; (3) Harihara *et al.* 1986; (4) Horai and Matsunaga 1986; (5) Brega *et al.* 1986a; (6) Bonné-Tamir *et al.* 1986; (7) Brega *et al.* 1986b; (8) Stoneking 1986; (9) Wallace *et al.* 1985.

Population	Number of Individuals	Number of Sites	Sequence Divergence	Reference
African				
Mixed African	20	363	0.47	(1)
Bantu	40	39	0.55	(2)
San (S.Africa)	34	39	0.59	(2)
Asian				
East Asian	34	363	0.35	(1)
East Asian	46	39	0.26	(2)
Ainu (Japan)	48	19	0.06	(3)
Japanese	74	19	0.32	(3)
Japanese	116	233	0.26	(4)
Tharu (Nepal)	91	51	0.21	(5)
Caucasian				
Mixed Caucasian	47	363	0.23	(1)
Mixed Caucasian	50	39	0.46	(2)
Israeli Arabs	39	42	0.35	(6)
Israeli Jews	39	42	0.38	(6)
Italians	95	50	0.32	(7)
Sardinians	134	50	0.24	(7)
South Pacific				
Australia	21	363	0.25	(1)
Papua new Guinea	119	371	0.21	(8)
New World				
Southwest U.S.	74	49	0.28	(9)
Venezuela	30	32	0.00	(2)

(Table 2.1 and Fig. 2.2), the conclusion from Figure 2.3 is that under-representation of mtDNA types is potentially a problem with their data.

MtDNA variability

The second line of evidence in support of an African origin – comparative levels of mtDNA variability – is summarized in Table 2.1. It is important to emphasize that the measure of variability used (average sequence divergence) is based on the actual number of mutations differentiating mtDNA types within a population, not simply on gene frequencies. Most mtDNA mutations either occur in noncoding regions or do not cause amino acid substitutions and hence are probably neutral (Cann *et al.* 1984; Stoneking 1986); statistical analyses demonstrate that the neutral model can account for most of the patterns of variability present in our data (Whittam et al. 1986).

If one accepts that mtDNA mutations are largely neutral, then their occurrence and accumulation are mostly a function of time: the more variability a population possesses, the older it is. Since African populations are the most variable (Table 2.1), it follows that they are the oldest. The

Table 2.2. Estimates of the rate of human mtDNA evolution derived from three populations.

Population	Age (Years)	% Divergence Per Million Years
Papua New Guinea	40,000	2.5–5.2
Australia	40,000	4.0–7.2
North America	12,000	4.2

measure of variability used in Table 2.1, average sequence divergence, is based on the number of mutations that have occurred along lineages. The stochastic accumulation of neutral mutations along mtDNA lineages should be directly influenced by time and the mutation rate, and little else. However, the survival of different mtDNA lineages in a population will be influenced by historical factors other than time; in particular, bottlenecks or admixture can alter the distribution of surviving lineages (and thus the amount of variability) in a population independent of its age (Wilson *et al.* 1985; Avise 1986). Furthermore, as shown in Table 2.1, the number of human populations examined for mtDNA variability is still quite small, and there are large differences between studies in the number of individuals and average number of restriction sites surveyed per individual, making it difficult to compare studies. Nevertheless, an African origin for human mtDNA provides the simplest explanation for the data in Table 2.1.

Dating the mtDNA ancestor

The genealogy in Figure 2.1 contains not only qualitative information on the branching order of the lineages leading to present-day mtDNA types, but also quantitative information on the amount of sequence divergence that has accumulated since any two mtDNA types last shared a common ancestor. An estimate of when the common ancestor lived can therefore be obtained directly from the amount of sequence divergence that has accumulated through the root of the tree, if the average rate of human mtDNA evolution is known.

We have described elsewhere how the rate of human mtDNA evolution can be estimated from the amount of mtDNA variation in human populations that colonized a specific region at a defined time and remained in relative isolation following colonization (Stoneking *et al.* 1986; Wilson *et al.* 1987). Table 2.2 presents estimates of the rate of human mtDNA evolution derived from human populations occupying three geographical regions (Australia, Papua New Guinea, and North America) that, to a first approximation, fulfill the above requirements. The rate of mtDNA evolution for a wide variety of vertebrates is 2% to 4% per million years (Brown *et al.* 1979; Wilson *et al.* 1985; Higuchi *et al.* 1987; Shields and Wilson 1987); the values in Table 2.2 are in approximate agreement, suggesting that the rate of human mtDNA evolution has remained constant at 2% to 4% per million years.

As Figure 2.1 shows, the average sequence divergence that has accumulated since the common mtDNA ancestor is 0.57%, which (using a rate of

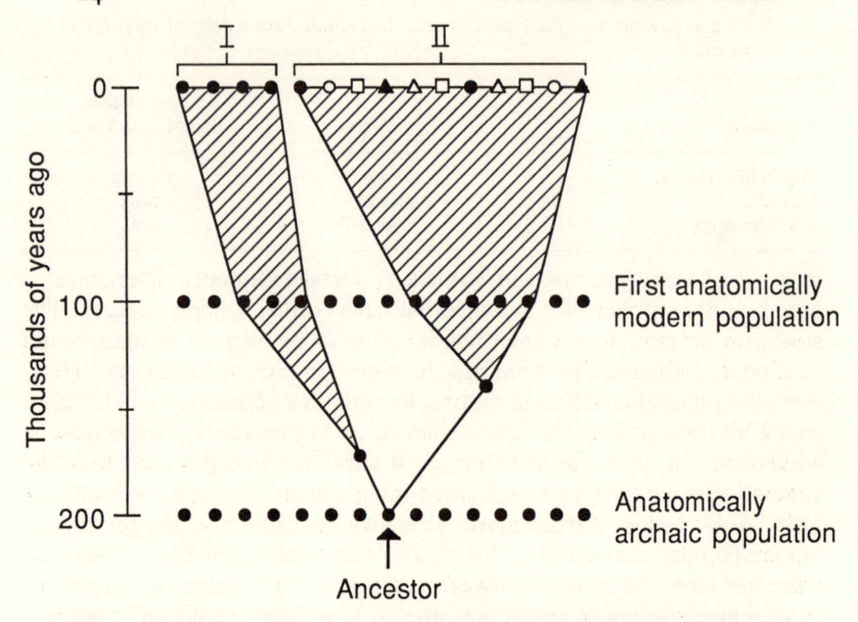

Figure 2.4. Scenario relating the mtDNA results to an African origin for modern humans. Solid circles represent African mtDNAs, and other symbols represent non-African mtDNAs, as in Figure 2.1; the shaded branches labelled I and II represent the two primary branches in the tree in Figure 2.1 and encompass the descendants of the common mtDNA ancestor. Assuming that this common ancestor lived about 200 000 years ago, from fossil evidence she would have been a member of an anatomically-archaic population. Descendants of this population would have undergone the transformation to anatomically-modern humans by 100 000 years ago, by which time the split between branches I and II had occurred, with some of the descendants of branch II later dispersing out of Africa. This figure is adapted from Wilson *et al.* (1987).

2% to 4% per million years) translates to 142 500 to 285 000 years ago. The scale of sequence divergence along the bottom of Figure 2.1 can similarly be used to estimate when the common ancestor of any mtDNA cluster lived.

How confident are we of this time for the common ancestor? Besides errors associated with the rate itself (discussed in Stoneking *et al.* 1986), the estimates of average sequence divergence are subject to both sampling and stochastic variances that are difficult to calculate but undoubtedly large (Nei 1985; Stoneking 1986). Thus, we can probably only state with certainty that the common ancestor was present at least 50 000 but less than 500 000 years ago. Additional comparative work will improve our estimates of the rate of human mtDNA evolution and the amount of sequence divergence that has accumulated since the common ancestor lived.

IMPLICATIONS FOR THE ORIGIN AND DISPERSAL OF MODERN HUMANS

It should be stressed that the transformation from archaic to modern humans is defined by the fossil and archaeological evidence. It is unlikely that mtDNA genes or the types of nuclear genes that have been studied in

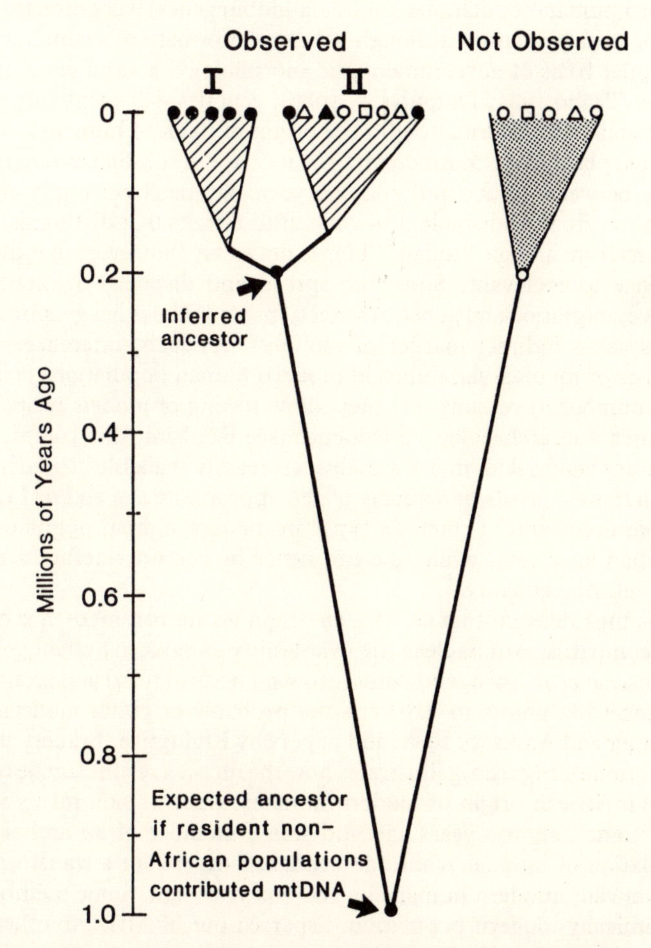

Figure 2.5. Expected genealogical relationship of human mtDNAs if resident populations in Asia and Europe contributed mtDNA to modern populations. Symbols are as in Figure 2.1, with diagonal shaded groups labelled I and II referring to the two primary branches observed in the tree in Figure 2.1. The other shaded branch shows that if the resident non-African populations that had diverged some 1 million years ago from the dispersing African population *did* contribute mtDNA to modern human populations, then we would expect to observe a third group of non-African mtDNA types approximately five times more divergent than any types heretofore observed.

modern human populations (e.g. beta-globin genes) were directly involved in this transformation (although as more information accumulates on the molecular basis of development and morphology, and on genes in ancient tissues (Pääbo 1985; Doran *et al.* 1986), genetics will eventually provide a direct source of information on the origin of modern humans).

Thus, placing our common mtDNA ancestor in sub-Saharan Africa somewhere between 50 000 and 500 000 years ago need not imply an African origin for the various biological and cultural traits that distinguish modern humans from archaic humans. This is not to say, however, that the mtDNA evidence is irrelevant. Since the spread and dispersal of mtDNA genes involves migration and genetic contact, mtDNA (like other genetic evidence) serves as an indirect marker of our past. As such, inferences based on patterns of mtDNA variability in modern human populations are valuable for a number of reasons: (1) they allow testing of models proposed from the fossil and archaeological evidence (see Rouhani, this volume); (2) the necessary material for mtDNA analyses is readily available in modern human populations – fossils or artifacts of the appropriate age and/or location are not required; and (3) mtDNA types in modern human populations must have had ancestors, while one can never be certain whether a particular fossil left descendants.

It is thus relevant that an African origin for human mtDNA is consistent with comparisons of nuclear DNA variability in modern human populations (Wainscoat *et al.* 1986) and with a growing body of fossil and archaeological evidence that points to Africa as the probable origin of modern humans (Stringer and Andrews 1988; and papers by Rightmire, Bräuer, and Klein, this volume). Figure 2.4 illustrates how the mtDNA results can be reconciled with an African origin of modern humans: the common mtDNA ancestor lived some 200 000 years ago and was a member of an archaic African population of humans whose descendants underwent a transformation to anatomically-modern humans by 100 000 years ago. Some members of the anatomically-modern population dispersed out of Africa to other parts of the world (branch II of the mtDNA tree), while others remained in Africa (branch I of the mtDNA tree).

An alternative view – the 'multiregional evolution' hypothesis – holds that the transformation to modern humans occurred more or less simultaneously in different parts of the world (Wolpoff *et al.* 1984; also Wolpoff, this volume). It is difficult to reconcile the molecular data (mtDNA plus nuclear DNA) with this scenario, since it is not clear how or why the observed greater antiquity in the gene pool of modern Africans would then arise. Furthermore, the primary observation that led to the multiregional evolution hypothesis – regional continuity in the Asian fossil record – is a source of some controversy (compare Bräuer 1984; Stringer *et al.* 1984; Wolpoff *et al.* 1984; Stringer and Andrews 1988; and papers in this volume by Bräuer, Habgood, Stringer, and Wolpoff). In any case, the African origin hypothesis can account for regional continuity by supposing that interbreeding occurred between the dispersing African population and the resident Asian populations.

Given the scenario depicted in Figure 2.4, what happened when the dispersing African population met the resident hominid populations in Asia (and Europe) that were descended from the *Homo erectus* populations that left Africa around one million years ago? Figure 2.5 illustrates the genealogical relationship expected among mtDNA types in modern human populations if the resident non-Africans had contributed mtDNA types to modern populations: specifically, we would expect non-African mtDNA types whose estimated time of divergence approaches one million years ago. To date mtDNA types have been determined for over 1000 non-African individuals (Table 2.1 and references therein) and extremely divergent mtDNA types have not been observed in any of these.

There are three reasons that could account for the lack of divergent non-African mtDNA types in modern human populations. First, such types may actually exist but have not yet been detected. Assuming that the populations listed in Table 2.1 provide a random sample of the mtDNA variability in the worldwide human population, then the frequency of unobserved types in this sample must be (with a 95% probability) less than 0.3%. As more information accumulates on mtDNA variability in modern human populations, either divergent non-African types will be found or the true frequency of such types will drop to essentially zero.

Second, it is possible that resident non-Africans did contribute mtDNA types to the dispersing African population, but they were subsequently lost by either selective or stochastic mechanisms. The likelihood of this is difficult to evaluate at present but merits consideration. If the mtDNA types in the dispersing population possessed a selective advantage relative to the resident types, then the resident types would eventually be lost. Although biochemical and statistical analyses, as discussed above, suggest that the bulk of mtDNA mutations now present in human populations are in fact neutral, this may not have been the case for all of the mutations that differentiated the dispersing mtDNA types from the resident types. Theoretical studies that estimate the magnitude of this putative selective advantage are required to assess the likelihood of selective loss within the timespan under consideration.

In addition, mtDNA types can be lost at random for reasons unrelated to the selective value of the mtDNA type (e.g. the loss of mtDNA types from females that leave no, or only male, offspring). Theoretical considerations suggest that the probability of stochastic loss is greatly influenced by the rate of population growth (Avise *et al.* 1984); this probability is essentially zero for even a very slowly-growing population but can be significant for stable or decreasing populations. Although the overall historical trend in human populations has been growth, suggesting that stochastic loss may not pose a problem, the effects of local fluctuations in population size and rate of growth on the probability of survival of mtDNA types need to be explored, using demographic parameters appropriate for early human populations.

The third explanation is that extremely divergent non-African mtDNA types are not found in modern human populations because they were never contributed by the resident populations. If this is the case, did the resident populations contribute any nuclear genes to modern populations? Given the lack of any direct evidence bearing on this question and the caveats mentioned above, such speculation is premature. Nevertheless, we would like to note that whenever a dispersing human population possessing some degree of technological and/or cultural advantage comes into contact with a resident population, gene flow usually involves invading males mating with resident females. These are precisely the conditions that would maximize the retention of the resident mtDNA types in the descendant 'hybrid' populations. Thus, *if further work rules out the first two explanations*, the rather staggering implication is that the dispersing African population replaced the non-African resident populations without any interbreeding.

SUMMARY

Consideration of all available information on mtDNA polymorphisms has led us to propose that the common ancestor of all existing mtDNA types in modern human populations lived in sub-Saharan Africa somewhere between 50 000 and 500 000 years ago. The mtDNA evidence is consistent with comparisons of nuclear DNA in modern human populations, and can easily be reconciled with one view of the fossil and archaeological record: Africa, home of the first split of our lineage from our nearest relatives, the African apes, was also the source of the last transformation in our ancestry – that to modern humans.

NOTE ADDED IN PRESS

Saitou and Omoto (*Nature* 329: 288, 1987) and Darlu and Tassy (*Nature* 329: 111, 1987) raised additional objections to our interpretation of our mtDNA results; these objection have been dealt with elsewhere by Cann *et al.* (*Nature* 329: 111-112, 1987).

ACKNOWLEDGEMENTS

We thank A. C. Wilson for his guidance and encouragement, and U. B. Gyllensten, H. Ochman, E. M. Prager, V. M. Sarich, and A. C. Wilson for advice and discussion. The research was supported by grants from the National Institutes of Health, the National Science Foundation, and the Foundation for Research into the Origin of Man.

REFERENCES

Anderson, S., Bankier, A. T., Barrell, B. G., de Bruijn, M. H. L., Coulson, A. R., Drouin, J., Eperon, I. C., Nierlich, D. P., Roe, B. A., Sanger, F., Schreier, P. H., Smith, A. J. H., Staden, R. and Young, I. G. 1981. Sequence and organization of the human mitochondrial genome. *Nature* 290: 457-465.

Avise, J. C. 1986. Mitochondrial DNA and the evolutionary genetics of higher animals. *Philosophical Transactions of the Royal Society of London* (Series B) 312: 325-342.

Avise, J. C., Neigel, J. E. and Arnold, J. 1984. Demographic influences on mitochondrial DNA lineage survivorship in animal populations. *Journal of Molecular Evolution* 20: 99-105.

Bonné-Tamir, B., Johnson, M. J., Natali, A., Wallace, D. C. and Cavalli-Sforza, L. L. 1986. Human mitochondrial DNA types in two Israeli populations – a comparative study at the DNA level. *American Journal of Human Genetics* 38: 341-351.

Bräuer, G. 1984. A craniological approach to the origin of anatomically modern *Homo sapiens* in Africa and implications for the appearance of modern Europeans. In F. H. Smith and F. Spencer (eds) *The Origins of Modern Humans: a World Survey of the Fossil Evidence*. New York: Alan R. Liss: 327-410.

Brega, A., Gardella, R., Semino, O., Morpurgo, G., Astaldi Ricotti, G. B., Wallace, D. C. and Santachiara Benerecetti, A. S. 1986a. Genetic studies on the Tharu population of Nepal: restriction endonuclease polymorphisms of mitochondrial DNA. *American Journal of Human Genetics* 39: 502-512.

Brega, A., Scozzari, R., Maccioni, L., Iodice, C., Wallace, D. C., Bianco, I., Cao, A. and Santachiara Benerecetti, A .S. 1986b. Mitochondrial DNA polymorphisms in Italy. I. Population data from Sardinia and Rome. *Annals of Human Genetics* 50: 327-338.

Brown, W. M., George, M. and Wilson, A. C. 1979. Rapid evolution of animal mitochondrial DNA. *Proceedings of the National Academy of Sciences (USA)* 76: 1967-1971.

Cann, R. L., Brown, W. M. and Wilson, A. C. 1984. Polymorphic sites and the mechanism of evolution in human mitochondrial DNA. *Genetics* 106: 479-499.

Cann, R. L., Stoneking, M. and Wilson, A.C. 1987. Mitochondrial DNA and human evolution. *Nature* 325: 31-36.

Doran, G. H., Dickel, D.N., Ballinger, W. E., Agee, O. F., Laipis, P. J. and Hauswirth, W. W. 1986. Anatomical, cellular and molecular analysis of 8000-yr-old human brain tissue from the Windover archaeological site. *Nature* 323: 803-806.

Greenberg, B. D., Newbold, J. E. and Sugino, A. 1983. Intraspecific nucleotide sequence variability surrounding the origin of replication in human mitochondrial DNA. *Gene* 21: 33-49.

Harihara, S., Hirai, M. and Omoto, K. 1986. Mitochondrial DNA polymorphism in Japanese living in Hokkaido. *Japanese Journal of Human Genetics* 31: 73-83.

Higuchi, R. G., Wrischnik, L. A., Oakes, E., George, M., Tong, B. and Wilson, A. C. 1987. Mitochondrial DNA of the extinct quagga: relatedness and extent of post-mortem change. *Journal of Molecular Evolution* 25: 283-287.

Horai, S. and Matsunaga, E. 1986. Mitochondrial DNA polymorphism in Japanese. II. Analysis with restriction enzymes of four or five base pair recognition. *Human Genetics* 72: 105-117.

Johnson, M. J., Wallace, D. C., Ferris, S. D., Rattazzi, M. C. and Cavalli-Sforza, L. L. 1983. Radiation of human mitochondria DNA types analyzed by restriction endonuclease cleavage patterns. *Journal of Molecular Evolution* 19: 255-271.

Nei, M. 1985. Human evolution at the molecular level. In T. Ohta and K. Aoki (eds) *Population Genetics and Molecular Evolution*. Tokyo: Japanese Scientific Societies Press: 41-64.

Nei, M. and Tajima, F. 1983. Maximum likelihood estimation of the number of nucleotide substitutions from restriction sites data. *Genetics* 105: 207-217.

Pääbo, S. 1985. Molecular cloning of ancient Egyptian mummy DNA. *Nature* 314: 644-645.

Shields, G. F. and Wilson, A. C. 1987. Calibration of mitochondrial DNA evolution in geese. *Journal of Molecular Evolution* 24: 212-217.

Stoneking, M. 1986. *Human Mitochondrial DNA Evolution in Papua New Guinea*. Unpublished Ph.D. Thesis, University of California, Berkeley.

Stoneking, M., Bhatia, K. and Wilson, A. C. 1986. Rate of sequence divergence estimated from restriction maps of mitochondrial DNAs from Papua New Guinea. *Cold Spring Harbor Symposia in Quantitative Biology* 51: 433-439.

Stringer, C. B. and Andrews, P. 1988. Genetic and fossil evidence for the origin of modern humans. *Science* 239: 1263-1268.

Stringer, C. B., Hublin, J. J. and Vandermeersch, B. 1984. The origin of anatomically modern humans in Western Europe. In F. H. Smith and F. Spencer (eds) *The Origins of Modern Humans: a World Survey of the Fossil Evidence*. New York: Alan R. Liss: 51-135.

Swofford, D. L. 1985. *Phylogenetic Analysis using Parsimony (PAUP), Version 2.4*. Illinois Natural History Survey, Champaign (Ill).

Wainscoat, J. S. 1987. Out of the garden of Eden. *Nature* 325: 13.

Wainscoat, J. S., Hill, A. V. S., Boyce, A. L., Flint, J., Hernandez, M., Thein, S. L., Old, J. M., Lynch, J. R., Falusi, A. G., Weatherall, D. J. and Clegg, J. B. 1986. Evolutionary relationships of human populations from an analysis of nuclear DNA polymorphisms. *Nature* 319: 491-493.

Wallace, D. C., Garrison, K. and Knowler, W. C. 1985. Dramatic founder effects in Amerindian mitochondrial DNAs. *American Journal of Physical Anthropology* 68: 149-155.

Whittam, T. S., Clark, A. G., Stoneking, M., Cann, R. L. and Wilson, A. C. 1986. Allelic variation in human mitochondrial genes based on patterns of restriction site polymorphism. *Proceedings of the National Academy of Sciences USA* 83: 9611-9615.

Wilson, A. C., Cann, R. L., Carr, S. M., George, M., Gyllensten, U. B., Helm-Bychowski, K. M., Higuchi, R. G., Palumbi, S. R., Prager, E. M., Sage, R. D. and Stoneking, M. 1985. Mitochondrial DNA and two perspectives on evolutionary genetics. *Biological Journal of the Linnean Society* 26: 375-400.

Wilson, A.C., Stoneking, M., Cann, R. L., Prager, E. M., Ferris, S. D., Wrischnik, L. A. and Higuchi, R. G. 1987. Mitochondrial clans and the age of our common mother. In F. Vogel and K. Sperling (eds) *Human Genetics: Proceedings of the 7th International Congress*. Berlin: Springer: 158-164.

Wolpoff, M. H., Wu, X. Y. and Thorne, A. G. 1984. Modern *Homo sapiens* origins: a general theory of hominid evolution involving the fossil evidence from East Asia. In F. H. Smith and F. Spencer (eds) *The Origins of Modern Humans: a World Survey of the Fossil Evidence*. New York: Alan R. Liss: 411-483.

3. Geographic Distribution of Alpha- and Beta-Globin Gene Cluster Polymorphisms

J. S. WAINSCOAT, A. V. S. HILL, S. L. THEIN,
J. FLINT, J. C. CHAPMAN, D. J. WEATHERALL,
J. B. CLEGG AND D. R. HIGGS

INTRODUCTION

Human population genetics has been traditionally studied by frequency comparison of blood group, protein and HLA polymorphisms (Nei and Roychoudhury 1982). More recently there has been an upsurge of interest in this subject due to the possibility of using techniques of DNA analysis for the detection of new molecular markers of human populations.

The optimism for the potential of DNA markers rests chiefly on the enormous numbers of these which must exist. Scattered along the genome every hundred bases or so are single base changes that represent polymorphisms between individuals and populations. The majority of these are in non-coding DNA and are thought to be selectively neutral. Many of these single base changes produce sequence differences that alter the recognition sequence for specific restriction endonucleases, and hence can be detected by single blot Southern analysis.

A large number of 'restriction fragment length polymorphisms' (RFLPs) have now been discovered, and their importance is enhanced by the linkage that exists between polymorphic sites of particular gene clusters, allowing the construction of haplotypes. It might be supposed that the linkage between sites would be inversely proportional to the distance that separates them. However, the situation is more complex and it seems that recombination takes place at defined hot spots rather than randomly throughout the genomes. Antonarakis and co-workers (1982) demonstrated that the polymorphic sites in the β- cluster could be divided into the two haplotypes groups (5' and 3'), and that although the sites within each group were in linkage disequilibrium, the groups themselves shared negligible linkage – that is there must be a hotspot for recombination between them. The existence of the two haplotype groups means that comparison of different populations is best undertaken by analysing the frequencies of one or both haplotype groups.

In addition to RFLPs resulting from single base changes there is another class, 'variable length polymorphisms', which result from differences in the copy number of a tandem repeat sequence. These are particularly useful as molecular markers both because they are, in principle, always detectable by Southern blot analysis, and because they are mutliallelic. It is likely they will become useful for population analysis although the problems of

Table 3.1 The frequency of β-globin gene haplotypes.
The numbers of chromosomes analysed in the population were:

United Kingdom	37	Thai	32	Polynesian	55
Mediterranean	132	North American Indian	34	Subsaharan African	26
Asian Indian	111	Melanesian	173	Nigerian	35

Haplotypes	United Kingdom	Mediterranean	Asian Indian	Thai	North American Indian	Melanesian	Polynesian	Sub-Saharan African	Nigerian
+−−−−	0.43	0.69	0.52	0.90	0.59	0.68	0.78	0.00	0.09
−+−++	0.40	0.22	0.25	0.00	0.18	0.17	0.11	0.08	0.11
−++−+	0.14	0.08	0.14	0.06	0.21	0.02	0.00	0.04	0.00
+−−++	0.00	0.00	0.00	0.00	0.00	0.06	0.00	0.00	0.00
−++−−	0.03	0.02	0.03	0.00	0.03	0.00	0.00	0.00	0.00
−++++	0.00	0.00	0.00	0.00	0.00	0.03	0.02	0.00	0.00
++−++	0.00	0.00	0.03	0.00	0.00	0.00	0.00	0.00	0.00
−−−−−	0.00	0.00	0.01	0.00	0.00	0.01	0.02	0.00	0.00
+++−+	0.00	0.00	0.01	0.00	0.00	0.00	0.00	0.00	0.00
−−−++	0.00	0.00	0.01	0.03	0.00	0.00	0.00	0.00	0.00
++−−−	0.00	0.00	0.01	0.00	0.00	0.00	0.00	0.00	0.00
−+−−−	0.00	0.00	0.00	0.00	0.00	0.00	0.00	0.04	0.03
−+−−+	0.00	0.00	0.00	0.00	0.00	0.02	0.07	0.23	0.17
−−−−+	0.00	0.00	0.00	0.00	0.00	0.01	0.00	0.62	0.60

interpretation they present must not be underestimated. Interestingly, the analysis of the human α-globin gene cluster has shown the association of a concentional restriction fragment length polymorphism and linked, hypervariable regions of DNA Higgs *et al.* 1986). The full analysis of this complex therefore offers the possibility of a comparison of the rate of change of the two types of DNA polymorphisms within populations.

Although new DNA polymorphisms are being described at an increasing rate, very few have been thoroughly studied in a range of ethnic groups. Most DNA polymorphisms described to date are found in the majority if not all of the various ethnic groups, and these are sometimes referred to as 'public' polymorphisms. Genetic distance analysis can be performed using either the frequencies of the polymorphisms or the frequencies of the haplotypes they constitute between populations. Perhaps more useful for population genetics are the polymorphisms which are population specific, sometimes called 'private' polymorphisms. It is likely that over the next ten years a large number of 'private' polymorphisms will be described which should enable a good picture to the built up of prehistoric migrations. In the present paper we present an analysis of the polymorphisms present in the α and β-globin gene clusters, and describe their geographical distribution.

β-GLOBIN GENE CLUSTER

The linkage of polymorphic sites along one gene cluster is referred to as a haplotype. The combination of the Hind II ε-globin, Hind III γ-globin and Hind III ψ-globin polymorphic sites constitutes the 5′ β-globin haplotype. The frequencies of the 5′ β-globin haplotypes in the β-globin gene cluster of normal β^A chromosomes of individuals from nine diverse populations are shown in Table 3.1.

One haplotype (+−−−−) is the most common haplotype in all non-African populations. The second most common haplotype (−+−++) is found in all the non-African populations with the exception of the Thai population (possibly due to the small sample size). A third haplotype (−++−+) accounts for the majority of the remaining haplotypes found in the non-African populations. These three haplotypes are present in African populations at low frequencies. In African populations two other haplotypes predominate (−−−+ and −+−−+). Two haplotypes (+−−++ and −++++) seem to be specific for Melanesian and Polynesian populations.

Table 3.2 shows the frequency of a Taq I polymorphic which is sited between the $^G\gamma$ and $^A\gamma$-globin genes. It is present at high frequencies in the three African populations but totally absent in the Eurasian populations tested.

α-GLOBIN GENE CLUSTER

Table 3.3 shows the frequency of the common polymorphisms in the α-globin gene cluster. These are defined as those in which the frequency of the less common allele is >0.05 in most populations. The polymorphisms listed in the Table are as illustrated in Figure 1 of Higgs *et al.* 1986. The three

Table 3.2 The frequency of the Taq I γ-globin polymorphism in the β^A chromosomes of different population groups
Data taken from Wainscoat *et al.* (1986)

Nigerian	0.47 (40)
South African Black	0.36 (22)
San	0.25 (32)
Mediterranean	0.00 (132)
Asian Indian (Non Tribal)	0.00 (11)
Asian Indian (Tribal)	0.00 (19)
British	0.00 (62)
South East Asia	0.00 (48)

African populations show remarkable similarity in the frequencies of all the polymorphisms. The remaining populations with the exception of Melanesia also show close similarities in the frequencies of the polymorphisms. The African and Eurasian groups show similar frequencies for several polymorphisms although some are clearly different. For example, the large IZ-HVR is absent in the three African populations.

As for the β-haplotype analysis it is probable that the populations would be more clearly differentiated by an analysis of the α-globin haplotypes rather than simply by the individual polymorphisms. Table 3.4 shows the α-globin haplotypes (as defined in Higgs *et al.* 1986) in the different population groups. It shows that we have only established the haplotypes of a relatively small number of African chromosomes, most remaining unclassified. This reflects the very small number of individuals homozygous for particular haplotypes in Africa and may imply a greater nucleotide diversity in Africans at the α-globin gene locus. However, it is clear that the haplotype Ia is common in Eurasian populations. The data also suggest the possibility that some haplotypes may be population specific, for example, IIe in Asians, and IIIf in Africans.

Some markers in the α-globin cluster are specific to population groups (Table 3.5). The Sph I and Pst I polymorphisms are African specific, whereas the Bgl I and Bgl II may be specific to Eurasian populations.

DISCUSSION

Our first DNA study of numerous world populations was of the restriction enzyme haplotype close to the β-globin gene (Wainscoat *et al.* 1986). We found that all non-African populations share a limited number of common haplotypes, whereas Africans have predominantly a different haplotype not found in other populations. A genetic distance analysis based on these nuclear DNA polymorphisms indicated a major division of human populations into an African and a Eurasian group.

Interestingly, none of the four common world haplotypes can be derived from any of the other three by either a single crossover or by a single base mutation causing the loss or acquisition of a restriction enzyme site. Most probably the haplotypes predate the racial divergence. Many of the rarer haplotypes can be derived from the four common ones by single crossovers.

Table 3.3 Common polymorphisms of the a-globin complex.
The polymorphisms are listed from top (Xbal) to bottom (Pst 1) as they are sited in the α-globin gene cluster in a 5′ to 3′ direction as described in Higgs et al (1986).

	Gambia	Nigeria	Zambia	United Kingdom	Italy	Saudi Arabia	India	Thailand	Island Melanisia
XbaI		0.50 (36)	0.53 (43)	0.63 (30)	0.53 (68)	0.47 (32)	0.61 (74)	0.41 (32)	0.63 (40)
SacI	0.38 (34)	0.42 (38)	0.53 (38)	0.76 (38)	0.80 (78)	0.47 (38)	0.78 (74)	0.69 (32)	0.08 (40)
BgII	0.08 (36)	0.03 (38)	0.11 (38)	0.17 (42)	0.13 (72)	0.16 (38)	0.11 (74)	0.19 (32)	0.38 (34)
IZHVR small	0.22 (36)	0.21 (38)	0.16 (38)	0.03 (40)	0.08 (78)	0.11 (38)	0.08 (74)	0.16 (32)	0.50 (40)
IZHVR medium	0.78 (36)	0.79 (38)	0.84 (38)	0.68 (40)	0.62 (78)	0.76 (38)	0.59 (74)	0.41 (32)	0.50 (40)
IZHVR large	0.00 (36)	0.00 (38)	0.00 (38)	0.30 (40)	0.31 (78)	0.13 (38)	0.33 (74)	0.44 (32)	0.00 (40)
Z/PZ, PZ	0.44 (34)	0.55 (38)	0.58 (38)	0.81 (42)	0.85 (78)	0.71 (38)	0.86 (74)	0.79 (29)	0.58 (40)
Acc I	0.75 (36)	0.09 (30)	0.68 (38)	0.79 (42)	0.84 (64)	0.67 (36)	0.85 (74)	0.72 (25)	0.58 (40)
Rsa I	0.14 (36)	0.29 (30)	0.21 (38)	0.48 (42)	0.45 (74)	0.32 (31)	0.43 (75)	0.31 (29)	0.03 (40)
Pst I	0.17 (38)	0.08 (38)	0.19 (38)	0.08 (38)	0.04 (72)	0.03 (21)	0.43 (40)		
Pst I	0.11 (36)	0.13 (38)	0.13 (38)	0.11 (38)	0.04 (72)	0.06 (18)	0.03 (60)	0.03 (32)	0.45 (20)

Table 3.4: α-globin haplotypes in different population groups.
Haplotypes as described in Higgs *et al* (1986)

Haplotypes	United Kingdom	Italy	India	Thailand	Jamaica	Nigeria	Zambia
Ia	16	28	26	11	3	0	1
IIa	4	7	6	11	0	0	0
IIe	0	0	2	4	0	0	0
IIIa	0	1	1	2	0	0	0
IIIb	3	0	1	0	1	0	0
IIIf	0	0	0	0	2	2	2
Others	3	6	8	3	6	0	1
Unclassified	6	20	18	22	26	26	30
Incomplete	12	16	12	17	0	10	4
Totals	44	78	74	70	38	38	38

Table 3.5 Population specific polymorphisms in the α-globin complex

	Gambia	Nigeria	Zambia	United Kingdom	Italy	India	Thailand	Island Melanisia
Bgl I	0.00 (38)	0.00 (38)	0.00 (38)	0.02 (44)	0.01 (72)	0.03 (74)	0.06 (34)	0.00 (35)
Bgl II		0.00 (50)		0.07 (68)	0.04 (241)	0.00 (97)		0.00 (1150)
Sph I	0.03 (32)	0.06 (35)	0.08 (38)	0.00 (40)	0.00 (70)	0.00 (37)	0.00 (29)	0.00 (40)
Pst I	0.08 (36)	0.16 (38)	0.05 (38)	0.00 (38)	0.00 (78)	0.00 (72)	0.00 (31)	0.00 (40)

In terms of different populations, some of the gaps in our original study are now being filled. We have found that the β-haplotype frequencies in a small population of Canadian Indians are similar to those of the Eurasian populations. Jenkins and Ramsay (1986) have studied South African populations, and they find very similar haplotype frequencies in the Bantu-speaking Africans to those of the Nigerians we have studied, whereas the frequencies are significantly different in the San ('Bushmen') group. Other studies have shown differences between the San and the neighbouring black African populations (Nurse *et al.* 1985). However, the Taq I γ-globin DNA polymorphism is found in both Bantu-speaking Negroes and San populations, indicating an ancient African origin (Wainscoat *et al.* 1986).

We now hope to confirm our general observations on human population affinities by the study of the polymorphisms and haplotypes of the α-globin gene cluster. In addition, since the α-globin haplotype potentially has more variability, it is possible that insight into the population affinities of more closely related groups and their prehistoric migrations might be gained.

As for the β-globin gene cluster it seems that the common polymorphisms and some common haplotypes in the α-globin gene cluster must have preceded human racial divergence, and have been inherited as stable linkage groups since that time. Inspection of the data on the frequency of the common polymorphisms shows a striking similarity between the three African populations on the one band, and between the Eurasian populations (excepting Melanesia) on the other. The frequencies of many polymorphisms in Melanesia are significantly different from that in Southeast Asia, presumably due to a founder effect.

There are many differences between the frequencies of the polymorphisms between the African and Eurasian groups. It is likely that, by analogy with the β-haplotypes, this African/Eurasian split would become more obvious from a comparison of the frequencies of the α-globin haplotypes rather than just the constituent polymorphisms. Unfortunately on account of the very high heterozygosity rates in Africans we have been unable to determine few of the African haplotypes. This is an indication perhaps of greater nucleotide diversity in African populations.

Many population-specific DNA polymorphisms have been described in Blacks. These include polymorphisms at the following well characterized loci: β-globin (Orkin and Kazazian 1984); growth hormone (Chakravarti *et al.* 1984); dihydrofolate reductase (Anagnou *et al.* 1984); and insulin (Bell *et al.* 1984). We have found two Black-specific polymorphisms in the α-gene cluster (SphI and PstI) but in addition have also found polymorphisms specific to other population groups. It seems likely that within a few years it will be possible to study population migrations with a battery of population specific polymorphisms.

Thus the two lines of evidence that are emerging from the DNA analysis of human populations are, firstly, genetic distance analyses, based on allele and haplotype frequencies, which show a distinct African lineage, and second, data indicating a greater nucleotide sequence diversity at particular loci in Africans (Cann *et al.* 1987). These observations are preliminary and require confirmation by analysis of other world populations and investigation of many other loci. The data are most consistent with the hypothesis that modern humans arose in Africa and subsequently spread to Eurasia and the Americas (Jones and Rouhani 1986; Wainscoat 1987).

ACKNOWLEDGEMENTS

This work has been supported by the Boise fund, the Rockefeller Foundation, and the Wenner-Gren Anthropological Foundation.

REFERENCES

Anagnou, N. P., O'Brien, S. J., Shimada, T., Nash, W. G., Chen, M.-J. and Nienhuis, A. W. 1984. Chromosomal organisation of the human dihydrofolate reductase genes: dispersion, selective amplification, and a novel form of polymorphism. *Proceedings of the National Academy of Sciences (USA)* 81: 5170-5174.

Antonarakis, S. E., Boehm, C. D., Giardina, P. J. V. and Kazazian, H. H. 1982. Nonrandom association of polymorphic restriction sites in the β-globin gene cluster. *Proceedings of the National Academy of Sciences (USA)* 79: 137-141.

Bell, G. I., Horita, S. and Karam, J. H. 1984. A polymorphic locus near the human insulin gene is associated with insulin-dependent diabetes mellitus. *Diabetes* 33: 176-184.

Cann, R. L., Stoneking, M. and Wilson, A. C. 1987. Mitochondrial DNA and human evolution. *Nature* 325: 31-36.

Chakravarti, A., Phillips, J. A., Mellits, K. H., Buetow, K. H. and Seeburg, P. H. 1984. Patterns of polymorphism and linkage disequilibrium suggest independent origins of the human growth hormone gene cluster. *Proceedings of the National Academy of Sciences (USA)* 88: 6085-6089.

Higgs, D. R., Wainscoat, J. S., Flint, J., Hill, A. V. S., Thein, S. L., Nicholls, R. D., Teal, H., Ayyub, H., Peto, T. E. A., Falusi, A. G., Jarman, A. P., Clegg, J. B. and Weatherall, D. J. 1986. Analysis of the human α-globin gene cluster reveals a highly informative genetic locus. *Proceedings National Academy of Sciences (USA)* 83: 5165-5169.

Jenkins, T. and Ramsay, M. 1986. β^S and β^A-globin gene cluster haplotypes in Southern African populations. *Human Genetics: Abstracts of the 7th International Congress of Human Genetics: 462.*

Jones, J. S. and Rouhani, S. 1986. Human evolution: how small was the bottleneck. *Nature* 319: 449-450.

Nei, M. and Roychoudhury, A. K. 1982. Genetic relationship and evolution of human races. *Evolutionary Biology* 14: 1-59.

Nurse, G. T., Weiner, J. S. and Jenkins, T. 1985. *The Peoples of Southern Africa and their Affinities*. Oxford: Oxford University Press.

Orkin, S. H. and Kazazian, H. H. 1984. The mutation and polymorphism of the human β-globin gene and its surrounding DNA. *Annual Review of Genetics* 18: 131-171.

Wainscoat, J. S., Hill, A. V. S., Boyce, A. L., Flint, J., Hernandez, M., Thein, S. L., Old, J. M., Lynch, J. R., Falusi, A. G., Weatherall, D. J. and Clegg, J. B. 1986. Evolutionary relationships of human populations from an analysis of nuclear DNA polymorphism. *Nature* 319: 491-493.

Wainscoat, J. S., Kulozik, A. E., Ramsay, M., Falusi, A, G. and Weatherall, D. J., 1986. A Taq-I γ-globin DNA polymorphism: an African specific marker. *Human Genetics* 74: 90-92.

Wainscoat, J. S. 1987. Out of the Garden of Eden. *Nature* 325: 13.

4. Evidence for the Paternal Ancestry of Modern Humans: Evidence from a Y-Chromosome Specific Sequence Polymorphic DNA Probe

G. LUCOTTE

INTRODUCTION

The human Y chromosome has considerable interest for research into human population genetics and the origins of our species. When fixed on the Y chromosome, mutations are transmitted through males only. Consequently, many variants have become fixed in this way, recording an evolutionary history through male individuals only, without recombinations at each generation. Examining Y chromosome DNA differences in human populations from different geographical regions thus has great potential for reconstructing recent events in human evolution, and can provide an interesting comparison with results of analyses using the maternally- inherited mitochondrial DNA (mtDNA).

THE Y-SPECIFIC PROBES STUDIED

In the present studies, several cosmidic Y-specific sequences have been used for the detection of restriction polymorphisms in a panel of unrelated DNAs (Ngo and Lucotte 1986). The results of these studies show that restriction polymorphism are detected only rarely (these results can be explained by the exemption of most of the Y chromosome from meiotic pairing and exchange). But the clone p49f constitutes a highly polymorphic probe, which detects several *Taq* I RFLPs ('Restriction Fragmentation Length Polymorphisms') on the Y chromosome (Lucotte and Ngo 1985). This probe detects about 16 Y-specific *Taq* bands corresponding to a low-copy number sequence family of genes located in the sub-region Yqll, which does not recombine with the X chromosome. Five of these bands, each representing a single DNA fragment, can be either present, absent, or variable in length (Fig. 4.1).

Familial segregation studies have shown that variations of these five fragments are inherited in a mendelian fashion, and are strictly Y-linked. A survey of numerous male individuals indicated that the five variable *Taq* I fragments detected by probes 49f and 49a can be considered as five independent allelic series, each series representing the different and mutually exclusive allelic forms observed for a single DNA fragment.

These *Taq* I restriction polymorphisms are not observed with other restriction digests, and have therefore been attributed to point mutations. An apparently independent reassortment of one of these series with respect

ALLELIC SERIES

	A					C		D			F		I	
Alleles	A1	A2	A3	A4	A0	C1	C0	D1	D2	D0	F1	F0	I1	I0

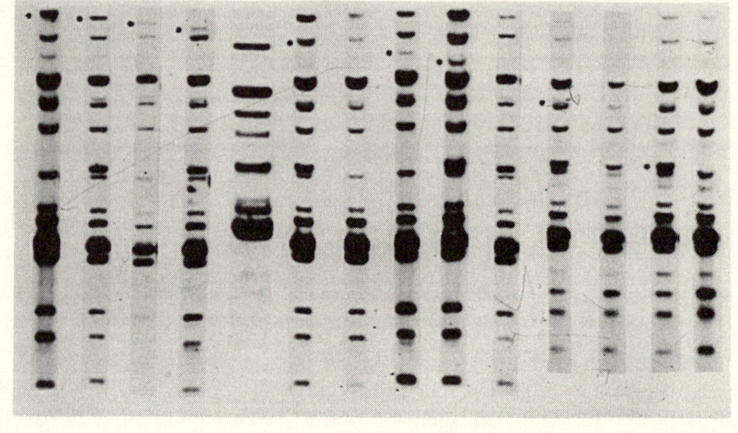

Figure 4.1. The five allelic series (A, C, D, F, and I) detected by probe 49f. *Taq* I hybridization patterns of different individuals each representing one allele (A1, A2, A3, A4, A0 for A; C1 and C0 for C; D1, D2 and D0 for D; F1 and F0 for F; I1 and I0 for I) are shown. Stars indicate the apparent variable allelic fragments.

to the others can be explained on the basis of mutations that occurred several times during the evolution of the Y chromosome.

SEVERAL RFLPS DEFINE MULTIPLE HAPLOTYPES OF THE HUMAN Y CHROMOSOME

A preliminary study had shown that the common haplotypes are present in all three of the main racial groups (Caucasoids, Negroids and Asiatic) and that no particular combination is specific to a particular race. A detailed analysis of the RFLPs was performed by a survey of about 120 Caucasoid unrelated males (Europeans, North Africans and East Asians). The 44 males retained for the study are only those for whom all the variable bands have been unambiguously observed.

Among the 44 individuals analysed, 16 different combinations of the five RFLPs have been found (Ngo *et al.* 1986) and these are indicated in Table 4.1, with the number of cases scored for each combination. Since all the elements of a combination are syntenic and separated on the chromosome by several kilobases at least, it can be considered that each combination defines a haplotype of the human Y chromosome.

This probe represents the most powerful tool so far developed for the

Table 4.1 Haplotypes of the human Y chromosome observed among 44 Caucasian unrelated individuals (haplotype XVII corresponds to Baruyas).

Haplotypes	Locus					Frequency
	A	C	D	F	I	(%)
I	A0	C0	D0	F1	I1	2.3
II	A0	C0	D1	F1	I1	2.3
III	A1	C0	D0	F1	I0	2.3
IV	A1	C0	D0	F1	I1	2.3
V	A2	C0	D0	F1	I1	4.6
VI	A2	C0	D1	F0	I1	2.3
VII	A2	C0	D1	F1	I0	6.8
VIII	A2	C0	D1	F1	I1	2.3
IX	A2	C1	D0	F1	I1	2.3
X	A3	C0	D0	F1	I0	4.6
XI	A3	C0	D0	F1	I1	4.6
XII	A3	C0	D1	F1	I0	18.2
XIII	A3	C0	D1	F1	I1	6.8
XIV	A3	C1	D1	F1	I1	11.4
XV	A3	C1	D2	F1	I1	22.7
XVI	A4	C0	D1	F1	I0	4.6
XVII	A2	C0	Db	F1	I1	—

analysis of patriarchal relationships in human populations. Because of the numerous mutation events the probe can detect, it is also of special interest in studying polymorphisms in small ethnic groups. *Taq* I polymorphism, revealed with probe 49f, was studied in Papuans of the Baruya tribe, living in the Wonenara valley. All the individuals screened are identical at the variable loci A, C, F and I, and fixed for the specific allele Db (Breuil *et al.* 1987). The deduced haplotype, number XVII (A2, CO, Db, F1,I1), is Baruya-specific.

TOWARDS A GENEALOGY OF THE Y CHROMOSOME

Taq I polymorphism can be considered as one aspect of CpG suppression in mammals, in which the high levels of DNA methylation accounts for the CpG deficiency. This implies that, in parallel to the evolutionary loss of CpGs, there should be a loss of *Taq* I sites, and therefore an increase in the size of *Taq* I restriction fragments. Hence, in most cases of *Taq* RFLPs, the smallest allele should represent the original form. As a corollary, restoration of *Taq* I sites must occur rarely, and most of the cases of apparent independent segregation mentioned above should be essentially the result of identical mutations that took place at the same site at different times. All the observed haplotypes could then be derived from another by single mutations.

This observation illustrates the possibility of establishing evolutionary relationships among the various haplotypes. We believe that in most cases

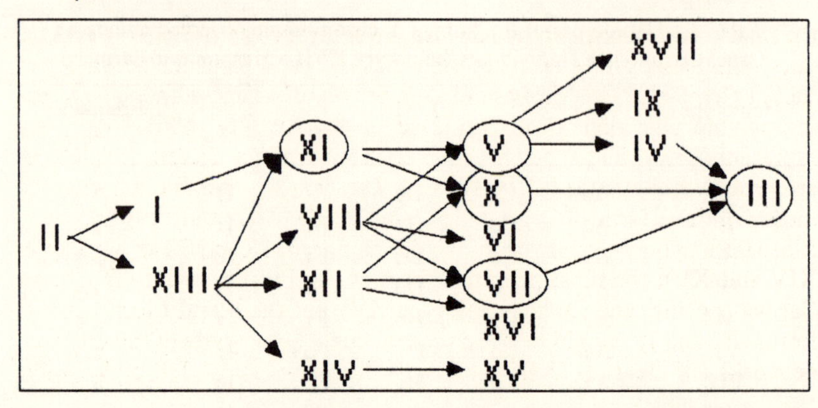

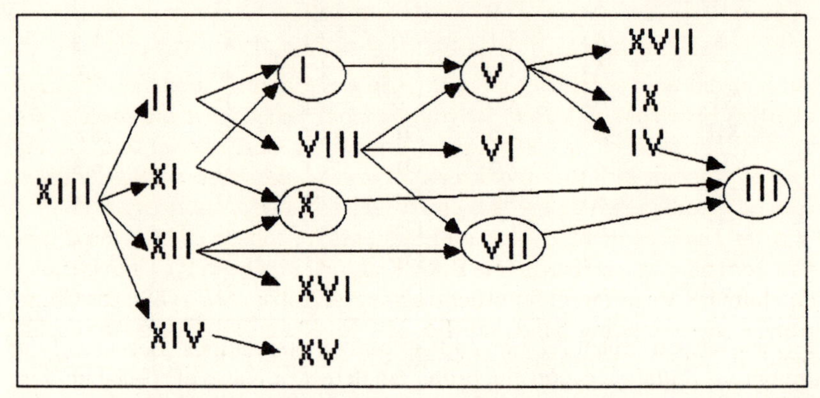

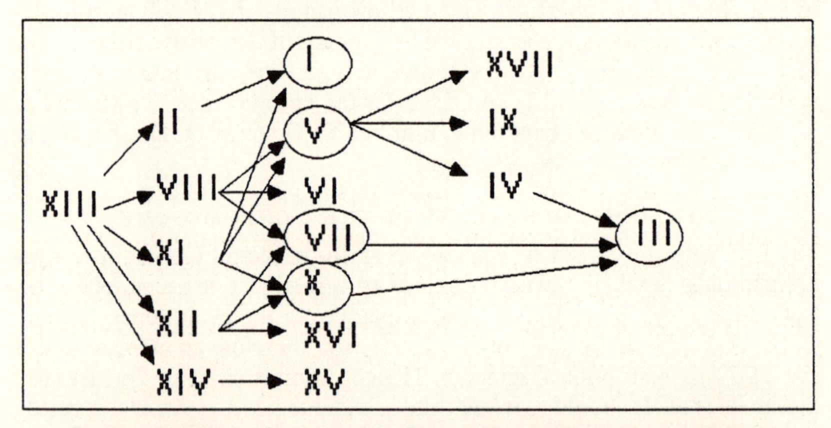

Figure 4.2. Networks for three of the four solutions retained. Filiations between the different haplotypes are shown, hybrids (haplotypes where several arrows converge) being encircled. From the ancestral haplotype (II, or more probably XIII), the different haplotypes are shown under successive columns toward the right of the figure, which indicate successive postulated mutational steps.

the smaller allelic forms at the *Taq* I fragments actually represent the original ones. Simulations have been are used to construct a plausible genealogy of these 17 haplotypes, on the basis of a limited set of basic hypotheses concerning polymorphic allele evolution (Hazout and Lucotte 1986). Figure 4.2 represents networks for the three solutions retained; the ancestral haplotypes are probably haplotype XIII (A3, CO, D1, F1, I1). For the A locus, the most primitive form is A3; for the C locus, CO → C1, but it is necessary to separate Cl between two forms (C1 for haplotype IX and C'l for haplotypes XIV and XV); the most primitive form at the D locus is D1 (D'1 being Baruya-specific, and D2 for haplotype XV); for the F and I loci, existing forms (F1 and I1) would be the ancestral ones. Some parts of the different networks are identical (XIII → XIV → XV; V → IV → III, or V → IV; V → XVII and V → IX).

MALE CHIMPANZEES RETAIN PRIMITIVE ALLELES

49f is anthropoid specific (Ngo *et al.* 1986), as established by hybridization of DNAs restricted with *Bam* HI coming from various primate species. An autoradiographic signal is obtained only with Chimpanzee and Gorilla species, and not with the other species (Baboons, Marmoset and *Microcebus murinus*) studied, which are less closely related to *Homo sapiens*.

After *Taq* I restriction, some 49f fragments are conserved between Chimpanzees and humans (fragments B, G, K, L, M, N, O and R). Chimpanzees and humans are different for other fragments (Abbas *et al.* 1988): the Chimpanzee corresponding bands are Eo, H', Jo, P' and Q'. For polymorphic bands in humans, Chimpanzee has A3, Co, F1 and I1 (Fig. 4.3). Chimpanzee is also polymorphic for bands F' and H, and one individual in the sample has AoD1 (after *Bcl* I restriction, this last individual has also polymorphic for bands 11-8,2 kb).

Consequently, four primitive alleles postulated by the model (A3, Co, F1 and I1) are fixed in Chimpanzee; the Chimpanzee is also polymorphic for D1.

HAPLOTYPE XIII IS THE PREDOMINANT HAPLOTYPE IN PYGMIES

Haplotype XIII, the most primitive one predicted by the model, is preponderant in Aka Pygmies, in which this haplotype represents more than half of the haplotypes present. The fact that this pattern has not been found in a limited sample of other African Blacks suggests that the preponderance of haplotype XIII may be Aka (Pygmy) specific (Guérin, Ruffié and Lucotte, in preparation). The polymorphic variation shown at the 49f locus among Pygmies establishes that at least two haplotypes are also Aka-specific, none of which has been observed so far in any samples from Africa or elsewhere. Confirmation of that will require study of several additional African groups. Although we could speculate on what alternative interpretations could be given depending on the results of these studies, it is clear that the pattern of variation involving the preponderance of haplotype XIII in Akas, and the existence of these new forms, will tell us much about the evolutionary relationships of African and non-African groups.

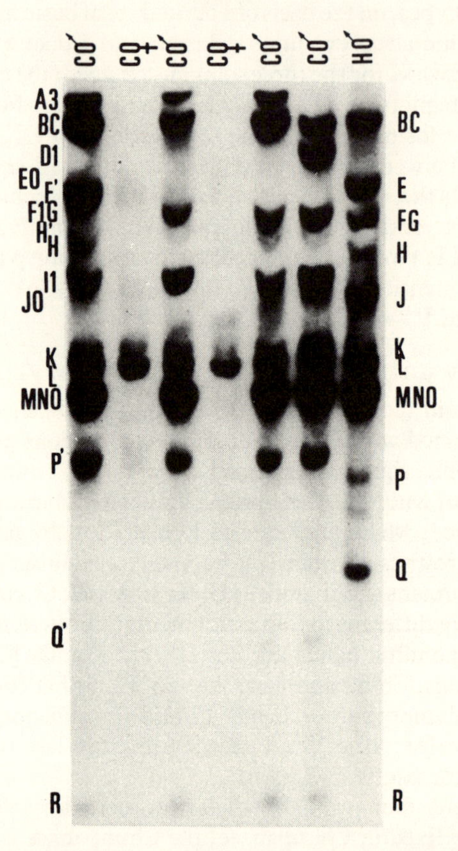

Figure 4.3. Hybridization of the 49f probe with one human DNA (H) and Chimpanzee DNAs (C) restricted with *Taq* I. The man studied here is Ao, Co, Do, F1, Io. *TAq* I bands are indicated for the Man, and corresponding bands for Chimpanzee. For variable bands, the Chimpanzee at left is F'H, and the Chimpanzee at right is AoD1. For other Chimpanzees, variable bands correspond to A3, F1 and I1 (all Chimpanzees are Do, Eo, H', Jo, P' and Q').

GENERAL CONSIDERATIONS

The work presented here represents a first step in the analysis of the paternal ancestry of modern human populations, with the use of Y-specific probes, which might enable a common paternal ancestor for modern humans to be identified. Some consensus is now developing that early forms of anatomically-modern *Homo sapiens* at sites in South Africa (such as the Border Caves, Natal, and at Klasies River Mouth near the Cape) may be older than any similar remains found so far in Eurasia (Rightmire 1982). This is sometimes explained by assuming an earlier split between African and Eurasian human groups. At the molecular level, enzyme data (Nei 1978; Nei and Roychoudhury 1982) favoured an initial split between African populations on the one hand, and all non-African populations on the other hand. Results of analysis of two types of DNA polymorphisms (mtDNA variants and genomic RFLPs in the β-globin region) have been described so far in several ethnic groups. Three major studies of mtDNA variation (Johnson *et al.* 1983; Wilson *et al.* 1985; Cann *et al.* 1987) have shown unexpectedly long branches for African populations. One set of data on polymorphisms in the β-globin region (Wainscoat *et al.* 1986) also indicates a longer branch for Africans, again in agreement with an early separation of African and Eurasian populations. Only by accumulating data over a large number of independent systems will an accurate reconstruction of phylogenies be possible. Samples of two aboriginal Pygmy populations have so far been studied in detail: the Bi-aka from the village of Bagandu, in the southwest corner of the Central African Republic, and the Mbuti from the Ituri forest in northeastern Zaire (Cavalli-Sforza *et al.* 1986). The comparison between Pygmies and Caucasians shows that the greatest differences (among the loci studied) between Africans and Caucasians are for loci like alcohol dehydrogenase 3, anti-thrombin III, and the anonymous locus D4S14.

The conclusion is that the data accumulated so far favour an early split between African and Eurasian populations, but are not sufficient, in isolation, to reach statistical significance. The differences observed, however, suggest that with three or four times as many probes as those studied here (Cavalli-Sforza *et al.* 1986) it may be possible to reach a statistically significant solution. The combined use of mtDNA, autosomal nuclear DNA, and Y-specific DNA markers has a great potential for the genetic analysis of human populations and reconstructions of human phylogeny.

REFERENCES

Abbas, N., Guérin, P., Bessières, P., Ruffié, J. and Lucotte, G. 1988. Y-specific sequences of humans present in Anthropoid Apes. *Biochemical Systematics and Ecology* 16: 105-109.

Breuil, S., Hallé, L., Ruffié, J. and Lucotte, G. 1987. Polymorphism of the p49 Y-specific probe in New Guinea Papuan Baruyas. *Annales de Génétique* 30: 209-212.

Cann, R. L., Stoneking, M. and Wilson, A. C. 1987. Mitochondrial DNA and human evolution. *Nature* 325: 31-36.

Cavalli-Sforza, L. L., Kidd, J. R., Kidd, K. K., Bucci, C., Bowcock, A. M., Hewlett, B. S. and Friedlaender, J. S. 1986. DNA markers and genetic variation in the human species. *Cold Spring Harbor Symposia in Quantitative Biology* 51: 411-417.

Hazout, S. and Lucotte, G. 1986. Towards a genealogy of the Y chromosome. *Annales de Génétique* 29: 246-252.

Johnson, M. J., Wallace, D. C., Ferris, S. D., Ratazzi, M. C. and Cavalli-Sforza, L. L. 1983. Radiation of human mitochondria DNA types analyzed by restriction endonuclease cleavage patterns. *Journal of Molecular Evolution* 19: 255-271.

Lucotte, G. and Ngo, K. Y. 1985. P49f, a highly polymorphic probe, that detects *Taq* I RFLPs on the human Y chromosome. *Nucleic Acids Research* 13: 82-85.

Nei, M. 1978. The theory of genetic distance and evolution of human races. *Japanese Journal of Human Genetics* 23: 341-352.

Nei, M. and Roychoudhury, A. K. 1982 . Genetic relationship and evolution of human races. *Evolutionary Biology* 14: 1-47.
Ngo, K. Y. and Lucotte, G. 1986. Strategies for detecting restriction polymorphisms of Y chromosome sequences. *Annales de Génétique* 29: 88-92.
Ngo, K.Y., Vergnaud, G., Johnson, C., Lucotte, G. and Weissenbach, J. 1986. A DNA probe detecting multiple haplotypes on the human Y chromosome. *American Journal of Human Genetics* 38: 407-418.
Ngo, K. Y., Ruffié, J. and Lucotte, G. 1986. Comparisons between restriction fragments Y-specific between Chimpanzee and Man, and also in other Primate species. *Biochemical Systematics and Ecology* 14: 141-148.
Rightmire, G.P. 1984. *Homo sapiens* in subsaharan Africa. In F. H. Smith and F. Spencer (eds) *The Origins of Modern Humans: a World Survey of the Fossil Evidence*. New York: Alan R. Liss: 295-325.
Wainscoat, J. S., Hill, A. V. S., Boyce, A. L., Flint, J., Hernandez, M., Thein, S. L., Old, J. M., Lynch, J. R., Falusi, A. G., Weatherall, D. J. and Glegg, J.B. 1986. Evolutionary relationship of human populations from an analysis of nuclear DNA polymorphisms. *Nature* 319: 491-493.
Wilson, A. C., Cann, R. L., Carr, S. M., George, M., Gyllensten, B., Helm-Bychowski, K., Higuchi, R. G., Palumbi, S. R., Prager, E. M., Sage, R. D. and Stoneking, M. 1985. Mitochondrial DNA and two perspectives on evolutionary genetics. *Biological Journal of the Linnean Society* 26: 375-400.

5. Molecular Genetics and the Pattern of Human Evolution: Plausible and Implausible Models

SHAHIN ROUHANI

INTRODUCTION

A recent proliferation of data on the human genome, together with recent advances in theoretical population genetics, provide us with a new tool for reconstructing mankind's evolutionary history. There now exist extensive data on various aspects of the human genome (Cavalli-Sforza and Edwards 1966; Nei and Roychoudhury 1982; Nei 1985; Brown 1980; Wainscoat *et al.* 1986; Cann *et al.* 1987; Ngo *et al.* 1986), which illuminate the pattern of genetic relatedness among present-day human populations, and also hint at the pattern of evolution of modern *Homo sapiens*. When combined with the fossil record, a clearer picture of human evolution emerges.

Palaeontologists have proposed two contrasting models for the origin of *H. sapiens*. The adherents of the rapid displacement model maintain that modern humans dispersed throughout the world from a single centre of origin, rapidly displacing their predecessors. However, the geographical location of this centre of origin is disputed (see, e.g. Howells 1976; Cavalli-Sforza and Bodmer 1971; Stringer 1988). The other view postulates the simultaneous emergence of modern humans across the world (Weidenreich 1947; Coon 1962; Van Valen 1966; Wolpoff *et al.* 1984). The supporters of the displacement scenario postulate a rapid dispersal of modern humans. An extreme view is that it took about 40 000 years to cover the Old World (Protsch 1975). This should have given rise to a discontinuity in the morphology of the fossil finds. Whether this is observed is disputed (Wolpoff *et al.* 1984; Stringer 1988).

Population genetics can contribute to this debate in two ways: firstly by ruling out implausible models on theoretical grounds, and secondly by probing the human genome.

Although theoretical population genetics does not have a definitive theory of speciation at its disposal, nevertheless most of the rival theories agree on a few common points. In the first section of this paper I shall argue, from this common ground, that the multiregional model of human evolution is theoretically implausible. It is evident that such a theoretical argument merely serves to decrease our confidence in a particular model rather than disprove it, but in this case theory favours one of the two competing scenarios.

The second prong of the argument is based on the interpretations of the genetic relationships among modern human populations. While the replacement model holds that the racial differences in modern humans have evolved relatively recently in response to regional differences and stochastic events, the multiregional model predicts an archaic origin for the variation among modern human populations. The genetic evidence supports a recent divergence of human races and populations, rather than an archaic origin. Furthermore there is some evidence that mankind has suffered a number of severe population bottlenecks, which implies that modern humans must have been localized in a single region of the world at some stage of their evolution. The molecular evidence implies that this may have been in Africa.

ORTHOGENESIS DOES NOT WORK

The multiregional model of the evolution of hominids was, in its original form (Weidenreich 1947; Coon 1962) based on the theory of orthogenesis, where evolution of isolated populations will take place in a single common direction. Proponents of the modern version of the multiregional model (Van Valen 1966; Wolpoff *et al.* 1984), have recognized that orthogenesis is an implausible mode of evolution (Wolpoff *et al.* 1984: 418), and have sought to bypass the difficulty by proposing a limited amount of gene flow.

Even under ecologically identical conditions – which is rarely the case in nature – geographically-isolated populations will diverge away from each other and eventually become reproductively isolated. Divergence of allopatric populations is believed to be governed by two major trends: stochastic forces arising from random fusions of gametes and mutation, and directional forces such as natural and sexual selection. Both tend to incorporate different genes in separate populations. Even in response to similar selection gradients, different genetical solutions may be found with the same phenotypic end product. A good example of this is convergent evolution, where similar morphological adaptations have evolved in totally different lineages (Futuyama 1979: 140). The large number of different ways in which a genome can change, in response to various evolutionary forces, may be appreciated by noting the large number of configurations which lie one mutational step away from any given configuration of the genome. Higher organisms have about 10^6 genes, and if each locus had at least 2 alleles, the number of alternative available configurations would be $2^{(10)}$, or approximately 10^{1000}! (Gillespie 1984). It is clearly highly improbable that evolution would take identical paths in this multi-dimensional landscape of possible configurations. Although development does impose constraints on evolutionary pathways and thus restricts the number of ways a genome may change, nevertheless an extremely large number of severe constraints would be needed before it would direct evolution along a highly predetermined path, leading to the evolution of the same biological species in separate allopatric populations. Finally, divergence may lead to the formation of co- adapted complexes and reproductive isolation. A co-adapted complex is a set of compatible chromosomes which are not compatible with other such sets. A cross between two populations with different complexes leads to infertility or death of the hybrid (Dobzhansky 1950).

For separate population demes of a species to undergo similar evolutionary changes, as argued by proponents of the multiregional scenario, and maintain compatible co-adapted complexes, it is necessary for them to exchange genes. This is argued to make a wholesale transformation of the demes of one species into another possible (Van Valen 1966; Wolpoff *et al.* 1984), but this depends on the magnitude of gene flow and the geographic range of the species. The multiregional and displacement models of human evolution require very different levels of gene flow among the continental populations. The amount demanded by the multiregional model may be unreasonably high.

GENE FLOW AND PHYLETIC SPECIATION

Exactly how much gene flow is required to suppress differentiation over the range of a continuously distributed species is a complicated problem which has received considerable attention and has recently been reviewed by Slatkin (1985). Gene flow affects all nuclear genes in the same way, but different loci may be under different selective regimes and have different mutation rates. Therefore all loci will not show similar patterns of differentiations and to discuss the problem, in relation to a possible global transformation of *Homo erectus* into *Homo sapiens*, estimates of a number of parameters regarding the demography of *Homo erectus* are needed. If the populations of *H. erectus* were similar in structure to present-day human hunter-gatherers, we could roughly estimate parameters such as population size, density and gene flow by using data available on these modern societies. Such demographic parameters of course depend on culture and language, and it is not clear how the estimated parameters from present-day societies differ from their ancestral values. However, to investigate the plausibility of the phyletic mode of speciation from *H. erectus* to *H. sapiens*, Weiss and Maruyama (1976) have estimated the values of the necessary demographic parameters, using data available at the time. Their estimates are reasonable and I shall use the parameter values used by them. My arguments will remain valid so long as these estimates are not far from their actual values by orders of magnitude.

It is generally believed that *H. erectus* occupied Africa and Southern Asia for roughly one million years, at a steady state of population size (Weiss 1984). The density of hunter-gatherers that an ecosystem can maintain is low. The total population size of *H. erectus* has been estimated at about 1 million individuals, which corresponds to a density of 0.1 individuals per square mile. This is consistent with the density of hunter-gatherers measured in recent times (Braidwood and Reed 1957; Birdsell 1968, 1972). Hunter-gatherers are usually organized into tribal units of roughly 500 individuals (Birdsell 1968). Tribal boundaries are complicated, but to maintain a density of 0.1, a tribe should on the average occupy an area of 5000 square miles. Thus taking the tribe as the unit deme of the *H. erectus* population, there must have been in the region of 10 000 *H. erectus* tribes scattered throughout Africa and Southern Asia. Weiss and Maruyama (1976) assume a rate of gene exchange between adjacent tribes of 5% per generation. This falls within the range of observed rates and seems reasonable, although slightly high. The *H. erectus* population therefore is modelled by an array of demes 30 wide and 300 long, with a population size of N = 500 and surface area of 5000 square miles. Each generation, every deme exchanges 5% (m = 0.05) of its inhabitants with its nearest neighbours. The average generation time for *H. erectus* was probably in the range 20-25 years (Weiss 1984). Within the constraints of this model I shall now discuss the expected pattern of differentiation for advantageous and neutral genes.

ADVANTAGEOUS GENES

Consider the spread of an advantageous mutant after it has been established in a deme. If we assume a diffusion model, this would happen via a wave of advance (Fisher 1937), governed by the diffusion equation:

$$\frac{\partial p}{\partial t} = \frac{\sigma^2}{2} \left(\frac{\partial^2 p}{\partial x^2} + \frac{\partial^2 p}{\partial y^2} \right) + sp(1-p)$$

Where p is the frequency of the advantageous allele, s is the selective advantage of the allele, x and y are the spatial coordinates, and σ is the standard deviation of the parent–offspring distance, $\sigma^2 = m\varepsilon^2$, where ε^2 is the area of the deme. In the *H. erectus* model, $\sigma^2 = 2.50$ square miles per generation.

The wave spreads at a speed of:

$$V = \tfrac{1}{2}\sigma\sqrt{s}$$

which for an allele with selective advantage s =0.01, is

$$V = \tfrac{1}{2}\sqrt{250 \times 0.01} = 0.8 \text{ miles per generation.}$$

With this speed of spread, an advantageous gene would take roughly 20 000 generations, or 400 000 years, to spread from South Africa to the coast of China. This assumes that the selective advantage of the gene remains the same irrespective of the varied environments, and that the rate of gene flow across geographical barriers (such as the Sahara desert) remains equal to the inter-tribal rate of 5%. In addition to these implausible assumptions, for the phyletic model to be possible, one has to add the further assumption that there were no gene interactions among the genes which characterize the differences between *H. erectus* and *H. sapiens*.

The rate at which advantageous mutants are established in a population depends on the rate of mutation and the selective advantages that they impart. The combination of these parameters may be represented by t, the expected time between establishment of advantageous mutants, anywhere in the population. Let us denote by T the time it would take for the wave of advance to spread throughout the population.

A population may accumulate advantageous genes as a coherent unit if $t \geqslant T$. This would mean that a new mutant is established and spreads throughout the range of the species before the next one appears, and that there will be no chance of gene interaction. For the *H. erectus* population, this requires $t \sim 4 \times 10^5$ years, which is far too long. As the *H. erectus* population probably was in a steady state for about 10^6 years (Weiss 1984), such a large value of t implies that there was sufficient time to accumulate only 3 or 4 advantageous alleles in transit from *erectus* to *sapiens*. This is far too few to be reasonable. On the other hand if $t \ll T$, a mosaic of different gene frequencies would form separated from each other by the crests of the waves of advance. But two genes, each of which are separately advantageous, may interact and be deleterious when both are present in an individual. This would then halt the waves of advance at the positions where they meet. This is even more likely to happen if the genes concerned are responsible for reproductive isolation. However, it is not clear whether *H. erectus* and *H. sapiens* were reproductively isolated species.

So far I have tacitly assumed that speciation occurs by gradual accumulation of advantageous genes and gradual divergence of populations. This may not necessarily be the case. However, other modes of speciation, such as the founder effect (Mayr 1942; Carson 1982), or speciation through chromosome rearrangement (White 1978) imply even more local modes of transition and cannot be reconciled with a wholesale transformation of a species with an extensive range.

The phyletic mode of transition is possible and plausible for a species with a limited range or very high rates of migration. The plausible demographic parameters of *H. erectus*, its global range of distribution, low density and low migration make a phyletic mode of speciation to *H. sapiens* implausible. As the range of *H. erectus* was global and spanned many different environments, it is reasonable to assume that clines were present separating the different races of *H. erectus* (Wolpoff *et al.* 1984). This would further hinder gene flow since clines act as barriers to neutral genes (Barton 1983). It is worth noting here that the expected scale of a cline is given by

$$\frac{\sigma}{\sqrt{2S}}$$

where σ is the dispersal rate and s is the selection coefficient (Slatkin 1985). This means that for reasonable selection coefficients (say $s = 0.01$) a cline would be narrow and would include a relatively small number of hybrid individuals. Therefore the chance of detecting such morphological hybrids in the fossil record is very small. The various sub-populations and races of *H. erectus* may have undergone considerable local differentiation and probably did not all coherently speciate into *H. sapiens*. This implies that racial variations observed in human populations must have occurred after speciation from *H. erectus* and therefore are relatively recent, which is in agreement with the findings of the genetic surveys of present-day human populations, as we shall see in the following sections.

THE GENETIC EVIDENCE

The distribution of gene frequencies in present-day human populations contains information on where and when the various human races and populations may have diverged from each other. Genes which are subject to strong selection are of little value to this kind of analysis. In response to natural selection, gene frequencies can change very rapidly, and thus cannot be used to extract information regarding the distant past of the population. Therefore the first step for a phylogenetic analysis is to isolate and survey neutral genes. A number of aspects of the human genome are thought to be under little or no selection. These are the variation in mitochondrial DNA (mtDNA); non-coding regions of nuclear DNA, including the Y chromosome; perhaps protein polymorphisms; and blood-group variation. In the next three sections I shall briefly discuss the implications of some of the findings based on the surveys of each of the above four genetic probes. I shall also discuss some of the disadvantages associated with each method. However, a number of theoretical issues are involved which should be discussed first.

Dendograms are pictorial representations of genetic distances measured from enzyme data, mtDNA, etc. Interpreting a dendogram as a phylogenetic tree, depicting the evolutionary descent of groups of organisms, requires a number of assumptions. The relevance of these assumptions to human populations has been questioned (Wolpoff *et al.* 1984). These authors claim that the use of a single root for a phylogenetic tree is an assumption. But the various groups within a phylogenetic tree are linked in some sense. If these groups are identifiable species, then by appealing to speciation we can postulate that a link must exist, and if the various groups are sub-populations of the same species, they must have been in contact with each other at an earlier stage of their evolution. Admittedly it is possible that the time of contact goes back few speciation events, as claimed by advocates of the phyletic speciation model of human evolution. In either case, only a single root need be postulated for a phylogenetic tree, and then the date of divergence may be estimated using well known techniques (Tajima 1983). Another assumption is that the rate of divergence of various sub-populations remains constant. This assumes a constant rate of mutations and also that severe bottlenecks in population size have not occurred. There is good evidence in support of a constant rate of mutation, as predicted by the molecular clock (Kimura 1983), although this has been questioned (Gillespie 1986). However, population bottlenecks probably have occurred (Brown 1980; Haig and Maynard-Smith 1972; Wainscoat *et al.* 1986; Jones and Rouhani 1986; Cann *et al.* 1987). The assumption of constant population size, where in fact a bottleneck had occurred, results in a longer branch in a genetic distance dendogram. A good parameter for dealing with bottlenecks is therefore divergence time (t) divided by effective population size N_e:

$$\beta = \frac{t}{2N_e}$$

Phylogenetic analysis is accurate and powerful when $\beta < 1$ (Tajima 1983). The races of a polytypic species are expected to have greater inter-racial than intra-racial variance if $\beta < 1$ (Tajima 1983). Indeed the pattern of variation observed within and among human races, using any genetic probe, is consistent with a small value of β. A very large component of the variation in human populations is due to within-population differences (Latter 1980; Jones 1981).

When analysing mitochondrial DNA (mtDNA), the relevant parameter is:

$$\beta mt = \frac{t}{N_f}$$

where N_f is the effective number of females. This is due to the haploid and maternal mode of inheritance of mtDNA. A further assumption concerns isolation of the groups within a phylogeny. It has been pointed out by a number of authors (Weiss and Maruyama 1976; Wolpoff *et al.* 1984) that gene flow must have existed between various human races and populations, and this would cause the phylogenetic analysis to underestimate the time of geographic divergence. Although this is undoubtedly true, for small amounts of gene flow ($2Nm < 1$) and relatively short times of divergence $\beta < 1$), the error introduced by ignoring gene flow is small (Wright 1941; Lande 1975; Weiss and Maruyama 1976). Given the extent of systematic error usually present in such analyses (Tajima 1983) this will not be a serious shortcoming, and populations which are separated by continents could be considered effectively isolated from each other. However the extent of inter-continental gene flow and its theoretical implications remain unknown.

I have simulated the effect of long-range gene flow on the genetic distance, on a computer. Consider two networks of population demes, in which each network has the same set of demographic parameters as used in the preceding section. But the two networks exchange genes only at their boundary. The evolution of a single neutral locus was simulated, through 1000 generations, in each of the populations, and their genetic distance was calculated using the Cavalli-Sforza and Edwards (1967) measure. By repeating the same simulation but with no gene flow between the two networks, we can see that a reasonable amount of inter-network gene flow does not seriously affect the genetic distance or divergence (Fig. 5.1). But gene flow by interdemic migration assumes that the global population is in equilibrium, and that no long-range migrations take place. However, long-range migrations do take place, although at a much lower rate (Slatkin 1985). Modelling such long-range gene movements is theoretically complicated and its effects remain unknown.

MITOCHONDRIAL DNA TREES

Mitochondria are cytoplasmic organelles with their own genetic coding system. Mitochondrial DNA is a single, circular strand of DNA, which is inherited maternally as the transmission takes place via the cytoplasm, and which shows no recombination. Mitochondrial DNA is generally believed to be under very mild selection, and it has a much higher mutation rate than nuclear DNA. It is therefore an ideal probe for constructing family trees. The disadvantage of mtDNA for exploring patterns of human migrations lies in its uniparental mode of inheritance, since it only reflects the pattern of migrations of females, which may give a false impression in the case of a highly asymmetrical pattern of migration. But a study of the Y chromosome variation, which has a paternal mode of inheritance, also points to Africa as the origin of modern humans (Hazout and Lucotte 1986). The Y chromosome study involved a sample of only 49 individuals which is too small to be conclusive, but it is consistent with the other surveys (see Lucotte *et al.*, this volume).

A number of surveys of human mtDNA, taken from wide-ranging populations, have been performed (Brown 1980; Johnson *et al.* 1983; Cann *et al.* 1987). All the surveys reveal a similar pattern of variation in mtDNA, but Johnson *et al.* (1983) have differed in their interpretation of the data.

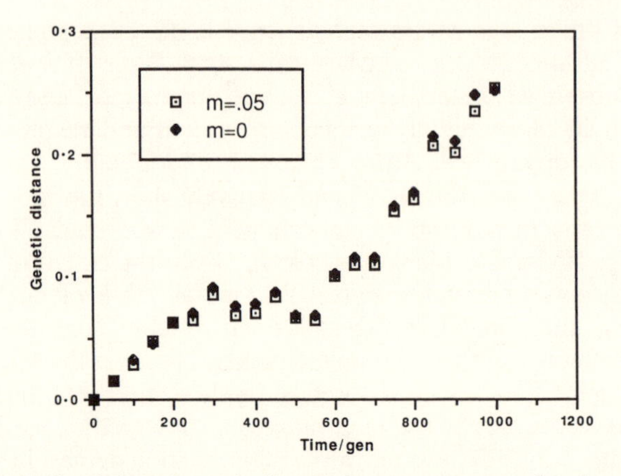

Figure 5.1. Simulation of genetic distance between two networks of populations in the presence of gene flow. The genetic distance is measured once every 100 generations using the Cavalli-Sforza and Edwards (1967) measure. A reasonable amount of gene flow (m = 0.05) results in roughly the same sort of distances as when there is no gene flow (m = 0).

The most comprehensive survey (Cann *et al.* 1987) involved 147 individuals from five geographically distinct populations. They sampled an average number of 370 sites and found 134 mtDNA types. They found that *inter*population variance was much smaller than the *intra*population variation. This finding is supported by almost all the other data sets, indicating that human races have recently diverged from each other (approximately 100 000 years since the Negroid-Mongoloid split: Nei 1985). This is in clear contradiction to the multi-regional theory of human evolution which predicts an ancient divergence of human races (Wolpoff *et al.* 1984).

A family tree of mtDNA types indicates that Africans form a separate branch, that all the major human groups contain many of the main branches of this tree, and that the mtDNA lineages in modern human populations are all descendants of a common female ancestor. According to Cann *et al.* (1987), our female ancestor lived 140-290 000 years ago in Africa. This is the most parsimonious assumption and minimizes the number of intercontinental migrations required to explain the resulting evolutionary tree. They interpret the multiple branches in each human group as a sign of multiple colonization, although it may also have come about by existing variance in the initial population that performed the first invasion. Their results point to two probable population bottlenecks in the course of human evolution. The first bottleneck probably occurred at the very early stage of human evolution – pointing to a single female ancestor of all mankind – and a second bottleneck probably occurred at the onset of invasion of the rest of the world. The finding of a single female ancestor is not in itself surprising. By chance, some female members of a population will not have female offspring, and hence these mtDNA types will be lost. This process will eventually lead to a population composed of a single mtDNA type together with its point mutations (Fig. 5.2a). In fact, the end result will not be different from a population that has suffered a severe bottleneck down to

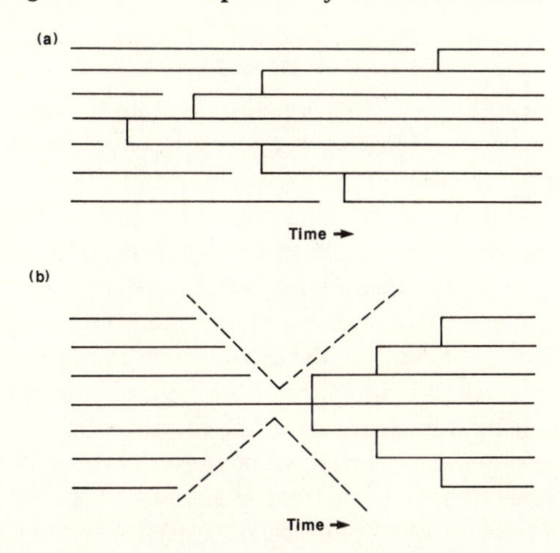

Figure 5.2. Mitochondrial DNA types may be lost in two ways: (a) through the action of drift, in which one type slowly replaces all others; (b) all mtDNA types except one are lost due to a drastic population bottleneck. The two end results would be indistinguishable, were it not for point mutations.

a single reproductive pair (Fig. 5.2b) (Wainscoat 1987; Lattore *et al.* 1986). But what is significant is the time at which our female ancestor lived. The expected time for fixation of a mtDNA type in a population of fixed size is $2N_f$ generations, where N_f is the effective number of reproductively active females. On the other hand, by calculating the rate of divergence of mtDNA and the amount of variation present we can also date the origin of divergence. This was done by Cann *et al.* (1987) and found to be roughly 200 000 years. Assuming a 20 year generation time for *H. sapiens* we can estimate the female populations of the world over the past 200 000 years to be about $N_f \simeq 5000$! This is clearly much less than the number expected (i.e. approximately 10^6 individuals). Therefore either our mtDNA type was fixed (or perhaps common) before the species line was crossed, or a bottleneck in population size occurred before womankind left her birth place. A correlated fixation of the same type in distant parts of the world, after migration to distant parts of the globe, would be highly unlikely. A similar population bottleneck is required to explain the existence of an ancestor to all non-African populations (ancestor 'c' in Cann *et al.* 1987). There is some evidence in support of both of these bottlenecks from other genetic surveys, (Brown 1980; Haig and Maynard-Smith 1967; Wainscoat *et al.* 1986; Jones and Rouhani 1986). A bottleneck at the onset of migration out of Africa is also consistent with a larger variation present in African populations ($\delta = 0.47$) as opposed to the smaller variation among Caucasians of $\delta = 0.23$ (less than half!).

Among the molecular data presently available, mtDNA is most convincingly supportive of the population displacement scenario. The larger amount of variation present in the African population, the rooting of the mtDNA tree in Africa, and a recent date for the divergence of the races, all support the displacement scenario with an origin in Africa.

NUCLEAR DNA TREES

Another window into genetic relationships among human populations is offered by studying restriction-fragment-length polymorphisms of the non-coding regions of the nuclear DNA. The phenotypic effect of the non-coding regions is unclear as they are probably not under direct selection and therefore are good candidates for phylogenetic analysis. However, linkage to nearby selected genes may cause a selective effect to bear on the non-coding regions. Also, unlike mtDNA, nuclear DNA recombines and therefore one must take care to study regions of low recombination rate. All of this makes phylogenetic analysis based on nuclear DNA less reliable than mtDNA, but it does not suffer from a uniparental mode of inheritance.

Restriction-site polymorphisms have been found throughout the human genome (e.g. Jeffreys 1979; Wainscoat *et al.* 1986; Casanova *et al.* 1985). Wainscoat *et al.* (1986) studied patterns of relatedness among eight human populations using five closely related linked restriction sites in the β-globin gene cluster. A striking geographical pattern emerges. Human populations are once again divided into two major groups – Africans and non Africans. Wainscoat *et al.* (1986) suggest that the pattern of distribution of the β-globin haplotypes is consistent with a spread of modern man from a centre of origin in Africa, and this passage to the rest of the world led to a drastic shift in haplotype frequencies because of a population bottleneck and sampling drift in the immigrant population. If the restriction sites are assumed to be unlinked to strongly selected genes (but see Hill and Wainscoat 1986), and thus are strictly neutral, it is then possible to estimate the size of the bottleneck necessary to give rise to the observed pattern of the haplotypes.

Let us concentrate on the most common haplotypes, assuming that the rare types have arisen since the divergence of various populations. There exist three common haplotypes in Eurasia which are not present in Africa, and the common African type is not present in Eurasia. Assuming that all four common haplotypes were present in equal frequencies among the ancestral African populations, the African type must have been lost in transit, due to intense drift resulting from a severe population bottleneck.

The expected time in units of population size, for this to take place is given by (Kimura and Ohta 1969):

$$\frac{t}{2N} = \frac{p \ln p}{1 - p} \simeq 0.46$$

This is consistent with an effective population size of four reproducing individuals for four generations, or roughly 80 years (Jones and Rouhani 1986). This leads to a total population size of about 20 individuals, if we assume that only a fifth of the population is reproductively active at any one generation. There are a number of problems with this interpretation. Firstly, the bottleneck is too severe to the believable, as small populations can become extinct. Secondly, the reverse scenario – i.e. an origin in Eurasia and migration to Africa (Ambrose and Giles 1986) – will also fit the data with a bottleneck parameter:

$$\frac{t}{2N} = 0.85$$

However, this is an even more severe bottleneck. The β- globin data cannot on its own be taken as evidence of migration out of Africa, but it is consistent with the other genetic surveys and theoretical expectations. Further studies of the restriction sites of nuclear DNA have been proposed (Higgs *et al.* 1981) and in particular a polymorphism of the Y chromosome (Casanova *et al.* 1985; Hazout and Lucotte 1986) may provide some information on the pattern of male migration.

ENZYMES AND BLOOD GROUPS

Surveys of spatial variation in enzyme polymorphisms and blood groups were the first to be used for phylogenetic analysis (Latter 1980; Nei and Roychoudhury 1982; Nei 1985; Jones 1986).

The most striking result is that 84 per cent of the variation present in protein polymorphism results from the differences among individuals of the same nationality or racial group (Latter 1980). This is a clear indication that racial and national divergences are recent events in human evolution, and therefore supports the displacement scenario. Furthermore, the number of loci involved in this analysis is large enough to make a dating of the divergence time feasible (Nei 1986). However, the blood group loci give a different picture to the protein loci: the latter suggest that the Negroid and the Caucasoid-Mongoloid groups diverged about 100 000 years ago, and this was followed by a divergence in the Caucasoid and Mongoloid line about 40 000 years ago. The blood group data on the other hand support an initial split of the Negroids from Mongoloids, with the Caucasians appearing later (Nei 1985). The blood group analysis seems to support a single centre of origin but places this in Asia (Cavalli-Sforza and Bodmer 1971). Much of this confusion may arise from the action of natural selection, which is known to be present on some human cell surface and enzyme variants (Mourant *et al.* 1978), and which will confuse their history.

DISCUSSION

I have attempted to show that the data on molecular divergence among human races, together with theoretically plausible models of human evolution, combine to increase our confidence in the 'rapid displacement' model for the emergence of modern humans. Undoubtedly a foolproof case for the rapid displacement theory and against the 'parallel evolution' scenario cannot be made, since construction of theoretical models and interpretation of the molecular data require demographic parameters for the *Homo erectus* population structure, which is no longer available for observation. Nevertheless, assuming that the *H. erectus* population structure was similar to that of modern hunter- gatherers, it is argued that the range of *H. erectus* was too large for the phyletic model to be plausible. The genetic surveys of different aspects of the human genomes produce similar genetic distances among human populations, and these have been argued to be consistent with the displacement theory and existence of population bottlenecks during the early stages of *H. sapiens* evolution. The other aspect of genetic surveys which supports the displacement scenario is the question of 'race'.

For political and cultural reasons, human racial differences have received undeserved attention for a long time. Attempts to define races in terms of a genetic basis have continually run into difficulty, clearly indicating that 'race' is an ambiguous notion when applied to *H. sapiens* (Latter 1980; Jones 1981). The failure to connect the apparent morphological differences to a more persuasive difference among human races in itself proves to be a clue to the mode of human evolution. The fact that individual differences far exceed national and racial differences is indicative of a recent time of divergence of racial groups, which is in turn consistent with the rapid displacement model.

The estimates of the time of divergence, of 40 000 years for the Caucasian-Mongoloid split and 110 000 years for the Mongoloid- Negroid split (Nei 1986) are based on the assumption that no migration took place between the groups over the past 100 000 years. This we know not to have been the case. According to the neutral theory, the great majority of evolutionary changes at the molecular level are caused by the action of random drift on neutral mutants (Kimura 1983; but also see Gillespie 1986). Random sampling drift, caused by random fusion of gametes, has a uniform effect on all nuclear genes. The effects of sampling drift are greater in smaller populations, so that for reasonable population sizes the effects of drift are only appreciable on neutral genes or genes which are under very mild selection. Under the action of drift, different neutral alleles are incorporated into geographically isolated populations which gradually diverge away from each other, but gene flow can pull them back towards each other again. Wright (1941) has shown that, provided two panmictic populations exchange only a few individuals, they will diverge away from each other. The critical amount of gene flow for this to be the case is $2 N_e m < 1$, where N_e is the effective population size and m is the fraction of migrants exchanged. If a fifth of the tribal population is at the reproductive age, this would give:

$$2 N_e m = 2 (100) (0.05) = 10.$$

This level of gene flow is sufficient to ensure uniformity among the demes of a well integrated tribal network (Weiss and Maruyama 1976), but it does not say anything about the extent of divergence between two tribal networks, whose only contact may be via a small number of interconnecting demes. However, my simulations indicate that with a reasonable amount of inter- population gene flow, and no long-range movements of populations, genetic distance can be a fairly good indicator of divergence time, over time scales shorter than the effective population size. Overall, the rapid displacement scenario (Cavalli-Sforza and Bodmer 1971; Stringer 1989) with a centre of origin in Africa, seems to be more plausible than the multiregional model (Coon 1962; Van Valen 1967; Wolpoff *et al.* 1984). The predictions of the rapid displacement model are consistent with both the molecular evidence (Nei 1985; Wainscoat *et al.* 1986; Cann *et al.* 1987), and the fossil record (Stringer 1989) and they fit a plausible mathematical model. One is hard pushed to make a similar case for multiregional models.

However, the problem is far from being resolved, and the extent of intercontinental gene flow and its implications remain unresolved. But, as information on the human genome is rapidly increasing, it will undoubtedly continue to sharpen our understanding of our origins.

ACKNOWLEDGEMENTS

I wish to thank Nick Barton, Steve Jones and Chris Stringer for many helpful comments and discussions. This work was financed by Science and Engineering Research Council grant GR/C/91529.

REFERENCES

Barton, N. H. 1979. Gene flow past a cline. *Heredity* 43: 333-339.

Birdsell, J. B. 1968. Some predictions for the Pleistocene based on equilibrium systems among recent hunter-gatherers. In R. B. Lee and I. Devore (eds) *Man the Hunter*. Chicago: Aldine: 229-249.

Birdsell, J. B. 1972. The problem of the evolution of human races: classification or clines? *Social Biology* 19: 136-162.

Braidwood, R. J. and Reed, C.A. 1957. The achievement and early consequences of food production: a consideration of archeological and natural historical evidence. *Cold Spring Harbour Symposia in Quantitative Biology* 22: 19-31.

Brown, W. M. 1980. Polymorphism in mitochondrial DNA of humans as revealed by endonuclease analysis. *Proceedings of the National Academy of Sciences (USA)* 77: 3605-3609.

Cann, R. L., Stoneking, M. and Wilson, A. C. 1987. Mitochondrial DNA and human evolution. *Nature* 325: 31-36.

Carson, H. L. 1982. Evolution of *Drosophila* on the newer Hawaiian islands. *Heredity* 48: 3-27.

Casanova, M., Leroy, P., Boucekkine, C., Weisenbach, J., Bishop, C., Fellous, M., Purrello, M., Fiori, G. and Siniscalco, M. 1985. A human Y-linked DNA polymorphism and its potential for estimating genetic and evolutionary distance. *Science* 230: 1403-1406.

Cavalli-Sforza, L. L. and Bodmer, W. F. 1971. *The Genetics of Human Populations*. San Francisco: Freeman.

Cavalli-Sforza, L. L. and Edwards, A. W. F. 1965. Analysis of human evolution. In S. J. Geerts (ed.) *Genetics Today: Proceedings of the XIth International Congress of Genetics:* 923-933.

Cavalli-Sforza, L. L. and Edwards, A. W. F. 1967. Phylogenetic analysis: models and estimation procedures. *American Journal of Human Genetics* 19: 233-257.

Coon, C. S. 1962. *The Origin of Races*. New York: Knopf.

Dobzhansky, T. 1950. Origin of heterosis through natural selection in populations of *Drosophila psuedoobscura*. *Genetics* 35: 288-302.

Fisher, R. A. 1937. The wave of advance of advantageous genes. *Annals of Eugenics* 7: 355-369.

Futuyama, D. J. 1979. *Evolutionary Biology*. Massachusetts: Sinauer Associates.

Giles, E. and Ambrose, S. H. 1986. Are we all out of Africa? *Nature* 322: 21-22.

Gillespie, J. 1984. Molecular evolution across the mutational landscape. *Evolution* 38: 116-129.

Gillespie, J. 1986. Rates of molecular evolution. *Annual Review of Ecology and Systematics* 17: 637-666.

Hazout, S. and Lucotte, G. 1986. Towards a genealogy of the Y-chromosome. *Annual of Genetics* 29: 246-252. (in French).

Higgs, D. R., Goodbourn, S. E. Y., Wainscoat, J. S., Clegg, J. B. and Weatherall, D. J. 1981. Highly variable regions of DNA flank the human alpha-globin genes. *Nucleic Acids Research* 9: 4213–4224.

Hill, A. V. S. and Wainscoat, J. S. 1986. The evolution of the α- and β-globin gene clusters in human populations. *Human Genetics* 74: 16-23.

Howells, W. W. 1967. Explaining modern man: evolutionists *versus* migrationists. *Journal of Human Evolution* 5: 477-496.

Haig, J. and Maynard-Smith, J. 1972. Population size and protein variation in man. *Genetics Research* 19: 73-89.

Jeffreys, A. J. 1979. DNA sequence variants in the G, A, δ and β-globin genes of man. *Cell* 18: 1-10.

Johnson, M. J., Wallace, D. C., Fernis, S. D., Rattazzi, M. C. and Cavalli-Sforza, L. L. 1983. Radiation of human mitochondrial DNA types analysed by restriction endonuclease cleavage patterns. *Journal of Molecular Evolution* 19: 255-271.

Jones, J. S. 1981. How different are human races? *Nature* 293: 188-190.

Jones, J. S. 1986. The origin of *Home sapiens*: the genetic evidence. In B. Wood, L. Martin and P. Andrews (eds) *Major Topics in Primate and Human Evolution*. Cambridge: Cambridge University Press: 317-330.

Jones, J. S. and Rouhani, S. 1986. Human evolution: how small was the bottleneck? *Nature* 319: 449-450.

Kimura, M. and Ohta, T. 1969. The average number of generations until fixation of a mutant gene in a finite population. *Genetics* 61: 763-771.

Latorre, A., Moya, A. and Ayala, F. J. 1986. Evolution of mitochondrial DNA in *Drosophila subobscura*. *Proceedings of the National Academy of Science (USA)* 83: 8649-8653.

Latter, B. D. H. 1980. Genetic differences within and between populations of the major human subgroups. *American Naturalist* 116: 220-237.

Lande, R. 1979. Effective deme size during long term evolution estimated from rates of chromosomal rearrangement. *Evolution* 33: 234-257.

Mayr, E. 1942. *The Growth of Biological Thought*. Harvard (Mass): Belknap.

Nei, M. 1985. Human evolution at the molecular level. In T. Ohta and M. Kimura (eds) *Population Genetics and Molecular Evolution*. Tokyo: Japan Scientific Press: 41-64.

Nei, M. and Roychoudhury, A. K. 1982. Genetic relationship and evolution of human races. In M. K. Hecht, B. Wallace and G. T. Prace (eds) *Evolutionary Biology* Vol. 14. New York: Plenum: 1-59.

Ngo, Y. K., Vergnaud, G., Johnsson, C., Lucotte, G. and Weissenbach, J. 1986. A DNA probe detecting multiple haplotypes of the human Y-chromosome. *American Journal of Human Genetics* 38: 407-418.

Protsch, R. 1978. *Catalogue of Fossil Hominids of North America*. New York: Fisher.

Slatkin, M. 1985. Gene flow in natural populations. *Annual Review of Ecology and Systematics* 16: 393-430.

Stringer, C. 1989. Documenting the origin of modern humans. In E. Trinkaus (ed.) *Patterns and Processes in Later Pleistocene Human Emergence*. Cambridge: Cambridge University Press. In Press.

Tajima, F. 1983. Evolutionary relationship of DNA sequences in finite populations. *Genetics* 105: 437-460.

Van Valen, L. 1966. On discussing human races. *Perspectives in Biology and Medicine* 9: 377-383.

Wainscoat, J. S., Hill, A. V. S., Boyce, A. L., Flint, J., Hernandez, M., Thein, S. L., Old, J. M., Lynch, J. R., Falusi, A. G., Weatherall, D. J. and Clegg, J. B. 1986. Evolutionary relationships of human populations from an analysis of nuclear DNA polymorphisms. *Nature* 319: 491-493.

Wainscoat, J. S. 1987. Human evolution: out of the Garden of Eden. *Nature* 325: 13.

Weidenreich, F. 1947. The trend of human evolution. *Evolution* 1: 221-236.

Weiss, K. M. 1984. On the number of members of the genus *Homo* who have ever lived, and some evolutionary implications. *Human Biology* 56: 637-650.

Weiss, K. M. and Maruyama, T. 1976. Archeology, population genetics and studies of human racial ancestry. *American Journal of Physical Anthropology* 44: 31-50.

White, M. J .D. 1978. *Modes of Speciation.* San Francisco: Freeman.

Wolpoff, M. H., Wu Xinzhi, and Thorne, A. G. 1984. Modern *Homo* sapiens origins: a general theory of hominid evolution involving the fossil evidence from East Asia. In F. H. Smith and F. Spencer (eds) *The Origins of Modern Humans: a World Survey of the Fossil Evidence.* New York: Alan R. Liss: 411-438.

Wright, S. 1941. On the probability of fixation of reciprocal translocations. *American Naturalist* 75: 513-522.

6. Multiregional Evolution: The Fossil Alternative to Eden

MILFORD H. WOLPOFF

INTRODUCTION

In 1976 W. W. Howells reviewed the evidence concerning the origin or origins of modern populations, and used the description 'Noah's Ark' hypothesis for one possible explanation – the contention that all living populations have a single recent origin from a source population that was already modern. In developing this hypothesis Howells proposed that once these modern populations appeared, they spread rapidly and replaced indigenous populations over the inhabited world. The 'Noah's Ark' hypothesis is not a classic punctuational scheme; the modern populations may not be a different species according to Howells' interpretation, and the population divergence may not have been at the ecological *or* geographic periphery of the human range, as the punctuational model requires. Howells suggested 'Noah's Ark' as one of several possibilities for the explanation of modern population origins, and was unsure as to which was correct.

In the past ten years, interpretations of new genetic evidence have been brought forward to modify the 'Noah's Ark' hypothesis into a much more focused statement of a rapid punctuational event at the origin of modern *Homo sapiens*. The interpretations of several different lines of genetic evidence are said to combine to suggest that modern populations originated within the last 200 000 years, all descending from a common mother who lived in Africa and gave rise to a small population of a *new species* (cf. Stringer and Andrews 1988) with severely reduced genetic variability in their mitochondrial DNA (mtDNA). This explanation of the limited mtDNA variation is based on an assumption of recent common genetic ancestry (see reviews by Cann *et al.* 1987; Wainscoat 1987; and Jones 1986) and depends on the assertion that the descendant populations spread throughout the inhabited world and replaced the indigenous inhabitants without admixture, which is why they must be considered a new species. Therefore, combining as it does the contention of African origins, the descent from a single Eve (the mitochondrial mother common to all modern populations) and the subsequent spread of modern peoples 'out of Africa', this interpretation of the genetic data might best be called the 'Garden of Eden' hypothesis.

The main elements of the Garden of Eden hypothesis are that modern humans *arose recently* from a small source population, approximately 200 000 years ago in Africa, and that this population was a *new species*. It is contended that this new species split into two main branches in Africa and that one of these passed through a bottleneck of small population size in the process of emigration from Africa, and rapidly replaced indigenous human populations in other parts of the world without mixture.

Cann *et al.* (1987: 35) claim that that there is one interpretation of the human fossil record that fits the Garden of Eden hypothesis, and similar contentions have been offered by other geneticists. The purposes of this paper are threefold: to critically examine these interpretations of the genetic data, and their alternatives; to present and discuss evidence from the human fossil record which indicates that the palaeontological data *do not* fit the Garden of Eden hypothesis; and to indicate alternative interpretations of the fossil record in the context of multiregional evolution theory and in accordance with a different consistent set of interpretations of the genetic evidence. Human fossils are absolutely critical in the ongoing research on modern population origins. The human fossil record has the potential to provide a valid independent basis for attempting to refute the contention that all living populations have a single recent common origin as a new species.

THE GARDEN OF EDEN HYPOTHESIS

Invasion without Admixture

For the Garden of Eden interpretation of the mtDNA variation to be correct, the absence of admixture must necessarily characterize the relations between the invading populations and the indigenous natives they presumably replaced. This interpretation requires that these populations cannot have interbred at all, or the older mitochondrial lines would be found in living populations. Presumably when modern humans left their place of origin, this population must have been reproductively isolated from the indigenous natives they were replacing in order to account for the lack of mixture between Eve's mtDNA line and the more ancient mtDNA lineages that is claimed in the Garden of Eden interpretation (Gould 1987). Since successful human invaders would be expected to incorporate at least the females of the native populations into their societies, the inability to have fertile offspring is a critical assumption in this interpretation. Therefore, the Garden of Eden interpretation requires that these invaders are a different species. Further, the observation of what was taken to be a unique African mitochondrial lineage has been interpreted to show a lack of gene flow after the ancestors of modern populations left the continent (otherwise this lineage would not be unique to Africa). If true, this would mean that any subsequent world-wide evolutionary changes in the mtDNA must have happened in parallel, and it is unlikely that there were many of these. *It follows that according to the Garden of Eden hypothesis Eve was in a population that was either directly and uniquely ancestral to modern Homo sapiens, or itself was the earliest modern Homo sapiens population.*

The evidence said to support the hypothesis that the appearance of modern populations required the origin of a new species and its subsequent rapid geographic dispersal is not based on genetic data alone. It also incorporates evidence from the fossil record for ascertaining the details of 'who, what, when, where, and why' for the earliest modern humans (cf. Stringer and Andrews 1988). This is because if modern humans are a new species with a recent origin, both the time and the place of this origin can be discovered by finding the earliest modern specimens. It is just such evidence that Cann *et al.* cite in support of their contentions. In particular, data concerning the 'when' and 'where' are critical for this interpretation.

When and Where are the Earliest Moderns?

A good deal of effort has been expended in seeking the *earliest* remains of any fossils with the modern human form because according to the Garden of Eden hypothesis they will be discovered at the place of origin of all modern populations – *a presumption which assumes the hypothesis to be tested, namely that modern populations have a single migrational origin.* The 'when', in other words determines the 'where' according to this hypothesis. Over the years, the candidate for the earliest modern human has changed considerably, and therefore so has the place of origin for modern *Homo sapiens.* Cautiously, in initially stating his 'Noah's Ark' hypothesis, Howells (1976) took no position as to where this source area might have been, but others have specifically suggested sub-Saharan Africa (Protsch 1975, 1978; Bräuer 1984a, 1984b; Stringer 1984a, 1984b, 1985, 1989; Wainscoat *et al.* 1986; Cann, Stoneking and Wilson 1987), the Levant (Vandermeersch 1970, 1981) or more generally (as the new Qafzeh dates suggest: Valladas *et al.* 1986) Western Asia (Howell 1951; Bodmer and Cavalli-Sforza 1976), China (Weckler 1957, Chang 1963; Macintosh and Larnach 1976; Denaro *et al.* 1981), Australia (Gribbin and Cherfas 1982), or even more unexpected places (Hogan 1977).

However, an important aspect of the Garden of Eden hypothesis is the specific evidence purported to reveal Africa as this place of origin. Some of this evidence is genetic (Wainscoat *et al.* 1986; Jones and Rouhani 1986a; Cann *et al.* 1987: but see Denaro *et al.* 1981; Johnson *et al.* 1983; Giles and Ambrose 1986). Morphological 'evidence' for an African origin relies on the presence of fossil human remains of modern *Homo sapiens* from this continent *with early dates* since there is no basis for the contention that there is anything plesiomorphic about the morphology of fossil or living Africans that would cause one to believe that they specifically reflect the ancestral condition for all living populations on the basis of their morphology alone. Unfortunately, the question of which African fossils are earliest is confused by what seem to be endless problems involved in the accurate determinations of how long ago the individuals concerned died.

There are many potential African Eves, or at least many fossil remains that are possible members of Eve's immediate family (Rightmire 1981), or mitochondrial lineage (see below). These fossils are said to demonstrate an African origin of modern populations because they represent the earliest appearing populations to closely resemble modern humans from any region. Yet the dates for these Africans are riddled with problems. In fact, of the sub-Saharan sites in question – Klasies River Mouth, Border Cave, and

Omo Kibish – *not one single specimen thought to be an 'early modern' has a defensible radiometric date*.

In the case of Border Cave it is not even clear that most of the adult specimens have a provenience! Stringer (1989), who generally supports the interpretation of modern populations migrating out of Africa, expresses caution about the dates claimed for the Border Cave specimens because of the provenience issue and the possibility of burial from a more recent level for the infant and adult mandible (see also Klein, this volume). As Rightmire, in his review of the situation at Border Cave, puts it (1979: 26):

> With the exception of a new adult mandible. . . all of the adult skeletal material was dug out of the cave by Horton and its original position in the deposits was not directly verified (p. 25) . . . The course of evolution outside of southern Africa cannot be determined from the evidence considered here.

As far as the Klasies River Mouth Cave specimens are concerned, the published dates are based on faunal correlations, attempts to relate the cave fauna to coastal faunas, and attempts to relate the coastal faunas to the oxygen-isotope-based sea-core chronology. These widely quoted 'dates' actually stand a good chance of being incorrect (Binford 1984, 1986) and the age of most of the specimens may be one half of what is generally assumed, and may correspond to the appearance of 'modern' populations in other regions. Moreover, it is important to consider exactly which specimens these dates are associated with. According to Stringer (1989) the more recent dates do not apply to three specimens: the mandible KRM 21776, the KRM 16425 frontal fragment, and a newly discovered maxilla. These three specimens are morphologically archaic. The frontal has a vertically thick although non-projecting superciliary arch that is not unlike that of Florisbad in its central portion, lack of nasal root depression, and low nasal profile. The mandible is extraordinarily robust for a 'modern' find from any geographic region, let alone from Africa where gnathic reduction is early. It is *not* 'modern' by any reasonable set of morphological criteria. Stringer describes the maxilla as edentulous and very robust. In sum, because this *earlier* sample from the Klasies cave *is not modern*, the older dates reported from the base of the Klasies sequence are not obviously relevant to the problem of *modern* population origins.

The Omo radiometric dates have been continuously disputed ever since their first publication because radiocarbon determinations based on shells are notoriously inaccurate, and recent Uranium/Thorium dates are problematic. Various faunal and stratigraphic 'dates' have been suggested as replacements for the radiometric estimates (Day 1972; Stringer 1989) and according to these the age of the three fossil humans could range between 40 000 and 130 000 years. However, which of the various date estimates may be correct cannot be established, and the fact is that there is no particular reason to accept any of them as valid!

There is one other aspect of the allegedly earliest modern human specimens from these three sites that must be taken into account with regard to the validity of the Garden of Eden hypothesis. Specimens from these sites are fundamentally *African* in their morphological details (Rightmire 1979, 1984a; Wolpoff 1980; Bräuer 1984b). The basis of this claim is in the numerous comparisons I have had the opportunity to examine (for instance, see Wolpoff 1980); I make this statement knowingly, and in spite of Day and Stringer's (1982) multivariate assessment of their Omo 1 reconstruction. I do not believe this multivariate assessment is valid. While the Day and Stringer analysis concluded that of the living populations they chose for comparison, Omo 1 is most like the Norse sample (although the difference from other comparisons was not judged significant), unfortunately – while there are published data – neither living nor subfossil East Africans (cf. Robbins 1974, 1980) were included within the comparative samples.

The important fact is that even if we assume the 'early' sub-Saharan sites are dated correctly and that the specimens associated with the dates can validly be regarded as modern *Homo sapiens* (but see Wolpoff 1986), the fact is that these three sites preserve the remains of the earliest modern *Africans*, morphologically as well as geographically. If they truly represent the earliest modern populations from any region, they establish the presence of an African morphological complex in moderns prior to their leaving Africa, as is required by the Garden of Eden hypothesis. The Eve of the genes, of course, need not have been in a modern *Homo sapiens* population. Yet, even if pre-modern, the fossil specimens themselves suggest that the population she presumably lived in must eventually have evolved into the African version of modern humanity *before* expanding to replace other populations within Africa and beyond (Stringer and Andrews 1988). No matter how long ago Eve lived, the 'no admixture' requirement of the Garden of Eden hypothesis requires that the migration out of Africa was *by a new species, of morphologically-modern Africans*.

This, however, creates a contradiction. Throughout the history of attempts to find the origin of the *earliest* modern *Homo sapiens* populations, in each place early modern remains are found their skeletal remains invariably have been recognized to exhibit the unique characteristics common to that region. Thus, for instance, in Australasia Niah and Mungo 1 closely resemble modern Australians; in China Liujiang specifically resembles modern south Chinese, and the similarities of the Zhoukoudian Upper Cave remains to modern North Asians are equally clear; and in Europe the Aurignacian Europeans such as the Mladeč specimens uniquely resemble modern Europeans. However, according to the Garden of Eden hypothesis and following from the discussion above, the earliest modern Africans were already distinctly African in their morphology (see Bräuer 1984a, 1984b). Therefore, because it is claimed that the ancestors of these modern populations left Africa as a new species, according to the Garden of Eden Hypothesis the characteristics of modern populations in each other region outside of Africa must have developed *in transit*, since by the time the populations arrived to presumably replace the indigenous inhabitants *they already were distinctly regional in morphology*.

Out of Africa: the Mitochondrial Connection

The most dramatic (and most highly publicized) statements of the Garden of Eden hypothesis are based on an analysis of mtDNA variation (Cann, Stoneking and Wilson 1987). From the study of 147 people, these geneticists concludes that most human fossils dated earlier than the late Pleistocene have nothing to to with human evolution, because all living people have a common African origin and left Africa 100-200 000 years ago, as discussed above. The basis for this 'out of Africa' hypothesis, with its migration date, is the interpretation that the existing mitochondrial lineages reflect the consequences of a tree-like divergence network from a single common mother (or a few mothers with the same mitochondrial type: see Gould 1987) that lived at this time. An unavoidable implication is that all indigenous (presumably *Homo erectus*) populations that were not ancestral to this African Eve were replaced by the new species without admixture. Since this would presumably include some of the earlier African fossil hominids and surely all of the fossil hominids found outside of Africa dated earlier than 200 000 years ago, the classic text by the late Wilfred Le Gros Clark, *The Fossil Evidence for Human Evolution*, could, on revision, be shortened dramatically.

The main interpretations said to be 'explained' by the hypothesis of total replacement from the Garden of Eden are based on assessments of population relationships as determined from trees of genetic or morphological information which are interpreted to show a common recent origin for human populations (Edwards 1971; Cavalli-Sforza *et al.* 1964; Cavalli-Sforza and Edwards 1965; Nei and Roychoudhury 1982; Howells 1973; Guglielmino-Matessi *et al.* 1979; Jones 1981). These trees are used to show relationships based on shared features, and account for variation as the result of population splits and divergences from a common ancestor, with subsequent differences between the branches a consequence of random independent mutations.

However, there is a fundamental question as to whether trees based on genetics or morphology are actually useful for determining the relationships between human populations, let alone relevant to ascertaining the timing of their divergences (Weiss and Maruyama 1976; Morton and Lalouel 1973; Harpending 1974; Howells 1976). In fact, it has been suggested that branching analyses do not provide any insight into the reconstruction of population history (Livingstone 1973; Wolpoff, Wu and Thorne 1984). This is because branching analysis necessarily assumes that population differences arose from a common ancestry through population splitting and continued isolation. For the analysis to be valid the branching must be recent (i.e. the bush must be shallow-rooted), the similarities on the branches must come from descent (i.e. must be shared with the stem population), and the branches can only be connected at their points of divergence (a prerequisite that is more likely to be valid when the branches are different species and not likely to be accurate when they are different populations within the same species). Unfortunately, just as a correlation analysis will provide a 'number' even when comparing apples to oranges, a branching analysis will provide branches to the form of the structure assumed to underlie population relationships whether or not these actually characterized population histories.

Interpretation of the branching pattern as clusters of relationships, and calibration of these associations in terms of evolutionary differences, assumes (1) that the differences are the consequence of constantly accumulating random mutations and drift, and (2) that gene flow did not occur. However, commonalities in selection may cause populations to appear more similar than their actual histories might suggest (Livingstone 1980, but see Cohan 1984). Even a small amount of gene flow between two populations will greatly reduce the observed magnitude of population differences and consequently minimize the time estimated to have passed since population splitting (Weiss and Maruyama 1976). Conversely, gene flow between one population in a recently-diverged sister pair and an 'outside' population will make the population pair less closely related genetically. However, are they less closely related *phylo*genetically? They will certainly appear to have diverged from each other earlier than was actually the case (Weiss 1986).

The obvious historic fact is that there have been numerous invasions, and a marked rate of gene flow between human populations since the end of the Pleistocene. These have affected every human population on the planet. All populations, therefore, should appear genetically and morphologically to be more closely related than they might actually be, if the analysis of the populational relationships assumes a splitting model of populational divergences. Another Holocene phenomenon that affects modern genetic variability in unknown ways is the demographic instability of the past two millennia, with numerous population replacements, ubiquitous admixture, and dramatic population expansions.

There are some additional problems in the genetic analyses supporting the Garden of Eden hypothesis. Geneticists discussing the palaeontological data have gleefully reported on the lack of agreement among the palaeoanthropologists, when claiming that at least one set of palaeoanthropological interpretations fit their own data. It is therefore appropriate here to point out that the disarray among the geneticists over the interpretation of the mtDNA data provides the same opportunity for a palaeoanthropologist! One set of contradictory interpretations emerges from a consideration of what should have been a much less ambiguous determination than modern human population divergences – the separation of human and chimpanzee lineages. The problem spotlighted by this analysis involves the *rate* of mtDNA divergence in different lineages. The 2%-4% divergence rate for mtDNA assumed in the Cann *et al.* analyses give a 1.4-2.8 million year divergence estimate (that is, a 2.1 million year mean divergence date) for human and chimpanzee lines (Saitou and Omoto 1987). Using similar techniques, Hasegawa, Kishino and Yano (1985) earlier derived an estimate for the chimpanzee-human split of 2.7 ± 0.6 million years ago. In attempting to explain their surprising result, Hasegawa *et al.* admit there might be some problems in this estimated splitting time because of the earlier dates known for *Australopithecus afarensis*. Instead of regarding this species as a 'dental hominid', as Sarich once described the australopithecines dated earlier than his divergence-date estimate, these authors propose that mtDNA passed across species boundaries between australopithecines and chimpanzee ancestors. This would account for the 'too recent' divergence determined by their method, since one consequence of interspecies mtDNA transfer is to make the species appear less diverged than they actually were. Hasegawa *et al.* note (1985: 171):

> If interspecies transfer of mtDNA between proto-human and proto-chimpanzee did indeed occur, it is tempting to speculate in which direction the transfer occurred. The lesser intraspecies polymorphism of human mtDNA compared to that of chimpanzees. . . suggests that the transfer occurred from proto-chimpanzee into proto-human.

Therefore, because of the potential for inter*species* mtDNA transfer, the rate determination may be incorrect, and this estimate of the chimpanzee-human split time may be too recent. But what is the real rate? The rate that gives an estimated date of 2.1 million years for the chimpanzee-human divergence is associated with other inconsistent or too-recent split dates. For instance, Cann *et al.* (1987: 33) determine their base substitution rate of 2%-4% per million years for modern populational divergences from 'known' dates of migration that include 30 000 years ago for the peopling of New Guinea, 40 000 for Australia (a surprising difference since these were the same continent at this time), and 12 000 for the New World. These dates, at the low end of the ranges presented by Stoneking, Bhatia and Wilson (1986), are far too recent even by conservative estimates, perhaps by as much as 50 per cent of the real value. Therefore, in the Cann *et al.* study the mutation rate has been overestimated, just as it is overestimated in the determination of the chimpanzee-human divergence at 2.1 million years. The reason for overestimation could be the same as for the chimpanzee-human split – i.e. gene flow between populations – but if so in this case the error is quite likely to have been much greater since the gene flow was among populations within the same species (the error in the chimpanzee-human divergence estimate is by a factor of 2-3 times).

Perhaps an error of this magnitude is to be expected. Divergence rates calculated per base-pair per million years vary between 0.5% and 2.0% in primate studies by different researchers (reviewed by Honeycutt and Wheeler 1987). The *fastest* of these is used as the *slowest* rate in the discussions of divergence times by Hasegawa, Cann and their colleagues.

In fact, there is a convergence in the estimates of how 'too recent' these chimpanzee-human split-time determinations may be. Based on palaeontological data I believe there is a maximum divergence date estimate of no more than 8 million years, and a minimum estimate of no less than 5 million years. Other estimates of this divergence are also quite different. Independently, Nei (1985, 1987) derives an mtDNA divergence rate of 0.71%, *between 2.8 and 5.6 times slower than the estimates used by Cann and her colleagues*. This slower rate suggests divergence times much more in line with the palaeontological (as well as other) estimates. Thus, as calculated from this rate, the Nei estimate gives a chimpanzee-human divergence determination of 6.6 million years, a date that markedly contrasts with the other mtDNA estimates (Hasegawa *et al.* 1985; Brown *et al.* 1982) but conforms to the palaeontological data and is very similar to that determined from nuclear DNA hybridization data (6.3 million years as ascertained by Sibley and Ahlquist 1984).

The base substitution mutation rate of 0.71% per million years suggests a divergence time for modern populations of approximately 850 000 years. At so early a time this divergence would be among *Homo erectus* populations. Mixture with indigenous populations would not be an issue, since this would probably represent the first hominid migrations out of Africa. Therefore, it need not have been a speciation event, and in any case at this date – at the beginning of the Middle Pleistocene – it could not possibly represent the origin of *modern Homo sapiens*.

If this estimated rate is a reasonable alternative to the 2%-4% rate of mtDNA divergence used by Cann and her colleagues, and others, it dramatically effects the dates widely reported in newspapers, popular magazines (Gould 1987) and scientific journals (Cann *et al.* 1987; Stringer and Andrews 1988) for modern human populational divergences. *One wonders how much more of an underestimate is involved in the mt*DNA *calibration of population-splitting within Homo sapiens* because of inter*population* mtDNA transfer?

According to the assumptions required by the mtDNA analysis discussed above, there was no interpopulation transfer at all, since the expanding population was a different species (see Stoneking and Cann, this volume). If this was the case, the rate differences that are seen when comparing the slower rate calculated from the widely accepted 5-8 million year chimpanzee-human divergence with the faster rate suggesting recent modern population origins (i.e. Eve) are inexplicable. But if the Garden of Eden hypothesis is incorrect because modern populations are *not* a new species, admixture could explain the differences in calculated rates. Moreover, a reasonable alternative interpretation of the mtDNA data presented by Nei also indicates a much slower divergence rate, and is supported by the nuclear DNA hybridization data. This slower rate indicates an estimated divergence time for modern human populations that approximates the Early-Middle Pleistocene boundary, a time when an increasing number of recent estimates (reviewed by Wolpoff and Nkini 1985) suggest *Homo erectus* might have first left Africa. This 'revised' estimate of a divergence time based on the mitochondrial data also fits a divergence estimate based on independent evidence for other faunal migrations into Eurasia (Turner 1984).

The more ancient estimate of human populational divergences is also in accord with the multiregional evolution hypothesis (Wolpoff, Wu and Thorne 1984) which proposes that human geographic variation dates to the initial emergence of *Homo erectus* from Africa. In fact, the only argument for not accepting the more ancient estimate for modern human populational divergences based on the slower rate of mtDNA evolution is that even at this age the time of divergence may be too recent for an accurate determination based on mtDNA analysis (Nei 1987; Honeycutt and Wheeler 1987). In sum, the slower (0.71% per base-pair per million years) mtDNA divergence rate determination is corroborated by other genetic data (nuclear DNA hybridization), and palaeontological data indicating a 5-8 million year chimpanzee-human divergence and a 0.75 to 1.0 million year date for the *Homo erectus* expansions out of Africa.

The contention that an actual divergence date can be determined from genetic data rests on two assumptions: that the main source of mtDNA lineage differences (i.e. variation) is mutations; and that the rate of mutation accumulation is constant (linear) for a finite period of time (presumably including the population divergences in question). But what direct evidence exists for rate constancy over this time? This 'evidence' may not be as good as is generally supposed. A study of mtDNA variation in Amerindians (Wallace, Garrison and Knowler 1985) revealed marked differences between Amerind tribes, important differences between the Amerindians and Asians, and the apparent retention of rare Asian variants in some of the Amerindian mtDNA lineages. Adhering to the assumption of a constant mutation rate, the authors account for these observations by presuming there were numerous founder events during the colonization of the New World, some of which established extremely rare Asian variants in Amerindian populations, *and that the living tribes studied each represent a separate migration from Asia some 20-40 000* BP. In my view, a much more parsimonious explanation would be an inconsistent and erratic mutation rate for this period, or the presence of selection acting on the mtDNA – *a possibility that cannot be discounted* given independent data that suggests that there is significant selection against some mtDNA variants (Hale and Singh 1986; Saitou and Omoto 1987).

In all, the assumption of rate-consistency for mtDNA evolution in human populations over this time span is very problematic. In particular, it is far from clear that a date estimate for human populational divergences is even possible using the technique, especially if these divergences were recent. Because of the slow rate and stochastic nature of base-pair substitutions, it is far more likely that the calibration of recent events will engender significant error than the calibration of more ancient events. Thus, of the two determinations that potentially do not agree, the calibration of recent migrations (such as used in the rate determinations by Cann, Stoneking and their colleagues) is far more problematic than rate determinations based on the divergence of chimpanzee and human lineages.

Apart from differences in divergence rate estimates, there is some independent evidence to support the contention that population mixing has significantly altered the interpretation of mtDNA variation. Mixing is ubiquitous in the historic record of human population movements, regardless of how different the physical characteristics or cultural variations of the populations might be. The effects of mixing are to invalidate branching analysis, and to confuse both the clusters of relationships and their magnitudes based on mtDNA variation. Thus, for instance, in an analysis of the mtDNA-based genetic distances presented by Cann and her colleagues, the New Guineans seem to be most closely related to Africans and only distantly related to Australians on the resulting phylogenetic trees (Saitou and Omoto 1987). Yet, by virtually every other measure (including nuclear DNA analysis) New Guineans and Australians are the same people, only separated since the end of the Pleistocene. Clearly, mixing takes place in non-human populations as well. In fact, it is the analysis of mtDNA itself that provided evidence of significant hybridization between two deer species, white-tailed and mule deer (Carr *et al.* 1986). A study of two closely related mice species with ranges that overlap in southern Denmark showed that while the nuclear DNA of each species penetrates only a few kilometres into the range of the other, the mtDNA variants of one are widespread within the other throughout Scandinavia (Ferris *et al.* 1983). A tree analysis based on mtDNA for these two species would be totally misleading (Jones 1986).

Moreover, one can ask whether *any* rate for recent populational divergence can be accurately measured. In a recent paper on rate determination based on the peoplings of New Guinea, Australia, and the Americas (Stoneking, Bhatia and Wilson 1986) the potential sources of error presented (and discounted) include three that in reality contribute considerable uncertainty to the determination because they reflect assumptions that are almost certainly in part or even as a whole incorrect (these assumptions are discussed in Bryan (ed.) 1986; Greenberg, Turner and Zegura 1986; Gruhn 1987; Kirk and Szathmary (eds) 1985; Kirk and Thorne (eds) 1976; Stewart 1974; Szathmary and Ossenberg 1978). Stoneking and his colleagues point out that the rate determination is based on the assumption that there were not multiple colonization events. However, the genetic, archaeological and linguistic evidence for multiple colonization events in the peopling of both greater Australia and the Americas is incontestable. A second assumption is that there was no appreciable back-migration from these areas. For North America, at least, this is unlikely to be correct. Finally, accurate colonization dates must be assumed. A review of the immense literature on this specific point suggests that the assumption of date accuracy is not supportable. My 25 years of reading the literature on this topic shows this assumption to be unrealistic, at best. A recent paper reviewing mtDNA variation concludes (Honeycutt and Wheeler 1987):

> The effectiveness of the molecular data to provide information on divergence times depends on a proper point of calibration as well as a demonstration that the molecules are evolving in a clock-like manner. When dealing with divergence times involving periods of less than one million years, the calibration must be accurate and errors small if meaningful estimates are desired. *Not only is there considerable disagreement as to the identification and date of key fossils, but the error or range of time estimates provided by both nuclear gene loci and mt*DNA *lead one to place little confidence in the dates* (my italics).

There are alternative interpretations of what the observed pattern of mtDNA variation reveals about evolutionary history. For instance, in a study of *Drosophila subobscura* (Latorre, Moya and Ayala 1986), a species whose colonization of the New World involves known times and places of origin, an analysis of the mtDNA variation 'provides no clue of the precise geographic origin of the colonizers'. Nor, in fact, does it provide a realistic estimate for the age of the Eve of the flies. It is possible that this is because there was no Eve, and that there is a different explanation for the observed variation.

> Today's world population of *D. subobscura* consists of many millions of individuals. It might well be the case that, a few hundred thousand years hence, all *D. subobscura* flies have mtDNAs derived from morph I. That would not mean that the mtDNA of the descendants derives only from one *D. subobscura* currently living – morph I is found in 44% of the living population. More importantly, the individuals living in that remote generation would count among their ancestors not only those females from which they inherited their mitochondria, but also innumerable other females and males from which they inherited their nuclear hereditary material (Latorre, Moya and Avise 1986: 8652-8653).

How could this be? It is clearly possible that differential lineage survivorship, rather than singular recent common ancestry for maternal lineages, is the cause of the limited mtDNA variation *Drosophila*, as well as *Homo* (Avise, Neigel and Arnold 1984). Probability models with stochastic survivorship assumptions show that virtually all existing mitochondrial lineages will become extinct, even in a stable population. With selection against some of the genome, and the variations in population sizes that almost certainly occurred, the process would be even more rapid. According to calculations by Avise *et al.*, a population *founded* by 15 000 *unrelated* females would have a 50 per cent chance of *appearing* to descend from a single female within 18 000 generations as a consequence of stochastic lineage extinctions in a stable population. After this time all of the *mitochondrial* lines would be traceable to a single female, although *contra* Gould (1987: 18), this much later population would have *multiple* nuclear DNA descent, possibly from most of the founders. There need not have been a mtDNA-nuclear DNA link through a 'killer' population spreading. It is interesting that 18 000 generations may represent perhaps as few as 300 000 years for humans, given the current estimates of a short lifespan for pre-modern humans (Trinkaus and Thompson 1987). This time-frame is similar to the Cann *et al.* estimate under a branching and replacement model. However, it is an overestimate if the likelihood of selection (Hale and Singh 1986; Whittam *et al.* 1986) and the virtual certainty that prehistoric human populations were much smaller (Weiss 1984) are taken into account. Therefore, a model of selection and differential lineage extinctions from an earlier population with equally limited mtDNA variation could account for today's limited DNA without recourse to assuming that there was a small original founding population of a new human species.

In a model much better fitting the fossil record, concordant with a number of interpretations of its evolutionary pattern (Weidenreich 1946; Wolpoff *et al.* 1984), female exchanges between indigenous populations of pre-modern humans, combined with a constant rate of stochastic lineage extinctions, may well have resulted in a homogeneous distribution of a few mtDNA lineages across widely spread human populations in the past (Saitou and Omoto 1987). This ancient distribution of genetic polymorphisms, perhaps itself the consequence of a previous long span of differential stochastic lineage extinctions, could have provided, by chance, a single surviving mtDNA line for descendant populations as the result of continued lineage extinctions. Therefore, the pattern of today may simply show random survivorship from what earlier was a single very common widespread mitochondrial lineage and would thereby not reflect a limited nuclear DNA source. A common origin for maternally-cloned cytoplasms can be completely independent of a common origin for nuclei, just as the descent of family names can be independent of genetic inheritance. The Late Pleistocene provides more than sufficient time to allow for the possibility that selection and stochastic lineage extinctions could account for the limited mtDNA variation reported by Cann *et al.* and others. If so, the dates estimated for population divergences from this variation, the discussions about a place of origin for 'Eve', and the evidence for bottlenecks in human evolution (see below), are all without obvious meaning.

Bottlenecking: a Different Foundation for the Garden of Eden Hypothesis

The second source of a genetic argument for population replacement comes from recent discussions of bottlenecking. The idea that a bottleneck occurred during the foundation of *Homo sapiens* developed in part as a consequence of these same mtDNA studies (Brown 1980; Wilson *et al.* 1985). The bottleneck involved a period of small population size and is presumably marked in living populations by reduced genetic variability. The data supporting this interpretation derive in part from the low level of mtDNA variability reported within several human populations, and a nuclear DNA analysis for the beta-globin gene cluster (Wainscoat *et al.* 1986, and this volume). The beta-globin data were interpreted to show a basic split between all African and non-African populations through a genetic distance analysis based on the 14 genotypes observed for this cluster. Moreover, the loss of what Wainscoat and his co-workers regard as the 'common' African genotype for this cluster in all non-African populations is regarded as evidence for genetic drift due to small population size (Wainscoat *et al.* 1986: 493) that presumably took place at the time that the populations split and *Homo sapiens* left Africa.

In a review article, Jones and Rouhani (1986a) bring together the mtDNA and beta-globin gene-cluster data to attempt an estimate of how small the bottleneck was. The size of the bottleneck is determined from the opportunity for drift and therefore is related to the length of time between leaving Africa and dispersing through the rest of the world. Jones and Rouhani provide us with an '*informed guess*' of 20 000 years. For the bottlenecked population (the ancestral group for all living non- African peoples) the estimated mean population size for a bottleneck of 20 000 years length is given as 600, or alternatively as 6 individuals for 200 years, or in the most blatantly stated Garden of Eden interpretation, a *single couple for 60 years*.

There are more difficulties with the bottlenecking argument than the biblical interpretation might suggest. Evidence now suggests that bottlenecks reduce fitness (Bryant, McCommas and Combs 1986) – hardly what one would expect in the population history of a new species that was so competitively superior to existing indigenous populations (of the old species) that it was able to rapidly replace them.

Moreover, the mtDNA bottleneck and the beta-globin bottleneck *cannot have been the same event*. The mtDNA bottleneck occurred during a speciation event and is used as evidence of modern *Homo sapiens* descending from Eve, or from a very small population. But if there was a split between a branch of the African and the non-African populations all of which remained within the same species, as the beta-globin evidence is taken to indicate, how could this split also be the *speciation event* required for the origin of modern populations by the Garden of Eden hypothesis? Therefore these events cannot be the same! The evidence for bottlenecking taken from the beta-globin analysis is often given as support for the recent African origin of modern humanity interpretation of mtDNA variation. This is incorrect. The two arguments contradict each other and therefore cannot both be valid, if in fact either is.

The beta-globin analysis also has its set of internal problems (Van Valen 1986). For instance, the divergence tree (and its calibration) is based on a genetic-distance analysis for ten populations in which three European populations (British, Cypriot and Italian) are analysed separately while two African populations (one a mix of East and West Africans, and the other Nigerians from three tribes) are lumped together. This lumping makes the African data seem more distinct from the rest of the world than might actually be the case. The ancestral haplotype for the five sites studied in the beta-globin gene cluster is yet to be identified, and therefore directionality of the changes cannot be determined (Honeycutt and Wheeler 1987). The African distinction is even more confused by Wainscoat *et al.*s treatment of the discovery of what they regard as the unique African genotype for the beta- globin gene cluster *outside of Africa*. In fact, *both* of the African genotypes also appear in the Melanesian sample where, it is asserted, they are homoplasies. But there are no data presented to show that these have the 'independent non-African origin' claimed for them, and evidence in this case is absolutely critical because with both of the so-called 'unique' African genotypes found *outside* of Africa (and, suspiciously, in the same population), there is no longer any evidence for a bottleneck, let alone the basis for an estimate of when it might have occurred.

In sum, the bottlenecking interpretation of the beta-globin data is beset with a number of factual and interpretative difficulties, and contradicts the mtDNA interpretation discussed above, even though proponents of each of these tend to quote the other for support. There are similar citational circularities between the mtDNA analysis literature and the palaeoanthropological literature (Eckhardt 1987), in which geneticists quote the conclusions of papers written by palaeoanthropologists about the time of modern populational origins while in these papers the palaeoanthropologists quote the same geneticists for support. It is quite possible that *neither* interpretation of genetic data is correct. Perhaps the geneticists Jones and Rouhani were premature in claiming (1986a: 449): 'the main lesson to be learned from palaeontology is that evolution always takes place somewhere else'.

REGIONAL CONTINUITY

The Fossil Evidence

In spite of all the above considerations, if the interpretations of the genetic data suggesting there was a fairly recent single origin for modern human populations are considered correct, then this explanation must apply to all *populations.* There can be no continuity between archaic and modern populations except in sub-Saharan Africa, where ironically the morphological evidence for continuity between archaic and more modern populations is poorest (Thorne 1981; Thorne and Wolpoff 1981; Rightmire 1976, 1981, 1984b). The Garden of Eden hypothesis requires the conclusion that the archaeological and morphological evidence for continuity in Australasia (Weidenreich 1946; Thorne and Wolpoff 1981; Jelínek 1982), North Asia (Weidenreich 1943; Wolpoff, Wu and Thorne 1984; Wolpoff 1985), North Africa (Ferembach 1979; Jelínek 1980, 1985), and Central Europe (Jelínek 1969, 1976, 1978, 1985; Smith 1982, 1984, 1985; Wolpoff 1982a) is incorrect. And indeed, some of those authors who support the interpretations of the genetic data discussed above recognize this and dismiss all these fossils (Cann *et al.* 1987; Gould 1987; Rightmire 1987; Stringer and Andrews 1988), although never with a discussion or refutation of the detailed morphological data presented by the fossil evidence which *supports* the regional evolution hypothesis.

However, because any convincing fossil evidence for morphological continuity outside of Africa refutes the Garden of Eden hypothesis (Eckhardt 1987), the fossil data especially at the peripheries provides the only potential refutation of the Garden of Eden hypothesis. At the same time, evidence of regional continuity at the peripheries would support the multiregional evolution hypothesis (Wolpoff, Wu and Thorne 1984; and see below) which is totally at odds with the Garden of Eden interpretation. Thus, there can be either a Garden of Eden interpretation for modern population origins, or a substantial non-African fossil record of Early and Middle Pleistocene hominids showing morphological evidence of continuity with living populations, but not both.

Continuity and Admixture

In this context, it is important to re-emphasize that evidence for admixture is evidence *against* complete replacement of one population by another (cf. Bräuer 1980, 1981, 1982), and therefore is evidence against the Garden of Eden hypothesis. Historically, *complete* replacement of one human population by another is an 'accomplishment' *that has not ever been possible,* even for invading populations as technologically 'advanced' as the Europeans in Tasmania, since a large number of individuals showing admixture persist on that island. If in spite of the European technological advantage replacement was incomplete in Tasmania, how could complete replacement be expected to have characterized the interaction of two different groups of hunter/gatherer populations spread widely across Europe? Whatever the case for the extreme west of Europe, the fact is that apart from a very few exceptions (Protsch 1975; Stringer 1984a, 1989; Stringer and Andrews 1988), among the palaeoanthropologists even the most ardent believers of the replacement hypothesis do not contend that the replacement in Central Europe was without significant mixture.

For instance, while Bräuer argues that modern populations arose first in southern Africa, and over the past 50 000 years spread through the rest of the world (1984a, 1984b), he also interprets the morphology of the remains of early 'modern' Europeans such as the Hahnöfersand frontal *as the consequence of 'hybridization'* between the local Neanderthals and the invaders. One might add that the persistence of this morphology even into the Holocene (see Schwalbe 1904; Weinert 1951) could lend additional confirmation to this interpretation. Bräuer is led to this interpretation because of the obvious transitional characteristics of Hahnöfersand, and the fact that such characteristics are logically either the result of in situ change (an unacceptable hypothesis for Bräuer (1982), or of hybridization between the indigenous Neanderthals and the invading populations.

Invasion *with* hybridization, however, is no longer the Garden of Eden hypothesis as described here because this would mean the invading and indigenous populations were in the same species. It is for this reason that I find it curious to observe that both Cann and Stoneking, in various publications, each quote Bräuer's palaeoanthropological interpretations in support of the Garden of Eden Hypothesis when in fact if this interpretation was correct it would actually *invalidate* the hypothesis. Admixture between invading and indigenous populations would invalidate the interpretation of the genetic evidence discussed above because it would result in the introduction of more ancient mitochondrial lineages into the ancestry of modern populations. The observation of more ancient mitochondrial lineages could then be regarded as evidence for a much more ancient and not necessarily African origin for modern populations. *So ancient a history of population divergences would firmly establish that when humans left their place of origin, the ancestors of the indigenous populations around the inhabited world were not modern.* Moreover, admixture would confuse any attempt to ascertain a date for modern population origins from the mtDNA data.

The Pattern at the Eastern Edge

I believe that because the compelling evidence for regional continuity in East Asia is ancient, beginning with the earliest inhabitants, this evidence disproves the Garden of Eden hypothesis no matter what date is assumed for *modern* populational origins. The regional interpretation of human evolution in East Asia is not new. The question of evolutionary continuity between archaic and modern populations outside of Africa was first approached by Weidenreich. By the end of the 1930s Weidenreich had accumulated first hand experience with the human fossil record of three regions. These were Europe (where he was trained and did his early work) North Asia (as the result of the rapidly accumulating Zhoukoudian finds from both the Lower and the Upper Cave); and Southeast Asia (as a consequence of communications and an exchange of casts with G. H. R. von Koenigswald, followed finally by an exchange of visits). Weidenreich was the only palaeoanthropologist of his time (and one of the very few of any time) with a detailed knowledge of three regions. As a scholar well educated in the morphological and evolutionary traditions of Central Europe, this put him in a unique position to appreciate the evidence for a world-wide pattern of human evolution. Weidenreich found such a pattern, and to explain it he developed his theory of polycentric evolution.

According to Weidenreich's polycentric interpretation (1939, 1943) all fossil hominids belong to a single species, *Homo sapiens* (this original 'single species hypothesis' first appears clearly stated in his 1943 monograph). He contended that there was no single centre of evolution from which new hominid types appeared from time to time to replace older ones. Instead, according to his interpretation there were at least four centres of origin: Asia Minor, Eastern or Southern Africa, north China, and the Sunda islands (perhaps today better referred to as the Sunda subcontinent because of the extensive land areas exposed during the glaciations). He argued that no fossil group or type could be excluded from the ancestry of recent hominids, and that racial differences (which he regarded as 'minor details') were as old as human evolution.

Weidenreich's different geographic lines shared a common ancestry and evolved in the same direction. The crux of the problem he faced was in how to explain this. Here he failed. In the mid 1940s he proposed an explanation based on orthogenesis (1947), just at the time when the foundations for the new evolutionary synthesis were being laid by Huxley, Mayr, Simpson, and others – a synthesis which denied any role for orthogenesis.

While the polycentric *interpretation* was widely dismissed when its orthogenic explanation failed, there remained his *observations of continuity* which have never been refuted when accurately considered (despite the claims made by Stringer and Andrews (1988), based on highly selective quotations of Weidenreich's observations). Today the fossil records of these regions are much better than Weidenreich could have imagined. Within these two areas, the Weidenreich interpretations of human fossil remains have been taken up by local scholars, intimately acquainted with the palaeontological materials of their regions and particularly concerned with the origins of the populations that inhabit them now.

Australasia: the Southern End

From the Australasian perspective, links between the Sangiran *Homo erectus* specimens, the later hominids from Ngandong, and finally the recent inhabitants of Kow Swamp and Coobool Crossing in Australia have been recognized by most authors, although there are exceptions, such as Rightmire (1987) and Stringer and Andrews (1988: 1267) who deny that the resemblances show regional links and instead interpret them as reflecting 'apparent evolutionary reversals'. The case for specific ancestral-descendent relations between hominid samples more than a half million years apart has improved dramatically since Weidenreich (1943) first proposed it for Sundaland (Thorne and Wolpoff 1981; Sartono 1982). Sambungmachan (Jacob 1976) is an excellent intermediary between the Ngandong hominids and the Sangiran *Homo erectus* remains (*including* Sangiran 17 which, contrary to the claims of number of authors, very clearly is a male *Homo erectus* specimen and not a male of a Ngandong-like population). While the Sambungmachan male is somewhat smaller than the Ngandong males, it resembles them in the development of an angular trigone at the lateral corner of the supraorbital torus, the reduction in basal pneumatization and the consequently higher position of the maximum cranial breadth, the flattening of the occipital plane of the occiput (and its vertical orientation), the doubled digastric sulci, and the tall, vertical posterior border of the temporal squama. Compared with the male *Homo erectus* fossils from the Kabuh levels, the Sambungmachan cranial capacity is larger, and the specimen is morphologically more similar to these remains than it is to the Ngandong hominids. Even without Sambungmachan, Weidenreich (1951) was able to link the Kabuh and Ngandong remains. With this specimen, the origin of the Ngandong hominids is firmly established.

Moreover the link between the Kabuh hominids and living Australians has long been evident even without consideration of the intermediate Ngandong sample. Recent discoveries further support this interpretation of evolutionary relationships in South Asia. A new reconstruction of Sangiran 17 (Thorne and Wolpoff 1981) – the most complete *Homo erectus* cranium and the only complete cranium of an adult male – allows comparisons of the face to be made in this region for the first time. This proved to be the anatomical area with the strongest evidence for regional continuity when the Sundaland and the latest Pleistocene and Holocene samples from Australia were compared.

From these comparisons, a number of regional (i.e. intraspecies clade) features were identified. Such features, of course, are not autapomorphic in the phylogenetic sense because they do not appear on genetically isolated lines. They are considered regional within the polytypic species because in combination they appear at higher frequencies in Australasia for much (or in some cases all) of human prehistory there, as compared with other regions of the world during equivalent time spans.

Features of the Sangiran 17 face, and of the other more fragmentary Sangiran faces, support the contention of a special relation between these and the living samples. Regional facial features include the marked ridge paralleling the zygomaxillary suture, the eversion of the lower border of the zygomatic, the rounding of the inferolateral orbital border, the lack of a distinct line dividing the nasal floor from the subnasal face of the maxilla, the curvature of the posterior alveolar plane of the maxilla that corresponds to the mandibular 'curve of Spee', the posterior position of the minimum frontal breadth, the relatively horizontal orientation of the inferior border of the supraorbital torus, and the marked and dramatic nature of the supraorbital or superciliary expression. Moreover, Australasian faces fundamentally differ from the faces of other regions in their combination of dramatic size (holding sex constant), lateral orientation of the maxilla, and marked subnasal prognathism. It is clear that the inhabitants of this southern region have had their own distinct morphology throughout the entire time that East Asia has been inhabited (Thorne and Wolpoff 1981; Jacob 1981; Jelínek 1982; Wolpoff, Wu and Thorne 1984).

At the same time there are equally firm links between the Ngandong hominids and the fossil and living Australians. Weidenreich (1943: 248-250) was able to relate the Ngandong specimens to the Australian Aborigines of today, even in the virtual absence of an Australian fossil record. He thus argued for the interpretation of an unbroken line of descent from the earliest hominids of Sundaland. Weidenreich recognized evidence for continuity in the occasional appearance among Australian Aborigines of a morphological complex that included a well- developed supraorbital torus which combines a discontinuity at glabella with the lack of a supratoral sulcus, and a long, flat, receding forehead. Additional evidence was found in prelambdoidal depressions, sharp angulations between the occipital and nuchal planes, and short or even non-existent sphenoparietal articulations in the region of pterion. Weidenreich's contention was quantified through the implications of a study conducted by Larnach and Macintosh over a decade ago (1974). These authors compared a number of Australian and New Guinea crania with Europeans and Africans, scoring them for the 18 characters that Weidenreich (1951) claimed were unique for the Ngandong hominids. Six of these were absent in all modern samples, while 9 of the 12 other features were found to attain their highest frequencies in the Australian and New Guinea natives and thereby fit the definition of 'regional' as indicated above. These are as follows: the large rounded zygomatic trigone; absence of a supraorbital sulcus; suprameatal tegmen; transverse squamo-tympanic fissure; angling of the petrous to tympanic in the petro-tympanic axis; lambdoidal protuberance; marked ridge- shaped occipital torus; external occipital crest emerging from the occipital torus; marked supratoral sulcus on the occiput. The authors did not consider frontal flattening. If they had, this would emerge as a tenth Ngandong character found at its highest frequency in modern Australians.

Therefore, from the Australian perspective, an Indonesian link (in the words of Macintosh, 'the mark of Java') has long been recognized for the living aboriginal populations. The comparisons of the Ngandong fossils with the living Australian Aborigines are further borne out by Australian fossil remains uncovered since Weidenreich and Larnach and Macintosh published their observations. Beginning with the temporally earliest of these, the WLH 50 hominid from the Willandra Lakes of Australia is a *very* convincing morphological and temporal intermediary between the Ngandong specimens and the recent and modern aboriginals of the continent (cf. Stringer 1989, but interpreted differently in Stringer and Andrews

1988) because of its pattern of robust features, the thickness of its vault, the position of the maximum cranial breadth, the form of the supraorbitals and the region superior to them, and a number of additional distinct morphological features (Thorne 1984; Flood 1983). The specimen is not especially like Jebel Irhoud 2 or Ngaloba (*contra* Delson 1985), and the lack of any unique resemblance to African specimens, archaic or modern, bodes ill for the Garden of Eden hypothesis. WLH 50 is ignored in Rightmire's (1987) analysis of evolution in the region, an analysis which concludes that 'solid evidence for evolutionary continuity is in fact not readily compiled', although he does not discuss most of the detailed evidence supporting the continuity interpretation in this region that has already been published for Australasia. This analysis is further flawed by a confusion of comparisons:

> Some of the traits said to link the Ngandong and Kow Swamp groups are either poorly expressed in the Sangiran fossils or differ explicitly. . . Indonesian *Homo erectus* displays many archaic features, whereas the Kow Swamp and Cohuna people are fully modern anatomically (Rightmire 1987).

The first point about links is irrelevant since the Sangiran folk are *earlier* than Ngandong and therefore would not necessarily be expected to have all of the special features shared by the Ngandong and sub- fossil Australian samples. With regard to the second point, about the differences between recent Australians and Indonesian *Homo erectus,* while the Kow Swamp sample is clearly distinguishable from living Australian aboriginal populations anatomically (Thorne and Wilson 1977), even if Kow Swamp and Coobool were exactly the same as the living populations it is unclear why a demonstration of evolution in the region over a span of more than half a million years disproves the regional continuity interpretation. Perhaps it is not surprising that what has come to be one popularly-quoted rebuttal (for instance quoted in both Stringer 1989 and Rightmire 1987) of the morphological evidence for regional continuity in Australasia is the paper by Kennedy (1984), which refutes claims of morphological continuity (if not identity) with *Homo erectus* for the postcranial remains of the subfossil Australians *that were never made!*

The Willandra Lakes 50 hominid is only the most recent addition to a substantial body of evidence linking hominid populations across the Pleistocene of the Sunda subcontinent. The Indonesian *erectus* connection is strengthened when the earlier discoveries of sub-fossil remains from Kow Swamp and Coobool Crossing are also taken into account, because so many of the evolutionary changes in Australia accelerated during the Holocene, after these populations lived. Because they were earlier, these late Upper Pleistocene/Holocene specimens retain a much higher frequency of features that are archaic in Sundaland than do the living populations of the continent, and thus reinforce the notion of morphological continuity.

In sum, if one firmly believed that modern populations had a single recent origin and replaced their predecessors throughout the world, the evidence discussed above would strongly suggest that Sundaland, if not greater Australia itself, comprised the region of origin. Not surprisingly, this has already been suggested (Gribbin and Cherfas 1982).

North China: the Opposite End

Of all those who would disagree with such a contention, the foremost would probably be among the scholars of the Institute for Vertebrate Palaeontology and Palaeoanthropology in Beijing (Wu and Zhang 1978; Wu and Lin 1985; Wu 1986). Since liberation there has been a continuous increase in the fossil record, with new discoveries of *Homo erectus* from Zhoukoudian, Lantian, Longgudong, and Hexian. New early (or archaic) *Homo sapiens* specimens include major remains from Dali, Maba, Yinnu Shan and Xujiayao, and a number of more fragmentary specimens. New individuals that are terminal Pleistocene or perhaps even Holocene in age include Chilinshan, Huanglong, Liujiang, Muchienchiao, Tzeyang, and a palate from south China found in a Hong Kong drugstore. Contrary to the assertions made by Stringer and Andrews (1988: 1266), these specimens confirm Weidenreich's interpretations, showing both morphological continuity within China and regional distinctions of the Chinese fossils from other areas with reasonably complete fossil records.

Working as he was in north China, Weidenreich (1939, 1943) was much more detailed in his discussions of morphological continuity for that region, in comparison with his rendering of the Australasian sequence. He described 12 features which he felt had particular importance in showing continuity between *Homo erectus* and the modern populations of north China. These were: mid- sagittal torus and parasagittal depression; metopic suture; Inca bones; 'Mongoloid' features of the cheek region; maxillary, ear and mandibular exostoses (the mandibular exostoses forming a mandibular torus); a high degree of platymerism in the femur; a strong deltoid tuberosity in the humerus; shovel-shaped upper lateral incisors; and the horizontal course of the nasofrontal and frontomaxillary sutures. While some of these, especially the postcranial features, are known to be generally characteristic of archaic populations everywhere, most of the others appear to be most frequent in combination in ancient and modern North Asians. Especially in combination, they characterize the region for a long period of time.

Over the decades since Weidenreich's death a number of additional observations have been added. Aigner (1976) pointed to: the profile contour of the nasal saddle and of the nasal roof; pronounced frontal orientation of the malar facies and the frontosphenoidal processes of the maxilla; and a rounded infraorbital margin, in line with the floor of the orbit. Wu (1981), in his description of the Dali cranium, was able to add a number of facial features which have great importance because the face seems to show more regional distinctions than does the cranial vault in north China, just as is the case in the Sundaland hominids (Thorne and Wolpoff 1981). Indeed, the Dali cranium (Wu 1981) and the morphologically similar, newly-discovered Yinnu Shan female (Wei 1984; Bunney 1984), have added much to an already very convincing case established by Weidenreich and elaborated upon the basis of more fragmentary remains recovered through the end of the last decade (Wolpoff, Wu and Thorne 1984). Besides facial flatness and the specific traits mentioned by Weidenreich and Aigner, we established additional regionally distinct features of these early *Homo sapiens* vaults which include the forehead profile (in particular the development of the frontal boss); the associated distinct angulation in the zygomatic process of the maxilla and anterior orientation of the frontal process of the zygomatic; lack of anterior facial projection and low degree of prognathism; rounded shape of the orbits; minimal nasal projection (the frontonasal and frontomaxillary sutures are virtually level); and a very low nasal profile. Other fragmentary remains of early *Homo sapiens,* especially those from Maba, Xujiayao, and Dingcun, also show these details when the appropriate parts are preserved. However, Dali and Yinnu Shan are clearly the centrepieces of the continuity argument in North Asia. The combination of regional features in these specimens so clearly link them specifically with the more recent and modern populations of north China that it is unlikely future discoveries will refute this interpretation. *With regard to the Garden of Eden hypothesis, it is ominous that none of these earlier specimens possess features which resemble archaic or modern Africans* in anything other than grade features that characterize all hominids of the time period.

Among the remains representing the earliest modern populations, Liujiang from south China and the Upper Cave specimens from Zhoukoudian have a number of features that are characteristically regional (Woo 1959; Wu 1961; Wolpoff, Wu and Thorne 1984; Wu and Wu 1985). They are very definitely *not* characteristically African, and therefore show an absence of any tendency for the earlier 'moderns' of the region to appear more African-like than the later populations (as would be expected according to the Garden of Eden hypothesis). Moreover, they are quite unlike other 'early modern' specimens such as Mladeč, Pavlov, WLH 50 or Mungo 1. Wu argued strongly against Weidenreich's 'Eskimo, Chinese, Melanesoid' trichotomy for the Upper Cave crania, instead showing that they all fall within the expected range of variation for north China. The early *Homo sapiens* remains have been particularly important in providing firm evidence for a morphological link between the Late Pleistocene/early Holocene specimens and the Zhoukoudian *Homo erectus* sample – a link that was lacking in Weidenreich's time, hampering his interpretation (1939) of the Zhoukoudian Upper Cave specimens.

In sum, the basis for claiming regional continuity in China, especially northern China, is at least as good as in Sundaland for three stages of human evolution: *Homo erectus*; archaic *Homo sapiens*; and, the earliest modern *Homo sapiens.* Continuity, however, is generally not found in different character states of the same traits that are regional in Sundaland, and characterizations such as 'homoplasy' or 'regional pleisomorphy' cannot possibly apply in attempting to discount what any observer can see. The differences between features showing continuity in these two ends of the eastern periphery were recognized as long ago as in Weidenreich's time and have continued to be recognized by scholars familiar with the fossil remains from both areas (Jacob 1981; Wu and Zhang 1978; Wu and Wu 1982; Wolpoff,

Wu and Thorne 1984). It was this evidence that led to Coon's (1962) chapter headings 'Pithecanthropus and the Australoids' and 'Sinanthropus and the Mongoloids'.

The pattern of hominid evolution at the eastern periphery still appears much as Weidenreich described it. It involves a morphological gradient spread over the eastern periphery between north and south ends that are unique and distinguishable as far back into the past as the earliest evidence for habitation can be found, but also involves geographically and morphologically intermediate specimens that show there never was populational isolation. Because of the marked differences in the morphological features that show continuity at the north and south ends of the eastern periphery, *this pattern clearly discounts complete replacement as a valid explanation of Pleistocene hominid evolution*.

Instead, a great deal of continuous evolutionary change (from the earliest inhabitants of East Asia to the modern populations of the region) can be documented that involves *no* speciation events. This creates a problem for those who insist on a cladistic definition for *Homo sapiens* (Delson 1985; Tattersall 1986). This is because while a great magnitude of gradual change is documented in the region, at the southern periphery the pattern of change is so continuous that no boundary can be drawn between the archaic and the modern *Homo sapiens* populations (Wolpoff 1986).

Considerable magnitudes of change without a speciation event and subsequent replacements is hardly a prediction of the Garden of Eden hypothesis. Thus, without the evidence for a complete replacement interpretation in East Asia, it seems to me that the Garden of Eden hypothesis fails the test, and those who ignore this evidence of the fossils do so at their peril.

MULTIREGIONAL EVOLUTION THEORY

The multiregional evolution hypothesis (Wolpoff, Wu and Thorne 1984) is a valid alternative to the Garden of Eden interpretation because it fits the data, especially in those places where the Garden of Eden hypothesis fails. It accounts for the appearance of modern populations throughout the hominid range without positing that they necessarily have a unique recent ancestry or represent the results of a recent speciation event. Specifically it proposes an evolutionary explanation for the observations of regional continuity at the northern and southern ends of East Asia, as discussed above, as well as elsewhere (Van Valen 1986). The observations which were contradictory in Weidenreich's explanatory scheme and were unconvincingly explained in Coon's, can be satisfactorily accounted for. This hypothesis does *not* account for modern populational origins by parallel evolution (Eckhardt 1987; *contra* Cann *et al.* 1987: 35; and Stringer and Andrews 1988), nor by simultaneous speciations in different parts of the world (*contra* Jones and Rouhani 1986b: 600). The parallelism interpretation of the regional continuity observations is as old as Howells' (1942) *mis*characterization of Weidenreich's 'trellis of human evolution' as a candelabra. The mischaracterization is as wrong now as it was then, and its continued restatement by supporters of the Garden of Eden hypothesis is misleading and counter-productive.

Fossil Hominids as a Polytypic Species

Explanations for the observation of regional continuity in long- lasting polytypic mammalian species do not abound in the palaeontological literature. This is because with only a very few exceptions (Freudenthal 1965; Martin 1970), fossil histories for polytypic species are virtually unknown. The hominids may prove to be an important exception to this (Van Valen 1966: 382):

> Our knowledge of the fossil history of non-human subspecies is almost negligible and *Homo* will probably furnish the first well studied case of the evolution of several subspecies in geological time.

The absence of information about the evolutionary histories of polytypic species is probably not because geographic variation within fossil species was rare, but rather because the tendency has been to define polytypic species out of existence, and interpret regional variation to be at the species level because it fits the morphospecies or evolutionary species criteria of phylogenetics.

A good example of this can be found in a recent symposium on Middle Pleistocene human evolution attended by a number of cladistically-minded scholars who, among other things, dealt at length with variation in *Homo erectus* (see Andrews and Franzen 1984). There emerged a consensus that the considerable geographic variation recognized in this hominid reflects the 'fact' that the *Homo erectus* sample was actually comprised of more than one species (Andrews 1984), and that perhaps in none of its taxonomic guises was it ancestral to living populations. However, it might be well to heed Stringer's cautionary statement (1984b: 141):

> We must beware of the position of saying on cladistic grounds that *H. sapiens* did not evolve from *H. erectus* but from a different species showing similar characteristics that lived at the same time!

Tattersall (1986), following Andrews' phylogenetic conclusions and incorrectly quoting me, characterizes *Homo erectus* as a grade and not a taxon, in a paper explicitly claiming that there are not enough taxa named in the human fossil record. And the regional variation in *Homo erectus* has been treated in even odder ways (Van Vark 1983)! It is clear to me that variation formerly regarded as at the subspecies level cannot and will not be recognized in the fossil record when phylogenetic criteria are uncritically applied, no matter how problematic the consequences of applying these criteria may be. In a curious way, by confusing what is almost certainly subspecies variation with species variation, these publications show by counter-example that in closely related groups long-lasting geographic variation in the past can only be recognized when the interbreeding concept of biological species is used to help generate criteria for distinguishing fossil species. When, instead, it is the morphotypic definition that is applied to closely related fossils, the observed variability in the Middle Pleistocene hominids cannot be reasonably interpreted at the subspecies level, and Middle Pleistocene hominid variants are misidentified as different species although they show variation no greater than that which characterizes comparisons between the regions today.

Moreover, ironically, there are other views that suggest that the multiplication of palaeospecies which is urged upon us (Tattersall 1986) may be more than inappropriate or a misapplication of otherwise useful approaches. It may be biologically misleading and incorrect. For instance, according to Templeton (1982: 117):

> Different groups of organisms are subjected consistently to certain speciation modes because of their genetic, population-structural, and ecological attributes. For instance. . . there are very consistent differences in population structure between frogs and mammals such that mammals are much more likely to display higher degrees of population subdivision. . . this implies that mammals are more subject to rapid population divergence. . . (and) that mammals are more prone to form polytypic species. The result is that. . . the number of fossil mammalian species could actually be overestimated.

Indeed, commonplace as they are among the living taxa, polytypic species are mainly discussed in the evolutionary literature with regard to peripatric speciation. In most other contexts, the analysis of geographic variation is simplified with the panmictic assumption. The evidence of marked divergences within continuously distributed species (Endler 1977) argues against the peripatric model of speciation and therefore tends to be ignored (Barton and Charlesworth 1984). I believe the genesis of this situation is not so much in a conviction that polytypic species can only change through peripatric speciation, as in a lack of realization that polytypic species pose an evolutionary problem. While the alternative to gradualism – punctuated equilibrium – provides a potentially valid but different explanation for evolutionary change in a polytypic species, it cannot account for regional continuity in evolving populations across the 'boundary' created by speciation events.

Clinal Distributions of Regional Morphology

Multiregional evolution theory (Wolpoff, Wu and Thorne 1984) attempts to explain three phenomena that characterize the Pleistocene evolution of the genus *Homo:* (1) the *initial* centre and edge contrast of a variable source population of early Middle Pleistocene African hominids, with a number of differing monomorphic peripheral populations (see Thorne 1981; Bryant, Coombs and McCommas 1986; Goodnight 1987); (2) the *early appearance* of features which show evolutionary continuity (i.e. regional features) with modern populations at the peripheries, contrasted with the much later appearance of features showing regional continuity at the centre (Thorne and Wolpoff 1981); (3) the *maintenance* of these regional contrasts and the persistence of regional continuity through most of the Pleistocene (Thorne 1981) until the population explosions (and dramatic movements) of the Holocene. These three phenomena are related to the development and maintenance of long-lasting gradients throughout the hominid range, balancing gene flow against opposing selection and (especially at the peripheries) perhaps also against the effects of drift.

The observations of regional continuity reflect the results of processes that lead to the *maintenance* of distinctly regional variants for long periods of time, during which general evolutionary trends characterize the entire polytypic species (Van Valen 1986; Wolpoff, Wu and Thorne 1984). This pattern of contrasting species-wide evolutionary trends and distinctly regional variations (especially for the peripheral populations of a polytypic species) is quite common (Mayr 1963, 1982). It is maintained through clines for multiple characters (Barton 1983), that can persist even in populations that have been in contact for very long periods of time (Clarke 1966; Endler 1977; White 1978).

Population differentiation is almost inevitable when species are widespread and subdivided (Wright 1943; Fisher 1950; Jain and Bradshaw 1966), a consequence of isolation by distance (Wright 1943). In the face of gene flow between and among such populations, the opposing selection may be at a very low level, or may even be replaced by drift in the maintenance of long-standing gradients (Barton and Charlesworth 1984). This is particularly likely when the contiguous populations along the gradients are small, as might be expected at the geographic or ecological peripheries of the range (Mayr 1982). The effects of drift are also maximized when populations are long lasting, providing the opportunity for fixations to occur (Slatkin 1987). Therefore the presence of clinal distributions is not necessarily a marker of corresponding selection differences over the range of the clines (Lande 1982). Moreover, clinal distributions involving small populations and a balance between gene flow and drift may be expected to produce (or maintain) more homogeneous populations at the geographic or ecological edge of the range of a polytypic species where population densities tend to be lower and the potential for drift is greater. The resulting morphological homogeneity at the peripheries provides the bases for the observations of regional continuities.

Gene Flow as a Source of New Alleles

Gene flow persists as the most *mis*understood aspect of multiregional theory. Gene flow plays several different roles in maintaining clines. Without gene flow, it is inevitable that there will be speciation. Thus, differential selection alone cannot maintain clines within a species for long spans of time and the very presence of long standing gradients for morphological features therefore indicates that gene flow was present, no matter what the form and magnitude that characterized it and whether or not there was also a gradient in selection.

Gene flow is commonly regarded as acting to spread new mutations among populations (Coon 1962; Ehrlich and Raven 1969) that may be promoted if they prove to be advantageous. Realistically, however, this probably represents only a small part of the creative role that gene flow normally plays in the evolutionary process (Slatkin 1987). In fact, a rather different assessment of the importance of new alleles in promoting morphological change follows from the implications of the extraordinary genetic similarity of humans and chimpanzees, recognized for more than a decade (King and Wilson 1975; Brown *et al.* 1982; Templeton 1983; Wilson *et al.* 1985). The contrast between the number of shared alleles and the divergent morphologies for these two living species has been interpreted in almost every possible way, but the most obvious is that *the notable morphological and behavioural differences between these two species are mainly a result of differing allele frequencies* (Lewontin 1974, 1984), and not of different alleles.

Studies of the nuclear DNA data from living humans have also revealed a surprising expression of genetic similarities among populations (Lewontin 1974, 1984; Nei and Roychoudhury 1982; Jones 1981). This similarity was also found in the mtDNA studies, and given a variety of interpretations (Cann, Brown and Wilson 1982, 1984; Brown *et al.* 1982). But if these data mainly indicate that virtually all of the genetic differences between populations that have evolved over the course of human history are the result of frequency differences, *it follows that gene flow must primarily be regarded to function in changing frequencies of existing alleles, and not usually in introducing new ones.*

Therefore, gene flow cannot usually be expected to have a 'stimulating' effect on human evolution by introducing new alleles for selection to act on. Instead, the usual, perhaps invariable, effect of gene flow is *change in frequencies of existing alleles* affected by the balance between gene flow and local selection and/or drift. When this balance persists, the role of gene flow in any locale could be envisioned as mainly providing an unending source of recombinatorial possibilities (rather than a source of new alleles) which are than subjected to selection.

The Magnitude of Gene Flow

The magnitude of gene flow is a closely related issue. Problems here involve how much gene flow actually exists, and the effects of various magnitudes of gene flow under different conditions. Consideration of clines and the gradients betweent them as two-dimensional networks magnifies the potential effects of gene flow on equilibria, providing a greater role for gene flow at low magnitudes than workers such as Ehrlich and Raven (1969) allowed. However, when a gradient in selection exists, it has a predominating effect on the resulting cline's form (Livingstone 1973; Endler 1977; Barton 1983) almost independently of how great the magnitude of gene flow might be. Under these circumstances, the main effects of differences in the magnitude of gene flow are to change the steepness of the gradient across the clines, and in some cases to displace the geographic positions of equilibria frequencies. These effects are magnified when the discontinuous distribution of individuals who live in groups is also taken into account. For these reasons, and because clines can exist when there are selection gradients even in the absence of gene flow, in most cases clines will be maintained within a species regardless of the magnitude of gene flow. It follows that the degree of differentiation along a cline, the steepness of the gradient, and the form that the gradient takes are all somewhat independent of the magnitude of genetic interchange.

I conclude that clines will invariably form in a polytypic species, regardless of the levels of gene flow. The particular attributes of these clines will be dictated by all of the relevant variables, including the magnitude and direction of gene flow (and the degree to which it is reciprocal) and the intensity and distribution of local selection, but also the distribution of populations, population sizes and breeding structures, the effect of drift, and the pattern of initial genetic variation. *I reject the notions that selection or drift is likely to 'override' the influence of gene flow, or that gene flow is likely to 'swamp' the effects of selection.* The idea that gene flow magnitudes can be too small or too large makes no sense in terms of the balance model of clines. Variation in the evolutionary forces underlying clines can alter their form, but will not erase the clines themselves.

This point is important for understanding the effects of the magnitude of gene flow between populations. One argument is that the magnitude must be high to account for simultaneous speciation when there is polytypism (Jones and Rouhani 1986b), while there is a different contention suggesting that this magnitude is rather low (Ehrlich and Raven 1969; Endler 1977) that is based on the observation that the expected effects of higher magnitudes cannot be found. Neither extreme view, unfortunately, is based on actual measurements of gene flow, and in fact actual magnitudes of gene flow between living animal populations remain largely unknown (Slatkin 1987). While the gene flow reported between adjacent living human populations is usually anything but minuscule, it is possible that this magnitude may be lower in other species, or in human prehistory (as might be suggested by the emerging understanding of how different Pleistocene archaeological sequences in various regions actually were).

However, I believe that when considered over geologic time, the question of whether gene flow was of sufficient magnitude to account for the spread of morphological features without the populations themselves moving might be of less importance than it may seem. The potential importance of the magnitude of gene flow has been misunderstood because comparisons among human populations or between humans and the African apes show that the actual number of *new alleles* whose spread is to be accounted for during the evolution of *Homo* is surprisingly small, and therefore that it is changing allele frequencies and not new alleles that must be accounted for.

The Role of Gene Flow

A final point of interest concerning gene flow is its role in a polytypic species. This can be discussed on several different levels. It is clear that this role is complexly dependent on 'both the geographic distribution of the species and on the importance of other evolutionary forces' (Slatkin 1987: 787). Without gene flow, polytypic species will eventually speciate (Mayr 1963, 1982). However gene flow has importance far beyond preventing speciation. At various times it can act as both a constraining force and a creative force in the evolutionary process. Opposing the Ehrlich and Raven (1969) perspective of species as internally subdivided with little or no internal gene flow is the 'species-unifying' concept for the role of gene flow. In this perception the gene flow between populations is seen as supporting cohesion and stability within the species as a whole. However a rather different problem arises from this concept – namely, how can gene flow both be responsible for maintaining stability in a homeostatic-like manner (small perturbations tend to be brought back to the mean) and at the same time be a mechanism of change?

One resolution to this potential contradiction is found in the precept that the genetic cohesion of species may not be maintained by gene flow alone. A rather different explanation for genetic cohesion is provided in the more recent discussions by Mayr (1982), Carson (1982), Lewontin (1984), and others. In these publications, the idea of co-adapted genetic systems representing the maximum fitness under balancing selection achieved by whole organisms has been envisioned as an internal restraint for change within a species, limiting how much a species can change without a complete reorganization of these genetic systems (Mayr's 'genetic revolution'). The co-adapted genetic system for a species is presumably not dependent on gene flow for its maintenance.

The problem I find with this concept is in the nature of normal subspecific or populational differences found within species. Such differences most often tend to reflect local adaptations. Consequently, the co-adapted genetic system model of species leads to a different contradiction. A single co-adapted genetic system is presumably established for an entire species because of the heterozygotic combinations promoted by balancing selection to maximize the fitness of individuals. Yet, at the same time the specific gene combinations in particular populations respond to local selection and change, often dramatically. To understand how this potential contradiction between co-adapted genetic systems at the species and subspecies (or population) levels is avoided, we must distinguish factors that maintain species cohesion and stability, and factors that maintain the phenetic similarity of populations within species. These are not the same phenomena (Barton and Charlesworth 1984; Charlesworth 1983). *In fact, a special explanation for phenetic similarity beyond the normal effects of shared ancestry and gene flow is hardly required in the face of the extremely small number of genetic differences distinguishing populations enclosed within the protected gene pool of a biological species.*

If the phenetic similarity within biological species is a phenomenon distinct from the sources of its genetic cohesion, one can envision the stability and cohesion-maintaining mechanisms in a somewhat different manner. It is likely that in polytypic species, co-adapted genetic systems are not species-wide, but rather can be expected to differ from population to population. Indeed, such differences would magnify the isolation by distance phenomenon (Wright 1943). This model (Wolpoff, Wu and Thorne 1984) resembles Wright's (1967, 1977, 1982) shifting balance theory, with morphologically distinct intraspecific groups reflecting different adaptive peaks maintained by stabilizing selection. The shifting balance model with its multiple adaptive peaks better fits what is known about polytypic species evolution in the absence of particular speciation events.

Gene flow is not so much the network that keeps the species whole, as it is the latticework that connects the populations of a polytypic species, represented by multiple adaptive peaks, through its participation in the support of clines and its contribution to the development of new recombinatorial possibilities. Therefore it can be suggested that the same mechanisms promoting stability and cohesion within a species can also provide for species-wide evolutionary changes; as the complex balances of gene flow and selection that produce clinal equilibria at the adaptive peaks vary, the clines change in response. Because the clines represent balances (albeit differing from area to area), evolutionary responses can be of much greater magnitude than the magnitude of changes in any of the equilibria-producing forces, just as a small shift in frequency can cause a great change in the interference pattern between two overlapping waves.

Evolutionary change in a species constituted as described here is neither directly related in magnitude to the amount of gene flow between populations, nor proportionately responsive to the magnitude of selection, nor directly limited by the inability to alter co-adapted genetic systems (if the distribution of these systems is the object of the clinal variation, as is likely since both unique genetic systems and the clines exist in polytypic species). One can envision the polytypic species as a *dynamic* system, externally bounded and protected by the limits to gene flow, and internally diversified by the evolutionary forces sustaining gradients between differentiated populations. The stability within such a species does not result in its stagnation.

Internal Diversification in Homo

With regard to the multiregional model of hominid evolution, it is particularly important to focus on the pattern of internal diversification. Clinal theory provides an explanatory basis for the maintenance of regional distinctions for populations at the periphery of the hominid range. Drift, due to small population effects, and differences in selection resulting from environmental variation as well as the existence of regionally distinct morphotypes, both characterized populations at the marginal or peripheral portions of the hominid range for long periods of time. The initially different gene pools of these populations were established during the process of first habitation (Thorne 1981). The pattern of regional variation was maintained throughout most of the Pleistocene by a balance between the local forces promoting homogeneity within populations and regional distinctions between them (selection, drift), and multidirectional gene flow which (*contra* Jones and Rouhani 1986b) for reasons discussed above need not have been particularly high (Slatkin 1987). As a consequence, a long-lasting dynamic system of morphological gradients came to characterize the multiregional distribution of our polytypic lineage.

If this was the normal pattern guiding the evolution of the genus *Homo*, the conflict assumed to exist between gene flow and local selection that underlies so many interpretative disputes should not actually exist. All evolutionary changes affect a balance between these potentially opposing forces and must be accounted for by shifts in this balance (except in the extreme cases of latest Pleistocene and Holocene large-scale population movements for which the multiregional evolution model does not apply). It makes no sense to argue about whether gene flow or selection or drift 'predominated' to account for a specific local evolutionary sequence because the multiregional evolution model requires all of these, and focuses on the balance between them in its explanation of evolutionary change. In this clinal model, shifts in the magnitudes of the opposing forces change the gradient of the cline, and it is this changing gradient, viewed over time in a specific region, that provides the data to characterize regional evolutionary change. This is not an alternative to the gradualist model, as applied to a polytypic species, this *is* the gradualist model.

The evidence for regional continuity in two areas of the eastern periphery has been reviewed, and it is as good as probably could be expected given the nature of the fossil record. But what of the evidence for gene flow, an important element in the evolutionary model proposed to account for it? To some extent the presence of common major evolutionary trends constitutes a valid source of evidence for gene flow. At the north and south ends of the eastern periphery there are brain-size expansions, reductions in muscularity, in sexual dimorphism, and in anterior tooth size. These, and a number of additional common evolutionary trends, result in many of the features shared by the modern populations of these areas and could be taken as evidence of gene flow. This is especially the case given the lack of technological and other forms of adaptive similarities. However, this evidence is insufficient because the data which the hypothesis was created to explain can hardly also be taken as independent support of the hypothesis itself. Moreover, it is likely that there are also some common elements of selection promoting these changes across the human range, particularly in the rapid spread of ideas which may be unique to humanity.

There is a growing body of direct evidence for gene flow along the eastern end of the human range. For the Upper Pleistocene this evidence was first recognized by Weidenreich (1945) during his comparisons of the Wadjak remains with Australian fossils such as Keilor. He argued that these specimens are virtually identical, and my observations also support the notion that there is a marked degree of similarity between them in size, proportions, and facial breadth and flatness. Moreover, the south China specimen from Liujiang shares many, perhaps most, of these features (Wolpoff, Wu and Thorne 1984). These and more fragmentary materials seem to provide a late-Pleistocene link between the Asian mainland, Sundaland, and greater Australia that indicates that there was persistent gene flow.

The evidence for gene flow begins much earlier, however. The Hexian *Homo erectus* cranium (Wu and Dong 1982, 1985) resembles the Zhoukoudian *erectus* remains in many details of the forehead profile as well as in the moderate frontal boss and the rounding of the superior orbital border. However, the weak expression of the supratoral sulcus on the frontal, the contrasting marked expression of the occipital's supratoral sulcus, the form of the basal pneumatization as seen in the posterior contour, the parietal angulation, and the expanded cranial breadths, all much more closely resemble the Indonesian hominids. Indeed, a case could be made that Hexian is a morphological as well as a geographical intermediary.

Yet it is interesting that to the west, the Narmada cranium (Lumley and Sonakia 1985) from Madhya Pradesh, Central India, closely resembles the later of the Chinese *Homo erectus* remains. Especially in the shape and configuration of the supraorbitals and the curvature of the frontal squama, its frontal is much like Hexian, while the round orbits resemble Maba. The temporal and occipital are similar to Zhoukoudian specimens (especially the H5 vault) in the form of the temporal squama, the expression and position of the angular torus, the curvature of the occipital in the sagittal plane, and the morphology of the nuchal torus. The much later Zuttiyeh frontofacial fragment, from even further west in Asia, also may reflect a strong East Asian influence in its morphology (Hublin 1976). Much like the Zhoukoudian hominids, its supraorbitals extend well anterior to the distinct frontal boss, creating a deep supratoral sulcus. Another resemblance lies in the very flat upper face and the anterior orientation of the orbital pillars, features shared with the later Skhūl/Qafzeh samples from the Levant as well. Evidently, on the Asian mainland, the features first recognized in China predominate far to the west during the Middle and the earlier Upper Pleistocene, even though intermediate specimens may be found, especially along the southern portion of its eastern periphery. In its totality the Asian evidence supports the interpretation of long-standing clines with significant gene flow between the north and south end regions. Unlike the lack of contrast between East and West Asia, North and South Asia has populations which are distinct and were distinct in the past. There is a fossil record showing distinguishable origins for the modern populations that inhabit the regions today, in spite of the evidence for Pleistocene and Holocene gene flow.

These patterns of diversification and change in the hominids are complex. Yet, in them are to be found the origins of modern populations. The critical role played by gene flow contradicts and invalidates the interpretations of the mtDNA data that derive these modern populations from a single recent African Eve through a speciation event. Indeed, the multiregional hypothesis predicts that there will be no single origin for these populations that can be located on genetic or morphological grounds. In contrast, interpretations of the mtDNA data that are based on differential lineage extinctions are fully compatible with the multiregional hypothesis. Differential mtDNA lineage extinction only works as a valid explanation if the mtDNA variants were widespread to begin with, a consequence of gene flow.

SUMMARY

According to the 'Garden of Eden' hypothesis, all modern human populations have a single recent origin as a new species. *Homo sapiens* is said to have originated in Africa approximately 200 000 years ago, and within the next 100 000 years some of the populations (representing one of the two main mitochondrial lineages that developed in the new species) went through a period of very small population size and migrated out of Africa to replace the indigenous human populations throughout the world, without mixing with them.

Evidence for an African origin is based on genetic and palaeontological inference. Neither of these bases are particularly strong. The genetic basis for claiming that Africa is the place of origin rests on the interpretation of the observation that there is more genetic diversity among African populations than among others. However, these genetic data can be interpreted in a different way. For instance, the trees of relationships developed from these data can be rooted on different continents, and the diversity arguments rest on as-yet unverified assumptions about which genetic variants are homologies and which are homoplasies. The palaeontological basis for claiming an African origin rests on the earlier dates claimed for modern *Homo sapiens* fossils in Africa than in other regions (which presupposes that there actually *was* a single unique place of origin for modern populations). However, the reality of the situation is that there are no firm dates for human fossils in the critical time span, and the early date for Qafzeh (if actually correct) is from *outside* the region (thousands of miles from where Eve's homeland is claimed to be) and applies to specimens described by Vandermeersch as 'Proto Cromagnoid' – hardly a reflection of perceived African resemblance (Valladas *et al.* 1988). In fact there are not even *equivocal* dates for the *African* specimens, so that it is far from clear that the earliest modern human populations first appeared there. Even without dates for the earliest modern humans, a different palaeontological problem arises from the claim that modern populations first evolved in Africa; the earliest modern human fossils from outside of Africa do not share any unique derived African features. Instead, in both North and South Asia, the earliest modern human fossil remains from each region show some of the unique features that characterized their predecessors in the same regions.

The genetic basis for the Garden of Eden hypothesis is most widely attributed to the interpretation of mtDNA variation. This interpretation rests on the reading of a genetic clock from the mitochondrial variation, which requires the assumptions that (1) the main source of mtDNA variation is mutations (selection and drift can be discounted); (2) the rate of mutations is constant everywhere, and mutation accumulation is linear with time; (3) the colonization events used in rate-determinations are accurately dated, singular, and unidirectional. It is unlikely that *any* of these assumptions are correct.

Moreover, rates of mtDNA evolution supporting the Garden of Eden hypothesis for modern populational origins are inconsistent with the mtDNA calibration of other more dramatic events, and inconsistent with the calibration of speciation events based on data from other genetic systems and from palaeontology. Using dates for the peopling of New Guinea, Australia, and the Americas, Cann and her colleagues determined a divergence rate of 2%-4% base substitutions per million years. It is the median rate in this range that is used in the estimate of 200 000 years for the divergence of modern human populations from their last common African source ancestor. However, when applied to other *species* divergences in the primate fossil record, this rate range gives a human-chimpanzee divergence date that averages only 2.7 million years ago by one estimate, and even less (2.1 million years ago) by another. Either date is clearly wrong, and is unambiguously disproved by the earlier presence of large samples of indisputable hominids in the fossil record, from Laetoli and Hadar.

Nei (1985, 1987) has shown that the length of time since these human migrations are supposed to have occurred is too short for the assumption of a consistent rate of base substitutions to apply. A combination of unknown earlier polymorphisms in the mtDNA, subsequent populational admixture (especially the effect of gene flow among recent populations), and the predominating effects of stochastic events over short periods of time, make it impossible to determine the time of the last common ancestor from a mitochondrial clock calibrated from recent migrations.

The average rate of base-pair substitutions is certainly much slower, perhaps only 0.71% per million years as Nei claims. This would give an estimate of the chimpanzee-human split at 6.6 million years, a date that closely approximates the estimate of 6.3 million years from nuclear DNA hybridization data, and that is also in conformity with estimates from the fossil record. *Estimates of modern populational divergence at 200 000 years based on a 2%-4% rate of base-pair substitutions per million years are therefore not likely to be correct.*

On the other hand, the 0.71% base-substitution-rate estimate indicates a divergence time for the ancestors of modern populations at about 850 000 years ago. At this date, near the beginning of the Middle Pleistocene, the appearance and spread of these populations out of Africa was probably not a speciation event; admixture with other populations is not an issue since there were probably no indigenous human populations to mix with. This date would suggest that *Homo erectus* was the first human species to be widely distributed around the world, and that the founders of many modern populations are to be identified in this species.

The mtDNA evidence said to support the Garden of Eden hypothesis requires that modern humans *be a new species*, reproductively isolated from all other populations that were contemporary with them at the time they originated. However, the bottlenecking evidence that is also said to support this hypothesis describes a population split *within a single species*. According to this interpretation the extra-African populations presumably underwent a period of small population size and substantial drift after they left Africa. This incident of bottlenecking could not have been the same as the speciation event because the descendant populations remain within the same species as the unbottlenecked (more variable) portion of *Homo sapiens* that remained in Africa. Therefore the bottleneck happened at a different, presumably later time. But if so, the bottleneck and not a recent time of origin could account for the lack of mtDNA diversity outside of Africa. *The fact is that these two sources of genetic data contradict rather than support each other.*

A rather different interpretation is much more likely for the small number of mtDNA lineages found among modern humans. The 'Eve' interpretation may be an artifact of tree analysis (which necessarily gives stems) and not an accurate reflection of population history. Population admixture, the migrations of recorded history, and the dramatic population expansions of the Holocene invalidates any tree analysis, even if populations actually diverged in a tree-like manner. The same pattern of mtDNA variation more probably reflects a history of stochastic differential mitochondrial lineage survivorship from numerous ancestors, especially if the small population sizes responsible for the bottlenecks described above actually characterized this portion of human evolution and if cytoplasmic variants were widespread because of gene flow. If so, a single mitochondrial line of cytoplasms would co-occur with populational descent from multiple nuclear DNA ancestors. Descent from a single Eve (or from a small population) would be an incorrect interpretation for the nuclear DNA evidence, which after all is the main focus of evolutionary studies. In terms of nuclear DNA descent, the idea of a single origin for the surviving lines would simply not apply.

Data from the human fossil record provide an independent basis for assessing the Garden of Eden hypothesis. This hypothesis suggests that the earliest dated specimens of modern *Homo sapiens* should be more similar to each other than are later ones, and that the later specimens should have more regional distinctions, as they would be more differentiated. The hypothesis also predicts that the earliest modern *Homo sapiens* from any region should have the most African characteristics of any sample in that region. *Fossil data do not conform to any of these predictions*. Instead, fossils from the peripheries of the human range show that the regional distinctions of today are as old as the earliest occupations of these regions. In fact, outside of Africa the regional distinctions precede the first appearance of modern *Homo sapiens* morphology. There definitely is no tendency for the earlier fossil specimens to diverge from a common morphological pattern (and certainly not from an African one). Moreover, in no region is modern *Homo sapiens* obviously distinct from earlier populations at a species level; indeed, in some places modern populations are not even particularly different from their Pleistocene predecessors at any taxonomic level.

A valid explanation of modern populational origins must account for two aspects of the human fossil record, recognized by Weidenreich a half century ago, both of which contradict the Garden of Eden hypothesis. These are the persistence of unique regional distinctions for long periods of time, even across species 'boundaries', and the commonalities of human evolution during the Middle and Later Pleistocene in different regions. Because of these evolutionary patterns, the fossil data indicate that – especially at the peripheries – a replacement model for the origin of modern *Homo sapiens* populations cannot be sustained. Instead, the origins of modern populations in different regions seem to be rooted in the earlier archaic populations of the regions, in most cases without evidence of a speciation event. Perhaps the most telling aspect of these interpretations is that they are rooted most firmly in different and independent morphological observations in samples from the northern- and southern-most ends of the eastern (Asian) periphery. There are different evolutionary patterns, based on complexes of different details and often of different character states for the same features, at the northern and southern ends of the eastern edge of the human range.

The Neanderthal problem of Europe, and the 'solutions' often accepted for it, might seem to support a replacement interpretation at the western margin of the human range. However, even if the European Neanderthals were replaced, it does not follow that the European situation necessarily supports the Garden of Eden hypothesis. This is because most workers, even those who generally believe that the Neanderthals were replaced by more modern populations, have come to accept the implications presented by intermediate-appearing specimens, suggesting that there was commonly hybridization between the Neanderthals and the invading populations. Hybridization, however, argues against the validity of the Garden of Eden hypothesis because it contradicts the necessary assumption that the immediate ancestor of modern populations was a new species.

The theory of multiregional evolution provides an alternative explanation for these patterns in the human fossil record. While the centre and edge hypothesis suggests how and why regional distinctions first appeared at the peripheries of the hominid range, regional continuity accounts for the persistence of these peripheral distinctions through the remainder of the Pleistocene. The pattern of central variability and regionally distinct peripheral monomorphisms was maintained by long-lasting gradients balancing gene usually from the centre toward the edges, against local selection and in some cases drift. This theory shares two things with the Garden of Eden hypothesis; it posits an African ancestry for all populations, and its validity relies on the widespread movements of genes. However, there are a number of notable differences between these hypotheses which allow one to determine whether data from various sources can refute one of them. In the multiregional perspective, all populations are potentially connected by gene flow, usually from the centre toward the edges, against local selection and in imply the movement of people (when there were migratory movements they involved small groups and led to admixture with existing indigenous populations rather than replacement without mixture) and speciation as populations rather than replacement without mixture, and speciation as described by this hypothesis was gradual (and was not an 'event') once the hominids became a geographically-dispersed polytypic species. The multiregional pattern and the precept that local human populations were probably fairly small for most of the Pleistocene, provides the basis for the conclusion that differential mtDNA lineage survivorships and the dramatic population increases of historic times combine as the most likely explanation for the pattern of limited mtDNA variation seen in living populations. This would replace the alternative explanation of a shallow root (i.e. recent divergence) for a single tree that is part of the Garden of Eden hypothesis.

I believe that the fossil record clearly and unambiguously contradicts the Garden of Eden hypothesis. If this fossil evidence has been correctly interpreted, as I believe it has, the Garden of Eden hypothesis can be considered disproved on the evidence of the hominid fossils. Much can be said in support of the contention that the fossil record is not an irrelevant source of information about human evolution.

There is an additional basis for this seemingly strong claim, to be found in the other interpretations of the mtDNA data that have been advanced, and in the alternative calibration of rates of mtDNA change that results in date estimates in concordance with split-times ascertained from both the fossil record and from nuclear DNA hybridization studies. If there were early Middle Pleistocene population divergences from a common African ancestor for the predecessors of modern, regionally-distinct groups, this would also contradict the Garden of Eden hypothesis. At so early a date the newly-divergent populations (and their immediate common ancestor) could not possibly be modern *Homo sapiens*. Therefore, if science can be said to progress by discovering refutations, in my opinion the Garden of Eden hypothesis can be said to be refuted.

Finally, both Cann *et al.* (1987) and Gould (1987, 1988) have indicated that the Garden of Eden hypothesis should be regarded as an important justification for accepting the reality of 'the underlying unity of all humans'. This is because the recent common-ancestry interpretation is said to be a means of showing that all human beings are closely related. But the contrary is not true, and the opposing hypothesis about multiregional human evolution does not support a different precept. In fact quite the opposite. The interpretation of human evolution as a persistent shifting pattern of population contacts and shared ideas may provide an even stronger biological basis for accepting the unity of all humanity. The spread of humankind and its differentiation into distinct geographic groups that persisted through long periods of time, with evidence of long-lasting contact and cooperation, in many ways is a more satisfying interpretation of human prehistory than a scientific rendering of the story of Cain, based on one population quickly, and completely, and most likely violently, replacing all others. This rendering of modern population dispersals is a story of 'making war and not love', and if true its implications are not pleasant. Of course, that an alternative interpretation is more satisfying does not make it true, but if valid for factual reasons there is no reason not to be more satisfied! A factual basis for supporting the multiregional evolution interpretation is provided by the prehistoric record itself.

ACKNOWLEDGEMENTS

I am very grateful to Alan Mann, Chris Stringer, David W. Frayer, Con Childress, Rachel Caspari, Karen Rosenberg and Kathy Stoner for their indispensable help in editing this paper, and for aiding me in rethinking some of the precepts in it. I am deeply indebted to Alan Thorne and Wu Xinzhi, my co-workers in developing the multiregional evolution hypothesis, for permission to examine the fossil human remains in their care, and for allowing me to join them in the development of these ideas about the pattern of human evolution in the regions they know so well. For the courtesy, cooperation, and kindness extended during my visits to their institutions, I thank the late G. H. R. von Koenigswald of the Senckenberg Museum; T. Jacob of the Gadjah Mada University Medical School; D. Kadar of the Bandung Geological Museum; C. B. Stringer of the British Museum (Natural History); I. Tattersall of the American Museum of Natural History; and P. Brown of the Department of Prehistory, University of New England. The research upon which this paper is based was supported by NSF grants BNS 75-21756 and BNS 75-82729, and grants from the National Academy of Sciences Committee for Scholarly Exchange with the People's Republic of China.

REFERENCES

Aigner, J. S. 1976. Chinese Pleistocene cultural and hominid remains: a consideration of their significance in reconstructing the pattern of human bio-cultural development. In A. K. Gosh (ed.) *Le Paléolithique Inférieur et Moyen en Inde, en Asie Centrale, en Chine et dans le sud-est Asiatique*. Paris: Centre Nationale de la Recherche Scientifique: 65-90.

Andrews, P. J. 1984. An alternative interpretation of the characters used to define *Homo erectus*. In P. J. Andrews and J. L. Franzen (eds) *The Early Evolution of Man, with Special Emphasis on Southeast Asia and Africa. Courier Forschungsinstitut Senckenberg* 69: 167-175.

Andrews, P. J. and Franzen, J. L. (eds) 1984. *The Early Evolution of Man, with Special Emphasis on Southeast Asia and Africa. Courier Forschungsinstitut Senckenberg* 69.

Avise, J. C., Neigel, J. E. and Arnold, J. 1984. Demographic influences on mitochondrial DNA lineage survivorship in animal populations. *Journal of Molecular Evolution* 20: 99-105.

Barton, N. H. 1983. Multilocus clines. *Evolution* 37: 454-471.

Barton, N. H. and Charlesworth, B. 1984. Genetic revolutions, founder effects and speciation. *Annual Review of Ecology and Systematics* 15: 133-164.

Binford, L. R. 1984. *Faunal Remains From Klasies River Mouth*. New York: Academic Press.

Binford, L. R. 1986. Reply to Singer and Wymer: 'On Binford on Klasies River Mouth: response of the excavators'. *Current Anthropology* 27: 57-62.

Binford, S. R. 1968. Early Upper Pleistocene adaptations in the Levant. *American Anthropologist* 70: 707-717.

Bodmer, W. F. and Cavalli-Sforza, L. L. 1976. *Genetics, Evolution and Man*. San Francisco: Freeman.

Bräuer, G. 1980 Die morphologischen affinitäten des jungpleistozänen stirnbeines aus dem Elbmündungsgebiet bei Hahnöfersand. *Zeitschrift für Morphologie und Anthropologie* 71: 1-42.

Bräuer, G. 1981. New evidence on the transitional period between Neanderthal and modern man. *Journal of Human Evolution* 10: 467-474.

Bräuer, G. 1982. A comment on the controversy 'Allez Neanderthal'. *Journal of Human Evolution* 11: 439-440.

Bräuer, G. 1984a. The 'Afro-European sapiens hypothesis' and hominid evolution in East Asia during the late Middle and Upper Pleistocene. In P. J. Andrews and J. L. Franzen (eds) *The Early Evolution of Man, with Special Emphasis on Southeast Asia and Africa. Courier Forschungsinstitut Senckenberg* 69: 145-165.

Bräuer, G. 1984b. A craniological approach to the origin of anatomically modern *Homo sapiens* in Africa and implications for the appearance of modern Europeans. In F. H. Smith and F. Spencer (eds) *The Origins of Modern Humans: a World Survey of the Fossil Evidence*. New York: Alan R. Liss: 327-410.

Bräuer, G. and Leakey, R. E. 1986. The ES-11693 cranium from Eliye Springs, West Turkana, Kenya. *Journal of Human Evolution* 15: 289-312.

Brown, W. M. 1980. Polymorphism in mitochondrial DNA of humans as revealed by restriction endonuclease analysis. *Proceedings of the National Academy of Sciences* (USA) 77: 3605-3609.

Brown, W. M., Prager, E. M., Wang, A. and Wilson, A. C. 1982. Mitochondrial DNA sequences of primates: tempo and mode of evolution. *Journal of Molecular Evolution* 18: 225-239.

Bryan, A. L. (ed.) 1986. *New Evidence for the Pleistocene Peopling of the Americas*. Orono (Me): Centre for the Study of Early Man, University of Maine.

Bryant, E. H., Combs, L. M. and McCommas, S. A. 1986. Morphometric differentiation among experimental lines of the housefly in relation to a bottleneck. *Genetics* 114: 1213-1223.

Bryant, E. H., McCommas, S. A. and Combs, L. M. 1986. The effect of an experimental bottleneck on quantitative genetic variation in the housefly. *Genetics* 114: 1191-1211.

Bunney, S. 1986. Chinese fossil could alter the course of evolution in Asia. *New Scientist* (11 September): 25.

Cann, R. L., Stoneking, M. and Wilson, A. C. 1987. Mitochondrial DNA and human evolution. *Nature* 325: 31-36.

Cann, R. L., Brown, W. M. and Wilson, A. C. 1982. Evolution of human mitochondrial DNA: a preliminary report. In B. Bonné-Tamir, P. Cohen and M. Goodman (eds) *Human Genetics: the Unfolding Genome*. New York: Alan R. Liss: 157-165.

Cann, R. L., Brown, W. M. and Wilson, A. C. 1984. Polymorphic sites and mechanisms of evolution in human mitochondrial DNA. *Genetics* 106: 479-499.

Carr, S. M., Ballinger, S. W., Derr, J. N., Blankenship, L. H. and Bickham, J. W. 1986. Mitochondrial DNA analysis of hybridization between sympatric white-tailed deer and mule deer in west Texas. *Proceedings of the National Academy of Sciences (USA)* 83: 9576-9580.

Carson, H. L. 1982 Speciation as a major reorganization of polygenic balances. In C. Barigozzi (ed.) *Mechanisms of Speciation*. New York: Alan R. Liss: 411-433.

Cavalli-Sforza, L. L. and Edwards, A. W. F. 1965. Analysis of human evolution. In S. J. Geerts (ed.) *Genetics Today: Proceedings of the XIth International Congress of Genetics*: 923-933.

Cavalli-Sforza, L. L., Barrai, I. and Edwards, A. W. F. 1964. Analysis of human evolution under random genetic drift. *Cold Spring Harbor Symposium on Quantitative Biology* 29: 9-20.

Chang, Kwangchi. 1963. *The Archeology of Ancient China*. New Haven: Yale University Press.

Charlesworth, B. 1983. Models of the evolution of some genetic systems. *Proceedings of the Royal Society of London* (Series B) 219: 265-279.

Charlesworth, B., Lande, R. and Slatkin, M. 1982. A Neo-Darwinian commentary on macroevolution. *Evolution* 36: 474-498.

Clark, W. E. LeGros 1964. *The Fossil Evidence for Human Evolution* (2nd edition). Chicago: University of Chicago Press.

Clarke, B. C. 1966. The evolution of morph-ratio clines. *American Naturalist* 100: 389-402.

Clarke, B. C. 1975. The causes of biological diversity. *Scientific American* 233: 50-60.

Clarke, R. J. 1985. A new reconstruction of the Florisbad cranium, with notes on the site. In E. Delson (ed.) *Ancestors: the Hard Evidence*. New York: Alan R. Liss: 301-305.
Cohan, F. M. 1984. Can uniform selection retard random genetic divergence between isolated conspecific populations? *Evolution* 38: 495-504.
Coon, C. S. 1962. *The Origin of Races*. New York: Knopf.
Day, M. H. 1972. The Omo human skeletal remains. In F. Bordes (ed.) *The Origin of Homo Sapiens*. Paris: UNESCO: 31-36.
Day, M. H. and Stringer, C. B. 1982. A reconsideration of the Omo Kibish remains and the *erectus-sapiens* transition. In M. A. de Lumley (ed.) *L'Homo erectus et la Place de l'Homme de Tautavel Parmi les Hominidés Fossiles:* Vol. 2. Nice: Centre National de la Recherche Scientifique / Louis-Jean Scientific and Literary Publications: 814-846.
Delson, E. 1985. Late Pleistocene human fossils and evolutionary relationships. In E. Delson (ed.) *Ancestors: the Hard Evidence*. New York: Alan R. Liss: 296-300.
Denaro, M., Blanc, H., Johnson, M. J., Chen, K. H., Wilmsen, E. and Cavalli-Sforza, L. L. 1981. Ethnic variation in Hpa I endonuclease cleavage patterns of human mitochondrial DNA. *Proceedings of the National Academy of Sciences (USA)* 78: 5768-5772.
Eckhardt, R. B. 1987. Evolution east of Eden. *Nature* 326: 749.
Edwards, A. W. F. 1971 Mathematical approaches to the study of human evolution. In R. Hodson, D. G. Kendall and P. Tauto (eds) *Mathematics in the Archaeological and Historical Sciences*. Edinburgh: Edinburgh University Press: 347-355.
Ehrlich, P. R. and Raven, P. H. 1969. Differentiation of populations. *Science* 165: 1228-1232.
Endler, J. A. 1973. Gene flow and population differentiation. *Science* 179: 243-250.
Endler, J. A. 1977. *Geographic Variation, Speciation, and Clines*. Princeton: Princeton University Press.
Ferembach, D. 1979. L'émergence du genre *Homo* et de l'espèce *Homo sapiens*. Les faits. Les incertitudes. *Biométrie Humaine* 14: 11-18.
Ferris, S. D., Sage, R. D., Huang, C. M., Nielsen, J. T., Ritte, U. and Wilson, A. C. 1983. Flow of mitochondrial DNA across a species boundary. *Proceedings of the National Academy of Sciences (USA)* 80: 2290-2294.
Fisher, R. A. 1950. Gene frequencies in a cline determined by selection and diffusion. *Biometrics* 6: 353-361.
Flood, J. 1983. *Archaeology of the Dreamtime*. Sydney: Collins.
Freudenthal, M. 1965 Betrachtungen über die Gattung *Cricetodon*. *Koninklijke Nederlandse Akademie Wetenschappen (Series B)* 68: 293-305.
Giles, E. and Ambrose, S. H. 1986. Are we all out of Africa? *Nature* 322: 21-22.
Goodnight, C. J. 1987. On the effect of founder events on epistatic variance. *Evolution* 41: 80-91.
Gould, S. J. 1987. Bushes all the way down. *Natural History* (June): 12-19.
Gould, S. J. 1988. A novel notion of Neanderthal. *Natural History* (June): 16-21.
Greenberg, J. H., Turner, C. G. and Zegura, S. L. 1986. The settlement of the Americas: a comparison of the linguistic, dental, and genetic evidence. Current Anthropology 27: 477-497.
Gribbin, J. and Cherfas, J. 1982. *The Monkey Puzzle: Reshaping the Evolutionary Tree*. London: Bodley Head.
Gruhn, R. 1987. On the settlement of the Americas: South American

evidence for an expanded timeframe. *Current Anthropology* 28: 363-365.
Guglielmino-Matessi, C.R., Gluckman, P. and Cavalli-Sforza, L.L. 1979. Climate and the evolution of skull metrics in man. *American Journal of Physical Anthropology* 50: 549-564.
Hale, L.R. and Singh, R.S. 1986. Extensive variation and heteroplasmy in size of mitochondrial DNA among geographic populations of Drosophila melanogaster. *Proceedings of the National Academy of Sciences (USA)* 83: 8813-8817.
Harpending, H.C. 1974. Genetic structure of small populations. *Annual Review of Anthropology* 3: 229-243.
Hasegawa, M., Kishino, H. and Yano, T. 1985. Dating of the human-ape splitting by a molecular clock of mitochondrial DNA. *Journal of Molecular Evolution* 22: 160-174.
Hogan, J.P. 1977. *Inherit the Stars*. New York: Ballantine.
Honeycutt, R.L. and Wheeler, W.C. 1987. Mitochondrial DNA: variation in humans and higher primates. In S.K. Dutta and W. Winter (eds) *DNA Systematics: Human and Higher Primates*. Boca Raton: CRC Press. In Press.
Howell, F.C. 1951. The place of Neanderthal man in human evolution. *American Journal of Physical Anthropology* 9: 379-416.
Howells, W. W. 1942. Fossil man and the origin of races. *American Anthropologist* 44: 182-193.
Howells, W. W. 1973. *Cranial variation in man: a study by multivariate analysis of patterns of difference among recent human populations*. Papers of the Peabody Museum of Archaeology and Ethnology 67. Cambridge (Mass): 1-259.
Howells, W. W. 1976. Explaining modern man: evolutionists versus migrationists. *Journal of Human Evolution* 5: 477-496.
Hublin, J.-J. 1976. *L'homme de Galilée*. Mémoire de DEA de Paléontologie. Paris: Université de Paris VI.
Jacob, T. 1976. Early populations in the Indonesian region. In R. L. Kirk and A. G. Thorne (eds), *The Origins of the Australians*. Canberra: Australian Institute of Aboriginal Studies: 81-93.
Jacob, T. 1981. Solo man and Peking man. In B.A. Sigmon and J.S. Cybulski (eds), *Homo erectus: Papers in Honor of Davidson Black*. Toronto: University of Toronto Press: 87-104.
Jain, S.K. and Bradshaw, A.D. 1966. Evolutionary divergence among adjacent plant populations. *Heredity* 21: 407-441.
Jelínek, J. 1969. Neanderthal man and *Homo sapiens* in Central and Eastern Europe. *Current Anthropology* 10: 475-503.
Jelínek, J. 1976. The *Homo sapiens neanderthalensis* and *Homo sapiens sapiens* relationship in Central Europe. *Anthropologie* (Brno) 14: 79-81.
Jelínek, J. 1978. Comparison of mid-Pleistocene evolutionary process in Europe and in South-East Asia. *Proceedings of the Symposium on Natural Selection, Liblice* 11978: 251-267.
Jelínek, J. 1980. Variability and geography: contribution to our knowledge of European and North African Middle Pleistocene hominids. *Anthropologie* (Brno) 18: 109-114.
Jelínek, J. 1982. The East and Southeast Asian way of regional evolution. *Anthropologie* (Brno) 21: 195-212.
Jelínek, J. 1985. The European, Near East and North African finds after australopithecus and the principal consequences for the picture of human evolution. In P. V. Tobias (ed.) *Hominid Evolution: Past, Present and Future. Proceedings of the Taung Diamond Jubilee International Symposium*. New York: Alan R. Liss: 341-354.
Jepsen, G. L., Simpson, G. G. and Mayr, E. (eds) 1949. *Genetics, Paleontology and Evolution*. Princeton: Princeton University Press.

Johnson, M. J., Wallace, D. C., Ferris, S. D., Rattazzi, M. C. and Cavalli-Sforza, L. L. 1983. Radiation of human mitochondrial DNA types analyzed by restriction endonuclease cleavage patterns. *Journal of Molecular Evolution* 19: 255-271.

Jones, J. S. 1986. The origin of *Homo sapiens*: the genetic evidence. In B. Wood, L. Martin and P. Andrews (eds) *Major Topics in Primate and Human Evolution*. Cambridge: Cambridge University Press: 317-330.

Jones, J. S. and Rouhani, S. 1986a. How small was the bottleneck? *Nature* 319: 449-450.

Jones, J. S. and Rouhani, S. 1986b. Mankind's genetic bottleneck. *Nature* 322: 599-600.

Kennedy, G. E. 1984. Are the Kow Swamp hominids 'archaic'? *American Journal of Physical Anthropology* 65: 163-168.

King, M. C. and Wilson, A. C. 1975. Evolution at two levels in humans and chimpanzees. *Science* 188: 107-116.

Kirk, R. L. and Szathmary, E. J. E. (eds) 1985. Out of Asia. *Journal of Pacific History*.

Kirk, R. L. and Thorne, A. G. (eds) 1976. *The Origin of the Australians*. Canberra: Australian Institute of Aboriginal Studies.

Lande, R. 1982. Rapid origin of sexual isolation and character divergence in a cline. *Evolution* 36: 213-233.

Larnach, S. L. and Macintosh, N. W. G. 1974. A comparative study of Solo and Australian Aboriginal crania. In A. P. Elkin and N. W. G. Macintosh (eds) *Grafton Elliot Smith: the Man and his Work*. Sidney: Sydney University press: 95-102.

Latorre, A., Mova, A. and Ayala, F. J. 1986. Evolution of mitochondrial DNA in *Drosophila subobscura*. *Proceedings of the National Academy of Sciences (USA)* 83: 8649-8653. 235; 1325-1327.

Lewontin, R. C. 1974. *The Genetic Basis of Evolutionary Change*. New York: Columbia University Press.

Lewontin, R. C. 1984. *Human Diversity*. San Francisco: Freeman.

Livingstone, F. B. 1973. Gene frequency differences in human populations: some problems of analysis and interpretation. In M. H. Crawford and P. L. Workman (eds) *Methods and Theories of Anthropological Genetics*. Albuquerque: University of New Mexico Press: 39-67.

Livingstone, F. B. 1980. Natural selection and random variation in human evolution. In J. H. Mielke and M. H. Crawford (eds) *Current Developments in Anthropological Genetics*. Vol 1: *Theory and Methods*. New York: Plenum Press: 87-110.

Lumley, M.-A. de and Sonakia, A. 1985. Première découverte d'un *Homo erectus* sur le Continent indien, à Hathnora, dans le Moyenne Vallée de la Narmada. *L'Anthropologie* 89: 13-61.

Macintosh, N. W. G. and Larnach, S. L. 1976. Aboriginal affinities looked at in world context. In R. L. Kirk and A. G. Thorne (eds) *The Origin of the Australians*. Canberra: Australian Institute of Aboriginal Studies: 113-126.

Magori, C. C. and Day, M. H. 1983. Laetoli hominid 18: an early *Homo sapiens* skull. *Journal of Human Evolution* 12: 747-753.

Martin, R. A. 1970. Line and grade in the extinct medium species of *Sigmodon*. *Science* 166: 1504-1506.

Mayr, E. 1963. *Animal Species and Evolution*. Cambridge (Mass): Belknap.

Mayr, E. 1982. Processes of speciation in animals. In C. Barigozzi (ed.) *Mechanisms of Speciation*. New York: Alan R. Liss: 1-19.

Morton, N. E. and Lalouel, J. 1973. Topology of kinship in Micronesia. *American Journal of Human Genetics* 25: 422-432.

Nei, M. 1985. Human evolution at the molecular level. In K. Aoki and T. Ohta (eds) *Population Genetics and Molecular Evolution*. Tokyo: Japan Science Society Press: 41-64.

Nei, M. 1987. *Molecular Evolutionary Genetics*. New York: Columbia University Press.

Nei, M. and Roychoudhury, A. K. 1982. Genetic relationship and evolution of human races. In B. Wallace and G.T. Prace (eds) *Evolutionary Biology* Vol. 14. New York: Plenum Press: 1-59.

Parsons, P. A. 1983. *The Evolutionary Biology of a Colonizing Species*. Cambridge: Cambridge University Press.

Protsch, R. 1975. The absolute dating of Upper Pleistocene sub- Saharan fossil hominids and their place in human evolution. *Journal of Human Evolution* 4: 297-322.

Protsch, R. 1978. *Catalogue of Fossil Hominids of North America*. New York: Fischer.

Rightmire, G. P. 1976. Relationships of Middle and Upper Pleistocene hominids from sub-Saharan Africa. *Nature* 260: 238-240.

Rightmire, G. P. 1978. Florisbad and human population succession in Southern Africa. *American Journal of Physical Anthropology* 48: 475-486.

Rightmire, G. P. 1979. Implications of the Border Cave skeletal remains for later Pleistocene human evolution. *Current Anthropology* 20: 23-35.

Rightmire, G. P. 1981. Later Pleistocene hominids of Eastern and Southern Africa. *Anthropologie* (Brno) 19: 15-26.

Rightmire, G. P. 1984a. *Homo sapiens* in sub-Saharan Africa. In F. H. Smith and F. Spencer (eds) *The Origins of Modern Humans: a World Survey of the Fossil Evidence*. New York: Alan R. Liss: 295- 325.

Rightmire, G. P. 1984b. Comparisons of *Homo erectus* from Africa and Southeast Asia. In P. J. Andrews and J. L. Franzen (eds) *The Early Evolution of Man, with Special Emphasis on Southeast Asia and Africa*. *Courier Forschungsinstitut Senckenberg* 69: 83-98.

Rightmire, G. P. 1987. L'Evolution des premiers hominidés en Asie du Sud-Est. *L'Anthropologie* 91: 455-465.

Robbins, L. H. 1972. Archaeology in the Turkana district, Kenya. *Science* 176: 359-366.

Robbins, L. H. 1974. *The Lothagam Site*. Michigan State University Museum Anthropological Series 1 (2).

Robbins, L. H. 1980. *Lopov: a Late Stone Age Fishing and Pastoral Settlement in the Lake Turkana Basin, Kenya*. Michigan State University Museum Anthropological Series 3 (1).

Saitou, N. and Omoto, K. 1987. Time and place of human origins from mtDNA data. *Nature* 327: 288.

Sartono, S. 1982. Characteristics and chronology of early men in Java. In M. A. de Lumley (ed.) *L'Homo erectus et la Place de l'Homme de Tautavel Parmi les Hominidés Fossiles:* Vol. 2. Nice: Centre National de la Recherche Scientifique / Louis-Jean Scientific and Literary Publications: 491-541.

Schwalbe, G. 1904. *Die Vorgeschichte des Menschen*. Braunschweig: Friedrich.

Sibley, C. G. and Ahlquist, J. E. 1984. The phylogeny of the hominoid primates, as indicated by DNA-DNA hybridization. *Journal of Molecular Evolution* 20: 2-15.

Singer, R. 1958. The Rhodesian, Florisbad, and Saldanha skulls. In G. H. R. Von Koenigswald (ed.) *Hundert Jahre Neanderthaler*. Köln: Bohlau: 52-62.

Slatkin, M. 1987. Gene flow and the geographic structure of natural populations. *Science* 236: 787-792.

Smith, F. H. 1982. Upper Pleistocene hominid evolution in South-Central Europe: a review of the evidence and analysis of trends. *Current Anthropology* 23: 667-703.

Smith, F. H. 1984. Fossil hominids from the Upper Pleistocene of Central Europe and the origin of modern Europeans. In F. H. Smith and F. Spencer (eds) *The Origins of Modern Humans: a World Survey of the Fossil Evidence*. New York: Alan R. Liss: 137-209.

Smith. F. H. 1985. Continuity and change in the origin of modern *Homo sapiens. Zeitschrift für Morphologie und Anthropologie* 75: 197-222.

Stewart, T. D. 1974. Perspectives on some problems of early man common to America and Australia. In A. P. Elkin and N. W. G. Macintosh (eds) *Grafton Elliot Smith: the Man and his Work*. Sydney: Sydney University Press: 114-135.

Stoneking, M., Bhatia, K. and Wilson, A. C. 1986. Rate of sequence divergence estimated from restriction maps of mitochondrial DNAs from Papua New Guinea. *Cold Spring Harbor Symposia on Quantitative Biology* 51: 433-439.

Stringer, C. B. 1984a. Human evolution and biological adaptation in the Pleistocene. In R. Foley (ed.) *Hominid Evolution and Community Ecology: Prehistoric Human Adaptation in Biological Perspective*. London: Academic Press: 55-83.

Stringer, C. B. 1984b. The definition of *Homo erectus* and the existence of the species in Africa and Europe. In P. J. Andrews and J. L. Franzen (eds) *The Early Evolution of Man, with Special Emphasis on Southeast Asia and Africa. Courier Forschungsinstitut Senckenberg* 69: 131-143.

Stringer, C. B. 1984c. The fate of the Neanderthals. *Natural History* (December): 6-12.

Stringer, C. B. 1985. Middle Pleistocene hominid variability and the origin of Late Pleistocene humans. In E. Delson (ed.) *Ancestors: the Hard Evidence*. New York: Alan R. Liss: 289-295.

Stringer, C. B. 1989. Documenting the origin of modern humans. In E. Trinkaus (ed.) *Patterns and Processes in Later Pleistocene Human Emergence*. Cambridge: Cambridge University Press. In Press.

Szathmary, E. J. E. and Ossenberg, N. S. 1978. Are the biological differences between North American Indians and Eskimos truly profound? *Current Anthropology* 19: 673-701.

Tattersall, I. 1986. Species recognition in human palaeontology. *Journal of Human Evolution* 15: 165-176.

Templeton, A. R. 1982. Genetic architectures of speciation. In C. Barigozzi (ed.) *Mechanisms of Speciation*. New York: Alan R. Liss: 105-121.

Templeton, A. R. 1983. Phylogenetic inference from restriction endonuclease cleavage site maps with particular reference to the evolution of humans and the apes. *Evolution* 37: 221-244.

Thoma, A. 1985 *Eléments de Paléoanthropologie*. Louvain-la-Neuve: Institut Supérieur d'Archéologie et d'Histoire de d'Art.

Thorne, A. G. 1981. The centre and the edge: the significance of Australian hominids to African palaeoanthropology. In R. E. Leakey and B. A. Ogot (eds) *Proceedings of the 8th Panafrican Congress of Prehistory and Quaternary Studies, Nairobi, September 1977*. Nairobi: TILLMIAP: 180-181.

Thorne, A. G. 1984. Australia's human origins: how many sources? *American Journal of Physical Anthropology* 63: 227.

Thorne, A. G. and Wilson, S. R. 1977. Pleistocene and recent Australians: a multivariate comparison. *Journal of Human Evolution* 6: 393-402.

Thorne, A. G. and Wolpoff, M. H. 1981. Regional continuity in Australasian Pleistocene hominid evolution. *American Journal of*

Physical Anthropology 55: 337-349.
Trinkaus, E. and Thompson, D. D. 1987. Femoral diaphyseal histomorphometric age determinations for the Shanidar 3, 4, 5, and 6 Neandertals and Neandertal longevity. *American Journal of Physical Anthropology* 72: 123-129.
Turner, A. 1984. Hominids and fellow travellers: human migration into high latitudes as part of a large mammal community. In R. Foley (ed.) *Hominid Evolution and Community Ecology: Prehistoric Human Adaptation in Biological Perspective*. London: Academic Press: 193-217.
Valladas, H., Reyss, J. L., Joron, J. L., Valladas, G., Bar-Yosef, O. and Vandermeersch, B. 1988. Thermoluminescence dating of Mousterian 'Proto-Cro-Magnon' remains from Israel and the origin of modern man. *Nature* 331: 614-616.
Van Valen, L. M. 1966. On discussing human races. *Perspectives in Biology and Medicine* 9: 377-383.
Van Valen, L. M. 1986. Speciation and our own species. *Nature* 322: 412.
Van Vark, G. N. 1983. Did our *Homo erectus* ancestors live in Eastern Asia? *Homo* 34: 148-153.
Vandermeersch, B. 1981. *Les Hommes Fossiles de Qafzeh (Israël)*. Paris: Centre National de la Recherche Scientifique.
Wainscoat, J. 1987. Out of the Garden of Eden. *Nature* 325: 13.
Wainscoat, J. S., Hill, A. V. S., Boyce, A. L., Flint, J., Hernandez, M., Thein, S. L., Old, J. M., Lynch, J. R., Falusi, A. G., Weatherall, D. J. and Clegg, J. B. 1986. Evolutionary relationships of human populations from an analysis of nuclear DNA polymorphisms. *Nature* 319: 491-493.
Wallace, D. C., Garrison, K. and Knowler, W. C. 1985. Dramatic founder effects in Amerindian Mitochondrial DNAs. *American Journal of Physical Anthropology* 68: 149-155.
Weckler, J. E. 1957. Neanderthal man. *Scientific American* 197: 89-97.
Wei Liming 1984. 200 000 year-old skeleton unearthed. *Beijing Review* 49 (December 3): 33.
Weidenreich, F. 1939. Six lectures on *Sinanthropus pekinensis* and related problems. *Bulletin of the Geological Society of China* 19: 1-110.
Weidenreich, F. 1943. The skull of *Sinanthropus pekinensis*: a comparative study of a primitive hominid skull. *Palaeontologia Sinica* (n.s. D) 10 (whole series 127). Beijing: Geological Survey of China.
Weidenreich, F. 1945. The Keilor skull: a Wadjak skull from southeast Australia. *American Journal of Physical Anthropology* 3: 21-33.
Weidenreich, F. 1946. *Apes, Giants and Man*. Chicago: University of Chicago Press.
Weidenreich, F. 1947. The trend of human evolution. *Evolution* 1: 221-236.
Weidenreich, F. 1951. Morphology of Solo man. *Anthropological Papers of the American Museum of Natural History* 43: 205-290.
Weinert, H. 1951. *Stammesentwicklung des Menschheit*. Braunschweig: Vieweg and Sohn.
Weiss, K. M. 1984. On the number of members of the genus *Homo* who have ever lived, and some evolutionary implications. *Human Biology* 56: 637-649.
Weiss, K. M. 1986. In search of times past: the roles of gene flow and invasion in the generation of human diversity. In N. Mascie-Taylor and G. Lasker (eds) *Biological Aspects of Human Migration*. London: Cambridge University Press.
Weiss, K. M. and Maruyama, T. 1976. Archaeology, population genetics and studies of human racial ancestry. *American Journal of Physical Anthropology* 44: 31-50.

White, M. J. D. 1978. *Models of Speciation*. San Francisco: Freeman.
Whittam, T. S., Clark, A. G., Stoneking, M., Cann, R. and Wilson, A. C. 1986. Allelic variation in human mitochondrial genes based on patterns of restriction site polymorphism. *Proceedings of the National Academy of Sciences (USA)* 83: 9611-9615.
Wilson, A. C., Cann, R. L., Carr, S. M., George, M., Gyllensten, U. B., Helm-Bychowski, K. M., Higuchi, R. G., Palumbi, S. R., Prager, E. M., Sage, R. D. and Stoneking, M. 1985. Mitochondrial DNA and two perspectives on evolutionary genetics. *Biological Journal of the Linnean Society* (London) 26: 375-400.
Wolpoff, M. H. 1980. *Paleoanthropology*. New York: Knopf.
Wolpoff, M. H. 1982. Comment on F. H. Smith 'Upper Pleistocene hominid evolution in South-Central Europe'. *Current Anthropology* 23: 693.
Wolpoff, M. H. 1982. The Arago dental sample in the context of hominid dental evolution. In H. de Lumley (ed.) *L'Homo erectus et la Place de l'Homme de Tautavel Parmi les Hominidés Fossiles:* Vol. 1. Nice: Centre National de la Recherche Scientifique / Louis-Jean Scientific and Literary Publications: 389-410.
Wolpoff, M. H. 1985. Human evolution at the peripheries: the pattern at the eastern edge. In P. V. Tobias (ed.) *Hominid Evolution: Past, Present and Future. Proceedings of the Taung Diamond Jubilee International Symposium*. New York: Alan R. Liss: 355-365.
Wolpoff, M. H. 1986. Describing anatomically modern *Homo sapiens*: a distinction without a definable difference. In V. V. Novotny and A. Miserová (eds) *Fossil Man: New Facts, New Ideas. Papers in Honor of Jan Jelínek's Life Anniversary. Anthropos* (Brno) 23: 41- 53.
Wolpoff, M. H., Wu Xinzhi and Thorne, A. G. 1984. Modern *Homo sapiens* origins: a general theory of hominid evolution involving the fossil evidence from East Asia. In F. H. Smith and F. Spencer (eds) *The Origins of Modern Humans: a World Survey of the Fossil Evidence*. New York: Alan R. Liss: 411-483.
Woo Jukang (Wu Rukang) 1959. Human fossils found in Liujiang, Kwangsi, China. *Vertebrata PalAsiatica* 3: 109-118.
Wright, S. 1931. Evolution in Mendelian populations. *Genetics* 16: 97-159.
Wright, S. 1940. Breeding structure of populations in relation to speciation. *American Naturalist* 74: 232-248.
Wright, S. 1943. Isolation by distance. *Genetics* 28: 114-138.
Wright, S. 1967. 'Surfaces' of selective value. *Proceedings of the National Academy of Sciences (USA)* 58: 165-172.
Wright, S. 1977. *Evolution and the Genetic of Populations*. Vol. 3: *Experimental Results and Evolutionary Deductions*. Chicago: University of Chicago Press.
Wu Rukang 1986. Chinese human fossils and the origin of Mongoloid racial group. In V. V. Novotny and A. Miserová (eds) *Fossil Man: New Facts, New Ideas. Papers in Honor of Jan Jelínek's Life Anniversary. Anthropos* (Brno) 23: 151-155.
Wu Rukang and Dong Xingren 1982. Preliminary study of *Homo erectus* remains from Hexian, Anhui. *Acta Anthropologica Sinica* 1: 2-13.
Wu Rukang and Dong Xingren 1985. *Homo erectus* in China. In Wu Rukang and J. W. Olsen (eds) *Paleoanthropology and Paleolithic Archaeology in the People's Republic of China*. New York: Academic Press: 79-89.
Wu Rukang and Lin Shenglong 1985. Chinese paleoanthropology: retrospect and prospect. In Wu Rukang and J.W. Olsen (eds) *Paleoanthropology and Paleolithic Archaeology in the People's Republic of China*. New York: Academic Press: 1-27.

Wu Rukang and Wu Xinzhi 1982. Comparison of Tautavel man with *Homo erectus* and early *Homo sapiens* in China. In H. de Lumley (ed.) *L'Homo erectus et la Place de l'Homme de Tautavel Parmi les Hominidés Fossiles:* Vol. 2. Nice: Centre National de la Recherche Scientifique / Louis-Jean Scientific and Literary Publications: 605-616.

Wu Xinzhi 1961. Study on the Upper Cave man of Choukoutien. *Vertebrata PalAsiatica* 3: 202-211.

Wu Xinzhi 1981. A well-preserved cranium of an archaic type of early *Homo sapiens* from Dali, China. *Scientia Sinica* 24: 530- 541.

Wu Xinzhi and Wu Maolin 1985. Early *Homo sapiens* in China. In Wu Rukang and J. W. Olsen (eds) *Paleoanthropology and Paleolithic Archaeology in the People's Republic of China*. New York: Academic Press: 91-106.

Wu Xinzhi and Zhang Yinyun 1978. Fossil man in China. In *Symposium on the Origin of Man*. Beijing: Science Press: 28-42.

7. Middle Stone Age Humans from Eastern and Southern Africa

G. PHILIP RIGHTMIRE

INTRODUCTION

Human remains from Middle Stone Age sites in Eastern and Southern Africa are rather scarce. Unlike the Neanderthals of Europe, these people of sub-Saharan Africa do not seem to have buried their dead in caves or shelters, and no very complete skeletons have been recovered. Nevertheless, the importance of the scattered fossils that have come to light is widely recognized. It was suggested some years ago that anatomically-modern people could be associated with the Middle Stone Age, and this claim has been substantiated. There is no doubt that some of the African skeletal material is of modern aspect, and there is good evidence that it is old, particularly at sites such as Klasies River Mouth. Other finds are still suspect, because of the fragmentary nature of the bones themselves or because of questions concerning archaeological context. It is likely that some of these problems will be resolved soon, as new work progresses.

I propose here to comment broadly on several assemblages, drawn from localities in Eastern and in Southern Africa. Most of the material can be associated at least tentatively with signs of Middle Stone Age activity. In keeping with the focus of this Symposium, I shall not emphasize any of the more archaic crania or jaws which have been discovered in Acheulian contexts. Although finds such as Ndutu, Broken Hill and Elandsfontein are usually referred to as *Homo sapiens,* it is in fact difficult to define clear links between these hominids and fully modern populations (Rightmire 1986). Fossils of more direct concern are known from the Omo, Laetoli and Florisbad, as well as from Upper Pleistocene sites where an association with Middle Stone Age occupation is more secure.

There is no need to touch on each of these specimens in detail, as all have been treated before, by myself and other workers. It will be enough to summarize the principal aspects of morphology, in an attempt to make some statements about change in sub-Saharan Africa. Although the record is still very poor, it is important to search for trends which characterize the evolution of these latest Middle Pleistocene and Upper Pleistocene populations. It will also be helpful to compare the picture emerging for Africa with that documented for Europe and Western Asia. Already it is apparent that there are differences in timing of the transition from archaic to more modern people. Other aspects of this transition as read so far from the Eurasian record may also have to be reinterpreted for Africa.

HOMINIDS FROM THE LATER MIDDLE PLEISTOCENE

Sites which are mentioned frequently in discussions of the origins of modern humans include both the Omo in Ethiopia and Laetoli in Tanzania. A partial skeleton, another cranium and fragments of a third individual were recovered from the Kibish Formation in the Omo Valley in 1967 (Leakey 1969). The skeleton of Omo 1 was found *in situ* at the base of Member 1, while the Omo 2 cranium was picked up on the surface a couple of kilometres away. Geological investigations at this second locality show that Omo 2 must be approximately contemporary with Omo 1. From time to time, uncertainty about the provenience of these individuals has been expressed, but such doubts have no firm basis (Butzer 1987). A date of 130 000 years obtained from shells in the deposits has been widely quoted but has still to be confirmed.

The hominid from Laetoli seems to be of about the same age. Most of a human cranium, some animal bones, and artifacts apparently linked with the Middle Stone Age were collected in the Ngaloba Beds at Laetoli in 1976 (Day, Leakey and Magori 1980). Tuff mineral content provides a basis for correlation with the stratigraphy at Olduvai Gorge, and the Laetoli deposits are estimated to be about 120 000 years old (Leakey and Hay 1982). This date is supported by Uranium-series determinations made recently at the site.

Of the Omo crania, Omo 2 is the more complete. This braincase is long and low in outline, and the back of the vault is strongly curved. There is general agreement that this individual is archaic in appearance, and many features recall the morphology of Broken Hill. At the same time, the frontal bone is relatively broad and flattened, and the supraorbital torus is appreciably less thickened than that of other specimens representing early *Homo sapiens*.

Omo 1 is said to exhibit more modern features. This is a fair interpretation of the cranial evidence, which consists mostly of an occiput and parietal fragments. Some of the frontal bone and a few pieces of the face are also available, and these have been worked into a reconstruction of Omo 1 by Day and Stringer (1982). Certainly the rear of this vault is not archaic in outline, and the occipital, which has a relatively long upper scale, is less sharply curved than that of Omo 2. The reconstruction suggests a long, flattened frontal, carrying a brow which is not massively developed.

Day and Stringer (1982) argue that Omo 1 must represent an anatomically-modern population, whereas Omo 2 is likely to be drawn from a distinct, archaic group. Other workers are reluctant to accept the view that different populations must be present, given the very incomplete state of the Omo 1 braincase. If the geological setting of the fossil localities provides no basis for separating these individuals in time, then perhaps the archaic aspects of Omo 2, which are brought out clearly in all comparisons, should be emphasized as a link between the Kibish people and populations represented by Ndutu or Broken Hill. It may be too soon to identify the Omo skeletons as related unequivocally to recent humans (Rightmire 1984).

Table 7.1. Measurements of crania recovered at selected later Middle Pleistocene and earliest Upper Pleistocene localities in Sub-Saharan Africa

	Broken Hill	Omo 1	Omo 2	Laetoli LH 18	Eliye Springs	Florisbad	Border Cave
Cranial length	205	–	–	204	–	–	–
Maximum cranial breadth	–	–	147	140	158	–	–
Biauricular breadth	140?	–	132	125?	>140	–	–
Maximum frontal breadth	118	–	121	115?	118	132?	–
Minimum frontal breadth	98	–	108	101	110	116?	108
Biorbital chord	125	–	–	–	–	124?	112?
Postorbital constriction index	78.4	–	–	–	–	93.5	96.4
Frontal sagittal chord	121	–	–	116	–	120	116
Frontal angle	141.5	–	–	143	–	137.5	122
Maximum biparietal breadth	145?	–	–	140	158	–	–
Parietal sagittal chord	113	–	119	116	117	–	–
Lambda-asterion chord	–	–	–	87	92	–	–
Biasterionic breadth	–	–	–	117	119	–	–
Occipital sagittal chord	89	101	106	–	93	–	–
Occipital angle	–	117	105	–	109	–	–
Lambda-inion	–	73	59	60	64	–	–
Inion-opisthion	–	47	74	–	48	–	–
Occipital scale index	–	64.3	125.4	–	75.0	–	–

The early Upper Pleistocene cranium from Laetoli offers more information. Day, Leakey and Magori (1980), Magori and Day (1983) and Bräuer (1984) suggest that this hominid (LH 18) displays primitive features, some of which are shared with Broken Hill. In my view, these claims are exaggerated. LH 18 seems to provide firmer evidence than does the Omo assemblage for a modern presence in East Africa. The flattened frontal bone shows practically no keeling in the midline. Although the supraorbital torus and temporal crest are damaged on both sides, it is clear that postorbital constriction is slight. The brow itself is only moderately thickened and differs markedly from that of Broken Hill. It may be comparable in size to the Omo tori, but there are resemblances in shape to later Upper Pleistocene specimens such as Lukenya Hill.

The parietals are large, and length of the bregma-lambda chord exceeds that of other archaic crania, including Broken Hill but not Omo 2 (see Table 7.1). Parietal bossing is apparent, and in rear view the Laetoli vault has a rounded profile. Greatest breadth falls higher on the parietal bones, rather than at the supramastoid crests. Biauricular width, taken at the cranial base, is relatively narrow. The mastoid portions of the temporal bones are not so laterally projecting as in Broken Hill and Omo 2, and there is little inturning of the mastoid tips. The small size of the mastoid process is said by Magori and Day (1983) to be a primitive character, but in fact projection of this process is subject to a lot of variation, both in archaic and in more recent humans.

Curvature of the occiput cannot be measured accurately, but this bone is less sharply flexed than that of Broken Hill. The upper scale presents a mound-like transverse torus. This torus has no clear upper margin, and swelling associated with it extends over much of the occipital plane, to give the rear of the LH 18 vault an inflated appearance. This is accentuated by hollowing of the nuchal area. The lower scale itself is damaged anteriorly but must be relatively short. On the right side, a little of the occipitomastoid junction is preserved, and there is some heaping up of bone on the medial aspect of the digastric incisure to produce a juxtamastoid eminence. Both glenoid cavities can be examined, and an articular tubercle is clearly defined. Here there is no departure from the anatomy characteristic of modern humans.

Another cranium picked up recently on the western shore of Lake Turkana, Kenya, also deserves comment, along with the hominids from Laetoli and the Omo. This new fossil was found in beach deposits near Eliye Springs, but its original geological provenience is unclear. So far, no dates are available. The cranium has been described by Bräuer and Leakey (1986). Although much of the facial skeleton is missing or heavily damaged, the braincase is quite well preserved. There is some distortion, which affects mostly the right side. The vault is large and must be nearly as long as LH 18. It is very broad, both at the supramastoid crests and higher on the parietal bones.

Despite damage, it is clear that the frontal is wide, and postorbital narrowing cannot have been pronounced. The frontal squama is rather less flattened than that of LH 18. Both sagittal and lambdoid chords taken on the parietal bone are slightly longer than comparable measurements of the Laetoli hominid. Bräuer and Leakey (1986) comment on the large size of the parietals but suggest nevertheless that the broad, low vault is archaic in overall shape. They point especially to prominence of the temporal lines and to the fact that some heaping up of bone occurs along the margins of the sagittal suture.

Breadth of the occipital taken at the asteria is great, relative to length measured from lambda to opisthion. At the same time, the occiput is (again) less sharply angled than that of Broken Hill, and the upper scale is relatively high. This portion of the occipital is rounded in profile. Bräuer and Leakey report that there is no transverse torus, but there is an area of increased swelling near the midline, above the superior nuchal lines. At the centre of this mound-like torus, there is a small depression, said to resemble the suprainiac fossa found in European Neanderthal crania. Below the superior lines, areas of muscle insertion are deeply excavated in the nuchal plane. The shape of the Eliye Springs occipital thus resembles that of LH 18 in many respects, and in both individuals there is noticeable projection of the rearmost part of the braincase.

Bräuer and Leakey conclude that the new Turkana cranium presents a mosaic of characters. The low vault and broad base recall *Homo erectus*, whereas frontal form, parietal size and occipital proportions align Eliye Springs with *Homo sapiens*. In fact, there can be little doubt that this individual is drawn from our own species, and a more pertinent question concerns the significance of the few archaic features that are expressed in a braincase otherwise of modern aspect. Bräuer and Leakey see affinities to both early and late 'grades' of archaic *Homo sapiens*. In my own opinion, there are some definite resemblances to more recent humans. The morphology of this large, short-faced cranium seems decidedly less archaic than that of Broken Hill, Elandsfontein, or the small hominid from Lake Ndutu. As with Laetoli 18, there are few similarities to Neanderthals, even in the occiput. The Eliye Springs occipital may show a faint suprainiac depression, but this is lacking in LH 18. Neither specimen has posterior vault proportions quite like those of archaic Western Europeans.

Additional human remains have been recovered from spring sands at Florisbad in South Africa. These consist of a frontal, parietal pieces and the incomplete right side of a face. The fossils were originally set in a plaster reconstruction by Dreyer (1935), but for some time it has been recognized that Dreyer's reconstruction was seriously flawed. The nasal bones, part of the maxilla and right zygomatic bone are fragmentary and show no secure contacts with one another or with the frontal. Nevertheless, Dreyer joined these bones to produce a face which was short with a peculiar flaring cheek, a deep canine fossa and substantial subnasal projection. When these faults are corrected, the orbit is seen to be enlarged, and outward flare of its lateral margin is much reduced. Infraorbital hollowing is also less pronounced, and the lower face exhibits little prognathism (Rightmire 1978). Ron Clarke (1985) who has recently cleaned the Florisbad facial fragments, notes also that the right and left sides of the palate as joined by Dreyer were too close together. The upper jaw and nasal cavity as newly reconstructed by Clarke are wider than before. These several adjustments give the face an archaic appearance. There are resemblances to Broken Hill, although the brows are less thickened, and the domed frontal is not greatly constricted behind the orbits. Unfortunately the Florisbad remains are so incomplete that comparisons with other fossils must be very limited.

Dating of the Florisbad deposits is still tentative, but the level from which the bones were recovered is clearly beyond the limits of radiocarbon dating. Butzer's (1984) analysis of the sediment stratigraphy permits some comparisons with alluvial sequences in the interior of South Africa, and his work, along with studies of the spring-vent fauna, suggests a late Middle Pleistocene age. Because of contamination problems, recent attempts to obtain Uranium-series dates directly for the earlier Florisbad peats and for associated bone samples have not proved successful.

Kuman and Clarke (1986) have described the excavation of a site at Florisbad containing Middle Stone Age artifacts, fauna and a hearth. The authors view this assemblage as resulting from multiple, short-term occupations by human hunters, who killed and butchered a relatively small number of game species. This material however occurs at a level substantially higher than that yielding the hominid cranium. Other tools found lower in the deposits both by Kuman and Clarke and by earlier excavators cannot be linked to Dreyer's original find with any certainty. At Florisbad, as at the East African localities already discussed, the association of human fossils with tools diagnostic of particular Stone Age lifeways is far from firm. Perhaps only at Laetoli is there a strong(er) indication that the hominid remains occur with a Middle Stone Age industry.

UPPER PLEISTOCENE ASSEMBLAGES FROM SOUTHERN AFRICA

At other sites in Southern Africa, more abundant signs of Middle Stone Age occupation are preserved. A number of caves and shelters have been investigated systematically, and studies of sediment sequences, animal and plant remains, and lithic artifacts have provided quite a lot of information. Questions concerning subsistence and ecology, adaptation in response to climatic shifts, tool typology and culture stratigraphy have been reviewed recently by Deacon *et al.* (1984, and this volume), Klein (1983, and this volume), Singer and Wymer (1982) and Volman (1984). Human bones have been recovered at several of the caves, located both on the Cape coast and in the interior of South Africa. Although these remains have been found mostly in heavily compacted deposits, and as a consequence are badly broken, they provide important clues. More than the Omo crania or Florisbad, these bones are recognizable as anatomically-modern.

Some of the assemblages are very small and consist primarily of teeth. This is the case at Die Kelders cave, which lies just to the west of Cape Agulhas. The oldest sediments in this cave predate the arrival of Middle Stone Age people, who inhabited the site during the early Last Glacial. Fauna from the period of Middle Stone Age occupation is well preserved, but isolated human teeth including a lower molar, a lower premolar and several deciduous specimens are not particularly informative. More teeth are known from Equus Cave, situated near Taung in the northern Cape. Sandy deposits in this cave contain large quantities of animal bones thought to have been accumulated by hyaenas. Excavations in the earlier levels at the site have produced Middle Stone Age artifacts (Beaumont *et al.* 1984). These levels are not well dated but may approach 90 000 to 100 000 years in age (Grine and Klein 1985).

Eight permanent teeth excavated at Equus Cave, along with a fragmentary mandible, have been described by Grine and Klein (1985). The jaw, which was picked up outside the cave but is probably derived from the Middle Stone Age deposits, has dimensions close to those of modern people. In most instances, lengths and breadths of the dental crowns also fall comfortably within the range of variation recorded for Southern African blacks. One lower molar and an upper canine are comparatively large and have mesiodistal diameters that lie at or slightly beyond the limits for modern males. Breadths for these teeth are not especially great. Pulp cavity size for the molars has been checked on radiographs, and none of the Equus Cave specimens can be considered to be more than marginally taurodont. There is no hint of the pulp chamber enlargement that is prevalent among Upper Pleistocene populations of Western Europe. Morphology of the molar crowns as described by Grine is similar to that of present day Southern Africans.

At Border Cave, located in northern Natal, there is a long record of Middle Stone Age occupation. Studies of the sedimentary sequence and cultural material, radiocarbon dates, and microanalytical data from bone samples have been discussed by Beaumont (1980), Beaumont, de Villiers and Vogel (1978), and by Butzer, Beaumont and Vogel (1978). Results of this work show that the site was first inhabited prior to the beginning of the Last Interglacial, and levels containing Middle Stone Age artifacts are all older than 49 000 years. Human remains from Border Cave consist primarily of cranial or mandibular parts of three adult individuals and the skeleton of an infant. Two of the adult individuals were discovered by guano diggers, and there is lingering doubt as to their exact provenience in the deposits. However, soil found attached to the partial cranium is best matched by that from levels identified by Butzer (1987) with oxygen-isotope stage 5. If the Border cave cranium is assumed to have come from this part of the sequence, then it may be close to 100 000 years old.

All of the human fragments from Border Cave are demonstrably modern in their anatomy. When measurements relating to frontal form, supraorbital development and projection of the nasal root are treated in discriminant analysis, the adult cranium approaches several recent populations and appears to fit best with Khoi or with San males (Rightmire 1979). Study of the distances separating the fossil from modern group centroids in the discriminant space confirms that these assignments have biological significance, in that Border Cave does fall within the range of variation expected for the recent populations (Rightmire 1981; Campbell 1984). De Villiers and Fatti (1982) have conducted a similar analysis and have also concluded that Border Cave is close to living Africans. Here the adult cranium is said to relate to Nguni males rather than to San, but all of the statistical findings suggest that the Border Cave people are entirely comparable to recent humans.

Perhaps the most important of the South African sites is Klasies River Mouth, on the Cape coast. At Klasies, cave deposits have yielded stone artifacts, a molluscan fauna, and bird and mammal bones. Studies of the sedimentary sequence (Butzer 1978; Deacon, this volume) together with oxygen-isotope analysis of shells obtained from the deposits place the start of the Klasies Middle Stone Age occupation at the beginning of the Last Interglacial. This age is supported by J. C. Vogel's Uranium disequilibrium results of greater than 110 000 years obtained for samples of a speleothem formation that is stratigraphically younger than the earliest Middle Stone Age levels.

One of the oldest and most interesting of the Klasies human remains is a lower jaw. This relatively complete mandible is rather damaged, and all of the anterior teeth are missing. The alveoli show evidence of extensive resorptive change, linked to age and dental disease. Because of these changes, measurements are difficult to take. The corpus is heavily built, and there is a strong chin. Internally, there is no development of an alveolar planum or superior torus. Neither here nor in the contour of the medial wall of the corpus is there any indication of archaic morphology. Singer and Wymer (1982) comment that this jaw represents fully modern *Homo sapiens,* and there is no reason to question this conclusion.

Other mandibles, which are not quite as old as the first, vary considerably in size. One is very small and gracile and may have belonged to a female. Others are more massive, and one specimen is especially robust. In this latter individual, much of the anterior corpus is preserved, and the bone is complete on the right side to the level of M_2. The corpus is relatively thick, and Singer and Wymer (1982) note that its upper and lower borders are approximately parallel. However, the anterior height of the body is difficult to measure accurately because of erosion of the tooth sockets. It is likely that the front of the jaw was deeper before this damage occurred. The symphyseal axis is nearly vertical. At the centre of this surface there is a blunt swelling which broadens below to provide evidence of chin formation. These traits tend to set the Klasies fossil apart from more recent jaws, but at the same time there is little development of an alveolar planum and no expression of a superior transverse torus. It seems fair to conclude that the Klasies jaw is within the range of variation to be expected in a modern population.

Additional remains are mostly fragmentary. One especially informative specimen consists of part of a frontal bone on which glabella and some of the orbital margin are preserved and to which the upper ends of both nasal bones are still attached. The superciliary eminence is not prominent, and the nasal root is broad and flat. This frontal can be compared to Florisbad, and the difference in robusticity is striking. In the latter face, glabella is projecting and the supraorbital torus is much more heavily constructed. Pieces of parietal and a zygomatic bone are also available, along with isolated teeth. One parietal fragment is thick and may have come from a relatively narrow vault. This bone could represent a Khoi or San-like cranium (Singer and Wymer 1982). The cheek bone is large in comparison to that of recent humans, but its anatomy is not noticeably archaic. With a few exceptions, all of the teeth fall within the size range expected for living African populations.

THE TRANSITION FROM ARCHAIC TO MORE MODERN POPULATIONS

The fossil evidence as reviewed here suggests that in sub-Saharan Africa, as in other regions of the Old World, archaic populations are followed by more modern people. Individuals from Broken Hill, Ndutu and Elandsfontein certainly differ in their morphology from Laetoli 18, Omo 1 and the cranium from Eliye Springs. Perhaps the incomplete Florisbad remains show fewer ties with this latter group, now that the face has been cleaned and reconstructed by Clarke. In any case, these people of the latest Middle Pleistocene are succeeded at sites such as Klasies River, Equus Cave and Border Cave by Middle Stone Age populations, apparently of anatomically-modern aspect.

This record can be interpreted in different ways. Some workers will wish to build a case for evolutionary continuity, in which even the more archaic specimens are viewed as the direct antecedents of recent Africans. Others will prefer to be cautious and to search for links to modern people only among the Upper Pleistocene assemblages. In fact, several questions concerning this material must be explored further, before population relationships can be assessed in an informed fashion. As a next step in the analysis, it may help to look for any trends or patterns of change that can be identified, in the evolution of cranial form, face shape or tooth size.

Cranial morphology has been emphasized in the preceding accounts of individual specimens. Characters which give Broken Hill an archaic appearance include very heavy brows, a slightly keeled and constricted frontal, a short parietal showing little bossing, an occiput which is sharply flexed with a relatively short upper scale, a prominent occipital torus, and expansion of the mastoid and supramastoid regions which are more laterally projecting than points higher on the parietal vault. Some of the same features, as expressed particularly in the rear and base of the braincase, can be documented for Omo 2. In other, latest Middle Pleistocene crania, there are clearer indications of change. The Laetoli 18 frontal is flat and less constricted behind brows which are reduced but still large by modern standards. The parietals are expanded, and in rear view this vault has a rounded profile. Occipital form is also more modern, although there is still substantial hollowing of muscle insertions on the nuchal area.

One point to be made here is that neither Broken Hill nor any of the other crania closely resemble Neanderthals, as I have emphasized before (Rightmire 1976, 1986). These African representatives of early *Homo sapiens* carry almost none of the features recognized as distinctive for Upper Pleistocene Europeans, even if LH 18 does exhibit some inflation or projection of the upper scale of the occipital bone. There are suggestions that the sub-Saharan populations were becoming more modern in overall shape of the braincase. However, it is difficult, certainly for Broken Hill and even in the case of Omo 1 or LH 18, to specify characters linking the Middle Pleistocene hominids to any later groups. Such resemblances become evident only at Klasies or at Border Cave, where both cranial bones and lower jaws can be matched in recent populations.

Essentially the same picture emerges from an examination of faces, which are less well represented as fossils. The Broken Hill face is large and moderately projecting, especially in its lower parts. The Florisbad cheek and upper jaw are so incomplete as to provide little information, but the Laetoli face shows definite hollowing of the maxillary wall to form a canine fossa. Pieces of frontal, nose and cheek from the Middle Stone Age localities are generally quite modern in appearance, although a fragmentary maxilla recovered recently from the main site at Klasies River is said to represent a young but very robust individual (Deacon, this volume). This upper jaw has not been described in detail. It may have to be considered, along with one of the more massive mandibles, to constitute evidence that there is substantial morphological variation among the Klasies humans.

Teeth are preserved for very few of the Middle Pleistocene hominids from sub-Saharan Africa. Although the Broken Hill maxillary dentition shows extensive pathology, it is clear that individual teeth are large. Incisor breadths are close to the means recorded by Trinkaus (1983) for European and Near Eastern Neanderthals. Areas of posterior tooth crowns are comparable to those expected for Neanderthals, or somewhat greater. Among the less archaic specimens likely to be of latest Middle Pleistocene age, only the Laetoli maxilla bears teeth, including P^3, P^4, M^1, and M^3. Limited comparisons with Broken Hill can be carried out. The Laetoli crowns are themselves quite badly damaged, but it is probable that all premolar and molar breadths are reduced, relative to Broken Hill. The length of M^1 is also less, although the lengths of P^3, and M^3, may equal or exceed the dimensions measured for the Zambian individual. It can readily be established that the Laetoli teeth, like many of those excavated at Equus Cave and Klasies River Mouth, are well within the modern size range.

This material is admittedly very sparse, but a trend toward tooth-size reduction seems to have begun well before the onset of the Upper Pleistocene. The story told by the more complete skulls and cranial fragments is less easily pieced together. The fossils are taken by many workers to show that a mix of archaic and modern characters is expressed in individuals from several of the latest Middle Pleistocene localities. However, if one considers especially the many modern features of the Laetoli vault, then it can be argued that even the earliest Middle Stone Age populations of Africa resemble recent people much more closely than do the Neanderthals of Europe. The emergence of modern-looking humans in sub-Saharan Africa perhaps 120 000 years ago stands in marked contrast to the picture in Europe, where such populations do not occur until after 40 000 years. In

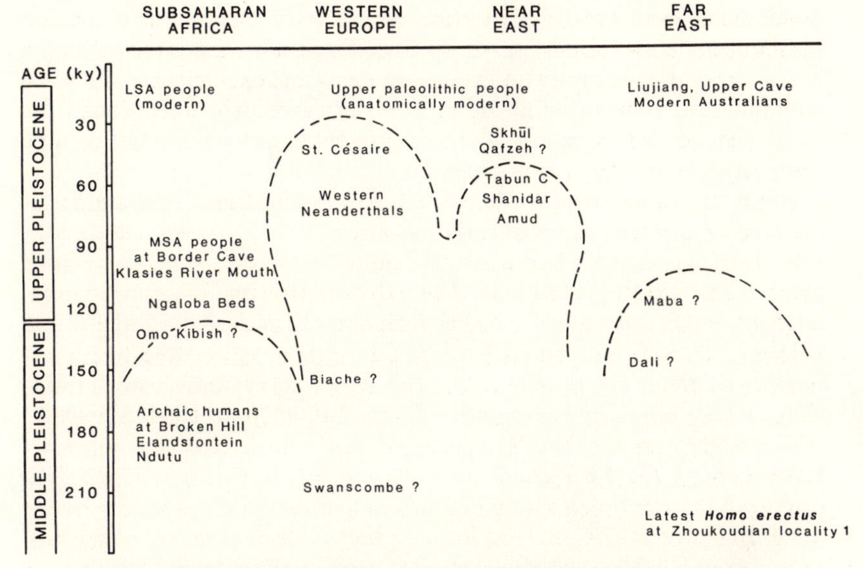

Figure 7.1. The evolution of Middle and Upper Pleistocene human populations, as documented in different geographic regions across the Old World. Remains judged to be representative of archaic forms of *Homo sapiens* are set apart by dashed boundaries from specimens which are more modern anatomically. It is clear that Neanderthals persist in the European record after the appearance of modern people associated with the Middle Stone Age in sub-Saharan Africa. Whether the populations sampled at Border Cave, Klasies River Mouth and the Ngaloba Beds at Laetoli are earlier than all anatomically-modern assemblages in Asia is still uncertain, pending resolution of dating questions at Qafzeh and some other localities. The record for the Far East remains relatively sparse and uninformative.

Western Asia, this transition may have taken place somewhat earlier than in Europe, although dates for important sites such as Qafzeh are disputed (see Bar-Yosef, this volume).

The transition in sub-Saharan Africa is documented by reduction in overall facial projection and brow size, some elevation of the frontal profile, increase in parietal length, and change(s) in the shape of the occipital. Although the record for Africa is incomplete and still subject to reinterpretation, these changes do not seem to parallel those occurring in Europe (Rightmire 1986; Trinkaus 1986). Nevertheless, it is almost as difficult to construct a case for evolutionary continuity from the African evidence as it is to derive later Upper Pleistocene Europeans directly from Neanderthal predecessors. The Laetoli cranium may share some frontal characters with the Omo individuals, but unique features linking LH 18 to archaic populations sampled at Lake Ndutu, Broken Hill, Elandsfontein and Florsibad are not easily identified.

This problem of linking earlier Upper Pleistocene humans to archaic antecedents is encountered not only in Africa and Europe but also in other geographic regions, including China and Southeast Asia (see Fig. 7.1). The fossil record, widely discussed at this Symposium, is richer for Western Europe than elsewhere, while serious gaps persist in the Far East. Where populations from this crucial time period have been sampled, albeit inadequately, there are increasingly firm suggestions that morphological change took place abruptly (in Europe) or even that Neanderthals co-existed with other Mousterian people of more modern aspect (in Southwest Asia). Such evidence is best interpreted to show replacement of archaic populations rather than local evolutionary continuity.

Given the antiquity of Middle Stone Age assemblages, Africa must be counted as a likely source of early modern people. However, neither the Laetoli nor the Omo 1 braincases fall quite within the range of variation exhibited by recent populations. Ties to living African groups are stronger only after the occupation of sites including Klasies and Border Cave, perhaps 100 000 years ago. These Middle Stone Age people were fully modern anatomically, even if they were not as skilled at fishing and hunting as their Late Stone Age successors (Klein, this volume). Deacon however (this volume) argues that Middle Stone Age populations did not differ behaviourally from later groups to the degree that is usually assumed. The origins of these people are uncertain. Whether they evolved locally from a more archaic stock, or moved into Southern Africa from other regions, cannot be established with any certainty from the present evidence.

ACKNOWLEDGEMENTS

Many thanks go to Chris Stringer and Paul Mellars, Symposium organizers, and to the National Science Foundation for research support.

REFERENCES

Beaumont, P. B. 1980. On the age of Border Cave hominids 1-5. *Palaeontologia Africana* 23: 21-33.

Beaumont, P. B., de Villiers, H. and Vogel, J. C. 1978. Modern man in sub-Saharan Africa prior to 49 000 years BP: a review and evaluation with particular reference to Border Cave. *South African Journal of Science* 74: 409-419.

Beaumont, P. B., van Zinderen Bakker, E. M. and Vogel, J. C. 1984. Environmental changes since 32000 BP at Kathu Pan, northern Cape. In J. C. Vogel (ed.) *Late Cainozoic Palaeoclimates of the Southern Hemisphere*. Rotterdam: Balkema: 329-338.

Bräuer, G. 1984. A craniological approach to the origin of anatomically modern *Homo sapiens* in Africa and implications for the appearance of modern Europeans. In F. H. Smith and F. Spencer (eds) *The Origins of Modern Humans: a World Survey of the Fossil Evidence*. New York: Alan R. Liss: 327-410.

Bräuer, G. and Leakey, R. E. 1986. The ES-1693 cranium from Eliye Springs, West Turkana, Kenya. *Journal of Human Evolution* 15: 289-312.

Butzer, K. W. 1978. Sediment stratigraphy of the Middle Stone Age sequence at Klasies River Mouth, Tsitsikama Coast, South Africa. *South African Archaeological Bulletin* 33: 141-151.

Butzer, K. W. 1984. Archeogeology and Quaternary environment in the interior of southern Africa. In R. G. Klein (ed.) *Southern African Prehistory and Paleoenvironments*. Rotterdam: Balkema: 1- 64.

Butzer, K. W. 1987. Comments on dating the hominid remains from Border Cave and Omo Kibish. Paper pre-circulated in the present Symposium, March 1987.

Butzer, K. W., Beaumont, P. B. and Vogel, J. C. 1978. Litho-stratigraphy of Border Cave, KwaZulu, South Africa: a Middle Stone Age sequence beginning *c.* 195 000 BP. *Journal of Archaeological Science* 5: 317-341.

Campbell, N. A. 1984. Some aspects of allocation and discrimination. In G. N. van Vark and W. W. Howells (eds) *Multivariate Statistical Methods in Physical Anthropology*. Dordrecht: Reidel: 177-192.

Clarke, R. J. 1985. A new reconstruction of the Florisbad cranium, with notes on the site. In E. Delson (ed.) *Ancestors: the Hard Evidence*. New York: Alan R. Liss: 301-305.

Day, M. H., Leakey, M. D. and Magori, C. 1980. A new hominid fossil skull (L. H. 18) from the Ngaloba Beds, Laetoli, northern Tanzania. *Nature* 284: 55-56.

Day, M. H. and Stringer, C. B. 1982. A reconsideration of the Omo Kibish remains and the *erectus-sapiens* transition. In H. de Lumley (ed.) *L'Homo erectus et la Place de l'Homme de Tautavel Parmi les Hominidés Fossiles*. Nice: Centre National de la Recherche Scientifique / Louis-Jean Scientific and Literary Publications: 814-846.

Deacon, H. J., Deacon, J., Scholtz, A., Thackeray, J. F. and Brink, J. S. 1984. Correlation of palaeoenvironmental data from the Late Pleistocene and Holocene deposits at Boomplaas Cave, southern Cape. In J. C. Vogel (ed.) *Late Cainozoic Palaeoclimates of the Southern Hemisphere*. Rotterdam: Balkema: 339-351.

Dreyer, T. F. 1935. A human skull from Florisbad, Orange Free State, with a note on the endocranial cast by C. U. Ariëns Kappers. *Koninklijke Akademie van Wetenschappen te Amsterdam, Proceedings* 38: 119-128.

Grine, F. E. and Klein, R. G. 1985. Pleistocene and Holocene human remains from Equus Cave, South Africa. *Anthropology* 8: 55-98.

Klein, R. G. 1983. The Stone Age prehistory of Southern Africa. *Annual Review of Anthropology* 12: 25-48.

Kuman, K. and Clarke, R. J. 1986. Florisbad: new investigations at a Middle Stone Age hominid site in South Africa. *Geoarchaeology* 1: 103-125.

Leakey, M. D. and Hay, R. L. 1982. The chronological position of the fossil hominids of Tanzania. In H. de Lumley (ed.) *L'Homo erectus et la Place de l'Homme de Tautavel Parmi les Hominidés Fossiles*. Nice: Centre National de la Recherche Scientifique / Louis-Jean Scientific and Literary Publications: 753-865.

Leakey, R. E. 1969. Early *Homo sapiens* remains from the Omo River region of south-west Ethiopia: faunal remains from the Omo Valley. *Nature* 222: 1132-1133.

Magori, C. C. and Day, M. H. 1983. Laetoli Hominid 18: an early *Homo sapiens* skull. *Journal of Human Evolution* 12: 747-753.

Rightmire, G. P. 1976. Relationships of Middle and Upper Pleistocene hominids from sub-Saharan Africa. *Nature* 260: 238-240.

Rightmire, G. P. 1978. Florisbad and human population succession in southern Africa. *American Journal of Physical Anthropology* 48: 475-486.

Rightmire, G. P. 1979. Implications of Border Cave skeletal remains for later Pleistocene human evolution. *Current Anthropology* 20: 23-35.

Rightmire, G. P. 1981. More on the study of the Border Cave remains. *Current Anthropology* 22: 199-200.

Rightmire, G. P. 1984. *Homo sapiens* in sub-Saharan Africa. In F. H. Smith and F. Spencer (eds) *The Origins of Modern Humans: a World Survey of the Fossil Evidence*. New York: Alan R. Liss: 295-325.

Rightmire, G. P. 1986. Africa and the origins of modern humans. In R. Singer and J. K. Lundy (eds) *Variation, Culture and Evolution in African Populations*. Johannesburg: Witwatersrand University Press:

209-220.
Singer, R. and Wymer, J. 1982. *The Middle Stone Age at Klasies River Mouth in South Africa*. Chicago: University of Chicago Press.
Trinkaus, E. 1983. *The Shanidar Neandertals*. New York: Academic Press.
Trinkaus, E. 1986. The Neandertals and modern human origins. *Annual Review of Anthropology* 15: 193-218.
Villiers, H. de and Fatti, L.P. 1982. The antiquity of the Negro. *South African Journal of Science* 78: 321-332.
Volman, T.P. 1984. Early prehistory of southern Africa. In R. G. Klein (ed.) *Southern African Prehistory and Paleoenvironments*. Rotterdam: Balkema: 169-220.

8. The Evolution of Modern Humans: a Comparison of the African and non-African Evidence

GÜNTER BRÄUER

INTRODUCTION

At the First International Congress of Human Palaeontology in Nice in 1982, I proposed a new model of *Homo sapiens* evolution called the 'Afro-European *sapiens* hypothesis'. Several comprehensive descriptions of this model appeared in 1984 (Bräuer 1984a, 1984b, 1984c). The reasons for advancing this model were the dating revisions of the African Stone Age of the early 1970s and the resulting redating of important hominids, as well as the increase in fossil *sapiens* discoveries in recent times. A new evolutionary framework appeared necessary to interpret all this evidence conclusively. Thus, the aim of my study was to analyse whether the newly-obtained datings of the African *sapiens* hominids – some of which were still quite uncertain – as a whole make sense from the viewpoint of phylogenetics.

The results led to a division of the evolution of *Homo sapiens* into three grades (Fig. 8.1). Such a division is, of course, only an artificial distinction, but it is useful in making the process of long-term morphological change within a species more graphic. It would, however, be even more artificial to introduce a larger number of species, such as Tattersall (1986) has done. The three grades are:

1. 'Early archaic *Homo sapiens*', which originated out of what is generally called 'developed *Homo erectus*', and which, along-side a certain brain enlargement, exhibits primitive features as well as derived features of *Homo sapiens* (Stringer 1984). Such hominids as Bodo, Hopefield, Broken Hill 1, Eyasi 1, and Ndutu are included in this grade.
2. 'Late archaic *Homo sapiens*', which comprises that phase of *sapiens* evolution which is intermediate between 'early archaic *Homo sapiens*' and early 'anatomically-modern *Homo sapiens*', and which gave rise to the latter. The specimens which may be counted among this more developed and essentially more modern grade include Laetoli H 18, Omo 2, Florisbad, and probably the new hominid ES-11693 from West Turkana.
3. 'Anatomically-modern *Homo sapiens*', which appears to have existed as early as about 100 000 years BP and which is represented especially by hominid 1 from Omo Kibish and the Klasies River Mouth remains.

If it is true that modern African humans are the result of such an evolutionary sequence and, moreover, existed in Africa at an earlier date than in any other continent, then the question arises as to whether these early moderns could also be the main source of the moderns in other parts of

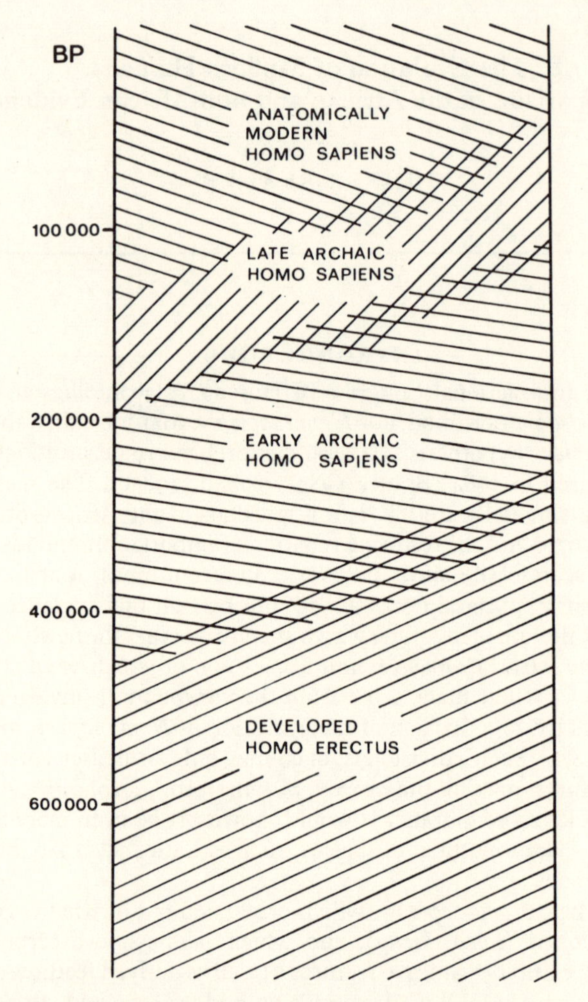

Figure 8.1. The evolutionary grades of *Homo sapiens* in Africa.

the world. The Afro-European *sapiens* hypothesis (Fig. 8.2) proposes this view with regard to Western Asia and Europe. Moreover, the existing evidence indicates that this may also be possible for the Far East as well. The causes of such an expansion out of Africa may primarily be climatic and environmental changes, as, for example, the increasing desertification in the area which now comprises the Sahara (Bräuer 1984a; Clark, this volume). Moreover, these expansions might have taken place quite slowly and through a mixing of gene pools to various degrees. Such mixed populations as, for example, the possible Afro-Southwest Asian hybrids, might have further expanded and mixed with other archaic *sapiens* populations. This process, which could be described as a 'hybridization and replacement model' (Bräuer 1984b) was certainly very complex, multicausal, different in various regions, and hardly rapid or complete. It does not exclude a

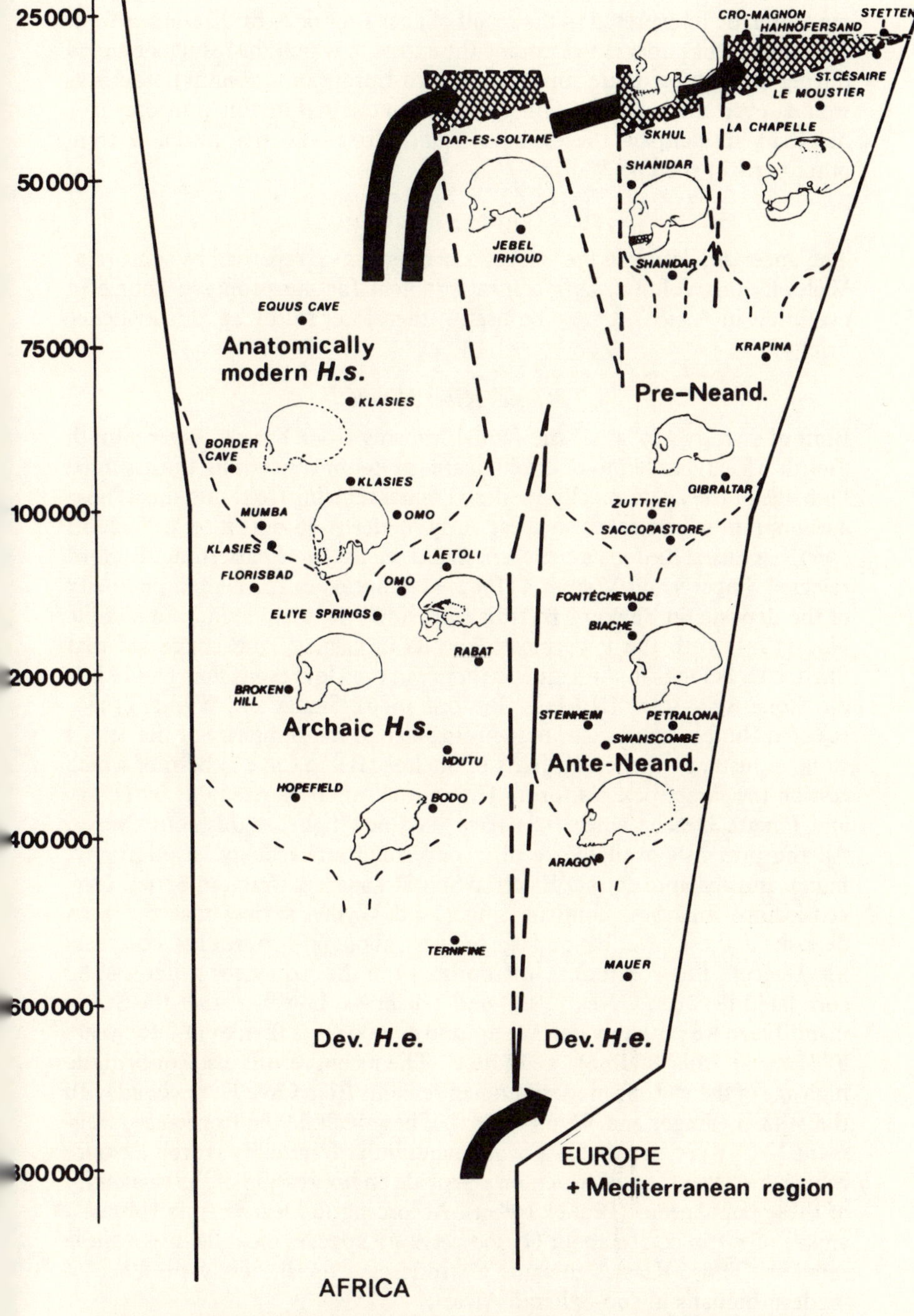

Figure 8.2. The Afro-European *sapiens* hypothesis (first proposed in Bräuer 1982). Revised schema reflecting 1987 evidence.

certain degree of regional continuity in the fossil record – although this may often be interpreted as the result of gene flow or hybridization as well. The aim of this paper is to examine the extent to which the results obtained during the last few years support the Afro-European *sapiens* hypothesis as well as even more global 'Out of Africa' models. For this purpose, it is necessary to compare the evidence from Africa with that available from other parts of the Old World.

THE AFRICAN EVIDENCE

The uncertain dating of the early modern remains is regarded by some (e.g. Wolpoff, this volume) as a central problem for the proposed course of evolution in Africa. It may be useful, therefore, to review this evidence briefly.

Klasies River Mouth

Binford's study (1984) of the faunal remains from Klasies River Mouth (South Africa) has led to a certain degree of doubt concerning the assumed high age of the anatomically-modern cranial remains from this site. Those authors into whose scenario these early moderns do not fit (e.g. Wolpoff 1989) have used Binford's criticism. Based on his faunal patterning, Binford rejected Singer's and Wymer's (1982) assumption of the contemporaneity of the deposits in Shelter 1B (which provided the early modern mandible No. 41815) with the lowermost deposits in Cave 1, and suggested that Shelter 1B belongs in the sequence between the Howiesons Poort and Middle Stone Age (MSA) III levels (Binford 1986). Singer and Wymer (1986) rejected Binford's criticism by pointing also to the similarity of the MSA I stone industry in the lowest parts of Shelter 1B and Cave 1, both of which rest on the same rock platform. The re-excavations carried out by Hilary and Janette Deacon since 1984 have shed new light on this controversy. All the presently-available results concerning archaeology, stratigraphy, fauna, and absolute dating (Electron Spin Resonance, Uranium-Series, Oxygen-isotope analyses) confirm Singer and Wymer's view that the MSA deposits in these caves have an age between about 60 000 and 125 000 years BP (Deacon, this volume). A discontinuity in the sequences which can be correlated between Caves 1, 1A, and 1B shows, however, that the 41815 mandible more probably dates to around 90 000 years than to 125 000 years BP (Deacon, this volume) (see Note 1). The recent results also confirm the high age of the various modern human remains from Cave 1 associated with the MSA II (Singer and Wymer 1982). These include the frontonasal fragment No. 16425 and a very gracile mandibular fragment, as well as more robust specimens, all of which may provide an impression of the variability of these populations (Bräuer 1984a). According to Deacon (this volume), an age of <100 000 to about 80 000 years BP appears most likely for these remains. Thus, Klasies remains a strong case for the early presence of modern humans in sub-Saharan Africa.

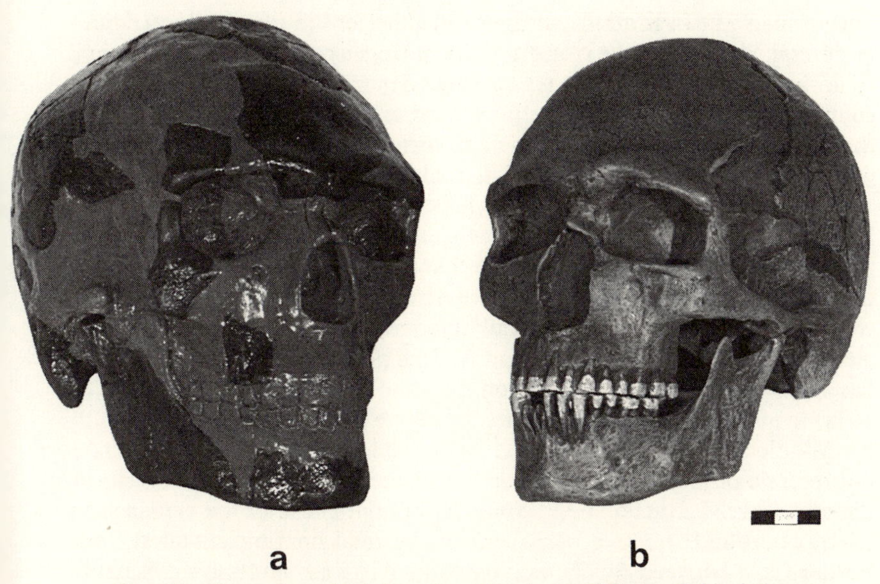

Figure 8.3. Comparison of (a) the early modern hominid Omo 1, Ethiopia (reconstruction by M. H. Day and C. B. Stringer); and (b) the Upper Palaeolithic specimen Předmost 3, Czechoslovakia. Scale in cm.

Omo Kibish 1

There is also good reason to assume a high age for the anatomically-modern Omo 1 hominid, which was found at the base of Member I of the Kibish Formation. The hominid level is situated much lower in the stratigraphy (cf. the section in Day and Stringer 1982) than an upper level of Member III for which a radiocarbon date of >37 000 BP was obtained. Butzer (pers. comm.) even assumes that the entire Member III lies beyond the range of conventional radiocarbon dating. Therefore, an age much greater than 40 000 years appears very reasonable. According to Butzer (pers. comm.), it is most probable that Members I, II, and III all belong to oxygen-isotope stage 5 (a non-glacial period), and are thus older than 75 000 years. Summarizing all the available evidence, including the Uranium/Thorium date of about 130 000 BP for Member I, an age of at least about 100 000 years seems well established for Omo 1. Wolpoff's claim (this volume) that there is no particular reason to choose between 40 000 and 130 000 years BP is thus not substantiated. Morphologically, Omo 1 represents a robust modern form (Fig. 8.3) which could well belong to the ancestral range of modern humans including those outside of Africa (Bräuer 1984a).

Further early modern remains

In addition to Klasies River and Omo Kibish, there are a number of further indications of the early presence of modern humans in Africa, especially the remains from the Border Cave, South Africa. In spite of intensive re-excavations at this site during the 1970s, the stratigraphic provenance of the cranium BC 1 and the mandible BC 2 found in 1941 still remains uncertain. Some indications, such as soil adhesions in the small interstices of the cranial vault (Cooke *et al.* 1945) and nitrogen content, point, however, to an association with early MSA layers (Beaumont 1980). Butzer (pers. comm.) assumes on the basis of the matrix of BC 1 an association between the hominid and his level 10, which he correlates with isotope stage 5b (*c.* 90 000 BP) While earlier multivariate analyses of the cranium have pointed to certain affinities to the Negroid-Khoisanoid spectrum (Rightmire 1979; De Villiers and Fatti 1982; Bräuer 1984a), recent analyses by Van Vark (1986, and pers. comm.) demonstrate that the hominid appears to be rather different from recent African and non-African populations and could thus well represent an early anatomically-modern type which was not yet as clearly differentiated towards recent African populations. Klein (1983: 34) has assumed that the infant skeleton BC 3 and the mandible BC 5 found in 1974, 'come from post-Middle Stone Age graves that were intrusive into the Middle Stone Age layers. This is suggested by a strong contrast in state of preservation between the human bones and animal bones that occur in the same levels'. Butzer (pers. comm.) does not consider the evidence so unequivocal and thinks a placement of the total hominid material from Border Cave between 65 000 and 95 000 BP is reasonable and consistent with the findings from Klasies River.

Small, but additional evidence for the presence of early modern humans comes from various dental remains from Equus Cave in the northern Cape Province (Grine and Klein 1985) and from Die Kelders Cave on the south coast (Klein, this volume); all the material appears to date from early in the last Glacial (*c.* 60 000-80 000 BP). The dental remains from the Mumba Rock Shelter in northern Tanzania may possess a somewhat greater, earliest Upper Pleistocene age (Bräuer and Mehlman 1988). All these teeth cannot be distinguished morphologically or metrically from those of recent populations in sub-Saharan Africa.

Summarizing the recent results with regard to the African early moderns, evidence of their early presence in sub-Saharan Africa has further increased. Yet what about the ancestors of these early moderns, the late archaic *Homo sapiens*?

Florisbad

One of the main representatives of this grade is the Florisbad cranium, the age of which has long been disputed. Since the 1950s there have been indications of an age around 40 000 BP. This, however, has always been seriously questioned as the cranium was found at a spring vent (Clark 1959). In the mid- 1970s, J. C. Vogel carried out new datings for the stratigraphically later Peat Layer II which yielded an age of >42 600 BP (Rightmire 1978).

Further clarification has been obtained only in the last few years by the new excavations of R. Clarke. New radiocarbon dates confirm that Peat Layer II has an age beyond the limits of the radiocarbon method. Datings of the Peat layers by the Uranium-Series method indicate a date well in excess of 100 000 years for Peat I (Clarke 1985). Based on the very different

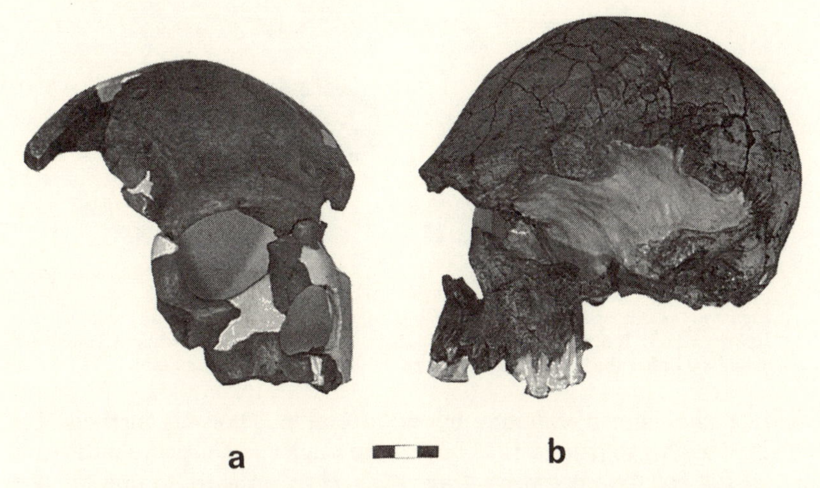

Figure 8.4. Late archaic *Homo sapiens* crania from (a) Florisbad, South Africa (reconstructed by R. Clarke); and (b) Laetoli (Hominid 18), Tanzania. Scale in cm.

states of preservation of bones found by Clarke in the levels above Peat I and those recovered by Dreyer from the spring vents, Clarke (1985: 305) sees 'little reason to doubt that Dreyer's bones and the hominid did originate from the general level of Peat I'. Clarke (1985: 305) further states that 'all indications suggest that the Florisbad cranium seems to belong to an archaic form of *Homo sapiens* that lived in Africa between 100 000 and 200 000 years ago. . .'. Butzer's (1984) analysis of the sedimentary units exposed in Clarke's new excavation as well as the fauna of Peat I also clearly point to a late Middle Pleistocene age (see also Kuman and Clarke 1986). The morphological analysis of the cranium (Fig. 8.4) newly reconstructed by Clarke (1985) have led him to the same conclusion which I had earlier reached (Bräuer 1984a), namely that Florisbad belongs to the same *Homo sapiens* grade as LH 18 and Omo 2, and that this stage of human evolution follows that to which such specimens as Hopefield and Broken Hill belong (early archaic *Homo sapiens*). The recent results thus provide further confirmation that Florisbad possesses a high age and may be among the direct forerunners of anatomically-modern humans.

Laetoli Hominid 18

The dating of this hominid was also still rather uncertain at the beginning of the 1980s. The hominid and associated artifacts of Middle Stone Age affinity come from around the middle of a 3-metre thick deposit of claystone and sandstone. The trachytic tuff below the beds which yielded the hominid is mineralogically similar to, and has been correlated with, the marker tuff in the lower unit of the Ndutu Beds at Olduvai Gorge (Leakey and Hay 1982). Its age has been estimated at 120 000 ± 30 000 BP on the basis of its relative stratigraphic position within the Ndutu Beds. This date has been used thus far as a preliminary estimate of the age of the LH 18 specimen. More recently, J. L. Bischoff has dated a giraffe vertebra from the bed

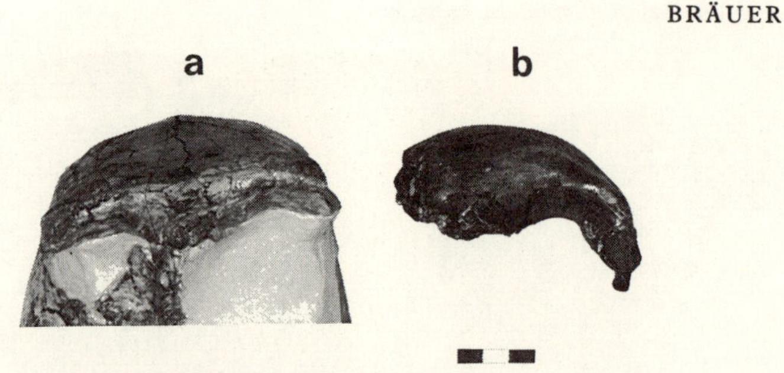

Figure 8.5. Comparison of the supraorbital regions of (a) Laetoli H 18; and (b) Lukenya Hill, Kenya. Lukenya Hill's shape is clearly modern, while LH 18 exhibits a rather thick torus-like structure. Scale in cm.

yielding the human cranium by the Uranium/Thorium method. He obtained a Th-230 date of 129 000 ± 4000 years and a concordant Pa-231 date of 108 000 ± 30 000 years (Hay 1987). Thus, an age of around 130 000 years can be regarded at present as rather well established. This is also in agreement with the most recent amino-acid date (isoleucine epimerization) of 100 000-200 000 years BP obtained by Bada (1987) on animal tooth enamel from the same level that yielded the hominid remains.

The various morphological studies of this hominid (Magori 1980; Magori and Day 1983; Bräuer 1984a) have shown (Fig. 8.4) that LH 18 as well as Florisbad and Omo 2, in spite of a number of archaic features, are close to the threshold of anatomically-modern *Homo sapiens*. In contrast to the results of these studies, Rightmire (1986) has claimed a nearly modern status for LH 18. In the supraorbital area, for example, he even sees close similarities to such late Upper Pleistocene specimens as Lukenya Hill (Fig. 8.5), for which the present author, however, cannot see any convincing evidence. Nevertheless, Rightmire's assessment of the morphology points to the close relationships between representatives of late archaic *Homo sapiens* and early modern humans, even if he does not seem to be fully convinced of an evolutionary continuity with preceding groups. However, what alternative phylogenetic interpretation could be more likely between nearly-modern humans and practically-modern humans than an ancestor-descendant relationship?

Eliye Springs (ES-11693)

The number of archaic *Homo sapiens* specimens has recently been increased by a quite well-preserved cranium from Eliye Springs, West Turkana. It comes from re-worked deposits of the beach and thus has no exact stratigraphic provenance. The associated fauna is modern but shows a degree of heavy mineralization similar to that of the cranium, thus suggesting that the fossils were part of the same unit. As there are not yet any concrete indications of its absolute age, only the morphological analysis can be used at present. A detailed description of the specimen has recently appeared (Bräuer and Leakey 1986a, 1986b) so that we shall here only briefly discuss those features which are most important for the classification of the cranium.

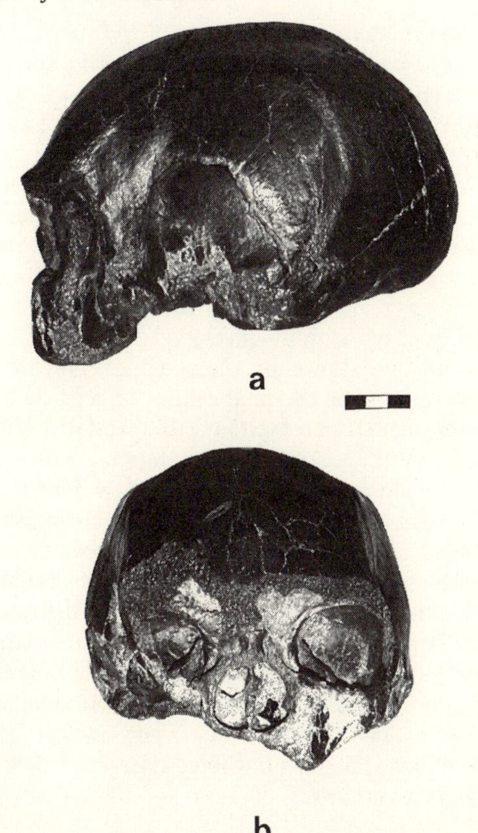

Figure 8.6. (a) Lateral, and (b) frontal views of the ES-11693 cranium from West Turkana, Kenya. Scale in cm.

There are few features in which the cranium exhibits certain marginal relationships to the *Homo erectus* range as described by Stringer (1984). The very low cranial vault of ES-11693 (Fig. 8.6) gives a height/length ratio lying quite close to the upper limits given for *Homo erectus*. The parietal walls converge superiorly, but only in their upper halves. The occipital angle (OCA) also falls within the limits of *Homo erectus*; this applies also to the ratio bidacryal breadth/biorbital breadth. In addition, ES-11693 exhibits numerous derived features which clearly show that the cranium can be classified as archaic *Homo sapiens*. The cranial capacity lies between 1300 and 1450 cc. The frontal and parietal angles are as small as those found in modern humans. The temporal squama is high and well curved. The other temporal features also fall within the range of *Homo sapiens,* although the tympanic plate is quite massively developed. While the occipital has a low and very strongly curved mid-sagittal profile, the occipital plane is considerably longer than the nuchal plane and there is no transverse torus.

It is interesting to consider to which of the two archaic *Homo sapiens* grades the hominid exhibits stronger affinities. Although the supraorbital region is almost completely missing, some of the surviving features, such as the presence of a supratoral sulcus, indicate that a torus or a torus-like structure was present. Yet based on the dimensions of the fracture, it can practically be ruled out that the torus was as massive as that of Broken Hill. With regard to the prominent temporal crests and the low degree of postorbital constriction, there are stronger similarities to the conditions of late archaic specimens, especially to Omo 2. A variable heaping up of bone or keeling on the frontal and on the parietals can be found in both early and late archaic *Homo sapiens*. The archaic outline of the parietals in the occipital view, however, is not found in the few known late archaic specimens. Closer affinities to the more developed and more modern grade, and especially to LH 18, are shown by the occipital. Nevertheless, the total morphological pattern of the occipital cannot be regarded as anatomically modern. Moreover, there is an occipitomastoid crest which is less prominent than that of Omo 2. With regard to some temporal features (glenoid fossa, tympanic plate), the closest affinities are to LH 18. Due to its fragmentary character, the face can only be roughly analysed. The most striking feature is its low and broad shape. Stringer (this volume) has been able to show similarities to the conditions in the Skhūl/Qafzeh hominids, the early Upper Palaeolithic Europeans, and even to modern Australian and Khoisan crania. The maxillary sinus appears to be heavily pneumatisized and more similar to the early archaic conditions. Summarizing the results of the various comparisons, only some of which could be mentioned here (see Bräuer and Leakey 1986b), ES-11693 exhibits a new mosaic of primitive and derived features which, taken as a whole, appears to be closer to the late archaic grade of *Homo sapiens*.

The process of sapiens evolution

The results of research over the past few years further confirm the existence of late archaic *Homo sapiens* and its phylogenetic position ancestral to early anatomically-modern humans. The hominids of this archaic grade exhibit mosaics of primitive and modern features. The transition to modern humans thus appears not to have been sudden, but has probably taken some tens of millennia (see also Trinkaus 1986b: 1042). As exact datings within this transitional period are generally not possible, a clear diachronic arrangement of the hominids is also not possible.

The specimens dated between *c.* 80 000 and 150 000 BP constitute a morphologically 'heterogeneous group' (Bräuer 1984b; Howells 1988) which reflects the transition from late archaic to early anatomically-modern *Homo sapiens*. Howells (1988: 225) has recently commented:

> The progression in Africa toward anatomically modern man as seen in the cranium is supported by evidence of the limb bones (Kennedy 1984): modern structure of these appears in robust form by 100 000 BP, while more archaic earlier specimens (e.g. Broken Hill) are modern in structure but retain the cortical thickness seen in *Homo erectus*. All of them differ morphologically from Neanderthal bones.

The ancestors of the late archaic *Homo sapiens*, the early archaic grade, can now be regarded as rather well documented. Thus, there are hardly any serious doubts that the main representatives of this grade (Bodo, Broken Hill 1, Hopefield, Ndutu and Eyasi) probably date between 200 000 and 400 000 BP (Leakey and Hay 1982; Partridge 1982; Vrba 1982; Klein 1983; White 1985; Mehlman 1987; Clark, this volume). This early archaic *Homo sapiens* grade will not be treated in detail here (see Bräuer 1984a). The evolutionary sequence from early archaic to late archaic and finally to anatomically-modern *Homo sapiens* is at present better documented than ever, and is thus also an important basis for any 'Out of Africa' hypothesis. Considering morphology, Smith (1985: 204) also sees continuity between the three grades, commenting that 'each group is a suitable ancestral group for the next younger group'. However, he regards this only as an indigenous African phenomenon.

Before turning to the evidence from other regions of the world, we shall consider the most recent results from molecular biology concerning the origins of modern humans.

DNA Analyses

The assumption that the roots of modern humans might lay in Africa has also gained strong support through recent DNA analyses. Cann *et al.* (1987) mapped the mitochondrial DNA (mtDNA) of 147 individuals from various continents by the high-resolution method, using 12 restriction enzymes. Their results showed that there is more sequence variation within the African sample than within the other groups from Asia, Europe, Australia, and New Guinea. Moreover, the average pairwise sequence difference between individuals who belong to two different groups showed that the Africans exhibit the largest divergence from the other four groups. Therefore, the Africans appear to have more mtDNA diversity than other populations, which would point to a longer existence of the human lineage in Africa.

The special position of the Africans is also shown by a cluster analysis relating the 134 different mtDNA types by the parsimony method (Cann *et al.* 1987). The dendrogram has two main clusters or branches, one exclusively African and one mixed, including African and non-African types (see Stoneking and Cann, this volume). The most simple explanation for this evidence is the assumption of a common African ancestor. On the basis of the best presently-available estimation of the rate of mtDNA divergence of 3% per million years, Wilson *et al.* (1987) assume that the 'common mother' lived about 200 000 years ago. Wainscoat (1987), however, also thinks it possible that the present kind of human does not go back to a single woman but to a larger number of women who were monomorphic for a particular mtDNA type. A certain ancestral mtDNA type could have become fixed by genetic drift.

In spite of these open questions, the results based on mtDNA with its special mode of inheritance and high divergence rate, have become an important source of support for the assumption of an African origin of modern humans. Wilson *et al.* (1987) think it unlikely that there was another evolutionary lineage from East Asian *Homo erectus* towards modern humans in East Asia which was independent from the African lineage (see also Rouhani, this volume). If this had been the case, one would expect there to be roughly five times as much divergence in the extant human mtDNA gene pool as has so far been observed. Such divergent mtDNA lineages have not yet been found among the more than 500 Asians studied (Wilson *et al.* 1987).

Assuming an age of about 200 000 years for the mtDNA origins of present-day humans, this would mean that the common ancestral 'Eve' belonged to African archaic *Homo sapiens*. Thus the common ancestor might not yet have been anatomically modern (Wilson *et al.* 1987; Stoneking and Cann, this volume).

This does not imply, however, that the populations which left Africa were archaic. Only some tens of millennia after the origin of modern humans in Africa, parts of this population left the continent – perhaps some 50 000 or even 90 000 years ago – and gave rise to the observed Eurasian mtDNA divergence.

A clear separation between African and non-African populations was also indicated by the studies by Wainscoat *et al.* (1986, and this volume) on nuclear DNA. Five linked polymorphic restriction sites of the beta-globin gene cluster were studied in approximately 600 individuals from eight populations of Europe, Asia, and Africa. Based on the combination of restriction sites, 14 different types could be differentiated. It could be shown that one of the four most frequent types was restricted exclusively to Africans while the other types were present in all non-African populations in similarly large frequencies, and less frequently in the Africans. Wainscoat *et al.* (1986) concluded that these four common types predate the racial divergence, and that a group of early modern humans left Africa and constituted the basic stock for all non-African populations. This group might have lost the type now present only in Africa through genetic drift – which would also favour the idea of a small founder population. Giles and Ambrose (1986) accept the differences between the African and the Eurasian populations but question the assumption as to which of the two lines represents the original population. However, the mtDNA studies mentioned above, as well as the fossil evidence, point to an older African line (Jones and Rouhani 1986).

The possibility that mixing occurred during the replacement period of the archaic populations could lead to disagreements between the evidence from palaeoanthropology and that of molecular biology. Under the assumption of strong mixing one would expect extremely divergent mtDNA types amongst, for example, the present-day Asians and Australians, which is not supported by the respective studies so far. Nevertheless, a certain degree of mixing between archaic and modern populations cannot be completely excluded. Archaic types of mtDNA could have been lost in the course of later evolution, or alternatively may not yet have been discovered. From the palaeoanthropological view, one must concede that hard evidence which could be interpreted as the result of mixing is relatively rare. The assumption of intensive mixing and gene flow is primarily inferred from the

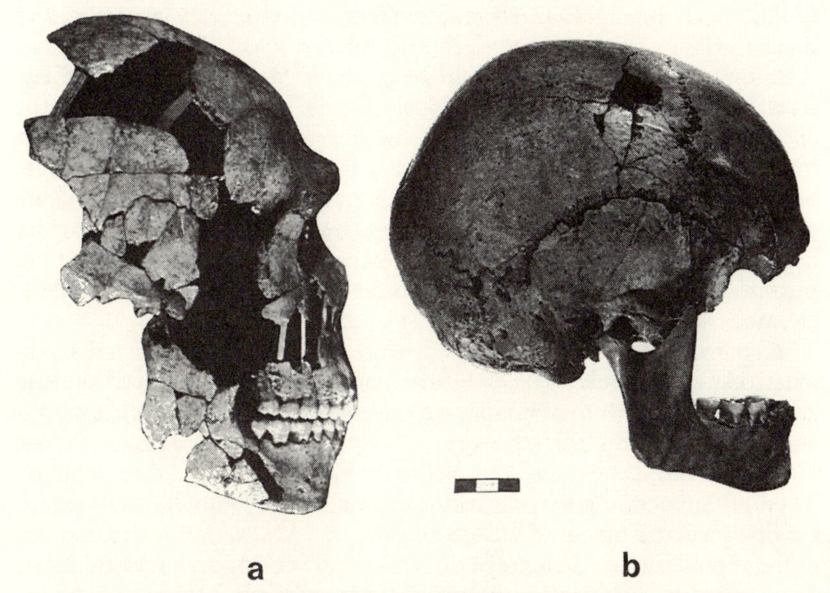

Figure 8.7. (a) The most recent Neanderthal from Saint-Césaire, France (photograph: B. Vandermeersch); and (b) early modern European Stetten 1, Germany (photograph: A. Czarnetzki). Both existed around 30 000-32 000 years BP. Scale in cm.

dynamics of population biology. There is, for example, little hard evidence of mixing between the European Neanderthals and the subsequent early moderns. The samples from Skhūl and Qafzeh may provide some indications of the effects of mixing. Some archaic features of various late Upper Pleistocene/early Holocene crania from Australia could perhaps also be due to some sort of hybridization in earlier times. The actual extent of mixing during the replacement period of archaic by modern populations, however, remains very unclear. Thus the present state of knowledge indicates that there is no fundamental contradiction between the evidence from DNA and that from the fossil record.

Both kinds of data, as well as other molecular evidence (see Rouhani, this volume), point to a recent origin of modern humans in Africa. Thus, I feel it is difficult to cast doubt on all this evidence and suggest that all of these various findings are false. If Africa was the main source of modern humans, this assumption should also be in agreement with the fossil evidence from the other parts of the world. Let us now consider whether this assumption is justified.

THE EUROPEAN EVIDENCE

Neanderthal-modern transition

There is a general consensus among the specialists with regard to the evolutionary steps in Europe up until the Neanderthals. The fossil remains show that the typical morphology of the large-brained Neanderthals evolved slowly and progressively from the range of the anteNeanderthals, to which such hominids as Arago, Steinheim, Petralona, Swanscombe, and Biache belong (Bräuer 1984c; Trinkaus 1986b: 1041). The Neanderthals *sensu stricto* first appeared in Europe at about the beginning of the last Interglacial (Smith 1985). There is less clarity, however, concerning the disappearance of the Neanderthals and the transition to early modern humans, an event which occurred in a span of only a few thousand years between *c.* 36 000 and 30 000 BP. The present controversy focuses on the question of whether the early modern Europeans originated primarily from immigrations from outside the region, or developed primarily out of the European Neanderthals. Both views, however, recognize that the Neanderthal-modern transition was an essentially multicausal process.

In spite of these clearly divergent points of view, there is general agreement that the Neanderthals and early modern Europeans differ from one another with respect to numerous significant features of the skull and postcranial skeleton (Smith 1984, 1985; Stringer *et al.* 1984; Trinkaus 1984; Howells 1988; Liberman, this volume; Vandermeersch, this volume). These differences made it difficult for Trinkaus (1986b: 1041) to regard the first modern humans of Europe and Western Asia as simply descendants of the Neanderthals. Stringer *et al.* (1984: 116) do not even recognize any fossil specimen as having an intermediate morphology between the Neanderthal and modern types (with the possible exception of the very fragmentary specimen from Hahnöfersand).

In addition to the considerable morphological differences between both groups there is also the fact that they are practically adjacent temporally and probably even co-existed in various regions. Such hominids as Saint-Césaire, Stetten (= Vogelherd 1), Hahnöfersand, and Velica Pecina support this assumption. An age of about 30 000 years has been well established through numerous radiocarbon dates for the completely modern and quite gracile skull Stetten 1 (Fig. 8.7) (Hahn 1986; Müller-Beck, pers. comm.). There are also interstratifications between Châtelperronian and Aurignacian industries in some deposits of southwestern France and Spain (Bordes 1972; Harrold, this volume) which support the view of a coexistence of both human types during the early Upper Palaeolithic, even if no human remains associated with the early Aurignacian have yet been discovered (Vandermeersch 1985: 100).

These chronological considerations have also led Demars and Hublin (1986) to conclude that a regional progressive evolution from one form to the other is very unlikely (see also Gambier, this volume). The numerous changes throughout the entire skeleton can be shown to have occurred within an extremely short time, whereas many tens of millennia would have been necessary for the evolutionary development of these features. The advocates of evolutionary continuity point, however, to certain changes – for example, a reduction in tooth breadth – which were already taking place in the period between the early (pre-Würm II) and the late (Würm II) Neanderthal samples (Wolpoff 1989). Mellars' (1986) new chronology of the French Mousterian could, however, make it necessary to rethink the composition of these samples. His results indicate that such hominids as

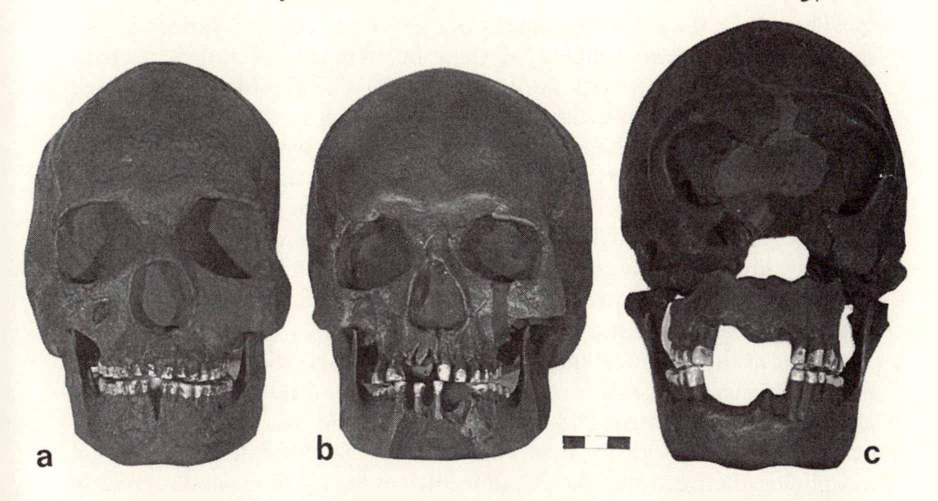

Figure 8.8. Modern Upper Palaeolithic female crania: (a) Brno 3 and (b) Předmost 4, Czechoslovakia, compared to (c) the probably female Neanderthal cranium La Quina H5, France (photograph: J.-L. Heim). Scale in cm.

La Ferrassie or La Quina probably date to the earliest part of Würm II (*c.* 70 000 BP), and one has to ask how far such changes would influence the assumed morphological trends.

The differences between the samples from Krapina and Vindija are regarded as especially good evidence of the changes from the early to late Neanderthals (Smith 1982, 1984). However, this evidence is not unequivocal, as the Vindija remains are very fragmentary. Therefore, it is possible that the observed trends from the older Krapina to the Vindija hominids, such as a reduction in mid-facial prognathism and thinner and less projecting supraorbital tori, are at least partly due to larger frequencies of female and juvenile individuals in the Vindija sample. But even if the differences represent real, diachronic, *in situ* changes within this regional group of Neanderthals, they by no means document a transition as complex as that towards the modern form. There thus remains a large morphological gap between the late Neanderthals and the early moderns in South-Central Europe as well as in the remainder of Europe, where such a pattern cannot be observed. What is more, the Vindija Neanderthals differ considerably in their total morphology from such early moderns as Mladeč and Brno (Smith 1982). This difference becomes even more evident if one also takes the female individuals of these early moderns into consideration (Fig. 8.8). No one would disagree that there is a considerable degree of variability within the Neanderthals, and that there may be certain recognizable trends; essential changes, however, cannot be observed subsequent to the early Würm (Trinkaus and Howells 1979; Howells 1988). Rak (1986: 163) also concludes from his recent analysis of the functional morphology of the Neanderthal face that 'it clearly represents a departure from the generalized fundamental architectural pattern that characterizes all the species of the genus *Homo*'. He fully agrees with Howells' (1975) view that the facial morphology of the *Homo* specimens preceding the classic Neanderthals is more similar to the morphology of those following the Neanderthals than either is to the Neanderthal morphology itself. This is particularly related to the architectural connection between the anterior teeth and the glabella. In the classic Neanderthals, the anterior buttresses terminate in two nasal apophyses extended up 'in the air', and not up against the glabella. Repeatedly, attention has been drawn to the presence of Neanderthal-like features among individual early modern specimens (e.g., Mladeč 5, Předmostí 3). For example, some specimens exhibit occipital buns (though these are less projecting than is the case with the Neanderthal condition) and quite large and projecting brow ridges (Smith 1985). Only a few of these features can, however, be interpreted as relating specifically to Neanderthals (Trinkaus and Le May 1982; Stringer *et al.* 1984). Most of them can also be found in early modern and archaic specimens of *Homo sapiens* from other continents. With regard to the considerable differences in the total morphology between such late Neanderthals as Saint-Césaire and such early moderns as Cro-Magnon or Stetten 1, the facts appear to indicate that the sporadic appearance of such similarities is due more to gene flow between Neanderthals and early moderns than to a direct evolution out of the Neanderthals (Demars and Hublin 1986; Stringer 1986). There are also clear differences in the post-cranial morphology and limb proportions of early modern specimens and Neanderthals (Stringer *et al.* 1984).

On the basis of the fossil evidence and the earlier presence of modern humans in Eastern and Central Europe, it appears more likely that early modern groups slowly expanded into Western Europe via Central Europe. They certainly coexisted with the Neanderthals for perhaps some thousand years (probably often in different territories), and also mixed with them to some degree (Bräuer 1982, 1984a; Vandermeersch 1984: 195). The modern humans, however, were obviously superior in many respects (e.g. physically, mentally, and demographically) and represented competitors for resources, which the Neanderthals could not match, notwithstanding their adaptation to the climate (Stringer 1984; Trinkaus 1986a). Under the assumption of interacting modern and Neanderthal populations, Zubrow (this volume) has demonstrated by demographic modelling that a small demographic advantage by the modern humans would have been sufficient to result in a rapid extinction of the Neanderthals in the course of only a thousand years.

Trinkaus (1986a: 198) sees a considerable increase in gene flow across Europe and Western Asia as the only possible explanation for the rapid and important morphological changes. He also thinks it likely that there was a population replacement with little or no genetic continuity in Western Europe. With regard to the source of the new anatomically-modern traits, he has remarked: 'since early modern humans appear to have evolved from preceding archaic human populations in sub-Saharan Africa sometime prior to 40 kyr BP, there was a potential source to the south of the Near East for the biobehavioral complex associated with early modern humans. . .'. Trinkaus (1986b: 1042) traced out a replacement model which is in good agreement with the Afro-European *sapiens* hypothesis: about 50 000 years ago, a modern sub-Saharan population expanded slowly, generation by generation, towards the north (and perhaps also the west) thereby absorbing Neanderthal groups as well. In other words, the ancestors of the inhabitants of Europe and Western Asia of some 30 000 years BP consisted of modern Africans with some admixture of Neanderthals. With regard to certain angles and indices which especially measure the facial projection, Stringer (this volume) has been able to demonstrate that the early Upper Palaeolithic hominids are considerably closer to the late archaic *Homo sapiens* of Africa than to their direct precursors, the Neanderthals.

Summarizing all of the recent evidence from Europe (including the archaeological evidence: see papers by Harrold, Allsworth-Jones, and Kozlowski, this Symposium) I see no hard evidence which would disagree with a replacement model in which the primary causes for the appearance of early modern humans were gene flow and population movements into Europe from outside. The extent to which other causes may have also played a role remains an open question.

THE SOUTHWEST ASIAN EVIDENCE

Neanderthal-modern transition

For quite a long time, the Near East has been viewed as a potential region of direct evolution to modern humans. Recent studies, however, clearly show that there is a considerable morphological gap between Neanderthals and early moderns in this region, comparable to that in Europe; this holds for the cranium as well as for the postcranial skeleton (Trinkaus 1981, 1984; Trinkaus and Smith 1985; Stringer, this volume; Tillier, this volume; Vandermeersch, this volume). The Neanderthals exhibit strong affinities with their European classic counterparts, although they seem to be somewhat less specialized (Stringer and Trinkaus 1981; Trinkaus 1983, 1984). In contrast, the samples from Skhūl and especially from Qafzeh are basically anatomically modern (Vandermeersch 1981). Stringer *et al.* (1984: 118) do not recognize any transitional fossils in this region either.

How far can one detect diachronic trends within the Neanderthal sample and towards modern humans? No diachronic trend with respect to size reduction can be observed in the Neanderthals. On the contrary, Trinkaus (1984) finds an increase of mid-facial prognathism from the early hominids of Zuttiyeh, and Shanidar 2 and 4 (perhaps via Amūd 1 and Tabūn C1) to the late Shanidar 1 and 5 specimens. The earlier group is characterized by the anterior positions of the anterior zygomatic roots, the angled zygomatic bones, and small or absent retromolar spaces (Trinkaus 1984: 281). The late Neanderthals are remarkably close in all respects to the Würm Neanderthals of Western Europe (Howells 1988). If one compares the Neanderthal sample (Shanidar, Tabūn, Amūd) to the Skhūl/Qafzeh sample with regard to tooth size, an increase towards the modern group can be observed (Stringer 1982). Trinkaus (1984: 274) confirms this trend with regard to the posterior teeth, while he finds a small decrease in breadth for the anterior teeth. As the differences are not statistically significant, however, there appears to be more similarity with regard to tooth size than any clear trend. However, in spite of this similarity in tooth size, all of the West Asian Neanderthals, exhibit heavy wear of the front teeth, while in the early moderns both the anterior and posterior regions of the dentition exhibit about the same rate of wear (Trinkaus 1984: 275). Although Trinkaus (1984) legitimately speculates about the biomechanical changes which could have led the modern type to develop out of the Neanderthal face, the facts appear to be in agreement with Stringer *et al.*'s (1984: 118) conclusion that there are no unequivocal directional trends towards modern humans evident in the Near Eastern Neanderthal sample as a whole.

Do the early moderns exhibit any features reminiscent of the Neanderthals? Vandermeersch (1982: 297) was unable to find any of the specialized Neanderthal characteristics within the Qafzeh hominids. Smith (1985: 207) has also described the Skhūl/Qafzeh sample as 'clearly modern, albeit robust, in total morphological pattern'. Nevertheless, some of the specimens, especially from the Skhūl sample, exhibit certain archaic features, for example in the supraorbital region (Skhūl 5 and 9). These features have been regarded by some as robust modern traits, but they could also be at least partially the result of mixing between modern and Neanderthal groups, an assumption which I think fits quite well with the total evidence in the Near East (Bräuer 1982, 1984a). In view of the considerable morphological gap and the small temporal difference of probably only about 5000 years, Trinkaus (1986a: 198) supports the idea of strong gene flow across Western Asia. He looks to sub-Saharan Africa as the potential source of the modern traits (Trinkaus 1986b: 1042), a view with which I can only agree. Smith (1985: 207) also concedes that under the assumption of regional continuity, one has to be ready to accept that the tempo of evolutionary change increased at the Neanderthal-modern transition. I would even say that there must have been a considerable increase in morphological change, and this in spite of the lack of any obvious change in the archaeological record. The close temporal succession of the Neanderthals and moderns has, however, been questioned in recent times. While an age of around 40 000 BP for the Skhūl Layer B remains can be regarded as likely on the basis of the correlation with Layer B of the neighbouring Tabūn Cave (Jelinek 1982), recent analyses of the Qafzeh microfauna point to a contemporaneity with Layer D at Tabūn, which could suggest an age of about 70 000 to 90 000 years (Bar-Yosef and Vandermeersch 1981). Jelinek (1982), however, doubts the correctness of this chronological correlation between Qafzeh and Tabūn. Nevertheless, recent TL dates support the suggested high age of the Qafzeh hominid layers (Bar-Yosef, this volume). From the perspective of population biology, however, the assumption of such a great age for Qafzeh might appear rather problematic, as Neanderthals and moderns would have coexisted during some tens of millennia in the small area of northern Israel, using the same cultural adaptive complex and yet remaining biologically distinct (Trinkaus 1984).

If this high age should be further confirmed, however, then the idea of regional continuity with regard to the Neanderthal-modern transition would no longer be tenable. Modern humans would have been present in this area as early as 80 000 or more years ago.

But what about the ancestors of these early moderns? Even such a great age would not contradict the idea that modern humans first evolved in Africa and later spread to the Near East. An exact reconstruction of the course of evolution in Southwest Asia is not currently possible owing to the remaining uncertainties in dating. The possibly high age of the Qafzeh remains, and the special morphology of the Zuttiyeh skull, have led Vandermeersch (1981) to propose still another model.

Zuttiyeh

While Trinkaus (1984, 1986b) and other authors place the early Upper Pleistocene cranial fragment from Zuttiyeh among the early Neanderthals of the Near East, Vandermeersch (1982: 198) regards this fossil as 'probably the most ancient *Homo sapiens sapiens* in the Near East'. Zuttiyeh indeed lacks the strong mid-facial prognathism and the retreating zygomatic profile typical of most Neanderthals, yet this pattern can also be found with the early Shanidar Neanderthals 2 and 4 (Trinkaus 1984; Smith 1985). Vandermeersch (this volume) describes other details in which Zuttiyeh is different from the Neanderthals. However, there remains the fact that the specimen consists of only a rather small part of the cranium. If one had access, for example, to only the same fragment of the quite similar-looking Steinheim cranium and, moreover, no other fossils dated to this period were known, then one would hardly place the Steinheim hominid into the lineage leading towards the Neanderthals. It thus appears that caution should be exercised when interpreting the Zuttiyeh hominid. To draw such a far-reaching conclusion, which would make the specimen practically the ancestor of a West Asian modern lineage which must have existed parallel to the Neanderthals, does not appear to be well founded in view of its unclear demarcation from the quite variable early Neanderthals and anteNeanderthals. Documentation of the direct evolutionary lineage assumed by Vandermeersch (1982, and this volume) from evolved *Homo erectus* through Zuttiyeh to Qafzeh and Skhūl remains to be demonstrated for most of the stages. Nevertheless, this model represents an interesting hypothesis about the difficult question of the origin of modern humans. Howels (1988) thinks it is more likely that early moderns moved along the Nile Valley and finally reached the Middle East some 40 000 or perhaps even 70 000 years ago, a view which agrees with the Afro-European *sapiens* hypothesis. The hypothesized spread from Africa would thus correlate with both the postulated early dates for Qafzeh and with the general evidence from the Near East.

THE EAST ASIAN EVIDENCE

Archaic Homo sapiens

The evidence for archaic *Homo sapiens* from East Asia consists of only a few hominids, especially from the People's Republic of China. These are the well-preserved cranium from Dali, the calotte from Maba, the vault fragments from Xujiayao, and the nearly complete cranium from Yingkou, which was discovered only recently. These hominids date to the late Middle Pleistocene, although Maba and probably also Xujiayao might have an early Upper Pleistocene age (Zhou Mingzhen *et al.* 1982; Wolpoff, this volume). Although the Dali cranium still exhibits quite close affinities to the *Homo erectus* remains from Zhoukoudian in some features, its total morphology appears to be more progressive and closer to that of early *Homo sapiens* (Wu Xinzhi 1981; Wu Rukang and Wu Xinzhi 1982). In a number of features, the Maba calotte shows similarities to Dali (Wu Rukang and Wu Xinzhi 1982). This also holds for the somewhat more gracile Yingkou cranium, at least as far as one can judge from photographs. Among the very fragmentary remains from Xujiayao, the occipital fragment particularly exhibits certain archaic features.

From Indonesia, we have the Ngandong sample which has been assigned to either archaic *Homo sapiens* or to late *Homo erectus* (Santa Luca 1980; Wolpoff 1980; Jacob 1981; Sartono 1982; Stringer 1984; Bräuer 1984b). These calvariae still exhibit close affinities to *Homo erectus*, although some of them clearly show progressive features which rather support an assignment to archaic *Homo sapiens*. Not only is the classification of these hominids unclear, but their age is also in question, and might range between 80 000 and 200 000 years.

Finally, the cranium found in 1982 at Hathnora, in the valley of the Narmada River, India, might date to the late Middle Pleistocene (Lumley and Sonakia 1985). Although the cranium has been classified by Lumley and Sonakia (1985) as 'more evolved *Homo erectus*', the authors also point to a number of similarities to East Asian archaic *Homo sapiens* and to the anteNeanderthals of Europe (see also Sonakia 1985; Stringer 1985). Altogether, the hominids referred to show that there are fossil representatives of the *erectus-sapiens* transitional period and of archaic *Homo sapiens* from India to China and down to Indonesia which date between about 80 000 and 250 000 years BP and which are morphologically very different from modern humans.

The gap before the early moderns

Between these archaic *sapiens* specimens and the earliest relatively well preserved anatomically-modern remains lies a period of at least some 50 000 years from which only a few fragmentary and uncertainly dated remains exist. These, however, do not allow any clear statements about the course of evolution. While there are no remains at all from Indonesia and India dating between Ngandong and Narmada, on the one hand, and the first Upper Pleistocene moderns, on the other, a few very fragmentary remains are available from China. These include in particular a maxillary fragment from Changyang which, according to Wolpoff *et al.* (1984), is too fragmentary to allow any statement with regard to morphological continuity. There are also three teeth as well as a juvenile parietal fragment from Dingcun which are said to be much less primitive than Peking Man (Zhou Mingzhen *et al.* 1982).

I cannot see that these few fragments can provide any certain indication as to how and when a regional transition from such types as Dali and Maba to modern humans occurred (Fig. 8.9), an assumption that has been made by various authors. Smith (1985: 208) has remarked that 'the earliest modern *Homo sapiens* from China are all undated, so it is not certain when the archaic- modern *H. sapiens* transition occurred'. There is certainly no justification to speak of fossil evidence of regional continuity, as Wolpoff (1985: 359) has claimed. Such an assumption is based primarily on the assumption of the presence of certain clade features in both the early modern humans and the Middle Pleistocene crania including *Homo erectus*. The relevance of these features, however, is by no means clear (see below).

Early modern humans

The early modern specimens from China may all be younger than 20 000 years (Zhou Mingzhen *et al.* 1982). The most important remains are those from Zhoukoudian Upper Cave, Liujiang, Ziyang, Chilinshan, Ordos, and Huanglong. The Upper Cave crania have now been dated to 10,500-18,300 BP, and the Ziyang cranium to *c.* 7000 BP. The others are still undated, although they are probably of latest Upper Pleistocene age.

Howells (1983: 298; 1988: 227) indeed assumes that some of these specimens are of a general Mongoloid morphology, but not of the strongly specialized form of recent Mongoloids. Due to the recent age of the specimens, this is not surprising. Multivariate analyses of cranium No. 101 from the Upper Cave have yielded the closest affinities to North American Plains Indians and the next closest to modern Europeans (Howells 1983). The Huanglong frontal showed close multivariate affinities to European Upper Palaeolithic specimens (Wang Linghong and Bräuer 1984). The Liujiang cranium also exhibits a quite robust morphology. This morphological spectrum, which can perhaps be described as 'Proto-Mongoloid', appears to be closer to a robust and rather unspecialized modern *Homo sapiens* type. Stringer *et al.* (1984: 121) point out: 'given the morphology and dating of the known material, it is still possible that the modern characters of the late Upper Pleistocene hominids of this area were ultimately derived from an exotic source such as Africa. . .'.

The assumption of regional continuity between East Asian *Homo erectus*, archaic *Homo sapiens*, and modern humans (Weidenreich 1939; Coon 1962; Thorne and Wolpoff 1981; Wolpoff *et al.* 1984) is primarily based on the presence of clade features which are said to be characteristic for this region. These include the presence of shovel-shaped teeth, mid-sagittal keeling, a rounded infra- orbital margin, a horizontal course of the naso-frontal suture, Inca bones, metopic suture, maxillary, ear and mandibular exostoses, and

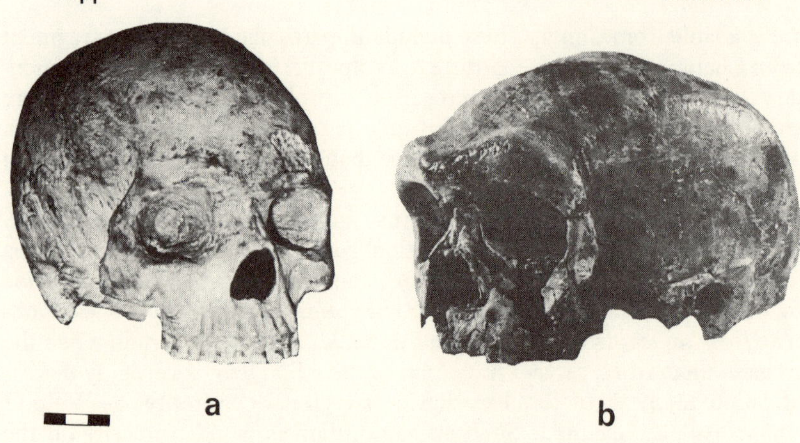

Figure 8.9. (a) The modern cranium Liujiang; and (b) the archaic *Homo sapiens* cranium Dali, from China (from Wu Xinzhi and Wu Maolin 1985). Scale in cm.

tori (Wolpoff *et al.* 1984: 425). The central problem with regard to these features, however, is that most of them also occur in recent and archaic populations from other areas, and are frequently primitive retentions. In other words, they do not provide unequivocal evidence of regional continuity. For example, shovel-shaped incisors are not only present in Dali and other Chinese specimens, but also in Arago and other European hominids (Wu Rukang and Wu Xinzhi 1982: 615). Sagittal keeling is also one of the common features of human fossils of such ancient age. Groves (this volume) analysed the various so-called clade features and concluded that there is little evidence for a special likeness of modern 'Mongoloids' to *Homo erectus pekinensis*. In the discussions during the present conference, Wolpoff emphasized that the East Asian clade features are not restricted to this area; they only appear more frequently in this area than in other regions. The latter assumption, however, has first to be demonstrated on the basis of large samples of anatomically-modern specimens from all parts of the world. Yet the problem remains that most fossil samples are so small that significant differences in the frequencies cannot really be proved.

As with Europe, a model involving varying amounts of gene flow from outside the area combined with local sources could also explain the variation present in the East Asian fossil hominids (Stringer *et al.* 1984). This would also be in better agreement with the recent molecular evidence.

Howells (1988: 227) has stated that specimens which could show a regional evolution to the Mongoloids are still meagre, and are not sufficient to yield a picture comparable to the archaic-modern transition in Africa: 'The best that can be said is that they do not contradict Weidenreich's hypothesis of Mongoloids, although perhaps calling for the eye of faith to see it in the morphology of all cases'.

A similar picture emerges further to the south, although the temporal gap in Indonesia between Ngandong and the few late Upper Pleistocene moderns is unclear, perhaps encompassing 50 000 to 200 000 years.

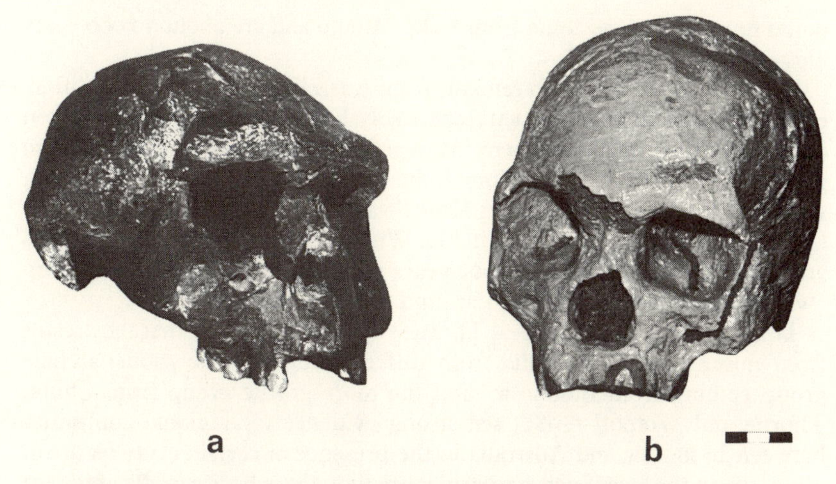

Figure 8.10. (a) The Homo erectus cranium Sangiran 17 (photograph: H. Meyer); and (b) the modern cranium Wadjak 1, from Indonesia. Scale in cm.

Nevertheless, Weidenreich (1943) also found a number of features in Australasia which he interpreted as evidence of a continuous line between the Ngandong crania and modern Australian crania. Groves (this volume) and Habgood (this volume) have shown, however, that these features can hardly be accepted as Australasian clade features. They are found frequently among African and European fossils as well. Groves (this volume) has also shown that published frequencies often have to be viewed with great caution.

The human remains from Niah Cave (Borneo) are generally regarded as the oldest anatomically-modern specimens from Southeast Asia. Their assumed age of 39 000 years, however, is heavily disputed (Brothwell 1960; Wolpoff 1980). According to Brothwell (1960), this skull of a juvenile female individual exhibits certain affinities to the Australoids. To this spectrum also seem to belong the Wadjak and Tabon cranial remains, the latter of which are dated to *c.* 23 000 years BP. Wolpoff *et al.* (1984) have pointed also to certain affinities between these specimens and those from Zhoukoudian Upper Cave and Liujiang. Santa Luca (1980) could not recognize any specific features of the Wadjak specimens which could prove a clear continuity with the Ngandong remains (Fig. 8.10). Due to the large gap in the fossil record between Ngandong and these early moderns, it is not possible to exclude regional evolution; it would be exaggerated, however, to claim that such a lineage is well documented. In order to find clearer evidence of continuity with Ngandong and even earlier Indonesians, Wolpoff *et al.* (1984: 441) have drawn attention to the human remains from Australia.

THE AUSTRALIAN EVIDENCE

Based upon the absolute dates of various archaeological sites, it can now be regarded as certain that Australia has been populated since at least 40 000 years ago (Pearce and Barbetti 1981; Jones, this volume). The oldest well-dated human remains come from Lake Mungo and are about 30 000 years old.

Viewing the fossil human remains from Australia as a whole, clear differences with regard to robusticity can easily be recognized. This has often led to a division of the specimens into a gracile/modern and a robust/archaic group. To the gracile group belong Lake Mungo, Keilor, and Lake Tandou, while the robust group includes Kow Swamp, Willandra Lakes, Cossack, Cohuna, Mossgiel, and Lake Nitchie. Whereas the more gracile specimens are documented as early as 30 000 years BP, the oldest of the robust group (WLH 50) appears to date from around 25 000 BP.

To explain these differences, Thorne (1976) has assumed that they result from migrations to Australia from different regions: the robust/archaic group coming from Indonesia, and the more gracile group from China. Thorne and Wolpoff (1981) see strong evidence for regional continuity between Indonesia and Australia in the presence of certain clade features, especially in the face. Such similarities are thought to be especially apparent between the 700 000 year-old *Homo erectus* cranium from Sangiran and some of the latest Pleistocene and Holocene specimens from Australia (see also Wolpoff, this volume). Groves (this volume) and Habgood (this volume) have, however, been able to show that most of these, as well as other assumed clade features, have a wide distribution and are of very dubious value. Certain similarities in one or two descriptive features of the face, the presence of which might not be restricted to Australasia (see Groves, this volume; Habgood, this volume), can hardly bridge a period of about 700 000 years in which facial remains are lacking. Thus, these can hardly be regarded as evidence of regional continuity.

Macintosh and Larnach (1976) have disputed the frequently-assumed division into 'robust' and 'gracile' forms and suggest that the observed differences reflect the general range of variation found in early modern humans. They also do not accept that the Australian population had its origin in the Ngandong lineage. Brown (1981) has shown that it is probable that some of the archaic-looking features, such as the frequently presence of flat and receding frontal bones as well as the pre-bregmatic eminence, were the result of artificial cranial deformations. Brown sees evidence for this practice in certain skulls from Kow Swamp, Coobool Creek, and Cohuna. A recent study by Brown of the very robust WLH 50 cranium has shown that this thick-walled cranium probably exhibits strong pathological alterations (Stringer, pers. comm.).

Recently, Habgood (1985) has found strong multivariate similarities between the robust and gracile crania, and he thus rejected the assumption of divergent origins for the Australian population. His results show that both Australian samples have considerably closer relationships to each other than to such Chinese specimens as Liujiang or Zhoukoudian Upper Cave. Wolpoff (1980: 330) has also pointed to the similarities between various individuals from Kow Swamp and Lake Mungo:

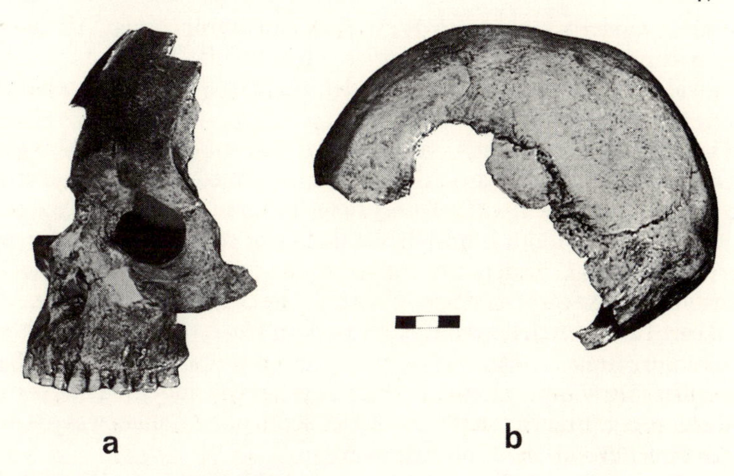

Figure 8.11. The modern Homo sapiens specimens (a) Kow Swamp 15; and (b) Mungo 3, from Australia. Scale in cm.

> I believe that the idea of contrasting groups has resulted from overemphasis on a few specimens, and the fact, that when the first Mungo specimen (the female, Mungo 1) was discovered, neither of the two fairly complete Kow Swamp females (4 and 16) had been reconstructed. Actually, Mungo 1 closely resembles Kow Swamp 4 and 16, while Mungo 3 (male) resembles Kow Swamp 14 in browridge development, and Kow Swamp 14 and 16 in frontal curvature. In other words, the range at a single site encompasses most of the known fossil material.

If specimens as gracile as Mungo 3 (Fig. 8.11) belonged to the populations which came from Indonesia and settled in Australia more than 30 000 years ago, as Habgood (1985) assumes, it appears very unlikely that evolutionary continuity existed between Ngandong and these gracile modern populations. The Ngandong sample cannot convincingly explain the line to either Mungo or to Wadjak.

As an explanation for the archaic elements in a number of Australian specimens, it appears quite possible that the replacement of the populations following the Ngandong type by modern populations coming from the north, was not as complete as appears to have been the case, for example in Western Europe. Thus in Indonesia, more archaic features were able to enter the gene pool of the early moderns, until they finally reached Australia. The Australian evidence would thus also be in agreement with a scenario of complex hybridization and replacement.

CONCLUSIONS

The main purpose of this paper has been to examine whether the fossil record of archaic and modern *Homo sapiens* in the various parts of the world supports the view of 'multiregional evolution' (Wolpoff, this volume), the view of an African archaic-modern evolutionary transition and a subsequent spread of modern humans to West Asia and Europe from Africa (as is proposed by the Afro-European *sapiens* hypothesis: Bräuer 1982), or an even wider global 'Out of Africa' model (Cann *et al.* 1987; Stoneking and Cann, this volume).

The evidence currently available from sub-Saharan Africa has strengthened the case for an early appearance of modern humans at around 100 000 years ago. Wolpoff's claim (1989, and this volume) that the dating of Klasies River Mouth is unreliable, no longer seems tenable. Moreover, there is substantial evidence of the late archaic yet rather modern-looking ancestors of these early moderns in Africa. The course of evolution in Africa from early to late archaic and then to modern *Homo sapiens* as proposed by this author (Bräuer 1982, 1984a) has received support from a number of specialists (see above). Howells (1988: 225) has concluded: 'Thus one may well argue (e.g. Bräuer 1984a) that Africa south of the Sahara was one scene of the direct evolution of modern man'.

The European and West Asian evidence discussed here does not show such an evolutionary sequence towards modern humans. Instead, this large area clearly documents a slow continuous evolution towards the highly specialized Neanderthals and their subsequent replacement by fully modern humans who already existed in Africa some tens of millennia before. The Afro-European *sapiens* hypothesis now appears to be even better supported by the facts than it was seven years ago. Based on the fossil and genetic evidence, it represents a well-supported model of *sapiens* evolution. Howells (1988: 226) remarks: 'one has to entertain the suggestion of Bräuer that moderns from Africa used the Nile Valley as a corridor out of Africa, hospitable at all times regardless of continental climates. . .'. Trinkaus (1986b) and others have also proposed a model with regard to *sapiens* evolution in Africa, Europe, and West Asia that is very close to the Afro-European *sapiens* hypothesis.

Finally, the question remains as to whether the evidence from East Asia and Australia supports the view of an evolution towards modern humans which was more or less independent from that in the western part of the Old World. If one considers the available fossil hominids, the claims for clear evidence of regional continuity in this area appear to be very problematic and hypothetical. Moreover, there is a large unbridged temporal and morphological gap between the representatives of archaic *Homo sapiens* and the early modern humans. A number of models are theoretically imaginable during this undocumented period – among which must surely be counted regional continuity – but conclusive evidence is missing. On the other hand, it has been shown that the model based on certain clade features is very uncertain. According to our present state of knowledge about the variability and distribution of these features, the conclusion is by no means justified that regional continuity is proven in the Far East without essential influence from outside.

Consequently, I cannot at present recognize any unequivocal facts which would disprove even a global 'Out of Africa' model, as has been suggested from recent genetic analyses. Nevertheless, the fossil record does not favour a radical replacement process, but rather a multicausal process of archaic-modern replacement which was quite variable in the various regions. Perhaps the next five or ten years will lead to an even clearer synthesis of the genetic and palaeoanthropologial evidence for the origin of modern humans.

NOTE

1. The new excavations at Klasies River Mouth also yielded a maxillary fragment from the very base of the sequence (>100 000 years BP) which has not yet been studied; it appears fairly robust (H. J. Deacon, pers. comm.).

ACKNOWLEDGEMENTS

In preparing my pre-Conference paper, I received helpful information from K. Butzer, J. and H. J. Deacon, R. L. Hay, W. W. Howells, M. J. Mehlman, H. J. Müller-Beck, G. N. van Vark, and A. C. Wilson; I am grateful to all of them. This final version of the paper profited from the papers presented during the Conference as well as from the discussion with many colleagues. Here, I would especially like to mention: P. Allsworth-Jones, O. Bar-Yosef, J. D. Clark, H. J. Deacon, D. Gambier, M. Green, C. Groves, P. Habgood, F. Harrold, R. Jones, R. Klein, J. Kozlowski, P. Lieberman, P. Mellars, S. Rouhani, F. H. Smith, C.B. Stringer, M. Stoneking, A.-M. Tillier, B. Vandermeersch, T. Volman, J. Wainscoat, M. H. Wolpoff, J. Wymer, and E. Zubrow. I would also like to thank John Baker for putting his final polish on my English. Finally, I am very grateful to the Hansische Universitätsstiftung in Hamburg for financially supporting my participation at this conference. This paper was written during my tenure as Heisenberg Fellow of the Deutsche Forschungsgemeinschaft.

REFERENCES

Bada J. L. 1987. Paleoanthropological applications of amino acid racemization for dating of fossil bones and teeth. *Anthropologischer Anzeiger* 45: 1-8.

Bar-Yosef, O. and Vandermeersch, B. 1981. Notes concerning the possible age of the Mousterian layers in Qafzeh Cave. In J. Cauvin and P. Sanlaville (eds) *Préhistoire du Levant*. Paris: Centre National de la Recherche Scientifique: 281-286.

Beaumont, P. B. 1980. On the age of Border Cave hominids 1-5. *Palaeontologia Africana* 23: 21-33.

Binford, L. R. 1984. *Faunal Remains From Klasies River Mouth*. New York: Academic Press.

Binford, L. R. 1986. Reply to Singer and Wymer's response. *Current Anthropology* 27: 57-62.

Bordes, F. 1972. Du Paléolithique moyen au Palólithique supérieur: continuité ou discontinuité? In F. Bordes (ed.) *The Origin of Homo sapiens*. Paris: UNESCO: 211-218.

Bräuer, G. 1982. Early anatomically modern man in Africa and the replacement of the Mediterranean and European Neandertals. In H. de Lumley (ed.) *L'Homo erectus et la Place de l'Homme de Tautavel*

Parmi les Hominidés Fossiles. Nice: Çentre National de la Recherche Scientifique / Louis-Jean Scientific and Literary Publications: 112.

Bräuer, G. 1984a. A craniological approach to the origin of anatomically modern *Homo sapiens* in Africa and implications for the appearance of modern Europeans. In F. H. Smith and F. Spencer (eds) *The Origins of Modern Humans: a World Survey of the Fossil Evidence*. New York: Alan R. Liss: 327-410.

Bräuer, G. 1984b. The 'Afro-European sapiens hypothesis' and hominid evolution in East Asia during the late Middle and Upper Pleistocene. *Courier Forschungsinstitut Senckenberg* 69: 145-165.

Bräuer, G. 1984c. Präsapiens-Hypothese oder Afro-europäische Sapiens-Hypothese? *Zeitschrift für Morphologie und Anthropologie* 75: 1-25.

Bräuer, G. and Leakey, R. E. 1986a. A new archaic *Homo sapiens* cranium from Eliye Springs, West Turkana, Kenya. *Zeitschrift für Morphologie und Anthropologie* 76: 245-252.

Bräuer, G. and Leakey, R. E. 1986b. The ES-11693 cranium from Eliye Springs, West Turkana, Kenya. *Journal of Human Evolution* 15: 289-312.

Bräuer, G. and Mehlman, M. J. 1988. Hominid molars from a Middle Stone Age level at the Mumba Rock Shelter, Tanzania. *American Journal of Physical Anthropology* 75: 69-76.

Brothwell, D. R. 1960. Upper Pleistocene human skull from Niah Caves, Sarawak. *Journal of the Sarawak Museum* 9: 323-349.

Brown, P. 1981. Artificial cranial deformation: a component in the variation in Pleistocene Australian aboriginal crania. *Archaeology in Oceania* 16: 156-167.

Butzer, K. W. 1984. Archaeogeology and Quaternary environment in the interior of Southern Africa. In R. G. Klein (ed.) *Southern African Prehistory and Paleoenvironments*. Rotterdam: Balkema: 1- 64.

Cann, R. L., Stoneking, M. and Wilson, A. C. 1987. Mitochondrial DNA and human evolution. *Nature* 325: 31-36.

Clark, J.D. 1959. Carbon-14 chronology in Africa south of the Sahara. In G. Mortelmans and J. Nenquin (eds) *Actes du IV Congrès Panafricain de Préhistoire et de l'Etude du Quaternaire:* 303-311.

Clarke, R. J. 1985. A new reconstruction of the Florisbad Cranium, with notes on the site. In E. Delson (ed.) *Ancestors: the Hard Evidence*. New York: Alan R. Liss: 301-305.

Cooke, H. B. S., Malan, B. D. and Wells, L. H. 1945. Fossil man in the Lebombo Mountains, South Africa: the 'Border Cave', Ingwawuma District, Zululand. *Man* 45: 6-13.

Coon C. S. 1962. *The Origin of Races*. New York: Knopf.

Day, M. H. and Stringer, C. B. 1982. A reconsideration of the Omo Kibish remains and the erectus-sapiens transition. In H. de Lumley (ed.) *L'Homo erectus et la Place de l'Homme de Tautavel Parmi les Hominidés Fossiles*. Nice: Centre National de la Recherche Scientifique / Louis-Jean Scientific and Literary Publications: 814-846.

Demars, P. Y. and Hublin, J. J. 1986. La transition Néandertaliens/hommes de type moderne en Europe: diffusion de caractères ou immigration? Paper presented to Colloquium on *L'Homme de Néandertal*, Liège, 1986. In Press.

De Villiers, H. and Fatti, L. P. 1982. The antiquity of the Negro. *South African Journal of Science* 78: 321-332.

Giles, E. and Ambrose, S. H. 1986. Are we all out of Africa? *Nature* 322: 21-22.

Grine, F. E. and Klein, R. G. 1985. Pleistocene and Holocene human remains from Equus Cave, South Africa. *Anthropology* 8: 55-98.

Habgood, P. J. 1985. The origin of the Australian Aborigines: an alternative approach and view. In P. V. Tobias (ed.) *Hominid Evolution: Past, Present and Future*. New York: Alan R. Liss: 367-380.

Hahn, J. 1986. *Kraft und Agression: die Botschaft der Eiszeitkunst im Aurignacien Süddeutschlands?*. Tübingen: Archaeologica Venatoria.

Hay, R. L. 1987. Geology of the Laetoli area. In M. D. Leakey and J. M. Harris (eds) *Results of the Laetoli Expeditions 1975-1981*. Oxford: Oxford University Press: 23-47.

Howells, W. W. 1975. Neanderthal man: facts and figures. In R. H. Tuttle (ed.) *Paleoanthropology: Morphology and Paleoecology*. Paris: Mouton: 389-407.

Howells, W. W. 1983. Origins of the Chinese People: interpretations of the recent evidence. In D. N. Keightley (ed.) *The Origins of Chinese Civilisation*. Berkeley: University of California Press: 297-319.

Howells, W. W. (1988). The meaning of the Neanderthals in human evolution. In Fondation Singer-Polignac (ed.) *L'Evolution dans sa Réalité et ses Diverses Modalités*. Paris: Masson: 221-239.

Jacob, T. 1981. Solo Man and Peking Man. In B. A. Sigmon and J. S. Cybulski (eds.) *Homo erectus: Papers in Honor of Davidson Black*. Toronto: University of Toronto Press: 87-104.

Jelinek, A.J. 1982. The Tabūn Cave and Paleolithic man in the Levant. *Science* 216: 1369-1375.

Jones, J. S. and Rouhani, S. 1986. How small was the bottleneck? *Nature* 319: 449-450.

Kennedy G. E. 1984. The emergence of *Homo sapiens:* the post-cranial evidence. *Man* 19: 94-110.

Klein R. G. 1983. The Stone Age prehistory of Southern Africa. *Annual Review of Anthropology* 12: 25-48.

Kuman, K. and Clarke, R. J. 1986. Florisbad: new investigations at a Middle Stone Age hominid site in South Africa. *Geoarchaeology* 1: 103-125.

Leakey, M. D. and Hay, R. L. 1982. The chronological position of the fossil hominids of Tanzania. In H. de Lumley (ed.) *L'Homo erectus et la Place de l'Homme de Tautavel Parmi les Hominidés Fossiles*. Nice: Centre National de la Recherche Scientifique / Louis-Jean Scientific and Literary Publications: 753-765.

Lumley, M.-A. de and Sonakia, A. 1985. Première découverte d'un Homo erectus sur le continent Indien à Hathnora, dans la moyenne vallée de la Narmada. *L'Anthropologie* 89: 13-61.

Macintosh, N. W. G. and Larnach, S. L. 1976. Aboriginal affinities looked at in world context. In R. L. Kirk and A. G. Thorne (eds) *The Origins of the Australians*. Canberra: Australian Institute of Aboriginal Studies: 113-126.

Magori, C. C. 1980. *Laetoli Hominid 18: Studies on a Pleistocene Fossil Human Skull from Northern Tanzania*. Unpublished Ph.D. Thesis, University of London.

Magori, C. C. and Day, M. H. 1983. An early *Homo sapiens* skull from the Ngaloba Beds, Laetoli, Northern Tanzania. *Anthropos* (Athens) 10: 143-183.

Mehlman, M. J. 1987. Provenience, age and associations of archaic *Homo sapiens* crania from Lake Eyasi, Tanzania. *Journal of Archaeological Science* 14: 133-162.

Mellars, P. A. 1986. A new chronology for the French Mousterian period. *Nature* 322: 410-411.

Partridge, T. C. 1982. The chronological positions of the fossil hominids of Southern Africa. In H. de Lumley (ed.) *L'Homo erectus et la Place de l'Homme de Tautavel Parmi les Hominidés Fossiles*. Nice: Centre

National de la Recherche Scientifique / Louis-Jean Scientific and Literary Publications: 617-675.

Pearce, R. H. and Barbetti, M. 1981. A 38 000 year old archaeological site at Upper Swan, Western Australia. *Archaeology in Oceania* 16: 173-178.

Rak, Y. 1986. The Neanderthal: a new look at an old face. *Journal of Human Evolution* 15: 151-164.

Rightmire, G. P. 1978. Florisbad and human population succession in Southern Africa. *American Journal of Physical Anthropology* 48: 475-486.

Rightmire, G. P. 1979. Implications of Border Cave skeletal remains for Later Pleistocene human evolution. *Current Anthropology* 20: 23-35.

Rightmire, G. P. 1986. Africa and the origins of modern humans. In R. Singer and J. K. Lundy (eds) *Variation, Culture and Evolution in African Populations*. Johannesburg: Witwatersrand University Press: 209-220.

Santa Luca, A. P. 1980. The Ngandong fossil hominids. *Yale University Publications in Anthropology* 78: 1-175.

Sartono, S. 1982. Characteristics and chronology of early men in Java. In H. de Lumley (ed.) *L'Homo erectus et la Place de l'Homme de Tautavel Parmi les Hominidés Fossiles*. Nice: Centre National de la Recherche Scientifique / Louis-Jean Scientific and Literary Publications: 491-541.

Singer, R. and Wymer, J. 1982. *The Middle Stone Age at Klasies River Mouth in South Africa*. Chicago: University of Chicago Press.

Singer, R. and Wymer, J. 1986. On Binford on Klasies River Mouth: response of the excavators. *Current Anthropology* 27: 56-57.

Smith, F. H. 1982. Upper Pleistocene hominid evolution in South-Central Europe: a review of the evidence and analysis of trends. *Current Anthropology* 23: 667-703.

Smith, F. H. 1984. Fossil hominids from the Upper Pleistocene of Central Europe and the origin of modern Europeans. In F. H. Smith and F. Spencer (eds) *The Origins of Modern Humans: a World Survey of the Fossil Evidence*. New York: Alan R. Liss: 137-210.

Smith, F. H. 1985. Continuity and change in the origin of modern *Homo sapiens*. *Zeitschrift für Morphologie und Anthropologie* 75: 197-222.

Sonakia, A. 1985. Early *Homo* from Narmada Valley, India. In E. Delson (ed.) *Ancestors: the Hard Evidence*. New York: Alan R. Liss: 334-338.

Stringer, C. B. 1982. Towards a solution to the Neanderthal problem. *Journal of Human Evolution* 11: 431-438.

Stringer, C. B. 1984. The definition of *Homo erectus* and the existence of the species in Africa and Europe. *Courier Forschungsinstitut Senckenberg* 69: 131-143.

Stringer, C. B. 1985. Evolution of a species. *The Geographical Magazine* 57: 601-607.

Stringer, C. B. 1986. The Origin of modern *Homo sapiens*. Paper presented to Colloquium on *L'Homme de Néandertal*, Liège, 1986. In Press.

Stringer, C. B., Hublin, J.-J. and Vandermeersch, B. 1984. The origin of anatomically modern humans in Western Europe. In F. H. Smith and F. Spencer (eds) *The Origins of Modern Humans: a World Survey of the Fossil Evidence*. New York: Alan R. Liss: 51-135.

Stringer, C. B. and Trinkaus, E. 1981. The Shanidar Neanderthal crania. In C. B. Stringer (ed.) *Aspects of Human Evolution*. London: Taylor and Francis: 129-165.

Tattersall, I. 1986. Species recognition in human paleontology. *Journal of Human Evolution* 15: 165-175.

Thorne, A. G. 1976. Morphological contrasts in Pleistocene Australians. In R. L. Kirk and A. G. Thorne (eds) *The Origin of the Australians*.

Canberra: Australian Institute of Aboriginal Studies: 95-112.
Thorne, A. G. and Wolpoff, M. H. 1981. Regional continuity in Australasian Pleistocene hominid evolution. *American Journal of Physical Anthropology* 55: 337-349.
Trinkaus, E. 1981. Neandertal limb proportions and cold adaptation. In C. B. Stringer (ed.) *Aspects of Human Evolution*. London: Taylor and Francis: 187-224.
Trinkaus, E. 1983. *The Shanidar Neandertals*. New York: Academic Press.
Trinkaus, E. 1984. Western Asia. In F. H. Smith and F. Spencer (eds) *The Origins of Modern Humans: a World Survey of the Fossil Evidence*. New York: Alan R. Liss: 251-293.
Trinkaus, E. 1986a. The Neandertals and modern human origins. *Annual Review of Anthropology* 15: 193-218.
Trinkaus, E. 1986b. Les néandertaliens. *La Recherche* 180: 1040- 1047.
Trinkaus. E. and Howells, W. W. 1979. The Neanderthals. *Scientific American* 241: 118-133.
Trinkaus, E. and Le May, M. 1982. Occipital bunning among Later Pleistocene hominids. *American Journal of Physical Anthropology* 57: 27-35.
Trinkaus, E. and Smith, F. H. 1985. The fate of the Neandertals. In E. Delson (ed.) *Ancestors: the Hard Evidence*. New York: Alan R. Liss: 325-333.
Vandermeersch, B. 1981. Les premiers *Homo sapiens* au Proche Orient. In D. Ferembach (ed.) *Les Processus de l'Hominisation*. Paris: Centre National de la Recherche Scientifique: 97-100.
Vandermeersch, B. 1982. The first *Homo sapiens sapiens* in the Near East. In A. Ronen (ed.) *The Transition from Lower to Middle Palaeolithic and the Origin of Modern Man*. Oxford: British Archaeological Reports International Series S151: 297-299.
Vandermeersch, B. 1984. A propos de la découverte du squelette néandertalien de Saint-Césaire. *Bulletins et Mémoires de la Société d'Anthropologie de Paris (série 14)* 1: 191-196.
Vandermeersch, B. 1985. Neanderthal Man and the origins of Modern Man. In *Homo: Journey to the Origins of Man's History*. Cataloghi Marsilio: Venise: 95-102.
Van Vark, G. N. 1986. More on the classification of the Border Cave 1 skull. *5th Congress of the European Anthropological Association*, Lisbon 1986 (abstract).
Vrba, E. 1982. Biostratigraphy and chronology, based particularly on Bovidae, of southern hominid-associated assemblages: Makapansgat, Sterkfontein, Taung, Kromdraai, Swartkrans; also Elandsfontein (Saldanha), Broken Hill (now Kabwe), and Cave of Hearths. In H. de Lumley (ed.) *L'Homo erectus et la Place de l'Homme de Tautavel Parmi les Hominidés Fossiles*. Nice: Centre National de la Recherche Scientifique / Louis-Jean Scientific and Literary Publications: 707-752.
Wainscoat, J. S. 1987. Out of the garden of Eden. *Nature* 325: 13.
Wainscoat, J. S., Hill, A. V. S., Boyce, A. L., Flint, J., Hernandez, M., Thein, S. L., Old, J. M., Lynch, J. R., Falusi, A. G., Weatherall, D. J. and Clegg. J. B. 1986. Evolutionary relationships of human populations from an analysis of nuclear DNA polymorphisms. *Nature* 319: 491-493.
Wang Linghong and Bräuer, G. 1984. A multivariate comparison of the human calva from Huanglong County, Shaanxi Province. *Acta Anthropologica Sinica* 3: 313-321.
Weidenreich, F. 1939. On the earliest representatives of modern mankind recovered on the soil of East Asia. *Peking Natural History Bulletin* 13: 161-174.

Weidenreich, F. 1943. The skull of *Sinanthropus pekinensis:* a comparative study of a primitive hominid skull. *Palaeontologia Sinica* (n.s. D) 10 (whole series 127). Pehpei: Geological Survey of China.

White, T. D. 1985. *Acheulian Man in Ethiopia's Middle Awash Valley: the Implications of Cutmarks on the Bodo Cranium.* Achtste Kroon-Voordracht Amsterdam. Haarlem: J. Ensched; en Zonen.

Wilson, A. C., Stoneking, M., Cann, R. L., Prager, E. M., Ferris, S. O., Wrischnik, L. A. and Higuchi, R. G. 1987. Mitochondrial clans and the age of our common mother. In F. Vogel and K. Sperling (eds) *Human Genetics: Proceedings of the 7th International Congress of Human Genetics.* Berlin: Springer: 158-164.

Wolpoff, M. H. 1980. *Paleoanthropology.* New York: Knopf.

Wolpoff, M. H. 1985. Human evolution at the peripheries: the pattern at the eastern edge. In P. V. Tobias (ed.) *Hominid Evolution: Past, Present and Future.* New York: Alan R. Liss: 355- 365.

Wolpoff, M. H. 1989. The place of the Neandertals in human evolution. In E. Trinkaus (ed.) *Patterns and Processes in Later Pleistocene Human Emergence.* Cambridge: Cambridge University Press. In Press.

Wolpoff, M. H., Wu Xinzhi and Thorne, A. G. 1984. Modern *Homo sapiens* origins: a general theory of hominid evolution involving the fossil evidence from East Asia. In F. H. Smith and F. Spencer (eds) *The Origins of Modern Humans: a World Survey of the Fossil Evidence.* New York: Alan R. Liss: 411-483.

Wu Rukang and Wu Xinzhi 1982. Comparison of Tautavel Man with *Homo erectus* and early *Homo sapiens* in China. In H. de Lumley (ed.) *L'Homo erectus et la Place de l'Homme de Tautavel Parmi les Hominidés Fossiles.* Nice: Centre National de la Recherche Scientifique / Louis-Jean Scientific and Literary Publications: 605-616.

Wu Xinzhi 1981. A well-preserved cranium of an archaic type of early *Homo sapiens* from Dali, China. *Scienta Sinica* 24: 530-539.

Wu Xinzhi and Wu Maolin. 1985. Early *Homo sapiens* in China. In Wu Rukang and J. W. Olsen (eds) *Palaeoanthropology and Palaeolithic Archaeology in the People's Republic of China.* Orlando: Academic Press: 91-106.

Zhou Mingzhen, Li Yanxian, and Wang Linghong, 1982. Chronology of the Chinese fossil hominids. In H. de Lumley (ed.) *L'Homo erectus et la Place de l'Homme de Tautavel Parmi les Hominidés Fossiles.* Nice: Centre National de la Recherche Scientifique / Louis-Jean Scientific and Literary Publications: 593-604.

9. The Evolution of Modern Humans: Recent Evidence from Southwest Asia

BERNARD VANDERMEERSCH

INTRODUCTION

One of the central problems posed by the study of human fossils is the origins of the populations which these fossils represent. Has there been local evolution (either gradual or rapid) from the previous population? Or must we invoke the hypothesis of movement of populations? The whole mechanism of human evolution is included in these questions, and in the answers we can give to them.

In the case of relatively recent remains (100 000 years old or less) the dating of the remains is, of course, as crucial to any discussion of these questions as is the biological character of the populations concerned.

Seen in these terms, the Near East is one of the most interesting areas for the study of modern human origins, since it has produced a relatively large sample of skeletal remains belonging to this period, and the remains display a range of morphological diversity which has given rise to a variety of competing interpretations. Over the past 20 years, however, a number of important advances have been made which allow a far better understanding of the character and potential relationships of the first modern populations of this area.

In the present discussion, I shall only take into account remains (principally skulls and skeletons) which are sufficiently complete to provide significant information on the biological characters of the human groups, and I shall only consider the remains of adults. The skeletal remains of children (especially very young ones) have a very different morphology from that of adults, which is related to both the order of appearance of particular skeletal characters during growth, and to the conditions controlling the development of body-build. The remains of young individuals are dealt with separately in the paper by A.-M. Tillier in the present volume.

HOMINID EVOLUTION IN SOUTHWEST ASIA

The available sample comprises the remains of 22 individuals, recovered from seven archaeological sites:

1. Amūd Cave (Israel) (Suzuki and Takai 1970)
2. Kebara Cave (Mount Carmel, Israel) (Arensburg *et al.* 1985)
3. Qafzeh Cave (Israel) (Vandermeersch 1981)

4. Skhūl Cave (Mount Carmel, Israel) (McCown and Keith 1939)
5. Shanidar Cave (Iraq) (Trinkaus 1983)
6. Tabūn Cave (Mount Carmel, Israel) (McCown and Keith 1939)
7. Zuttiyeh Cave (Israel) (Keith 1927)

All these remains derive from levels which are archaeologically 'Mousterian', with the exception of the partial skull from Zuttiyeh. The latter is almost certainly older, as Gisis and Bar-Yosef (1974) have shown. The latter authors showed that the archaeological material recovered from the single archaeological level recognized in the excavations by Turville-Petre (1927) was in fact a mixture of two archaeological industries, and that the fragments of breccia that still remained in place in several parts of the cave showed a succession of an Acheulian of Yabrudian facies, followed by Mousterian. The Zuttiyeh skull most probably came from the Acheulian level. It is therefore older than was previously thought, and could be more than 100 000 years old.

The first remains to be discovered in a Mousterian context came from the Mount Carmel caves of Tabūn and Skhūl. In a preliminary publication, Keith and McCown (1937) separated these remains into two distinct populations – the one from Tabūn resembling the European Neanderthals, and the other from Skhūl showing a more modern morphology. In the final publication, however (McCown and Keith 1939), they combined these remains into a single group, which they attributed to the genus '*Palaeoanthropus*' (a term derived from the palaeoanthropian stage proposed in 1916 by Elliot Smith). According to this definition, the '*Palaeoanthropus*' remains were those previously attributed to the Neanderthals, and other morphologically similar finds. It would appear that this change in interpretation was prompted (at least in part) by the pre-last-glacial age attributed to the finds.

McCown and Keith suggested that the Mount Carmel populations might show evolutionary trends in the direction of both Neanderthals and modern human forms with, however, a prevalence towards the former. The discoveries made by Neuville at the site of Qafzeh (between 1933 and 1935) had not been studied at this time, and were therefore not taken into account in these discussions.

From 1939 onwards the concept of hybridization was introduced into these discussions (Coon 1939; Ashley-Montagu 1940; Dobzhansky 1944). Viewed from this perspective, the various hominids from Mount Carmel were considered to be the result of cross-breeding between separate populations of Neanderthals and *Homo sapiens sapiens*. More recently, this interpretation has been well defended in publications by A. Thoma (1957-58, 1962, 1963). According to this author, the fossils from Skhūl should be seen as the result of hybridization between a Neanderthal population (as represented at Tabūn) and some other population which had not yet been discovered in the area. This interpretation therefore allowed (as in the initial interpretation by McCown and Keith) a separation between the remains from Tabūn and Skhūl.

At the same time, F. Clark Howell (1959) proposed that the chronology of the Mousterian industries of the Near East was essentially synchronous with that of the Mousterian in Europe – i.e. coinciding with the earlier part of the last glaciation. He accepted the morphology of the Skhūl and Qafzeh hominids as essentially modern, and proposed to classify them as 'Proto-Cro-Magnon'. This interpretation therefore required the existence of two populations – one Neanderthal and one of modern form – both dating from the earlier part of the last glaciation within the Near Eastern region. Further discoveries and publications served only to confirm this interpretation.

At present, therefore, we have at our disposal two series of important human remains deriving from the Near Eastern Mousterian, each represented by several skeletons:

1. The Neanderthal forms – represented by the finds from Shanidar, Tabūn, Amūd and Kebara; and
2. *Homo sapiens sapiens* forms (i.e. the 'Proto-Cro-Magnons' according to the terminology of Clark Howell and myself) – represented by the remains from Skhūl and Qafzeh.

The remainder of the present discussion will focus on the second group of finds, with the aim of clarifying the origin of anatomically modern populations within the Near East.

THE ORIGINS OF ANATOMICALLY MODERN POPULATIONS IN SOUTHWEST ASIA

In this context, two separate, alternative, hypotheses must be considered. Either

1. The appearance of *Homo sapiens sapiens* forms in the Near East is a result of purely local evolution from pre-existing populations in the same area. Or,
2. These early *Homo sapiens sapiens* forms evolved in some other area, and arrived as immigrants into an area already occupied by Neanderthal populations.

The former hypothesis can in turn be split into two alternatives:

a. The early *Homo sapiens sapiens* forms in the Near East evolved from local Neanderthal populations in the same areas. Or,
b. These forms derive from some other population of archaic *Homo sapiens* form.

The first theory can only be discussed in the context of the remains from the Near East themselves. If it is possible to demonstrate that there could not have been any direct phylogenetic link between the Proto-Cro-Magnons and their immediate predecessors in the same region, then the second hypothesis becomes, *ipso facto*, the only plausible interpretation. In order to evaluate either interpretation, however, we must have information on the anatomical characteristics of the supposedly ancestral population, and its area of origin.

Viewed from the perspective of purely local evolution, we have three groups of human remains to take into account. Two of these are of Mousterian age (i.e. the Proto-Cro-Magnons and the Neanderthals), while the third is of Acheulian age (i.e. the fragmentary Zuttiyeh skull).

The possibility of a direct phylogenetic link between the Near Eastern

Neanderthals and the local Proto-Cro-Magnon forms has sometimes been argued on chronological grounds (Jelinek 1982a, 1982b). In order to maintain this theory, however, there is an essential chronological requirement: within the local sequence, the Neanderthals must be older than the oldest Proto-Cro-Magnon forms. At present there is still no definitive chronology for the Middle Palaeolithic sequence of the Near East, and a major programme on *The evolution of the cultures and human types in the Near East from the end of the Lower Palaeolithic to the beginning of the Upper Palaeolithic* has recently been initiated by myself and O. Bar-Yosef in an attempt to resolve these issues (Bar-Yosef *et al.* 1986). At present, there is certainly no unequivocal evidence to support the chronological precedence of the Neanderthals over the Proto-Cro-Magnon forms. On the contrary, there are strong arguments against this view. At Qafzeh, for example, there are indications of major disturbance in the uppermost part of the Mousterian levels. The final Mousterian levels have been eroded by water, and there is a clear break between these layers and the overlying Upper Palaeolithic levels. In fact, the most recent Middle Palaeolithic human remains from this site date from well before the end of the Mousterian (Bar-Yosef and Vandermeersch 1981) and one of the Proto-Cro-Magnon skeletons derives from layer XXII, at the very base of the deposits. In order to support Jelinek's chronology, we would therefore have to propose that the entire Mousterian sequence at Qafzeh is younger than the Mousterian levels at Tabūn, Amūd and Kebara which have produced Neanderthal remains.

The latter interpretation can however be contested on at least two grounds:

1. Palaeontological analysis of the micro-mammalian remains from Qafzeh supports an early date for these levels (Haas 1972; Tchernov 1984) (see Note 1);
2. Preliminary results of thermoluminescence dating of the Mousterian levels at Kebara (Valladas *et al.* 1987) show that these deposits have an age of c. 50-60 000 years. The recently discovered Neanderthal skeleton from this site therefore dates from a relatively late stage in the overall Mousterian succession of the Near East.

In any event, it seems to me that chronological arguments are of limited relevance in this context. Dating is of course essential to determine the relative sequence of the various Neanderthal and modern human remains within the region, but these arguments are neither necessary nor sufficient to determine the phylogenetic relationships between these forms. Chronological succession does not automatically imply phylogenetic affiliation. If the anthropological data are opposed to such a relationship, then chronology itself cannot establish it.

The real problem is strictly anthropological and is linked to the way in which specific Neanderthal characters are interpreted. These characters are not confined to the superstructures of the skull or to other detailed anatomical features. The whole shape of the Neanderthal skull is different from that of modern humans, and the studies of Trinkaus (1977, 1983, 1984 etc.) have similarly emphasized the pronounced specializations of the postcranial skeleton. This strong morphological specialization can be shown to have been acquired gradually over the course of approximately 300 000 years of evolution in Europe (Stringer, Hublin and Vandermeersch 1984). This pattern of evolution also explains why archaic *Homo sapiens* forms (from which the Neanderthal lineage was derived) have a general morphology which is closer to that of modern humans than to the typical Neanderthal forms dating from the Würm glaciation.

If we postulate a direct transition from Neanderthals to modern humans we are forced to propose an extremely rapid evolutionary development during the course of the Würm II/III interstadial. A further problem concerns the lack of evidence for Neanderthal origins outside of Europe. There are in fact no obvious ancestral Neanderthals known from elsewhere.

In addition, the Neanderthals from the Near East differ from those of Europe in a number of special features and, in the former group, the specific Neanderthal characters are often less clearly pronounced. On the other hand, Condemi's work (1985) has emphasized the many and close resemblances between the Near Eastern Neanderthals and the hominids from Saccopastore (Italy). In the light of all this it is possible to suggest that the Neanderthals reached the Near East during the Last Interglacial or at the very beginning of the Würm glaciation. Populations must have been established on a substantial scale, as representatives have been found ranging from Israel to Uzbekistan. Subsequently, connections with the European Neanderthals may have been broken, so that the two populations evolved separately. Differences in climatic and environmental conditions would have led to slightly different patterns of development in the two regions, but the basic anatomical characters were retained.

In Western Europe we now know that the Neanderthals survived until the beginning of the Upper Palaeolithic period, and occur in association with a Châtelperronian industry at Saint-Césaire (Charente-Maritime) (Levêque and Vandermeersch 1980). We unfortunately have no human remains which can be reliably attributed to the earliest stages of the Aurignacian, but from the available indications it seems highly unlikely that these were also Neanderthals (see Gambier, this volume). Hence it is likely that the latest Neanderthal populations and the earliest populations of anatomically-modern humans in Western Europe were (at least in part) contemporaneous.

If our current understanding of the evolutionary significance of Neanderthal features is correct, it is very difficult to accept an evolution from Neanderthals to modern humans on the basis of purely anthropological criteria. In addition, current evidence from both archaeology and geochronology (despite their present limitations) make this theory even more unlikely. For all these reasons, I find it difficult to accept any direct phylogenetic relationship between Neanderthals and modern humans either in the Near East or in Europe.

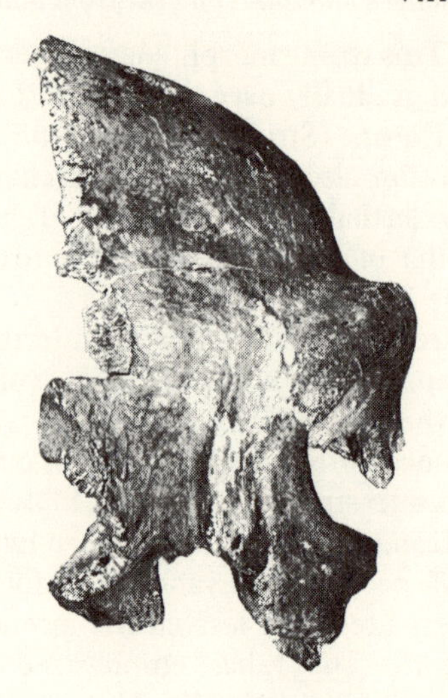

Figure 9.1. The Zuttiyeh skull. Lateral view.

The Zuttiyeh Skull

So what was the origin of the Proto-Cro-Magnon population in the Near East? The character of the earlier hominid from Zuttiyeh is clearly crucial in this context, since this represents the only evidence at present available for the pre-Mousterian, Acheulian populations in the Levant. Any reconstruction based on this skull will inevitably be incomplete, as the skull is represented only by the frontal right zygomatic and part of the right sphenoid (see Fig. 9.1). Nevertheless the morphology of this find deserves close examination.

The *squama frontalis* is fairly long (125 mm). The nasion-bregma chord reaches 113 mm, which gives an index of 90.4. The height above the cord is 21 mm, but we must allow for the fact that the nasion is not recessed. The convexity of the squama is within the range of modern populations as well as that of Neanderthals. More important is the orientation of the squama, which is very difficult to estimate on such an incomplete specimen. It is possible, as Keith did (1927), to use the *sututa frontozygomatica,* but as this is not absolutely straight, the result is only approximate. It is also possible to take into account the plane of the orbits, which cannot differ much from the vertical: for a skull like Zuttiyeh which has a strong supraorbital brow-ridge, this plane cannot be oriented towards the top and backwards. Whichever feature is being used, and in spite of imprecision, the *squama frontalis* of Zuttiyeh is always set vertically, as Piveteau (1957) had already noted.

The brow-ridge has a double arched shape. A slight glabellar depression, as well as a strong supraglabellar depression, can be seen. The *arcus superciliaris* is much thicker than the *arcus supraorbitalis*. The separation between these two structures is very clear, especially on the right, showing that the process of dissociation between the supraorbital structures was occurring.

The brow-ridge is marked, as in Neanderthals, some of which show a lateral thinning. But none have both constituents so distinct. This is reminiscent of what can be observed in Skhūl IV or Předmostí III, although the structure is more robust in Zuttiyeh.

The brow-ridge and *squama frontalis* are separated by a *sulcus supraciliaris*, which is both wide and hollow. The torus extends on both sides beyond the *squama*. The part of the brow-ridge under the glabella is almost flat, which implies that the nasion is set back very little from the glabella, and the nasal notch is very weak. This feature is very different from what can be observed on Neanderthal skulls.

The minimum frontal width is much lower than in Neanderthals (except for Tabūn), and very close to that of Skhūl V, Předmostí I and the French Cro-Magnons. The maximum frontal width is lower than in the Neanderthals, as well as in the Proto-Cro-Magnons and Cro-Magnons (except for Qafzeh 9 and Skhūl V).

The measurements of this skull can only provide rather imprecise information, but the morphology of the skull allows some more specific observations. The brow-ridge, with its heavy shape and expansion on both sides of the *squama frontalis*, is incontestably archaic. But on the frontal bone comparisons can be made with the the Neanderthals for the morphology of the brow-ridge, position of the nasion, and heightening of the squama, which is a gradual feature.

A further point must be noted: the frontal sinus has neither the shape nor the thickness of that of the Neanderthals (Keith and McCown 1937). The nasal bones are flat, and not in a roof shape as in Neanderthals, and they slope obliquely backwards rather than being set upright. But at the same time the orbits are rectangular, a little wider than high. The zygomatic bone is robust, with a high body and a *processus frontalis* which is rather narrow, more so than in the Neanderthals from Amūd and Shanidar. The posterior side is regularly structured, again differing from Amūd I and Shanidar 1 and 2, without a *processus marginalis*. But the most important character is the anteriorly facing position of the bone body, which makes it very different from the Neanderthals.

One of the most revealing features of the facial architecture of the Neanderthals – which is already apparent in the pre-last glaciation Preneanderthals – is therefore missing from Zuttiyeh. The *ala major* is retained. It is thick and large in an anteroposterior direction (39 mm) and is included in the macrosphenoid type as described by McCown and Keith (1939). We have here an archaic feature which can also be observed in Skhūl V.

Thus on the skull from Zuttiyeh we can observe a set of archaic features: a strong brow-ridge and *sulcus supraciliaris*, wide *ala major*, large *facies temporalis* of the zygomatic bone etc. Some of these features may be common in Neanderthals, but distinctive Neanderthal features cannot be observed in Zuttiyeh. The latter shows, by contrast, a facial architecture which is similar to that of modern humans, except for the strong brow-ridge. It can be compared with that of Djebel Irhoud I, with the following differences: the brow-ridge from Djebel Irhoud I is larger in the supraorbital part;

laterally it extends more widely beyond the *squama frontalis*; and it shows no signs of dissociation. It is also less arched above the orbit, but the torus curvature in anterior aspect varies more or less in every fossil population. The *squama frontalis* is wider in Djebel Irhoud I. In contrast, the orbits and zygomatic bones of both fossils are very close.

Assessed on the basis of these features, is it possible to postulate a direct phylogenetic relationship between the Zuttiyeh hominid and the Near Eastern Neanderthals? Clearly, we must make allowance for the very incomplete nature of the Zuttiyeh skull. The European Pre-Neanderthals show that the apomorphic features of the lineage have actually evolved in a 'mosaic' manner. For example, we could suggest that the occipital bone from Zuttiyeh, had it been preserved, might have had Neanderthal features. This argument cannot be accepted, however, as the mosaic features of the Neanderthal lineage can only be documented for the oldest remains, and for only a few of them (notably Arago and Petralona). In these early European fossils the distinctive features of the Neanderthal face can already be observed. Furthermore, the typical morphology of the Neanderthal skull had already been established (in all its features) by the middle of the Riss glacial period – and there is no evidence to suggest that the Zuttiyeh skull could be older than this. In proposing a phyletic link between Zuttiyeh and the Neanderthals we would have to admit two independent lineages – one in Europe and the other in the Near East, both of which would have given rise to Neanderthals. This hypothesis is scarcely defensible.

Can we propose an evolution from the Zuttiyeh hominids to the Proto-Cro-Magnons? Whilst there are no specific Cro-Magnon features on the preserved portion of the Zuttiyeh skull, the morphological changes that would be necessary to approach the morphology of the crania from Qafzeh, Skhūl V or even Předmostí III are not great. These would include:

1. Reduction of the brow-ridge: this feature is still very strong in Qafzeh 6, Skhūl IV and V and Předmostí III, and it shows a clear separation of its elements in Předmostí III and Qafzeh 9. This separation is less pronounced in other skulls.
2. Increase of the squama size: in fact this feature is not much larger in Skhūl V than that of Zuttiyeh.
3. Modification of the shape of the orbits, which moves from square to rectangular and low.
4. Reduction in overall robusticity: but with the exception of the torus, the differences in bone thickness between Zuttiyeh and certain Proto-Cro-Magnons are not great.

The changes that would be needed to progress from the morphology of the Zuttiyeh skull to that of the Proto-Cro-Magnons are therefore relatively unimportant, and could have taken place in a simple evolutionary pattern. These changes need not have involved any basic modifications, and can be compared to the patterns of early *Homo sapiens* evolution documented in South and East Africa (Bräuer 1984). Therefore the transformation in the Near East of an archaic *Homo sapiens* population – represented by the skull from Zuttiyeh – into a Proto-Cro-Magnon population is probably the sounest hypothesis in the present state of knowledge.

NOTE

1. Whilst the present article was in press, the Mousterian levels in the Qafzeh cave have been dated by thermoluminescence dating to *c.* 92 000±5000 BP – confirming the early age predicted for these levels by the micro-mammalian studies: see Valladas *et al.* 1988.

REFERENCES

Arensburg, B., Bar-Yosef, O., Chech, M., Goldberg, P., Laville, H., Meignen, L., Rak, Y., Tchernov, E., Tillier, A.-M. and Vandermeersch, B. 1985. Une sépulture néandertalienne dans la grotte de Kébara (Israël). *Comptes-Rendus de l'Académie des Sciences de Paris (Série D)* 300: 227-230.

Ashley-Montagu, J. R. 1940. Review of T. D. McCown and A. Keith: *The Stone Age of Mount Carmel*. Vol 2: *The Fossil Remains of the Levalloiso-Mousterian*. *American Anthropologist* 42: 518-522.

Bar-Yosef, O., Vandermeersch, B., Arensburg, B., Goldberg, P., Laville, H., Meignen, L., Rak, Y., Tchernov, E. and Tillier, A.-M. 1986. New data on the origin of Modern Man in the Levant. *Current Anthropology* 27: 63-64.

Bräuer, G. 1984. A craniological approach to the origin of anatomically modern *Homo sapiens* in Africa and implications for the appearance of modern Europeans. In F. H. Smith and F. Spencer (eds) *The Origins of Modern Humans: a World Survey of the Fossil Evidence*. New York: Alan R. Liss: 327-410.

Condemi, S. 1985. *Les Hommes Fossiles de Saccopastore (Italie) et Leurs Relations Phylogénétiques*. Unpublished Ph.D. Thesis, University of Bordeaux I.

Coon, C. S. 1939. *The Races of Europe*. New York: Macmillan.

Dobzhansky, T. 1944. On the species and races of living and fossil man. *American Journal of Physical Anthropology* 2: 251-265.

Gisis, I. and Bar-Yosef, O. 1974. New excavations in Zuttiyeh cave, Wadi Amūd, Israel. *Paléorient* 2: 175-180.

Haas, G. 1972. The microfauna of Djebel Qafzeh Cave. *Paleovertebrata* 5: 261-270.

Howell, F. C. 1959. Upper Pleistocene stratigraphy and early man in the Levant. *Proceedings of the American Philosophical Society* 103: 1-65.

Jelinek, A. 1982a. The Tabun Cave and Paleolithic man in the Levant. *Science* 216: 1369-1375.

Jelinek, A. 1982b. The Middle Palaeolithic in the Southern Levant, with comments on the appearance of modern *Homo sapiens*. In A. Ronen (ed.) *The Transition from Lower to Middle Palaeolithic and the Origin of Modern Man*. Oxford: British Archaeological Reports International Series S151: 57-104.

Keith, A. 1927. A report on the Galilee skull. In F. Turville-Petre (ed.) *Researches in Prehistoric Galilee: 1925-1926*. London: British School of Archaeology in Jerusalem: 53-106.

Keith, A. and McCown, T. D. 1937. Mount Carmel man: his bearing on the ancestry of modern races. In G. G. MacCurdy (ed.) *Early Man*. Philadelphia: E.B. Lippincott: 41-52.

Lévêque, F. and Vandermeersch, B. 1980. Découverte de restes humains dans un niveau castelperronien à Saint-Césaire (Charente- Maritime). *Comptes-Rendus de l'Académie des Sciences de Paris (Série D)* 291: 187-189.

McCown, T. D. and Keith, A. 1939. *The Stone Age of Mount Carmel.* Vol. 2: *The Fossil Human Remains from the Levalloiso-Mousterian.* Oxford: Clarendon Press.

Piveteau, J. 1957. *Traité de Paléontologie.* Vol 7: *Primates, Paléontologie Humaine.* Paris: Masson.

Stringer, C. B., Hublin, J.-J. and Vandermeersch, B. 1984. The origin of anatomically modern humans in Western Europe. In F. H. Smith and F. Spencer (eds) *The Origins of Modern Humans: a World Survey of the Fossil Evidence.* New York: Alan R. Liss: 51-135.

Suzuki, H. and Takai, F. 1970. *The Amud Man and his Cave Site.* Chicago: Chicago University Press.

Tchernov, E. 1984. Faunal turnover and extinction rate in the Levant. In P. S. Martin and R. G. Klein (eds) *Quaternary Extinctions: a Prehistoric Revolution.* Tucson: University of Arizona Press: 528-552.

Thoma, A. 1957-58. Essai sur les hommes fossiles de Palestine. *L'Anthropologie* 61: 470-502; 62: 30-52.

Thoma, A. 1962. Le déploiement évolutif de l'*Homo sapiens. Anthropologia Hungarica* 5: 1-179.

Thoma, A. 1963. La définition des Néandertaliens et la position des hommes fossiles de Palestine. *L'Anthropologie* 69: 519-534.

Trinkaus, E. 1977. A functional interpretation of the axillary border of the Neandertal scapula. *Journal of Human Evolution* 6: 231-234.

Trinkaus, E. 1983. *The Shanidar Neandertals.* New York: Academic Press.

Trinkaus, E. 1984. Western Asia. In F. H. Smith and F. Spencer (eds) *The Origins of Modern Humans: a World Survey of the Fossil Evidence.* New York: Alan R. Liss: 251-293.

Turville-Petre, F. (ed.) 1927. *Researches in Prehistoric Galilee: 1925-1926.* London: British School of Archaeology in Jerusalem.

Valladas, H., Joron, J. L., Valladas, G., Arensburg, B., Bar-Yosef, O., Belfer-Cohen, A., Goldberg, P., Laville, H., Meignen, L., Rak, Y., Tchernov, E., Tillier, A.-M. and Vandermeersch, B. 1987. Thermoluminescence dates for the Neanderthal burial site at Kebara in Israel. *Nature* 330: 159-160.

Valladas, H., Reyss, J. L., Joron, J. L., Valladas, G., Bar-Yosef, O. and Vandermeersch, B. 1988. Thermoluminescence dating of Mousterian 'Proto-Cro-Magnon' remains from Israel and the origin of modern man. *Nature* 331: 614-616.

Vandermeersch, B. 1981. *Les Hommes Fossiles de Qafzeh (Israël).* Paris: Centre National de la Recherche Scientifique.

10. New Skeletal Evidence Concerning the Anatomy of Middle Palaeolithic Populations in the Middle East: the Kebara Skeleton

B. ARENSBURG

INTRODUCTION

During the summer of 1983, a substantial part of a well preserved human skeleton was discovered in the course of the combined Franco-Israeli excavations in the Kebara Cave on Mount Carmel (see Bar-Yosef *et al.* 1986). The skeleton was found clearly stratified within the Middle Palaeolithic occupation levels, and was overlain by a total of 8 metres of later Middle and Upper Palaeolithic deposits. Recently, the age of this skeleton has been estimated on the basis of thermoluminescence dating of burnt flint samples in the region of 60 000 BP (Valladas *et al.* 1987; see also Valladas *et al.* 1988).

As a result of four years of intensive study, it is now clear that these remains present a number of highly significant anatomical features which cast important new light on many aspects of the character of the human populations associated with Middle Palaeolithic industries in the Middle Eastern region. (For reasons which will be explained later, we prefer for the present to refer to all of these hominids recovered from Middle Palaeolithic contexts simply as 'Mousterian hominids', and to avoid more interpretive labels such as 'Neanderthals'.) The remains available for the Kebara skeleton include the mandible, the whole thoracic cage and vertebral column, complete upper limbs, the pelvis, and the hyoid bone – all in a remarkably good state of preservation. Unfortunately, the skull and most parts of the lower limbs of the skeleton are missing. The present communication will focus briefly on three main aspects of the anatomy of this individual, namely, the vertebral column, the pelvis, and the hyoid bone. This of course is not meant to imply that the remains from other parts of the skeleton are any less significant in anatomical terms, and these features will be described more fully in later publications.

The vertebral column. All 24 presacral vertebrae are represented, although the first five cervical vertebrae are lacking parts of their dorsal aspect. All the vertebrae are morphologically similar to, and metrically within the range of variation of, vertebrae in modern populations. As among some Neanderthals from Europe and the Middle East, but also found in some modern individuals, the inclination of the spinous process of the sixth and seventh cervical vertebrae of the Kebara skeleton tends to be horizontal. This feature (as already discussed by Arambourg (1955) and Straus and

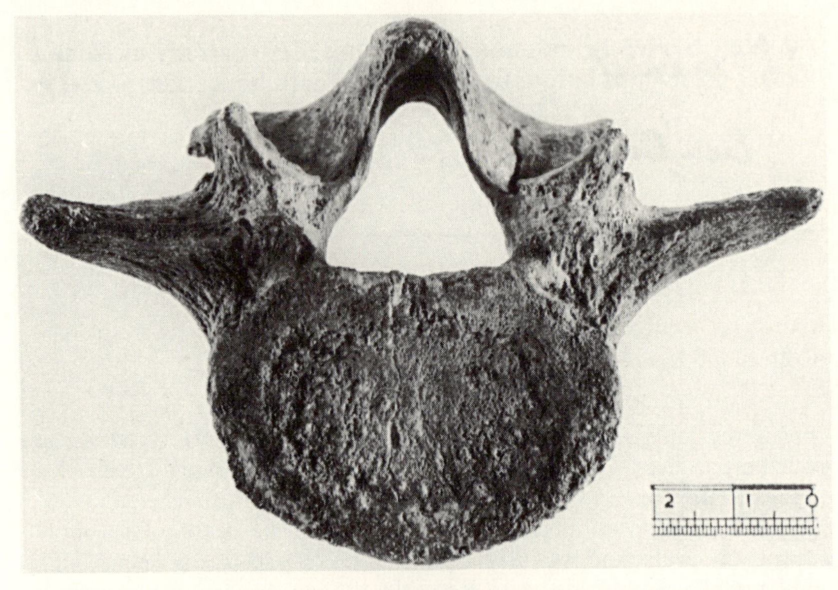

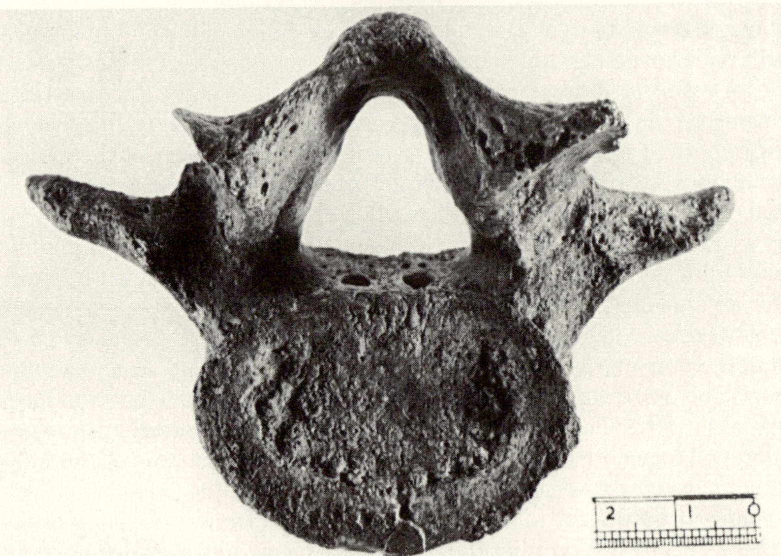

Figure 10.1. Fifth lumbar vertebra of the Kebara skeleton. Upper = superior view; lower = inferior view. Scale = 2 cm.

Cave (1957)) most probably has no taxonomic value, and may be related to functional factors such as squatting or the equilibrium of the head during load bearing.

The cervical column of the Neanderthals has generally been considered (following Boule's monograph on the La Chapelle-aux- Saints skeleton) as being 'short and stiff' (Boule and Vallois 1952). Both earlier and modern studies (Trinkaus 1983) on Mousterian remains have generally avoided any reference to the intervertebral discs when the length of the column was estimated. The cervical discs in modern man account, however, for 22% of the total length of the neck, 20% of the thoracic part, and 33% of the lumbar, according to De Palma and Rothman (1970). Thus the calculation of the actual length (including disc heights) of the different segments of the vertebral column of some Middle Eastern Mousterian hominids would indicate that the neck length of the Kebara and Skhūl V fossils were, most likely, between the 55th and 75th percentile of modern man's neck length, while the neck length of Shanidar II was in the higher percentile. The incomplete La Chapelle-aux-Saints cervical segment seems to have belonged to a column similar to that of Shanidar II.

The sum of all the heights of the vertebral elements from the second cervical vertebra to the last lumbar vertebra (regardless of whether or not the disc heights are counted) indicates in the Kebara man a shorter cervical segment relative to the thoracic or lumbar parts when compared with the proportions in modern man. Yet, the total length of the column of this fossil undoubtedly falls within the modern human range. In the absence of a complete fossil record on human vertebral columns, these results are difficult to interpret.

A further interesting point concerning the vertebral column from Kebara relates to the shape and size of the last lumbar vertebra (see Figure 10.1). This bone presents three very modern features: first, a larger superior than inferior surface of the body; second, a higher ventral than dorsal aspect of the body; and third, a horizontal, cranio-caudal compressed pedicle of L5. All these features occur only among bipedal humans, whereas non-human primates show a completely different morphology. The human adaptation of the last lumbar vertebra seems to be directly related to a redistribution of the load above the lumbo-sacral joint due to bipedalism, a change that is not present in semi- erect, knuckle-walking or brachiating primates.

The pelvis. The pelvis of the Kebara skeleton is the most complete pelvis of a Mousterian hominid so far recovered (see Figure 10.2). Hence it may well contribute to a better understanding of questions concerning gestation among ancient humans (see Trinkaus 1984; Rak and Arensburg l987).

The message of the Kebara pelvis is simple: it confirms that among some Mousterian remains the superior, ilio-pubic ramus is elongated, a typical morphology which is not found in modern humans. Nevertheless the pelvic inlet is unequivocally modern in size, very similar to that of present-day humans. Despite the fact that we have here a male, it is suggested that the pelvic inlet was fundamentally similar in the female. In contrast to the opinion of Trinkaus (1984), a longer gestation period than in modern

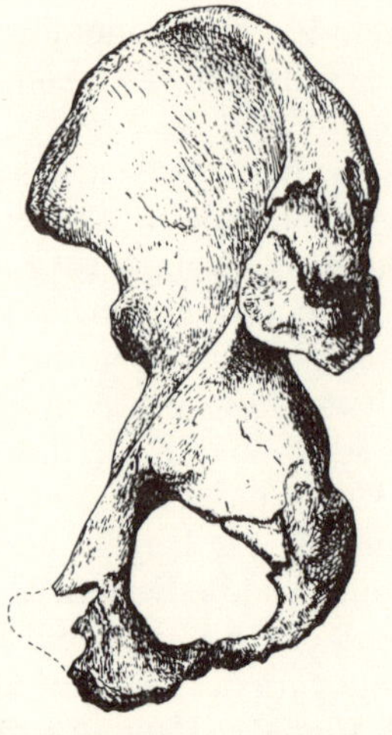

Figure 10.2. Internal view of the right innominate bone of the Kebara skeleton.

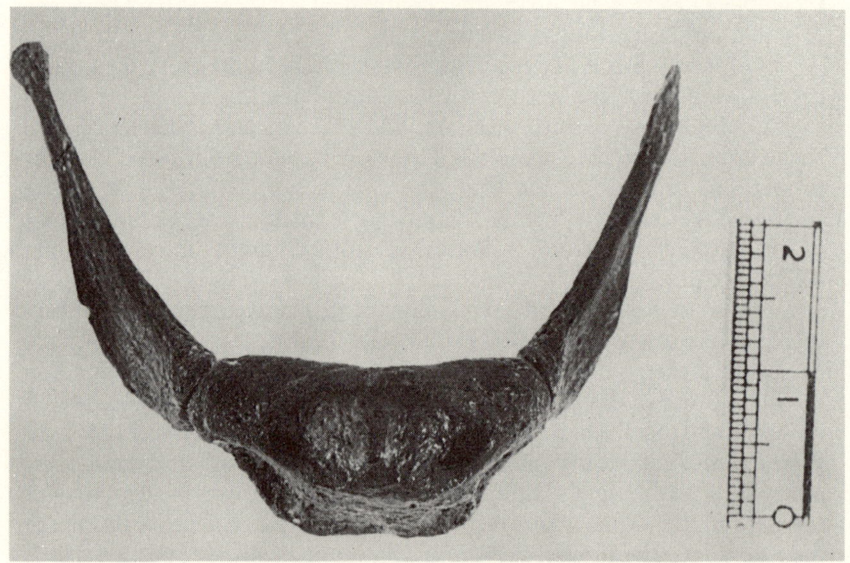

Figure 10.3. Hyoid bone of the Kebara skeleton: antero-superior view. Scale = 2 cm.

women, which could permit further development of the head and brain in the uterus, was *not* achieved in this fossil population. The elongated superior, ilio-pubic rami of the Mousterians seem to correspond to a special disposition of the pelvic inlet osseous components without affecting the actual size of the inlet, since a lateral rotation of the hip bone places the sacrum in a more anterior position than in modern man. The distance between the promontorium in an anteriorly placed sacrum and the symphysis pubis, necessary to maintain the minimum normal size of the birth canal, is achieved by an elongated superior ilio-pubic ramus. These features of the ramus that are typical of the European Neanderthal pelvis and in the Shanidar and Kebara hominids, are missing in other Middle Eastern specimens such as those from Skhūl and Qafzeh, which present a more 'modern' morphology.

Indeed it is suggested here that postural factors and not gestation period are responsible for the lateral rotation of the hips and the subsequent elongation of the superior ilio-pubic ramus, as observed in the Kebara specimen and in other Mousterian human remains.

The hyoid bone. As far as we are aware, this is the first hyoid bone so far reported from a fossil hominid. The bone was found in the prevertebral area of the neck, between the two mandibular gonia. It is well preserved, and the absence of the lesser horns suggests that these structures were cartilaginous (see Figure 10.3).

The Kebara hyoid resembles that of modern man both in its configuration and size. The body presents in its anterior surface two distinct areas: a superior one excavated by a depression at each side of the midline, and an inferior one, flat in the central part and excavated at the sides. A transverse line divides the upper and lower areas. The attachment for the geniohyoid, omohyoid, mylohyoid and other muscles are clear. The greater horns are slender, with a sharp, everted inferior border, a large facet for the articulation with the body and a blunt, button-like distal end for the thyrohyoid ligament.

This essentially modern human morphology of the Kebara hyoid bone stands in sharp contrast to the archaic shape and large size of the mandible. These two bones – the hyoid and the mandible – are closely related by many muscles and by common functions such as speech, deglutition, etc. It is therefore surprising to find a combination in the same individual of a modern-shaped hyoid bone related to a morphometrically massive mandible. The mandible differs in many features not only from modern mandibles, but also from mandibles of other Mousterian specimens. Yet the mechanism responsible for the general reduction in size of the masticatory apparatus during the last 60 000 years apparently did not affect the hyoid bone nor any component of the visceral skeleton. Thus the hyoid bone, middle ear ossicles, and most probably the cartilaginous elements related to the larynx, appear to be the most conservative in terms of evolutionary changes of all the skeletal components (Arensburg *et al.* 1981).

The newly-discovered hyoid bone from Kebara casts totally new light on the speech capability of this individual. It has been suggested in earlier publications (Lieberman and Crelin 1971; Laitman 1978) that a high position of the hyoid and related larynx – as present in newborn humans, in apes and supposedly in Neanderthals – prevented a fully modern capability for speech in the Mousterian hominids. The discovery of the hyoid from Kebara shows no different anatomical relationships of this bone with the mandible, and in fact strongly suggests a low position in the neck of the hyoid and of the subjacent larynx. Viewed in anatomical terms, it would seem that the Mousterian man from Kebara was just as capable of speech as modern man.

CONCLUDING REMARKS

In sum, the morphology of the skeleton from Kebara, as briefly described here, shows a mosaic of characteristics that may be found in modern populations, but also to some extent in Middle Palaeolithic or even earlier groups. Thus, 'typical' Neanderthal traits in the metacarpal bones, the pelvis or the cervical vertebrae appear concurrently with modern features in the vertebral column, hyoid, ribs, sternum, together with seemingly more archaic morphological traits in the mandible. The overall morphology of the Kebara skeleton is, generally speaking, so robust that the robust skeletons described by Trinkaus (1983) from Shanidar seem almost gracile by comparison.

In the light of the preceding discussion, one must now pose a central question: to what extent is it appropriate to refer to the Kebara fossil as 'Neanderthal'?

Following the studies of McCown and Keith (1939), Endo and Kimura (1970), Vandermeersch (1981) and others, it is now clear that the human remains recovered from Middle Palaeolithic contexts in the Middle East comprise a variety of relatively diverse human morphological 'types'. Unfortunately, there is still considerable uncertainty concerning the relative and absolute chronology of many of these remains, which appear to span a period ranging from at least 80-90 000 BP to *c.* 35 000 BP. At present, the precise relationships between the morphology and chronology of the different hominids remains speculative – although there are strong indications that a combination of modern and archaic anatomical features were present within these remains throughout this long period.

In my opinion it is safer, in the present state of research, to avoid the use of compromising terminology such as 'Neanderthal Man' in referring to the Middle Palaeolithic hominids in the Middle East – partly because of the high degree of anatomical variability apparent within the Middle Eastern samples. It may well be that this variability reflects a wide range of factors including evolution from the local populations of *Homo erectus,* combined with varying degrees of gene flow, population migrations, and associated evolutionary changes, related at least in part of changing environmental conditions.

For these reasons it seems safer to refer to all of these hominids simply under the descriptive terms of 'Mousterian Men' – although of course in the clear recognition that this is a strictly cultural label, and is not meant to imply any formal biologically taxonomic status. The use of more specific labels such as *Homo sapiens* or *Homo neanderthalensis* is best avoided until the precise relationships between these different hominids have been clarified by the discovery and analysis of more fossil material.

REFERENCES

Arambourg, C. 1955. Sur l'attitude, en station verticale, des Néanderthaliens. *Comptes-Rendus de l'Académie des Sciences de Paris (Série D)* 240: 804-806.

Arensburg, B., Harell, M. and Nathan, H. 1981. The human middle ear ossicles: morphometry, and taxonomic implications. *Journal of Human Evolution* 10: 199-205.

Bar-Yosef, O., Vandermeersch, B., Arensburg, B., Goldberg, P., Laville, H., Meignen, L., Rak, Y., Tchernov, E. and Tillier, A.-M. 1986. New data on the origins of Modern Man in the Levant. *Current Anthropology* 27: 63-64.

Boule, M. and Vallois, H. V. 1952. *Les Hommes Fossiles*. Paris: Masson.

De Palma, A. F. and Rothman, R. H. 1970. *The Intervertebral Disc*. Philadelphia: W. B. Saunders.

Endo, B. and Kimura, T. 1970. Post cranial skeleton of the Amud Man. In H. Suzuki and F. Takai (eds) *The Amūd Man and His Cave Site*. New York: Academic Press: 231-406.

Laitman, J. T., Heimbuch, R. C. and Crelin, E. S. 1978. Developmental change in the basicranial line and its relationship to the upper respiratory system in living primates. *American Journal of Anatomy* 152: 467-482.

Lieberman, P. and Crelin, E. S. 1971. On the speech of Neanderthal man. *Linguistic Inquiry* 2: 203-222.

McCown, T. D. and Keith, A. 1939. *The Stone Age of Mount Carmel*, Vol. 2: *The Fossil Human Remains from the Levalloiso-Mousterian*. Oxford: Clarendon Press.

Rak, Y. and Arensburg, B. 1987. The Kebara 2 Neanderthal pelvis: first look at a complete inlet. *American Journal of Physical Anthropology* 73: 227-231.

Straus, W. L. and Cave, A. J. E. 1957. Pathology and posture of Neanderthal man. *Quarterly Review of Biology* 32: 348-363.

Trinkaus, E. 1983. *The Shanidar Neandertals*. New York: Academic Press.

Trinkaus, E. 1984. Neandertal pubic morphology and gestation length. *Current Anthropology* 25: 509-514.

Valladas, H., Joron, J. L., Valladas, G., Arensburg, B., Bar-Yosef, C., Belfer-Cohen, A., Goldberg, P., Laville, H., Meignen, L., Rak, Y., Tchernov, E., Tillier, A.-M. and Vandermeersch, B. 1987. Thermoluminescence dates for the Neanderthal burial site at Kebara in Israel. *Nature* 330: 159-160.

Valladas, H., Reyss, J. L., Joron, J. L., Valladas, G., Bar-Yosef, O. and Vandermeersch, B. 1988. Thermoluminescence dating of Mousterian 'Proto-Cro-Magnon' remains from Israel and the origin of modern man. *Nature* 331: 614-616.

Vandermeersch, B. 1981. *Les Hommes Fossiles de Qafzeh (Israël)*. Paris: Centre National de la Recherche Scientific.

11. Geographic Variation in Supraorbital Torus Reduction during the Later Pleistocene (c. 80 000-15 000 BP)

FRED H. SMITH, JAN F. SIMEK AND MARIA S. HARRILL

INTRODUCTION

The issue of modern human origins is one of the oldest in the study of human evolution, having emerged with the controversy surrounding interpretation of the Feldhofer Neanderthal during the last century. Perspectives on this issue have changed considerably over the intervening years (Spencer and Smith 1981; Smith and Spencer 1984), but despite the accumulated knowledge of more than a century, the nature of modern human origins remains one of the most intensely debated topics in modern Palaeoanthropology. Recent years saw something of a compromise approach to this issue. Rather than stressing the extreme views of classic monocentrism or polycentrism, many researchers agreed that both local genetic continuity and gene flow of some form and magnitude were involved in the emergence of modern humans throughout the Old World (Bräuer 1984; Smith 1985; Stringer *et al.* 1984; Trinkaus and Smith 1985; Wolpoff *et al.* 1984). Though the relative importance of continuity *versus* gene flow to the emergence of modern humans continued to be a subject of considerable disagreement, it seemed obvious that the phenomenon of modern human origins was not the result of simple population dynamics.

Also, however, a recent tendency has emerged to stress the existence of qualitative morphological and behavioural differences between archaic *Homo sapiens* and modern *Homo sapiens*, particularly in Europe (Binford 1985, 1989; Cartmill *et al.* 1986; Eldredge and Tattersall 1982; Klein 1983; Pilbeam 1986; Stringer 1989; Tattersall 1986). Based on genetic (Cann *et al.* 1987), anatomical and archaeological evidence, an argument has emerged that most archaic *Homo sapiens* (especially Eurasian Neanderthals) were reproductively isolated from emergent modern *Homo sapiens*. Following this perspective, the appearance of modern humans in areas other than their region of initial origin must be explained ultimately as the result of an influx or physical migration of modern peoples. Virtually no genetic input from localized archaic *Homo sapiens* populations is proposed to have occurred. Indeed in order to accept certain genetic arguments for this pattern of modern human emergence (Cann *et al.* 1987), any significant genetic exchange between localized archaic *Homo sapiens* and incoming moderns must be ruled out (Smith and Paquette 1989).

While this explanation for modern humans may be attractive to some for its simplicity, we believe it unlikely that the population dynamics which resulted in modern humans in most regions were as simple as complete replacement. Such an argument totally ignores an impressive body of evidence for varying degrees of morphological, and thus presumably genetic, continuity across the archaic-modern *Homo sapiens* transition in several areas of the Old World. This evidence is most convincing for Eastern Asia (Thorne and Wolpoff 1981; Wolpoff *et al.* 1984; Wolpoff 1985) and Central Europe (Frayer 1986, Frayer *et al.*, in press; Smith 1982, 1984; Smith and Ranyard 1980), but reasonable cases can also be made in other regions as well (Smith 1985; Wolpoff 1980, 1989). In this paper, we further document morphological continuity in Central Europe through statistical analysis of one structure -- the supraorbital torus. Furthermore, we show that the pattern and timing of evolution in the supraorbital region differ between Central and Western Europe. This implies that the process leading to modern humans may not have been the same in both regions. This, in turn, suggests that a simple replacement model is not adequate to explain the emergence of modern Europeans.

THE SUPRAORBITAL TORUS

The supraorbital torus in European Neanderthals is ". . . an osseous bar projecting from the frontal squama at the inferior border of the frontal bone and arching from glabella laterally over each orbit to the frontozygomatic suture" (Smith and Ranyard 1980: 591). This structure has been recognized as a characteristic aspect of the Neanderthal total morphological pattern since the first Neanderthal was recognized (e.g. Schwalbe 1901), and our choice of the supraorbital torus as the focus of this study is partially due to the fact that loss or at least marked reduction of the torus is one of the most characteristic differences between Neanderthals and modern Europeans. But there are other reasons. First, the torus in Neanderthals and comparable regions in modern humans are well represented in the late Pleistocene human fossil record, insuring adequate sample sizes for statistical evaluation. Second, several recent studies have demonstrated that the supraorbital region is structurally integrated with several aspects of craniofacial form and function. Specifically, supraorbital morphology reflects facial forwardness (Ravosa, in press; Smith and Ranyard 1980) and loading of the anterior dentition (Rak 1986; Russell 1985; Smith 1983; Smith and Paquette 1989). Thus, although it is true that focusing on any single structure in assessing relationships requires caution, we believe that the supraorbital region is one of the most suitable single structures for such analysis, because it can be considered a general indicator of *overall* craniofacial form in late Pleistocene hominids.

Two general aspects of the supraorbital region must be considered in any metric analysis of the torus and structures derived from it. First, the projection of the torus relative to the brain case is important, primarily as a reflection of facial forwardness. This antero-posterior dimension is measured as a chord from the most anterior point on the internal surface of the frontal's inferior aspect to the most anterior point on the external surface of the torus. Second, the thickness of the torus is measured as a chord from the inferior to the superior surface of the projecting portion of the torus. Both projection and thickness dimensions are necessary for a complete study of changes in the supraorbital region.

Similarly, the torus is not evenly thick, nor does it project to the same degree at glabella, as it does over the orbit or at its lateral-most extent. Thus, thickness and projection measurements must be taken at differing points along the torus. We utilize three points (see Smith and Ranyard 1980: 596-597): (1) the *medial* point is just lateral to glabella and corresponds to the highest point in the arch of the torus over the orbit; (2) the *lateral* point is the thickest portion of the torus lateral to a perpendicular plane at frontotemporale; (3) the *midorbital* point is defined as the point of minimum thickness between the medial and lateral point. This definition may seem vague, but this point invariably corresponds to the flattened portion of the supraorbital trigone (as defined by Schwalbe 1901, and Cunningham 1908) and is thus always found at the same relative position along the torus.

Finally, Tattersall (1986) has recently chided that analysis of morphological continuity in temporally contiguous 'species' is ". . . extraordinarily uninformative if one wishes to know if the character states being compared are primitive or derived. . ." (p. 170). Based on comparisons with earlier species of the genus *Homo,* we would assert that the supraorbital torus is a primitive feature in the evolution of the genus. Thus our study is based on a plesiomorphic feature, which Tattersall notes is common for such studies. This means that one must consider the absence of a true torus to be an apomorphy in modern *Homo sapiens* compared to earlier members of the genus. In this regard, demonstration that the Neanderthal supraorbital torus grades into the early modern European condition, both morphologically and metrically (Frayer 1986; Frayer *et al.*, in press; Smith 1982, 1984; Smith *et al.* 1985; Wolpoff *et al.* 1981), is interesting theoretically as well as phylogenetically, since it demonstrates the emergence of an apomorphy from its primitive forerunner.

THE SAMPLE

The fossil human specimens examined for this study are presented in Table 11.1, along with the dates assigned to each individual specimen or site for purposes of this analysis. The location of specimens in Western or Central Europe follows Smith (1984). The dates assigned are determined from information in Laville *et al.* (1980), Oakley *et al.* (1971), Smith (1984), and Stringer *et al.* (1984). Where specific dates are unavailable and only gross temporal assignment is made (e.g. 'Würm II', 'Riss-Würm Interglacial'), the mid-point of the generally-accepted dates for that temporal period are employed for this study.

ANALYTIC PROCEDURES

For each of the six metric variables discussed above, at least two (and usually three) distinct regression procedures will be applied to fossils from

Table 11.1. Fossil human specimens utilized for this study. Date assignments are derived from sources listed in the text.

Western Europe		Central Europe	
Specimen	Date (Ky BP)	Specimen	Date (Ky BP)
La Chapelle	47.5	Krapina[1]	75
La Ferrassie 1	47.5	Vindija[2]	42.4
La Quina 5	40.0	Hahnöfersand	36.3
Gibraltar	75	Podbaba	32
Spy 1 & 2	57.5	Kelsterbach	31.2
Saint-Césaire	33	Paderborn	27.4
Cro-Magnon (1-4)	30	Stetten 1 & 2	30
La Madeleine	12.64	Velika Pečina	33.85
Abri Pataud	21.54	Mladeč (1,2,5)	35
Englis 2	21	Brbo 2	26.32
Oberkassel 1 & 2	15	Dolní Vestoniče	25.82
		Pavlov	26.62
		Zlaty Kun	35

Notes:
1 Krapina specimen numbers: 3, 4, 6, 28, 37-1, 37-3, 37-4, 37-5, 37-6, 37-7, 37-8, 37-10, 37-11.
2 Vindija specimen numbers: 202, 260, 261, 262, 284, 305

each geographic area. In all regressions, fossil date will constitute the predictor (or independent) variable, with the metric torus measurements as response (dependent) variables. In multiple regression analyses (see below), geographical area will be included in calculations as a second predictor. First, within each region a variety of regression models are fitted to the data to assess whether abrupt or gradual change over time characterizes the variable of interest. For the most part, quadratic and simple linear models will be assessed, but cubic models will be applied if quadratic functions are shown to be significant to determine if even higher order functions might be appropriate. Second, based on these analyses, change in each variable will be characterized as (1) not related to time (i.e. no change across the transition in a region); (2) a simple linear function of time (i.e. gradual, with no breaks in the pattern across the transition); or (3) a polynomial function of time (i.e. showing a change or shift in the pattern across the transition).

Once patterns in morphometric change have been defined for each variable by region, the patterns will be compared between the two geographic areas. However, comparisons between regressions are required only when *both* regions show the same kind of statistically significant trend. Thus, if both geographic regions have simple linear or polynomial relationships for the same variable, comparative regressions will be performed. If the two regions are characterized by different relationships for the same variable, then no comparison is necessary because the processes of change are distinct. If either or both regions have no trend indicated, then comparison would not be useful. In these cases, only the first two regression steps are performed.

As will be seen, regression comparisons will be required only for variables where simple linear models provide the best fit in both geographic regions. Thus, only the method employed for this kind of test will be described here. The first task involves determining if the two regression slopes are equivalent. This is accomplished by examining interaction effects between geographic region and date using common multiple regression of all fossils in the sample. If regression slopes are distinct (i.e. the interaction is significant), then the rates of change are different between the two regions. If the interaction effects are not significant, then the rates of change are similar in both areas, and the regressions must be tested for equivalence of intercept values. Here, multiple regression is again performed, testing for the effects of geographic region on the combined data sets, but interaction effects between predictor variables (area and date) are removed from the calculations. If the effects of geographic region are significant, then the intercepts are distinct and the changes observed, while proceeding at similar rates, occurred at different times in the two regions. If the geographic region effects are not significant, the processes of change in the two regions must be considered identical in form, rate, and time. Here, a single common regression formula may be derived to characterize the transition. The reader is referred to any standard statistical text for more detailed discussion of these straightforward procedures (e.g. Neter *et al.* 1985).

RESULTS OF ANALYSIS

The first stage of analysis involves describing the nature of change in the metric variables within each of the two European areas. In these studies, fossils will not be differentiated as to their morphological type, since it is the chronological transition from Neanderthal to modern human forms that is of concern. First, results obtained for Central European fossils will be described. Second, results for West European hominids will be presented. Finally, necessary comparisons between patterns in these two geographical regions will be discussed.

Central Europe

Table 11.2 shows the result of fitting a quadratic function to lateral thickness measurements for Central European fossils. As the addition of a power term does not produce a significant fit, a simple linear regression was performed; results of that analysis are also presented in Table 11.2. This regression shows a slope parameter significantly greater than 0, indicating that a trend in lateral thickness is present over time. The trend can be characterized by the following formula:

$$\text{Lateral thickness} = 6.01345 + .00008694(\text{Date}). \quad (1)$$

This regression relationship accounts for over 60% of the variability in lateral thickness ($r = 0.779887$) within the Central European sample. Figure 11.1a shows the regression line plotted over the observed data point distribution.

Table 11.3 gives results obtained by fitting quadratic and simple linear models to lateral projection measurements for Central European specimens.

Table 11.2. Results of Regression Analyses for Central Europe: Lateral Thickness. Significant parameters are marked by "**".

Variable	Coefficient	*T*-value	Probability (>*T*)
Quadratic Solution			
Intercept	3.03699	0.8	0.4325
Date	0.00022	1.32	0.1992
Date²	−0.00000	−0.79	0.4332
Simple Linear Solution			
Intercept	6.01345	8.34	0.0000**
Slope	0.00009	6.47	0.0000**

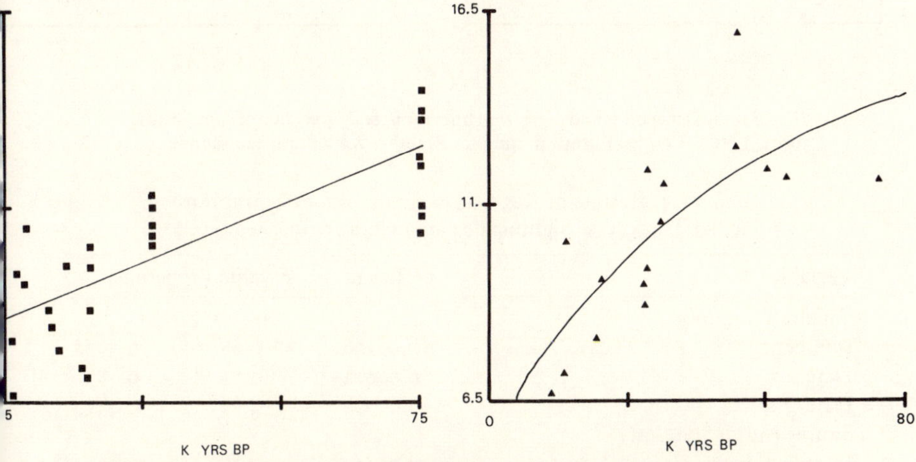

Figure 11.1. Data points and regression lines for lateral thickness of supraorbital torus. Left = Central European sample; Right = West European sample.

Table 11.3. Results of Regression Analyses for Central Europe: Lateral Projection. Significant parameters are marked by "**".

Variable	Coefficient	*T*-value	Probability (>*T*)
Quadratic Solution			
Intercept	14.32458	2.07	0.0504
Date	0.00022	0.76	0.4574
Date²	−0.00000	−0.44	0.6635
Simple Linear Solution			
Intercept	17.3218	13.59	0.0000**
Slope	0.00009	3.89	0.0009**

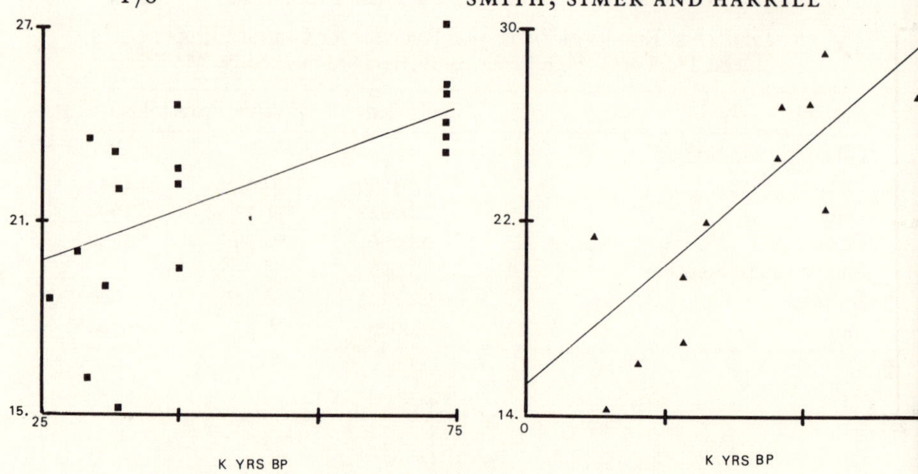

Figure 11.2. Data points and regression lines for lateral projection of supraorbital torus. Left = Central European sample; Right = West European sample.

Table 11.4. Results of Regression Analyses for Central Europe: Medial Thickness. Significant parameters are marked by "**".

Variable	Coefficient	*T*-value	Probability (>*T*)
Quadratic Solution			
Intercept	10.90106	0.88	0.3928
Date	0.00022	0.71	0.487
Date²	−0.00000	−0.81	0.4317
Simple Linear Solution			
Intercept	20.7344	9.59	0.0000**
Slope	−0.00005	−1.1	0.2899

Again, the quadratic term is not significant. However, simple linear regression produces a model with a slope significantly greater than 0. The trend in lateral projection over time defined for the Central European fossils can be described by the following model:

$$\text{Lateral projection} = 17.3218 + .0000939(\text{Date}). \quad (2)$$

The associated r statistic for this regression is 0.646787, indicating that over 41% of the variability in lateral projection can be accounted for by time. Figure 11.2a shows the defined regression plotted with the actual data values.

Table 11.4 presents results of regression performed on medial thickness measurements for Central Europe. The squared term is not significant, and, in contrast to the preceding variables, analysis of variance for simple linear regression parameters indicates that the slope for medial thickness is not significantly different from 0. This indicates that there is no significant trend through time in this measurement within the Central European sample.

Table 11.5 shows regression results for medial projection. The quadratic

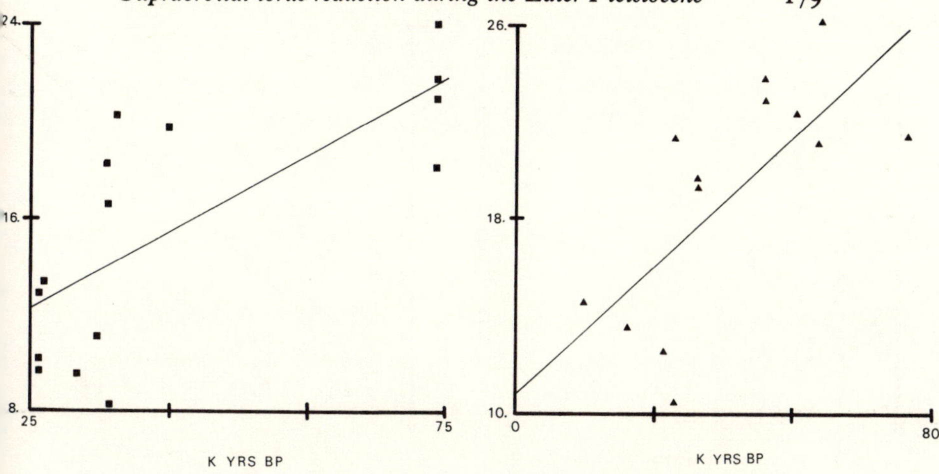

Figure 11.3. Data points and regression lines for medial projection of supraorbital torus. Left = Central European sample; Right = West European sample.

Table 11.5. Results of Regression Analyses for Central Europe: Medial Projection. Significant parameters are marked by "**".

Variable	Coefficient	*T*-value	Probability (>*T*)
Quadratic Solution			
Intercept	−12.99328	−0.9	0.3874
Date	0.00109	1.71	0.1124
Date^2	−0.00000	−1.44	0.1753
Simple Linear Solution			
Intercept	7.5792	2.99	0.0123**
Slope	0.00018	3.49	0.0051**

is not significant, while results for a simple linear regression show a regression slope significantly greater than 0. The defined trend in medial projection can be characterized by the formula:

$$\text{Medial projection} = 7.57921 + .000175789(\text{Date}). \quad (3)$$

Temporal change accounts for 52.3% of the observed variability in medial projection among the Central Europe specimens ($r = 0.724396$). Figure 11.3a illustrates the data points with the derived regression line.

Fitting a quadratic model to measurement of midorbital thickness for the Central Europe fossils fails to produce significant results. A simple linear model, however, again defines a significant trend for this measurement. Table 11.6 gives parameters for midorbital thickness predicted by time. The following linear formula is derived by regression:

$$\text{Midorbital thickness} = 3.39283 + .000098817(\text{Date}). \quad (4)$$

This model accounts for 65.2% of the variability in midorbital thickness within the region's sample ($r = 0.807159$). Figure 11.5a illustrates the data points with the line defined by this regression formula.

As for the other five torus variables, midorbital projection measurements

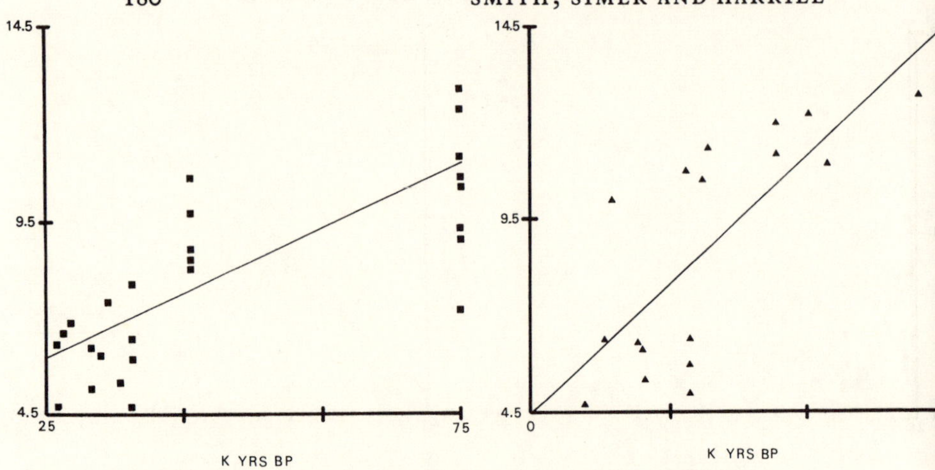

Figure 11.4. Data points and regression lines for midorbital thickness of supraorbital torus. Left = Central European sample; Right = West European sample.

Table 11.6. Results of Regression Analyses for Central Europe: Midorbital Thickness. Significant parameters are marked by "**".

Variable	Coefficient	*T*-value	Probability (>*T*)
Quadratic Solution			
Intercept	−2.57493	−0.7	0.4862
Date	0.00036	2.28	0.0294**
Date²	−0.00000	−1.66	0.106
Simple Linear Solution			
Intercept	3.39283	4.68	0.0001**
Slope	0.0001	7.49	0.0000**

show no polynomial trend over time in Central Europe. Again, a simple linear model does produce a slope value significantly greater than 0, indicating patterning change over time (Table 11.7). The line defined by regression has the following form:

$$\text{Midorbital projection} = 10.9672 + .000173184(\text{Date}). \quad (5)$$

An r value of 0.789205 is associated with this regression, which thus accounts for over 62% of the sample variation. Figure 11.5a illustrates the midorbital projection data and regression line for Central European fossils.

In summary, results of regression analyses for Central European specimens show that all torus measurements but one (medial thickness) have significant patterns of change over time and across the Neanderthal-modern human boundary. All variables showing temporal trends have simple linear relationships with time, indicating direct and gradual change in torus morphology. We will discuss the single variable having no trend indicated in concluding this paper. In general, this analysis confirms the results of earlier studies (Smith and Ranyard 1980; Wolpoff *et al.* 1981; Smith 1987), which showed that early modern Central European supraorbital form derived gradually from the Neanderthal condition. The lack of any abrupt change

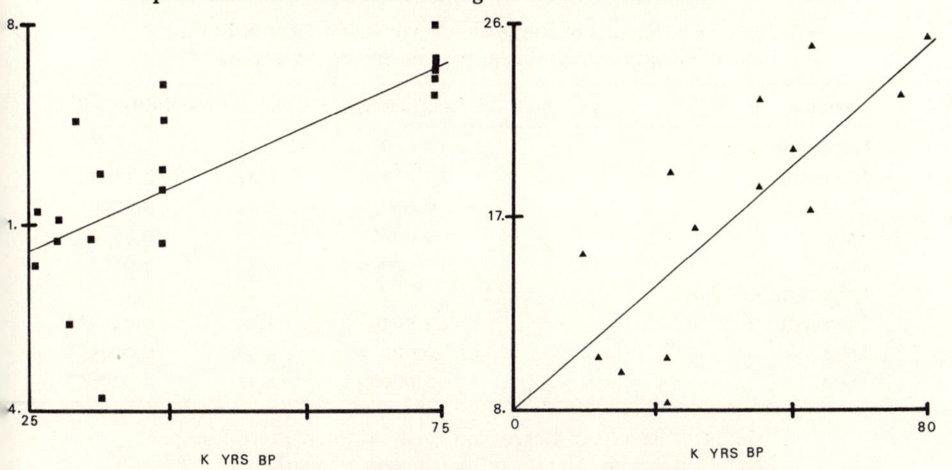

Figure 11.5. Data points and regression lines for midorbital projection of supraorbital torus. Left = Central European sample; Right = West European sample.

Table 11.7. Results of Regression Analyses for Central Europe: Midorbital Projection. Significant parameters are marked by "**".

Variable	Coefficient	*T*-value	Probability (>*T*)
Quadratic Solution			
Intercept	9.30177	1.14	0.2648
Date	0.00025	0.71	0.4858
Date^2	−0.00000	−0.21	0.8371
Simple Linear Solution			
Intercept	10.9672	7.35	0.0000**
Slope	0.00017	6.43	0.0000**

in this sample's supraorbital region, either metrically or morphologically, suggests that no replacement of archaic *Homo sapiens* population by an influx fully developed modern ones occurred in Central Europe.

Western Europe

Table 11.8 shows the result of fitting a quadratic model to lateral thickness measurements for the West European fossil sample. As can be seen, the quadratic term provides a significant improvement in fit over the simple linear model. Table 11.8 also shows the fit of a cubic model to the data; the cubed term is not significant, and the quadratic model is retained. The curvilinear formula that characterizes the derived relation is:

$$\text{Lateral thickness} = 3.594 + .003(\text{Date}) - .000000002(\text{Date}^2). \quad (6)$$

Over 65% of the variability in the sample is accounted for by this formula ($r = 0.66153$). Figure 11.1b shows the Western European lateral thickness data plotted with the regression curve produced by this analysis.

Table 11.9 presents the fit of a quadratic model to lateral projection measurements. The squared term here is not significant. Therefore, a simple linear regression was performed, and that model provided a significant

Table 11.8. Results of Regression Analyses for Central Europe: Lateral Thickness. Significant parameters are marked by "**".

Variable	Coefficient	*T*-value	Probability (>*T*)
Cubic Solution			
Intercept	5.27764	1.44	0.1747
Date	0.00016	0.49	0.6299
$Date^2$	−0.00000	0.18	0.8572
$Date^3$	−0.00000	−0.52	0.6102
Quadratic Solution			
Intercept	3.59397	2.09	0.0555**
Date	0.00032	3.38	0.0045**
$Date^2$	−0.00000	−2.42	0.0298**

Table 11.9. Results of Regression Analyses for Western Europe: Lateral Projection. Significant parameters are marked by "**".

Variable	Coefficient	*T*-value	Probability (>*T*)
Quadratic Solution			
Intercept	12.73163	2.86	0.0169**
Date	0.00032	1.37	0.2008
$Date^2$	−0.00000	−0.64	0.5383
Simple Linear Solution			
Intercept	15.1762	6.93	0.0001**
Slope	0.00017	3.49	0.0051**

Table 11.10. Results of Regression Analyses for Western Europe: Medial Thickness. Significant parameters are marked by "**".

Variable	Coefficient	*T*-value	Probability (>*T*)
Quadratic Solution			
Intercept	14.73624	4.73	0.0003**
Date	0.00016	0.96	0.3537
$Date^2$	−0.00000	−0.57	0.5745
Simple Linear Solution			
Intercept	16.30525	11.13	0.0001**
Slope	0.00007	1.86	0.0831

Table 11.11. Results of Regression Analyses for Western Europe: Medial Projection. Significant parameters are marked by "**".

Variable	Coefficient	*T*-value	Probability (>*T*)
Quadratic Solution			
Intercept	5.58137	0.96	0.3642
Date	0.00049	1.68	0.1277**
$Date^2$	−0.00000	−1.00	0.3416
Simple Linear Solution			
Intercept	10.6966	3.74	0.0038**
Slope	0.0002	3.2	0.0094**

fit to the data. The indicated trend can be described formally as:

$$\text{Lateral projection} = 15.18 + .00017(\text{Date}). \quad (7)$$

Figure 11.2b shows the data plotted with the derived regression line. This relationship accounts for over 52% of the sample variation in lateral projection ($r = 0.72476$).

Table 11.10 gives results of quadratic and simple linear regressions for medial thickness measurements. The quadratic term is not significant, so simple linear regression was performed for this variable. As was the case in the Central European fossils, medial thickness measurements have a slope in this regression that cannot be distinguished from zero. Thus, in both geographic areas, medial thickness seems to show no time-related trend across the Neanderthal-modern human transition.

Table 11.11 shows regression results for medial projection among the West European fossils. The quadratic is not significant, but simple linear regression has a slope parameter significantly greater than 0. The temporal trend in medial projection can be defined as follows:

$$\text{Medial projection} = 10.697 + .0002(\text{Date}). \quad (8)$$

The correlation coefficient associated with this regression formula is 0.7116, indicating that over 50% of the sample variability in medial projection is accounted for by the derived model. Figure 11.3b shows the data and regression line.

Table 11.12 shows that quadratic regression for midorbital thickness is unsuccessful. A simple linear model, however, provides a slope parameter significantly greater than 0. The linear formula produced by this regression is:

$$\text{Midorbital thickness} = 4.06 + .000126(\text{Date}). \quad (9)$$

Over 58% of the Western European sample variation in midorbital thickness can be accounted for by chronological change ($r = 0.76575$). Figure 11.4b shows the derived and observed relations between midorbital thickness and date.

Table 11.13 shows the results of quadratic regression for the final variable, midorbital projection. The squared term is not significant in this model. A slope significantly greater than 0 is produced by simple linear regression of these same data. The relationship can be expressed formally as follows:

$$\text{Midorbital projection} = 12.857 + .00018(\text{Date}). \quad (10)$$

An associated r value of 0.74542 indicates that about 55% of the variability in midorbital projection is accounted for by the regression relationship. Figure 11.5b illustrates the derived regression line along with the observed data values for West European fossils.

Results of regression analyses for West European fossil supraorbital measurements reveal both similarities and differences compared with those derived for Central Europe. Lateral thickness in the western sample shows a quadratic relationship to time, indicating an accelerating rate of change across the transition from Neanderthal to modern forms of *Homo sapiens*. Such a pattern did not appear in an identical analysis of Central European fossils. This difference suggests that the process of change in lateral torus

Table 11.12. Results of Regression Analyses for Western Europe: Midorbital Thickness. Significant parameters are marked by "**".

Variable	Coefficient	*T*-value	Probability (>*T*)
Quadratic Solution			
Intercept	2.95552	1.27	0.2255
Date	0.00019	1.51	0.1522**
$Date^2$	−0.00000	−0.54	0.5981
Simple Linear Solution			
Intercept	4.05731	3.71	0.0021**
Slope	0.00013	4.61	0.0003**

Table 11.13. Results of Regression Analyses for Western Europe: Midorbital Projection. Significant parameters are marked by "**".

Variable	Coefficient	*T*-value	Probability (>*T*)
Quadratic Solution			
Intercept	13.04252	2.97	0.0142**
Date	0.00017	0.73	0.4812**
$Date^2$	0.00000	0.05	0.9619
Simple Linear Solution			
Intercept	12.85661	6.05	0.0001**
Slope	0.00018	3.71	0.0000**

thickness was more abrupt in Western than in Central Europe. As was also the case in Central Europe, no relationship between time and medial thickness could be derived for the West European sample. The other four variables all show simple linear patterns with respect to time in Western Europe, which are similar to the patterns for the corresponding variable in Central Europe. Before arguing that the patterns of transition in these dimensions are the *same* in both areas, however, the regression results must be compared to determine if the derived patterns totally coincide.

Comparing Central and West Europe

Comparisons between the two regions need only be carried out for those variables that have significant time-related trends defined by specific regression *and* also have the same *kind* of relationship to time in formal terms. Thus, two variables can be eliminated from comparative analysis because they do not meet these criteria: lateral thickness and medial thickness. Pattern comparison would be irrelevant in these cases.

Four variables – lateral projection, medial projection, midorbital thickness, and midorbital projection – show simple linear relationships with time in both geographically-defined samples. These data are amenable to examination between regions using techniques designed to establish equivalence of regression lines. Such assessments involve two steps. First, the paired regression lines are compared for equivalence in slope parameters. Similar slopes indicate similar rates of change, and the regressions must be

Table 11.14. Results of Comparative Regression Analyses between Central and West Europe: Lateral Projection. Note that Type I sum of squares are used to test for slope equivalence, and Type III sum of squares are used to test for intercept equivalence. Significant parameters are marked by "**".

Variable	Sum of Squares	*F*-value	Probability (>*F*)
Test for Equivalence of Slopes (Rate)			
Model	189.14968	10.3	0.0001**
Area	0.0665	0.01	0.9177
Date	165.94387	27.11	0.0001**
Area x Date	23.13932	3.78	0.061
Test for Equivalence of Intercepts (Timing)			
Model	166.01036	12.48	0.0001**
Area	8.15397	1.23	0.2765
Date	165.94387	24.94	0.0001**

Table 11.15. Results of Comparative Regression Analyses between Central and West Europe: Medial Projection. Note that Type I sum of squares are used to test for slope equivalence, and Type III sum of squares are used to test for intercept equivalence. Significant parameters are marked by "**".

Variable	Sum of Squares	*F*-value	Probability (>*F*)
Test for Equivalence of Slopes (Rate)			
Model	316.43438	8.71	0.0007**
Area	47.32042	3.91	0.0621
Date	265.70561	21.93	0.0001**
Area x Date	3.40835	0.28	0.6017
Test for Equivalence of Intercepts (Timing)			
Model	313.02602	13.38	0.0002**
Area	83.17345	7.11	0.0145**
Date	265.70561	22.71	0.0001**

tested further to determine if intercept parameters also coincide.

Table 11.14 shows the results of multiple regression performed on lateral projection measurements for the pooled fossil sample. Here, interaction effects between geographic region and date are not significant; thus, the slopes of the two area regressions coincide ($B_1=B_2$). Change in lateral projection of the torus occurs at about the same rate (*c.* 0.00017 mm per year reduction) in both regions. With variable interaction removed, the coefficient for geographic region is not significant, indicating no difference between the two regions in intercept. Comparison between Central and Western European samples suggests that trends in lateral projection of the torus do not differ in the two regions.

Table 11.15 compares regressions between the two areas for medial projection. Interaction between geographic region and date does not produce a significant parameter; this indicates equivalent regression slopes between the area samples. Medial projection of the supraorbital torus reduces *c.* 0.000199 mm per year over Europe generally in this period. Interaction

Table 11.16 Results of Comparative Regression Analyses between Central and West Europe: Midorbital Thickness. Note that Type I sum of squares are used to test for slope equivalence, and Type III sum of squares are used to test for intercept equivalence. Significant parameters are marked by "**".

Variable	Sum of Squares	*F*-value	Probability (>*F*)
Test for Equivalence of Slopes (Rate)			
Model	199.44042	24.4	0.0001**
Area	0.03487	0.01	0.9104
Date	196.17217	71.99	0.0001**
Area x Date	3.23338	1.19	0.2819
Test for Equivalence of Intercepts (Timing)			
Model	196.20704	35.85	0.0001**
Area	25.97753	9.49	0.0035**
Date	196.17217	71.7	0.0001**

Table 11.17 Results of Comparative Regression Analyses between Central and West Europe: Midorbital Projection. Note that Type I sum of squares are used to test for slope equivalence, and Type III sum of squares are used to test for intercept equivalence. Significant parameters are marked by "**".

Variable	Sum of Squares	*F*-value	Probability (>*F*)
Test for Equivalence of Slopes (Rate)			
Model	396.59715	21.47	0.0001**
Area	0.94821	0.15	0.6971
Date	394.17846	64.01	0.0001**
Area x Date	1.47048	0.24	0.6281
Test for Equivalence of Intercepts (Timing)			
Model	395.12667	32.78	0.0001**
Area	23.47447	3.9	0.0492**
Date	394.17846	65.4	0.0001**

effects from multiple regression are then removed to determine if regression intercepts are the same for the two regions. Here, the resulting coefficient for geographic area is significant, indicating that change in medial torus projection occurs at different times in the two areas, although the rate of change is similar. The intercept parameter for West European fossils is larger than that for the Central European sample, and this means that changes occurred later in the west than in the east.

Table 11.16 assesses equivalence of slopes for midorbital thickness between the samples. Again, interaction effects are not significant, indicating similar rates of change over the European continent. The rate of midorbital thickness reduction, defined by common regression, is of the order of 0.000123 mm per year. When regression is performed to examine equivalence of intercept values between regions, the coefficient for geographic area is significant, indicating that changes took place at different times in the two areas. Again, the intercept for western fossils is greater than for the Central European sample; changes occurred later in the west.

Finally, Table 11.17 shows the results of regression to compare slopes for midorbital projection between Central and Western Europe. Interaction effects are not significant. A similar rate of reduction in this feature (*c.* 0.000179 mm per year) characterizes both geographic regions. With the interaction removed, the coefficient for area is, again, significant, and from this we can conclude that changes occurred at different times. As was the case for the two preceding variables, the intercept for midorbital projection is greater in Western Europe, indicating that change occurred later there than in Central Europe.

Comparisons among these variables between Western and Central Europe over time have produced some interesting results, which can be summarized as follows. Lateral thickness of the torus reduces in both regions over the time-period of concern, but the process of change is distinct in each area. Change is gradual in Central Europe but abrupt and accelerating in the west. Reduction in three measurements (medial projection, midorbital thickness, and midorbital projection) occurs at the same rate in both areas, but change occurs later in Western Europe than in the Central part of the continent. Lateral projection of the torus reduces at the same time and rate in both regions. Finally, medial thickness shows no patterned relation to time in either of the two geographic regions considered.

DISCUSSION AND CONCLUSIONS

These results have some important implications for the models of change across the Neanderthal-modern human transition discussed earlier. We have demonstrated that variation in the nature and/or timing of supraorbital reduction exists between two geographic areas of Europe during the later Pleistocene. While explanation of that variation might be considered somewhat equivocal at the present time, the very existence of variability lends credence to notions that evolutionary processes may have differed in the two regions. In concluding this paper, we will discuss some of the implications of our studies for the models for change outlined earlier, and we will discuss some of the results in terms of functional evolution in the supraorbital region.

Perhaps the most obvious implication of these analyses is the empirical need for a shift in the rate of torus evolution some time *prior* to the period of concern. Simple linear trends were defined for 9 of the 12 measurements considered, all of which have a positive slope (i.e. variable values increase the further back in time one considers). These regressions can be used to predict the relevant variable values in the more-remote past. For example, formula (1) would predict that Central European hominids at *c.* 300 000 BP had lateral thickness measurements of *c.* 32.1 mm. This value is clearly too great given known values for roughly contemporary hominids (e.g. Steinheim at 8.1 mm or Broken Hill at 16.3 mm). In other words, the observed trend for lateral thickness must represent an accelerated rate of change that began somewhere between 300 000 and 80 000 years ago. Similar exercises could be carried out for the other torus measurements analysed here, and these would yield similar empirical discrepancies between predicted and observed values for earlier hominids. The point is that acceleration in the pattern of supraorbital reduction generally occurs *prior to,* or *during* the early phases of the emergence of European Neanderthals and *not* so much with the appearance of modern humans.

Certain patterns in the data are rather easily interpretable on functional/morphological grounds. For example, all projection variables in both geographic regions exhibit simple linear patterns of reduction over time. This is not unexpected, since it is clear that supraorbital projection is a function of facial forwardness or prognathism (see discussions in Smith and Ranyard 1980, and Ravosa, in press). Increased degree of facial forwardness is characteristic of Neanderthals compared to early modern humans (Stringer *et al.* 1984; Tattersall 1986). But especially in Central Europe, a series of cranial features show that facial size and projection were reduced in later Neanderthals compared to earlier Neanderthals and in early moderns compared to late Neanderthals (Wolpoff *et al.* 1981; Smith 1982, 1984; Smith *et al.* 1985). Thus since there is evidence of a non-abrupt pattern of facial reduction over the transition, it is not surprising that projection of the supraorbital region reflects a similar pattern of change over the same period.

Patterns of change in thickness dimensions are somewhat more complex to explain. Generally, we consider increased thickness as a functional reflection of extensive vertical occlusive forces, since the supraorbital region is the major anchor for the facial superstructures onto the cranial vault, and these superstructures are necessary for generating and dissipating such forces. Reduction in thickness thus reflects lessening of the need for such an anchoring structure, brought about by substantial decreases in generation of vertical occlusive forces and the supportive structure related thereto.

As mentioned previously, medial thickness demonstrates no significant pattern of change over time in either Central or Western Europe. We feel this lack of change is an artifact of the invasion of the lower frontal squama by the frontal sinus in early modern humans, a condition not found in Neanderthals (Vlček 1967). Exactly why this occurs is unknown, but it is associated with a distinct bulging in many modern human skulls. This bulging is essentially the superciliary arch (Cunningham 1908) of modern human supraorbital regions. Since these regions are invariably well-developed in early modern Europeans, and since the mode of measurement employed here would measure this entire 'bulge' (Smith and Ranyard 1980: 596), the lack of difference with Neanderthals is not puzzling. Whether these specific measurements are functionally equivalent in these two groups is not clear.

Thicknesses in the lateral and mid-orbit regions are not complicated by the presence of sinuses; and in both geographic regions, both dimensions decrease over the time period presented here. We believe this pattern of reduction reflects decreased functional demands relating to force disruption and anchoring of the face onto the cranial vault that characterize modern humans in contrast to Neanderthals. Furthermore, because of the distinctive infra-orbital and lateral orbital morphology of Neanderthals (Smith 1983; Rak 1986), we believe that a substantial amount of the force generated through their habitual anterior dental loading was dissipated laterally through the lateral orbital margins onto the cranial vault, rather than through the inter-orbital region as is the case with modern humans (cf. Russell 1985). Thus, changes in thickness in the midorbital and lateral regions of the torus are, in our opinion, more accurate reflections of functional changes than changes in the medial region.

Interestingly, midorbital and lateral thickness in Central Europe follow exactly the same linear patterns of reduction over time as projection dimensions do. In Western Europe, however, the situation is more complex. Midorbital thickness follows a linear pattern, but lateral thickness does not. It exhibits a quadratic relationship with time, suggesting an accelerated change with the origin of modern humans in this region. In other words, there is evidence here of an abrupt change with the appearance of modern humans in Western Europe, rather than the continuous reduction demonstrated in Central Europe.

Evidence that the population dynamics responsible for the appearance of modern humans differ in Western and Central Europe is also seen in comparisons of the linear regression for medial and midorbital projection and midorbital thickness. These comparisons suggest that the process of reduction in these regions of the torus occurred later in the west. In evolutionary terms, this implies that the selective forces which maintained the torus and related morphological complex were reduced or removed later in Western than in Central Europe. Taken together with the abrupt change noted in lateral thickness, we interpret this pattern to suggest that population replacement may have been more a factor in the emergence of modern humans in West than in Central Europe.

Some recent, albeit tentative, support for this interpretation can be found in the archaeological record of southern and central France, if one is willing to accept that all Châtelperronian is the product of Neanderthals (as it apparently is in the two sites where we have human fossils in Châtelperronian contexts – Grotte du Renne at Arcy-sur-Cure (Leroi-Gourhan 1959) and Saint-Césaire (Vandermeersch 1984)) and that all Aurignacian is the product of modern humans (although there are *no* fossil human associations with *early* Aurignacian in the west). Leroyer and Leroi-Gourhan (1983) have shown that the Aurignacian appears first in the southern portion of France at a time when the Châtelperronian is widespread to the north. Gradually, the Aurignacian expands into the Périgord region of Aquitaine and replaces the Châtelperronian.

Furthermore, detailed stratigraphic analyses of two sites in southwestern France (Le Piage and Roc de Combe) demonstrate the presence of Châtelperronian components stratigraphically *above* early Aurignacian ones, rather than the more common relationship (Bordes and Labrot 1967; Champagne and Espitalié 1967). Chronostratigraphic analysis on a regional scale also supports the inference of Aurignacian and Châtelperronian contemporaneity (Laville *et al.* 1980). If these complexes are human-type- specific, then we have evidence that Neanderthals and early modern humans were contemporaneous in this region of France for a time. If the archaeological complexes are *not* human-type-specific, then the 'transition' process is exceedingly complex – indeed so complex as to invalidate most of the simplistic models for relationships between biological and cultural change that have been proposed recently.

In Central Europe, there is no evidence which implies a possible replacement of Middle by Upper Palaeolithic populations, and the patterns of supraorbital evolution in this region can be interpreted as the result of localized change without external influence. However, we do recognize that lack of any external influence on this process is unlikely, primarily because of the great probability that modern humans existed earlier in Africa and Western Asia than in Europe (Bräuer 1984; Rightmire 1979; Smith 1985; Stringer *et al.* 1984). Thus it is more reasonable to think that the genetic basis of modern human morphology entered Europe through gene flow rather than that it evolved independently there. However, there is absolutely *no* evidence in Central Europe that this gene flow involved any substantial population replacement. Significant genetic changes may well have been introduced into Central Europe, but not by an influx of fully-developed modern humans from some adjacent geographic region.

The fact that such population replacement may have been operative in Western Europe, while being completely non-operative in Central Europe, uncovers what we feel is the most important point to be made from our analysis: that the population dynamics involved in the origin of modern humans is complex and does not conform to any simplistic model. The re-emergence of total replacement as a world-wide model for the origin of modern humans is simply not supported by the bulk of evidence from the human fossil record, not even in Europe. Moreover, the archaeological record also fails to support models for radical change (Rigaud 1989; Simek and Price, in press, this Symposium vol. 2; Simek and Snyder 1988). It might be possible to argue that West European Neanderthals were replaced by incoming moderns – although probably not without some degree of interbreeding. But the source for these modern populations would certainly have been in Central Europe, where no form of replacement is compatible with existing data. Thus, some European Neanderthals probably had a significant role in the origin of modern Europeans. It is clear, then, that a total- replacement approach to the origin of modern humans in general, and modern Europeans in particular, should not be resurrected from its grave in the pages of history.

REFERENCES

Binford, L. R. 1985. Human ancestors: changing views of their behavior. *Journal of Anthropological Archaeology* 4: 292-327.

Binford, L. R. 1989. Isolating the transition to cultural adaptations: an organizational approach. In E. Trinkaus (ed.) *Patterns and Processes in Later Pleistocene Human Emergence.* Cambridge: Cambridge University Press. In Press.

Bordes, F. and Labrot, S. 1967. La stratigraphie du gisement du Roc de Combe (Lot) et ses implications. *Bulletin de la Société Préhistorique Française* 64: 29-34.

Bräuer, G. 1984. A craniological approach to the origin of anatomically modern *Homo sapiens* in Africa and implications for the appearance of modern Europeans. In F. H. Smith and F. Spencer (eds) *The Origins of Modern Humans: a World Survey of the Fossil Evidence*. New York: Alan R. Liss: 327-410.

Cann, R. L., Stoneking, M. and Wilson, A. C. 1987. Mitochondrial DNA and human evolution. *Nature* 325: 31-36.

Cartmill, M., Pilbeam, D. and Isaac, G.Ll. 1986. One hundred years of paleoanthropology. *American Scientist* 74: 410-420.

Champagne, F. and Espitalié, R. 1967. Le stratigraphie du Piage: note préliminaire. *Bulletin de la Société Préhistorique Française* 64: 35-40.

Cunningham, D. J. 1908. The evolution of the eyebrow region of the forehead, with special reference to the excessive supraorbital development in the Neanderthal race. *Transactions of the Royal Society of Edinburgh* 46: 243-310.

Eldredge, N. and Tattersall, I. 1982. *The Myths of Human Evolution*. New York: Columbia University Press.

Frayer, D. W. 1986. Cranial variation at Mladeč and the relationship between Mousterian and Upper Paleolithic hominids. *Anthropos* 23: 243-256.

Frayer, D. W., Jelínek, J., Minigh, N., Oliva, M., Seitl, L., Smith, F. H. and Wolpoff, M.H. (in press). *Upper Pleistocene Human Remains from the Mladeč Cave,* Moravia. In Press.

Klein, R. 1983. What do we know about Neanderthals and Cro-Magnon Man? *American Scholar* 52: 386-392.

Laville, H., Rigaud, J.-P. and Sackett, J. 1980. *Rock Shelters of the Périgord*. New York: Academic Press.

Leroyer, C. and Leroi-Gourhan, A. 1983. Problèmes de chronologie: le castelperronien et l'aurignacien. *Bulletin de la Société Préhistorique Française* 80: 41-44.

Leroi-Gourhan, A. 1959. Etude des restes humains fossiles provenant des Grottes d'Arcy-sur-Cure. *Annales de Paléontologie* 44: 87-147.

Neter, J., Wasserman, W. and Kuther, M. H. 1985. *Applied Linear Statistical Models*. Homewood (Ill): Richard D. Irwin.

Oakley, K., Molleson, T. and Campbell, B. 1971. *Catalogue of Fossil Hominids*. Vol 2: *Europe*. London: British Museum (Natural History).

Pilbeam, D. R. 1986. The origin of *Homo sapiens:* the fossil evidence. In B. Wood, L. Martin, and P. Andrews (eds) *Major Topics in Primate and Human Evolution*. London: Cambridge University Press: 331-338.

Rak, Y. 1986. The Neanderthal: a new look at an old face. *Journal of Human Evolution* 15: 151-164.

Ravosa, M. J. (in press). Browridge development in *Cercopithecidae:* a test of two models. *American Journal of Physical Anthropology*. In Press.

Rigaud, J.-P. 1989. From Middle to Upper Paleolithic: transition or convergence? In E. Trinkaus (ed.) *Patterns and Processes in Later Pleistocene Human Emergence*. Cambridge: Cambridge University Press. In Press.

Rightmire, P. 1979. Implications of the Border Cave skeletal remains for later Pleistocene human evolution. *Current Anthropology* 20: 23-35.

Russell, M. D. 1985. The supraorbital torus: 'a most remarkable peculiarity'. *Current Anthropology* 26: 337-360.

Schwalbe, G. 1901. Der Neanderthalschädel. *Bonner Jahrbücher* 106: 1-72.

Simek, J. and Snyder, L. M. 1988. Patterns of change in Paleolithic archaeofaunal diversity. In H. Dibble and A. Montet-White (eds) *The Pleistocene Prehistory of Western Eurasia*. Philadelphia: University of Pennsylvania Press: 321-332.

Smith, F. H. 1982. Upper Pleistocene hominid evolution in south-central

Europe: a review of the evidence and analysis of trends. *Current Anthropology* 23: 667-703.
Smith, F. H. 1983. Behavioral interpretation of changes in craniofacial morphology across the archaic-modern *Homo sapiens* transition. In E. Trinkaus (ed.) *The Mousterian Legacy: Human Biocultural Change in the Upper Pleistocene*. Oxford: British Archaeological Reports International Series S164: 141-163.
Smith, F. H. 1984. Fossil hominids from the Upper Pleistocene of central Europe and the origin of modern Europeans. In F. H. Smith and F. Spencer (eds) *The Origins of Modern Humans: a World Survey of the Fossil Evidence*. New York: Alan R. Liss: 137-209.
Smith, F. H. 1985. Continuity and change in the origin of modern *Homo sapiens*. *Zeitschrift für Morphologie und Anthropologie* 75: 197-222.
Smith, F. H., Boyd, D. C. and Malez, M. 1985. Additional Upper Pleistocene human remains from Vindija Cave, Croatia, Yugoslavia. *American Journal of Physical Anthropology* 68: 375-383.
Smith, F. H. and Paquette, S.P. 1989. The adaptive basis of Neanderthal facial form, with some thoughts on the nature of modern human origins. In E. Trinkaus (ed.) *Patterns and Processes in Later Pleistocene Human Emergence*. Cambridge: Cambridge University Press. In Press.
Smith, F. H. and Ranyard, G. C. 1980. Evolution of the supraorbital region in Upper Pleistocene fossil hominids from south-central Europe. *American Journal of Physical Anthropology* 53: 589-610.
Smith, F. H. and Spencer, F. (eds) 1984. *The Origin of Modern Humans: a World Survey of the Fossil Evidence*. New York: Alan R. Liss.
Spencer, F. and Smith, F. H. 1981. The significance of Alesv Hrdlička's "Neanderthal Phase of Man": a historical and current assessment. *American Journal of Physical Anthropology* 56: 435-459.
Stringer, C. B. 1989. Documenting the origin of modern humans. In E. Trinkaus (ed.) *Patterns and Processes in Later Pleistocene Human Emergence*. Cambridge: Cambridge University Press. In Press.
Stringer, C. B., Hublin, J.-J. and Vandermeersch, B. 1984. The origin of anatomically modern humans in western Europe. In F. H. Smith and F. Spencer (eds) *The Origins of Modern Humans: a World Survey of the Evidence*. New York: Alan R. Liss: 51-135.
Tattersall, I. 1986. Species recognition in human palaeontology. *Journal of Human Evolution* 15: 165-175.
Thorne, A. G. and Wolpoff, M. H. 1981. Regional continuity in Australasian Pleistocene hominid evolution. *American Journal of Physical Anthropology* 55: 337-349.
Trinkaus, E. and Smith, F. H. 1985. The fate of the Neandertals. In E. Delson (ed.) *Ancestors: the Hard Evidence*. New York: Alan R. Liss: 325-333.
Vandermeersch, B. 1984. A propos de la découverte du squelette néandertalien de Saint-Césaire. *Bulletins et Mémoires de la Société d'Anthropologie de Paris* (série 14) 1: 191-196.
Vlček, E. 1967. Die sinus frontales bei europäischen Neandertalern. *Anthropologischer Anzeiger* 30: 166-189.
Wolpoff, M. H. 1980. *Paleoanthropology*. New York: A. A. Knopf.
Wolpoff, M. H. 1985. Human evolution at the peripheries: the pattern at the eastern edge. In P. V. Tobias (ed.) *Hominid Evolution: Past, Present, and Future*. New York: Alan R. Liss: 355-365.
Wolpoff, M. H. 1989. The place of Neandertals in human evolution. In E. Trinkaus (ed.) *Patterns and Processes in Later Pleistocene Human Emergence*. Cambridge: Cambridge University Press. In Press.
Wolpoff, M. H., Smith, F. H., Malez, M., Radovčić, J. and Rukavina, D. 1981. Upper Pleistocene human remains from Vindija Cave,

Croatia, Yugoslavia. *American Journal of Physical Anthropology* 54: 499-545.
Wolpoff, M. H., Wu, X. Z. and Thorne, A.G. 1984. Modern *Homo sapiens* origins: a general theory of hominid evolution involving the fossil evidence from East Asia. In F. H. Smith and F. Spencer (eds) *The Origin of Modern Humans: a World Survey of the Fossil Evidence*. New York: Alan R. Liss: 411-483.

12. Fossil Hominids from the early Upper Palaeolithic (Aurignacian) of France

DOMINIQUE GAMBIER

INTRODUCTION

In France, the earliest remains of anatomically-modern humans derive from twenty sites situated in the area between the Loire and the Pyrénées. Arcy-sur-Cure (Yonne) is the only site located to the north of the Loire.

In general, these remains have not been well documented in the earlier literature, principally because of the fragmentary condition of the remains. Moreover, most of the earlier interpretations have concentrated on identifying typological differences, rather than on studying morphological variation and evolution. The recent debates surrounding the origins of modern populations in Europe, and their relationship to the Neanderthals, have led to increased interest in these early Upper Palaeolithic remains. The first part of this paper will be devoted to a description and analysis of cranial variation in the early modern humans in France. In the second section, their relationship with the European Neanderthals will be assessed.

THE AVAILABLE MATERIAL

If we accept the attributions given in the *Catalogue of Fossil Hominids* (Oakley *et al.* 1971), there are many French sites which have produced human remains in association with cultures of the early Upper Palaeolithic. Unfortunately, most of these sites were excavated around the end of the 19th century or the beginning of the present century, when methods of both excavation and recording were poorly developed. As a result, the stratigraphic position of the remains is frequently imprecise, and for many of them the dating evidence is very uncertain. As we shall see, the age of some of the specimens is so uncertain that they are better omitted from the present discussion.

For the present study we shall only consider remains which, if not certainly, then at least probably, come from the early Upper Palaeolithic. Most of the specimens derive not from the very beginning of the Upper Palaeolithic but from levels which, in archaeological terms, belong to the later stages of the Aurignacian.

As noted above, most of this material is in a fragmentary condition. There are few complete individuals, and bones from the postcranial skeleton are rare. Most of the remains derive from the skull, teeth and mandible.

If we consider the material as a whole, we have remains of around 30 individuals including about 10 children. Of these, the remains of the skull and mandible employed for the present study total around 12 specimens, including three children.

Combe-Capelle (Dordogne)

This rock-shelter produced an almost complete adult skeleton. The discovery was made in 1909 by labourers working under the direction of O. Hauser. According to Hauser and Klaatsch (1910), the burial occurred at the interface between a Mousterian and an 'early Aurignacian' (i.e. Châtelperronian) level, in contact with the rock floor. The thin Châtelperronian level was covered by a thin sterile layer. Above it occurred successive layers of Aurignacian, Périgordian and Solutrean industries, each separated by further sterile layers. The whole sequence, according to the authors, reached a maximum depth of 2.8 metres. Later studies by Peyrony (1943) and Sonneville-Bordes (1960) have shown that the Aurignacian industries belonged to Aurignacian I and II. The supposedly Châtelperronian age of the skeleton, as well as the authenticity of the skeleton itself, have been the subject of many discussions. For Peyrony (1943), Bordes (1972) and Genet Varcin (1979) this age is well established. Asmus (1964), Frayer (1978) and Vandermeersch (1984) on the other hand suggest a younger age, while Thoma (1972) has contested the authenticity of the skeleton itself. The question is clearly of considerable importance, since an acceptance of the Châtelperronian age of the skeleton would make this one of the oldest representatives of modern humans in France.

To discuss this point we shall refer to the stratigraphic section of the site recorded by Peyrony before 1909 (during the excavations of M. Villeréal – published in 1943), the indications provided by Hauser himself (1910-1911), the available information concerning the morphology of the skeleton (Klaatsch 1910; Genet Varcin 1969), and recent data relating to the other skeletal remains found in association with the Châtelperronian.

There is no drawing available showing the skeleton *in situ*, as it was originally discovered, and no stratigraphic section was ever recorded in the area immediately surrounding the burial. The only available documents are Hauser's photographs, which seem to demonstrate that the skeleton was indeed discovered in the site.

While Peyrony (1943) accepts the Châtelperronian age of the skeleton, he expressed some reservations concerning the interpretation of Hauser's stratigraphy. In particular he disputes the presence of a Mousterian level at the base of the stratigraphic sequence. The remainder of Peyrony's section however, conforms with that described by Hauser, and reveals no interruptions in any of the overlying layers.

More recently, Asmus (1964) has provided a reanalysis of the data published by Hauser (1910), and has drawn attention to the following points:

1. The general lack of rigour in the observations recorded by Hauser, and the existence of certain contradictions surrounding the conditions of the discovery, due to the methods of excavation employed. The records of

stratigraphy and the position of the skeleton would appear to have been based on the evidence of workmen, as Hauser himself was rarely present on the site. The stratigraphy of the site was evidently more complex than that recorded on the published section (Hauser and Klaatsch 1910: 274). Hauser mentions variations in the thickness of the layers, and notes that the Aurignacian and Périgordian layers even disappeared at certain spots.

2. The shallow thickness of the Châtelperronian layer (c. 30 cm) and the sterile layer above it (15 cm). According to Asmus, this thickness would have been incompatible with the survival of the skeleton, which we know was in an excellent state of preservation. Clearly, this observation provides an argument against the Châtelperronian age of the Combe-Capelle skeleton. The recent discovery of a characteristically Neanderthal skeleton in a Châtelperronian level at Saint-Césaire in Charente-Maritime (Vandermeersch 1984), further supports this conclusion.

The possibility of an Aurignacian age for the Combe-Capelle skeleton seems equally dubious. As Asmus (1964) points out, to accept an Aurignacian age with any confidence it would be necessary to demonstrate that the sterile layer overlying the Aurignacian level was intact. In fact, the methods of excavation and the available stratigraphic evidence (including the evidence for the interruption of some of the overlying layers) do not allow us to assert that the layers were intact above the burial. Under these conditions we think that an Aurignacian age cannot be proved.

Two other observations would seem to argue in favour of a post-Aurignacian age (or even post-Palaeolithic age) for the Combe-Capelle skeleton:

1. Later excavations on the site revealed the presence of human remains in association with much more recent archaeological material (Billy, pers. comm.).

2. The available information on the morphology of the Combe- Capelle skeleton (as documented by the studies of Klaatsch (1910) and Genet Varcin (1969)) indicates that this skeleton is very different from other skeletons of Aurignacian age. It would therefore stand (in morphological terms) as an isolated specimen in Western Europe – and the supposed parallels with the specimen from Brno (Vlček 1971) do not appear evident to us. Taking into account all of the available data, the most likely explanation is that the Combe-Capelle skeleton represents a relatively recent, intrusive burial into the site, which penetrated through the Upper Palaeolithic levels.

Finally, Thoma (1972) has emphasized the uncertain nature of the published metric and morphological data, since the reconstruction of the face is open to question. As it is impossible for us to check these data (since the skeleton has now been lost or destroyed), and given the uncertainties as to its age, it seems reasonable to exclude this specimen from the Upper Palaeolithic sample under discussion here (see Note 1).

Roche-Courbon (Charente-Maritime)

In 1956 the Bouil Bleu cave (or cave 164) of the site of Roche-Courbon produced an adult skeleton associated with Aurignacian tools (Geay and

Colle 1956). According to Vallois (1957: 155), the excavations were "faites au hasard et sans aucune méthode". Most probably this represents a Neolithic burial which intruded into the Upper Palaeolithic levels (Perpère 1971; Debénath 1974). Once again, this specimen is best omitted from the present discussion.

Cro-Magnon (Les Eyzies de Tayac, Dordogne)

In 1868 the skeletons of four adults and four very young children were discovered in the rock-shelter of Cro-Magnon (Lartet 1868). The adult remains consist of both cranial and postcranial bones. According to Sonneville-Bordes (1959) the human remains were contemporary with level J, and were associated with a late Aurignacian industry.

Subsequently, Movius (1969) suggested an age of *c.* 30 000 BP for these remains, based on the dating of comparable late Aurignacian levels at the neighbouring site of Abri Pataud. Further confirmation of the dating of these specimens, and their contemporaneity, is however required.

The Skulls. Cranium one (CM1) and cranium two (CM2) are well preserved. They are voluminous and the general dimensions are large (see Table 12.1). They have a long and high vault, with lambdoidal depression and posterior projection of the occiput. The maximum cranial width (M8) is directly above the mastoid process. The frontal region rises vertically, and the parietal dimensions and vaulting are modern. In *norma occipitalis* the cranial shape is "*en maison*" as in modern humans. The occipital reliefs are very pronounced on CM1. Both show a marked external occipital protuberance. The temporal bone has a high squama and the auditory meatus is situated under the zygomatic process root. The mastoid process is well developed. The face is very broad (Table 12.1) and short, particularly on CM1, and does not project. A well marked canine fossa is present. The supraorbital region is not very pronounced and is divided into superciliary and supraorbital arches. The nasal aperture is long and narrow with projecting nasal bones. The orbits are low and rectangular.

Cranium two (CM2) is smaller and more gracile than the others, which indicates that it is probably female.

Cranium three (CM3) preserves the frontal, the parietal and the upper part of the occipital bone. Wolpoff (1980) has claimed the presence on this specimen of some supposedly Neanderthal features, such as as a low frontal bone, a very pronounced supraorbital torus, a markedly projecting occipital bone and a flattened cranial base. The vault contour is similar to that observed on the other skulls. The frontal region rises vertically and the vault height (Table 12.1) is more pronounced on CM3 than the other specimens. The lambdoidal depression is marked and the posterior projection of the occipital is substantial. But the morphology of the occipital bun is very different from that of Neanderthal specimens (Ducros 1967). The flattening of the cranial base does not seem more marked than on CM1 or CM2, and it is very difficult to observe this feature on CM3 because the inferior part of the occipital bone is missing. The supraorbital region is very robust but is divided into two parts as in modern humans. The super-

Table 12.1. Cranial dimensions for west European Neanderthal and early Upper Palaeolithic groups. The "M" dimensions listed in the left hand columns are those defined by Martin and Saller (1956-1966). Specimen numbers are as follows: (1) La Chapelle-aux-Saints; (2) La Ferrassie; (3) Spy 1; (4) Spy 2; (5) Neanderthal; (6) Monte Circeo; (7) La Quina; (8) Cro-Magnon 1; (9) Cro-Magnon 2; (10) Cro-Magnon 3; (11) Les Cottés; (12) La Crouzade. Letters below the specimen numbers indicate the sources of data: (a) Vallois and Billy 1965; (b) Patte 1954; (c) Gambier, personal observations; (d) Vandermeersch 1981. All dimensions are in millimetres.

Dimensions:	Neanderthals							Early Upper Palaeolithic				
Martin & Saller 1956-1966	No. 1 (d)	No. 2 (d)	No. 3 (d)	No. 4 (d)	No. 5 (d)	No. 6 (d)	No. 7 (d)	No. 8 (a, c)	No. 9 (a, c)	No. 10 (a, c)	No. 11 (b)	No. 12 (c)
M1	208.0	207.5	200.4	200.00	199.2	204.0	203.0	202.0	192.0	202.0	192.0	—
M2	196.5	194.5	198.0	—	199.0	198.0	194.0	200.0	183.0	191.5	—	—
M2a	190.5	187.0	193.0	188.0	189.0	194.0	190.0	196.0	—	184.0	—	—
M8	156.0	158.0	144.3	153.2	146.7	155.0	138.0	149.5	138.0	152.0	138.0	—
M8/M1x100	75.0	76.1	71.9	76.6	73.6	75.9	67.9	74.0	71.9	75.2	71.9	—
M9	109.0	109.0	101.1	107.9	105.0	106.0	101.3	102.5	97.5	96.5	99.0	104.0
M9/M8x100	69.9	69.0	70.1	70.4	71.6	68.4	72.7	68.6	70.7	63.5	—	—
M10	122.0	121.0	—	125.9	122.3	127.0	108.3	126.0	120.0	123.0	—	122.0
M9/M10x100	89.3	90.1	—	85.7	85.8	83.5	92.6	81.3	81.2	78.4	—	85.2
M12	130.5	125.0	121.2	131.2	—	124.0	109.3	112.0	—	—	—	—
M12/M8x100	83.6	79.1	84.0	85.6	—	80.0	79.2	—	—	—	—	—
M20	111.0	114.0	111.2	114.0	—	111.0	111.5	122.5	115.0	—	—	—
M20/M1x100	53.3	54.9	55.4	57.0	—	54.4	54.9	60.6	59.8	—	—	—
M20/M8x100	71.1	72.1	77.6	74.4	—	71.6	80.8	81.9	83.3	—	—	—
M22a	90.0	93.0	81.0	87.0	80.5	88.5	79.5	98.0	—	108.0	—	—
M22a/M2x100	40.5	47.8	40.9	44.4	40.4	47.7	40.1	49.0	—	56.4	—	—
M25	357.0	367.0	—	—	—	361.0	—	403.0	—	—	—	—
M26	121.0	135.0	110.0	—	133.0	131.0	120.0	147.0	132.0	148.0	126.0	140.0
M27	121.0	120.0	126.0	115.0	109.0	117.0	112.0	130.0	133.0	132.0	126.0	—
M28	115.0	112.0	—	—	—	113.0	—	126.0	—	—	—	—
M28-1	44.0	69.0	58.5	55.0	57.2	—	62.0	53.5	72.5	—	—	—
M29	110.0	116.0	102.8	—	117.4	117.0	106.4	125.0	115.0	126.0	—	116.0
M30	111.0	112.0	114.9	109.0	102.9	109.0	102.9	118.5	122.0	121.0	—	—
M30/M27x100	91.7	93.3	91.2	94.8	93.7	93.2	96.2	91.1	91.7	91.7	—	—
M29/M26x100	90.9	85.9	93.4	—	88.3	89.3	88.7	85.0	87.1	85.1	—	82.9
M48	86.0	88.0	—	—	—	87/92	—	69.0	70.0	—	—	—
M45	153.0	148.5	—	—	—	147.0	—	142.0	—	—	—	—
M48/M45x100	56.2	59.2	—	—	—	59.2	—	48.6	—	—	—	—
M54	34.0	34.0	—	—	—	36.0	—	24.0	26.0	—	—	—
M55	61.0	62.0	—	—	—	66.0	—	51.0	54.0	—	—	—
M54/M55x100	55.7	54.8	—	—	—	54.5	—	47.0	48.1	—	—	—
M72	83.0	83.0	—	—	—	83.0	83.5	88.0	92.0	—	—	—
M73	83.0	83.5	—	—	—	83.5	—	95.0	94.5	—	—	—
M74	88.0	82.0	—	—	—	81.5	—	68.0	83.0	—	—	—

ciliary arches are very pronounced.

In *norma occipitalis* the cranial shape is "*en maison*". The occipital relief is very pronounced and there is an external occipital protuberance, as on CM1 and CM2. In fact, all the modern features exhibited by CM1 and CM2 are present on CM3. Overall, CM3 is simply more robust than the other specimens. Cro-Magnon 4 is very fragmentary (see Note 2). Fragments of the maxilla, occipital, parietal and temporal bones are preserved. These remains indicate a robust specimen. The fragment of occipital bone shows a large external occipital protuberance, as in modern humans.

The Mandibles. Three fragments of mandible are preserved (nos. 4253, 4256, 4258), and all are robust. The best preserved (no. 4253) has a markedly projecting chin and the muscle insertions are strongly pronounced. None show a retromolar space, but the mental foramen on the first mandible is located below the first molar as in Neanderthal mandibles. On the second (no. 4256), the mental foramen is below P2/M1. The digastric fossae are deep and broad on mandibles 4253 and 4256, and they are orientated slightly downwards.

The Cro-Magnon sample – both skulls and mandibles – exhibit some variation, but the total morphological pattern of each specimen is modern.

Les Cottés (Saint-Pierre-du-Maillé, Vienne)

In 1881, R. de Rochebrune excavated the incomplete skeleton of an adult from the Les Cottés cave, associated with 'Aurignacian' tools. The skeleton consisted of a calotte, a mandible fragment, limbs bone and a few ribs and vertebrae. Its association with the Aurignacian was contested by Cartailhac (1881). However, Breuil (1906) accepted the age of this skeleton (Rochebrune having affirmed that the levels had not been disturbed), and Cartailhac subsequently (1912) accepted this opinion.

The later excavations carried out by L. Pradel outside the cave entrance (between 1951 and 1958) confirmed the existence of an Aurignacian level, and revealed a succession of four archaeological levels, separated by sterile layers – i.e. Mousterian, 'Périgordian II', Aurignacian I, and early Upper Périgordian of 'La Gravette' type (Pradel 1961). However, can the skeleton be reliably associated with any of these levels? In his excavation report, Rochebrune wrote: "A deux metres environ au dehors de la grotte, dans un humus noir et friable nous apercûmes tout à coup le sommet d'un crâne. . ." (Rochebrune 1881: 489). For Breuil (1906) this "black and crumbly humus" corresponded to the black layer, rich in bones and Aurignacian flints, which was said to have been found during the 1881 excavations two metres outside the cave entrance. Perpère however (1973) does not accept this correlation, on the grounds that the later excavations had not in fact extended onto the terrace in front of the cave, but were restricted to a single trench located at the cave entrance. In these circumstances, the skeleton could derive from a superficial level, without any connection with the Upper Palaeolithic levels. In a slightly later publication, however, Rochebrune (1883: 426) wrote: "Enfin à l'entrée de la caverne, que j'achevais de déblayer à la fin de mes fouilles je trouvais un cadavre humain. . .". This revised description of the location would of course be consistent with Pradel's observation (1961) and with an Upper Palaeolithic age for this skeleton.

On present evidence it is hardly possible to be certain about the real antiquity of this skeleton. Even if we accept an Upper Palaeolithic age, it is difficult to know whether the skeleton derived from the Aurignacian level or from the overlying early Périgordian level, as Rochebrune did not recognise this latter level. The observations made by Pradel would perhaps favour an Aurignacian age, since according to Pradel (1961), the Périgordian II layer was very thin near the cave entrance, and only attained a substantial thickness in front of the cave – probably beyond the limits of de Rochebrune's excavations.

Direct dating of this fossil is clearly necessary. If its association with the Aurignacian could be was confirmed, it would be one of the oldest representatives of modern humans in France.

Unfortunately, it is at present impossible to examine the skull. The information quoted here is taken from the published study by Patte (1954). The specimen from Les Cottés consists of a well preserved calotte and fragmentary mandible. On the basis of size (Table 12.1) and the lack of marked development of cranial relief, the skull is similar to that of Cro-Magnon 2. The frontal rises vertically, while the occipital bone is not projecting (there is no occipital bun). The maximum cranial width is directly above the mastoid process. In *norma occipitalis* the cranial shape is "*en maison*" as in modern humans. On the occipital bone there is an external occipital protuberance, but it is not very marked. The supraorbital region is modern, with a rather weak superciliary arch.

The mandible is broken in two pieces and the symphysis is not preserved. It is robust in size, but the muscle insertions are not pronounced. There is no retromolar space, and the mental foramen is under the first premolar.

A detailed study of this specimen is necessary, but there is some evidence that the skull and mandible are of modern form and do not show any Neanderthal features.

La Crouzade (Gruissan, Aude)

This site was excavated by Helena in 1928. Level F produced a frontal bone and a fragment of right maxilla with P2, M1, M2, M3 in place.

The dentition indicates a young adult. According to Sacchi (1973), the associated tools indicate an early Aurignacian industry. The La Crouzade frontal bone is almost complete. It is wider than that of the Cro-Magnon specimens. The sagittal convexity is marked (Table 12.1). The measurable dimensions of the arch are among the largest of the early modern human sample. The supraorbital region is modern, and the superciliary and supraorbital arches are weak. There is no frontal sinus, in contrast to the Neanderthals which have an extensive pneumatisation of the torus (Tillier 1977). The maxilla is very fragmentary, but it does not show any Neanderthal features, and is anatomically-modern.

Isturitz (Pyrénées-Atlantiques)

Between 1930 and 1940, excavations by R. and S. de St. Périer in the cave of Isturitz produced an incomplete mandible of an adult. According to Sonneville-Bordes (1959), the tools which associated with the mandible indicate an Aurignacian I level.

This is a left hemimandible. The chin region, the condyle and the coronoid process are broken, but I2, C, and P1 are present. This mandible is robust and the muscular insertions are very marked. There is no retromolar space and the mental foramen is under P2. The left *fossa digastrica* is deep and broad, and is oriented downwards and backwards.

Overall, this specimen is modern, although it is more robust than mandibles of present-day humans.

Fontéchevade (Orgedeuil, Charente)

The Fontéchevade remains were recovered during the excavations of Durousseau-Dugontier between 1902 and 1912, and include a parietal bone of an adult and a right hemimandible of a child. The whole associated industry has been describe by Sonneville-Bordes (1959) as Aurignacian.

The parietal: The anterior part is broken. The morphology and size (Table 12.3) of this bone present modern characteristics. The lambdoid convexity is more developed than that of Neanderthal parietal bones, and the *tuber parietale* is marked. The child's mandible: The body is broken at the level of the right canine socket, and the upper part of the ascending branch is not preserved. The second deciduous molar, the germ of the permanent premolar, and the first and second molar are present. We can estimate the dental age as somewhere between 4 and 5 years (after Ubelaker 1978). The robustness of this mandible (Table 12.2) is substantial but it is not very different from that of some recent children. The mandibular angle is rounded and the masseteric crest is not differentiated. On the inner surface, the *lingula mandibulae* is broken. The crest for the attachment of the inner pterygoid is not well marked. On the lateral surface of the body, the *prominentia lateralis* is clearly differentiated, and on the inner surface the *linea mylohyoidea* is well marked. The mental foramen at the mid-height of the body is below the socket for dM1. This mandible is not very different from that of a recent child of the same dental age.

Les Rois (Mouthiers en Boeme, Charente)

The Les Rois juvenile mandibles were discovered during excavations between 1948 and 1952 by P. Mouton and R. Joffroy. The human remains derive from levels A2 and B, and were found in direct association with the Aurignacian I and II industries on the site.

Mandible A: This consists of the left and right parts of the body. The ascending *ramus* is not preserved, and the body is broken at the level of the second permanent molar socket. The second and first deciduous molar, the first permanent molar and the canine are present. We can estimate the dental age at somewhere between 10 and 11 years. Compared to recent

Table 12.2. Mandibular dimensions (in mm.) for West European Neanderthal, early Upper Palaeolithic, and recent children. Specimen numbers are as follows: (1) Les Rois A; (2) Fontéchevade; (3)(4) recent; (5) La Chaise; (6) Archi; (7) Roc de Marsal; (8) Gibraltar; (9) Combe Grenal; (10) Malarnaud; (11) Montgaudier. Letters under the specimen numbers indicate the sources of data: (a) Tillier 1983; (b) Tillier and Genet-Varcin 1980; (c) Gambier, personal observations; (d) Vandermeersch and Duport 1976; (e) Tillier 1984; (f) Tillier 1983 and personal communication.

	Early Upper Palaeolithic		Recent		Neanderthal						
	No. 1 (c)	No. 2 (c)	No. 3 (b)	No. 4 (c)	No. 5 (b)	No. 6 (a)	No. 7 (a)	No. 8 (f)	No. 9 (b)	No. 10 (e)	No. 11 (d)
Age at death (years)	10-11	5	4-5	10	4-5	4-5	3	5	6-7	13-14	—
Bimental. Breadth	45.3	—	—	40.0	—	—		—	—	—	—
Bicanine Breadth	26.4	—	—	19.5	—	—		—	—	—	—
Bi-M1 Breadth	60.2	—	—	41.0	—	—		—	—	—	—
Bi-M1 Long.	31.8	—	—	31.0	—	—		—	—	—	—
Symphysial Height	27.6	—	22.8	25.1	22.1	21.0	20.4	21.2	—	25.0	18.5
Height dm2/M1	25.2	—	—	22.0	—	—	—	—	—	—	—
Height dm1/dm2	—	18.2	20.0	—	20.5	20.0	17.0	22.8	26.9	—	—
Proj. thickness											
—symphysial	15.4	—	12.0	13.0	12.8	13.5	12.5	12.6	—	12.5	12.8
—dm2/M1	16.8	—	—	15.0	—	—	—	—	—	—	—
dm1/dm2	—	12.5	11.4	—	12.5	11.8	13.1	13.6	13.8	—	—
Robusticity											
—Symphysial	55.8	—	52.6	51.8	57.9	59.0	61.3	59.4	—	50.0	60.2
—dm2/M1	66.7	—	—	68.2	—	—	—	—	—	—	—
—dm1/dm2	—	68.7	57.0	—	60.9	59.0	77.0	59.6	51.3	—	—
Symphisial perimeter	67.0	—	—	56.0	—	—	—	—	—	—	—

Table 12.3. Dimensions (in mm.) of parietal bones for West European Neanderthal and early Upper Palaeolithic fossils (from Gambier, personal observations, and (a) Vandermeersch 1981).

	Fontéchevade	Cro-Magnon 1	Cro-Magnon 2	Les Cottés (a)	La Chaise (a)	La Chapelle (a)	Spy 1 (a)	Spy 2
Lambdoid arc (1)	116.0	115.0	104.5	—	—	—	—	
Lambdoid Chord (2)	098.4	100.0	094.0	—	—	—	—	—
I=2/1x100	84.8	86.9	89.9	—	87.2	86.8	—	—
Thickness at parietal tuberosity	—	—	—	6.0	—	—	—	—
Thickness at Lambda	6.1	9.0	—	9.0	—	—	10.0	8.0

children of the same dental age, it is a very robust specimen. The general size and thickness of the body (Table 12.2) are substantial, and the posterior reduction of the height of the body is slight. On the inner surface of the body, the *linea mylohyoidea* is very strong while the *fovea submandibularis* is very discrete. The *foramen mentale* is below P1 and the anterior surface of the body is vertical in lateral view, with incisors and canines which are not in a frontal arrangement. The chin is not well marked but the *tuber symphyseos* and *tubercula lateralia* can be observed.

On the posterior surface, below the alveolar margin, there is an incipient *planum alveolare*. The digastric fossae are large, deep, and orientated downwards. The Les Rois mandible differs strongly from the mandibles of Neanderthal children (such as Montgaudier, Malarnaud and Teshik-Tash), where only a slight tuber symphyseos is differentiated (Tillier 1984).

Mandible B is very fragmentary. The alveolar margin with I2, C, and P1 are preserved. We can estimate the dental age somewhere between 10 and 11 years. It is a robust specimen like mandible A. The juvenile mandible of Les Rois A shows modern characteristics such as the decrease of the height of the body and the morphology of the chin. But the internal surface of the symphysis displays primitive features, such as an incipient alveolar planum and the direction of the digastric impressions. The most striking feature of this mandible is its substantial size and robusticity, compared to the mandibles of recent children of the same dental age. This is also true for Mandible B.

Other Sites

Three other sites – La Quina X-Y, La Chaise de Vouthon (Charente) and Les Roches (Indre) – have produced human remains associated with Aurignacian material. But these remains are extremely fragmentary and can provide little useful anatomical information.

DISCUSSION

The overall morphology of the cranium and mandible of the oldest French Upper Palaeolithic modern humans is thus very similar to that of modern humans. Essentially, differences are confined to a greater robusticity, a longer cranial vault, and the persistence of a few archaic features. Comparisons with Neanderthals show apparent shared features in some cases, such as an occipital bun, the presence of an alveolar plane, the orientation of the digastric fossae, posterior positioning of the mental foramen, and the general higher degree of robusticity. However, these characters are not consistently present, and in this respect Upper Palaeolithic crania and mandibles show significant variation. Moreover, none of the Upper Palaeolithic fossils show the most diagnostic Neanderthal apomorphies, nor are 'transitional' morphologies present.

Observations on other Western European crania are comparable (Stringer *et al.* 1984). Only the specimen from Hahnöfersand (West Germany), represented by its frontal bone, could be considered distinctive, although the morphology of the supraorbital region is modern. From its robusticity (especially cranial thickness), degree of glabella projection, and extent of vault flattening, Bräuer (1981) considered it to be related to the Neanderthals. However, neither cranial robusticity nor glabella projection are specific Neanderthal characters, and the degree of flattening can only be inferred, not demonstrated, in an isolated frontal bone. In my opinion, this fossil represents a very robust modern human.

Data concerning the teeth show that there is a trend in the reduction of crown diameters between the early and late Upper Palaeolithic samples (Frayer 1978; Gambier, in preparation). This trend seems most marked in the anterior teeth. Comparison with the teeth of Neanderthals suggests, as Frayer (1978) has shown, that there is also a reduction in crown diameters moving from Neanderthal to Aurignacian samples. However, the samples are very small, differences in dimensions are slight, and their interpretation is difficult. Given existing sample variability, a few additional teeth could completely alter the picture. Also it should be borne in mind that the gracilization is a general feature of later human evolution, and therefore the possibility of parallelism in this feature must be considered carefully before drawing phylogenetic conclusions.

Finally, as regards the postcranial skeleton, early Upper Palaeolithic material is comparable in morphology to that of recent humans, although it is generally more robust. However, none of the most distinctive Neanderthal features are present (Trinkaus 1983).

CONCLUDING REMARKS

Considering the data discussed above, it is now possible to turn to the problem of the actual origin of the Upper Palaeolithic peoples, and their relationship to the Neanderthals. At the moment, arguments about these matters centre on two different hypotheses. According to one, the Neanderthals played an insignificant, or even non-existent, role in the emergence of modern humans in Europe. Modern humans had a single origin in Africa (Bräuer 1984, and this volume; Stringer 1985, and this volume) or Southwest Asia (Vandermeersch 1970, and this volume), and then replaced the Neanderthals by 30 000 years ago during a radiation into Europe. The other hypothesis proposes that most of the Neanderthals were ancestral to modern humans (Frayer 1978; Smith 1982, 1985; Wolpoff 1980). Trinkaus (1986) defends an intermediate position: certain Neanderthal groups would have been absorbed by modern populations. In the first case, highly specialized characters of the Neanderthals are said to distinguish them completely from the first modern humans. In the second case, a morphological continuity between Neanderthals and early modern humans in various areas of Europe is accepted although such continuity is less apparent in Western Europe (Smith 1982, 1983, 1984, 1985). Trinkaus (1986) notes that a partial morphological continuity between some Neanderthals (e.g. La Quina) and certain early modern humans (Hahnöfersand, Mladeč and Brno).

For proponents of local continuity, two kinds of data are quoted to justify this model:

1. Early Upper Palaeolithic crania often show Neanderthal-like features,

such as an occipital bun and supraorbital robusticity (Smith 1982, 1985, and this volume; Wolpoff 1980).

2. Aurignacian humans show an overall skeletal robusticity which places them in an intermediate position, morphologically as well as chronologically, between late Neanderthals and samples from the later Upper Palaeolithic. Thus they are part of a process of gracilization which continues during the Upper Palaeolithic (Riquet 1971; Billy 1972; Frayer 1978).

Certainly such parameters can be observed on some of the Upper Palaeolithic specimens I have just discussed, but their phylogenetic significance is questionable. The occipital bun, the incipient alveolar plane and the orientation of the digastric impressions apparent on some Aurignacian mandibles are not specific Neanderthal characters. They can be found on more ancient non-Neanderthal specimens, and are thus archaic (primitive) characters. Features which would indicate a strong genetic continuity between Neanderthals and modern humans are the distinctive derived characters of Neanderthals (Stringer *et al.* 1984).

However, I can recognize neither derived Neanderthal features nor appropriate intermediate morphologies in the Aurignacian sample. The only exception might be the posterior position of the mental foramen on Cro-Magnon mandible 4253. But in this case there is no accompanying retromolar space, as would be found in Neanderthals. Furthermore, the overall architecture of the associated crania is quite different from that of Neanderthals, as is generally true for the whole Aurignacian sample (Vandermeersch, this volume; Stringer, this volume).

Characteristics such as general robusticity, supraorbital morphology and dental dimensions do not appear to show a special relationship between Neanderthals and early modern humans. Such features can just as easily indicate links with samples such as the one from Qafzeh (Israel), where Vandermeersch (1981) has reported many characters which evoke the 'Cro-Magnons'.

Some authors (e.g. Frayer; Smith 1985) have emphasised the limitations of the Upper Palaeolithic sample in Western Europe. It is true that the remains are not necessarily dated to the earliest Upper Palaeolithic, but this also means that they may not be ideal for demonstrating morphological continuity. However, if the dating of the Hahnöfersand frontal to *c.* 36 000 BP (Bräuer and Protsch 1980) could be confirmed, this specimen would provide interesting evidence. However, despite its claimed antiquity, this specimen has no specific Neanderthal features, and can be interpreted as a very robust modern human. Moreover, even if we accept Bräuer's (1981) interpretation that it represents a Neanderthal-early modern hybrid, this specimen would indicate contemporaneity between the two populations, not an evolutionary development from one to the other.

Other data also support this hypothesis of contemporaneity between the two groups in Western Europe. This includes the discovery of a Neanderthal in a Châtelperronian level at Saint-Césaire in Charente (Lévêque and Vandermeersch 1981), as well as the interstratification of the Châtelperronian and the Aurignacian in two French sites – Le Piage and Roc de Combe

(Bordes 1972). The synchroneity of these two cultures is further supported by palynological and sedimentological analyses (Laville 1975; Leroyer and Leroi-Gourhan 1983). In the absence of human remains associated with the earliest Aurignacian, it impossible to confirm that it was invariably modern humans who produced this culture. But the cultural unity of the Aurignacian and the fact that the Aurignacian I is systematically associated in Western Europe with modern humans (as for example, at Les Rois, Isturitz) suggests that the initial stages of the Aurignacian were similarly the product of modern humans.

Finally one further observation argues against a local evolution from the Neanderthals to anatomically-modern humans in Western Europe. This is the lack of morphological evolution within the latest Neanderthals from this area. The skeleton from Saint-Césaire is little different from earlier Neanderthals, and does not exhibit any features suggesting an evolution towards modern humans (Vandermeersch 1984). The reduction trend for Neanderthal teeth, claimed by Frayer (1978), has been questioned by Stringer (1982), on the basis of the composition of the analysed samples (but compare Frayer 1984). It seem to me that the fossils from the French Aurignacian present a robust morphology which, while being indisputably modern, still shows plesiomorphic retentions. They demonstrate that there was a certain degree of variability among these early modern populations, but they exhibit no metrical or morphological character that brings them closer to the Neanderthals, and provide no strong evidence to support the hypothesis of a local evolution from the Neanderthals. In this area, at least, the idea of a rapid replacement of the Neanderthals is the hypothesis which agrees best with available palaeoanthropological and cultural data.

According to Smith and Simek (this volume), population replacement is possible in Western Europe but impossible in Central Europe. Certainly, the immediate source of modern populations would have been Central Europe, and in this case Central European Neanderthals might have a significant place in the origin of modern Europeans. But the claimed link between Central European Neanderthals and Central European modern humans is based on archaic retentions rather than shared derived features. For example, the Mladeč specimens show a modern morphology combined with some archaic characters – such as an occipital bun and a very developed supraorbital region (Frayer 1986: 249, 252). According to Frayer (1986: 242) the female specimens "provide little evidence for continuity between the two samples" but the Mladeč males present more archaic features than the Mladeč females. But neither the females nor the males seem to show the apomorphic conditions described on Neanderthals.

The Mladeč specimens are more robust than those from Cro-Magnon, but their skull architecture does not seem very different. According to Stringer (this volume), "it is not at all Neanderthal-like". The difference between the earliest modern human populations in Central Europe and those in Western Europe, in my view, can be explained as individual variations without phylogenic significance. Under these circumstances, I think that even in Central Europe the palaeoanthropological data supports the hypothesis of population replacement, as in Western Europe, but here there is better evidence for some interbreeding during the replacement phase.

NOTES

1. Amboise and Bouvier (1973) have described a fragment of a human frontal bone from Combe-Capelle, preserved in the National Museum of Prehistory at Les Eyzies. In view of the small size of this fragment and the lack of data concerning its origin, this can hardly provide any new evidence concerning the age of the Combe-Capelle finds, and cannot be used to confirm the antiquity of the original discovery.

2. The fourth skull described by Wolpoff (1980), and by Stringer *et al.* (1984) is a recent skull. It does not come from the Cro- Magnon rock-shelter.

REFERENCES

Ambroise, P. and Bouvier, J. M. 1973. Un fragment crânien humain de Combe-Capelle (Dordogne). *Bulletins et Mémoires de la Société d'Anthropologie de Paris* 10: 413-419.

Asmus, G. 1964. Kritishe Bemerkungen und neue Gesichtspunkte zur jungpalaolithischen Bestatting von Combe-Capelle, Périgord. *Eizeitalter und Gegenwart* 15: 181-186.

Bouvier, J. M., Debénath, A., Delpech, F. and Duport, L. 1969. Les restes humains de la grotte Duport à La Chaise de Vouthon (Charente) dans leur contexte stratigraphique et paléontologique. *Bulletin de la Société d'Anthropologie du Sud Ouest* 5: 32-46.

Bordes, F. 1972. Du Paléolithique moyen au Paléolithique supérieur: continuité ou discontinuité? In F. Bordes (ed.) *Origine de l'Homme Moderne*. Paris: UNESCO: 211-218.

Breuil, H. 1906, Les Cottés. Une grotte du vieil âge du renne à St. Pierre de Maillé (Vienne). *Revue de l'Ecole d'Anthropologie* 16: 47-62.

Billy, G. 1972. L'évolution humaine au Paléolithique supérieur. *Homo* 72: 2-12.

Bräuer, G. 1981. New evidence on the transitional period between Neanderthal and modern man. *Journal of Human Evolution* 10: 467-474.

Bräuer, G. and Protsch, R. 1980. New Upper Pleistocene hominids with neanderthaloid affinities from northern and eastern Germany. *American Journal of Physical Anthropology* 52: 207.

David, P. 1956. Les gisements préhistoriques de la Chaise de Vouthon (Charente). *Congrés Préhistorique de France, XVième Session* (Poitiers-Angoulême): 148-152.

Debénath, A. 1974. *Recherches sur les Terrains Quaternaires Charentais et les Industries qui leur sont Associées*. State Doctoral Thesis, University of Bordeaux 1.

Ducros, A. 1967. Le chignon occipital, mésure sur le squelette. *L'Anthropologie* 90: 89-106.

Duport, L. and Vandermeersch, B. 1976. La mandibule moustérienne de Montgaudier (Montbron, Charente). *Comptes-Rendus de l'Académie des Sciences de Paris* 283: 1161-1164.

Frayer, D. W. 1978. *Evolution of the Dentition in Upper Paleolithic and Mesolithic Europe*. University of Kansas, Publications in Anthropology 10.

Frayer, D. W. 1984. Biological and cultural change in the European late Pleistocene and early Holocene. In F. H. Smith and F. Spencer (eds)

The Mousterian Legacy: Human Biocultural Change in the Upper Pleistocene. Oxford: British Archaeological Reports International Series S164: 211-250.

Frayer, D. W. 1986. Cranial variation at Mladeč and the relationship between Mousterian and Upper Paleolithic Hominids. In *Fossil Man: New Facts, New Ideas. Anthropos* (Brno) 23: 243-256.

Geay, P. 1957. Sur la découverte d'un squelette aurignacien? en Charente-Maritime. *Bulletin de la Société Préhistorique Française* 54: 193-197.

Genet Varcin, E. 1979. *Les Hommes Fossiles*. Paris: Boubée.

Héléna, P. 1926-1927. La stratigraphie de la grotte de La Crouzade (Commune de Gruissan, Aude). *Bulletin de la Commission Archéologique de Narbonne* 17: 49-94.

Henri-Martin, G. 1936. Nouvelles constatations dans une station aurignacienne de La Quina (Charente). *Bulletin de la Société Préhistorique Française* 31: 177-203.

Howells, W. W. 1974. Neanderthals: names, hypotheses and scientific methods. *American Anthropologist* 76: 24-38.

Hublin, J.-J. 1978. *Le Torus Occipital Transverse et les Structures Associées: Evolution dans le Genre Homo*. Unpublished Thesis, Université Pierre et Marie Curie, Paris.

Klaatsch, H. and Hauser, O. 1910. *Homo Aurignacensis Hauseri. Prähistorische Zeitschrift* 1: 273-338.

Lartet, L. 1968. Une sépulture des troglodytes du Périgord (Crânes des Eyzies). *Bulletin de la Société d'Anthropologie de Paris* 3: 335-349.

Laville, H. 1975. *Climatologie et Chronologie du Paléolithique en Périgord*. Laboratoire de Paléontologie Humain et de Préhistoire, Université de Provençe, Mémoire 4.

Leroyer, C. and Leroi-Gourhan, A. 1984. Problèmes et chronologie: le Castelperronien et l'Aurignacien. *Bulletin de la Société Préhistorique Française* 80: 41-44.

Martin, R. and Saller, K. 1956-1966. *Lehrbur der Anthropologie* 4. Stuttgart: Fischer.

Mouton, P. and Joffroy, R. 1958. *Le Gisement Aurignacien des Rois à Mouthiers (Charente)*. 9th Supplement to *Gallia Préhistoire*.

Movius, H. L. 1969. The Abri of Cro-Magnon, Les Eyzies (Dordogne) and the probable age of the contained burials on the basis of the nearby Abri Pataud. *Anuario de Estudio Atlanticos* 15: 323-344.

Oakley, K. P., Campbell, B. G. and Molleson, T. I. 1971. *Catalogue of Fossil Hominids*. Vol. 2: *Europe*. London: British Museum (Natural History).

Patte, E. 1954 Le crâne aurignacien des Cottés. *L'Anthropologie* 58: 470-471; 59: 39-61.

Perpère, M. 1971. *L'aurignacien en Poitou-Charentes (Etude des Collections d'Industries Lithiques)*. Unpublished Doctoral Thesis, University of Paris.

Perpère, M. 1973. Les grands gisements aurignaciens du Poitou. *L'Anthropologie* 77: 683-716.

Peyrony, D. 1943. Le gisement du roc de Combe-Capelle. *Bulletin de la Société Historique et Archéologique du Périgord:* 158-173.

Pradel, L. 1961. La grotte des Cottés, commune de St. Pierre de Maillé (Vienne). *L'Anthropologie* 65: 229-258.

Riquet, R. 1970. La race de Cro-Magnon: abus de langage ou réalité objective? In G. Camps and G. Olivier (eds) *L'Homme de Cro-Magnon: Anthropologie et Archéologie*. Centre de Recherches Anthropologiques, Préhistoriques et Ethnographiques. Art et Métiers Graphiques: Paris: 37-57.

Rochebrune, R. de. 1881. *Les Troglodytes de la Gartempe, Fouille de la Grotte des Cottés*. Fontenay le Comte: Imprimerie Caurit.

Rochebrune, R. de. 1881. Seconde fouille de la grotte des Cottés. *Matériaux pour l'Histoire Primitive et Naturelle de l'Homme* 16: 487-489.

Rochebrune, R. de. 1883. La grotte des Cottés. *Bulletins et Mémoires de la Société d'Anthropologie de Paris* 6: 423-426.

Sacchi, D. 1973. Les civilisations du würmien récent dans le Narbonnais. Paper presented to the XLVième Congrès de la Fédération Historique du Languedoc mediterranéen et du Rousillon. *Narbonne Archéologie et Histoire:* 2-28.

Smith, F. H. 1982. Upper Pleistocene hominid evolution in South Central Europe: a review of the evidence and analysis of trends. *Current Anthropology* 23: 667-703.

Smith, F. H. 1985. Continuity and change in the origin of modern *Homo sapiens. Zeitschrift für Morphologie und Anthropologie* 75: 197-222.

Sonneville-Bordes, D. de. 1959. Position-stratigraphique et chronologique relative des restes humains du Paléolithique supérieur entre Loire et Pyrénées. *Annales de Paléontologie* 45: 19-51.

Sonneville-Bordes, D. de. 1960. *Le Paléolithique Supérieur en Périgord*. Bordeaux: Delmas.

Stringer, C. B. 1974. Population relationships of later Pleistocene hominids: a multivariate study of available crania. *Journal of Archaeological Science* 1: 317-342.

Stringer, C. B. 1982. Toward a solution to the Neanderthal Problem. *Journal of Human Evolution* 11: 431-438.

Stringer, C. B., Hublin, J.-J., and Vandermeersch, B. 1984. The origin of anatomically modern human in Western Europe. In F. H. Smith and F. Spencer (eds) *The Origins of Modern Humans: a World Survey of the Fossil Evidence*. New York: Alan R. Liss: 51-135.

Thoma, A. 1972. L'origine des Cro-Magnoides. In Fondation Singer-Polignac (ed.) *Les Origines Humaines et les Epoques de l'Intelligence*. Paris: Masson: 261-271.

Tillier, A.-M. 1977. La pneumatisation du massif cranio-facial chez les hommes actuels et fossiles. *Bulletins et Mémoires de la Société d'Anthropologie de Paris* 4: 287-316.

Tillier, A.-M. 1983. L'enfant néandertalien du Roc de Marsal (Campagne du Bugue, Dordogne). Le squelette facial. *Annales de Paléontologie* 69: 137-149.

Tillier, A.-M. 1984. L'enfant Homo 11 de Qafzeh (Israël) et son apport à la compréhension des modalités de la croissance des squelettes moustériens. *Paléorient* 10: 7-47.

Tillier, A.-M. and Genet Varcin, E. 1980. La plus ancienne mandibule d'enfant découverte en France dans le gisement de la Chaise de Vouthon (Abri Suard) en Charente. *Zeitschrift für Morphologie und Anthropologie* 71: 196-214.

Trinkaus, E. 1983. Neandertal postcrania and the adaptative shift to modern humans. In E. Trinkaus (ed.) *The Mousterian Legacy: Human Biocultural Change in the Upper Pleistocene*. Oxford: British Archaeological Reports International Series S164: 165-200.

Trinkaus, E. 1986. Les Néandertaliens. *La Recherche* 17: 1040-1047.

Twiesselmann, F. 1941. Méthode pour l'évaluation de l'épaisseur des parois craniennes. *Bulletin du Musée Royal d'Histoire Naturelle de Belgique* 17: 2-33.

Vallois, H. V. 1952. Les restes humains. In R. de Saint Périer and S. de Saint Périer (eds) *La Grotte d'Isturitz*. Vol. 3: *Les Solutréens, les Aurignaciens, et les Moustériens*. Paris: Archives de l'Institut de Paléontologie Humaine 25.

Vallois, H. V. 1957. Nouvelles découvertes d'hommes fossiles. *L'Anthropologie* 61: 154.

Vallois, H. V. 1958. Les restes humains de la grotte des Rois. In P. Mouton and R. Joffroy (eds) *Le Gisement Aurignacien des Rois à Mouthiers (Charente)*. 9th Supplement to *Gallia Préhistoire:* 118-137.

Vallois, H. V. and Billy, G. 1965. Nouvelles recherches sur les hommes fossiles de l'abri de Cro-Magnon. *L'Anthropologie* 69: 47-74; 249-272.

Vandermeersch, B. 1981. *Les Hommes Fossiles de Qafzeh (Israël)*. Paris: Centre National de la Recherche Scientifique.

Vandermeersch, B. 1984. A propos de la découverte du squelette neandertalien de St. Césaire. *Bulletins et Mémoires de la Société d'Anthropologie de Paris* 35: 191-196.

Vandermeersch, B. and Lévèque, F. 1980. Découvertes de restes humains dans un horizon castelperronien à St. Césaire (Charente-Maritime). *Comptes-Rendus de l'Académie des Sciences de Paris (Série D)* 291: 187-189.

Vlček, E. 1967. Relation morphologique des types humains fossiles de Brno et Cro-magnon au Pléistocene supérieur d'Europe. *Folia Morphologica* 15: 214-221.

Wolpoff, M. H. 1980. *Paleoanthropology* New York: Knopf.

13. The Demographic Modelling of Neanderthal Extinction

EZRA ZUBROW

INTRODUCTION

The evolutionary fate of the 'Neanderthal' populations of Europe and the Middle East has been a topic of debate for over 80 years. Although current opinions remain sharply divided on certain aspects of this question, there is an increasing consensus that in at least certain areas of Europe and Western Asia, Neanderthal populations not only came into direct contact with intrusive populations of *Homo sapiens sapiens*, but were eventually replaced by the new populations, with few if any indications of interbreeding. Clearly, these questions raise interesting issues of the nature of demographic relationships between the Neanderthal and Sapiens populations, which are of central concern to the theme of the present Symposium.

For several years, I have been interested in modelling demographically the extinction of Neanderthals. Allow me to assume that there were contemporaneous interacting populations of *Homo sapiens sapiens* and *Homo sapiens neanderthalensis*. If so, I believe that only a small demographic advantage is necessary for the modern forms to grow rapidly, and for the archaic forms to become extinct. This superiority may be as paltry as a one per cent difference in mortality, and the extinction may be as rapid as 30 generations. In other words, Neanderthals could have become extinct in a single millennium.

I wish to make a disclaimer. Unlike many of the other papers in this volume, this study is not based upon substantive evidence. Instead, it is based upon theory, logic, and admitted speculation. Its merit is that it restricts the range of possibilities, which is a sufficient rationale in itself. First, certain generally assumed scenarios of *Homo sapiens sapiens* and *Homo sapiens neanderthalensis* are impossible. Conversely, what was thought impossible is indeed credible and perhaps even probable. Second, by analysing the range of outcomes for differing input conditions, it is possible to obtain a more precise impression of both the type and the scale of possible causes. The realisation that small differences in competitive advantage can be transformed into large changes in mortality and life expectancy may provide insights into how one conducts substantive field research. Third, the chain between cause and eventual effect may be long and very complicated. Often the results run counter to intuitive expectations. For example, increased mortality in certain age groups may actually *decrease* the probability of extinction for the Neanderthal population. Fourth, there are heuristic values in attempting to formalize the relationship among late Middle Palaeolithic and early Upper Palaeolithic populations. It compels one to think about aspects of Palaeolithic life in a manner that one otherwise would not.

ARCHAEOLOGICAL BACKGROUND

The substantive archaeological background to Neanderthal existence and extinction is a complicated picture made up of brushstrokes from human palaeontology, prehistory, archaeological interpretation and climatology. An impressionistic portrait of Neanderthal adaptation may be painted. The broad outlines are clear but the minutiae are difficult to discern. The details which need to be interpreted are based on a very incomplete record.

There are more than 100 sites containing Neanderthal material that have been excavated since the first skeleton was found near Düsseldorf in 1856. It appears that the Neanderthals subsisted as hunters and gatherers with a primary emphasis probably on the hunting or scavenging of large game. There were small populations living throughout the unglaciated regions of Western, Central and Eastern Europe (Allsworth-Jones 1986; Svoboda 1986; Oliva 1986) and extending into the Middle East (Jelinek 1982; Vandermeersch 1985). It was not only an extensive adaptive radiation but a successful one. Extending from the time of the Last Interglacial approximately 125 000 years ago to *c.* 35 000 years ago, when they became extinct, the Neanderthals triumphantly survived major climatic change, predation, and competition. Testimonials to their presence may be found in a variety of environments and ecological zones. Evidence of their occupations may be unearthed in open-air, rock-shelter, and cave sites. They made a wide variety of stone tools using Mousterian techniques. Whether these cultures are called Mousterian or Micoquian, there was remarkable consistency in both form and function of their tools. This undisguised sophistication in technology was extended to other arenas. For example, they were sufficiently advanced as to bury their dead, suggesting a widespread awareness of the significance of death for both the individual and society. Use of colouring materials (red ochre and black manganese dioxide) similarly attests at least an incipient capacity for abstract, symbolic thought.

Early *Homo sapiens sapiens* populations are associated exclusively with the Upper Palaeolithic period in Europe. Earliest occurrences date from approximately the same time as the extinction of the Neanderthals. There are extensive debates concerning the possible contemporaneity of the two sub-species. I will return to these in a moment. From the Middle-Upper Palaeolithic transition onwards there was an increasing variety in the form and function of tool kits. When amassed at an aggregated level, these lithic assemblages forming the Aurignacian, Périgordian, Solutrean and Magdalenian phases attest to increasing diversity and more rapidly occurring differentiation of culture.

As hunters and gatherers, *Homo sapiens sapiens* populations adapted to climates of the last glaciation and then expanded rapidly in the warming that followed. Analogies have been drawn to ethnographic bands of hunters and gatherers on the margins of modern society. The inference is life in small bands of perhaps 25 to 50 individuals. Sometimes these groups may have combined into larger 'regional bands' or 'tribes' of perhaps several hundred individuals. These larger groups were not very stable and would dissolve back into extended family units. In the broad perspective, they were not limited by resources. However, at specific times and at specific places, the Malthusian shoe may have fit very tightly indeed. In any case, there was sufficient labour efficiency and surplus resources that effort could be allocated to the development of art and ceremonial (Conkey 1983). In short, the evidence points to the rates of change being very different from that of the Neanderthals. If the adaptive radiation of Neanderthals was a leisurely stroll through Europe, then the adaptive radiation of *Homo sapiens sapiens* was at a jogger's pace which rapidly became an all-out sprint after the Neolithic revolution.

There are several views of the transition from the Middle to the Upper Palaeolithic and from *Homo sapiens neanderthalensis* to *Homo sapiens sapiens*. They can be paired into two groups. One is the 'replacement/acculturation hypothesis', the other is a 'multi-regional evolution' process. Smith (1985) has suggested that nowhere in Western Europe is there clear synchrony of archaic and modern humans. By synchrony I mean both temporal and geographic overlap. Smith's view as restated by Allsworth-Jones (1986) is that Neanderthals may have lived as late as 38 000 years ago in Central Europe and sometime about 33-31 000 years ago in Western Europe. Earliest *Homo sapiens sapiens* appear in Central Europe about 36-34 000 years ago and in Western Europe less than 30 000 years ago. The gap in time between the two forms of *Homo sapiens* is approximately 2000 years for both Central and Western Europe.

Recent evidence supports a new synchronism. In this context the discovery of a typical Neanderthal associated with a Chatelperronian industry at Saint-Césaire in western France, and dated to *c.* 35-34 000 years ago is crucial. It provides a temporal overlap with the Aurignacian and (by implication) with the earliest populations of *Homo sapiens sapiens* in western France (see papers by Harrold and Gambier, this volume). There is actual interstratification of Chatelperronian and Aurignacian at Le Piage and Roc de Combe. Allsworth-Jones (1986) suggests a similar picture for Central Europe, based on new information from Das Geissenklosterle and Boch-Kine.

Given the above, I will assume in this paper that there was both temporal and geographic synchrony for these two sub-species. I will set my range inclusively. Thus, I will suggest when modelling that *Homo sapiens sapiens* and *Homo sapiens neanderthalensis* have overlapped in both Western Europe and Central Europe for a period ranging from 10 000 years to 1000 years.

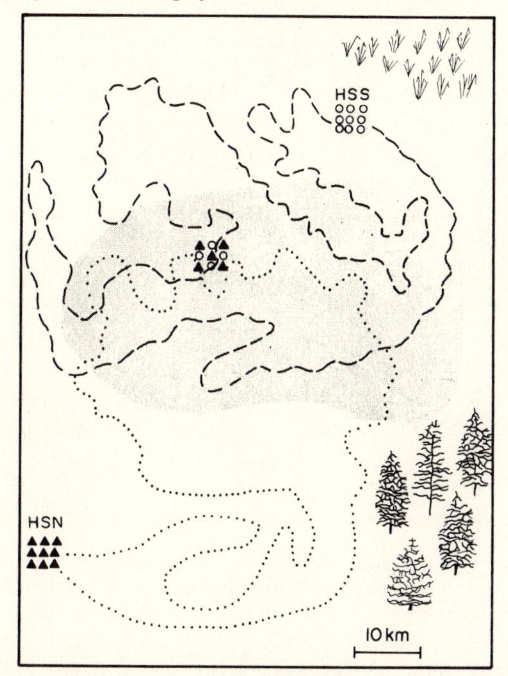

Figure 13.1. The scenario for three models: bands of modern humans and Neanderthals meet occasionally while pursuing foraging strategies in a Pleistocene landscape.

DEVELOPMENT OF DEMOGRAPHIC MODELS

It may be useful to provide a brief background to my previous attempts to model the Neanderthal and *Homo sapiens sapiens* interaction. Afterwards, I will discuss the nature of stable populations which is the basis of my new model. My first model examined entire populations independent of age structure or geographic location. The second model examined these populations from the perspective of the spatial location of the sites (Zubrow 1986). My new model considers the populations from the perspective of age structure and relative fertility and mortality rates.

All three models are based upon the same scenario. Imagine two bands of hominids moving through a Pleistocene landscape following their respective game animals. The sun rises and falls on their respective camps. As the seasons pass, each traverses a route through their territory. These routes are established by the schedule of harvesting wild plants, game routes, predators, and the location of water. They are also determined by a variety of imponderables – volition, religion, idiosyncratic personality and simple chance.

These populations are not static. They grow and decline; they break up and reaggregate. This depends on many factors – the local environment, the skill of the subsistence gatherers, disease and the fertility of the child bearers.

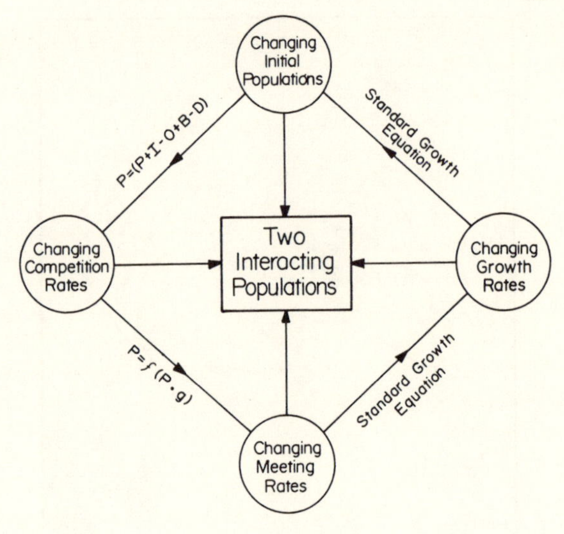

Figure 13.2. The initial parameters for a model of interactive growth in which the growth equations operate on entire populations. (For explanation of the growth equations, see text.)

These roving bands inhabit areas of very low density, so low, it is almost inconceivable to the modern urban dweller. The idea of walking for two weeks and never seeing another individual is true solitude. If you did, it would be a member of your immediate household or local band. However, even these small populations are not completely isolated. Occasionally, one of these populations meets another (Fig. 13.1). When this happens a complex set of interactions takes place. There may be (a) immediate withdrawal; (b) competition for resources; (c) warfare; and (d) trade and exchange. My models address all but the first alternative. Each alternative is a type of interaction and thus I call my models 'models of interactive growth'.

In the paper entitled *An Interactive Growth Model Applied to the Expansion of Upper Paleolithic Populations* (Zubrow 1986), I reported on the first two models of population interaction which simulated the growth of *Homo sapiens neanderthalensis* and *Homo sapiens sapiens* based upon standard growth equations which operated on the entire populations. The growth of each population was partially dependent on the initial size and growth rate of each population. These were inputs into the model. Within the confines of this model it was possible for the two populations to grow independently of each other (the multi-centre hypothesis). On the other hand, it was also possible that one population was dependent upon the resources of the other (the partially competitive hypothesis) or that both were dependent on each others' resources (the fully competitive hypothesis). The model allowed one to enter the initial probability that members of one population met another, and the initial replacement rate. Replacement might be complete, partial, or non-existent. In other words, when members of each population met, they might totally or partially replace the ones who did not succeed in acquiring resources. Alternatively, there may be withdrawal and no replacement at all (Fig. 13.2).

This model was initially run some 50 times. In more than 60 per cent of the cases one of the two populations (almost always *Homo sapiens sapiens*) survived more than 100 generations. However, no matter how one varied the parameters it was almost impossible for the Neanderthal population to survive. The sizes of the initial populations were varied with *Homo sapiens neanderthalensis* usually ten or more times the size of *Homo sapiens sapiens*. This was the conservative assumption for the Neanderthals, since they have the significant advantage of numbers for survival.

A second set of 250 runs showed that under many different demographic and interactive variations the continuation of Neanderthals was impossible. The demographic window which could have made survival possible was quite improbable. It required unreasonably low sizes and growth rates for the populations of *Homo sapiens sapiens*. Even this was insufficient. Survival would have required very low replacement and interaction rates. This model showed that in general, Neanderthal continuation was more prolonged in competitive situations where both populations were small. Neanderthal survival was actually enhanced if both populations were extremely small initially – i.e. below 20. As soon as either population was larger, the probability of *Homo sapiens neanderthalensis* extinction increased dramatically. It would appear that the one advantage the modern forms had was the ability to more rapidly reach a form of 'stable' growth.

The second model was based upon competition in a regional context. The locations in the matrix corresponded to the location of the sites. The rules were: (1) a site with no neighbour dies of isolation; (2) a site with 1-3 neighbours survives; (3) a site with 4 or more neighbours dies from competition; and (4) 3 neighbours surrounding an empty location creates growth of a new site by budding (Fig. 13.3). I simulated growth for Neanderthal and Cro Magnon sites separately and together. In both groups peripheral sites were rapidly stripped off. However, after a few generations the sites containing *Homo sapiens sapiens* rapidly divided into spatially separate and stable groups. Extinction followed the lack of stability for the sites occupied by *Homo sapiens neanderthalensis*. Frequently, the cause of extinction was the lack of neighbouring sites. Thus in both of these models, the extinction of *Homo sapiens neanderthalensis* was probable.

My new model is based upon stable population models. In order to understand stable population models, it is necessary to briefly consider the life table. Life tables are the combinations of mortality rates for the different ages and sexes of a population into a single schedule. They may be viewed from two different perspectives. One is the 'photographic' or 'population' approach in which the life table represents the mortality experience of an entire population across a very short period of time such as a year. It is a snapshot of death. The other is the 'processual' or 'generational' viewpoint. In this table one follows the mortality record of a single age-sex cohort throughout their existence, from the time the first member is born to the time the last dies. It is not a photograph but the trail of death.

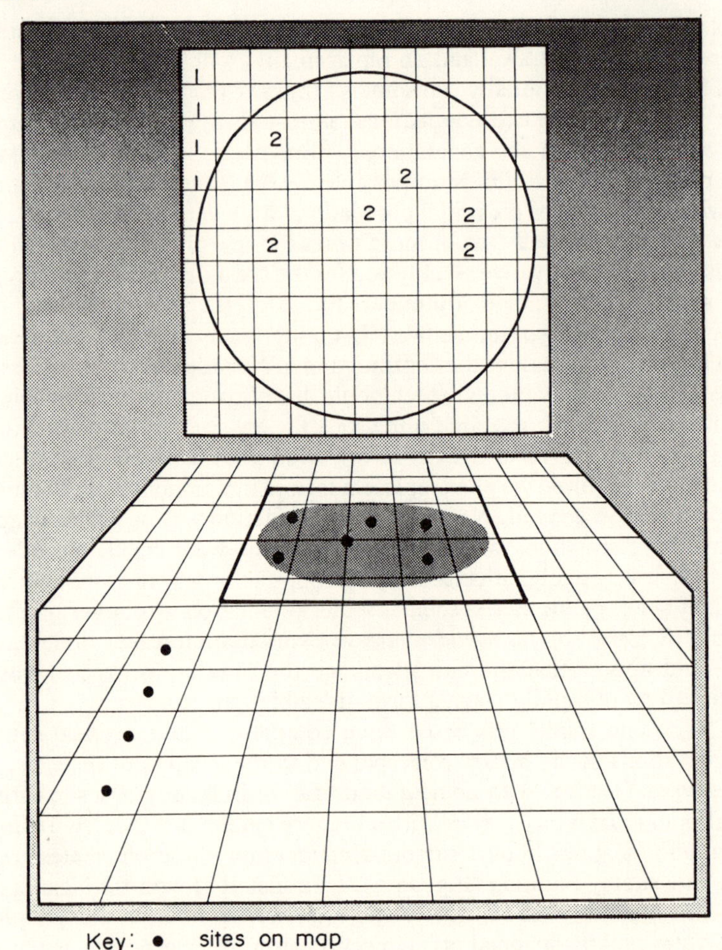

Figure 13.3. A geographic model based upon competition in a regional context, in which the site locations are positions in a matrix.

A life table consists of seven functions easily calculated and interpreted. They are:

$x \rightarrow x+n$ This is the age cohort or the time interval – i.e. $20 \rightarrow 20+5$ means the individuals who are between the ages of 20 and 25. The next cohort is similarly incremented by 5. It is thus $25 \rightarrow 25+5$ or the individuals between 25 and 30.

l_x This figure is the number of people alive at the beginning of a particular age interval, x.

${}_nq_x$ This value is the rate of death for the people alive at the beginning of the cohort who will die during the next n years. For example

Table 13.1 Stable Life Table

Table Type: south 7 general stable model
sex: female

Age-Sex Cohorts $x \rightarrow x+n$	% Dying $x \rightarrow x+n$ ${}_nq_x$	No. Living at x l_x	No. Dying $x \rightarrow x+n$ ${}_nd_x$	No. in $x \rightarrow x+n$ ${}_nL_x$	No. in this X to 85+ T_x	Life Expectancy $e^o x$
0-1	0.1990	100000.00	19900.00	91045.00	849742.53	8.50
1-5	0.3640	80100.00	29156.40	278123.22	758697.53	9.47
5-10	0.3950	50943.60	20122.72	205417.33	480574.31	9.43
10-15	0.4110	30820.88	12667.38	123069.31	275156.97	8.93
15-20	0.4330	18153.50	7860.46	71509.35	152087.67	8.38
20-25	0.4600	10293.03	4734.80	39864.92	80578.32	7.83
25-30	0.4870	5558.24	2706.86	21159.38	40713.40	7.32
30-35	0.5140	2851.38	1465.61	10666.14	19554.03	6.86
35-40	0.5410	1385.77	749.70	5092.08	8887.88	6.41
40-45	0.5680	636.07	361.29	2295.19	3795.81	5.97
45-50	0.5950	274.78	163.49	973.34	1500.62	5.46
50-55	0.6710	111.29	74.67	373.48	527.28	4.74
55-60	0.7300	36.61	26.73	117.58	153.79	4.20
60-65	0.8020	9.89	7.93	30.00	36.21	3.66
65-70	0.8820	1.96	1.73	5.56	6.21	3.17
70-75	0.9490	0.23	0.22	0.62	0.65	2.81
75-80	0.9840	0.01	0.01	0.03	0.03	2.66
80 and +	0.9850	0.00	0.00	0.00	0.00	4.48

if $_nq_x$ is 0.088 for the cohort 20-25 which has 1000 people in it at the beginning of the cohort, 88 will die during the next five years.

$_nd_x$ is the number of persons who will die within the age cohort or the time interval($x \rightarrow x+n$). In other words, the 88 individuals mentioned above.

$_nL_x$ refers to the number of person-years lived during the time interval by those persons alive during the age cohort or time interval ($x \rightarrow x+n$).

T_x This is the total number of person-years lived from the beginning of the indicated time interval x until all members of the cohort are dead.

$e^o x$ This figure is the life expectancy or the average remaining life span for a person who survives to x.

An example of an abbreviated life table is given in Table 13.1. It is based upon data which will be relevant to our later discussions. One may interpret a life table as a stationary population. This term refers to a population whose total number and whose age and sex distribution do not change with time. Of course, such a population is hypothetical. If the fertility and mortality rates remain constant over a long period of time, births would equal deaths and the distribution of age and sex would remain constant.

In 1907 Lotka developed the concept of a stable population. He assumed no migration. Given age-specific fertility and mortality rates applied for an indefinite period of time, he demonstrated that the age composition of the population would assume a fixed distribution. Later, Dublin and Lotka (1936) proved that a closed population with fixed age-specific mortality and fertility rates would eventually have a constant rate of natural increase which they labelled the 'true rate of increase'. This line of research was continued by Coale (1968) who began to investigate the length of the period that differing age, mortality, and fertility structures would take to reach stability. There is a surprising range in the periods. From the context of this demographic perspective, a life table is a stationary population as well as a stable population with a natural increase of zero.

These tables have been grouped according to their characteristic mortality and fertility rates by demographers into four different regions – 'West', 'North', 'East', and 'South'. The groups are partially based on the actual demographic attributes of such populations in various parts of the world. Thus, the lowest fertility and mortality rates are represented in the Western group and the highest in the Southern tables. Within each group the tables are ranked according to fertility and mortality from one to twenty. Table twenty has the lowest fertility and mortality rates; table one the highest. With this short introduction we can turn to my most recent model.

THE NEW MODEL

My new model consists of linked stable and stationary population models. The easiest path to understanding is to take the reader through the creation process. This procedure consisted of three steps. The first step was to create

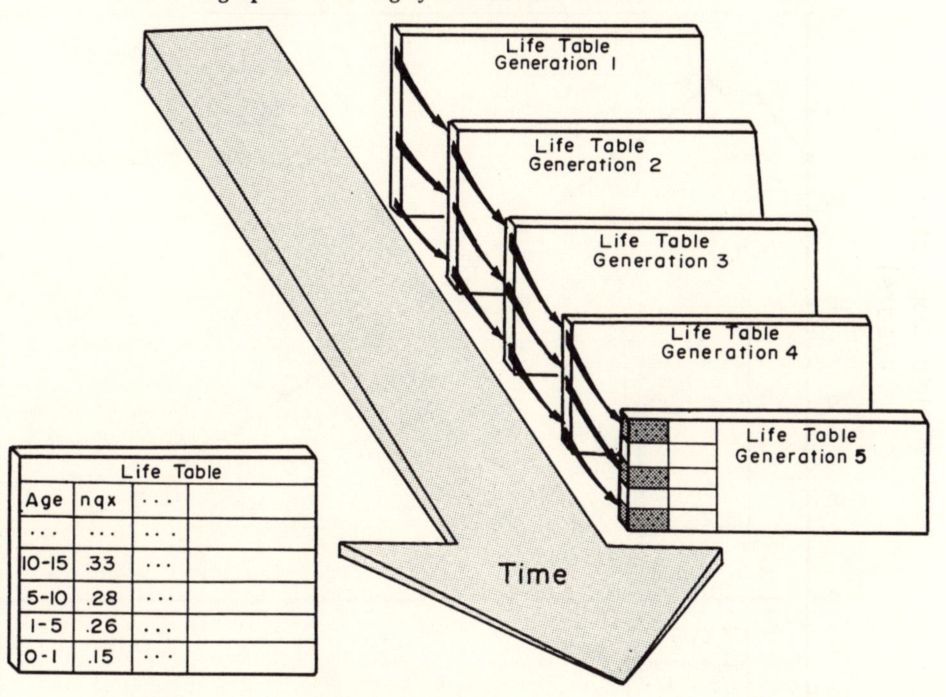

Figure 13.4. The new model in very simplified form. In this version, multi-generational life tables are linked through mortality for each cohort across time. The arrows show the linking from generation to generation of the age-specific mortality. In other words, the cohort-specific mortality at one point of time (the n+1 generation) is not only a standard life table function but is dependent upon the cohort-specific mortality at another point in time (the nth generation).

a *linked* set of multi-generational life tables (Fig. 13.4). As in the case of all populations, as time passed mortality was applied to each cohort and the ageing cohort was placed in the next life table. In other words, one generational life table outputted its newly changed mortality and population structure to the next life table. These were tested for their stationary and stable characteristics. If the population was stationary and/or stable, then the age structure and life expectancy remained the same. For example, Figures 13.5a and 13.5b show simulated life expectancies by cohort. The life expectancies for the zero, third, sixth, ninth, twelfth, and fifteenth generations are computed and shown. They are the result of the linked multi-generational life tables for two stable models – West 11 and South 11 respectively. The stability is confirmed across the generations. The life expectancies vary by cohort. However, the life expectancies for the same cohort from each generation are exactly equal. Thus, the observations and their consequent connecting lines overlie each other. Although both are stable, please note the great difference in life expectancy for each cohort. For comparative purposes one might note that Trinkaus (1986) has

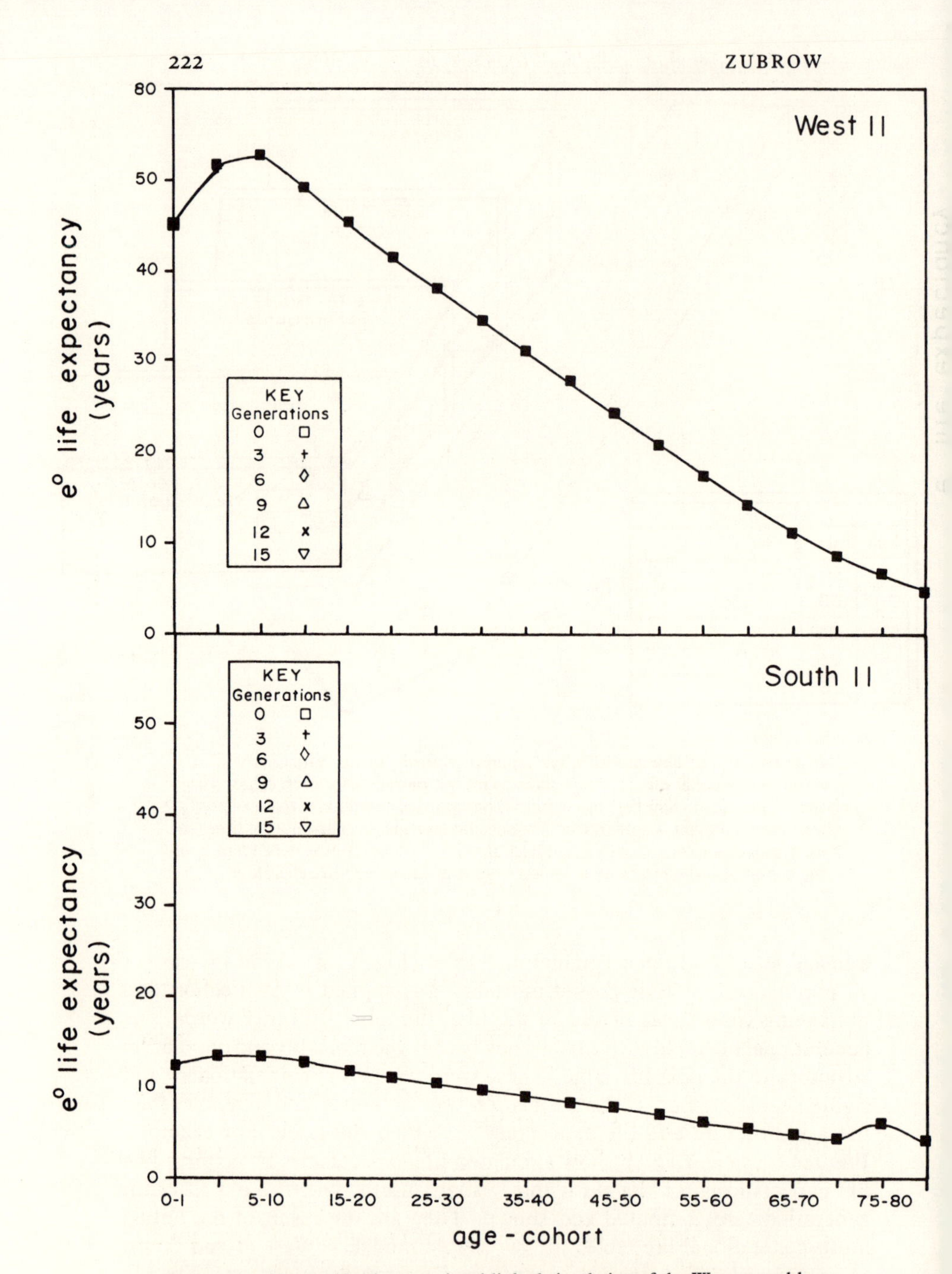

Figure 13.5. Upper: Multi-generational linked simulation of the West II stable population model. The stability is demonstrated by each generation having the same cohort-specific life expectancies. Thus, each observation and the consequent lines overlay each other. The other dependent rates are also the same. Lower: Multi-generational linked simulation of the South II stable population model. The stability is demonstrated by the same as the above. However, note the significant differences in life expectancy by cohort.

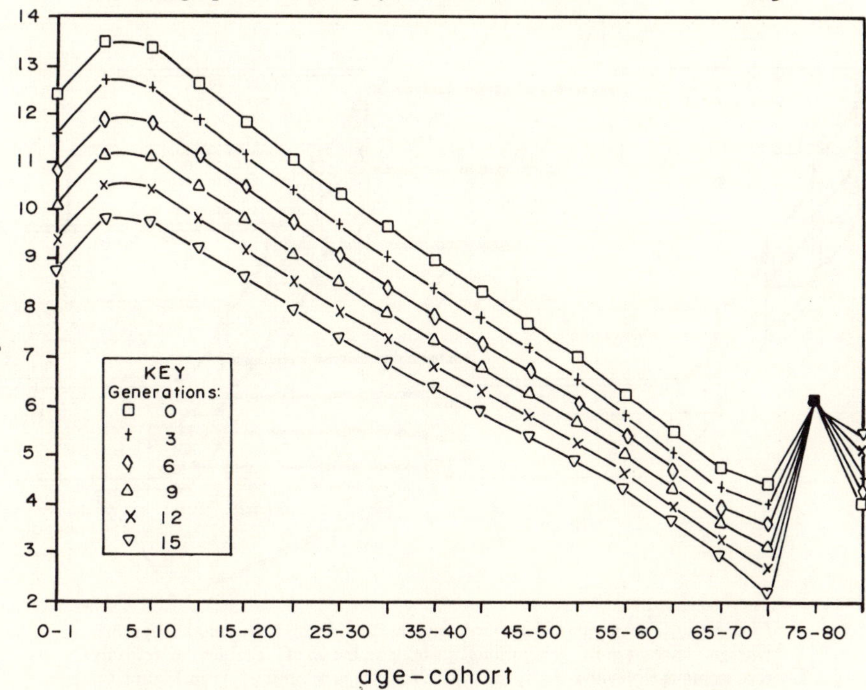

Figure 13.6. Life expectancy from multi-generational linked life tables. The initial mortality for the first generation of early hominids was set at the values for stable population model South 11. Mortality was allowed to increase by 2 percent to k = 1.02. The initial generation is indicated by square boxes, the third by crosses, the sixth by diamonds, the ninth by triangles, the twelfth by x's, and the fifteenth by inverted triangles.

suggested that Neanderthal life spans might reach the late thirties or early forties.

Having demonstrated their stability, I decided to test what changes in life expectancy would occur if there were changes in the mortality pattern. Initially, I did not care what caused the changing mortality. Any external source such as disease, chance mortality, or competition for food would be more than adequate. Figure 13.6 shows an example. There is an increase in mortality of 2 per cent which is applied equally to all cohorts in the first generation. As one would expect, life expectancy for the next resulting generation diminishes. This process continues generation by generation. The life expectancies of each subsequent generation are less than the one preceding it. The figure shows the initial generation, the third, sixth, ninth, twelfth and fifteenth. Although the initial increase in mortality is applied equally, the resulting life expectancy diminishes unequally. It has different consequences cohort by cohort and generation by generation. The increases in mortality have a greater effect on the immediately subsequent generations and on the younger cohorts. The one exception to the above generalization are infants.

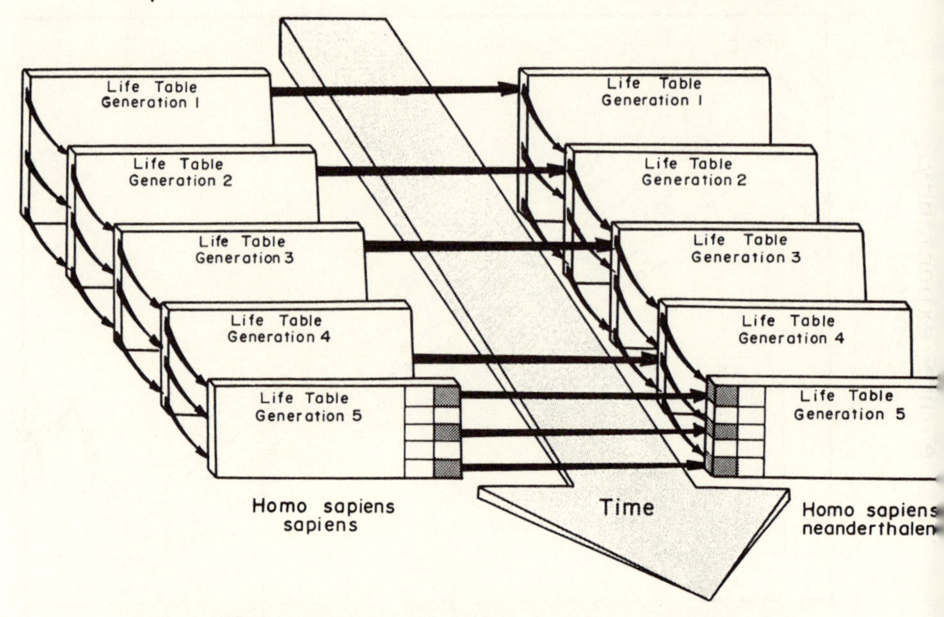

Figure 13.7. The new model in a more sophisticated form. In this model there are two sets of linked multi-generational life tables. One set of life tables theoretically represents the Neanderthals, the other *Homo sapiens sapiens*. As in Figure 13.4, the cohort-specific mortality for any generation is dependent upon the cohort-specific mortality at previous times. The important new addition is that *Homo sapiens sapiens* age-specific mortality affects the Neanderthals' cohort mortality but not *vice versa*. This is shown by the arrows which connect the *Homo sapiens sapiens* life tables to the Neanderthal life tables.

The second step was to model one-way or unidirectional interaction. This is a type of limited interaction between the two populations. This was done by setting two initial populations side by side and creating for each their multi-generational linked life tables. Figure 13.7 shows an overview of this process. The mortality rates of the independent population, *Homo sapiens sapiens,* affect the rates of the dependent population, *Homo sapiens neanderthalensis,* but not *vice-versa*.

One must initialize each population. I will use one of the stable models which I believe is reasonable. I might assume that demographically there was an inherent superiority about *Homo sapiens sapiens*. For example, they may have had a lower mortality rate due to greater resistance to disease. If this was so I would initialize the *Homo sapiens sapiens* population with a stable South 7 population, while *Homo sapiens neanderthalensis* had a stable South 3 population. This would give the demographic advantage to the modern forms. They would have slightly lower fertility and mortality as well as considerably longer life expectancy.

On the other hand one might assume that the two populations had almost the same demographic characteristics. In this case, one would initialize both populations with the same stable model. Both types of analyses have been completed. In order to conserve space I will only present the more

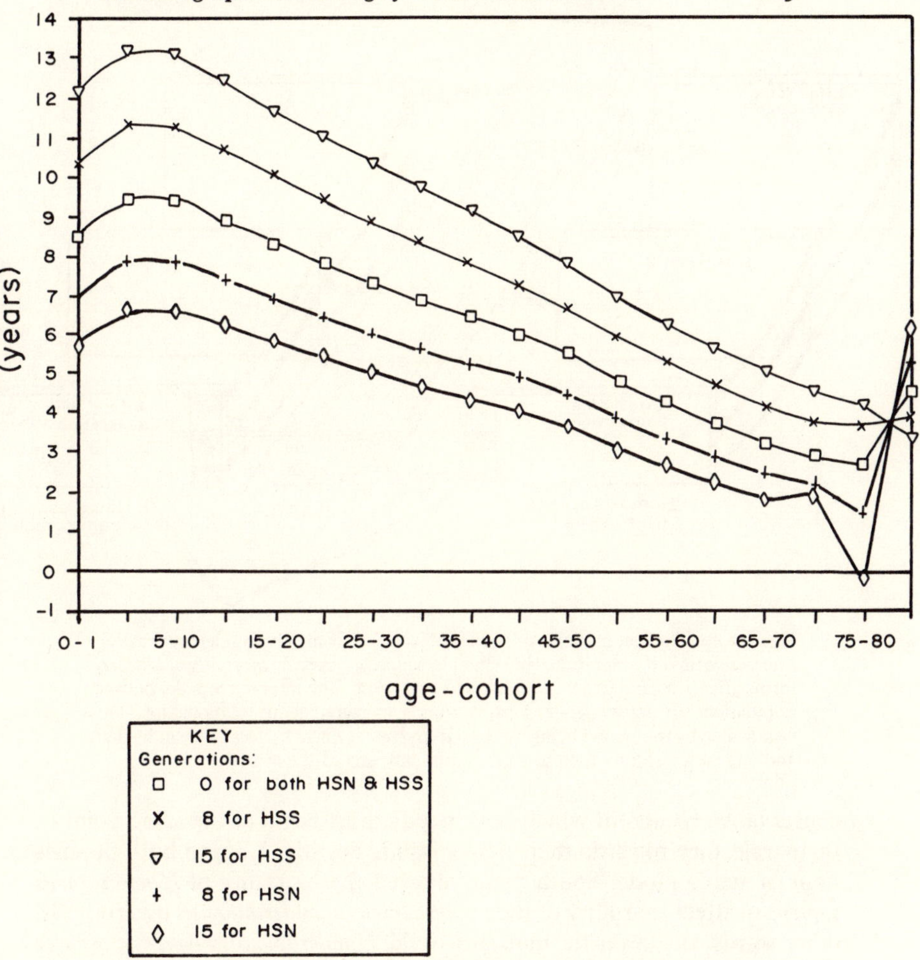

Figure 13.8. Limited interactive growth model of *Homo sapiens sapiens* and *Homo sapiens neanderthalensis* using interactive multi-generational linked life tables. Both populations begin with the same initial configuration, stable model 7. The initial generation for both forms of humans is indicated by boxes. With a small demographic advantage of less than two percent functional interrelationship in mortality, the modern human forms have improved their life expectancies by approximately two years in eight generations. This is indicated by the line of x's. The Neanderthals suffered a decreased life expectancy by approximately one year in the same eight generations, indicated by the crosses. By the fifteenth generation, life expectancy for modern human forms has increased by four years (inverted triangles) while Neanderthals have suffered another year decline in their life expectancy (diamonds).

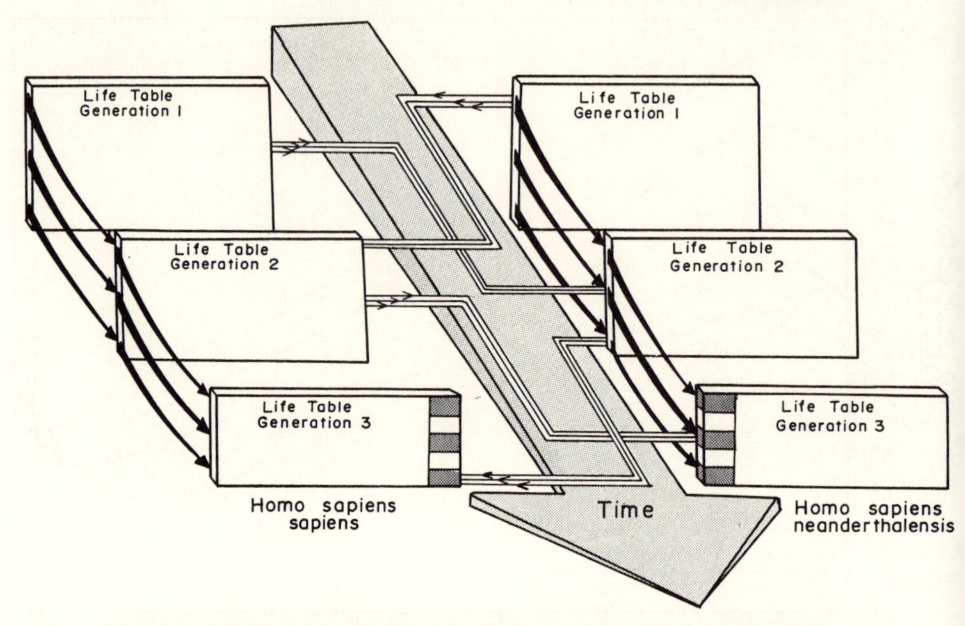

Figure 13.9. The new model in its most sophisticated form. This model uses entirely interactive linked generational life tables in which age-specific mortality of modern forms affects Neanderthal mortality and *vice versa*. The interactive links between populations are across generations, causing a staggered set of relationships. This is indicated by the arrows being directed from *Homo sapiens sapiens* to Neanderthals and *vice versa,* and by the cross generation crossing of the arrows.

conservative results in which both populations begin at the same point in their trajectory towards their demographic destiny. I began both populations at stable model South 7 and allowed the mortality of *Homo sapiens sapiens* to affect mortality of the *Homo sapiens neanderthalensis* inversely. In other words, the lower the mortality of the *Homo sapiens sapiens* the greater the effect upon the mortality of the Neanderthal population. This resulted in an increasing life expectancy for *Homo sapiens sapiens* and a decreasing life expectancy for *Homo sapiens neanderthalensis*.

Figure 13.8 presents the results in life expectancies for two simulated populations, one Sapiens and one Neanderthal. Both are initialized at South 7. Their early demographic history is similar. The life expectancies are equal for the first generation (g=0). They appear as boxes on the graph. The eighth and fifteenth generations of the Neanderthal population are also shown. They are the two lines below the zero generation (g=0) baseline. The eighth and the fifteenth generations of *Homo sapiens sapiens* are the two lines above the g=0 line.

Two aspects of this figure are important. The growth and the decline in life expectancy for *Homo sapiens sapiens* and *Homo sapiens neanderthalensis* respectively are largest in the younger cohorts. This is exactly where the change will cause the greatest long-term consequences. In other words, there is a type of multiplier effect seen here. It is being applied positively

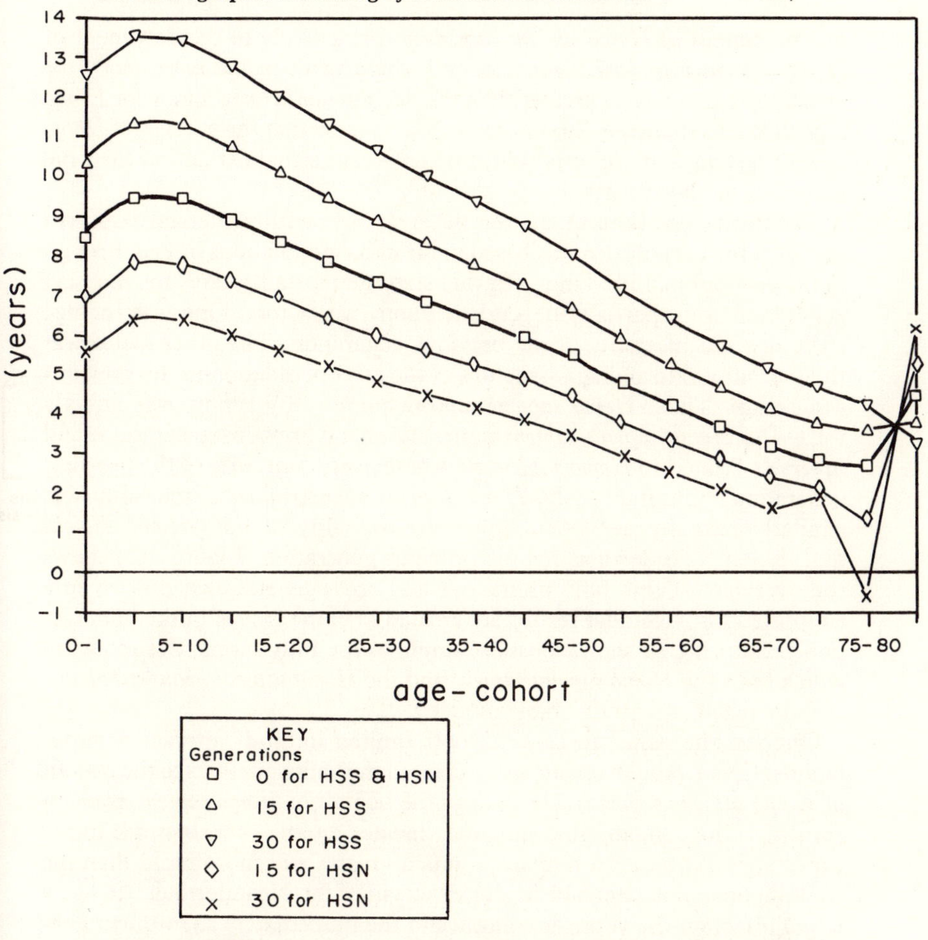

Figure 13.10. The fully interactive multi-generationally linked simulation for *Homo sapiens sapiens* and for *Homo sapiens neanderthalensis*. Each population is partially dependent upon the mortality rates of the other in the previous generation. This interrelationship is less than two and a half percent of the mortality rates. *Homo sapiens sapiens* are indicated by boxes in the initial generation, crosses in the fifteenth generation, and inverted triangles in the thirtieth generation. Neanderthals are indicated also by boxes in the initial generation, diamonds in the fifteenth and x's in the thirtieth. One sees the same pattern as in limited interaction and growth shown in Figure 13.7, but in a more restrained manner. Note the difference in the number of generations being simulated. The lengthening of the life span of *Homo sapiens sapiens* takes place more rapidly than does the shortening of the life span of the Neanderthals. In other words it takes less time for *Homo sapiens sapiens* to increase life expectancy than it takes for *Homo sapiens neanderthalensis* to have their life expectancy decreased.

for the benefit of *Homo sapiens sapiens* and negatively to the detriment of *Homo sapiens neanderthalensis*. Second, the growth in life expectancy for *Homo sapiens sapiens* is greater than the decline in life expectancy for *Homo sapiens neanderthalensis*. In other words, it is clear that the success of *Homo sapiens sapiens* is more important for the Neanderthal extinction than the failure of the Neanderthals.

The third step of this model was the creation of a fully interactive growth model. This version also was based upon stable populations linked through multi-generational life tables. In this step the mortality rates for the next generation were partially dependent upon two factors. One was the age structure and mortality of the previous generation. The other factor was the mortality rate and age structure of the *other* population in the previous generation. Thus, *Homo sapiens sapiens* morality by cohort was directly related to *Homo sapiens sapiens* mortality of the previous generation, and inversely related to *Homo sapiens neanderthalensis* mortality of the previous generation. Simultaneously, *Homo sapiens neanderthalensis* mortality was similarly partially dependent upon the mortality of both *Homo sapiens sapiens* and Neanderthals for the previous generation. Figure 13.9 shows the overview of this fully interactive and cross-generational process in a simplified form. Similar results are created as in the case of limited interaction. Figure 13.10 shows a case example of the fully interactive model in which both the *Homo sapiens sapiens* and the *Homo sapiens neanderthalensis* are initialized at a South stable model 7.

One sees the same processes for both limited and fully interactive populations. Given a slight advantage, increased mortality operates to the benefit of *Homo sapiens sapiens* more than to the benefit of *Homo sapiens neanderthalensis*. The demographic situation changes rapidly. The increase in life expectancy for modern humans is much greater and more rapid than the contemporaneous decrease in life expectancy for Neanderthals. It has a larger effect on the younger cohorts and the immediately subsequent generations. If one compares the limited interaction step to the fully interactive one, one sees that the advantage that *Homo sapiens sapiens* has is slightly decreased as the Neanderthals provide some demographic resistance. However, the effect is the same.

If one increases the initial increment in mortality, one can see the catastrophic effect it has on the Neanderthal population. At first the life expectancy decreases, and then the population is destroyed (Fig. 13.11). The wildly oscillating graph is indicative of no life expectancy. The program continues to compute with negative values and is therefore showing erroneous peaks.

I have systematically modelled these populations following them from one to 200 generations. Clearly extinction has begun by 15 generations and in most cases is completed in 30. The increment in mortality need only be between one and two per cent. If one has an initial increment of seven per cent, extinction within 15 generations is assured. If one is talking of a band of 50 people, it only requires that four individuals die. The death of four people is easy to imagine when *Homo sapiens neanderthalensis* and *Homo*

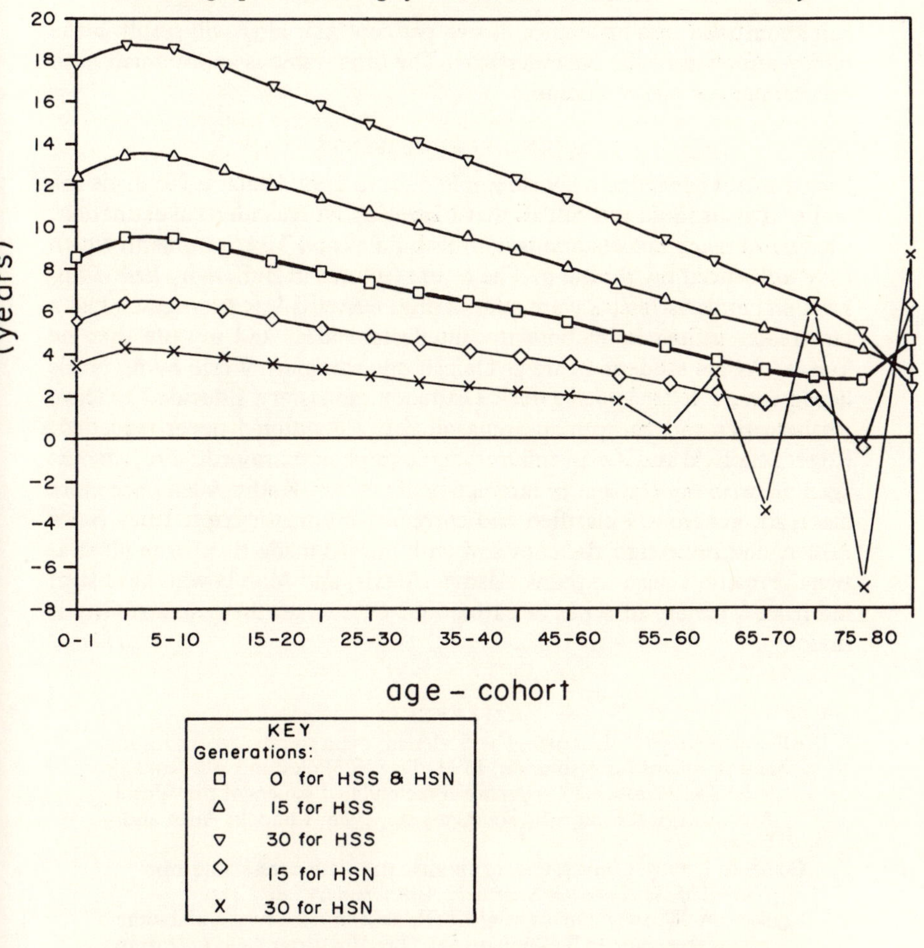

Figure 13.11. The same as Figure 13.10 with the exception that the demographic advantage of *Homo sapiens sapiens* has increased. The functional interrelationship of the cohort-specific mortality rates has increased up to five percent from two and a half percent mortality. The symbols remain the same. Not only is the pattern similar, but the wildly oscillating life expectancies in the thirtieth generation indicate the beginning of actual extinction of Neanderthal populations.

sapiens sapiens came into contact. It is even easier to envisage when the contact occurs not once but repeatedly.

CONCLUSIONS

This paper makes two assumptions. First, that there were interacting populations of *Homo sapiens sapiens* and *Homo sapiens neanderthalensis*. Second, that these populations may be modelled demographically with stable populations. Given these assumptions, a small demographic advantage in the neighbourhood of a difference of two per cent mortality will result in the rapid extinction of the Neanderthals. The time frame is approximately 30 generations, or one millennium.

ACKNOWLEDGEMENTS

I wish to acknowledge a considerable debt to Paul Mellars. He made me realise that I should rethink all that I knew about Neanderthal extinction. I have had many conversations with Rob Foley and Ted Steegmann which have influenced my thinking. The recent lectures in Buffalo by Erik Trinkaus on similar subjects were stimulating. Bernard Vandermeersch had a far greater influence on both my intellectual ideas and my life than he knows. As his student I dug at Qafzeh one summer. While living in his headquarters at an Israeli Greek Orthodox monastery I decided to leave mathematics and become an archaeologist, a decision I never regretted. Roger Schofield and Kevin Schurer, demographers extra-ordinaire, encouraged me with my concept of interactive life tables. Kathy Allen once more has read, generously clarified and corrected my manuscript. Lucy Balos Miller took my rough sketches and with insight made them true illustrations. Finally, I wish to thank Alanya, Alexis, and Marcia who have kept me from a variety of types of extinction. Of course, the responsibility is theirs.

REFERENCES

Allsworth-Jones P. L. 1986. The Szeletian: main trends, recent results, and problems for resolution. In M. Day, R. Foley and Wu Rukang (eds) *The Pleistocene Perspective*. Precirculated papers of the World Archaeological Congress, Southampton, 1986. London: Allen and Unwin.

Coale, A. J. 1963. Convergence of a human population to a stable form. *Journal of the American Statistical Association* 63: 395- 435.

Conkey, M. W. 1983. On the origins of Paleolithic art: a review and some critical thoughts. In E. Trinkaus (ed.) *The Mousterian Legacy: Human Biocultural Change in the Upper Pleistocene*. Oxford: British Archaeological Reports International Series S164: 201-277.

Delson, E. 1984. *Ancestors: the Hard Evidence*. New York: Alan R. Liss.

Dublin, J. and Lotka, A. J. 1936. *Length of Life: the Study of the Life Table*. New York: Roland.

Jelinek, A. J. 1982. The Tabūn Cave and Paleolithic Man in the Levant. *Science* 216: 1369-1375.

Lotka, A. J. 1907. Relation between birth rates and death rates. *Science* 26: 21-22.

Oliva, M. 1986. From the Middle to the Upper Palaeolithic: a Moravian Perspective. In M. Day, R. Foley and Wu Rukang (eds) *The Pleistocene Perspective*. Precirculated papers of the World Archaeological Congress, Southampton, 1986. London: Allen and Unwin.

Smith, F. H. 1985. Continuity and change in the origin of modern *Homo sapiens*. *Zeitschrift für Morphologie und Anthropologie* 75: 197-222.

Svoboda, J. 1986. Origins of the Upper Palaeolithic in Moravia. In M. Day, R. Foley and Wu Rukang (eds) *The Pleistocene Perspective*. Precirculated papers of the World Archaeological Congress, Southampton, 1986. London: Allen and Unwin.

Trinkaus, E. 1983. *The Shanidar Neandertals*. New York: Academic Press.
Trinkaus, E. 1986. The Neandertals and modern human origins. *Annual Review of Anthropology* 15: 193-218.
Trinkaus E. and Smith, F. H. 1985. The fate of the Neandertals. In E. Delson (ed.) *Ancestors: the Hard Evidence*. New York: Alan R. Liss: 325-343
Vandermeersch, B. 1985. The origin of the Neandertals. In E. Delson (ed.) *Ancestors: the Hard Evidence*. New York: Alan R. Liss: 306-309.
Zubrow, E. B. W. 1986. An interactive growth model applied to the expansion of Upper Paleolithic populations. In M. Day, R. Foley and Wu Rukang (eds) *The Pleistocene Perspective*. Precirculated papers of the World Archaeological Congress, Southampton, 1986. London: Allen and Unwin.

14. The Origin of Early Modern Humans: a Comparison of the European and non-European Evidence

C. B. STRINGER

INTRODUCTION

While most workers now agree that the Middle-early Late Pleistocene hominid fossil record of Europe documents the evolution of the Neanderthal lineage over a time span probably exceeding 200 000 years, there is still little agreement as to whether this evolutionary lineage continued with the appearance of the first anatomically-modern humans in the area about 35 000 BP, or was replaced, with or without accompanying gene flow (for example, compare the contributions in this volume by Bräuer and by Smith *et al.*). The following study of measures of face shape and face-vault relations in European and non-European crania from this time period represents an attempt at testing the models of Neanderthal-early modern continuity or discontinuity by identifying Neanderthal and non-Neanderthal characters, and determining their presence or absence in the crucial early Upper Palaeolithic European sample. A number of non-European early modern crania are also studied to document their similarities to, or differences from, earlier, penecontemporaneous and later samples from various parts of the world (see Table 14.1).

Discussion of Sample Composition

Crania which were complete enough to provide at least one facial index or angle were arranged into samples on chronological or geographical grounds as listed in Table 14.1. Whether these specimens should all have been arranged in this way could itself require a whole paper for discussion, but in general the most controversial inclusions (such as Forbes' Quarry in the NEA group rather than ENEA and Sala and Zuttiyeh in ENEA) make little difference to the relevant group means when rearranged. The only case where this is not so is for MIDPL, where the damaged Steinheim cranium might be excluded because of distortion affecting the data. Removing Steinheim does somewhat alter each MIDPL mean value in the direction of NEA values (the SSR/GOL ratio becomes 0.524; BNL/GOL becomes 0.528; NLH/EKB 0.511; NFA 141°; and SSA 120°, respectively). A further point worth raising is the uncertainty now attached to the grouping of the Skhūl and Qafzeh samples. New thermoluminescence dating determinations (Valladas *et al.* 1988) place the Qafzeh Mousterian at about 90 000 years, whereas

Table 14.1. Hominid samples studied in the present analyses. Bracketed figures show in how many analyses specimens have been used. Measurements were taken on original specimens except where indicated otherwise. Some Neanderthal data are taken from Trinkaus (1983)

Middle Pleistocene (MIDPL)

Bodo (3)	Petralona (5)
Broken Hill (5)	Arago 21 (3)
	Steinheim cast (5)

Middle/early late Pleistocene Africa (EAF)

Irhoud 1 (4)	ES-11693 cast (1)
Irhoud 2 (1)	Singa (2)
Ngaloba (2)	Florisbad cast (3)

Early Neanderthals (ENEA)

Saccopastore 1 (3)	Shanidar 2 (1)
Saccopastore 2 (3)	Shanidar 4 (1)
Krapina C (1)	Šala (1)
Krapina E (1)	Zuttiyeh (1)
Ehringsdorf 9 (1)	

Neanderthal (NEA)

La Chapelle-aux-Saints (5)	La Quina 5 (1)
La Ferrassie (5)	Neanderthal (1)
Guattari 1 (5)	Forbes' Quarry (5)
Saint-Césaire (3)	Spy 1 (1)

Asian Neanderthal (ASNEA)

Shanidar 1 (4)	Tabūn 1 (1)
Shanidar 5 (3)	Amud 1 cast (4)

Skhūl-Qafzeh (SQ)

Skhūl 4 cast (1)	Qafzeh 6 (5)
Skhūl 5 (2)	Qafzeh 9 (5)

Early Upper Palaeolithic (EUP)

Cro-Magnon 1 (5); 2 (4); 3 (1)	Pavlov 1 cast (2)
Abri Pataud (5)	Dolni Vestoniče 3 (4)
Grotte des Enfants 6 cast (5)	Brno 3 (4)
Mladeč 1 (5)	Předmostí 3 cast (5)
Mladeč 5 (1)	Předmostí 4 cast (4)

Early Australian (EAU) (all casts)

Cohuna (4)	Kow Swamp 15 (3)
Keilor (5)	Talgai (3)
Kow Swamp 5 (3)	

Upper Cave Zhoukoudian (UCZ) (all casts)
101 (5); 102 (3); 103 (5)
Liujiang cast (LJ) (5)
Maba cast (MABA) (1)
Dar-es-Soltane 5 (DES) (1) (data from Ferembach 1975)
Border Cave 1 cast (BC) (1)
Modern European (MEU): 110 Norse; 99 Zalavar; 109 Berg (5 analyses each; data from Howells 1973)
Modern African (MAF): 83 Teita; 101 Dongon; 102 Zulu (5 analyses each; data from Howells 1973)
Modern 'Bushman' (MBU): 90 (5 analyses each; data from Howells 1973)
Modern Australasia (MAU) : 101 South Australia; 86 Tasmania; 110 Tolai (5 analyses each; data from Howells 1973)
Modern 'Asia' (MAS): 100 Mokapu; 109 Buriat; 108 Eskimo (5 analyses each; data from Howells 1973)
Modern Amerindian (MAM): 69 Arikara; 110 Peruvian (5 analyses each; data from Howells 1973)

the Skhūl Mousterian remains uncertainly dated. Given morphological similarities noted elsewhere (e.g. Vandermeersch 1981; Stringer and Trinkaus 1981; Trinkaus 1984) they will be retained as a single group for the present analyses.

ANALYSES

The samples and specimens have been studied using five different angles and indices reflecting face shape and the relationship of the face to the cranial vault. Some of these angles and indices have previously proved useful in distinguishing certain Pleistocene samples (Stringer 1974, 1978, 1983, 1987; Stringer and Trinkaus 1981), and they also separate, to varying extents, sample means of recent human geographical variants. These are the related transverse angles which measure midfacial projection at the top of the nose (nasio-frontal angle, NFA) and base of the nose (subspinale angle, SSA) (Howells 1973), and the indices which measure nasal height/biorbital breadth (NLH/EKB), basion-nasion length/skull length (BNL/GOL) and subspinale radius/skull length (SSR/GOL) (measurements taken as defined by Howells 1973). The transverse angles are generally correlated with each other (Trinkaus 1983), yet they do display different patterns in other hominoid taxa studied (Stringer 1986). It could also be argued that using GOL as a measure of skull length biases the results where the supraorbital torus is a significant contributor to total GOL; however, as will be seen, this does not prevent archaic hominids with large brow-ridges from showing a large degree of variation when compared with each other and with small-browed anatomically-modern crania. Indeed, this use of GOL might be expected to minimize contrasts between the Neanderthal and early modern samples in the indices using GOL.

Extreme midfacial projection is a feature known to distinguish Neanderthals from other hominids (Howells 1974; Stringer 1974, 1978, 1986; Stringer and Trinkaus 1981; Trinkaus 1983), and to a lesser extent it also separates modern Europeans (relatively low mean NFA values) from non-Europeans, and modern Australasians (relatively low mean SSA values) from non-Australasians. High NLH/EKB values similarly distinguish Neanderthals from non-Neanderthals, while in recent populations this is to a lesser extent characteristic of Asian and Amerindian samples. Neanderthals display high SSR/GOL ratios, as is also true to a lesser extent of Australasian and Asian crania, while high BNL/GOL indices characterize both Neanderthal and modern Asian samples. This summary of average differences is not meant to imply that similarities in these indices can be used to classify specimens or groups together – these are not discriminant functions, and there is considerable individual variation. However, because the Neanderthals are so distinctive, we should at least expect there to be some similarities with early Upper Palaeolithic specimens, if these are descended from such unusual ancestors. Similarly, where certain modern populations are *on average* special, we may be able to pick up such characters in possible antecedent early modern crania.

If local evolution was predominantly responsible for early modern vari-

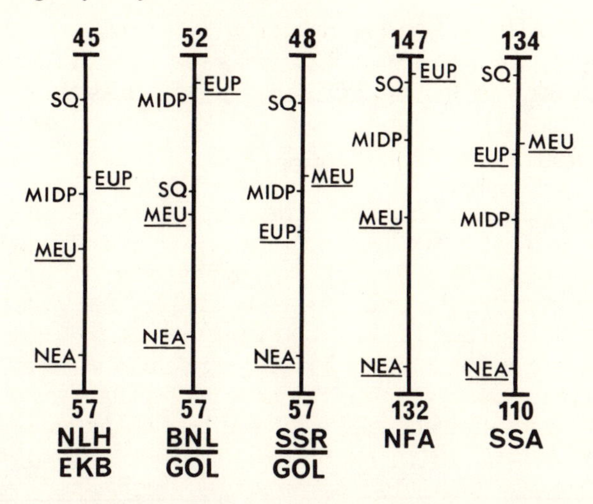

Figure 14.1. Diagrammatic representation of mean values of European Neanderthal (NEA), early Upper Palaeolithic (EUP), modern European (MEU), Skhūl-Qafzeh (SQ) and Middle Pleistocene (MIDP) samples, for the three cranial indices (x 100) and two transversal facial angles (in degrees) studied. The NEA, EUP, and MEU sample means are underlined for ease of comparison (see Tables 14.1 and 14.2 for details of samples and measurements).

ation, then differences between the early modern samples should already be pronounced, and we might expect the European early Upper Palaeolithic sample to resemble both ancestral (European Neanderthal) and descendant (modern European) groups, and the Skhūl-Qafzeh sample to resemble its supposed ancestral group (Asian Neanderthals). If the single origin model of a recent (early late Pleistocene?) African origin for modern human variation is appropriate, we should expect to find a progressively more similar morphology in the earlier modern human crania as they approach more closely to the hypothesized African time of origin – i.e. later samples should look more like their present geographical counterparts (*if* there was a close relationship between them), while older samples should appear more similar to each other and to early African specimens (if those specimens resemble the actual ancestral morphology).

EUROPEAN DATA

Figure 14.1 displays the group means for European Neanderthal (NEA), early Upper Palaeolithic (EUP) and modern European (MEU) samples. For comparison, the Middle Pleistocene (MIDP) and Skhūl-Qafzeh (SQ) groups are added. As can be seen, the NEA and SQ groups are markedly contrasted in four of the analyses, while the EUP group is even more distinct from NEA in two of the analyses. MEU is most like NEA in three analyses, EUP is most similar to NEA in one analysis (SSR/GOL), and MIDP in one analysis (SSA). The distribution of the group means emphasizes the distinctiveness of NEA, and it can be seen that the EUP sample is certainly more like the SQ group than the NEA group in four analyses, and is in fact more like

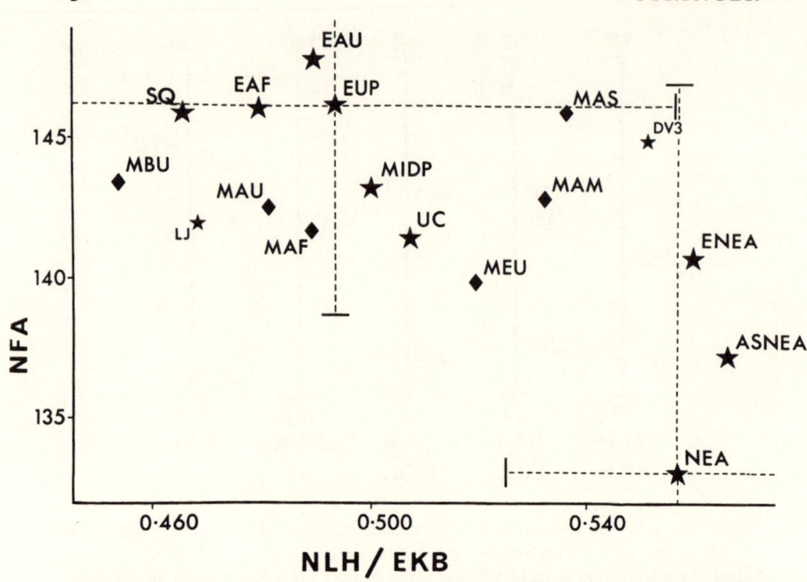

Figure 14.2. Scatter diagram of sample mean and individual values for nasio-frontal angle (NFA) against nasal height/biorbital breadth index (NLH/EKB). Note the distinctiveness of the three Neanderthal groups, and the wide range of early Upper Palaeolithic (EUP) values, marked by standard deviation bars. In this case, the positioning of the early modern fossils supports a close relationship between them.

MIDP than the NEA group in all five analyses! From these analyses, the NEA sample seems a better ancestor for *recent* Europeans than it does for the earliest modern Europeans!

Table 14.2 provides a complete listing of the group means and individual specimen means for comparison. Some comments follow: The early and late Neanderthal samples from Europe (ENEA and NEA) and Western Asia (ASNEA) show similarly distinct values in the analyses, although the ENEA sample (dominated by the Saccopastore specimens) is generally less extreme. Considering the position of the Neanderthals and their supposed special relationship to recent Europeans, it is surprising to find that the MEU sample is the closest modern group in only one analysis (NFA).

The early Upper Palaeolithic sample (EUP) is unexceptional in three analyses and extreme *in a direction opposite to the Neanderthals* in two analyses (BNL/GOL, NFA). Considering that it is generally agreed that the EUP sample might be representative of the ancestors of modern Europeans, and that they are considered by many workers to already look 'European', it is surprising to note that the MEU sample is only the nearest modern neighbour in one analysis (SSA, joint nearest neighbour with MAM).

The Skhūl-Qafzeh (SQ) sample is consistently very different from both Neanderthals and Asian Neanderthals in all the analyses, which does not support models of supposed ASNEA-SQ evolutionary continuity or gene flow.

Figures 14.2 and 14.3 show graphically how the group means and certain

Table 14.2. Complete listing of sample and individual specimen means for the five cranial indices and angles employed in the present analyses. The indices and angles measure the following parameters: SSR/GOL = subspinale radius/glabello-occipital length; BNL/GOL = basion-nasion length/glabello-occipital length; NLH/EKB = nasal height/biorbital breadth; NFA = nasio-frontal angle; SSA = subspinale angle (measurements and angles taken as defined in Howells 1973; see text for further discussion). The composition of the samples analysed is given in Table 1. The European Neanderthal (NEA) and early Upper Palaeolithic (EUP) positions are underlined for ease of comparison.

SSR/GOL	*BNL/GOL*	*NLH/EKB*	*NFA (°)*	*SSA (°)*
.596 ENEA	.570 EAU	.566 ASNEA	133 *NEA*	112 ASNEA
.560 *NEA*	.568 LJ	.560 ENEA	137 ASNEA	112 *NEA*
.541 ASNEA	.566 MAS	.557 *NEA*	140 MEU	121 MIDPL
.540 MAU	.562 *NEA*	.536 MAS	141 ENEA	122 ENEA
.539 UCZ	.562 ASNEA	.532 MAM	142 UCZ	123 MAU
.536 MAS	.557 ENEA	.518 MEU	142 LJ	125 EAU
.536 EAU	.557 UCZ	.507 UCZ	142 MAF	127 EAF
.529 *EUP*	.555 MAM	.499 MIDPL	143 MAU	127 *EUP*
.527 EAF	.552 MAF	.493 *EUP*	143 MAM	128 UCZ
.527 MAM	.548 EAF	.489 EAU	143 MIDPL	128 MAM
.520 MAF	.544 MEU	.488 MAF	143 MBU	128 MEU
.516 MIDPL	.540 MAU	.480 MAU	146 MAS	131 MAF
.512 MEU	.540 SQ	.479 EAF	146 SQ	133 SQ
.504 MBU	.533 MBU	.468 LJ	146 EAF	133 MBU
.500 LJ	.527 *EUP*	.465 SQ	146 *EUP*	133 MAS
.492 SQ	.526 MIDPL	.453 MBU	148 EAU	142 LJ
		.392 DES	151 BC	
			153 MABA	

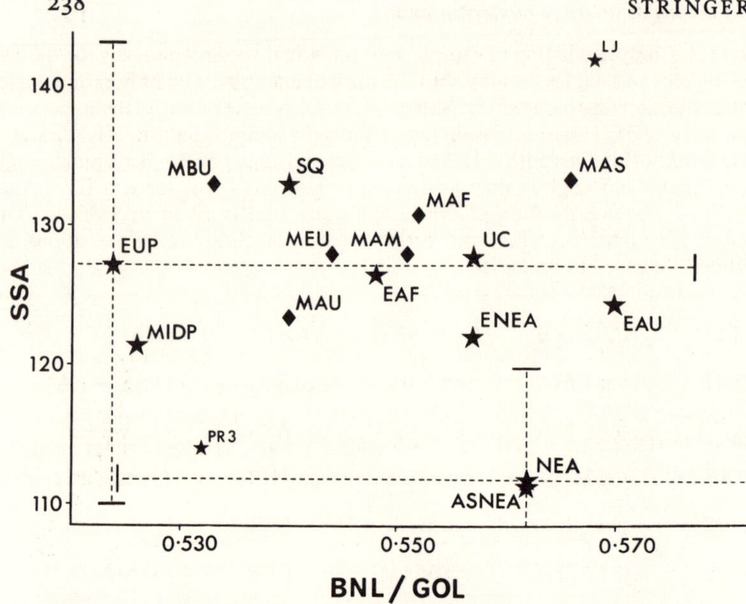

Figure 14.3. Scatter diagram of sample mean and individual values for subspinale angle (SSA) against basio-nasion length/glabello-occipital length index (BNL/GOL). Note the much closer positioning of the Liujiang (LJ) and modern Asian (MAS) groups compared with Figure 14.3, and the position of Předmostí in the area of overlap of Neanderthal (NEA) and early Upper Palaeolithic (EUP) ranges (marked by two standard deviation bars).

specimen means appear when plotted for four of the angles and indices. Two standard deviation bars are added for the EUP and NEA groups to show the high variation (considerably larger for EUP than for any single modern sample). These figures show by how much the Neanderthal groups fall away from the other groups analysed, and it can be seen that for the scatter diagram of NFA versus NLH/EKB (Figure 14.2), the expectations of the single origin model are well met, since the early modern groups *are* more closely matched than recent groups, and moreover the possibly ancestral EAF sample is also similar. LJ, and to a lesser extent UC, resemble the other early modern groups rather than their possible recent counterparts (MAM and MAS). However, in the plot of SSA *versus* BNL/GOL (Figure 14.3), the early modern group means are more dispersed than the means of their modern counterparts, with the EUP, LJ and EAU groups forming outliers. LJ displays even more marked lower facial flattening than MAS (high SSA value), while the EAU group (represented only by Keilor) has a very high BNL/GOL value, resembling the MAS and NEA groups. Were it not for this extreme position of EAU (Keilor), the expectations of the single origin model would be met to a greater extent.

GENERAL RELATIONSHIPS

I have attempted to analyse the results in two further ways: by the use of nearest neighbour relationships, and by a numerical cladistic analysis using

PAUP. Since I hope that the cladistic analyses will be discussed in more detail elsewhere, I will present only the most parsimonious cladograms here, and make a few general comments.

(i) *Nearest neighbour analysis considering the fossil groups only*. For the three Neanderthal groups, there are 17 nearest fossil neighbours in the 5 analyses, and 12 of these positions are occupied by one of the other Neanderthal groups. In other words, *the Neanderthal groups are primarily similar to each other rather than to other groups* (the remaining five positions are occupied by UCZ in three cases, MIDPL and LJ).

For the early modern groups (i.e. SQ, EUP, UCZ, LJ and EAU), there are 31 nearest neighbour positions, and these are predominantly occupied by EAF (8 positions) and SQ and EUP (5 positions each). Other nearest neighbours are UCZ (3), LJ (4), EAU (2), MIDPL (2), ENEA (1) and ASNEA (1). These results certainly seem to support a single African origin model in that they emphasize the similarities of the early modern groups to each other, and especially to the EAF group.

The isolation of the Neanderthals from non-Neanderthal groups is further shown by the nearest neighbour positions for the EAF and MIDPL groups. For EAF there are 6 nearest neighbour positions (3 occupied by EUP, 2 by SQ and 1 by EAU) while for MIDPL the nearest neighbours are EUP (2), EAF (1), UCZ(1), LJ (1) and ENEA (1).

(ii) *Numerical cladistic analysis (PAUP)*. By dividing the whole range of values for each index and angle on a scale from 0 to 8, scores were assigned to each specimen or group. PAUP was then used to analyse the data, employing the FARRIS and branch and bound options (see Swofford 1985, and Stringer 1987). Various alternative rootings were employed but the most appropriate choice of outgroup and rooting (the MIDPL sample) is illustrated here. However, in considering these preliminary results it is as well to note the cautionary comments of Pimentel and Riggins (1987) about the use of continuous characters in cladistic analyses, especially indices (and cranial angles of the kind employed here are also really disguised indices). Additionally, the ordered character option used here favours gradual sequential changes through the cladograms, rather than more 'punctuational' changes.

Figure 14.4 shows a typical pattern from 22 equally parsimonious PAUP alternatives, with only the anatomically-modern groups placed as terminal taxa. The consistency indices of these 22 best trees were low (C.I. = 0.486, tree length 72) indicating considerable homoplasy. The structure of the cladograms suggests a relatively primitive face shape for the EUP and MAU groups, especially reflected in BNL/GOL, NLH/EKB and NFA values. EUP formed the sole outside sister group to the rest in 11 cases, and MAU in 4 cases. The EAF group was also relatively primitive, sharing the first (outermost) sister group position in 5 cases, and the second sister group position in another 9 cases, mostly with either the EUP or MAU groups. The apparent relatively primitive SQ and MBU groups were consistently linked in all cases (via low SSR/GOL, BNL/GOL and NLH/EKB, and high NFA and SSA values), with LJ as next most closely related in 18 trees. The more derived UCZ

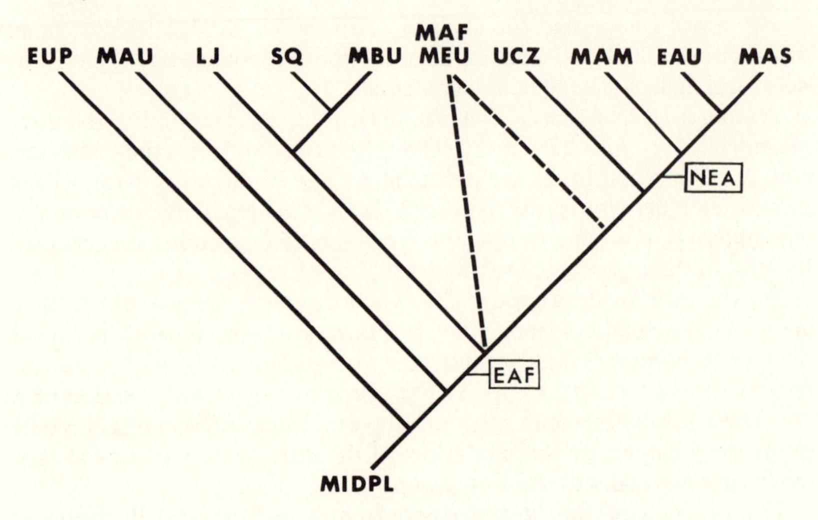

Figure 14.4. Typical cladogram derived from 22 most parsimonious solutions provided by PAUP for an analysis of the five cranial indices and transverse facial angles studied (C.I. = 0.486, tree length 72). Outgroup and rooting via the middle Pleistocene (MIDPL) sample. Anatomically-modern groups have been positioned as terminal taxa, with typical relative positions of the Middle/early Late Pleistocene African sample (EAF) and Neanderthal clade (ENEA/NEA/ASNEA) indicated. The modern African (MAF) and modern European (MEA) groups did not form a clade, but often exchanged positions, sometimes grouping with the LJ/SQ/MBU clade and sometimes with NEA/UCZ/MAM/EAU/MAS.

and MAM samples were consistently linked with the EAU and MAS groups through relatively high BNL/GOL and NLH/EKB values and more moderate SSR/GOL, NFA and SSA values, while the ENEA, NEA and ASNEA groups always formed a clade characterized by high values for each index and low values for both angles. This Neanderthal clade was always most closely related to the MAM/EAU/MAS clade.

Since, to my knowledge, no worker other than Thoma (1973) has recently supported a specific Neanderthal-Mongoloid phylogenetic relationship, and such a relationship is contradicted by much data from the rest of the cranium and skeleton, the structure of this cladogram is unlikely to reflect the actual phylogenetic relationships of these groups. Instead, like Thoma's metrical comparisons, this appears to provide evidence of parallel change in certain of these indices and angles (most notably in an increase in SSR/GOL, BNL/GOL and NLH/EKB) between the Neanderthal clade and certain anatomically-modern populations of Asian origin. However these relationships are interpreted, it is evident that the Neanderthal clade does not display primitive characters, which are apparently still retained in the EAF, EUP and SQ groups. In any scheme of regional continuity these latter two groups could be expected to be the most closely related to the Neanderthals, given their geographic and temporal proximity to them, but this is certainly not reflected in these facial angles and indices.

GENE FLOW FROM NEANDERTHALS?

The analyses have demonstrated that in these particular angles and indices, late Neanderthals from Europe or Asia are improbable ancestors for early modern humans in the same areas. As far as can be determined from preserved parts, this is as true for what may be the youngest known Neanderthal (Saint-Césaire) as for earlier specimens. However, the EUP sample does contain some anomalies which need to be explained. While most of the sample, including the Cro-Magnon and Mladeč specimens, is not at all Neanderthal-like, Dolní Vestonice 3 and Předmostí 3 *are* more like Neanderthals in analyses of NLH/EKB (DV 3 = 0.551, PR 3 = 0.534) and SSR/GOL (DV 3 = 0.554, PR 3 = 0.566), while Brno 3 appears atypical in displaying a high BNL/GOL index (0.568). And, as has been noted before (Howells 1974; Stringer 1978), Předmostí 3 has an extremely projecting lower midface (SSA value *c.* 114°).

So while an ancestry for the EUP population from outside Europe appears very likely (both on the grounds of these analyses and others which include limb proportions: Trinkaus 1981), these anomalies could indeed indicate that some gene flow from Neanderthals did occur in Eastern Europe, unless they are examples of individual EUP variation paralleling some Neanderthal features. However, even if there was some degree of hybridization, one must then consider whether those genes necessarily continued into modern European populations, since these analyses show that the EUP sample is different both from earlier and later European populations (another parallel with the limb proportion data, as well as matching a retrodiction of the single origin model). Modern Europeans certainly had ancestors, but that does not mean we must accept, without further proof, that the whole EUP sample represents those ancestors. Furthermore, despite the possibility of some low level of hybridization, this cannot be equated with a significant ancestor-descendant relationship between the Neanderthals and early modern humans, since all the important novel morphological features of the early modern humans were apparently derived from non-Neanderthal sources.

GENERAL CONSIDERATIONS

Given the characteristics of the Middle Pleistocene sample used here, and what little is known of earlier Pleistocene facial morphology from specimens like Sangiran 17 and KNM-ER 3733, it is possible to reconstruct the primitive (ancestral) condition which presumably gave rise to the derived condition found in Neanderthals and the variation found in the early modern groups. Overall, the primitive condition was for low values of NLH/EKB, BNL/GOL and SSR/GOL, and high values of NFA and SSA. The MIDPL sample used here has an intermediate value for NLH/EKB and a much higher value for SSA than the inferred ancestral condition, but it seems that the EAF group was more similar to the ancestral condition, either through primitive retentions or evolutionary reversals. The short, broad and flat upper face found in the EAF group not only looks 'primitive' but is also present in the SQ, EUP, EAU and LJ groups. As far as can be determined it was also present in Border Cave I (high NFA) and Dar-es-Soltane 5 (very low NLH/EKB).

A link, but a rather unsatisfactory one based primarily on primitive retentions rather than shared derived characteristics, can thus be made between the EAF and early modern groups. While late Middle Pleistocene/earliest Late Pleistocene Asian hominids are so poorly known, it remains possible that they also displayed this short, broad and flat upper facial morphology. Zuttiyeh and Maba certainly both display flat upper faces (NFA values *c.* 150°) and it will be valuable to obtain comparable data for the more complete Dali and Yinkou crania from China. Without such data, I can only conclude that the EAF sample is the most plausible ancestral one for early modern humans known from the appropriate time period. This does not mean that these specimens, drawn from various parts of Africa, must represent the actual ancestral population for early modern *Homo sapiens*. However, they appear morphologically and metrically close to the hypothesized common ancestor.

Further, and independent, data which support the model of an African origin for modern humans can also be added. First, Africa has plausible (and fairly complete) morphological intermediates between 'archaic' and 'modern' human morphologies (such as Irhoud 2 and Omo Kibish). Second, from Klasies River Mouth Caves, Border Cave, Equus Cave, Mumba Cave, Omo Kibish I and East Turkana (KNM-ER 3884) we may have the oldest known fossils approximating the modern human anatomical pattern (see, for example, papers in this volume by Bräuer, Deacon, Klein, and Rightmire). However, absolute dating of the Qafzeh Mousterian suggest that modern humans were already present in Southwest Asia by about 90 000 years ago (Valladas *et al.* 1988; Stringer 1988), and it remains to be seen whether any of the known African fossils of early modern humans are actually older than this. Nevertheless, the Levant is adjacent to Africa and a close relationship between the Qafzeh and African samples can be adduced from various analyses, including those conducted here. Limb proportion data certainly also support a tropical/subtropical derivation for the Qafzeh fossils (Trinkaus 1981).

If, as now suggested from the fossil data (Gowlett 1987; Lewin 1987a) and supported from the genetic data (Lewin 1987b; Stringer and Andrews 1988), modern humans did originate in Africa during the late Middle or early Late Pleistocene, it follows that the extensive data available on Eurasian Middle-Upper Palaeolithic behavioural changes are irrelevant to that origin. However, such data become very relevant to the subsequent radiation of early modern humans across the Old World (see Mellars 1989). But if behavioural factors do lie behind the evolution of the first modern humans, those factors probably remain to be deciphered from the early part of the Middle Stone Age/Middle Palaeolithic archaeological record of Africa.

ACKNOWLEDGEMENTS

I would like to thank the many workers and institutes who have allowed me to collect data on their fossil material, and Robert Kruszynski and the British Museum (Natural History) photographic studio for the preparation of figures.

REFERENCES

Ferembach, D. 1975. Les restes humains de la Grotte de Dar-es-Soltane 2 (Maroc): Campagne 1975. *Bulletin et Mémoires de la Société d'Anthropologie de Paris* 3: 183-193.

Gowlett, J. A. J. 1987. The coming of Modern Man. *Antiquity* 61: 210-219.

Howells, W. W. 1973. *Cranial Variation in Man*. Cambridge (Mass): Papers of the Peabody Museum 67.

Howells, W. W. 1974. Neanderthal Man: facts and figures. *Yearbook of Physical Anthropology* 18: 7-18.

Lewin, R. 1987a. Africa: cradle of modern humans. *Science* 237: 1292-1295.

Lewin, R. 1987b. The unmasking of mitochondrial Eve. *Science* 238: 24-26.

Mellars, P. A. 1989. Major issues in the emergence of modern humans. *Current Anthropology* 30: 349-385.

Pimentel, R. A. and Riggins, R. 1987. The nature of cladistic data. *Cladistics* 3: 201-209.

Stringer, C. B. 1974. Population relationships of later Pleistocene hominids: a multivariate study of available crania. *Journal of Archaeological Science* 1: 317-342.

Stringer, C. B. 1978. Some problems in Middle and Upper Pleistocene hominid relationships. In D. J. Chivers and K. Joysey (eds) *Recent Advances in Primatology*. Vol. 3: *Evolution*. London: Academic Press: 395-418.

Stringer, C. B. 1983. Some further notes on the morphology and dating of the Petralona skull. *Journal of Human Evolution* 12: 731-742.

Stringer C. B. 1986. The credibility of *Homo habilis*. In B. Wood, L. Martin and P. Andrews (eds) *Major Topics in Primate and Human Evolution*. Cambridge: Cambridge University Press: 266-294.

Stringer, C. B. 1987. A numerical cladistic analysis for the genus *Homo*. *Journal of Human Evolution* 16: 135-146.

Stringer, C. B. 1988. The dates of Eden. *Nature* 331: 565-566.

Stringer, C. B. and Trinkaus, E. 1981. The Shanidar Neanderthal crania. In C. B. Stringer (ed.) *Aspects of Human Evolution*. London: Taylor and Francis: 129-165.

Stringer, C. B. and Andrews, P. 1988. Genetic and fossil evidence for the origin of modern humans. *Science* 239: 1263-1268.

Swofford, D. L. 1985. *PAUP: version 2.4*. Champaign (Ill): Natural History Survey.

Thoma, A. 1973. New evidence for the polycentric evolution of *Homo sapiens*. *Journal of Human Evolution* 2: 529-536.

Trinkaus, E. 1981. Neanderthal limb proportions and cold adaptation. In C. B. Stringer (ed.) *Aspects of Human Evolution*. London: Taylor and Francis: 187-224.

Trinkaus, E. 1983. *The Shanidar Neandertals*. New York: Academic Press.

Trinkaus, E. 1984. Western Asia. In F. H. Smith and F. Spencer (eds) *The Origins of Modern Humans: a World Survey of the Fossil Evidence*. New York: Alan R. Liss: 251-293.

Valladas, H., Reyss, J. L., Joron, J. L., Valladas, G., Bar-Yosef, O. and Vandermeersch, B. 1988. Thermoluminescence dating of the Mousterian 'Proto-Cro-Magnon' remains from Israel and the origins of modern man. *Nature* 331: 614-616.
Vandermeersch, B. 1981. *Les Hommes Fossiles de Qafzeh (Israël)*. Paris: Centre National de la Recherche Scientifique.

15: The Origin of Anatomically Modern Humans in Australasia

PHILLIP J. HABGOOD

INTRODUCTION

To quote Clark Howell from the Introduction of a recent volume dealing with the origins of modern humans:

> There is now a near consensus among students of human evolutionary biology that the origins of our own species, *Homo sapiens*, is somehow intimately linked with the first intercontinental ancient hominid, *Homo erectus*. However, neither the transformation of *erectus* to *sapiens* nor the transformation of ancient (archaic) populations of *Homo sapiens* to their anatomically modern succeedents (*H. s. sapiens*) are matters of agreement in this scientific fraternity (1984: xiii).

Most discussions of this topic concentrate on the hominid material from Europe and Africa with only cursory mention of that from Australasia. This is surprising in that three of the most modern-appearing crania older than 20 000 years BP (the 'Deep Skull' from Niah, and Lake Mungo 1 and 3) have been recovered from this region. The present paper will discuss the evidence for the evolution of modern humans in Australasia and how it relates to data from other parts of the world. At present, there are two major explanations for the origin of anatomically-modern humans (Wolpoff's terms (1986) "modern populations" or "recent and modern populations" could also be used here); the *Rapid Replacement* hypothesis, and the *Regional Continuity* hypothesis. The Rapid Replacement hypothesis suggests a single origin of anatomically-modern *Homo sapiens* within a restricted geographical region, and then an outward migration and rapid replacement of existing archaic populations throughout the rest of the occupied world (Bräuer 1984b; Howells 1976a; Stringer 1984a; Stringer *et al.* 1984). Currently, the most commonly held source area for anatomically-modern *Homo sapiens* is sub-Saharan Africa (Bräuer 1984a, 1984b; Rightmire 1984; Stringer 1984a; Stringer *et al.* 1984). The Regional Continuity hypothesis proposes that anatomically-modern *Homo sapiens* evolved in a number of geographical regions from already differentiated ancestral populations (Coon 1962; Thorne and Wolpoff 1981; Weidenreich 1943, 1947; Wolpoff 1985, and this volume; Wolpoff *et al.* 1984). This explanation proposes morphological continuity within the differentiated geographical regions, and suggests the presence of distinct regional morphological features. As Howells (1976a: 492) has stated in relation to these two hypotheses: "One must wrong in its conclusions".

Which of these two explanations of the origin of anatomically- modern *Homo sapiens* best explains the data from Australasia? To answer this question, the cranial sample from this region will be examined so as to ascertain if there is evidence of 'regional continuity' which, if established, would discount the possibility of the large-scale movement of anatomically-modern populations into the area.

Many authors have seen a morphological link between the *Homo erectus* remains from Indonesia and both prehistoric and modern crania from Australia (Coon 1962; Habgood 1986a; Thorne 1976, 1977, 1980, 1984; Thorne and Wolpoff 1981; Weidenreich 1943, 1947; Wolpoff 1980, 1985, and this volume; Wolpoff *et al.* 1984), with Macintosh (1965: 59) stating, in relation to the Australian fossil crania, that "the mark of Ancient Java is on all of them". If this suggested 'regional continuity' can be proven 'beyond a reasonable doubt' it would demonstrate that anatomically-modern *Homo sapiens* evolved within the Australasian region and did not migrate into the area. The morphological link between Indonesian and Australian hominids is based on the occurrence of suggested 'regional features' on both archaic (*Homo erectus*) and more modern (prehistoric and recent) crania (Coon 1962; Thorne 1976; Thorne and Wolpoff 1981; Weidenreich 1943; Wolpoff 1985, and this volume; Wolpoff *et al.* 1984). It assumes that the Australian Aborigines are phylogenetically closer to the Javan *Homo erectus* material than they are to *Homo erectus* and/or archaic *Homo sapiens* fossils from other geographical areas (Africa, Europe, and China), and that they are phylogenetically closer to the Javan *Homo erectus* material than are modern groups from other regions. It is, however, necessary to examine these features so as to ascertain if they are actually 'regional', and so are not found on *Homo erectus* and/or archaic *Homo sapiens* crania from other geographical regions. It has been contended by some scholars (Groves, this volume; Rightmire 1987; Stringer 1984b, 1985; Stringer *et al.* 1984) that the suggested 'regional features' are actually primitive retentions, and so do not document any special 'clade' relationship between Australian and Indonesian crania. Stringer *et al.* state that:

> "identification of supposed 'clade' characters that can be traced back from late Pleistocene Australian fossils to early Middle Pleistocene Indonesian specimens. . . must be viewed with caution until it has been determined whether or not such characters might be plesiomorphies found more widely in early anatomically-modern *H. sapiens*" (1984: 121).

Few detailed morphological evaluations of the validity of the suggested 'regional clade features' for Australasia have been attempted (see Groves, this volume, for an alternative evaluation of these features). This paper will attempt such an evaluation. The features that have been proposed will be individually assessed so as to ascertain if they are 'regional' in distribution and so demonstrate 'regional continuity', or if they have a wider distribution. This will be done by an examination of published descriptions, and examinations of the originals and casts of the relevant skeletal material.

INDONESIA AND AUSTRALIA

In the 1940s Weidenreich stated that:

> ". . . the ancient Javanese forms, *Pithecanthropus* and *Homo soloensis*, agree in typical but minor details with certain fossil and recent Australian types of today so perfectly that they give evidence of a continuous line of evolution leading from the mysterious Java forms to the modern Australian bushman" (1946: 83).

And that:

> ". . . the original, special feature of the *Pithecanthropus-Homo soloensis* forehead. . . has undergone very little change despite the fact that the skull as a whole has been transformed into the modern-human pattern" (1943: 249).

He did, however, caution that:

> ". . . at least one line leads from *Pithecanthropus* and *Homo soloensis* to the Australian aborigines of today. This does not mean, of course, that I believe all the Australians of today can be traced back to *Pithecanthropus* or that they are the sole descendants of the *Pithecanthropus-Homo soloensis* line" (1943: 249-50).

Weidenreich did not give a detailed list of features that document this morphological link. Instead, he highlighted a number of general similarities between the Ngandong crania and Australian Aborigines (Weidenreich 1943: 249). These are:

1. Well-developed superciliary ridges. These are a primitive character for hominids (Andrews 1984) and are found on all *Homo erectus* and archaic *Homo sapiens* crania. The supraorbital torus is, on average, thicker on Asian *Homo erectus* crania than on those from Africa, although they are matched by those on Olduvai H9. Archaic *Homo sapiens* from Africa, such as Bodo 1, Kabwe 1 and Saldanha 1, however, have as thick and prominent brow ridges as does the Dali cranium from China, and so this earlier contrast between the two regions has disappeared. Early modern crania from Europe, North Africa and sub-Saharan Africa also have large brow ridges.

2. A flat receding forehead. This is a general character of hominoids (Andrews 1984) and is typical of all *Homo erectus* crania (Howell 1978), being especially evident on Olduvai H9. Archaic *Homo sapiens* crania, such as Kabwe 1 and Laetoli H18, also have flat receding frontals.

3. A pre-lambdoid depression. Archaic *Homo sapiens* crania including Skhūl 4, 5, 9, Jebel Qafzeh 6, and Kabwe 1, Neanderthal crania such as Teshik-Tash, and anatomically-modern crania from Jebel Sahaba (Anderson 1968) clearly have depressions anterior to the lambda, as does the *Homo erectus* crania KNM-ER 3883. This feature, therefore, is not limited to Australasian crania.

4. A sharply-angled occipital bone with a torus-like demarcation line between the occiput and the nuchal plane. All *Homo erectus* (Howell 1978) and archaic *Homo sapiens* crania have an angulated occipital bone with a

prominent occipital torus, as do African apes and some australopithecines (Andrews 1984), making this feature a primitive retention.

5. The pterion region has a short sphenoparietal, or 'H type', articulation, measuring 5-8 mm. Kabwe 1 has an 'H type' sphenoparietal articulation measuring 5.25 mm long (Baker 1968). Many of the Zhoukoudian *Homo erectus* crania similarly display a short sphenoparietal articulation at pterion. For example, Sinanthropus 11 has an 'H type' articulation measuring 7 mm, while Sinanthropus 12 has one measuring 3-4 mm (Weidenreich 1943). Neanderthal and modern crania also have an 'H type' sphenoparietal articulation (Baker 1968). Weidenreich drew attention to this feature even though "in the Ngandong skulls the conditions are obscure" (1943: 199). His later examination of the Ngandong material (Weidenreich 1951) demonstrated that the most common form of articulation at the pterion in that sample was the 'I type', or frontotemporal articulation pattern. This pattern is not common on Australian Aboriginal crania (Larnach and Macintosh 1966, 1970). Therefore, this feature is of no use in demonstrating a morphological link between Indonesian and Australian crania.

6. A deep and narrow infraglabellar notch. In a later publication Weidenreich (1951) observed that there is no distinct nasion notch on any of the Ngandong crania. In any case, a deep and narrow infraglabellar notch may be observed on some australopithecines, on African *Homo erectus* crania, and on archaic *Homo sapiens* crania, such as Arago 21, Petralona and Skhūl 5.

It is obvious from this discussion that the features cited by Weidenreich as linking the Ngandong crania from Indonesia with the Australian Aborigines have a much wider distribution than can be accepted if they are Australasian "regional features" They can be found on most *Homo erectus* and archaic *Homo sapiens* crania, and many appear to be retained primitive features.

Weidenreich thought that Wajak 1 (formerly 'Wadjak') was a morphological link between the early Indonesian material and the Australian Aborigines, stating that there ". . . is an almost continuous phylogenetic line leading from the *Pithecanthropus* group through *Homo soloensis* to the Wadjak Man, and from there to the Australian aboriginal of today" (1945: 30).

He felt that Wajak 1 was similar in overall size, shape and proportions to the Keilor cranium from Australia, stating that "the likeness could not be greater if the skulls belonged to identical twins" (Weidenreich 1945: 27). These similarities, however, are so general as not to reveal any special relationship between the crania from the two regions. Also, as previously noted, the Wajak material now appears to be much younger (Holocene) than thought by Weidenreich.

Coon (1962) added little to Weidenreich's list of features documenting a morphological link between Indonesian *Homo erectus* and Australian crania, and again used features, such as large brow-ridges, flat frontal bones, and prognathic faces, which may be found on all *Homo erectus* and archaic *Homo sapiens* crania. He too saw the Wajak specimens as an extension of the evolutionary sequence in Java, suggesting that the Wajak "brain cases resemble those of *Pithecanthropus* and Solo in a familial way" (Coon 1962: 402), in that they still had an angular cranial contour, large mastoid processes, temporal lines that follow the contour of the vault, no supratoral fossa or ophryonic groove, and a large bizygomatic diameter. All of these features, however, may be found on *Homo erectus* and archaic *Homo sapiens* crania from outside of Java. Although Coon thought that the resemblance between Wajak 1 and Keilor was not as great as stated by Weidenreich, he still suggested that even if they were "not identical twins, they could have been brothers" (Coon 1962: 407). The relationship between Wajak 1 and and prehistoric Australian crania, such as Keilor, will be discussed in more detail later.

In the last few decades the number of late Pleistocene skeletal remains from Australia has greatly increased, and new studies of older material have been conducted. Thorne (1976, 1977, 1980; Thorne and Wilson 1977) has proposed the existence of two morphological types, one 'gracile' and the other 'robust' within this expanded sample – a suggestion that has not been accepted by all workers (Habgood 1985, 1986a, 1986b; Macintosh and Larnach 1976). Thorne hypothesised that his two morphological types were the result of separate migrations to Australia by morphologically distinct groups – the 'robust' type from Indonesia and the 'gracile' type from China. He states: "as Macintosh (1965) noted, the 'mark of ancient Java' is on the early Australians. So, too, is the stamp of ancient China" (Thorne 1980: 40).

Thorne, in collaboration with a number of other authors, has provided a detailed list of features to demonstrate the morphological link between Australian and East and Southeast Asian crania (Thorne and Wolpoff 1981; Wolpoff *et al.* 1984). They regard these features as 'clade features' in that they "uniquely characterise a geographic region" (Wolpoff 1986: 42) and document a 'morphological clade' – a term they use "to refer to a fossil sequence showing both continuity over time and differentiation from other contemporary morphological clades" (Thorne and Wolpoff 1981: 342).

Thorne and Wolpoff (1981) propose twelve 'clade' features which display a general resemblance between the Sangiran hominids from Indonesia, particularly Sangiran 17, and prehistoric Australian crania, especially those from Kow Swamp, which typify Thorne's 'robust' type. They state that: "Continuity. . . can be shown in any lineage. For this reason we wish to emphasise those elements of continuity which are either unique to or especially characteristic of this region [i.e. Australasia]" (1981: 342).

Wolpoff's reconstruction of Sangiran 17 (Thorne and Wolpoff 1981) allowed comparisons of the facial skeleton between early Indonesian and Australian hominids for the first time. Thorne and Wolpoff contend that the "evidence of local morphological continuity can be found in the vault and face. The facial region better expresses the clade relation" (1981: 345).

Thorne and Wolpoff (1981; Wolpoff *et al.* 1984; Wolpoff 1985) contend that they are dealing with the frequency of the features they are suggesting are 'regional', yet they base most of their comparisons on a single individual, Sangiran 17. This provides a frequency of 100% for the features in

Table 15.1. The occurrence of proposed 'Regional Features' (Thorne and Wolpoff 1981; Wolpoff *et al.* 1984) in fossil hominids. The numbering of the features follows the text. 'X' = present; '–' = absent; '?' = questionable identification or area not preserved.

	FEATURES											
	1	2	3	4	5	6	7	8	9	10	11	12
HOMO ERECTUS												
KNM–ER 3733	–	–	–	–?	X	X	–	–	–	X	?	?
KNM–ER 3883	X	–	–	–	X	?	–	–?	–	?	?	
Olduvai H9	X	–	X	?	X	?	?	?	?	?	?	?
Sangiran 17	X	X	X	X/–	X	X	X/–	X	X/–	X	?	X
Solo 12	X	X	X	X	X	?	?	?	?	?	?	?
Zhoukoudian reconstruction	X/–	–	–	–	X	X	–	–	X	X	–	X
AUSTRALIAN												
WLH 50	X	X	X	–	X	?	X	X	X	?	?	?
Kow Swamp 1	X	X	X	–	X	X	X	X	X	X	X	X
Kow Swamp 5	X	X	X	?	X	X	X	X	X	X	?	X
Kow Swamp 15	–	X?	X	?	?	X	X	X	X	X	X?	X
Cohuna	X	X	X	X	X	X	X	X	X	X	X	X
Talgai	X	X?	X?	?	X	X	?	X?	X?	X	X?	X
Cossack	X	X	?	?	X	X	X	?	?	?	X	X
Mossgiel	X	X	X	–	X	X?	X	X	X	?	X?	X
Lake Nitchie	–	–	X	–	X	X	X	X	X	X	X	X
Keilor	–	–	X	–	X	X	X?	X	X	X	?	X
Lake Mungo 1	–	–	X	–	–?	?	–	?	?	?	?	?
Lake Mungo 3	–	–	X	–	X	?	?	?	?	?	?	?
Lake Tandou	–	–	X	–	X?	X	X?	X	X	?	X	X
SOUTHEAST ASIA												
Niah	–	?	?	–	?	X	–	–	X	X	?	?
Wajak 1	–	–	X?	–	X	X	–	–	X	X	?	?
Wajak 2	?	?	X?	?	?	?	?	?	?	X	–	X
EAST ASIA												
Maba	–	–	–	?	?	?	?	?	?	?	?	?
Liujiang	–	–	X?	–	–	X	–?	–	X	X	–	X
Upper Cave 101	–	–	X	–	X	X	–?	–	X	X	–	–
Upper Cave 102	X	–	X?	X	X	X	–?	–	X	X	?	?
Upper Cave 103	–	–	X	–	X	X	–?	X	X	X	?	?
SUB-SAHARAN AFRICA												
Bodo 1	–?	–	X	?	?	X	–	X	X	X?	?	?
Ndutu	?	–	–	?	X	X	?	–	?	?	?	?
Kabwe 1	X	–	X	–	X	X	X	–	X	X	X	X
Laetoli H18	X	–	X	–	X	X	?	?	?	X	?	?
Omo 1	–	–	?	–	?	?	–?	?	X	?	?	?
Border Cave 1	–	–	–	–	?	?	X?	?	X	?	?	?
Florisbad	–	–	–	–	?	?	–?	?	X	–	?	?
NORTH AFRICA AND WESTERN ASIA												
Jebel Irhoud 1	–	–	–?	–	X	X	–	–	X	X	?	?
Wadi Halfa Sample	X/–	–	–	–	X/–	X	X	X	X?	?	–	X
Zuttiyeh	–	–	–	–	?	?	–?	–	–	?	?	?
Skhūl 4	–	–	?	?	X	X	–	–	X	X	–	X
Skhūl 5	–	–	X	–	–	X	–	–	X?	?	–	X
Jebel Qafzeh 9	–	–	X	?	–	X	–?	–	X	X	–	X
Amūd 1	X	–	–	–	–	X	–?	–	–	–?	–	X
EUROPE												
Petralona	X	–	X	–	X	X	–?	–	X	–	–	X
Arago 21	X	X	–	–	X	X	–	–	X	–	–	X
Steinheim	X	–	–	–	X	X	–	–	X	?	–?	X
La Ferrassie 1	X	–	–	–	–	X	–	–	X	–	–	X
Cro–Magnon 1	–	–	X	–	–	–	?	–	X	?	?	?
Oberkassel 1	–	–	X	–	–	–	X	X	X	?	–	?
Oberkassel 2	–	–	X	–	–	–	–	–	–	?	–	X
Dolni Vestoniče 3	–	–	X	–	–	–	–	–	–	?	X?	X
Předmost 3	–	–	–	–	–	–	–	–	–	–	–	X
Předmost 4	–	–	–	–	–	–	–	–	–	–	–	X
Mladeč 1	–	–	–	–	–	–	–	–	–	–	?	?

Table 15.2. Percentages of the occurrence of proposed 'Regional Features' (Thorne and Wolpoff 1981; Wolpoff *et al.* 1984) in Australian Aboriginal populations. The numbering of the features follows the text. Note that Feature 1 here refers to marked frontal recession, and not actually frontal flatness, whereas Feature 2 refers to marked postorbital constriction, and not posterior position of the minimum frontal breadth.
M = Male; F = Female.

	Fenner 1939 Total Series		Larnach & Macintosh 1966 New South Wales			Larnach & Macintosh 1970 Queensland		
Features	M	F	Total	F	M	Total	F	M
1	45.0	28.0	39.5	23.1	53.2	40.0	34.0	44.5
2	–	–	–	37.5	67.2	–	39.1	77.9
3	53.0	72.0	–	–	–	–	–	–
4	–	–	–	–	–	–	–	–
5	–	–	–	–	–	–	–	–
6	–	–	74.8	76.0	73.7	78.0	81.8	75.2
7	43.0	7.0	45.9	2.1	78.1	54.0	17.4	80.0
8	92.0	95.0	54.9	40.5	66.7	50.5	27.3	69.1
9	55.0?	21.0?	80.7	83.7	78.5	79.6	71.7	85.0
10	37.0	40.0	73.6	69.4	77.2	67	56.8	73.8
11	–	–	–	–	–	–	–	–
12	–	–	–	–	–	–	–	–

Indonesia, but it is not a very good 'sample' on which to base an argument for 'regional continuity'.

The occurrence of the morphological traits suggested to be regional features by Thorne and Wolpoff on Middle Pleistocene, Upper Pleistocene, and prehistoric and modern Australian Aboriginal crania are presented in Tables 15.1 and 15.2.

The features Thorne and Wolpoff highlighted are:

1. The flatness of the frontal in the sagittal plane. Brown (1981) has demonstrated that some of the late Pleistocene crania from Australia suffered cranial deformation that produced a number of typical cranial features, one being a flat frontal. Even when disregarding those crania that are definitely or possibly deformed, a flat frontal is still thought to be an important 'clade feature' (Wolpoff *et al.* 1984). This feature, however, is a general character of hominoids (Andrews 1984), is typical of *Homo erectus* (Howell 1978; Rightmire 1984; Stringer 1984a; Wood 1984), and may be found on the crania of archaic *Homo sapiens*, such as Laetoli H18, Omo 2 and Kabwe 1.

2. A posterior position of the minimum frontal breadth well behind the orbits. All *Homo erectus* crania have marked postorbital constriction and frontal bones that narrow anteriorly (Howell 1978; Wood 1984). Both of these features are characteristic of hominoids (Andrews 1984) and may be found on most archaic *Homo sapiens*. Sangiran 17, however, parallels the Kow Swamp configuration more closely than the other hominids examined.

3. A relatively horizontal orientation of the inferior supraorbital border. Neither Sangiran 17 nor any of the late Pleistocene Australian cranial sample (except possibly WLH 50) have what can be called 'horizontal' inferior supraorbital borders, in that they all have obliquely-angled superior orbital borders. Fenner, in relation to the orbital shape of a sample of Australia Aboriginal crania, states that "the axis of the orbits varied in obliquity between horizontal and strongly oblique, but the commonest condition was a slight obliquity of the axis" (1939: 269).

The axis of the orbit generally equates with the orientation of the inferior supraorbital border, and so most Australian Aboriginal crania would have slightly oblique inferior supraorbital borders. This feature has, therefore, been taken to mean a relatively straight (not necessarily horizontal) inferior supraorbital border. Weidenreich stated that on the Ngandong calvaria:

> "the contour of the supraorbital margin viewed from in front shows its edge as a long straight line. . . that curves downward fairly abruptly only towards the lateral ends. . . Conditions are similar in *Sinanthropus,* and they must also have been so in *Pithecanthropus*. . . The same condition is found in the Rhodesian skull. . ."(1951: 252).

The inferior supraorbital borders of some archaic *Homo sapiens,* such as Bodo 1, Petralona, and Laetoli H18, are, like those of Kabwe 1, relatively horizontal, as are those of many early anatomically-modern *Homo sapiens* crania from Europe, Western Asia, and China. Those on Olduvai Hominid 9 are also relatively straight, whereas on Narmada, KNM-ER 3733 and KNM-ER 3883 they appear to be more rounded.

4. Presence of a distinct pre-bregmatic eminence. Sagittal thickening, especially near the bregma, is said by Howell (1978) to be typical of *Homo erectus* in general. Most of the Asian *Homo erectus* crania have well-defined metopic, coronal, and sagittal ridges, and a bregmatic eminence (Santa Luca 1980; Stringer 1984a; Weidenreich 1943, 1951). These prominences may unite at the bregma to form a cruciate configuration, or the metopic ridge may remain separated from the bregmatic eminence. Santa Luca (1980) does not describe a pre-bregmatic eminence for any of the Asian *Homo erectus* crania he examined. He describes the bregmatic eminence as being either continuous with the metopic ridge, as on the Zhoukoudian crania and the early Indonesia *Homo erectus* material including Sangiran 17, or separated from it as on the Ngandong hominids. Examination of a cast of Sangiran 17 did not reveal a pre-bregmatic eminence but Wolpoff (pers. comm.) has reaffirmed the presence of such a structure on the original specimen. Brown (1981) has demonstrated that the pre-bregmatic eminence seen on some prehistoric Australian crania may have been the result of cranial deformation. This is consistent with the presence of a pre-bregmatic

eminence on the cranially-deformed Zhoukoudian Upper Cave 102 crania (Weidenreich 1939; Coon 1962). The bregmatic region of Olduvai Hominid 9 is missing, while on KNM-ER 3883 the region is preserved but there is no evidence of bregmatic swelling (Rightmire 1984). The bregma is not preserved on KNM-ER 3733, but the bone anterior to it does display some thickening. It is impossible on a cast, however, to ascertain if this represents a pre-bregmatic eminence, a metopic ridge, or postmortem damage. Leakey and Walker (1985: 148) state that "there is a very large (25.0 mm diameter) defect over bregma" on KNM-ER 3733. Kabwe 1 has a metopic ridge continuous with a bregmatic eminence that is accentuated by a marked postbregmatic depression (Santa Luca 1980). Wu Xinzhi (1986) contends that the Ziyang calvaria from China has a pre-bregmatic eminence, whereas in an earlier publication (Wu Xinzhi and Zhang Zhenbiao 1985) he suggests it has a cruciate-shaped eminence at bregma that extends both posteriorly and anteriorly.

5. A low position of maximum parietal breadth (at or near the parietal mastoid angle). This feature is a shared primitive character found on all *Homo erectus* and archaic *Homo sapiens* crania (Andrews 1984; Howell 1978). Brown (1981) has also suggested that the cranial deformation of some prehistoric Australian crania may have caused the development of this feature.

6. Marked facial prognathism. This is a primitive feature that is found on all *Homo erectus* and archaic *Homo sapiens* crania (Andrews 1984; Howell 1978; Stringer 1984a). Thorne and Wolpoff (1981) have tried to demonstrate that Sangiran 17 is more prognathic than KNM-ER 3733, which they take to be representative of African *Homo erectus*, and so feel that marked prognathism is one of the most striking 'clade features' for Australasia. The recently discovered KNM-WT 15000 male *Homo erectus* skeleton from Kenya, however, appears more prognathic than the female KNM-ER 3733 (Brown *et al.* 1985), which may negate Thorne and Wolpoff's (1981) conclusion. There are also problems with Thorne and Wolpoff's argument (Stringer, pers. comm.). The face of Sangiran 17 has been crushed up under the vault so that it is difficult to ascertain the exact degree of prognathism. Thorne and Wolpoff (1981) provide the basion-nasion measurement (111.2 mm) and the gnathic index (121.9 mm) for Sangiran 17 which allows the calculation of the basion-prosthion value (135.6 mm). If one tries to position callipers, set on 135.6 mm, on a cast of Sangiran 17 it seems impossible for the face to be pulled out far enough for the basion-prosthion chord to reach this value (Stringer, pers. comm.). The degree of facial prognathism of Sangiran 17 would, therefore, appear to have been over-estimated by Thorne and Wolpoff (1981). Archaic *Homo sapiens* crania from the sites of Mugharet es-Skhūl (McCown and Keith 1939) and Jebel Qafzeh (Vandermeersch 1981) in Western Asia, and Dar-es-Soltane 2 in North Africa (Ferembach 1976) are very prognathic, as are anatomically-modern *Homo sapiens* crania from North African sites (Anderson 1968; Greene and Armelagos 1972).

7. Presence of a malar (zygomaxillary) tuberosity. Howell (1978: 195) has stated that *Homo erectus* crania have "malar facies, sometimes with

zygomaxillary tuberosity", yet this feature does not appear to be evident on KNM-ER 3733, nor KNM-ER 3883, and is not commonly found on the Zhoukoudian material (Leakey and Walker 1985; Weidenreich 1943). However, Weidenreich (1943: 83) does state that on a malar fragment, which probably belonged to Skull 10, ". . . the entire anterior area shows a distinct prominence. . . which projects also beyond the lower margin of the bone" (1943: 83). He refers to this structure as a malar tuber, and suggests that it "may also occur in modern man, but certainly cannot be considered a common feature" (1943: 83).

Examination of a cast of Sangiran 17 did not reveal the presence of a distinct zygomaxillary tuberosity, in that the entire malar was evenly curved, although Wolpoff (pers. comm.) contends that it is clearly evident on the original. Kabwe 1 has a malar projection that may be classified as a tuberosity. Border Cave 1 also has a ridge on its malar which again may be a tuberosity. The anatomically-modern Oberkassel 1 cranium from Germany has a clear ridge running perpendicular to the zygomaxillary suture. Many crania from the Mesolithic sites of Jebel Sahaba and Wadi Halfa in North Africa display zygomaxillary tuberosities (Anderson 1968; Greene and Armelagos 1972). Thorne (1976) found that 50 per cent (6 individuals) of his sample from Kow Swamp had marked malar tuberosities. Wu Xinzhi (1986) states that all Chinese Upper Pleistocene crania have developed malar tuberosities, although these were not evident on casts of this material.

8. Eversion of the lower border of the malar. This feature is not evident on KNM-ER 3733 nor on any remains from Zoukoudian (Leakey and Walker 1985; Weidenreich 1943), but appears to be present on the Bodo 1 hominid from Ethiopia (Conroy 1980). KNM-ER 3883 does display eversion of its right malar, but this appears to be due to postmortem damage. The anatomically-modern crania from Wadi Halfa and Jebel Sahaba, North Africa, display marked malar eversion (Anderson 1968; Greene and Armelagos 1972) as do Oberkassel 1 and Zhoukoudian Upper Cave 103.

9. Rounding of the inferolateral border of the orbit. This feature would appear to be typical of *Homo erectus* in general (Howell 1978), with Weidenreich (1943: 212) stating that in *Sinanthropus* "the infraorbital margin is rounded and is at the same level as the even floor of the orbit". This same trait is also used to document continuity in China (Weidenreich 1943). Rightmire (1987) suggests that this feature is common in many crania of archaic *Homo*. It is found on archaic *Homo sapiens* crania including Kabwe 1, Bodo 1, Omo 1, Border Cave 1, Florisbad, Jebel Irhoud 1, Petralona, Arago 21, Steinheim, and Neanderthal crania such as La Ferrassie 1. The Cro-Magnon 1, Oberkassel 1, Skhūl 4, and Jebel Qafzeh 9 cranai also have it, as do some of the crania from Wadi Halfa (Greene and Armelagos 1972).

10. The lower border of the nasal aperture lacks a distinct line dividing the nasal floor from the subnasal face of the maxillae. This is a hominoid condition (Andrews 1984), and is found on some *Homo habilis*, and most *Homo erectus* crania (Howell 1978; Stringer 1984a). Weidenreich (1943: 208) stated that in Sinanthropus "the nasal floor is even and separated from the clivus nasoalveolaris by a simple margo limitans". Leakey and Walker (1985: 148)

suggest that on KNM-ER 3733 the "lateral border of the aperture is rounded inferiorly". Archaic *Homo sapiens* crania such as Kabwe 1, Bodo 1, Laetoli H18, and Jebel Irhoud 1 also demonstrate this pattern, as do Skhūl 4 and Jebel Qafzeh 9.

11. Marked expression of curvature of the posterior alveolar plane of the maxillae. This feature is hard to assess, as the maxillary region is not very well preserved on African and Chinese (and for that matter Indonesian) *Homo erectus* material, and archaic *Homo sapiens* crania. Kabwe 1, however, appears to display a similar configuration to that seen on Sangiran 17, as do some of the crania from Jebel Sahaba (Anderson 1968), and possibly Dolní Vestoniče 3 from Czechoslovakia.

12. The degree of facial and dental (especially posterior) reduction. Thorne and Wolpoff state "...reduction in the face and posterior dentition seems to have been characteristic of both this region [Australasia] and mainland Asia during the Middle and Upper Pleistocene" (1981: 345), and that "in both regions, facial heights and posterior tooth sizes in the Middle Pleistocene had already reduced to what might be thought of as the late Upper Pleistocene (but not modern) condition" (1981: 345) – admitting that the faces and teeth of Middle Pleistocene and modern crania from Australasia are larger than those from mainland Asia. Yet, Wolpoff *et al.* (1984) cite the 'maintenance' of large posterior dentition as a 'clade feature' (Rightmire 1987). In either case, a large dentition and orofacial complex is a retained primitive feature found on all *Homo erectus* and archaic *Homo sapiens* crania. Reduction of these features is a world-wide phenomenon and not typical of any particular geographical region.

In a later publication (Wolpoff *et al.* 1984), these twelve characters are reaffirmed as 'clade features', and other more specific resemblances between prehistoric Australian crania and the Ngandong hominids are proposed, These are:

1. *Cohuna:* The temporal has a straight superior border and a low posterior border. This configuration is a retained primitive feature (Andrews 1984), and is typical of all *Homo erectus* crania (Howell 1978). Larnach and Macintosh (1974) found that such a pattern was more common on Caucasoid crania than on Australian Aborigines.

The temporal line forms a ridge spanning the entire length of the frontal. KNM-ER 3733 and KNM-ER 3883 also display this pattern (Rightmire 1980, 1984), as do many of the crania from the Mesolithic sites of Wadi Halfa and Jebel Sahaba (Anderson 1968; Greene and Armelagos 1972).

The coronal suture closely approaches the anterior angle of the temporal squama. This is the pterion in the 'H narrow type' of articulation, but as mentioned earlier, the fronto-temporal, or 'I type', is the most common articulation form within the Ngandong sample, while the 'H type' is commonly found on the Zhoukoudian *Homo erectus* material, as well as on Neanderthal, and modern crania (Larnach and Macintosh 1966, 1970; de Villiers 1968).

Maximum cranial breadth at the mastoid angle of the parietals. The primitive nature of this feature has been discussed previously (Andrews

1984).

A *torus angularis* at the lambdoidal suture superior to the asterionic region. As discussed previously, an angular torus is typical of most Asian *Homo erectus* crania (Andrews 1984; Santa Luca 1980; Stringer 1984a; Weidenreich 1943), and is evident on a number of archaic *Homo sapiens* crania, such as Bodo 1, Castel di Guido 5 and Arago 47 (Asfaw 1983; Grimaud 1982; Mallegni *et al.* 1983; Stringer 1984a). As mentioned before, Cohuna does not appear to have a typical (cf. Santa Luca 1980) *torus angularis*.

2. *Kow Swamp 5:* The entire articular surface anterior to the roof of the glenoid fossa forms a vertical (anterior) wall. The glenoid fossa in its entirety is deep and the sagittal narrow. This pattern is not confined to the Ngandong hominids in that it is also seen on Olduvai Hominid 9 (Hublin 1986).

3. *Kow Swamp 7:* A broad sulcus on the frontal joins lateral sulci paralleling the temporal ridges producing a continuous sulcus outlining the entire superior portion of the frontal on three sides. This suggested resemblance is interesting in that a lack of a supratoral sulcus, which typifies the Ngandong crania (Santa Luca 1980; Weidenreich 1943, 1951), is also proposed as a 'clade feature' (see below).

4. *Kow Swamp 9:* The lateral supraorbital corners form a posteriorly facing triangle (lateral frontal trigone) whose apex is the temporal ridge. The supraorbital region of this specimen appears fragmentary with only a small portion of the right lateral segment evident on the available photograph (Thorne 1975). It is difficult to ascertain if the form of this fragment is close to the characteristic form seen on the Ngandong crania. Macintosh and Larnach (1972) also consider a lateral supraorbital wing as typical of *Homo erectus* in general.

A nuchal torus with a marked posterior projection, combined with a distinct separation of the torus from the occipital, with superior and inferior sulci and an inferior dip at inion. But, as Weidenreich states with regard to the occipital torus of the Ngandong hominids: "The general pattern of the torus. . . displays interesting variations. . . some characteristic differences in the shaping of the torus particularly involve its middle portion and the method of connection with the occipital planum" (1951: 255).

The preserved portion of the occipital torus of Kabwe 1 would appear to be similar to the pattern described above (Santa Luca 1978, 1980).

It should be noted that these more specific resemblances between the Ngandong hominids and Australian Aboriginal crania are variable within both the Ngandong and Australian samples (Fenner 1939; Larnach and Macintosh 1966, 1970; Santa Luca 1980; Thorne 1976; Weidenreich 1951). Some of the features also differ from the state that Thorne and Wolpoff (1981) describe for Sangiran 17. For example, Sangiran 17 lacks a supratoral sulcus above the occipital torus and a posteriorly elongated triangle at the lateral supraorbital corner, and has only a weak angular torus which, at its most posterior position, is still anterior to the lambdoidal suture. These features are clearly quite variable, and so their usefulness as 'regional features' must be doubted.

Wolpoff *et al.* (1984) also accept as 'clade features' nine features thought to be typical of the Ngandong crania by Larnach and Macintosh (1974), and which the latter had, in an analysis of modern crania, found to have their highest frequency on Australian Aboriginal crania, or crania from New Guinea. These features are:

1. A very large rounded zygomatic trigone. This feature, however, was scored as present in only 0.4% of the sample of 207 Australian Aboriginal crania (Larnach and Macintosh 1974: Table 2), and so its significance must be questioned.

2. Absence of a distinct ophryonic groove (supratoral or supraglabellar sulcus). This feature was commonly found within the entire modern sample (Larnach and Macintosh 1974: Table 2). Five out of twelve crania from Kow Swamp, however, are said to combine a distinct supraglabellar fossa and an ophryonic groove (Thorne 1976). Howell (1978) suggests that some *Homo habilis* crania may lack a supratoral groove, and that *Homo erectus* crania may or may not have a supratoral sulcus. Olduvai Hominid 9, Hominid 12, KNM-ER 3733, and KNM-ER 3883 all have shallow supratoral sulci, which differ from the more prominent ophryonic groove found on the Zhoukoudian crania (Leakey and Walker 1985; Rightmire 1984; Weidenreich 1943). As mentioned earlier, Wolpoff *et al.* (1984) cite the presence of a supratoral sulcus on Kow Swamp 7 as a specific resemblance to the Ngandong crania. The absence of this character, therefore, does not seem to imply a special relationship between the Ngandong and Australian crania. It does, however, differentiate the Ngandong and Zhoukoudian sample.

3. Presence of a suprameatal tegmen. In an earlier publication Macintosh and Larnach (1972) proposed that this is a characteristic of *Homo erectus* in general. This feature was not present in any of the Australian Aboriginal crania, and was present in only 1.2% of the 80 crania from New Guinea studied (Larnach and Macintosh 1974: Table 2).

4. The course of the squamo-tympanic fissure is transverse. Again, in an earlier publication Macintosh and Larnach (1972) found this pattern to be characteristic of *Homo erectus* in general. This feature was also found to be more common in the samples of Mongoloid and New Guinean crania than in the sample of Australian Aborigines (Larnach and Macintosh 1974: Table 2).

5. Angling of the petrous to the tympanic in the petro-tympanic axis. Macintosh and Larnach (1972) also regarded this feature as being typical of *Homo erectus* in general. The New Guinean sample had the highest frequency of this feature (5%), while only 1.9% of the Australian Aboriginal sample had it (Larnach and Macintosh 1974: Table 2).

6. A lambdoidal protuberance. This character, which is evident on KNM-ER 3733 and on crania from Zhoukoudian, appears to be a common *Homo erectus* feature. It was only present on 1.4% of the Australian Aboriginal crania studied (Larnach and Macintosh 1974: Table 2).

7. Presence of a marked ridge-shaped occipital torus. Some archaic *Homo sapiens* crania such as Kabwe 1, Saldanha 1, and Omo Kibish 2, have

occipital tori that can be regarded as ridge- shaped. This form of occipital torus was found on only 3.4% of the Australian Aboriginal crania studied (Larnach and Macintosh 1974: Table 2).

8. An external occipital crest emerging from the occipital torus. This character is not found on all of the Ngandong sample, while it is present on Sangiran 4 (Macintosh and Larnach 1972; Santa Luca 1980). Although Macintosh and Larnach (1972) consider absence of an external occipital crest as being typical of *Homo erectus,* Howell (1978) mentions it as being present on *Homo erectus* crania. Hublin (1986) believes that no *Homo erectus* crania have a true external occipital protuberance, which he feels is only found on relatively recent *Homo sapiens* crania. All samples of modern crania studied by Larnach and Macintosh (1974: Table 2) had some individuals with this feature present, although it was most common in the Australian Aborigines, being present in 13% of the sample. None of the Zhoukoudian Upper Cave crania have this feature (Weidenreich 1939; Wu Xinzhi 1961).

9. Presence of a marked occupital *sulcus supratoralis*. Macintosh and Larnach (1972) previously regarded the presence of this sulcus to be typical of *Homo erectus,* while Weidenreich (1943) found it on the Zhoukoudian material, and Howell (1978) suggested that it was usually present on *Homo erectus* crania. Only 8.7% of the Australian Aboriginal sample had it, while it was also present on 5% of the Mongoloid sample (Larnach and Macintosh 1974).

Thus, most of the characters used by Larnach and Macintosh (1974) would appear to be commonly found on *Homo erectus* crania in general and not unique to the Ngandong hominids. They therefore do not demonstrate any specific relationship between the Ngandong hominids and prehistoric or modern Australian Aboriginal crania. Many of the traits did have a higher frequency in the Australian Aboriginal crania, but they were still found on a very low percentage of the total sample studied. Larnach and Macintosh felt that from their study ". . . it is difficult to point to any particular racial group as a substantial inheritor of Solo genes" (1974: 100), and that there is ". . . a great morphological difference between the Solo crania we know and the crania of any modern group" (1974: 101).

Hence the usefulness of these features as 'clade features' must be seriously doubted.

As mentioned earlier, Brown (1981) found that some late Pleistocene Australian crania displayed evidence of cranial deformation, and he suggests that ". . .there is a fine gradation from the crania that are obviously deformed (such as Kow Swamp 5 and Cohuna) into those which show no evidence of deformation" (1981: 165).

Brown does not believe, however, that his study negates the concept of 'regional continuity' stating that "this [cranial deformation] does not detract from the general argument for regional continuity which is supported by the size and morphology of the orofacial complex. . ." (1981: 166).

The features that Brown (1981) mentions as supporting 'regional continuity', such as great symphyseal height and thickness, marked gonial eversion, great masseteric fossae, and a negative chin with little incurvature, are typical of *Homo erectus* mandibles (Howell 1978; Rightmire 1980; Weidenreich 1936), while others, such as large dentitions and broad palates, are general primitive characters. Mandibles from the North African Mesolithic sites of Mechta-el-Arbi, Jebel Sahaba and Wadi Halfa display pronounced gonial eversion (Anderson 1968; Briggs 1950; Greene and Armelagos 1972). This feature also appears to be present on the Oberkassel 1 and the Zhoukoudian Upper Cave 101 mandibles. Larnach and Macintosh (1970), studying mandibles from eastern Australia, found high symphyses on 50.6% of males and 8.9% of females (34% of the total sample), gonial eversion on 40.9% of males and 3.5% of females (26.2% of the total sample), depressed masseteric fossae on 43.5% of males and 31.3% of females (38.5% of the total sample), and a negative chin on 33.7% of males and 43.6% of females (37.9% of the total sample).

From this discussion it is evident that all of the characters proposed by Thorne and Wolpoff (1981), Wolpoff *et al.* (1984), and Brown (1981) to be 'clade features' linking Indonesian *Homo erectus* material with Australian Aboriginal crania, are retained primitive features present on *Homo erectus* and archaic *Homo sapiens* crania in general. Many are also commonly found on the crania and mandibles of anatomically-modern *Homo sapiens* from other geographical regions, being especially prevalent on the robust Mesolithic skeletal material from North Africa.

NORTH ASIA AND AUSTRALIA

As previously discussed, Thorne (1977) has identified two morphological types within the corpus of late Pleistocene Australian crania which he relates to two separate migrations to Australia. This conclusion is supported by Wolpoff *et al.*, who state that: "We conclude that a very specific case can be made linking features in some of the Pleistocene Australians with the Ngandong fossils. At the same time the range of variation and the morphological details of that variation indicate that the ancestry of the Australians is more complex than a simple unique line of descent from the Ngandong folk" (1984: 445). They go on to outline similarities between East Asian and Australian crania.

Wolpoff *et al.* (1984) place Wajak 1 within their East Asian morphological type, and reaffirm Weidenreich's (1945) and Coon's (1962) suggested similarity between it and the Keilor cranium in size, proportion and facial flatness. They do not consider Keith's comment that ". . . to make the Wadjak type into the Australian needs an extensive reduction in all parts, and when this is done, there remain so many points of difference that we cannot regard their relationship as more than a cousinship. . . . Proto-Australoid is not the right name for the Wadjak man" (1925: 455).

Wolpoff *et al.* (1984) also find Keilor to be similar to the Liujiang cranium from China in overall size and proportion, and in having a wide interorbital area, an indistinct nasal margin, thick lateral supraorbital corners, low square orbits and a similar degree of facial flatness. The parallels between Keilor, Wajak 1, and Liujiang, however, do not demonstrate any particular relationship between the crania, in that most of the features are very general and broadly descriptive. They could be used to suggest similarities with numerous crania from various localities. A similarity in overall cranial size and proportion does not necessarily indicate a special relationship. Neither does the presence of low square orbits or a wide interorbital area, as these may be found on many crania, including those of some archaic and early modern *Homo sapiens*. An indistinct nasal margin and thickened lateral supraorbital corners were also cited by Wolpoff *et al.* (1984) as features that demonstrated a link between Indonesian *Homo erectus* and Australian crania. However, these features have been shown to be primitive characters, and so of little use in demonstrating any morphological links. No data were presented to demonstrate a similarity in facial flatness, and this similarity appears to have been overstated, in that Keilor is more like other Australian crania in having a less flat upper and middle face than crania from China. These crania do, as Wolpoff *et al.* (1984) acknowledge, have major differences, but they feel that these are outweighed by the similarities. This conclusion is hard to understand, in that the Liujiang cranium, although robust, is generally 'East Asian' in overall morphology, while Keilor is typically 'Australian' in form (Habgood 1985, 1986b; Howells 1976b; Wu Xinzhi and Zhang Zhenbiao 1985). Liujiang lacks most of the features thought to be typical of Australian Aboriginal crania (Larnach and Macintosh 1966, 1970), while Keilor does not present the features that Wu Rukang (1986) proposes are characteristic of 'Mongoloid' crania. Wajak 1 is a robust cranium which displays both 'Australoid' characters, such as a dolichocephalic cranium with thick vault bones, a sagittal keel, occipital protrusion, alveolar prognathism, indistinct lower nasal margin, large teeth (Jacob 1976; Wolpoff *et al.* 1984), and 'Mongoloid' characters, including a flat broad face with laterally projecting malars, a broad flat nasal root, a non-depressed nasion, poorly expressed canine fossae and broad low rectangular orbits (Jacob 1976; Wolpoff *et al.* 1984). Many of these features, however, are not uniquely 'Australian' or 'Chinese', and may be found on robust modern crania from other areas such as Europe. Wolpoff *et al.* (1984) see marked similarities between Wajak 1 and the Zhoukoudian Upper Cave 101 cranium, and to a lesser extent Liujiang, in that they share a combination of suggested 'Australoid' and 'Mongoloid' features. The proposed 'Australoid' features of this material would appear, generally, to be the result of robusticity, and not due to any special relationship with Australian Aborigines. The Liujiang and Upper Cave crania appear to be robust early 'Mongoloid' crania that have not yet developed the modern 'Mongoloid' configuration (Wu Rukang 1986; Wu Xinzhi 1961; Wu Xinzhi and Zhang Zhenbiao 1985), whereas Wajak 1 (depending on its date) may reflect more recent genetic contact with Australia and/or New Guinea. Multivariate analyses also suggest that Wajak 1, Liujiang and the Zhoukoudian Upper Cave crania are more similar to each other than they are to the Keilor cranium and other prehistoric Australian material (Habgood 1985, 1986b).

The dating of the Wajak 1 remains also has a bearing on their particular morphological relationships. Wajak 1 is often suggested to be Late Pleistocene in age, but could actually date from the Holocene period. The fauna collected by Dubois from Wajak is essentially modern, with only two species that are now extinct on Java, but these may be found on Sumatra and the Malay Peninsula (van den Brink 1982). The Wajak fauna is similar to that from the Sampung cave (Java) which is considered to be Mesolithic or Neolithic and has an estimated age of 3000-6000 years BP (van den Brink 1982). Other caves from the region of the Wajak cave (which was not destroyed as stated by van den Brink (1982), Howells (1973), and others, but has recently been relocated (de Vos, pers. comm.)) have provided fauna that is similar in composition and degree of fossilisation to that from Wajak and Sampung (van den Brink 1982; de Vos, pers. comm.). Three of these caves, Hukgrot (Hoekgrot)-Eastern Corner Cave (which also had bone tools similar to those from Sampung), Gua Cimbe (Goea Djimbe) and Gua Kecil (Goea Ketjil), have produced fragmentary human remains which included teeth that have the same morphology and size as those from Sampung and Wajak (de Vos, pers. comm.). This evidence seems to confirm von Koenigswald's (1956) suggestion that the Wajak material was Mesolithic or Neolithic in age. The probable Holocene date of Wajak 1 makes its usefulness as a morphological link between Late Pleistocene crania from China and Australia very limited, as its morphology may be the result of more recent population movements or gene flow.

Similarities are also suggested by Wolpoff *et al.* (1984) between the Lake Mungo 1 calotte from Australia and the Ziyang cranium from China. These are: the small overall size, general proportions, the degree of central supraorbital development, presence of a distinct supratoral sulcus, and the vertical separation of the internal and external occipital protuberances. Again, these features are quite general and broadly descriptive, and so cannot demonstrate morphological continuity between Chinese and Australian crania. Also, Wolpoff *et al.* (1984) state that the central supraorbital region of Lake Mungo 1 is vertically thicker and more anteriorly projecting than that of Ziyang, yet they still contend that the development of this area is similar in the two specimens. As discussed previously, both the lack of, and the presence of, a supratoral sulcus have been used as 'regional features' linking Indonesian and Australian crania (Wolpoff *et al.* 1984).

It is worth noting that the comparisons outlined earlier were made with Indonesian Middle Pleistocene *Homo erectus* material, whereas those with East Asia utilize late Pleistocene and Holocene crania.

In summary, Wolpoff *et al.* (1984) have clearly failed to identify any 'clade features' common to Australian and East Asian crania. All they have demonstrated is that there are broad and general similarities between some crania from the two regions, but these do not demonstrate any close phylogenetic relationship.

DISCUSSION

The question that must now be asked is: Are any of the features that have been discussed above of use as 'clade features'? On an individual basis they are not. As proposed by several authors (Groves, this volume; Rightmire 1987; Stringer 1984b, 1985; Stringer *et al.* 1984), all of the features have been shown to be retained primitive characters in that they can be found on *Homo erectus* and archaic *Homo sapiens* crania in general. None of these features, therefore, can be used as 'clade features' in that they do not "uniquely characterise a geographic region" (Wolpoff 1985: 42). The occurrence of the traits used by Wolpoff *et al.* (1984) to identify an Australasian 'morphological clade' on hominid crania from other parts of the world was not considered because

> ". . . the validity of non-metric analysis is limited by the diversity of the groups analysed. Comparisons are likely to be misleading or invalid unless they are between populations that are fairly closely related. In sum, we believe that by limiting the geographic region over which these comparisons are made...observations such as those made by Weidenreich are phenetically valid" (1984: 426).

This is a rather circular argument in that it assumes a close relationship between the groups under study, yet this is what the analysis should be attempting to demonstrate. Another example of circular argument employed by the above authors is exemplified by the starting point for their 'clade features' – namely, the use of Sangiran 17 as the cranium with which they compared the Australian crania. The archaic *Homo sapiens* Kabwe 1 cranium, which is chronologically closer to the Australian skeletal material, shares numerous morphological features with Australian Aboriginal crania including many that Thorne and Wolpoff (1981) and Wolpoff *et al.* (1984) use as 'clade features', as outlined above. If they had started their investigation with Kabwe 1, they would have been able to document a fair degree of morphological continuity between this fossil and later Australian material. This continuity would have been further strengthened with the addition of other sub-Saharan African archaic *Homo sapiens* material, such as Bodo 1.

As outlined on the preceding pages, all of the traits used by Thorne and Wolpoff (1981) and Wolpoff *et al.* (1984) can, in a broad sense, be found on *Homo erectus* and archaic *Homo sapiens* crania in general. The particular configurations of a number of the features, however, cannot be matched outside of Australasia. Facial prognathism is greater in Australian Aborigines than in any other modern regional group, and Thorne and Wolpoff (1981) contend that Sangiran 17 is more prognathic than the older KNM-ER 3733 crania. However, as discussed previously, this suggestion is problematic. The long, sagittally-flat frontal bone, with a posterior position for the minimum frontal breadth well behind the orbits, as seen on Kow Swamp 1 (a non-deformed cranium: Brown 1981), has its closest match on Sangiran 17. The presence of a medium-to-marked malar tuberosity is quite common within the Kow Swamp sample (Thorne 1976), and on Australian Aboriginal crania (Table 15.2). It is said to be present on Sangiran 17 (Thorne and Wolpoff 1981), but is lacking on most of the Zhoukoudian material (Weidenreich 1943), and on KNM-ER 3733 and KNM- ER 3883 (Leakey and Walker 1985). A malar tuberosity may be present on Kabwe 1 and Border Cave 1, and is evident on Oberkassel 1 and the cranium from Wadi Halfa and Jebel Sahaba. The only archaic *Homo sapiens* crania that has eversion of the lower border of the malar would appear tc Bodo 1, although it can be seen on the anatomically- modern crania from Wadi Halfa and Jebel Sahaba, on Oberkassel 1, and Zhoukoudian Upper Cave 103 (Table 15.1). These features, therefore, could be evidence of 'regional continuity' within Australasia, but they are a shaky base upon which to construct a 'morphological clade', for the individual features are not unique to this particular geographical region.

Thorne and Wolpoff state that

> "we recognize that many of these features found in Sangiran 17 can be found independently in other fossil crania of the genus *Homo*. Yet, in no other region can a specimen be found that combines so many features that seem unique, or at least of high frequency, in Pleistocene Australians" (1981: 345).

A similar view is expressed by Wood who proposes that:

> ". . . a definition of a taxon should include all its autapomorphies, but a definition is not limited to such features. For, while symplesiomorph and shared-derived characters are (by definition) not unique to a taxon, what may be unique is a particular combination of such character states" (1984: 104).

A combination of some of the traits suggested by Thorne and Wolpoff (1981) and Wolpoff *et al.* (1984) (but not all, because many are found in combination on *Homo erectus* and/or archaic *Homo sapiens* crania in general) may be evidence of 'regional continuity' in Australasia. They cannot, however, be documenting a 'morphological clade', for they are not unique to the Australasian region. A cranium that has a long and sagittally- flat frontal bone with a posteriorly placed minimum frontal breadth, very prognathic face (?) and malars with everted lower borders and prominent zygomaxillary tuberosities, could be regarded as having an 'Australianness' about it, in that all of these features are commonly found on both prehistoric and modern Australian crania (Tables 1 and 2). These features, however, could be functionally related, and may reflect a large orofacial complex. If this is correct, one must explain why all crania with large orofacial complexes do not have this combination of features. As discussed above, these features are also found on Sangiran 17 (Thorne and Wolpoff 1981), but are not matched, in combination, on *Homo erectus* or archaic *Homo sapiens* crania from other geographical areas. Some, especially those on the malar bone, are, however, found in combination on a number of anatomically-modern *Homo sapiens* crania such as those from Wadi Halfa, Jebel Sahaba, and Oberkassel 1 (Table 15.1). All of these crania have large orofacial complexes. It is difficult to ascertain if this patterning negates the possibility of 'regional continuity' in Australasia, since only anatomically-modern *Homo sapiens* crania regularly display the combination of features. It does, however, demonstrate that a combination of these features is not unique to Australasia. If this combination of features is a primitive retention, it should be commonly found on *Homo erectus* and archaic *Homo sapiens* crania, which is not the case.

CONCLUSION

One problem in attempting to document a 'morphological clade' in Australasia is the lack of skeletal material dating from between the late *Homo erectus* Ngandong material (which appears to be at least 200 000 years old: Habgood 1986c; Santa Luca 1980), and the earliest Australian skeletal material from Lake Mungo dating to approximately 25 000-30 000 years ago (White and Habgood 1985; White and O'Connell 1982). It is unlikely that a Ngandong-type cranium will ever be found in Australia, for it is very probable (if not essential) that the earliest inhabitants of the Sahul continent were fully modern *Homo sapiens*. It should be remembered, however, that a *Homo erectus* cranium will always be more similar to other *Homo erectus* crania than it will be to *Homo sapiens* crania, and *vice versa*. For this reason it will be difficult to identify a 'morphological clade' within Australasia – if indeed one exists – until hominids are found that fill the morphological and chronological gap between the latest Indonesian *Homo erectus* material and the earliest Australian crania. A number of remains are, however, thought to date to this intervening period, although to the latter end of it. A fragmentary human cranium and some fragmentary postcranial material, including a left talus, were found near the western mouth of the Great Cave of Niah in Sarawak (Harrisson 1959). The 'Deep Skull' of Niah is suggested to date to approximately 40 000 years BP (Coon 1962; Howells 1973; Kennedy 1979; Oakley *et al.* 1975) on the basis of radiocarbon dates of 39,820± 1012 BP (Harrisson 1958; Oakley *et al.* 1975), 39,600±1000 BP (Harrisson 1958) and 32,630±700 BP (Harrisson 1958). The burnt bone and charcoal used for the dating came from some 20 cm higher in the excavation than the skeletal material, and so are not directly associated with it. It is stated that the dated charcoal came from directly above the skull (Harrisson 1958). Most of the faunal remains from the site are of extant species (Harrisson *et al.* 1961). It has been suggested that the Niah skull may represent an intrusive burial into older layers (Bellwood 1985; Wolpoff 1980). Fluorine, uranium and nitrogen analyses, however, indicate that the cranium was probably contemporaneous with *Chiroptera* bones from adjacent layers (Oakley *et al.* 1975). This supports the contention that the 'Deep Skull' was not intrusive, but rather a possible secondary deposition which may have been incompletely cremated (Kennedy 1979; Howells 1973). Bellwood also questions the overall reliability of the stratigraphic sequence in the cave, stating that:

> "Harrisson's reconstructions of the cultural sequence at Niah were based partly on the idea that depths and ages could be correlated regularly across the site. However, the site has an uneven surface, and arbitrary levels of excavation up to 24 inches thick, plus a set of partially contradictory radiocarbon dates recorded only by depths below surface, clearly do not encourage much confidence in the finer details of the 'Niah area phaseology'. . ." (1985: 176).

The exact dating and stratigraphical relationships of the 'Deep Skull' of Niah are far from settled, and so we must await the results of dating some of the postcranial fragments by the accelerator radiocarbon technique (Stringer pers. comm.). A date of approximately 40 000 years BP is, however, not incompatible with the early dates for human occupation of Sahul – the combined landmass of New Guinea, Australia and Tasmania (Groube *et al.* 1986; White and Habgood 1985). The cranium, which is thought to be that of a female, is poorly preserved, fragmentary, and distorted (Birdsell 1979; Coon 1962). Based on the state of development of the unerupted third molar, Brothwell (1960) thought the skull to be that of an adolescent aged approximately 15-17 years, although Bulbeck (1981) has suggested that it had a retarded eruption of the third molars. Birdsell (1979), based predominantly on basilar suture closure, suggests that the individual was a young adult aged 20-30 years. The reconstructed cranium has thin vault bones, a relatively steep supraglabellar region behind which the frontal gently recedes, slight supraorbital development, square orbits, indications of a deep nasal root, prominent parietal bosses, well marked temporal lines, a rounded occipital, a short, broad and prognathic face, a broad nasal aperture that lacks a sharply-defined lower margin, a rudimentary anterior nasal spine, moderately developed canine fossae, and a large, shallow palate (Brothwell 1960; Kennedy 1979). Many of these morphological features are commonly found on Australian Aboriginal crania (Larnach and Macintosh 1966, 1970). Brothwell (1960) also found the Niah specimen to be closest to Australian crania.

Another interesting group of hominid remains came from Tabon Cave on Palawan Island in the Philippines (Fox 1968; Oakley et al. 1975). The remains of at least five hominids, including cranial, mandibular and postcranial elements, have been recovered, and are said to derived from deposits dated to about 23 000 years BP (Oakley *et al.* 1975). The area where the remains were found, however, had been disturbed by Megapode birds which destroyed the stratigraphy (Fox 1968; Howells 1973). There appears to be no evidence that the remains had been buried, and the frontal has provided fluorine and uranium values consistent with those obtained from associated animal bones (Oakley *et al.* 1975). The dating of these remains is, therefore, far from satisfactory at present, but human occupation of the Tabon Cave extends beyond 30 000 years BP, and so the presence of 23 000-year-old human bones is quite possible. An adult frontal and mandible, which are collectively referred to as 'Tabon 1', are the most widely discussed remains from the site. The frontal, which Howells (1973) suggests is from a male, is quite gracile in appearance, with thin vault bones, little postorbital constriction, and a relatively steep supraglabellar region, behind which the squama gently recedes towards the bregma. It displays a moderate supraorbital region, prominent glabella, broad shallow supratoral sulcus, marked temporal lines, broad interorbital region, relatively horizontal inferior supraorbital borders, slightly depressed nasion, and prominent, relatively narrow nasal bones that are pinched with a central ridge (Wolpoff et al. 1984). The damaged mandible lacks the upper halves of both rami, and part of the left gonial angle. The only teeth remaining are the right canine, first and second premolars, and the left first molar – although the roots of the left lateral incisor, canine and first premolar are also present. The mandible has a three-rooted left molar and congenital absence of both third molars (Barker 1978). The (possibly male) mandible was moderately long and broad with relatively narrow rami, medium symphyseal height, a slightly positive chin, and no trace of a mandibular torus (Macintosh 1978). The frontal displays a number of features (such as a moderate supraorbital region with prominent glabella, relatively horizontal inferior supraorbital borders, and prominent temporal lines) that are commonly found on Australian Aboriginal crania (Larnach and Macintosh 1966, 1970). Some features of the mandible, such as the three-rooted molars and agenesis of the third (Barker 1978), suggest Mongoloid affinities, although Macintosh believes that "at least the evidence derived from the specimen itself appears to be quite consistent with the hypothesis that the Tabon mandible belonged to a racial stock closely related to Australian Aborigines" (1978: 160).

The Niah and Tabon 1 remains seem to indicate an Australoid presence in Wallacea from at least approximately 40 000 years BP, although there are problems in the dating of both specimens. They do not (due to their fragmentary nature and the possible young age of the Niah individual) shed light on the 'regional continuity' debate. A relatively recent find of a calvaria and some fragmentary postcranial material on the surface near Lake Garnpung (north of Lake Mungo, and part of the Willandra Lakes System: Flood 1983; Thorne 1984) is suggested to reinforce the idea of regional continuity in Australasia (Habgood 1986a, 1986c; Thorne 1984; Wolpoff 1985). This hominid, which has not yet been fully published, is designated 'Willandra Lakes Hominid' 50 ('WLH 50'). It is 'opalized' in that all the normal phosphate in the bone had been replaced by silicates (Flood 1983). WLH 50, which is fully sapient in overall morphology and 'Australian' in nature, is very large and robust with marked postorbital constriction and thick vault bones involving enlargement of the diploe (Delson 1985; Flood 1983; Thorne 1984). The calvaria has a flat and receding frontal with a posterior position of minimum frontal breadth, marked temporal crests, a protruding occipital with a well-developed transverse torus, prominent brow-ridges, and a maximum parietal breadth located towards the parietal mastoid angle (Flood 1983). The transverse arc is said to fall close to the Ngandong mean value for this measurement (Thorne, pers. comm,). A left malar fragment displays eversion of its lower border, rounding of the inferolateral border of the orbit, and a zygomaxillary tuberosity (Wolpoff, pers. comm.). The date of WLH 50 has not yet been conclusively established. Since the material was a surface find, it is difficult to ascertain which geological layer it should equated with. Quoting unpublished papers written by A. G. Thorne, Flood (1983: 67) records that "radiocarbon and trace-element analysis indicate a minimum age of 25 000 to 30 000 BP, but the remains are probably much older", while Delson (1985: 298) states that "using an experimental electron-spin-resonance approach, the oldest specimen, WLH 50, is far older than 30 000, perhaps something like 60 000 years old". These dates are, of course, preliminary estimates and have not been substantiated in print. It is, however, unlikely that WLH 50 is older than 45 000 years BP, as this is the estimated date for the base of the Mungo Unit, and as yet no archaeological remains have been recovered from the underlying Golgol Unit (Bowler 1976).

At present, the evidence for regional continuity within Australasia is limited, due to the lack of skeletal material dating between the Ngandong material from Indonesia, and the hominids from Niah Cave, Tabon Cave and the Willandra Lakes – the earliest anatomically-modern *Homo sapiens* crania from the region. As discussed above, however, there are a number of morphological features which, when found in combination, appear to document continuity between the early Indonesian material and some prehistoric and modern Australian crania. There is a major problem with the idea that the combination of features, outlined above, document 'regional continuity'. This is the fact that the earliest modern skeletal material from the region (Lake Mungo 1, Lake Mungo 3, and Niah), although fragmentary, does not have these features, either individually or in combination (Table 15.1). A malar fragment from Lake Mungo 2 (Bowler *et al.* 1970), however, displays a malar tuberosity (Brown, pers. comm.). There is, then, a major discontinuity in the sequence. This gap may be partially filled by WLH 50, which appears to have the combination of features (Table 15.1). Even if it is demonstrated that WLH 50 is older than the Lake Mungo and Niah crania, it does not explain why the combination of possible 'regional features' are not present. This material does, however, share morphological features with prehistoric Australian crania that display the combination of possible 'regional features' (Table 15.1). The Lake Mungo material does have an 'Australianness' about it, in that it displays most of the features typical of Australian Aboriginal crania (Larnach and Macintosh 1966, 1970).

If one wishes to advocate the 'Rapid Replacement' hypothesis to explain the origin of modern humans in Australasia, one must explain why there are some features that appear to document 'regional continuity', and why WLH 50 resembles a possible morphological intermediate between the Ngandong material and later prehistoric and modern Australian Aboriginal crania. Alternatively, those who wish to advocate the 'Regional Continuity' hypothesis to explain the evolutionary sequence in Australasia must explain the lack of regional features on the Lake Mungo 1 and 3 and the Niah crania. Thorne's (1976, 1977, 1980; Thorne and Wilson 1977) 'Dual Source' hypothesis for the origin of the Australian Aborigines combines both continuity and migration into the region, and has been used to explain away the Lake Mungo and Niah material (Wolpoff *et al.* 1984). Wolpoff himself does not totally accept Thorne's explanation (Wolpoff 1980), and highlights the morphological similarities between Thorne's migrating 'gracile' group and his local 'robust' group. Recent multivariate analyses have also demonstrated that Thorne's 'robust' and 'gracile' groups are more similar to each other than either is to material from other geographic regions (Habgood 1985, 1986b). A recent demographic and genetic explanation as to the reasons for the large morphological variation within the corpus of late Pleistocene crania from Australia has also been formulated (Habgood 1986c) which can be used to explain the morphological dichotomy within Australia, and why the Lake Mungo material does not display the combination of possible 'regional features'.

The late Pleistocene archaeological record from Australia does not seem to support the 'Rapid Replacement' hypothesis. As previously mentioned, no artifacts have been conclusively associated with *Homo erectus* remains from Indonesia. There is no evidence from Australia of an influx of new archaeological elements during the late Upper Pleistocene, with the art and stone artifacts being classified into pan-continental assemblages (Flood 1983; White and Habgood 1985; White and O'Connell 1982).

To conclude, the present skeletal sample from Australasia is not adequate to allow a clear distinction between the two competing explanations as to the origins of modern humans in the region, in that there is evidence that could be used to support both the 'Rapid Replacement' and 'Regional Continuity' hypotheses. The present investigation has, however, demonstrated that there is a combination of at least four morphological features that seem to be indicating a relatively high degree of morphological continuity within the region.

ACKNOWLEDGEMENTS

Discussions with many people have greatly helped to develop the ideas expressed in this paper. Those I would especially like to acknowledge are C. B. Stringer, M. H. Wolpoff, G. P. Rightmire, C. P. Groves, A. G. Thorne, D. Hodgson and M. Green. I would also like to thank C. B. Stringer, R. Kruszynski, J. Zias, J.-L. Heim, J. Brink, Q. B. Hendy, P. V. Tobias, K. Parsons and H. de Lumley for allowing me to examine original specimens and/or casts of hominid crania in their care. This research was supported by grants from the Australian Institute for Aboriginal Studies, Canberra, and the Carlyle Greenwell Research Fund, Department of Anthropology, University of Sydney.

REFERENCES

Anderson, J. E. 1968. Late Paleolithic skeletal remains from Nubia. In F. Wendorf (ed.) *The Prehistory of Nubia*. Dallas: Southern Methodist University Press: 996-1040.

Andrews, P. 1984. On the characters that define *Homo erectus*. In P. Andrews and J. L. Franzen (eds) *The Early Evolution of Man with Special Emphasis on Southeast Asia and Africa. Courier Forschungsinstitut Senckenberg* 69: 167-175.

Asfaw, B. 1983. A new hominid parietal from Bodo, Middle Awash Valley, Ethiopia. *American Journal of Physical Anthropology* 61: 367-371.

Baker, J. R. 1968. Observations on the cranium of Broken Hill Man *Homo rhodesiensis* Woodward. *Zeitschrift für Morphologie und Anthropologie* 60: 121-127.

Barker, B. C. 1978. Dental features of the Tabon mandible. *Archaeology and Physical Anthropology in Oceania* 13: 160-166.

Bellwood, P. 1985. *Prehistory of the Indo-Malaysian Archipelago*. Sydney: Academic Press.

Birdsell, J.B. 1979. A reassessment of the age, sex and population affinities of the Niah cranium. *American Journal of Physical Anthropology* 50: 419 (Abstract).
Bowler, J. M. 1976. Recent developments in reconstructing late Quaternary environments in Australia, In R. L. Kirk and A. G. Thorne (eds) *The Origin of the Australians*. Canberra: Australian Institute of Aboriginal Studies: 55-77.
Bowler, J.M., Jones, R., Allen, H. and Thorne, A,G. 1970. Pleistocene human remains from Australia: a living site and human cremation from Lake Mungo, western New South Wales. *World Archaeology* 1: 39-60.
Bräuer, G. 1984a. A craniological approach to the origin of anatomically modern *Homo sapiens* in Africa and implications for the appearance of modern Europeans. In F. H. Smith and F. Spencer (eds) *The Origin of Modern Humans: a World Survey of the Fossil Evidence*. New York: Alan R. Liss: 327-410.
Bräuer, G. 1984b. The Afro-European *sapiens*-hypothesis, and hominid evolution in Asia during the late Middle and Upper Pleistocene. In P. Andrews and J. L. Franzen (eds) *The Evolution of Man with Special Emphasis on Southeast Asia and Africa. Courier Forschungsinstitut Senckenberg* 69: 145-166.
Briggs, C L. 1950. On three skulls from Mechta-El-Arbi, Algeria: a re-examination of Cole's adult series. *American Journal of Physical Anthropology* 8: 305-313.
Brink, L.M, van den. 1982. On the mammal fauna of the Wajak Cave, Java (Indonesia). *Modern Quaternary Research in Southeast Asia* 7: 177-193.
Brothwell, D.R. 1960. Upper Pleistocene human skull from Niah Caves. *Sarawak Museum Journal* 9: 323-349.
Brown, Harris, J., Leakey, R. and Walker, A. 1985. Early *Home erectus* skeleton from west Lake Turkana, Kenya. *Nature* 316: 788- 792.
Brown, P. 1981. Artificial cranial deformation: a component in the variation in Pleistocene Australian aboriginal crania. *Archaeology in Oceania* 16: 156-167.
Bulbeck, F. D. 1981. *Continuities in Southeast Asian Evolution since the Late Pleistocene*. Unpublished M.A. Dissertation, Department of Prehistory and Anthropology, Australian National University, Canberra.
Conroy, G. 1980. New evidence of Middle Pleistocene hominids from the Afar desert, Ethiopia. *Anthropos* (Brno) 7: 96-107.
Coon, C. S. 1962. *The Origin of Races*. New York: Knopf.
Delson, E. 1985. Late Pleistocene human fossils and evolutionary relationships. In E. Delson (ed.) *Ancestors: the Hard Evidence*. New York: Alan R. Liss: 296-300.
Fenner, F. J. 1939. The Australian Aboriginal skull: its non- metrical morphological characters. *Transactions of the Royal Society of South Australia* 63: 248-306.
Ferembach, D. 1976. Les restes humains de la grotte de Dar-es-Soltane 2 (Maroc), Campagne 1975. *Bulletins et Mémoires de la Société d'Anthropologie de Paris* 13: 183-193.
Flood, J. 1983. *Archaeology of the Dreamtime*. Sydney: Collins.
Fox, R.B. 1968. The prehistory of the Philippines. *Hemisphere* 12 (10): 10-16.
Greene, D. L. and Armelagos, G. 1972. *The Wadi Halfa Mesolithic Population*. Research Report 11, Department of Anthropology, University of Massachusetts, Amherst.
Grimaud D. 1982. Le parietal de l'Homme de Tautavel. In H. de Lumley (ed.) *Actes du Premier Congrès International de Paléontologie Humaine*.

Nice: Centre Nationale de la Recherche Scientifique: 62-88.
Groube, L., Chappell, J., Muke, J. and Price, D. 1986. A 40 000 year-old human occupation site at Huon Peninsula, Papua New Guinea. *Nature* 324: 453-455.
Habgood, P. J. 1985. The origin of the Australian Aborigines: an alternative approach and view. In P. V. Tobias (ed.) *Hominid Evolution: Past, Present and Future*. New York: Alan R. Liss 367-380.
Habgood, P. J. 1986a. A late Pleistocene prehistory of Australia: the skeletal material. *Physical Anthropology News* 5: 1-5.
Habgood, P. J. 1986b. The origin of the Australians: a multivariate approach. *Archaeology in Oceania* 21: 130-137.
Habgood, P. J. 1986c. Aboriginal fossil hominids: evolution and migrations. In M. Day, R. Foley and Wu Rukang (eds) *The Pleistocene Perspective*. Precirculated papers of the World Archaeological Congress, Southampton, 1986. London: Allen and Unwin.
Harrisson, T. 1958. Carbon-14 dated palaeoliths from Borneo. *Nature* 181: 792.
Harrisson, T. 1959. New archaeological and ethnographical results from Niah Cave, Sarawak. *Man* 59: 1-8.
Harrisson, T., Hooijer, D. A. and Lord Medway. 1961. An extinct giant pangolin and associated mammals from Niah Cave, Sarawak. *Nature* 189: 166.
Howell, F. C. 1978. Hominidae. In V. J. Maglio and H. B. S. Cooke (eds) *Evolution of African Mammals*. Cambridge (Mass): Harvard University Press: 154-248.
Howell, F. C. 1984. Introduction. In F. H. Smith and F. Spencer (eds) *The Origins of Modern Humans: a World Survey of the Fossil Evidence*. New York: Alan R. Liss: xiii-xxii.
Howells, W. W. 1973. *The Pacific Islands*. Wellington: Reed.
Howells, W. W. 1976a. Explaining modern man: evolutionists *versus* migrationists. *Journal of Human Evolution* 5: 477-496.
Howells, W. W. 1976b. Physical variation and history in Melanesia and Australia. *American Journal of Physical Anthropology* 45: 641- 650.
Hublin, J.-J. 1986. Some comments on the diagnostic features of *Homo erectus*. In *Fossil Man: New Facts, New Ideas*. Anthropos (Brno) 23: 175-185.
Jacob, T. 1976. Early populations in the Indonesian region. In R. L. Kirk and A. G. Thorne (eds) *The Origin of the Australians*. Canberra: Australian Institute of Aboriginal Studies: 81-93.
Keith, A. 1925. *The Antiquity of Man* (2nd Edition). London: Williams and Norgate.
Kennedy, K. 1979. The Deep Skull of Niah: an assessment of twenty years of speculation concerning its evolutionary significance. *Asian Perspectives* 20: 32-50.
Koenigswald, G. H. R. von. 1956. The geological age of the Wadjak man from Java. *Proceedings Koninklijk Nederlands Akademie van Wetenschappen* (Series B) 5: 455-457.
Larnach, S. L. and Macintosh, N. W. G. 1966. *The Craniology of the Aborigines of Coastal New South Wales*. Oceania Monographs 13.
Larnach, S. L. and Macintosh, N. W. G. 1970. *The Craniology of the Aborigines of Queensland*. Oceania Monographs 15.
Larnach, S. L. and Macintosh, N. W. G. 1974. A comparative study of Solo and Australian Aboriginal crania. In A. P. Elkin and N. W. G. Macintosh (eds) *Grafton Elliot Smith: the Man and His Works*. Sydney: Sydney University Press: 95-102.
Leakey, R. E. F. and Walker, A. C. 1985. Further hominids from the Plio-Pleistocene of Koobi Fora, Kenya. *American Journal of Physical*

Anthropology 67: 135-164.

Macintosh, N. W. G. 1965. The physical aspect of man in Australia. In R. M. Berndt and C. H. Berndt (eds) *Aboriginal Man in Australia*. Sydney: Angus and Robertson: 29-70.

Macintosh, N. W. G. 1978. The Tabon cave mandible. *Archaeology and Physical Anthropology in Oceania* 13: 143-159.

Macintosh, N. W. G. and Larnach, S. L. 1972. The persistence of *Homo erectus* traits in Australian Aboriginal crania. *Archaeology and Physical Anthropology in Oceania* 12: 1-7.

Macintosh, N. W. G. and Larnach, S. L. 1976. Aboriginal affinities looked at in World context. In R. L. Kirk and A. G. Thorne (eds) *The Origin of the Australians*. Canberra: Australian Institute of Aboriginal Studies: 113-126.

Mallegni, F., Mariani-Costantini, R., Fornaciari. G., Longo, E. T., Giacobini, G. and Radmilli, A. M. 1983. New European fossil hominid material from an Acheulian site near Rome (Castel di Guido). *American Journal of Physical Anthropology* 62: 263-274.

McCown T. D. and Keith, A. 1939. *The Stone Age of Mount Carmel*. Vol. 2: *The Fossil Human Remains from the Levalloiso-Mousterian*. Oxford: Clarendon Press.

Oakley K. P., Campbell, B. G. and Molleson, T. I. 1975. *Catalogue of Fossil Hominids*. Vol. 3: *Americas, Asia, Australia*. London: British Museum (Natural History).

Rightmire, G. P. 1980. Middle Pleistocene hominids from Olduvai Gorge, northern Tanzania. *American Journal of Physical Anthropology* 53: 225-241.

Rightmire, G. P. 1984. Comparisons of *Homo erectus* from Africa and South Asia. In P. Andrews and J. L. Franzen (eds) *The Early Evolution of Man with Special Emphasis on Southeast Asia and Africa. Courier Forschungsinstitut Senckenberg* 69: 83-98.

Rightmire, G. P. 1987. L'Evolution des premiers hominidés en Asia du Sud-Est. *L'Anthropologie* 91: 455-465.

Santa Luca, A.P. 1978. A re-examination of presumed Neandertal-like fossils. *Journal of Human Evolution* 7: 619-636.

Santa Luca, A. P. 1980. *The Ngandong Fossil Hominids*. Yale University Publications in Anthropology 78.

Stringer, C. B. 1984a. The definition of *Homo erectus* and the existence of the species in Africa and Europe. In P. Andrews and J. L. Franzen (eds) *The Early Evolution of Man with Special Emphasis on Southeast Asia and Africa. Courier Forschungsinstitut Senckenberg* 69: 131-143.

Stringer, C. B. 1984b. Human evolution and biological adaptation in the Pleistocene. In R. Foley (ed.) *Hominid Evolution and Community Ecology: Prehistoric Human Adaptation in Biological Perspective*. London: Academic Press: 55-84.

Stringer, C. B. 1985. Middle Pleistocene hominid variability and the origin of Late Pleistocene humans. In E. Delson (ed.) *Ancestors: the Hard Evidence*. New York: Alan R. Liss: 289-295.

Stringer, C. B., Hublin J.-J. and Vandermeersch, B. 1984. The origin of anatomically modern humans in Western Europe. In F.H. Smith and F. Spencer (eds) *The Origins of Modern Humans: a World Survey of the Fossil Evidence*. New York: Alan R. Liss: 51-135.

Thorne A. G. 1975. *Kow Swamp and Lake Mungo: Towards an Osteology of Early Man in Australia*. Unpublished Ph.D. Dissertation, Department of Anthropology, University of Sydney.

Thorne, A.G. 1976. Morphological contrasts in Pleistocene Australians. In R. L. Kirk and A. G. Thorne (eds) *The Origin of the Australians*. Canberra: Australian Institute of Aboriginal Studies: 95-112.

Thorne, A. G. 1977. Separation or reconciliation: biological clues to the development of Australian society. In J. Allen, J. Golson and R. Jones (eds) *Sunda and Sahul: Prehistoric Studies in Southeast Asia, Melanesia and Australia*. London: Academic Press: 197-204.

Thorne, A. G. 1980. The longest link: human evolution in Southeast Asia and the settlement of Australia. In J. J. Fox, R. G. Earnaut, P. T. McCawley and J. A. C. Maukie (eds) *Indonesia: Australian Perspectives*. Canberra: Australian National University Research School of Pacific Studies: 35-43.

Thorne, A. G. 1984. Australia's human origins: how many sources? *American Journal of Physical Anthropology* 63: 227 (Abstract).

Thorne, A. G. and Wilson, S. R. 1977. Pleistocene and recent Australians: a multivariate comparison. *Journal of Human Evolution* 6: 393-402.

Thorne, A. G. and Wolpoff, M. H. 1981. Regional continuity in Australasian Pleistocene hominid evolution. *American Journal of Physical Anthropology* 55: 337-349.

Vandermeersch, B. 1981. *Les Hommes Fossiles de Qafzeh (Israël)*. Paris: Centre National de la Recherche Scientifique.

Villiers, H. de. 1968. *The Skull of the South African Negro*. Johannesburg: Witwatersrand University Press.

Weidenreich, F. 1936. The mandible of *Sinanthropus pekinensis:* a comparative study. *Palaeontological Sinica* (n.s. D) 7. Pehpei: Geological Survey of China: 1-169.

Weidenreich, F. 1939. On the earliest representatives of modern mankind recovered on the soil in East Asia. *Peking Natural History Bulletin* 13: 161-174.

Weidenreich, F. 1943. The skull of *Sinanthropus pekinensis:* a comparative study of a primitive hominid skull. *Palaeontologia Sinica* (n.s. D) 10 (whole series 127). Pehpei: Geological Survey of China.

Weidenreich, F. 1945. The Keilor skull: a Wadjak type from southeast Australia. *American Journal of Physical Anthropology* 3: 21-32.

Weidenreich, F. 1947. Facts and speculations concerning the origin of *Homo sapiens*. *American Anthropologist* 49: 187-203.

Weidenreich, F. 1951. Morphology of Solo Man. *Anthropological Papers of the American Museum of Natural History* 43: 205-290.

White, J. P. and Habgood, P. J. 1985. La préhistoire de l'Australie. *La Recherche* 167: 730-737.

White, J. P. and O'Connell, J. F. 1982. *A Prehistory of Australia, New Guinea and Sahul*. Sydney: Academic Press.

Wolpoff, M. H. 1980. *Paleoanthropology*. New York: Knopf.

Wolpoff, M. H. 1985. Human evolution at the peripheries: the pattern at the eastern edge. In P. V. Tobias (ed.) *Hominid Evolution: Past, Present and Future*. New York: Alan R. Liss: 355- 365.

Wolpoff, M. H. 1986. Describing anatomically modern *Homo sapiens:* a distinction without a definable difference. *Anthropos* (Brno) 23: 41-53.

Wolpoff, M. H., Wu Xinzhi and Thorne, A. G. 1984. Modern *Homo sapiens* origins: a general theory of hominid evolution involving the fossil evidence from East Asia. In F. H. Smith and F. Spencer (eds) *The Origins of Modern Humans: a World Survey of the Fossil Evidence*. New York: Alan R. Liss: 411-483.

Wood, B. 1984. The origin of *Homo erectus*. In P. Andrews and J .L. Franzen (eds) *The Early Evolution of Man with Special Emphasis on Southeast Asia and Africa. Courier Forschungsinstitut Senckenberg* 69: 99-111.

Wu Rukang. 1986. Chinese human fossils and the origin of Mongoloid racial group. *Anthropos* (Brno) 23: 151-155.

Wu Xinzhi. 1961. Study on the Upper Cave man Zhoukoudian. *Vertebrata PalAsiatica* 3: 202-211.

Wu Xinzhi. 1986. Upper Palaeolithic man in China and their relation with populations of neighbouring areas. In M. Day, R. Foley and Wu Rukang (eds) *The Pleistocene Perspective*. Precirculated papers of the World Archaeological Congress, Southampton, 1986. London: Allen and Unwin.

Wu Xinzhi and Zhang Zhenbiao. 1985. *Homo sapiens* remains from Late Pleistocene and Neolithic China. In Wu Rukang and J.W. Olsen (eds) *Palaeoanthropology and Palaeolithic Archaeology in the People's Republic of China*. Orlando: Academic Press: 107-133.

16: A Regional Approach to the Problem of the Origin of Modern Humans in Australasia

COLIN P. GROVES

INTRODUCTION

The problem of the origin of any given people can be looked at in two ways: what relationship, if any, do they bear to the people who preceded them in the same area; and what relationship do they bear to their contemporaries in neighbouring areas?

As far as Australasia is concerned, the first of these questions can be approached from several angles. Do modern *Homo sapiens* populations, whether in Australia, Melanesia or Southeast Asia, have some special likeness to *Homo erectus* populations in the region, such that one can envisage some sort of ancestor- descendant relationship? Or can such special likenesses, if they exist, be explained in some other way? Again, there are the prehistoric *Homo sapiens* of the same region: do they, or do they not, bear a total or partial ancestral relationship to modern peoples? Are today's (traditional) inhabitants of an area derived, in the main, from unitary stock, or are they the product of blending between two other peoples?

The second question is related to the first. Is one of the living populations of the region in some ways more 'primitive' than the others, such that it might even be a relatively unchanged ancestral stock? Has one population totally replaced another, unrelated to it, or are all the populations of the region associated at the same end of the spectrum of genetic variation that is modern *Homo sapiens?* Are there, within the region, any peoples that are genuinely intrusive? Have formerly more distinct populations migrated here and there, intermixing on the way with other groups, or are regional differences the result mainly of *in situ* diversification, with gene-flow playing at most a marginal role? Finally, what role has been played by environmental factors, either as challenges to the power of natural selection, or as barriers to gene flow and migration?

HOMO ERECTUS AND HOMO SAPIENS

The idea that there has been regional continuity within Southeast Asia from some period well back in human evolution is due to Weidenreich (1939). Pointing to a dozen morphological character states which are shared in common between *Homo erectus pekinensis* (as we would now call it) and modern 'Mongoloid' populations, and to a less-well-specified or enumerated number of traits in common between *H. e. erectus* and modern 'Australoids', Weidenreich postulated ancestor-descendant sequences. While he was not entirely clear on the exact course of events, his scenario seems to be one whereby the precursor species had evolved *en masse* into modern humanity, via a number of grades: a pithecanthropine (i.e. *H. erectus*) grade, a palaeanthropic (more or less 'Neanderthal') grade, and an archaic and modern sapient grade. While the characters that survived through the 'Mongoloid' lineage were more clear-cut, the 'Australoid' lineage was the more complete, going from *Pithecanthropus erectus* (i.e. *Homo erectus erectus*) via Ngandong (the 'Far Eastern Neandertaler') to Wajak (archaic *sapiens*) and so on to Australian Aborigines (modern *sapiens*). At a pinch, *Pithecanthropus robustus* could be added on to the beginning of this sequence; but that is another story.

Coon (1962) supported this scheme, extending the Regional Continuity concept to African and European populations as well. His model, however, is confused: at one point he seems to exclude any gene flow between his five 'evolving subspecies', while at another he denies that such an exclusion is possible. On a minimum of evidence he postulated a very early crossing of the *erectus-sapiens* boundary in some regions (China, Europe), but a very late transition in others (Africa, Southeast Asia). As with Weidenreich, but more strikingly because Coon's model was global and not just regional, the only actual traits offered as evidence for regional continuity were those of the 'Mongoloid lineage'.

With a few exceptions, most palaeoanthropologists turned their backs on Regional Continuity models. Larnach and Macintosh (1974) for the first time listed traits linking Ngandong and Australoids, although they specifically excluded *H. erectus* proper from consideration. Aigner (1976) added to the list of traits of the Mongoloid lineage. Wu (1981) described a fossil, the Dali skull, which he allocated to a position in the Mongoloid lineage analogous to that of Ngandong in the Australoid one. Now, at last, the Mongoloid and Australoid lineages could be placed on the same footing: both could be said to have character traits defining them, and a continuum of fossils along them.

Most recently Wolpoff *et al.* (1984) have surveyed the evidence as far as Eastern Asia is concerned, coming out in strong support of Regional Continuity and proposing an evolutionary model to account for it.

The evidence for regional continuity needs to be looked at closely. First, the regional traits said to characterize the Mongoloid and Australoid lineages will be discussed, and then the fossils cited as links between the Middle Pleistocene and the modern ends of the chain will be examined.

THE REGIONAL TRAITS

The Mongoloid lineage

(1) Mid-sagittal 'crest' (Weidenreich 1939) and parasagittal depression. This morphology is more correctly characterized as a sagittal ridge-like thickening emphasized by bilateral depressions. Wolpoff *et al.* (1984) note that this

becomes fainter through time, and is characteristic of the Australoid lineage as well. Among modern 'Mongoloid' populations, some – perhaps most – lack a ridge at all, and those that have it (such as Eskimos) lack the parasagittal depressions, making the cranial vault somewhat 'tent-shaped'. The same is true of its form in those Australian skulls that possess the ridge. Although keeling of the vault occurs in 93% of New South Wales crania and 94.8% of Queensland crania, associated longitudinal parasagittal depressions are distinct in only 2.6% of the New South Wales crania and 0.9% of the Queensland crania (Larnach and Macintosh 1966, 1970; Larnach 1978). Whether this can be truly said to be the same morphology is a moot point; even if it is, its total absence in many Australoids, and in whole Mongoloid populations, means that some evolution must be considered to have occurred in it.

(2) High frequency of metopic suture (Weidenreich 1939). This character is no longer part of Wolpoff *et al.*'s list; indeed, its highest modern frequency is found outside the region (in Arabia and India, where it occurs in 6-7% of skulls).

(3) High frequency of Inca bones. The concept of an Inca bone is not an exceptionally well-defined one, but the general tendency to develop extra bones on and around Lambda is of more evident significance than the impossibly subdivided specifications of Meyer (1877). Here again, Wolpoff *et al.* (1984) note that the frequency of occurrence of Inca bones diminishes through time. In modern populations the frequency appears to be highest in Polynesia (60%), but is nearly as high in Melanesia (58%), and varies between 40% and as much as 65% in Australian samples. East Asian and American samples show frequencies between 30% and 54%. Outside the region, Inca bones occur notably in KNM-ER 1813, a Lower Pleistocene skull from East Africa; and two European Middle Pleistocene or Pliocene specimens, Petralona and Vértesszöllös, show them, while they are not unknown in Neanderthalers. To maintain the Regional Continuity significance of this character, one must thus envisage it travelling out from an East Asian centre, retracting into that centre again, spreading southeast from that centre (subsequent to the Middle Pleistocene, for it does not occur in any *H. e. erectus* cranium in which sutures are visible), and finally achieving even higher frequencies in the Pacific region than in the original heartland. Other explanations are possible, but none that fit the 'Regional Continuity of Mongoloid Characters' model.

(4) Mongoloid features of the cheek region (Weidenreich 1939) – i.e. Facial Flatness (Wolpoff *et al.* 1984). Weidenreich (1943: Fig. 165) provides a series of figures to illustrate this similarity; from these figures it would appear that it characterizes not only Zhoukoudian *H. erectus* and modern Mongoloids, but apes as well! In other fossils with a well enough preserved cheek region to demonstrate this character, the flatness feature certainly occurs in SK847, a Plio-Pleistocene specimen from Swartkrans; and, despite the distortion, probably also in Sangiran 17, the only Indonesian *H. erectus* skull in which the facial skeleton is preserved. Facial flatness would therefore seem to be not a regional link, but a primitive *Homo* trait which has been retained to a greater degree in Mongoloids than in other

modern peoples: and not in all Mongoloids equally, for in the figures for the Zygomaxillary Index collected by Bulbeck (1981: Table 6-16) some 'Mongoloid' populations (Buriats, Eskimos) have extreme facial flatness (index 18.8, 19.0), and others (Japan) much less (index 24.6), giving a face less flat than many sub-Saharan Africans and even a few European samples.

(5) Mandibular exostoses, (6) ear exostoses, (7) maxillary exostoses (Weidenreich 1939). Wolpoff *et al.* (1984) say that jaw exostoses still characterize Mongoloids although, as in other characters, the high frequency found in the Zhoukoudian sample has dropped. Mandibular exostoses are, truly, rare outside the 'Mongoloid lineage', but a fine maxillary exostosis occurs in SK847 from Swartkrans. As for auditory exostoses, it now seems dubious whether they are genetic in origin at all (Kennedy 1986).

(8) Femoral platymeria (Weidenreich 1939). This feature is part of a complex of femoral morphology characterizing all the known Lower and Middle Pleistocene *Homo*, no trace of which persists in any modern population (Kennedy 1980, 1983). As such, it is rightly excluded from Wolpoff *et al.*'s list.

(9) Strong deltoid tuberosity of humerus (Weidenreich 1939). This feature is at present imponderable; too little fossil material exists to plot its former distribution, and no special study appears to have been made to determine its relative representation in modern groups.

(10) Shovel-shaped upper incisors, notably laterals (Weidenreich 1939). This is perhaps the most commonly cited similarity between Zhoukoudian *erectus* and modern Mongoloids; it stands at the head of Wolpoff *et al.*'s (1984) list. In fact, shovelling frequencies of central and lateral mandibular incisors do not differ consistently from those of central and lateral maxillary incisors, so the two can be treated together here.

Among modern peoples, shovel-shaped incisors in the upper jaw are most frequent among some 'Mongoloid' peoples. Bulbeck (1981: Table 3-1) gives the following frequencies for the combined Full and Semi grades (the two top grades of shovelling):

Northeast Asians	90-99%	Bhutan	61.5%
Thailand	50.5%	Indonesia	35.3-36.0%
Ainu and Jomon	28.0-32.1%	Polynesia	5.6-42.6%
Melanesia	4.0-16.9%	Andamanese	17.3%

As the frequencies for some European populations exceed 30% (Cadien 1972), this means that not all 'Mongoloid' populations have higher rates than all other populations.

Australian figures are hard to come by. Birdsell (1967) gives figures for the three quasi-racial groups recognized by him, dividing them into 'Absent', 'Traces', 'Medium', 'Large' and 'Pronounced', while admitting that the categories are not comparable to the standards laid down by Hrdlička. Birdsell gives data for 'whites' also, and comparing these with the data compiled by Cadien (1972) we get the following comparisons for I^1:

Cadien (ex. Hrdlička)		*Birdsell*	
		Absent	37
Absent	68	Traces	34
Trace	23	Medium	22
Semi	6	Large	7
Full	2	Pronounced	0

The figures are thus most comparable if Birdsell's 'Absent' plus 'Traces' are together equivalent to Cadien's and Hrdlička's 'Absent'. The standard 'Semi plus Full' will now be equivalent to Birdsell's 'Large' plus 'Pronounced'. So, the Australian figures (i.e. Birdsell's three types) to compare with Bulbeck's figures will be:

Barrineans	(N = 53) 8%
Carpentarians	(N = 66) 6%
Murrayians	(N = 12) 42%

Moreover, shovel-shaped incisors are, contrary to the usual assumption, the norm, not the exceptional state, among protohominids. Moderate shovelling (apparently, judging from the illustrated example, of 'Semi' grade) is the usual condition in australopithecines (Robinson 1956). The little-worn lateral incisors of Sangiran 4 are well-shovelled (von Koenigswald 1978: 356). It would seem from this that, like facial flatness, incisor shovelling is a primitive trait which has been retained more in certain modern 'Mongoloid' peoples than in other modern groups.

As an added item of interest, Bulbeck (1981) finds that when modern peoples and their prehistoric (but sapient) precursors in the same regions are compared, the percentage of 'Full' plus 'Semi' incisor shovelling is generally found to have increased over time: from 78% in Bao Ji to 99% in modern Chinese; from 20-27% in Non Nok Tha and Ban Kao to 51% in modern Thais; from 19-36% in Leang Cadang, Gilimanuk and Gua Kepah (though 43% in Gua Cha) to 36-44% in modern and early Metal Age Malays and Indonesians. Fluctuations in frequencies of polymorphic traits are the expected finding in genetics; so why not in palaeoanthropology too?

(11) A horizontal course of the nasofrontal and fronto-maxillary sutures (Weidenreich 1939). This is again a feature which varies in modern samples, but is usual not only in Zhoukoudian but in all Middle Pleistocene *Homo* (for example, Kabwe). It could be postulated that the high frequency of the trait in 'Mongoloids', if true, is a third example of a primitive retention; but the prominent glabella of the archaic taxa may influence the form of the suture, in which case it would not be a genuine resemblance at all, but an epiphenomenon of different conditions.

(12) Rounded profile of nasal saddle and nasal roof (Aigner 1976). The absence of the median nasal angulation or even ridge commonly seen in non-Mongoloid human skulls is connected to facial flatness, and is thus plausibly another primitive retention.

(13) Rounded infraorbital margin (Aigner 1976). This again appears to be part of the facial flatness complex, and is distributed accordingly.

(14) Posterior teeth are reduced earlier in time in the Mongoloid heartland

than elsewhere: Zhoukoudian posterior molars are already smaller than those of modern Australians (Wolpoff *et al.* 1984). This is not completely true: in *Homo sapiens heidelbergensis,* and *Homo sp.* from Koobi Fora and Swartkrans, molar size is approximately the same as in *H. erectus pekinensis;* only the molars of *H. e. erectus* being different from (but larger than!) other Lower/Middle Pleistocene *Homo.* This might therefore be thought a similarity between Java *erectus* and modern Australoids (molar enlargement) rather than between Chinese *erectus* and modern Mongoloids (molar reduction); but many modern sub-Saharan populations have extremely large teeth as well, so this will not really work either.

(15) High frequency of M3 agenesis (Wolpoff *et al.* 1984). This may or may not be the logical (and genetic?) extreme of molar reduction, but its occurrence, being easy to trace, needs to be documented separately. M3 agenesis occurs in the Lantian jaw; but even earlier in a jaw from Omo. The frequency of M3 absence (on one or both sides) rises to 40% in Eskimos (Cadien 1972) and 41% in Mokapu Hawaiians (Bulbeck 1981), but in other Mongoloid-affiliated populations is not necessarily higher than in non-Mongoloids: Anyang Chinese 25.7%; Papuans 25%; some European samples 24%; Nicobarese ('Mongoloid') about 13%; Eastern Melanesia 14.3% (figures from Cadien 1972, and Bulbeck 1981).

(16) Small frontal sinuses (Weidenreich 1939), restricted to the interorbital area (Wolpoff *et al.* 1984). Frontal sinuses are noticeably smaller in all East Asian *H. erectus* than in their European contemporaries such as Petralona. They are often small in modern peoples and may, contrary to the Middle Pleistocene specimens, be absent altogether. Weidenreich (1945) notes that the sinus was absent in 19 out of 22 Buriats (i.e. 86%), but also in 37% of Melanesians and 30% of Australians; on the other hand it is large, of 'archaic Europe' proportions, in a late Middle Pleistocene specimen from China (Maba).

This survey of specific traits demonstrates that there is little evidence for special likeness of modern 'Mongoloids' to *Homo erectus pekinensis.* What there is can be put down to two causes: (1) a similarity of Zhoukoudian to some populations, but not others, of modern 'Mongoloids', and not always the same ones at that; and (2) perhaps, even in these cases, retention in some Mongoloid groups of primitive character states. It is possible that many authors have failed to recognize these because of the strongly expressed neoteny (surely a derived condition) of 'Mongoloids', but the mosaic nature of evolution has been demonstrated often enough, and one must expect to find a mosaic of primitive and derived traits in all populations.

The Australoid Lineage

(1) The posterior position of the minimum frontal breadth (well behind the orbit) (Wolpoff *et* al. 1984). This does tend to unite *H. e. erectus* (including Ngandong) and Australoids, but is also seen in such Lower Pleistocene fossils as KNM-ER 3733 and 1813, and consequently is plausibly a primitive retention.

(2) Flatness of frontal in the sagittal plane (Wolpoff *et al.* 1984). Even given

that the most flattened frontals, those of the Kow Swamp series, are now supposed to be due to artificial deformation (Brown 1981), this feature is said to characterize modern Australoids and Java *erectus* specimens. Clearly, this is true (though not of all 'Australoids'); but it occurs equally in some Pliocene-Lower Pleistocene specimens (notably KNM-ER 1813), and sporadically in African Middle Pleistocene skulls (Omo II, Saldanha and, to a lesser degree, Kabwe). Thus, while it may be a genuine reflection of regional continuity, it can also be looked at as a primitive retention: certainly its opposite – frontal convexity – takes different forms in the various groups where it occurs (Zhoukoudian; Arago, Petralona; Neanderthal; modern Mongoloids; modern European/African peoples), as if it had arisen independently.

(3) Horizontal orientation of the lower border of the supraorbital region (Wolpoff *et al.* 1984). This in fact characterizes all *Homo erectus* (in the restricted sense, i.e. the East Asian forms), and could thus be regarded as either indicating continuity between China *and* Java with Australoids, or as something to do with the type of supraorbital buttressing, which is still poorly understood.

(4) Distinct prebregmatic eminence (Wolpoff *et al.* 1984). This characterizes not only all East Asian *erectus* fossils, but other skulls too, such as KNM-ER 3733. It is best interpreted as a primitive retention.

(5) Marked prognathism (Wolpoff *et al.* 1984). This is admitted to depend on Sangiran 17 alone, i.e. the only Java *erectus* specimen with a complete facial skeleton. It is surely a primitive retention, without going into detail on the precise definition of the different types of prognathism.

(6) Maintenance of large posterior teeth (Wolpoff *et al.* 1984). This character as presented, being an admitted primitive retention, would not be of itself constitute evidence for regional continuity; but there are other aspects to it, already discussed in (14) above.

(7) Persistence of a zygomaxillary ridge (Wolpoff *et al.* 1984). This again depends entirely on Sangiran 17. Its occurrence in other fossil specimens, such as Swartkrans SK847, suggests that it is either a primitive retention or something purely connected with prognathism.

(8) Eversion of lower border of malar (Wolpoff *et al.* 1984). This again occurs in other fossils (SK847, KNM-ER 3733), and would reflect the size of the masseter.

(9) Rounding of inferolateral border of orbit (Wolpoff *et al.* 1984). This is a variant of a proposed Mongoloid-lineage feature discussed in Section (13) above, and is part of a facial specialization: in this case not facial flatness as such (though some degree of this condition may occur in Australoids) but an enlarged zygomatic trigone, a reflection of masseter development.

Apart from the above, Wolpoff *et al.* (1984) also quote with approval the finding of Larnach and Macintosh (1974) that nine features of the Ngandong fossil sample are found in highest frequency in modern Australoids:

1. Large rounded zygomatic trigone (see no. 9 above).
2. Absence of supraorbital sulcus.

3. Suprameatal tegmen.
4. Transverse squamo-temporal fissure.
5. Angling of petrous to tympanic in petro-tympanic axis.
6. Lambdoidal protuberance.
7. Marked ridge-shaped occipital torus.
8. External occipital crest emerging from occipital torus.
9. Marked occipital supratoral sulcus.

This list looks convincing until one refers to the actual figures involved: feature [2] occurs in 84.5% of Australians and 62.2% of New Guineans, compared to only 50% in Mongoloids and Caucasoids; feature [5] occurs in 5% of New Guineans and 1.9% of Australians, but in no others; feature [6] occurs in just 1.4% of Australians, 1.2% of New Guineans, but no Mongoloids or Caucasoids; feature [9] occurs in 8.7% of Australians, 6.2% of New Guineans, compared to 5% of Mongoloids, and no Caucasoids. In the above list, only feature [2] shows interpopulation variation that is in any way large enough to be convincing; but it must be admitted that in all four characters both Australians and New Guineans show frequencies greater than the other samples. This is not however the case with the other five features, in which either Australians (in [1], [7] and [8]) or New Guineans (in [3]), but *not both,* exceed other populations' frequencies, or else (in [4]) the New Guinean frequency equals – but does not exceed the Mongoloid frequency, that for Australians being lower. It would seem in this list that the case for *erectus*/Australoid affinities rests at least in part on the construction of a 'catch-all' Australoid category, a concept that suffers from the same strictures as that of an all-purpose Mongoloid 'race', as discussed above. The only really plausible similarity lies in character no. [2], which is of course the same as Wolpoff *et al.*'s (1984) second character (frontal flatness).

Larnach and Macintosh's purpose was to list the diagnostic features of the Ngandong sample, and see to what extent they occurred in modern series. They therefore – perfectly fairly – listed two others in which 'Australoids' were actually less like Ngandong than one or both of the other two samples: feature [10] low, unarched squamous suture (3.8% Caucasoids, 1.2% New Guinea, 0.5% Australoids, no Mongoloids); and feature [11] presence of a juxtamastoid ridge (100% Caucasoids, 81-85% all three others).

From the 'shared characters' point of view, the Regional Continuity model lacks much real substance. What of the 'fossil continuum' aspect?

THE FOSSIL LINKS

The then-to-now sequence in China runs: Lantian- Zhoukoudian/Hexian-Dali/Maba-Liujiang/perhaps Zhoukoudian Upper Cave-modern Mongoloids. It seems that there are two breaks in this sequence which cannot be overlooked: between stages two and three, and between stages three and four. It is argued here, building upon the work of Wood (1984), Stringer (1984), and Andrews (1984), that *Homo erectus* is a species with its own uniquely derived traits; that, as far as the Middle Pleistocene is concerned at least, the species is known only from China and Java; that the Middle Pleistocene fossils of Europe and Africa lack the derived traits of *H. erectus* and begin to show those of *H. sapiens;* and that these archaic *sapiens* forms first appear in China in the late Middle Pleistocene (Dali and Maba), where they appear to replace *H. erectus,* and are later themselves replaced by *H. sapiens* of modern type which had first appeared in Northeast Africa. In Java the picture is again replacement: Ngandong is *H. erectus* (indeed the Ngandong series differs only on average values from the earlier Sangiran and Trinil *erectus* series) and is replaced by *H. sapiens,* but this time directly by *sapiens* of modern type, there being no trace of the archaic form.

These uniquely derived traits of *Homo erectus* include the development of the supraorbital ridges into a true torus, with lateral 'wings'; the large occipitomastoid crest; the low frontal angle, below 45°; the angulated occiput; strong occipital and angular tori, and supramastoid crest; platycephaly; the mastoid/petrosal fissure; the thick tympanic plate. None of these traits occur in any populations of *Homo sapiens.*

But the case will not be re-argued here; suffice it to say that the cladistic assortment of the characters of the various samples suggests that *H. erectus,* thus narrowly defined, had no noticeable part to play in the ancestry of modern humans, whose origin is instead convincingly demonstrated by Bräuer (1984) further to the west.

THE DISPERSAL OF HOMO SAPIENS IN THE INDO-PACIFIC REGION

Once again we are in the presence of competing models: replacement or *in situ* evolution. As in the previous case, population replacement (in a broad Australoid-replaced-by-Mongoloid scheme, sometimes varied with Negritos and Melanesioids, and division of 'Mongoloids' into Proto- and Duetero-Malays) was virtually assumed until successive challenges by Hooijer (1952), Jacob (1967) and others raised serious questions.

Hooijer (1952) had noticed that reduction in general size, in particular dental size, had occurred in many mammalian lineages (rhino, tapir, orangutan, siamang, leaf-monkeys etc.) in Southeast Asia, and asked the question, "Why should it not have occurred in humans also?" The supposed replacement of 'Austromelanesians' by immigrant Malay-like peoples might, in fact, be better seen as simply another example of dental size reduction.

The prize exhibit of the 'replacement' model has always been the Wajak skull, supposedly an unchallengeable representative of the Austromelanesoids that preceded the immigrant Mongoloids. Jacob (1967) did finally challenge this assessment: for him, Wajak was much more a robust 'southern Mongoloid' than had been admitted. Bulbeck (1981) described a specimen – undated, but certainly pre-metal – from Leang Buidane in Sulawesi which strikingly recalls Wajak, and differs from Metal Age burials from the same site. Arguing for Wajak, Leang Buidane, Niah, Tabon, Gua Kepah and the Gua Cha Hoabhinian sample as representative of a general Sundaland/Wallacean pre-metal (probably early Holocene, or even latest Pleistocene) population, Bulbeck compares these to Metal Age

(Leang Buidane jar burials, Gilimanuk, Leang Cadang etc.) and modern samples from the same region. The trends that seem to occur are of reduction of the masticatory apparatus, associated with general gracilization, and brachycephalization. Throughout the Southeast Asian sequence, typical 'Sundadont' morphology is maintained: moderate degrees of incisor shovelling, high levels of lower molar complexity, short faces (in these characters differing from Northeast Asian or 'Sinodont' morphology, and resembling Australoids), high frequencies of M^3 hypocone reduction, fairly marked facial flatness, little or no prognathism, relatively broad – at least mesocephalic – braincase, and no malar eversion (in these characters resembling Sinodont morphology and differing from Australoids).

It is important, Bulbeck concludes, not to see Australoid morphology as simply 'primitive' *Homo sapiens,* but to disentangle what is genuinely primitive from what is 'cladistically Australoid'. When this is done, the pre-metal Southeast Asians can be seen as Australoid in *grade* characters only: in *clade* characters they are as Southeast Asian (or Sundadont) as their present-day successors.

It is important to recall that Sinodont-Sundadont differentiation is clinal, not sharp. The cline continues east along the Lesser Sunda (Nusatenggara) chain to New Guinea (Bijlmer 1929), although it is obviously a considerably steeper one than the North-South Asian cline. One way of looking at the changes in Southeast Asian *Homo sapiens* is to see the total cline as simply shifting southward (and southeastward), while becoming constricted by the Wallacean barrier (Bulbeck 1981). Such a view would in effect give the Hooijerian dental reduction trend, for humans at least, a centre of origin and dispersal.

The Negrito problem remains. Who are these pygmoid, black- skinned, woolly-haired peoples in the Philippines, West Malaysia, and the Andamans? Following Howells (1973), Bulbeck suggests dissociating the Andamanese from the Semang and Aeta; the latter are, cranially, much more Southeast Asian than is commonly realised, and (*contra* Coon, 1962) totally lack any trace of Australomelanesian robusticity. They are rainforest-living people, hunter-gatherers but with important reciprocal economic links to their neighbours, and a strong awareness of their own cultural distinctness which, until recently at least, had the effect of reinforcing marriage rules and so enhancing and sharpening their distinctive appearance.

The case is clearly far from closed. At present, however, it does appear that the characters appealed to in the Regional Continuity model are more logically analysed in other ways, and that more rigorous polarity determinations are necessary before they can be used to support an hypothesis of genetic continuity between present-day populations and other predecessors. There is a certain irony in this conclusion, in that a Replacement model within anatomically-modern *Homo sapiens* in the region seems increasingly hard to maintain as Late Pleistocene and Early Holocene fossil remains become better known.

REFERENCES

Aigner, J. S. 1976. Chinese Pleistocene cultural and hominoid remains: a consideration of their significance in reconstructing the pattern of human bio-cultural development. In A. K. Ghosh (ed.) *Le Paléolithique Inféreur et Moyen en Inde, en Asie Centrale en Chine et Dans le Sud-est Asiatique*. Paris: Centre National de la Recherche Scientifique: 65-90.

Andrews, P. 1984. An alternative interpretation of the characters used to define Homo erectus. In P. Andrews and J. L. Franzen (eds) *The Early Evolution of Man with Special Emphasis on Southeast Asia and Africa. Courier Forschungsinstitut Senckenberg* 69: 167- 175.

Bijlmer, H. J. T. 1929. *Outlines of the Anthropology of the Timor Archipelago*. Indischer Comite voor Wetenschappen, Onderjook III, Weltevreden.

Birdsell, J. B. 1967. Preliminary data on the trihybrid origin of the Australian Aborigines. *Archaeology and Physical Anthropology in Oceania* 11: 100-155.

Bräuer, G. 1984 The 'Afro-European *sapiens* hypothesis', and hominoid evolution in East Asia during the late Middle and Upper Pleistocene. In P. Andrews and J. L. Franzen (eds) *The Early Evolution of Man with Special Emphasis on Southeast Asia and Africa. Courier Forschungsinstitut Senckenberg* 69: 145-165.

Brown, P. 1981. Artificial cranial deformation: a component in the variation in Pleistocene Australian Aboriginal crania. *Archaeology in Oceania* 16: 156-167.

Bulbeck, F. D. 1981. *Continuities in Southeast Asian Evolution since the Late Pleistocene: some new Material Described and some Old Questions Reviewed*. Unpublished M.A. Thesis, Department of Prehistory and Anthropology, Australian National University, Canberra.

Cadien, J. D. 1972. Dental variation in man. In S.L. Washburn and P. Dolhinow (eds) *Perspectives on Human Evolution:* Vol. 2. New York: Holt, Rhinehart and Winston: 199-222.

Coon, C. J. 1962. *The Origin of Races*. New York: Knopf.

Hooijer, D. A. 1952. Austromelanesian migrations once more. *Southwestern Journal of Anthropology* 8: 472-477.

Howells, W. W. 1973. *The Pacific Islanders*. Wellington: Reed.

Jacob, T. 1967. *Some Problems Pertaining to the Racial History of the Indonesian Region*. Unpublished Ph.D. Thesis, University of Utrecht.

Kennedy, G. E. 1980. *Palaeoanthropology*. New York: Melwar Hill.

Kennedy, G. E. 1983. Some aspects of femoral morphology in *Homo erectus*. *Journal of Human Evolution* 12: 587-616.

Kennedy, G. E. 1984. Are the Kow Swamp hominids archaic? *American Journal of Physical Anthropology* 65: 163-168.

Kennedy, G. E. 1986. The relationship between auditory exostoses and cold water: a latitudinal analysis. *American Journal of Physical Anthropology* 71: 401-415.

Koenigswald, G. H. R. von. 1978. The palate of *Pithecanthropus* modjokertensis. In P. M. Butler and K.A. Joysey (eds) *Development, Function and Evolution of Teeth*. London: Academic Press: 353-357.

Larnach, S. L. 1978. *Australian Aboriginal Craniology*. Oceania Monographs 21. Sydney: University of Sydney.

Larnach, S. L. and Macintosh, N. W. G. 1966. *The Craniology of the Aborigines of Coastal New South Wales*. Oceania Monographs 13. Sydney: University of Sydney.

Larnach, S. L. and Macintosh N. W. G. 1970. *The Craniology of the Aborigines of Queensland*. Oceania Monograph 15. Sydney: University of Sydney.

Larnach, S. L. and Macintosh, N. W. G. 1974. A comparative study of Solo and Australian Aboriginal crania. In A. P. Elkin and N. W. G. Macintosh (eds) *Grafton Elliot Smith: the Man and his Work*. Sydney: Sydney University: 95-102.

Meyer, A.B. 1875, 1877, 1878. Uber Hundert Fünf und Dreissig Papua-Schädel von Neu-Guinea und der Insel Mysore (Geelvinksbai). *Mitteilungen aus dem Koniglischen Zoologischen Museum zu Dresden* 1: 59-84, Tafel II-IV; 2: 163-204, Tafel VIII-X; 3: 383-411, Tafel XXXI-XXXV.

Robinson, I. T. 1956. *The Dentition of the Australopithecinae*. Transvaal Museum, Memoir 9.

Stringer, C. B. 1984. The definition of *Homo erectus* and the existence of the species in Africa and Europe. In P. Andrews and J. L. Franzen (eds) *The Early Evolution of Man with Special Emphasis on Southeast Asia and Africa*. *Courier Forschungsinstitut Senckenberg* 69: 131-143.

Weidenreich, F. 1939 On the earliest representatives of modern mankind recovered on the soil of East Asia. *Peking Natural History Bulletin* 13: 161-174.

Weidenreich, F. 1943. The skull of *Sinanthropus pekinensis:* a comparative study of a primitive hominid skull. *Palaeontologia Sinica* (n.s. D) 10 (whole series 127). Pehpei: Geological Survey of China.

Wolpoff, M.H., Wu Xinzhi and Thorne, A.G. 1984. Modern *Homo sapiens* origins: a general theory of hominid evolution involving the fossil evidence of East Asia. In F. H. Smith and F. Spencer (eds) *The Origins of Modern Humans: A World Survey of the Fossil Evidence*. New York: Alan R. Liss: 411-483.

Wood, B. A. 1984. The origin of *Homo erectus*. In P. Andrews and J. L. Franzen (eds) *The Early Evolution of Man with Special Emphasis on Southeast Asia and Africa*. *Courier Forschungsinstitut Senckenberg* 69: 99-111.

Wu Xinzhi 1981. A well-preserved cranium of an archaic type of early *Homo sapiens* from Dali, China. *Scientia Sinica* 24: 530- 541.

17: The Evolution of Modern Humans: Evidence from Young Mousterian Individuals

ANNE-MARIE TILLIER

INTRODUCTION

The hominids recovered from the Skhūl and Qafzeh caves in Israel are generally regarded as early representatives of anatomically-modern humans in Southwest Asia dating from the earlier part of the Upper Pleistocene (McCown and Keith 1939; Howell 1958; Vandermeersch 1981). They are sometimes called 'Proto-Cro- magnoids'. In addition to the important samples of adult remains, these two sites have also provided substantial numbers of immature individuals, which provide a unique opportunity to study the growth processes for early modern humans.

The Near East has also yielded a number of fossils closely aligned with, but not strictly identical to, the Western European Neanderthals, in both cranial and post-cranial morphology. These eastern Neanderthals found in the Tabūn, Amūd, Kebara and Shanidar caves appear to have been contemporaneous with the European ones. The Teshik-Tash child is probably the most eastern representative of the eastern group. Juvenile Neanderthals are more numerous than Proto-Cro-magnoid ones, with a wide distribution extending from Western Europe to Southwest Asia.

If there is general agreement about the presence of two morphologically distinct types of populations which share the same cultural context (i.e. Mousterian technology, burial practices etc.), the question of their origins and/or their contemporaneity in Southwest Asia is still debated (see papers by Bar-Yosef, Bräuer, Smith *et al.*, and Stringer, this volume). For some authors (Jelinek 1982; Trinkaus 1982, 1986; Wolpoff 1989), the Neanderthals may predate the early anatomically-modern humans, the most recent Neanderthals in Western Asia being much older than 50 000 years BP. For others (Bar-Yosef and Vandermeersch 1982; Bar-Yosef 1989), the Qafzeh and Skhūl hominids are older, and the Neanderthals are relatively late in the Mousterian sequence. Recently, substantial support for the latter interpretation has been provided by the application of TL dating to the sites of Kebara and Qafzeh. Whereas the robust Neanderthal skeleton recently excavated from the Kebara Cave has been dated to *c.* 60 000 BP, the anatomically-modern remains from Qafzeh have been dated to *c.* 92 000 BP (Valladas *et al.* 1987, 1988). Clearly, this supports the idea that anatomically-modern populations were already established in the area at an early stage in the Upper Pleistocene and that the Neanderthals were latecomers in the Levant. The chronological ordering of the Upper Pleistocene hominids from the Southwest Asia is of course directly relevant to phylogenetic considerations about the origins of modern humans. In the available sample of Neanderthals – especially in this area – there are no apparent evolutionary trends in the direction of modern humans, and there are no transitional fossils.

NEANDERTHAL MORPHOLOGY AND DEVELOPMENT

The Near Eastern Neanderthals show close morphological affinities to the Western European sample (Stringer and Trinkaus 1981; Trinkaus 1983), even if some evolutionary trends are admitted for the former. During the past few decades, progress in phylogenetic analysis has enabled us to identify on the cranial skeleton (cranial bones and mandible) derived features of Neanderthals (autapomorphies), which provide evidence for excluding them from the ancestry of modern humans (Sergi 1947; Vallois 1969; Howells 1975; Hublin 1978; Trinkaus 1983; Stringer *et al.* 1984; Rak 1986). The identification of Neanderthal apomorphic features in the post-cranial skeleton remains less evident. Although some features are well defined with regard to modern European samples (notably by Trinkaus 1981, 1983), their polarity is unclear when considered in relation to the older fossil record. Part of these features may be plesiomorphies (Vandermeersch 1981; Stringer *et al.* 1984; Tillier 1986), which were present on *Homo erectus* or other archaic *Homo sapiens* in addition to Neanderthals (Rosenberg 1986; Hublin *et al.* 1987). In this case, their recognition at an early developmental stage in immature Neanderthals, as pointed out by Vlček (1973), Heim (1982), Trinkaus (1983), and Tompkins and Trinkaus (1987) would not be surprising. In attempts to understand the place of the Neanderthals in hominid phylogeny and the significance of their total morphological pattern ('Neanderthalization'), a new interest in juvenile Neanderthals has emerged in the last ten years. The purpose of such studies, conducted primarily on the cranial skeleton, is to distinguish between plesiomorphic retentions, Neanderthal autapomorphies, and juvenile characters. Among the last group, features related to incomplete ossification, shared by all *Homo sapiens* juvenile specimens, are clearly distinguishable from Neanderthal features which are not yet fully developed.

Cranial Data. Accurate data are still lacking for the first two years of life for Neanderthals. The cranial remains are fragmentary and no specimen is sufficiently complete to provide evidence that the Neanderthals and the first anatomically-modern humans are markedly contrasted in cranial size at birth. In 1982 Trinkaus asserted that "young Neanderthals, approximately one year of age, are similar to infants of modern humans" (*in* Ronen 1982: 315). This hypothesis would appear to be confirmed by the La Ferrassie 4, 5, L'Hortus I/I *bis*, Kebara 1, and Shanidar 7 individuals (Lumley 1973; Trinkaus 1983; Tillier, personal observations). But the cranial remains are very fragmentary and nothing can be said with certainty about the recognition of Neanderthal traits.

More information is available for the older stages of growth, i.e. from approximately two years of age. The analysis of immature Neanderthals demonstrates that, among the autapomorphic Neanderthal features (well defined on the adult individuals), a distinction can be made during growth between two categories. Some features appear quite early in development (recognizable from about two years old), while others are identified later, principally related to the permanent dentition (Hublin 1980; Tillier 1981, 1982, 1983a, 1983b, 1986, 1987). If the fact that derived characters appear late in ontogeny may not be surprising, the early occurrence of some of them has to be recognized. In this latter group are found features which mainly involve the form of the cranial vault (occipital, parietal and temporal bones) and also the nasal protrusion (see Tillier 1983b, 1986). Most of the relatively retarded characters relate to the face, especially in the *extreme* mid-facial projection typical of the Neanderthals. This observation may be connected to data known from facial models of growth in modern humans.

Special attention has been paid recently to the maturation of juvenile Neanderthals. For instance, Dean, Stringer and Bromage (1986) propose the idea that the rate of development among Neanderthals may have been accelerated relative to that of modern humans. The claim for a different rate of cranial growth within *Homo sapiens* is based – for Neanderthals – on a new ageing of the Devil's Tower remains. This re-evaluation supports the idea that these hominids achieved rapid brain growth before birth. This new interpretation has given rise to some controversy (Tillier, in press). In addition, the acceptance of growth acceleration requires further support from other fossils and an accurate estimation of age at death. An increase in the duration of childhood dependency associated with the emergence of modern man has also been postulated by Trinkaus (1986). From this short review it is clear that there is now growing interest in ontogeny, and by implication in adaptive questions, among fossil representatives of *Homo sapiens*. Morphological and phylogenetically oriented studies, on Neanderthals and on early anatomically-modern humans, would appear to have produced as many questions as answers, where juvenile individuals are concerned.

THE MORPHOLOGY AND DEVELOPMENT OF EARLY ANATOMICALLY-MODERN HUMANS

Our current understanding of the earliest *Homo sapiens sapiens* in Southwest Asia is based on 13 adult specimens from the Skhūl and Qafzeh caves in Israel. The analysis of the post-cranial skeleton (axial bones, upper and lower limbs) shows that a large proportion of modern features are represented, as compared with the Neanderthals, and in common with the Upper Palaeolithic hominids and their successors in Europe (McCown and Keith 1939; Vandermeersch 1981). Beside these features, a few characters which are shared with the Neanderthals can be recognized. As already noted, the polarity of these features is unclear, however, and some of them may be plesiomorphies, but more accurate data on the predecessors of

Middle Palaeolithic hominids are still required. There have been many attempts to explain the distinctive aspects of post-cranial Neanderthal morphology and limb proportions, in terms of adaptive hypotheses (Brace 1964; Trinkaus 1981, 1983; Trinkaus and Smith 1985). But for early modern humans, biomechanical requirements and the environment were probably more similar to those of the Neanderthals than to those of later Upper Palaeolithic hominids. As pointed out by Bar-Yosef (1989), these variations in hominid morphology do not appear to be reflected in the typology of the associated Mousterian industries.

Cranial Data. As for the Neanderthals, a more accurate analysis is available for the cranial and mandibular remains of the Proto-Cro-magnoids. A list of true derived features has been composed, strictly apart from some metric and non-metric plesiomorphic retentions. Within the last category, some traits were first classified by McCown and Keith as "intermediate" for the Skhūl sample alone, as distinct from the features shared by the Tabūn-Skhūl sample. These features were used to support the attribution of the Skhūl remains to the 'Paleoanthropic genus'. Among metric traits, the large biorbital, bifrontal and biasterionic breadths are now thought of as primitive retentions (Vandermeersch 1981). In the same group, are included morphological characters such as the more or less developed supraorbital area, the robust zygomatic process which strongly deviates from the squamous part of the temporal bone, the styloid process which is mesially placed, and the short lateral extension of the tympanic plate. Finally, there is the general robustness of the mandible, with a *fossa genioglossa* and large teeth. It should be noted that among the Skhūl-Qafzeh hominids, some degree of slight variability can be recognized. Having regard to the total morphological pattern, however, these archaic-looking features cannot be used as evidence for evolutionary continuity with Neanderthals.

From the detailed description of the adult specimens, an evaluation of the morphological pattern exhibited by the earliest modern humans in Southwest Asia can be obtained. The important series of immature individuals found in the Skhūl and Qafzeh caves – representing a total of 11 individuals – should provide further new insights into the origin of modern humans in this particular area.

The Skhūl and Qafzeh children are important for several reasons (see Table 17.1). They constitute the only available sample of anatomically-modern humans so far discovered within a clearly Mousterian context; they retain portions of all anatomical regions; they display several stages of development; and they can be compared closely to the adult sample. These remains provide an opportunity for several lines of research: to examine the development of modern autapomorphies; to make comparisons between ontogenetic processes in Mousterian and living modern humans; to distinguish between features shared with the adults and those which are restricted to juveniles; and to compare ontogenetic models among different Middle Palaeolithic populations of the same species, *Homo sapiens*.

Table 17.1. List of immature human remains from the sites of Skhūl and Qafzeh. Age at death has been estimated by reference to modern standards and is therefore approximate (indicated by ± prefix). Anthropological arguments, together with evidence of association, explains the combination of Qafzeh 4a/21 and Qafzeh 4/22 (see Tillier, in press). (+) = present; (–) = absent; (f) = fragmentary; (ff) = very fragmentary.

Immature Individuals	Skull	Mandible	Teeth	Post-cranial skeleton	Age at death
Qafzeh 13	(f)	+	+	+	new-born
Qafzeh 14	(ff)	–	+	–	± 6 months
Qafzeh 4a	(ff)	–	–	(ff)	± 3 years
Qafzeh 21	(f)	–	+	(f)	
Skhūl I	+	+	+	+	± 4 years
Qafzeh 12	+	–	+	(f)	± 5 years
Qafzeh 10	+	+	+	+	± 6 years
Skhūl X	–	(ff)	+	–	± 6 years
Qafzeh 4	(ff)	+	+	–	± 7 years
Qafzeh 22	–	–	–	(ff)	
Skhūl VIII	–	–	–	(f)	± 8 years
Qafzeh 15	+	+	+	(ff)	±9 years
Qafzeh 11	+	+	+	(f)	± 13 years

Data from the Skhūl Juveniles

From Mugharet es-Skhūl, three juvenile specimens were described by McCown and Keith (1939): Skhūl I, X and VIII, respectively about 4, 6 and 8 years old. In their survey of the post-cranial bones (mainly those of Skhūl I and VIII), the authors emphasized the predominantly modern features in these infantile stages of development. In particular, the limb proportions and limb segments correspond with those of modern specimens. However they noticed that when compared with those of a modern child, the bones showed a few characters "which have a paleanthropic (neandertalian) character". As a result of recent progress in phylogenetic analysis and the recent extensive research on Neanderthalian children (not available at the time of McCown and Keith's work) these features cannot now be accepted as specifically Neanderthalian characteristics. This is the case, for example, for the width of the distal epiphysis of the humerus and the robustness of the foot bones.

Most of the data relating to the cranial morphology come from the Skhūl I child. The cranial characters, both metric and non- metric, conform on the whole to the modern type. Only the large biasterionic breadth and the molar pattern recall plesiomorphous traits. McCown and Keith also mentioned two other features (the less rounded frontal eminences, and rudiments of an early developmental supra-orbital torus) which cannot now be maintained (personal observation). The Skhūl I mandible lacks part of the symphysial region. Nevertheless the bony framework is more robust, in the width of the arch and the slope and thickness of the preserved upper symphysial part, than in modern children of equivalent age.

The Skhūl X mandible belongs to an older child but is even less complete. It consists of the symphysial region and a fragment of the body. The mental elements (*tuber symphyseos, tubercula lateralia, fossae mentales*) are recognizable on the anterior face of the symphysis, but are not strongly marked. The fourth mental element, the *incurvatio mandibulae,* is slightly curved and this is partly due to the unerupted permanent incisors. The chin eminence is slightly prominent. McCown and Keith described a wide shallow *fossa genioglossa* on the posterior symphysial face, as on the two adult mandibles Skhūl IV and V. This plesiomorphous trait on Skhūl X cannot be confirmed, because this region is now damaged (Tillier 1979).

On Preneanderthal and Neanderthal juvenile mandibles, some of the same elements may be recognized (*tuber symphyseos, fossae mentales*, a rough shape of an *incurvatio mandibulae*), but the chin development never approaches the modern configuration (Tillier 1981). During the growth of the bone, the *tuber symphyseos* moves from a strictly basal position to an upper central one, but the trigonum mentale never occurs. On the posterior symphysial face the presence of plesiomorphous retentions (*planum alveolare, fossa genioglossa, torus transversus*) is variable, since the symphysis is either still receding or it is not. Unlike those of the Neanderthals, the Skhūl X mandible shows a chin development almost that of modern children.

Data from the Immature Qafzeh Sample: Preliminary Results

The juvenile sample from the Qafzeh cave, recently enlarged to 8 individuals (Tillier, in press), includes several stages of development from new-born to about 12-13 years. The sample is noteworthy for the relatively large numbers of both cranial and postcranial remains, although much of the material is weathered as a result of the processes of fossilisation and (above all) breccification during the formation of the cave deposits. Since the study of these remains is still in progress, full information is not yet available on all aspects of the Qafzeh juveniles. Nevertheless, a number of observations can be made with regard to the cranial and mandibular remains.

Five mandibles, from new-born to 12-13 years (Qafzeh 13, 10, 4, 15, 11) provide information about bone growth. As on Skhūl X, the mental elements are present but the rate of development of the maximal chin eminence seems slightly retarded, compared to growth processes in modern mandibles, even though the comparison is limited to European sample. The comparison with equivalent stages of dental development shows that the fully-developed chin type is present on the oldest specimen, Qafzeh 11 (Tillier 1984). This last individual displays the most 'modern' configuration

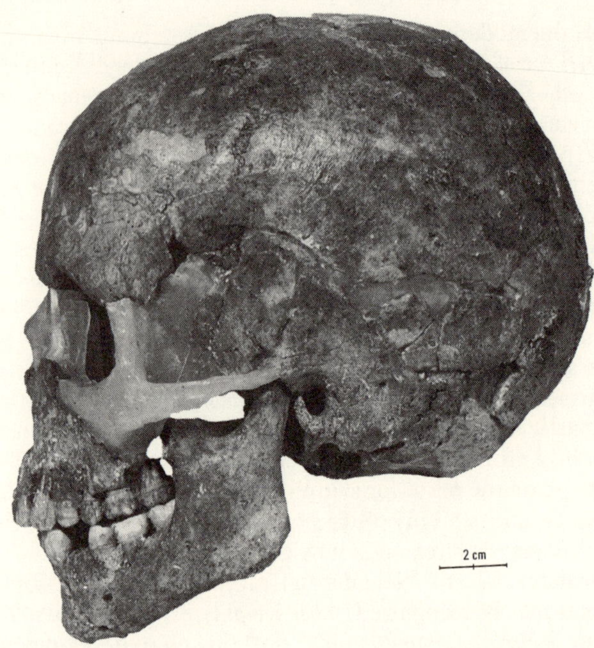

Figure 17.1. The Qafzeh 11 immature skull (approximately 13 years of age) in *norma lateralis*, left.

within the Skhūl-Qafzeh population as a whole, along with the adult Skhūl V. Certain plesiomorphous characters can also be recognized: a slight *planum alveolare* on Qafzeh 4 (Tillier 1979), a *fossa genioglossa* on Qafzeh 4, 10, 15, and large teeth on Qafzeh 4, 10 and 15. Most of these features have already been identified on the adult mandibles from Qafzeh and Skhūl.

Special attention has been paid to the cranial morphology of Qafzeh 11 (Tillier 1984), since the age of this specimen is closest to that of the adults from the site. As on the Skhūl I child (as well as on all the adults) autapomorphies of modern man predominate on this skull (see Figure 17.1). In addition to these features, other features which are specific to juveniles can be observed – as, for instance, the accentuated sagittal frontal curvature, with marked frontal eminences, and the slight external occipital structures. The Qafzeh 11 skull also displays some plesiomorphous retentions, such as the flexed maxilla with a canine pit (also displayed by Qafzeh 4), the weak juxta-mastoid eminence (see Note 1), and the lateral (or transverse) shortness of the tympanic plate. The glabellar eminence and the supraorbital morphology are more developed than on modern European children, but clearly distinct from those of Neanderthals. These archaic-looking features are shared with the adults from Qafzeh, as described by Vandermeersch (1981), and cannot be used as indications of evolutionary continuity with Neanderthals or as the results of gene flow. The three primitive metric features quoted by Vandermeersch on the adults (large biorbital, bifrontal and biasterionic breadths) are lacking on Qafzeh 11. But Qafzeh 11 displays an archaic feature which is missing on the adults – namely the form of the tympanic plate, which is divided in two unequal parts (anterior and posterior) by a slight crest separated from the mastoid process. This feature, claimed to be characteristic of Neanderthal bones (Vallois 1969; Vandermeersch, 1981), can be observed on some older fossil hominids, as well as on Skhūl specimens (McCown and Keith 1939). It is also present in some juvenile bones from fossil and living populations (personal observation). Qafzeh 11 displays a few features shared with Neanderthal children. Among these are a large inter-orbital breadth, the porion-bregma height (but still within the modern range), a frontal arc longer than the parietal one (but also found on some modern children), the configuration of the tympanic plate already noted, and the robusticity indices of the mandibular body between the first and second molars.

Apart from the tympanic configuration, other features described on Qafzeh 11 are missing on the adult skulls; the prevalence of the frontal arc over the parietal one (shared with the adult Skhūl IX) and the slenderness of the mandible with a salient chin and small teeth (with the exception of the canines and premolars). An earlier study of the Qafzeh 4 child (Tillier 1979) provided evidence that the morphology of the jaw and the teeth were closer to those of the adults. From the survey of the other juveniles from the site we can expect more data to confirm this variability, or to reinforce the peculiarities of the immature individuals within the Qafzeh sample.

In his description of the adult skulls, Vandermeersch (1981) noted that on two specimens (Qafzeh 7 and 9) a slender cranial vault was associated with a rather robust mid-facial pattern. However, a strictly opposite morphological combination was displayed by Qafzeh 5. Only Qafzeh 6 showed a balanced morphological pattern. This variability has to be recognized as far as a close relationship between large teeth and facial robustness is concerned. Among the children, Qafzeh 11 more closely recalls Qafzeh 6 in the whole cranial aspect, while Qafzeh 15 seems closer to Qafzeh 9.

CONCLUSIONS

Major advances have been made in recent years in the study of immature remains among Neanderthals. Analyses of their cranial morphology have emphasized the distinctiveness of the Neanderthal autapomorphies during growth, which display heterochrony. The basicranial traits appear earlier in development than most of the facial traits (see below), the latter being more related to the eruption of the permanent teeth. Most aspects of Neanderthal facial morphology are generally regarded as reflections of biomechanical factors, due to a heavy use of the anterior permanent teeth, which are strongly worn (Brace 1964; Trinkaus 1983). However, it remains uncertain to what extent anterior dental enlargement and anterior tooth wear can be attributed to dietary as opposed to non-dietary use. Furthermore, few of the Neanderthals shows extreme anterior tooth wear. It should be recalled that other Middle Palaeolithic hominids display large teeth in both Southwest Asia (Vandermeersch 1981) and in North Africa (Hublin and Tillier 1981), but lack this mid facial prognathism. But a generalized dental enlargement is characteristic of this African sample.

As in the Neanderthals, a mosaic of cranial features can also be identified among the immature Proto-Cro-magnoids. These include (1) juvenile features which are common to all children within *Homo sapiens;* (2) primitive retentions shared with archaic *Homo sapiens* (or at least Neanderthals); (3) primitive retentions shared with both Neanderthals and modern children; (4) modern autapomorphies displayed by immature and adult Proto-Cro-magnoids; and (5) modern features present on immature Proto-Cro-magnoids and immature modern humans.

More osteologically oriented studies, related to growth patterns, are still required on different living populations to clarify the significance of some features (as for example on the temporal bone) which are shared between all children within *Homo sapiens* and the adult Neanderthals. The available data relating to mandible growth patterns – especially for the chin eminence – need to be increased. In the original description of the Staroselye child, Roginskij (1954) emphasized the modern features of this Mousterian fossil. He concluded that this child from the Crimea was closely related to the same population as the Skhūl individuals. As also pointed out by Howell (1958), Alexeiev (1976) and Tillier (*in* Ronen 1982: 3l5), the attribution of this young child of about one year old to *Homo sapiens sapiens* seems certain. If the association of this fossil with a Mousterian context is well documented (as argued by Jelínek 1969 and Alexeiev 1976), the jaw shows a more accelerated chin growth than that of the immature Proto-Cro-magnoids. The large size of the anterior teeth on some specimens from Qafzeh is not sufficient to explain the chin growth form, as the mandible from the Mousterian site of Jebel Irhoud, Morocoo (Irhoud 3) clearly shows (Hublin and Tillier 1981).

Clearly, there are many problems inherent in attempting to construct a model of skeletal growth peculiar to Middle Palaeolithic hominids (including Neanderthals, other archaic *Homo sapiens,* and Proto-Cro-magnoids) on the basis of information derived from living human populations – especially in terms of growth rates. Clearly, we must beware of applying circular arguments, which apply *modern* standards for ageing these specimens, and then employ these standards to argue for *different* rates of growth in the fossil specimens!

Many studies have focused on behavioural features which are peculiar to Neanderthal populations to explain their typical morphology (especially in the face and post-cranial bones). In considering the evolutionary origins of early anatomically-modern humans recovered from Middle Palaeolithic contexts, several questions emerge – mainly relating to functional hypotheses concerning the post-cranial peculiarities of the Neanderthals. In South-west Asia, the great morphological contrasts between the Skhūl-Qafzeh samples and the Asian Neanderthals do not accord with the model of regional continuity, whatever the date of the former samples may be. This point has been argued from the characteristics of the adult specimens by both Stringer and Vandermeersch (this volume) and is supported by the present analysis of ontogenetical processes. Even if some archaic features are still recognizable in these hominids, the most important question which remains is the source of their modern autapomorphies.

ACKNOWLEDGEMENTS

The anatomical remains from Skhūl and Qafzeh discussed in the present paper were studied by courtesy of the Department of Antiquities, Rockerfeller Museum, Jerusalem. My thanks are due to Chris Stringer for his review of the final version of the paper.

NOTE

1. The juxta-mastoid eminence is distinct from the occipito-mastoid crest, as noted by Hublin (1982: 349).

REFERENCES

Alexeiev, V. P. 1976. Position of the Staroselye find in the hominid system. *Journal of Human Evolution* 5: 413-421.

Bar-Yosef, O. 1989. Upper Pleistocene human adaptations in Southwest Asia. In E. Trinkaus (ed.) *Patterns and Processes in Later Pleistocene Human Emergence*. Cambridge: Cambridge University Press. In Press.

Bar-Yosef, O. and Vandermeersch, B. 1982. Notes concerning the possible age of the Mousterian layers in Qafzeh Cave. In J. Cauvin and P. Sanlaville (eds) *Préhistoire du Levant*. Paris: Centre National de la Recherche Scientifique: 281-285.

Brace, C. L. 1964. The fate of the 'Classic' Neanderthals: a consideration of hominid catastrophism. *Current Anthropology* 5: 3-43; 7: 210-214.

Dean, M. C., Stringer, C. B. and Bromage, T. D. 1986. Age at death of the Neanderthal child from Devil's Tower, Gibraltar, and the implications for studies of general growth and development in Neanderthals. *American Journal of Physical Anthropology* 70: 301-309.

Heim, J. L. 1982. *Les Enfants Néandertaliens de La Ferrassie*. Paris: Masson.

Howell, F. C. 1958. Upper Pleistocene Men of the Southwest Asian Mousterian. In G. von Koenigswald (ed.) *Hundert Jahre Neanderthaler*. Köln, Böhlau: 185-198.

Howells, W. W. 1975. Neanderthal Man: facts and figures. In R. H. Tuttle (ed.) *Paleoanthropology, Morphology and Paleoecology*. The Hague: Mouton: 389-407.

Hublin, J. J. 1978. *Le Torus Occipital Transverse et les Structures Associées: évolution dans le genre Homo*. Unpublished PhD. Thesis, University of Paris VI.

Hublin J. J. 1980. La Chaise (Suard), Engis 2 et La Quina 18: développement de la morphologie occipitale externe chez l'enfant pré-néandertalien et néandertalien. *Comptes-Rendus de l'Académie des Sciences de Paris (Série D)* 291: 669-672.

Hublin, J. J. 1982. Les antenéandertaliens: présapiens ou prénéandertaliens? *Phylogénie et Paléogéographie*. Géobios Special Mémoire 6, Lyon: 345-357.

Hublin, J. J. and Tillier, A.-M. 1981. The Mousterian juvenile mandible from Irhoud (Morocco): a phylogenetic interpretation. In C. B. Stringer (ed.) *Aspects of Human Evolution*. London: Taylor and Francis: 167-185.

Hublin, J. J., Tillier, A.-M. and Tixier, J. 1987. L'humérus d'enfant moustérien (Homo 4) du Jebel Irhoud (Maroc) dans son contexte archéologique. *Bulletins et Mémoires de la Société d'Anthropologie de Paris* 4: 115-142.

Jelinek, A. 1982. The Middle Paleolithic in the Southern Levant, with comments on the appearance of Modern *Homo sapiens*. In A. Ronen (ed.) *The Transition from Lower to Middle Palaeolithic and the Origin of Modern Man*. Oxford: British Archaeological Reports International Series S151: 57-104.

Jelínek, J. 1969. Neanderthal Man and *Homo sapiens* in Central and Eastern Europe *Current Anthropology* 10: 475-504.

Lumley, M. A. de. 1973. *Antenéandertaliens et Néandertaliens du Bassin Méditerranéen Occidental Européen*. Marseille: Etudes Quaternaires, Université de Provence.

McCown, T. D. and Keith, A. 1939. *The Stone Age of Mount Carmel*. Vol. 2: *The Fossil Human Remains from the Levalloiso-Mousterian*. Oxford: Clarendon Press.

Rak, Y. 1986. A new look to an old face. *Journal of Human Evolution* 15: 151-164.

Roginskij, Y. Y. 1954. Morphological features of the skull of the child from the Late Mousterian level of the Cave of Staroselye. *Sovietskaia Etnografia* 1: 27-39 (in Russian).

Ronen, A. (ed.) 1982. *The Transition from Lower to Middle Palaeolithic and the Origins of Modern Man*. Oxford: British Archaeological Reports International Series S151.

Rosenberg, K. R. 1986. *The Functional Significance of Neanderthal Pubic Morphology*. Unpublished Ph.D. Thesis, University of Michigan.

Sergi, S. 1947. Sulla morfologia della 'facies anterior corporis maxillae' nei paleantropi di Saccopastore e del Monte Circeo. *Rivista di Antropologia* 35: 401-408.

Smith, P. and Arensburg, B. 1977. A Mousterian skeleton from Kebara Cave. In O. Bar-Yosef and B. Arensburg (eds) *Eretz-Israel* 13: 164-176.

Smith, P. and Tillier, A.-M. (in press). Additional infant remains from the Mousterian strata, Kebara Cave (Israel) In O. Bar-Yosef and B. Vandermeersch (eds) *Prehistoric Investigations in Southern Levant*. Oxford: British Archaeological Reports International Series. In Press.

Stringer, C. B. and Trinkaus, E. 1981. The Shanidar Neanderthal crania. In C.B. Stringer (ed.) *Aspects of Human Evolution*. London: Taylor and Francis: 129-165.

Stringer, C. B., Hublin, J. J. and Vandermeersch, B. 1984. The origin of anatomically modern humans in Western Europe. In F. Smith and F. Spencer (eds) *The Origins of Modern Humans: a World Survey of the Fossil Evidence*. New York: Alan R. Liss: 51-135.

Tillier, A.-M. 1979. Restes craniens de l'enfant moustérien Homo 4 de Qafzeh (Israël): la mandibule et les maxillaires. *Paléorient* 5: 67-85.

Tillier, A.-M. 1981. Evolution de la région symphysaire chez les *Homo sapiens* juvéniles du Paléolithique moyen: Pech de l'Azé, Roc de Marsal et La Chaise 13. *Comptes-Rendus de l'Académie des Sciences de Paris* 293: 725-727.

Tillier, A.-M. 1982. Les enfants néanderthaliens de Devil's Tower (Gibraltar). *Zeitschrift für Morphologie und Anthropologie* 73: 125-148.

Tillier, A.-M. 1983a. Le crâne d'Engis 2: un exemple de distribution des caractères juvéniles, primitifs et néanderthaliens. *Bulletins de la Société Royale Belge d'Anthropologie et de Préhistoire* 94: 51-75.

Tillier, A.-M. 1983b. L'enfant néanderthalien du Roc de Marsal, Campagne du Bugue, Dordogne: le squelette facial. *Annales de Paléontologie* 69: 137-149.

Tillier, A.-M. 1984. L'enfant Homo 11 de Qafzeh (Israël) et son apport à la compréhension des modalités de croissance des squelettes moustériens. *Paléorient* 10: 7-48.

Tillier, A.-M. 1986. Quelques aspects de l'ontogenèse du squelette cranien des Néanderthaliens. In V. V. Novotny and A. Mizerova (eds) *Fossil Man: New facts – New Ideas*. Brno: *Anthropos:* 207-216.

Tillier, A.-M. 1987. L'enfant néanderthalien de La Quina H 18 et l'ontogénie des Néanderthaliens. In B. Vandermeersch (ed.) *Préhistoire de Poitou-Charentes: Problèmes Actuels*. Paris: Comité des Travaux Historiques et Scientifiques: 201-206.

Tillier, A.-M. (in press). La place des restes de Devil's Tower (Gibraltar) dans l'ontogenèse des Néanderthaliens. *Bulletins et Mémoires de la Société d'Anthropologie de Paris*. In press.

Tillier, A.-M. (in press). Les enfants Proto-Cro-Magnons de Qafzeh (Israël): mise au point. In O. Bar-Yosef and B. Vandermeersch (eds) *Prehistoric Investigations in Southern Levant*. Oxford: British Archaeological Reports International Series. In Press.

Tompkins, R. L. and Trinkaus, E. 1987. La Ferrassie 6 and the development of the Neanderthal pubic morphology. *American Journal of Physical Anthropology* 73: 232-239.

Trinkaus, E. 1981. Neanderthal limb proportions and cold adaptation. In C. B. Stringer (ed.) *Aspects of Human Evolution*. London: Taylor and Francis: 187-224.

Trinkaus, E. 1982. Evolutionary continuity among archaic *Homo sapiens*. In A. Ronen (ed.) *The Transition from Lower to Middle Paleolithic and the Origin of Modern Man*. Oxford: British Archaeological Reports International Series S151: 301-314.

Trinkaus, E. 1983. *The Shanidar Neandertals*. New York: Academic Press.

Trinkaus, E. 1986. The Neandertals and modern human origins. *Annual Review of Anthropology* 15: 193-218.

Trinkaus, E. and Smith, F. H. 1985. The fate of the Neandertals. In E. Delson (ed.) *Ancestors: the Hard Evidence*. New York: Alan R. Liss: 325-333.

Valladas, H., Joron, J. L., Valladas, G., Arensburg, B., Bar-Yosef, O., Belfer-Cohen, A., Goldberg, P., Laville, H., Meignen, L., Rak, Y., Tchernov, E., Tillier, A.-M., Vandermeersch, B. 1987. Thermoluminescence dates for the Neanderthal burial site at Kebara in Israel. *Nature* 330: 159-160.

Valladas, H., Reyss, J. L., Joron, J. L., Valladas, G., Bar-Yosef, O. and Vandermeersch, B. 1988. Thermoluminescence dating of Mousterian 'Proto-Cro-Magnon' remains from Israel and the origin of modern man. *Nature* 331: 614-616.

Vallois, H. V. 1969. Le temporal néandertalien H27 de La Quina. Etude anthropologique. *L'Anthropologie* 73: 365-400, 525-544.

Vandermeersch, B. 1981. *Les Hommes Fossiles de Qafzeh (Israël)*. Paris: Centre National de la Recherche Scientifique.

Vlček, E. 1973. Post-cranial skeleton of a Neandertal child from Kiik-Koba, USSR. *Journal of Human Evolution* 2: 149-180.

Wolpoff, M. H. 1989. The place of the Neandertals in human evolution. In E. Trinkaus (ed.) *Patterns and Processes in Later Pleistocene Human Emergence*. Cambridge: Cambridge University Press. In Press.

18: The Ecological Conditions of Speciation: a Comparative Approach to the Origins of Anatomically-Modern Humans

ROBERT FOLEY

THE EVOLUTIONARY CONTEXT OF ANATOMICALLY-MODERN HUMAN ORIGINS

The appearance of anatomically-modern humans (AMH) represents the evolution of a taxon that has dominated the world more than any other single *species*. It is, however, an event that has long been the subject of considerable debate, and, as this volume shows, one in which there is still no consensus view.

Several reasons for this may be suggested. One is that, contrary to the usual complaint of palaeontologists and archaeologists, there is a relative wealth of information. This is not to say that the fossil record is perfect, but rather that because both morphological and behavioural change can be observed in some detail, conflicting lines of evidence, or, perhaps more accurately, lines of evidence that cannot easily be integrated, produce a confused picture (Pilbeam 1985). Palaeontologists study the shifting patterns of species appearance and disappearance over long periods of time (10^5 and 10^6 years), and so seldom deal with precise evolutionary events occurring at the level of sub- species and over periods of 10^4 and 10^5 years. There is consequently little expectation of what should occur at the boundary between macro- and micro-evolution. Furthermore, because it is a specifically human evolutionary event, explanation has largely rested on specifically human characteristics, especially that of 'culture'. What is particularly confusing here is that culture has been treated in these explanations as both the cause and the consequence of becoming a modern human. By fulfilling both these roles, the use of culture as an all embracing and all pervading explanation has probably done most to obscure the processes by which modern humans evolved (Foley, in press).

Through this anthropocentric perspective, the origin of AMH has been treated as a unique event, outside the context of the ordinary processes of evolutionary biology, and incorporating significant elements of non-biological disciplines (archaeology, cultural anthropology). Given the fact that this approach has not led to any general agreement, understanding of the origins of AMH can perhaps now best be served not by stressing the uniqueness of this evolutionary event, but by looking at it in a comparative evolutionary and ecological framework. This paper will focus on the process of speciation in the context of (a) general factors in evolutionary ecology that affect speciation; and (b) primate and earlier hominid analogues. Expectations about the nature of speciation arising from these considerations can then be applied to the specifics of AMH origins.

Before exploring these areas it is necessary to clarify certain problems concerning the transition to AMH. First to show that it is evolutionarily significant (rather than just being the final gasp of continuous hominid evolution); and second, to clarify its position in time and space.

The Evolutionary Significance of the Appearance of AMH

The classic nomenclature of hominid phylogeny recognizes two species of middle and later Pleistocene hominid – *Homo erectus* and *Homo sapiens*. It is broadly accepted that there is an ancestor-descendent relationship between the two (see Day 1986). However *Homo sapiens* is a polytypic species, with distinct sub-species recognized (Table 18.1), the evolutionary relationships of which have been debated for a long time (Howell 1984). Of these sub-species, it has generally been assumed that the archaic *sapiens* populations are ancestral to the subspecies of AMH *(Homo sapiens sapiens)*, although the relative position in this transition of the other major group, the neanderthals *(Homo sapiens neanderthalensis)* still remains unclear. Whatever the precise relationships, the assumption underlying the nomenclature and systematics is that the transition from *Homo erectus* to *Homo sapiens* is more significant in evolutionary terms than is the transition from archaic forms of *Homo sapiens* to AMH.

There are, though, relatively few features that support this view. The anatomical distinctions between *Homo erectus* and *Homo sapiens* are not great, and tend to be continuous in character (Stringer 1985). Many of these relate to the overall increase in cranial capacity and reduction of facial proportions that occurs in later hominid evolution. Confirmation of this morphological continuity is perhaps provided by the difficulty of assigning particular specimens to one or other taxon. For example, the cranium from Petralona has been assigned to both *Homo erectus* (Hemmer 1982) and archaic *Homo sapiens* (Stringer 1983). Similar problems of defining the boundary between these taxa exist in other parts of the world: in Java the distinction between *Homo erectus* and *Homo sapiens* is far from clear; in China a new specimen from Yinkou demonstrate continuity in archaic features (Wu Rukang, in press), and in Africa, the Kabwe and Bodo crania are equally ambiguous (Day 1986).

In comparison to this, the distinction between AMH and archaic representatives of *Homo sapiens* is normally clear cut. In China, in Europe and in Southeast Asia, fully modern forms appear relatively suddenly, with little continuity with earlier forms. Only in Africa, and to a lesser extent the Middle East, do specimens showing continuity exist (Stringer *et al.* 1984). Furthermore, the change from archaic to modern forms of *Homo sapiens* is more than just a minor anatomical shift, but involves a radical alteration in direction – a move away from the robusticity associated with

Table 18.1. Alternative taxonomic nomenclature of the genus *Homo*. The one on the right proposed here emphasizes the relative importance of the shift from archaic to modern hominids. It would however fall foul of the Rules of Zoological Nomenclature on the grounds that the name *Homo neanderthalensis* has priority over that of *Homo erectus* (Stringer, *pers. comm.*).

Campbell	Tattersall 1986	Nomenclature implied by this paper
H. habilis	*H. habilis*	*H. habilis*
H. erectus erectus	*H. erectus*	*H. erectus erectus*
pekinesis		*pekinensis*
olduvaiensis		*olduvaiensis*
		heidelbergensis
H. sapiens heidelbergensis	*H. heidelbergensis*	*soloensis*
rhodesiensis		*neanderthalensis*
soloensis		
capensis		
neanderthalensis	*H. neanderthalensis*	
sapiens	*H. sapiens*	*H. sapiens*

thick cranial and post-cranial bones, accompanied by a reorganization of cranial and facial proportions with the raising and shortening of the cranial vault and the tucking under of the face.

These anatomical considerations are reinforced by the archaeological evidence. The transition from *Homo erectus* to *Homo sapiens* is not marked by any major technological change. Bifaces continue throughout this period, accompanied by a gradual increase in the proportion of prepared-core technique employed (Table 18.2). Apart from expansion in geographical distribution, few other major changes are apparent with the appearance of the 'Middle Palaeolithic'. In contrast to this, the appearance of AMH is marked in many contexts by an extremely rapid transition in technology to blade production, a sharp increase in the complexity and variability of tool forms, and the development of regional diversity and temporal instability (Isaac 1972). Furthermore, new forms of behaviour appear, including the development of bone technology, and ultimately artistic forms of representation (Mellars 1973).

Overall, therefore, both morphological and behavioural considerations lead to the conclusion that the transition from archaic forms of *Homo sapiens* to modern forms is of greater evolutionary (i.e. genetic and adaptive) significance than the shift from *Homo erectus* to *Homo sapiens*. As Stringer (1985) has suggested, 'sapienization' is not a single process, but rather two distinct ones. The first is a general trend among archaic hominids within existing morphological frameworks towards encephalization; the second is cranial reorganization and a move away from the robusticity of archaic hominids.

Nomenclature is not the focus of this paper, but provisionally it might be suggested that this pattern of hominid evolution is better represented by confining the term *Homo sapiens* to anatomically-modern forms, and considering the 'archaic' sapiens forms to be subgroups within *Homo erectus* or, as Tattersall (1986) has suggested, by recognizing several distinct species among the earlier representatives of what are traditionally referred to as '*Homo sapiens*' populations. This revision of nomenclature would better reflect the significance of the transition to AMH.

Evidence for the Origins of Anatomically-Modern Humans

Historically, the Middle East has been the favoured location for the origins of AMH, but attention has recently turned to sub-Saharan Africa (see papers in this volume). The earliest-known specimens with anatomically-modern features are from sub-Saharan Africa (see Rightmire 1984, 1985 and Bräuer 1984, and this volume, for a full review of this evidence) – principally the specimens from Border Cave (95-120 000 BP), Klasies River Mouth (>80 000 BP), Laetoli LH-18 (120 000 BP) and Omo 1 (>100 000 BP). Archaeological evidence, particularly the very early appearance of blade industries in South Africa, provides some support to this fossil evidence for an African origin. While many of these dates have considerable errors attached to them, they exceed by a substantial margin the earliest evidence from most parts of the world, and only the recently published dates for the Qafzeh hominid approach them (Valladas *et al.* 1988). Furth-

Table 18.2. Characteristics of archaic hominids (= *Homo erectus*; see Table 18.1) and *Homo sapiens* (anatomically modern humans). See also Figure 2 in Stringer 1985).

Archaic Hominids	Modern Hominids
Heavy musculature	Lighter musculature
Robust skeleton	More gracile skeleton
Thick cranial bones	Thinner cranial bones
Long low crania	High and rounded cranial vault
Relatively flat basi-cranium	Flexed basi-cranium
Moderate to large brain	Large brain
Large face and dentition	Reduced face and dentition
Extremely widespread and stable stone-tool assemblages (continental distribution and duration > 50 Kyr.	Variable and unstable tool assemblages (regional distribution and duration ≤ 10 Kyr)
Repetitive assemblages (technological modes 1-3)	Highly structured assemblages (technological modes 4-5)
Tropical and warm-temperate distribution; intermittent colonization of cold habitats	Global distribution
No art	Art, etc.
Language?	Language

ermore, in Europe and in Asia the Upper Pleistocene hominid fossil record shows a discontinuity; this is particularly marked in Western Europe, where full Neanderthals are known from as late as 31-34 000 BP (i.e. Saint-Césaire (Lévêque and Vandermeersch 1980; Stringer *et al.* 1984)).

Recent work on human mitochondrial DNA (mtDNA) has also indicated an African origin for modern humans (Cann *et al.* 1987). MtDNA, which evolves more rapidly than nuclear DNA, and is inherited through the female line only, shows a pattern of diversity suggesting that African forms are ancestral to those found elsewhere in the world. This suggests a *single* origin for modern humans, and a dispersal from Africa in the late middle or early Upper Pleistocene. While these data are still controversial (Wainscoat 1987), on account of the limited African sample and uncertainties that arise from estimates of the rate of mutation in mtDNA, they perhaps provide further circumstantial evidence for an African origin for AMH.

THE EVOLUTIONARY ECOLOGY OF SPECIATION

Two points have been established so far. The first is that the appearance of AMH is evolutionarily significant, involving a major change from earlier archaic hominids such that it is not a simple case of continuously gradual change. The second is that this change seems to have occurred first in sub-Saharan Africa during the early Upper Pleistocene. These conclusions will form the underlying assumptions for the remainder of this paper.

The basic question that this poses is 'why Africa?' as the location of this speciation event. Various answers can be proposed to this question. Perhaps the simplest of these is that this is where the mutations or innovations that made AMH adaptations possible occurred. Various key 'mutations' have been suggested – for example reorganization of the brain, shortening of the gestation period, extension of the juvenile stage, invention of blade technology, and so on. Such mutations are random, and therefore it is merely an accident where those combinations liable to lead to AMH will occur. This seems, though, to be a rather pointless explanation, reducing all patterns in evolution to molecular stochasticism. It is also an incomplete explanation; mutations, cultural or genetic, may be random, but they only supply the source of variation on which selection may operate. To understand why there may be selection for novel adaptations requires also a consideration of the factors that are likely to promote speciation. The remainder of this paper will concentrate on this issue by looking at what factors have promoted speciation in primate evolution and during the earlier phases of hominid evolution. This comparative approach is adopted since hominoid and hominid evolution has not been a unilineal process, but appears to consist of a series of branching (speciation) events, resulting in between 5 and 9 species being recognized in sub-Saharan Africa between 5 and 1.5 million years ago (Foley 1987). The questions surrounding the origins of AMH should be seen in the comparative context of this earlier adaptive radiation of hominids.

The factors likely to promote speciation can be considered at a series of levels: (a) large-scale biogeography; (b) environmental dynamics; (c) spatial ecology; and (d) the evolutionary ecology of adaptive divergence. These will be considered in turn.

Biogeography

The biogeographical scale of evolutionary change is relevant to a question that has persisted in discussions of the origins of AMH. Is speciation more likely to occur across a broad geographical area? Or will increasing variability between regions result in different evolutionary trajectories, and so to new species appearing locally and then expanding, either through colonization or gene flow?

An examination of primate and earlier hominid evolution would suggest that multiple origins are relatively unlikely. Recent work has increasingly shown that geographical variation often lies at the root of evolutionary problems. For example, the diversity of form in Miocene hominoids seems to be best explained by continental-wide patterns of distribution. The great apes, despite evolutionary parallels, are best seen as independent Asian and African evolutionary progressions – with the African apes and humans representing a clade relative to the Asian apes. Fossil apes from the Miocene seem also to follow this distinction, with *Sivapithecus* and *Gigantopithecus* belonging to the Asian clade, while *Kenyapithecus* and the early hominids belong to the African one (see papers in Ciochon and Corruccini 1983). The evolution of living cercopithecoids also indicates the importance of geography in evolutionary patterns – the emergence of cercopithecines in Africa, radiation within Africa of colobines and cercopithecines, their expansion out of Africa, followed by parallel radiations of *Macaca* and *Presbytis* in Asia, and *Papio, Cercopithecus,* and *Colobus* in Africa. Biogeographical patterns such as these can be found in many widely distributed mammalian species. The inference to be drawn from this is that populations that disperse across broad geographical areas also diverge; they do *not* continue to have a single evolutionary trajectory. Local factors (selection and isolation) predominate over other broad phylogenetic characteristics.

Divergence, however, is not the only process operating. While evolutionary separation may be continuing (i.e. genetic differences are accumulating), functional and adaptive parallels may still occur. Similar animals will face similar problems, despite the fact that they are geographically isolated. Selection is likely to favour *similar* but not *identical* adaptive solutions to these problems, thus producing a pattern of parallel evolution – that is, genetic divergence but morphological similarity as a result of local selection in response to similar selective pressures. Such parallelisms are well known within the primates. The catarrines and platyrrhines are a classic example. More specifically, within the catarrhines, *Presbytis* and *Colobus,* despite evolutionary divergence in response to geographical separation, show similarities as folivorous primates that are not solely due to a common ancestor. More speculatively, parallels may be suggested within the Hominoidea, between australopithecines and *Gigantopithecus*. These belong to different families, but show similar, megadontic adaptations to coarse plant foods. Within earlier hominid evolution the problem of parallel evolution is now being given considerable attention, including suggestions that the South African *(Australopithecus robustus)* and East African *(Australopithecus boisei)* australopithecines may in fact be the products of parallel evolution (Wood and Chamberlain 1986; Foley 1987).

The continental pattern of primate evolution leads to two expectations – regional divergence of lineage combined with adaptive parallelisms. This is not dissimilar to the pattern seen in later hominid evolution. First, late Middle Pleistocene hominids display a semi-global distribution; during the Middle Pleistocene, hominids had colonized large parts of the tropical and sub-tropical world. Secondly, within that total range, despite broad morphological similarities of grade, they show regional variability that perhaps indicates local populations, and, according to some authorities (Andrews 1984; Wood 1984), separate evolutionary clades (Figure 18.1). This indicates evolutionary divergence based on wide geographical separation.

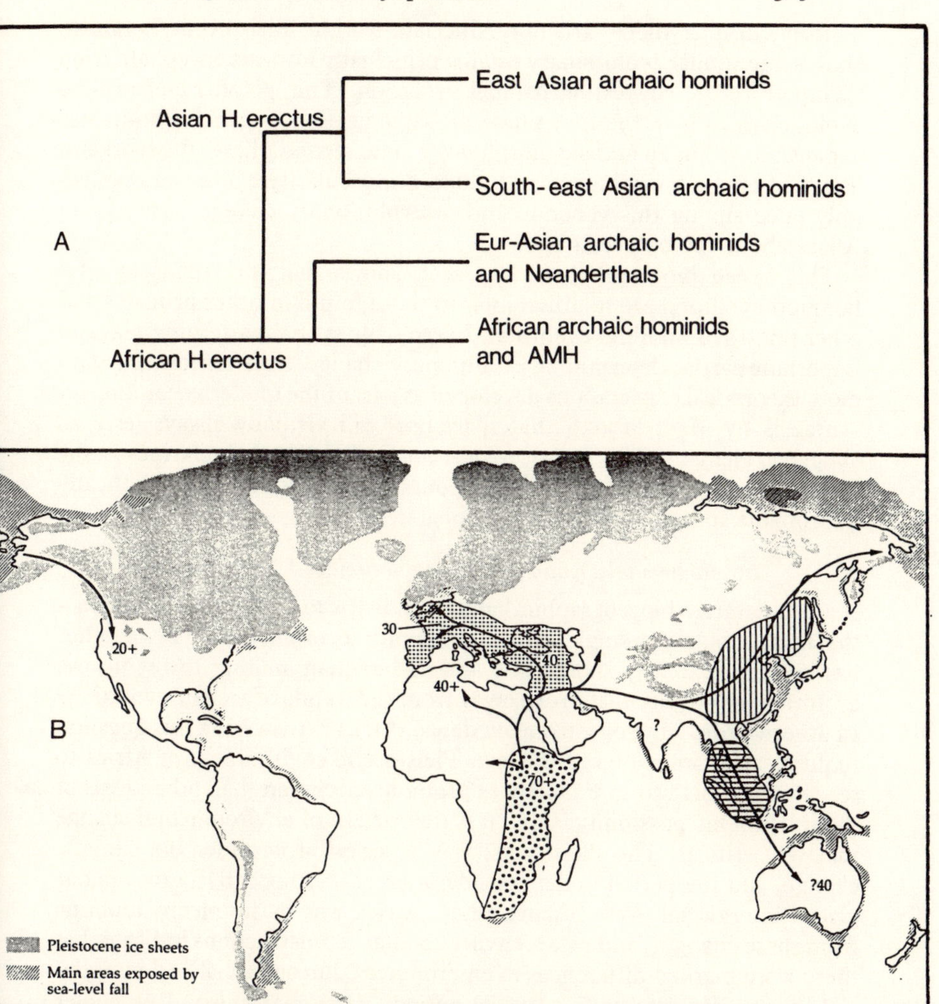

Figure 18.1. The phylogenetic and spatial patterning of later hominids. (A) The phylogenetic relationships of the hominids of the Pleistocene, based on Andrews (1984). This phylogeny indicates that only the African hominids of the late Middle Pleistocene are ancestral to anatomically-modern hominids (AMH). (B) Distribution of these clades through the Old World. The vertical shading shows archaic Chinese populations (e.g. Yinkou); the horizontal shading shows the archaic hominids of South-east Asia (e.g. Solo). Fine stippling represents the Neanderthals of Europe and the Middle East, and the coarse stippling the African populations from which AMH arose. The Middle East may well have been an area of overlap of the Neanderthal and African populations. Arrows indicate dispersal of AMH from Africa, with approximate dates in thousands of years shown.

Thirdly, despite their variability, the late Middle Pleistocene hominids show some similar evolutionary trends, principally towards encephalization (Wolpoff 1983), suggesting parallel evolution. This parallel and general evolution relates to the first phase of 'sapienization' – general brain size expansion within an archaic morphology. The second phase (the reorganization of cranial morphology and reduced muscularity), however, evolves only once among this general band of evolutionary change – that is, in Africa about 100 000 years ago.

This broad biogeographical approach shows that the trends in later hominid evolution are not dissimilar to those found in other primates and other phases of hominid evolution. Large scale spatial patterning plays an important part in determining evolutionary change, thus conforming with more theoretical expectations developed as part of the modern evolutionary synthesis by Mayr (1963), that speciation will virtually always have an allopatric element. This perspective accounts for global similarities in the pattern of later hominid evolution without depending upon the theoretically improbable special explanation of global evolution.

Environmental Dynamics and Opportunities for Speciation

If a comparative biogeographical approach helps to answer one question – that there should be a single, rather than multiple, origin of AMH – another has been raised: why should one area rather than another throw up the evolutionary novelty that broke away from the trends of archaic evolution? In other words, why, on current evidence, Africa? Answering this question requires an examination of the later Pleistocene environment of Africa to see whether this was more likely to promote speciation than other areas of the world. One possibility is that it is the pattern of environmental change that was critical. The Pleistocene was a period of considerable climatic change, and the period under consideration (150-20 000 BP) saw a major glacial-interglacial cycle (Figure 18.2). Africa was by no means immune from these changes, and palaeoenvironmental reconstructions indicate that there were marked differences in environment during glacial and interglacial phases (Roberts 1984). Glacial periods were characterized by cooler but drier conditions. Savanna (grass-dominated understory) conditions were more widely distributed, and rainforest reduced and confined to a few refuge areas. Interglacials (such as the present one) saw a reversal of this pattern, with more extensive rainforest and restricted woodland/grassland.

This pattern of environmental change, creating habitat distributions that would have isolated populations, has had marked evolutionary consequences. Isolation produces both founder effects and local selective pressures which result in divergence. When habitats are reunited, either inter-breeding is difficult or hybrids have reduced viability. A close relationship between Pleistocene environmental change and speciation in Africa has been shown for various groups, including invertebrates, fish, birds, mammals and primates (see Roberts 1984 for discussion and references).

Two primate examples are particularly relevant here. The first is the

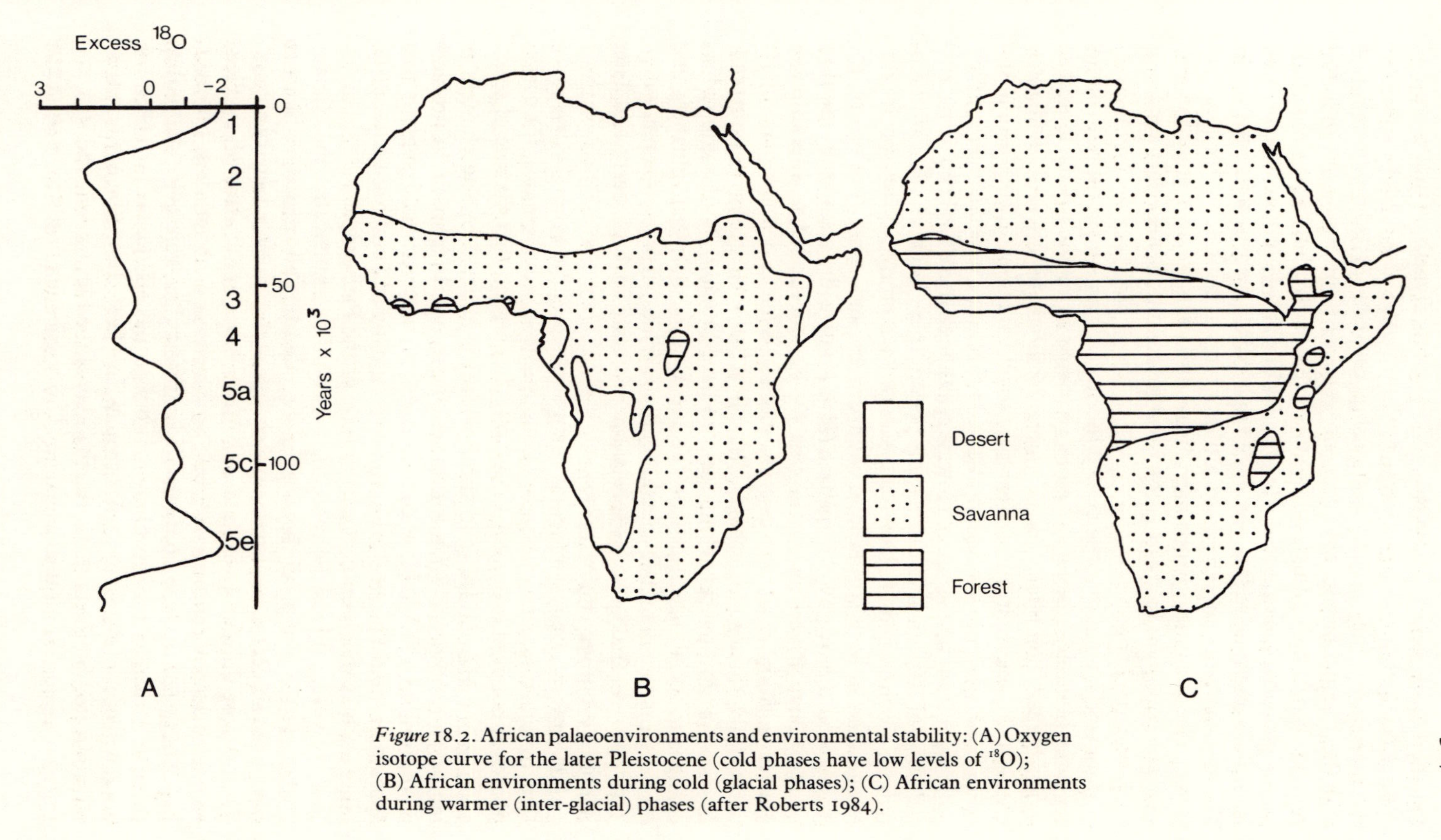

Figure 18.2. African palaeoenvironments and environmental stability: (A) Oxygen isotope curve for the later Pleistocene (cold phases have low levels of ^{18}O); (B) African environments during cold (glacial phases); (C) African environments during warmer (inter-glacial) phases (after Roberts 1984).

radiation of the genus *Cercopithecus*. This genus diversified into at least 16 species occupying a range of areas in Central Africa. Kingdon (1982) has recently shown the detailed relationship between this adaptive radiation and the fluctuations in environment, especially the formation of forest refugia. What makes this evolutionary event particularly relevant here is that it probably took place during the Middle and Late Pleistocene – that is, at the same time as it is postulated that AMH were evolving in Africa. For the cercopithecines the distribution of forest is critical, but for hominids it may have been the mirror-image distribution of more open environments that was important in creating the isolating conditions necessary for major evolutionary change.

It may be argued that the cercopithecines are not an appropriate model for later hominids, being far more restricted in ranging behaviour than larger-bodied hominids, and so more likely to be affected by habitat changes. Furthermore, it is clear that small variations in coat colour and pattern can promote speciation in cercopithecines – a mechanism unlikely to be significant in hominids! However, although there has been no parallel radiation of African apes during this period, there has been some differentiation. Environmental change appears to have produced sub-species differentiation in both gorillas and chimpanzees, and, more specifically, the isolation of the pygmy chimpanzee *(Pan paniscus)* by changes in the flow of the Congo River during the (early?) Pleistocene led to speciation or greater isolation in already existing species. What is perhaps particularly exciting about the *paniscus/troglodytes* split is that small changes in morphology and in feeding behaviour seem to have brought about some radical changes in social organization – the break up of male-male alliances and the development of female-female bonding (White, in press; Susman 1984). This may be analogous to speciation processes in both early and later hominid evolution (Foley 1987).

The pattern of environmental change in the Pleistocene of Africa may thus have had major evolutionary consequences. Furthermore, for earlier phases of hominid evolution Vrba (1985) has argued that the rate of speciation in the Hominidae (and the Bovidae) is affected by the changes in temperature observable in the deep-sea core record. In terms of the problem of the origins of AMH, a comparative perspective would suggest that speciation should occur during times of climatic and environmental instability such as that seen in the later parts of the Pleistocene.

If, however, this is the case, then it could be argued that the climatic changes at the root of all this were global, and that all areas were likely to have been subject to environmental change and so to evolutionary change. This is more difficult to examine, as Africa is one of the areas where Pleistocene palaeoenvironments are best documented. Some studies, however, have suggested that the rainforest of Southeast Asia were more stable than those of Africa, and so less likely to produce the conditions in which speciation will occur (Pope 1983). Africa may have been the region of the tropics most affected by the climatic fluctuations of the Pleistocene, and so a continent of major evolutionary change. This suggestion needs to be tested by examining the comparative rates of endemic evolutionary change in Asia and Africa.

Spatial Ecology

While Africa may be the area of the tropics most likely to be the source of AMH, what about other areas that were subject to major environmental change during the Pleistocene? The mid-latitudes, occupied by hominids from the beginning of the Middle Pleistocene, saw even more drastic environmental change, so if environmental instability promotes speciation, it is in Eurasia that we should perhaps expect AMH to have arisen.

However, from an evolutionary perspective, environmental change does not necessarily result in speciation and the development of new adaptations. It should be remembered that other responses are possible, not the least of which is extinction. Whether a population responds to evolutionary change by adaptive shifts or by becoming extinct would depend upon the scale and nature of the change in habitat. Different levels of change will prompt different levels of response, and so the explanatory apparatus of 'environmental change' that is so crucial to evolutionary biology cannot be used in an undifferentiated way.

The concept of an 'environmental gradient' may be useful here. Changes between habitats and environments, whether through time or space, can be continuous and smooth, such that the resources and conditions available to an animal may not vary markedly over short distances or periods of time. This situation may be described as a 'smooth' environmental gradient. In contrast, some habitat shifts are extremely sudden, with major differences in resource character and availability occurring across a temporal or spatial boundary. In this case the gradient of environmental variability may be described as 'steep' (Figure 18.3).

Differences in habitat gradient may be significant in determining responses to environmental change. The prediction may be made that where one habitat is replaced (especially rapidly) by another that is radically different, it is far more likely that those animals already adapted to these conditions elsewhere will take over through dispersal and colonization. *In situ* evolution is unlikely to occur. In contrast, where the differences between habitats are not marked, then there is a greater chance that evolutionary and adaptive changes within the existing populations will be able to accommodate the environmental change. It may be expected that this latter type of spatial ecology, with smooth environmental gradients, is more likely to produce innovative evolutionary change. In summary, environmental gradient, spatially and temporally, will determine whether faunal change occurs through migration and replacement or through local evolution.

This theoretical discussion leads to certain expectations about Pleistocene evolutionary history. Changes in environment during a glacial/interglacial cycle in higher latitudes are likely to be very marked and quite rapid – i.e. involving very steep environmental gradients. This may explain why the faunal history of Eurasia over the last million years is largely one of changing distributions of animals (Stuart 1982). In contrast, tropical environments

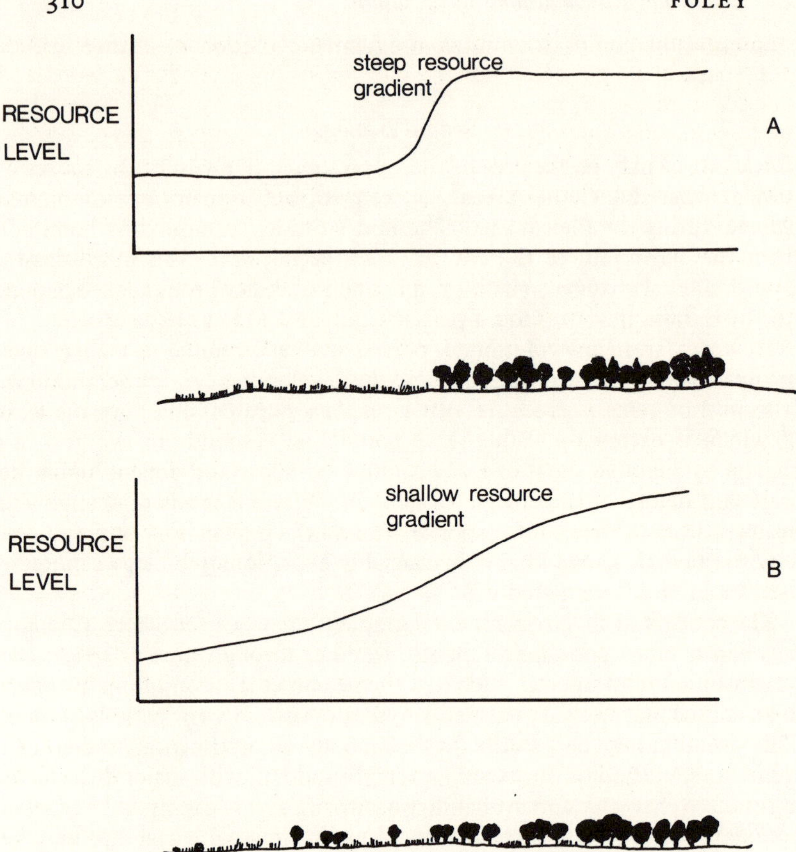

Figure 18.3. Environmental gradients: the change between habitats in terms of resource structure may be sharp (A) or shallow (B). In the former, migration and population replacement is expected. In the latter, local adaptive changes should occur more frequently. High latitudes are characterized by steep environmental gradients; tropical woodland/grassland by shallow gradients.

have a more mosaic and continuous spatial and temporal patterning, with one habitat grading into another – a situation in which local evolution is comparatively more likely to occur. This would perhaps be one reason why major evolutionary changes within the Hominidae appear to have taken place predominantly in Africa, and why the archaic hominid colonization of Europe was intermittent (Gamble 1986).

Evolutionary Ecology of Adaptive Divergence

So far we have seen that speciation is more likely to occur locally than globally, and that it is more likely to occur in changing environments where the gradient of environmental change is not too great, and so in the tropics rather than the mid- latitude regions. This remains a rather general discussion, though, with little mention of the details of the African environments in which the later hominids were living. Can these environments be specified more closely?

An examination of distributions in the *Atlas of African Prehistory* (Clark 1967) indicates a preference among Pleistocene hominids for the woodland/grassland mosaic (see Foley 1987 for a discussion). There may be a taphonomic bias inherent here, but it is generally accepted that these environments have been important in the development of hominids. This complex of habitats, colloquially known as 'woodland' and 'savanna' conditions, has several interesting characteristics from an evolutionary perspective. Principal among these is their graded and mosaic nature. As discussed above, they do not constitute clear environmental bands, but are fragmented. This pattern reflects the local climatic and geological conditions, so that such traits as, for example, tree cover and seasonal variability, vary continuously. In the spirit of the comparative approach adopted here, it may be interesting to examine how the evolutionary diversity of non-human primates is a response to this environmental variability.

The predominant primate of the savanna (other than hominids) is the baboon group (Papionini). With the expansion of savanna conditions this group has diversified during the same period that the hominids evolved, so that a variety of species and sub- species now occupies the full range of habitats. To some extent this diversity reflects geographical factors, but Dunbar (1984 and references therein) and others have shown through an 'eco- correlate' approach that the variability in feeding behaviour and social organization between and within different species of baboons is related to habitat characteristics. An example of this is the way that predation by baboons increases with aridity, as does group size and home and day range. This is not the place to examine the work in full, but the patterns that emerge are of some interest to the problem of AMH origins.

The critical variables determining the socio-ecology of the Papionini are food quality, distribution (patchiness) and predictability. The differences between the various species of baboon can be accounted for by the interaction of these variables (Figure 18.4). For example, hamadryas baboons (*Papio hamadryas*) have certain characteristics (large group size, infra-group structure, etc.) that reflect an environment in which food quality is low, patchily distributed, but nonetheless predictable. *Theropithecus gelada*, on the other hand, is adapted to evenly distributed, low quality and predictable food supplies.

Two points are worth stressing from this work. The first is that Dunbar's results do show that quite small differences in environment can produce evolutionary diversity and quite marked differences in behavioural adaptation. The second is that the variables producing evolutionary novelty can be specified quite precisely, and their relative significance assessed. Of the three variables shown here, a more predictable environment will result in increases in group size and in the amount of structuring within those groups; increased patchiness will result in increases in the size of home and day ranges. Furthermore, patchy environments may produce tightly bonded groups, while less patchy ones will favour rather looser associations (fission/fusion) (Wrangham 1980). And finally, an increase in food quality will lead to greater feeding selectivity.

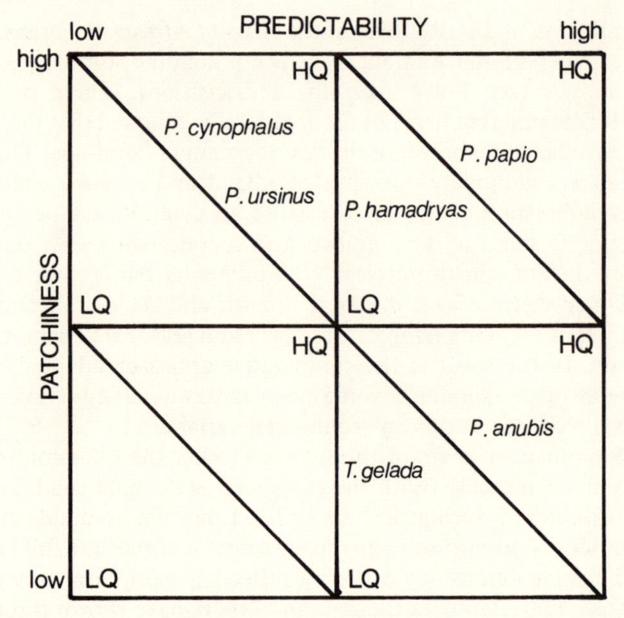

Figure 18.4. Evolutionary divergence of the Papionini in relation to ecological factors. High levels of predictability promote larger group size and structuring; patchiness produces larger home and day ranges; and high quality food leads to dietary selectivity (based on Lee 1988). The ecological factors associated with AMH (see Table 3 and text) are most likely to be a response to the combination of patchy, high quality and predictable resources. (HQ = high quality food; LQ = low quality food).

These results enable us to identify at least some of the environmental factors that are likely to promote speciation and diversification. More precisely, they may also help us to relate the characteristics of AMH to their ecological circumstances. In other words, what conditions are likely to promote the characteristics of AMH?

Table 18.3 suggests some ecological apomorphies of AMH (prior to food production) – that is, ways in which they may be thought to have differed from archaic hominids. These must of necessity be somewhat speculative as we do not know how many of these characteristics go back to the earliest AMH. However, AMH may be considered to have had very widespread ranging behaviour, to exist in large groups with considerable (kin-based) sub-structure, and to be highly selective in their feeding behaviour (i.e. a preference for high-quality foods such as meat, fruit, nuts, etc.) Meat is particularly critical here as it also tends to come in large package sizes.

An examination of Figure 18.4 shows that these characteristics are most likely to occur in patchy environments where the food is both of high quality and predictable. I would argue that this is the context in which we should seek the origins of AMH. Where is it likely to occur? The answer to this lies partly in the type of environment (i.e. the finely graded woodland/grassland mosaic discussed above) and partly in changes in the species ability

Table 18.3. Ecological apomorphies of anatomically modern humans.

1. Very large home range
2. Very large day range
3. Large group size
4. High degree of sub-structure within groups
5. Dietary selectivity

to exploit those environments. Principally this means that changes in the foraging behaviour of the hominids of the later Pleistocene will change the relative patchiness, predictability and food quality available in specific environments. In this case, it is probably an increase in hunting ability that may be most significant. Meat is a high quality food for any primate, and furthermore is very patchily distributed in the environment. An increase in hunting would certainly account for the AMH ecological apomorphies.

The principal problem lies with the predictability of meat as a resource. Compared to plant foods, animals are by and large less predictable in their distribution. The key to this problem, though, lies not so much in a change in environment as a change in the characteristics of the hominids. The predictability of a resource lies not only in its location in time and space, but also the chances of acquiring it once located. Herein lies the unpredictability of carnivory, for animals, unlike plants, can easily escape capture even after location. Anything that would reduce this element of the unpredictability of a resource could have a very great influence on the socioecological characteristics of a species. The critical variable here might well be the development of a technology that allows predation from a distance – i.e. hunting projectiles. In a woodland or bushland environment, even over short distances, this would be a major advantage. The very early appearance in parts of Africa of a blade technology may well be indicative of this.

Predictability is an element of risk management (Mithen 1988), and this is currently considered to be one of the major adaptive problems facing animals. Behavioural ecologists recognize two broad strategies – 'risk averse' and 'risk prone'. The former consists of minimizing the occurrence of risk, the latter of attempting to maximize payoffs. Increasing the predictability of resource acquisition is in effect allowing a greater degree of risk-prone behaviour. Increasing planning depth, technological efficiency, foraging organization etc. will reduce the unpredictability of the environment, and will be enhanced both by increased 'intelligence' and other elements of social and ecological behaviour associated with AMH.

SOCIAL VERSUS ECOLOGICAL MODELS FOR THE EMERGENCE AND DISPERSAL OF AMH

We have seen that a series of general evolutionary and ecological principles can help to pinpoint both the location in time and space and the processes of speciation in later hominids, and so help to account for the emergence of AMH. At this stage this is a question of general modelling using biological concepts and processes. However, it also allows for prediction, which may help to identify future lines of research. On the basis of the discussion here it would be particularly interesting to know more about the responses to environmental change in tropical *versus* high latitudes – continuous in the former, more abrupt in the latter. It would also be predicted that the late Middle and early Upper Pleistocene archaeological sites of Africa should show the development of more efficient and intensive hunting behaviour than those of Eurasia and Southeast Asia. Furthermore, if the origins of AMH are local to Africa, then dispersal would seem to be the mechanism for the appearance of AMH in the rest of the world. In this context, both the fossil and archaeological evidence tend to show the sudden appearance of modern forms from 45 000 to 12 000 BP throughout the non-African world. It is interesting to note that rapid dispersal is a well known phenomenon in the biological world, both historically (e.g. the spread of rabbits in Australia, grey squirrels in Europe, domestic ungulates in Africa) and palaeontologically (many examples of rapid colonization, such as *Hipparion*). What these cases show are the advantages of *exogeny* – species entering a new habitat often have a considerable selective advantage, probably due to their generalized exploitation of the environment relative to local specialists. Frequently their impact upon these environments is catastrophic, with higher rates of extinction and environmental change. The rapid dispersal of modern hominids through the world, and (in many parts of the world) high rates of late Pleistocene extinction, are perhaps evidence for the exogenous nature of AMH outside Africa.

The patterns of dispersal of AMH can perhaps also throw light on the question of social *versus* ecological factors in their origins. Many of the key characteristics of modern humans are social and 'cultural' rather than ecological, and so the primacy of these in their origin is a distinct possibility. However, the model proposed here is one of ecological primacy. To some extent ecological and social changes may be expected to develop in phase. However, the consequences of social development (the human social 'revolution'/the origins of 'culture'?) are likely to be dramatic and self-reinforcing. Evolutionary change as a response to social changes may be expected to be rapid. The origins of AMH, though, seem to fall into two phases – a relatively slow and intermittent development in Africa between 140 000 and 50 000 BP, followed by rapid dispersal (Fig. 18.5). It could be argued that this pattern reflects the relatively slow changes in morphology and adaptive ecological behaviour leading to social changes, which then results in the explosive spread of AMH throughout the world. Clearly the mechanisms of this need to be explored, but if the changes in ecology

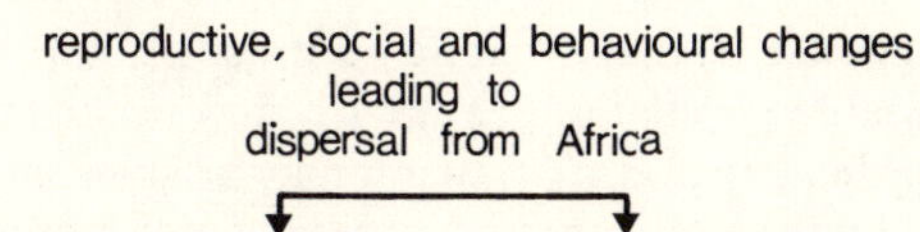

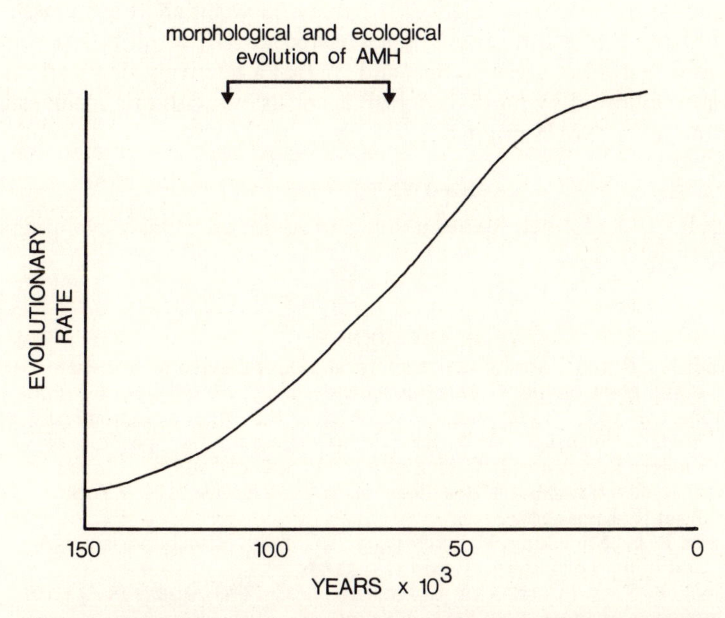

Figure 18.5. Model of the origins of AMH. The earlier phases of this process are characterized by relatively slow evolutionary change in response to ecological conditions. This is followed by more rapid changes in social and reproductive behaviour leading to dispersal out of Africa.

resulted in larger group sizes with greater internal structuring, as argued above, then this new social context could provide the context for more intense selective pressures for social and reproductive changes (parental care, kin-bonding, communication, inter-population differentiation, etc.). Ecological conditions, therefore, provide the basis for evolution in social behaviour.

SUMMARY

The purpose of this paper has been to place the emergence and dispersal of anatomically-modern humans into a broader, comparative framework. That framework has been provided by principles of evolutionary ecology and the patterns of primate and earlier hominid evolution. A series of expectations can be derived from this analysis which may help to resolve some of the issues in the AMH origins debate.

It was argued that the evolution of modern humans was a speciation event occurring in Africa around 100 000 years ago. Consideration of speciation events and the pattern of speciation in primates and earlier hominids suggests: (1) that it should occur locally within a widespread and regionally-variable population; (2) that it may occur in association with considerable parallel evolution; (3) that it may be associated with environmental change; (4) that the rate and magnitude of environmental change should not be too great; (5) that the direction of environmental change, in terms of the patchiness, predictability and quality of food processes, should determine the direction of evolutionary change; (6) that the woodland/bush/grassland areas of late Middle and early Upper Pleistocene Africa fulfil these conditions, and that other areas do not; and (7) that a relatively slow pattern of evolution, followed by rapid dispersal, is consistent with this ecologically-based model of AMH origins.

ACKNOWLEDGEMENT

I thank P.C. Lee for helpful discussion and advice on primate evolutionary ecology.

REFERENCES

Andrews, P. 1984. An alternative interpretation of the characters used to define *Homo erectus*. *Courier Forschungsinstitut Senckenberg* 69: 167-175.

Bräuer, G. 1984. A craniological approach to the origin of anatomically modern *Homo sapiens* in Africa and implications for the appearance of modern Europeans. In F. H. Smith and F. Spencer (eds) *The Origins of Modern Humans: a World Survey of the Fossil Evidence*. New York: Alan R. Liss: 327-410.

Cann, R. L., Stoneking, M. and Wilson, A.C. 1987. Mitochondrial DNA and human evolution. *Nature* 325: 31-36.

Ciochon, R. and Corruccini, R. (eds) 1983. *New Perspectives in Ape and Human Ancestry*. New York: Plenum.

Clark, J. D. 1967. *Atlas of African Prehistory*. Chicago: University of Chicago Press.

Day, M. H. 1986. *Guide to Fossil Man*. London: Cassell.

Dobzhansky, T. 1963. Possibility that *Homo sapiens* evolved independently five times is vanishingly small. *Scientific American* 208: 169-172.

Dunbar, R. 1984. *Reproductive Decisions*. Princeton: Princeton University Press.

Foley, R. 1987. *Another Unique Species: Patterns in Human Evolutionary Ecology*. London: Longman.

Foley, R., in press. The inadequacy of the culture concept in palaeoanthropology. In M. Day, R. Foley and Wu Rukang (eds) *The Pleistocene Perspective: Hominid Evolutionary Behaviour and Dispersal*. London: Allen and Unwin.

Gamble, C. 1986. *The Palaeolithic Settlement of Europe*. Cambridge: Cambridge University Press.

Isaac, G. Ll. 1972. Chronology and the tempo of cultural change in the Pleistocene. In W. W. Bishop and J. Miller (eds) *Calibration of Hominoid Evolution*. Edinburgh: Scottish Academic Press: 381-430.

Hemmer, H. 1982. Major factors in the evolution of hominid skull morphology. Biological correlates and the position of the anteneanderthalers. In H. de Lumley (ed.) *L'Homo erectus et la Place de l'Homme de Tautavel Parmi les Hominidés Fossiles*. Nice: Centre National de la Recherche Scientifique / Louis-Jean Scientific and

Literary Publications: 339-354.

Howell, F. C. 1984. Introduction. In F. H. Smith and F. Spencer (eds) *The Origins of Modern Humans: a World Survey of the Fossil Evidence*. New York: Alan R. Liss: xiii-xxii.

Lee, P. C. 1988. Ecological constraints and opportunities: interactions, relationships and social organization in primates. In J. E. Fa and C. Southwick (eds) *The Ecology and Behavior of Food-Enhanced Primate Groups*. New York: Alan R. Liss.

Lévêque, F. and Vandermeersch, B. 1980. Découverte de restes humains dans un niveau castelperronien à Saint-Césaire. *Comptes-Rendus de l'Académie des Sciences de Paris (Série D)* 291: 187-189.

Kingdon, J. S. 1980. The role of visual signals and face patterns in African forest monkeys (guenons) of the genus *Cercopithecus*. *Transactions of the Zoological Society of London* 35: 425-475.

Mehlman, M. 1984. Archaic *Homo sapiens* at Lake Eyasi, Tanzania: recent misrepresentations. *Journal of Human Evolution* 13: 487-501.

Mellars, P. A. 1973. The character of the Middle-Upper Palaeolithic transition in southwestern France. In C. Renfrew (ed.) *The Explanation of Culture Change*. London: Duckworth: 255-276.

Mayr, E. 1963. *Animals, Species and Evolution*. Cambridge (Mass): Harvard University Press.

Mithen, S. 1988. *Hunter Gatherer Decision-Making*. Unpublished Ph.D. Thesis, University of Cambridge.

Pope, G. G. 1983. Evidence for the age of the Asian Hominidae. *Proceedings of the National Academy of Sciences (USA)* 80: 4988-4992.

Pilbeam, D. R. 1985. The origin of *Homo sapiens*. In B. Wood, L. Martin and P. Andrews (eds) *Major Topics in Primate and Human Evolution*. New York: Alan R. Liss: 295-328.

Roberts, N. 1984. Pleistocene environments in time and space. In R. Foley (ed.) *Hominid Evolution and Community Ecology: Prehistoric Human Adaptation in Biological Perspective*. London: Academic Press: 25-54.

Stringer, C. B. 1985. Middle Pleistocene hominid variability and the origin of late Pleistocene humans. In E. Delson (ed.) *Ancestors: the Hard Evidence*. New York: Alan R. Liss: 289-295.

Stringer, C. B., Hublin, J.-J. and Vandermeersch, B. 1984. The origin of anatomically modern humans in Western Europe. In F. H. Smith and F. Spencer (eds) *The Origins of Modern Humans: a World Survey of the Fossil Evidence*. New York: Alan R. Liss: 51-135.

Stuart, A. J. 1982. *Pleistocene Vertebrates in the British Isles*. London: Longman.

Susman, R. (ed.) 1984. *The Pygmy Chimpanzee*. New York: Plenum.

Tattersall, I. 1986. Species recognition in human palaeontology. *Journal of Human Evolution* 15: 165-175.

Valladas, H., Reyss, J. L., Joron, J. L., Valladas, G., Bar-Yosef, O. and Vandermeersch, B. 1988. Thermoluminescence dating of Mousterian 'Proto-Cro-Magnon' remains from Israel and the origin of modern man. *Nature* 331: 614-616.

Vrba, E. 1985. Ecological and adaptive changes associated with early hominid evolution. In E. Delson (ed.) *Ancestors: the Hard Evidence*. New York: Alan R. Liss: 63-71.

Wainscoat, J. S. 1987. Out of the garden of Eden. *Nature* 325: 13.

White, F. (in press). The socioecology of the pygmy chimpanzee. In V. Standen and R. Foley (eds) *Comparative Socioecology of Mammals and Man*. Oxford: Blackwell Scientific Publications. In Press.

Wolpoff, M. 1983. Evolution in *Homo* erectus: the question of stasis. *Paleobiology* 10: 389-406.

Wood, B. 1984. The origins of *Homo erectus*. *Courier Forschungsinstitut Senckenberg* 69: 99-111.
Wood, B. and Chamberlain, A. T. 1986. The nature and affinities of the "robust" australopithecines. *Journal of Human Evolution* 16: 625-642.
Wrangham, R. 1980. An ecological model of female-bonded groups. *Behaviour* 75: 262-300.

SECTION II

Behavioural Change

19. How Different was Middle Palaeolithic Subsistence? A Zooarchaeological Perspective on the Middle to Upper Palaeolithic Transition

PHILIP G. CHASE

INTRODUCTION

Because the way in which an organism goes about obtaining nourishment is one of the most important aspects of its adaptation, the investigation of subsistence behaviour is a logical starting point for studying differences and similarities in hominid adaptation at the Middle-Upper Palaeolithic boundary. Such a study can contribute to our understanding of the subject in two ways. First, we may find evidence of a change in subsistence practices that will help to explain other phenomena associated with the transition. For example, S. R. Binford (1968) has suggested that much of the change involved in the evolution of anatomically-modern *Homo sapiens* can be attributed to the initiation of cooperative hunting of large, migratory herd animals. Second, a close examination of subsistence behaviour may provide evidence concerning other aspects of Middle Palaeolithic adaptation that are of relevance to the question at hand. For example, L. R. Binford (1984, 1985) has suggested that Middle Stone Age peoples of Southern Africa and the Middle Palaeolithic peoples of Europe were, unlike their modern successors, incapable of hunting large game, and that the faunal evidence indicates an absence of long-range planning, of cooperation and of sharing. If this is so, it means that the behavioural and, presumably, neurological changes involved in the Middle-Upper Palaeolithic transition were much more drastic than would be the case if Middle Palaeolithic subsistence strategies were essentially the same as those of the earlier Upper Palaeolithic.

What follows will be a survey of what is known of Middle and early Upper Palaeolithic subsistence practices in Europe, directed toward determining how great a change in adaptation was involved in the Middle-Upper Palaeolithic transition, and what aspects of adaptation were involved. Such a survey is, unfortunately, limited largely to a study of faunal data. At the present time, few data are available concerning the use of plant materials. Although incomplete, however, an investigation of subsistence behaviour based on the faunal evidence can nevertheless shed light on the nature of the transition.

In any study of this transition one must keep in mind that the problem involves a set of phenomena that apparently arose at the boundary between the two time periods. This means that, as Fish and Dibble (1982) have

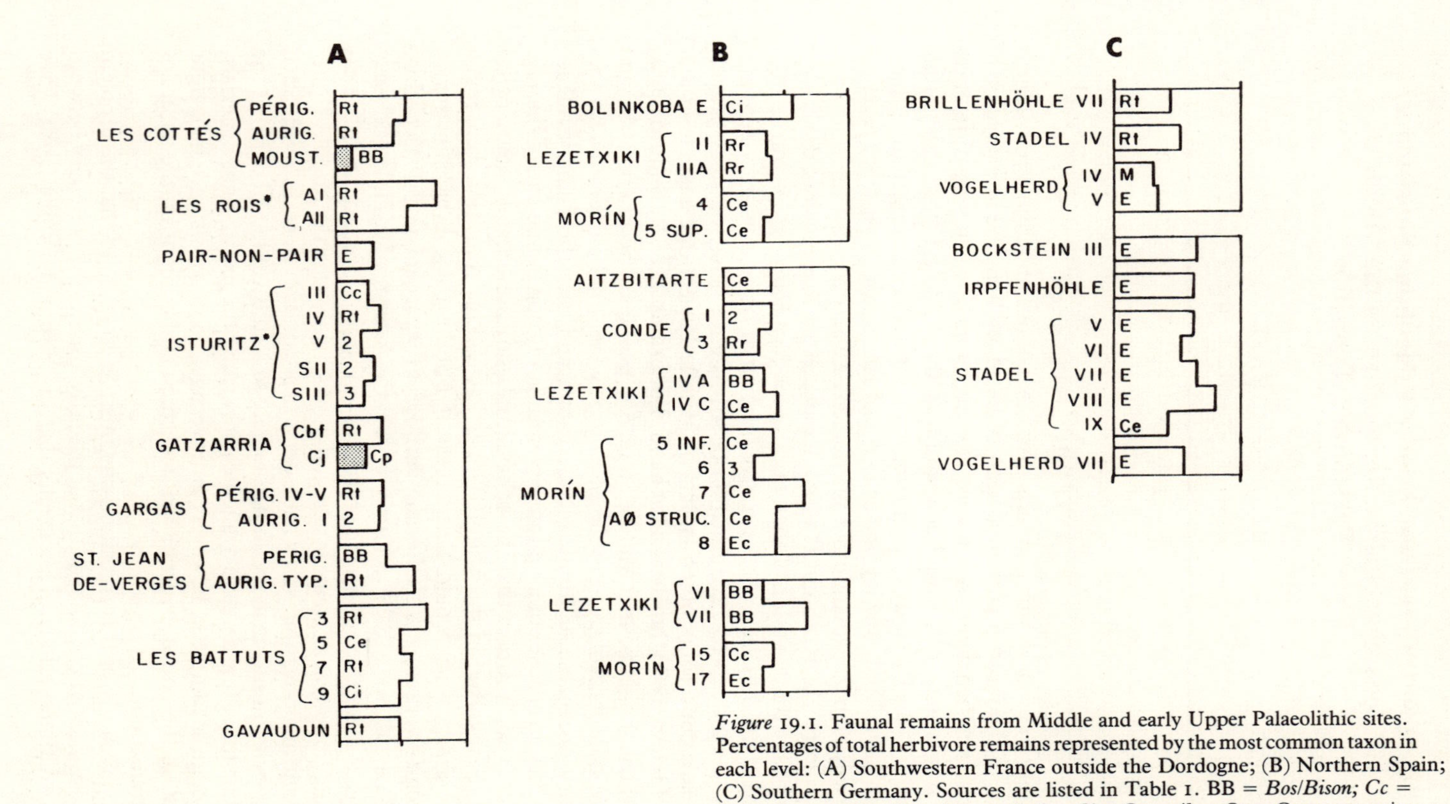

Figure 19.1. Faunal remains from Middle and early Upper Palaeolithic sites. Percentages of total herbivore remains represented by the most common taxon in each level: (A) Southwestern France outside the Dordogne; (B) Northern Spain; (C) Southern Germany. Sources are listed in Table 1. BB = *Bos/Bison; Cc = Capreolus capreolus;* Ce = *Cervus elaphus;* Ci = *Capra ibex;* Cp = *Capra pyrenaica;* E = *Equus;* Ec = Equus caballus; M = *Mammuthus;* Rr = *Rupicapra rupicapra;* Rt = *Rangifer tarandus*. *2, 3* = two or three species equally common.

pointed out, we cannot use the late Upper Palaeolithic as a basis for contrast with the Middle Palaeolithic. Many characteristics of the later Upper Palaeolithic may be the result of developments during the course of the Upper Palaeolithic. While such developments are of interest in themselves, they cannot be used to explain changes in hominid morphology or changes in behaviour that occurred much earlier, at the very beginning of the Upper Palaeolithic.

In the discussion that follows, it will be assumed that early Upper Palaeolithic populations were essentially modern, and that any changes in behaviour from that time to the present can be attributed to cultural rather than to biological evolution. Therefore, no attempt will be made to investigate such variables as the capacity for foresight or cooperation among Upper Palaeolithic peoples. Upper Palaeolithic subsistence will be studied only (1) for the purpose of finding changes that may help to explain the transition; or (2) as a base-line of comparison for evaluating Middle Palaeolithic adaptation.

The following survey of the evidence from Europe is organized topically. First, we shall discuss whether or not the transition from the Middle to the Upper Palaeolithic was accompanied by a shift to specialized hunting. Next, we shall test S. R. Binford's (1968) hypothesis that the cooperative hunting of herds of large migratory herbivores was the key factor in the evolution of anatomically-modern *Homo sapiens*. Next, the evidence for foresight and planning as a part of Middle Palaeolithic adaptation will be evaluated. Finally, the problem of hunting techniques in the Middle Palaeolithic will be discussed.

SPECIALIZED HUNTING

Paul Mellars (1973), in a detailed comparison of the Middle and Upper Palaeolithic archaeological record in the Dordogne region of southwest France, suggested that Upper Palaeolithic peoples, unlike their predecessors, appear to have been specialized hunters, concentrating very heavily upon a single species – namely reindeer (*Rangifer tarandus*). Certainly by the later Upper Palaeolithic, there is evidence that at least some populations may have been specialized hunters in this sense. Moreover, it would not be unreasonable, on the basis of the data from the earlier Palaeolithic of the Dordogne region, to conclude that specialized reindeer hunting may have begun somewhere near the Middle-Upper Palaeolithic boundary.

The archaeological evidence from the Dordogne is unusual in this respect however. Even in southwestern France outside the Dordogne area, the faunal assemblages from the earlier Upper Palaeolithic show no signs of heavy reliance upon a single species (Fig. 19.1a). The same is true of the data from northern Spain (Fig. 19.lb) and from southern Germany (Fig. 19.1c), the only other two areas in Europe with a reasonable sample of quantitatively reported faunal assemblages. If hunting in the Dordogne was indeed specialized, then it was apparently an exception. It may be that the sample of sites from the Dordogne is somewhat biased. At all times in the sequence, there are sites where species other than reindeer predominate

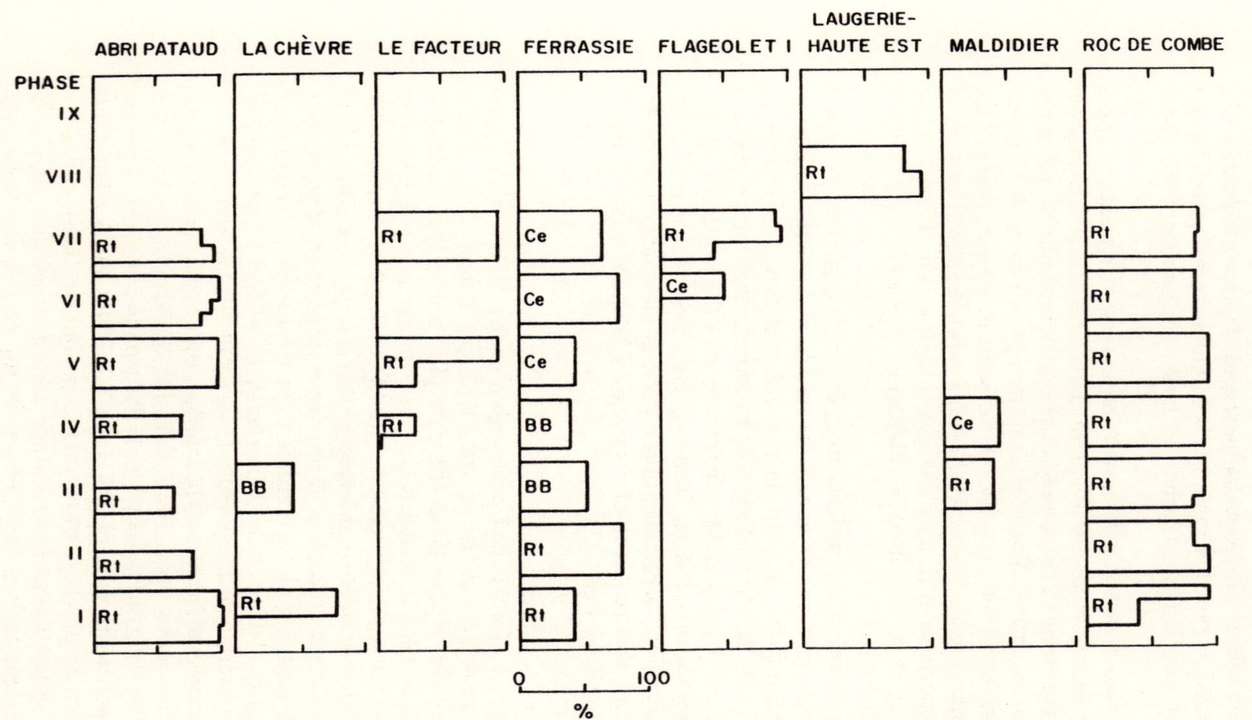

Figure 19.2. Herbivore remains from early Upper Palaeolithic sites in the Dordogne, arranged according to Laville, Rigaud and Sackett's (1981) chronology. Percentages of remains of ungulates represented by the most common taxon in each level. Sources are listed in Table 1. BB = Bos/Bison; Cc = Capreolus capreolus; Ce = *Cervus elaphus;* Ci = *Capra ibex;* Cp = *Capria pyrenaica;* E = *Equus;* Ec = *Equus caballus;* Rr = *Rupicapra rupicapra;* Rt = *Rangifer tarandus.* 2, 3 = two or three species equally common.

(Fig. 19.2). Some sites, notably La Ferrassie, regularly yielded diverse assemblages. The specialized appearance of the data is due primarily to the heavy predominance of reindeer at two sites, the Abri Pataud and the Roc de Combe. Since at the Abri Pataud the percentage of *Graminaea* pollen is consistently extremely low (Donner 1975), it is possible that that site, and perhaps both, were simply locations where reindeer were always more abundant than other ungulates, and where the faunal remains reflect this fact rather than specialization as a subsistence strategy.

If we are to infer specialization from such Upper Palaeolithic sites as the Abri Pataud and the Roc de Combe, then the existence of some specialized subsistence systems cannot be ruled out for the Middle Palaeolithic. Sites with very heavy concentrations upon a single species are known from the Middle Palaeolithic of Europe (Chase 1986a, 1987; see below). It does not necessarily follow, however, that Middle Palaeolithic peoples were specialized hunters, since such assemblages may reflect only a single aspect of a strategy involving the exploitation of different species at different locations (or during different seasons of the year). The data from the early Upper Palaeolithic of the Dordogne may be interpreted in the same way: the Dordogne as a whole may have served as a seasonal reindeer-hunting area involved in part of a more eclectic subsistence system.

It may be, therefore, that specialized hunting did not occur until later in the Upper Palaeolithic, in which case it is of no relevance to our understanding of the Middle-Upper Palaeolithic transition. If it did occur in the early Upper Palaeolithic, it was apparently the exception rather than the rule, and is therefore of doubtful importance in the transition, especially since isolated specialized hunting groups may have existed in the Middle Palaeolithic.

COOPERATIVE HUNTING OF MIGRATORY HERDS

S. R. Binford's hypothesis (1968) is one of the few real attempts to explain rather than simply describe the transition from Middle to Upper Palaeolithic and the origins of anatomically-modern humans. Based on the faunal data from the Levant, she concluded that the hunting of herds of large herbivores during their seasonal migrations between the coastal plains and the highlands began near the end of the Middle Palaeolithic. She believed that, unlike earlier hunting patterns, this strategy would have required the cooperation of large numbers of adult males, and therefore the aggregation of sizeable bands. This cooperation and aggregation would have led to the conditions necessary for genetic change in a population.

Her work in the Near East is hampered by the nature of the faunal data. Most of it is from old excavations, for which quantitative information is not available. The only areas in the World where usable samples of quantitative data are available are in Southern Africa and in Europe, and in the former an adequate sample is not available for the early Later Stone Age.

In Europe, it seems likely that the hunting of migratory herds of large herbivores was practised in the Upper Palaeolithic. However, there is also evidence for such a pattern in the Middle Palaeolithic. At the Abri Pataud, the highest minimum number of individuals calculated by Spiess (1979: 215) was 40 reindeer in layer 3:main. Much higher numbers of individuals are known from the Middle Palaeolithic. At Mauran, in the French Pyrenees, Girard and David (1982) reported the remains of a minimum of 108 large bovines (*Bos* and *Bison*). At Starosel'e in the Crimea the remains of 287 steppe ass (*Equus asinus*) were recovered (58,900 bones) (Klein 1969) (see Note 1). Gábori (1979) estimated that Il'skaya in the Kuban Basin contained the remains of perhaps 1200 bison. While Middle Palaeolithic sites are difficult to date, Il'skaya at least appears to date from early in the Middle Palaeolithic (Gábori 1976: 139). It would appear, then, that if large numbers of herd animals indicate cooperative hunting, such a practice appeared long before the transition from the Middle to the Upper Palaeolithic, and cannot be used to explain the change.

FORESIGHT AND SHARING

Lewis Binford (1984) has argued that the faunal data from the Klasies River Mouth sites (South Africa) indicate a striking lack of foresight on the part of Middle Stone Age peoples there. He bases this conclusion on two observations. First, the way in which small animals at the site were butchered indicates to him that the best parts were consumed at the point of procurement. Second, the food packages consumed were generally small – either the animal itself was small, or the parts exploited mere small. This latter argument is used to turn what would normally be taken as evidence of foresight and planning into evidence of lack of foresight:

> Filleting of both the tenderloin and the upper-front quarter may well betray the preparation of *biltong*, or air-dried stripped meat, for future consumption. If so, this suggests very short-term planning and a lack of sharing, because these were very small animals indeed. If sharing had been widespread, such small animals would likely have been shared out for consumption, with no surplus available for drying (Binford 1984: 182).

It is difficult to understand how the preparation of meat for future consumption is indicative of a lack of planning. (Binford's conclusion concerning lack of sharing is based on the argument that only small packages of meat were sought out because small amounts of meat were required, since hunters were killing only for their own individual and immediate needs (*ibid.*: 194) Elsewhere in the same work, however, he attributes the use of small game and of the least meaty portions of large animals not to choice but to an inability to kill large animals!) Nor is it necessarily true that the exploitation of small animals implies lack of foresight or sharing. Many foods that come in small packages, such as fish, rabbits, and grains, may be exploited cooperatively, shared, and preserved for future use.

Be that as it may, the evidence in Europe for deliberate foresight on the part of Middle Palaeolithic peoples is strong. In order to appreciate it, we must first consider what one would expect to find in the archaeological record in the absence of foresight.

Animals would be killed as they were encountered and, for the most part, consumed on the spot. Most faunal assemblages would therefore represent kill sites. While very small faunal assemblages, representing the killing of one or a very few animals, might appear specialized, large assemblages would contain, more or less, a cross-section of the game available in the vicinity. Moreover, most of the assemblages would contain (allowing for taphonomic bias) all parts of the skeleton, since transportation of choice parts for consumption elsewhere would be rare, or at least a minor factor.

In fact, what we find is quite different. While many sites have produced a generalized fauna, in many of them one species outnumbers all the rest. At Combe Grenal in southwest France, this is true of red deer in layers 5, 40, 42-43, 50A, 50, 52 and 54; of reindeer in layers 4, 23, 24, 25, and 27; and of *Equus* in layer 14 (Chase 1986a). Elsewhere in France, this is true of ibex (*Capra ibex*) in layers 10A3 through 21B, 28A, and 29 of Hortus (Lumley 1972); of reindeer in layers 1 and 4 at Hauteroche (Bouchud 1959: 120-122); and of large bovids at Mauran (Girard and David 1982). In Germany, reindeer dominate the fauna of Salzgitter-Lebenstedt (Tode *et al.* 1953: 180-181) and of Heidenschmiede (Bosinski 1967: 144); while horse dominate those of Vogelherd (Lehmann 1954: 115) and of Kogelsteinklufte (Bosinski 1967: 149). In Austria, ibex outnumber all other ungulates in both the grey sandy layer and the red-brown layer at Repolust in Steiermark (Chase 1986a: 107; Mottl 1960: 130). In the Soviet Union, where we have already mentioned the sites of Il'skaya and Starosel'e, Volgograd, too, is specialized, with 78% of the bones those of *Bison* (Gábori 1976: 205).

It is possible that some of these sites represent the activities of groups heavily dependent upon a single species of game. It is more likely that most of them represent the activities of peoples who, while exploiting a range of resources, tended to go to a specific location in order to hunt a specific species, probably at a particular time of year. (A more detailed discussion of this point is contained in Chase 1986a.) However, it is very unlikely that such sites were produced simply by random, opportunistic killing. It is true that the immediate surroundings of some of these sites may have favoured hunting of one kind of animal. For example, Girard and David (1982) describe the location of Mauran as perhaps ideal for killing large bovids grazing in the vicinity. However, the huge number of animals, killed there indicates either (1) that it was visited repeatedly for the purpose of exploiting these animals; or (2) (if the bones there represent only a few hunting episodes) that the amount of meat produced at one time was far more than would be needed for immediate consumption by just a few individuals. Either of these two forms of accumulation runs contrary to Binford's notions.

Moreover, the data indicate that Middle Palaeolithic sites are not simple kill-and-consumption sites. The Binfords (L. R. Binford and S. R. Binford 1966) were among the first to make this point, arguing on the basis of stone tools that Middle Palaeolithic sites differed in terms of the economic activities carried out in each location. Although their argument has been criticized (Bordes 1967, 1973; Bordes and de Sonneville-Bordes 1970;

Bordes, Rigaud and de Sonneville-Bordes 1972; Mellars 1970; Rolland 1981) and Lewis Binford has apparently abandoned the argument (L. R. Binford 1982: 27), there is good faunal evidence to indicate that such differences in site functions did in fact exist (although these apparently do not account for most of the variability in stone tool industries: see Gábori 1979; Chase 1986b).

At Combe Grenal, during the Würm I (or phases I through X of the Early Würm following to the terminology of Laville, Raynal, and Texier 1984) and during the first part of the Würm II (or phases XI through XVIII), both red deer and reindeer are represented by the portions of the carcass poorest in meat, with some bias within this category toward those bones richest in marrow (Chase 1986a). This implies either (1) that the animals were scavenged, or (2) that they were killed near the site and that the hunters were preparing the best portions of the carcasses for consumption elsewhere, eating, in the meantime, the best of the poorer parts. Since the data indicate that these animals were not scavenged (Chase 1988; see below), this implies that Combe Grenal served over a long period of time as a specialized hunting/processing site that was part of a larger subsistence system which also included sites with other functions. Later in the Würm II, reindeer were evidently consumed at the site. However, since the meatiest parts and those with the highest quality marrow are over-represented relative to those of lesser value, it appears that now reindeer were being killed at some distance from the site, were being butchered where they died, and that the best parts were being transported to Combe Grenal for consumption there. This indicates that, while the site's function changed, the people of Combe Grenal regularly, throughout the occupation of the site, engaged in a pattern of subsistence that included the deliberate hunting of animals at one location with the clear intention of consuming them elsewhere.

This pattern is repeated at other Middle Palaeolithic sites where data on the relative frequencies of different parts of the skeleton are available. At San Agostino in southwestern Italy, the hides of boar, roe deer, and fallow deer were brought into the site more frequently than the meat of these animals. By contrast, red deer and ibex apparently provided much of the meat at the site, and red deer hides appear to have been removed from the site (Chase 1986a; Tozzi 1970). Lumley (1972) and his colleagues, on the basis of a variety of evidence, determined that the site of Hortus in southern France was at different times a kill/butchery site, a hunting camp, and a base camp. All three of these functions imply the existence of other sites with complementary functions, with meat transported out of the kill/butchery sites and hunting camps and into the base camps.

The existence of complementary functions among sites, especially when it persists over long periods of time, indicates that activities were planned in advance to fulfil future needs at other locations. Archaeologically, the Middle Palaeolithic pattern thus resembles in many ways that described by Binford (1978) for the modern Nunamiut Eskimos, who engage in a complex system of food-processing, transportation, and caching, all

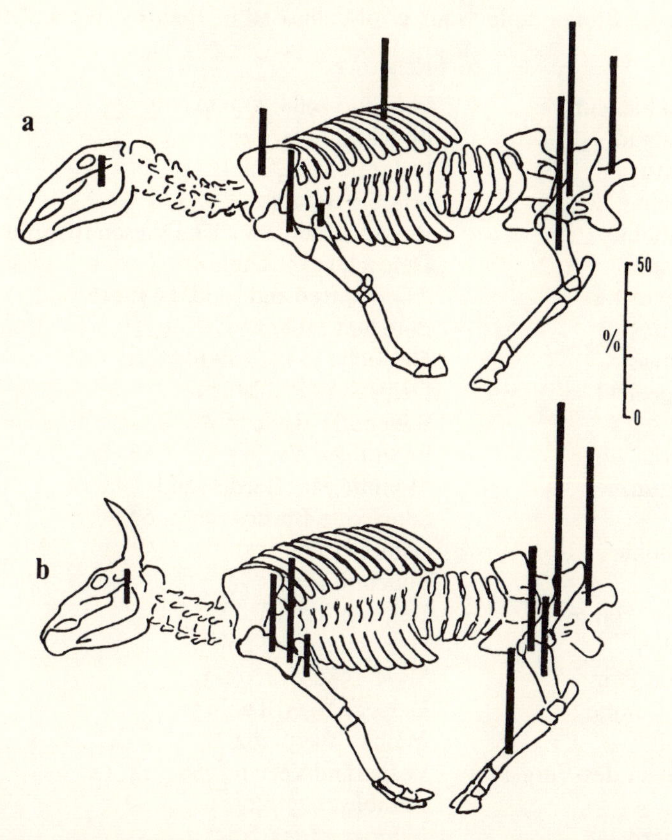

Figure 19.3. Percentage of each portion of the skeletons of equids (upper) and of large bovids (lower) in the Mousterian of Combe Grenal bearing tool marks. The location of each vertical bar indicates the portion of the skeleton. The height of each bar represents the percentage. Percentages for long bones are calculated separately for proximal and distal ends and for shafts.

designed to fulfill future needs. It is therefore hard to believe that foresight and planning did not comprise a key part of the adaptation of Middle Palaeolithic peoples.

However, it is not only possible but quite likely that *ad hoc* and unplanned food-procurement, combined with on-the-spot consumption also occurred in the Middle Palaeolithic. In fact, such behaviour is also common among modern hunter-gatherers.

Whether the Middle Palaeolithic pattern implies sharing of food is perhaps less clear. The regular transportation of food certainly implies regular sharing. In addition, the killing of large game, perhaps in large numbers, sometimes perhaps by driving (see below), implies cooperation on the part of a number of hunters. It seems likely that such cooperation would have taken place in the context of other forms of cooperation. For example, different parties might hunt from and return to the same camp

Table 19.1 Bibliographic sources of data used in Figures 19.1 and 19.3

Site	*Reference*
Abri du Facteur	Bouchud 1968: 113-121
Abri Pataud	Bouchud 1975: 120
Les Battus	Delpech 1975, Table 14
Bockstein	Gamble 1979: 47
Brillenhöhle	Boessneck and von den Driesch 1973: 66
Caminade	Delpech 1975: Table 8
La Chèvre	Arambourou and Jude 1964: 116
Les Cottés	Bouchud 1961: 259
La Ferrassie	Delpech 1975: Table 6
Le Flageolet I	Delpech 1975: Table 10
Gargas	Bouchud 1959: 177
Gatzarria	Bahn 1983: Table 5
Gavaudun	Momméjean, Bordes, and Sonneville-Bordes 1964: 262
Irpfenhöhle	Gamble 1979: 47
Isturitz	Bouchud 1959: 173
Laugerie-Haute	Delpech 1975: Table 16
Maldidier	Delpech 1975: Table 9
Pair-non-Pair	Musil 1980-81 1: 32-33
Roc de Combe	Delpech 1975: Table 5
Les Rois	Mellars 1973: 262
Saint-Jean-des-Verges	Vézian and Vézian 1966: 124, 127
Stadel	Gamble 1979: 40-41
Vogelherd	Lehmann 1954: 115

as a way of increasing the probability that at least one of them would have been successful; or different groups might engage in complementary activities (e.g. hunting *versus* gathering, or procuring *versus* processing) and share the results of their labour. This inference is strengthened by the skeletal evidence from Shanidar (Trinkaus 1983: 422-423) that injured and partially disabled individuals were able to survive – an unlikely phenomenon without sharing and cooperation. The very existence of burials in the Middle Palaeolithic implies affective bonds among individuals that would be hard to understand without at least the emotional capability for sharing and cooperation.

HUNTING TECHNOLOGY

There are two ways of investigating the technology of subsistence. The first is to infer technological competence from our conclusions concerning the subsistence strategies in use. For example, Lewis Binford (1984, 1985) has concluded, on the basis of faunal data, that Middle Palaeolithic peoples were incapable of killing large animals and were forced to scavenge game, such as equids, *Bos*, and *Bison*. The second method is to study the artifacts used in procuring and processing foods. Luis Orquera (1984) has followed this course, attributing many of the differences between the Middle and the Upper Palaeolithic to increased efficiency in exploiting animals, an efficiency reflected in the increased specialization of food-processing tools.

Binford's conclusions that Middle Palaeolithic hunters lacked the skill or technology to kill the kinds of large animals that Upper Palaeolithic hunters indubitably killed on a regular basis could have important consequences for our understanding of the transition from the Middle to the Upper Palaeolithic. However, as I have argued elsewhere (Chase 1988), this conclusion is not sustained by the data from Europe. On the one hand, it is hard to see how huge concentrations of large game, such as those found at Mauran, Il'skaya, and Starosel'e, can be attributed to scavengers, who would by definition be dependent upon non-human predators and other natural agents for the carcasses they scavenged. For this reason, one would expect the faunal assemblages left by hominids incapable of hunting large game to be made up primarily of small- to medium-sized game, with some admixture of large game they were fortunate enough to be able to scavenge. By the same token, the age profiles of the large animals recovered should reflect the ages of animals that die from natural attrition. However, the ages of both the equids from Combe Grenal (Levine 1983) and the elands (*Taurotragus oryx*) from Klasies River Mouth (Klein 1976, 1986) indicate that hominids of this period were perfectly capable of killing animals of prime age in proportion to their frequencies in the population. Finally, Binford's work at Klasies River Mouth was based on two assumptions. First, he reasoned that since only the poorest portions of the carcass would be left to scavengers by the predators responsible for the kills, both the relative frequencies of different parts of the skeleton and the locations of tool marks on the bones would indicate a concentration upon these parts of the body. Second, since hunters may leave behind assemblages dominated by meat-poor bones, he used a second and necessary criterion for recognizing scavenging: evidence of the butchering of partially-desiccated, stiffened carcasses. This evidence consists primarily of traces on the bones of heavy-handed hacking rather than of cutting and slicing through soft flesh. Yet at Combe Grenal none of these criteria characterizes the remains of the large animals Binford believes were scavenged – horses, aurochs, and bison. Both equids and large bovids are represented more heavily by the meat-bearing bones of the proximal limbs and trunks than by those of the distal limbs (Chase 1986a, 1988). Tool marks are concentrated upon these meat- bearing bones (Fig. 19.3), and almost all of the tool marks are cuts, none are the result of hacking (Chase 1986a, 1988). One must also take into account the discovery of a wooden spear with the skeleton of an elephant (*Hesperoloxodon antiquus*) at the site of Lehringen from the Riss-Würm interglacial of northern Germany (Movius 1950; Adam 1951). Taken together, all these facts make it impossible to accept that hominids of the Middle Palaeolithic were incapable of killing large animals on a regular basis. It would thus appear that if they were technologically less competent than modern *Homo sapiens,* this incompetence did not affect their ability to adapt by means of big-game hunting.

The direct evidence for how such hunting was carried out is very scarce. Bone implements are usually extremely crude; there does not appear to have been any widespread pattern of making projectile points from bone or antler. Nor is there any real evidence for the manufacture of stone projectile points. Some of Bordes' (1979) tool types (notably Levallois and Mousterian points) might be taken as intentional projectile points. However, Bordes (1980) himself concluded that Levallois points were often by-products of blade production. Mousterian points differ from convergent scrapers only in being more pointed. Given that Dibble (1984, 1986, 1987a) has shown that convergent scrapers are probably simply the result of resharpening double scrapers, it seems likely that Mousterian points are just one end of a continuum of scraper reduction, rather than the result of deliberate manufacture of projectile points (Dibble 1987b). This is reinforced by the fact that in most European industries the bulb of percussion is not removed, so that hafting a Mousterian point, or even a Levallois point, would have been difficult.

The best artifactual evidence from the Middle Palaeolithic consists of the discovery of the wooden spear under the carcass of the elephant at Lehringen. Because wood is so poorly preserved, it is conceivable that this was the primary raw material for weapons. Perhaps our best hope for testing this possibility lies in future lithic use-wear studies (e.g. Beyries 1987).

It is also possible that driving was one of the primary means of killing game. At Combe Grenal, Levine (1983) found that the age distributions of equids from all three layers she studied fit best with either a 'life assemblage' or 'family group' model. Since animals of all ages were being killed in proportion to their frequencies in a living group, this implies catastrophic death and therefore, most probably, driving. In fact, the abundance of Middle Palaeolithic sites in the Dordogne may be in part due to the nature of the terrain, with its rolling plateaus cut by steep, often cliff-lined valleys, which makes it ideal country for driving game. It is perhaps suggestive that the site of La Quina, which has the richest faunal assemblage in the Charente, is at the base of a cliff ideally situated for driving game, in a region where such locations are rare (Jelinek, Debénath and Dibble 1986). Further study of the relative frequencies of different parts of the skeleton and of the age profiles of animals at sites like La Quina will help to clarify this problem.

Orquera's (1984) notion that Upper Palaeolithic tools were more specialized in nature than Middle Palaeolithic tools, and that this increased specialization implies greater efficiency, is worthy of further study, especially in view of Dibble's (*op. cit.*) findings that many Middle Palaeolithic types represent not desired end-products but rather segments of a continuum of reduction and resharpening. Since Upper Palaeolithic tools seem to fall into distinct categories that represent intentional classes of tools, it is quite possible that these represent a greater specialization and consequent efficiency of function, as Orquera suggests. It may be, however, that the standardization of tool morphology observed by Orquera is due to the fact that even utilitarian artifacts were now doubling as bearers of symbolic information. It is at the beginning of the Upper Palaeolithic in Europe that representative art first appears, and this art very likely has a significant socio-economic function (see, for example, Gamble 1983; see also Chase and Dibble 1987). On the basis of what we know now, it seems possible that either or both of these hypotheses may be true. Certainly this is one area in which at present we have no clear answers.

Orquera also points out that the absence of a highly developed bone and antler industry may indicate that Middle Palaeolithic peoples specialized less heavily upon large animal resources and more on a variety of resources (such as wood). However, it is difficult to determine whether this represents greater efficiency, either in the sense of killing more animals per man-hour of labour or per person, or in the sense of producing more goods per animal, per person, or per man-hour. Certainly Middle Palaeolithic peoples appear to have been perfectly competent hunters of big game. Given the nature of taphonomic considerations, and given that the incomplete nature of the sample of sites known from the Middle Palaeolithic makes inferences about either demography or animal consumption per person difficult (see Chase 1986a: 3), it seems unlikely that we will be able to measure efficiency with any precision.

CONCLUSIONS

On the basis of the evidence reviewed above, it would seem that Middle Palaeolithic subsistence differed little in its overall nature from that of the earlier Upper Palaeolithic. Specialized hunting cannot be shown to have appeared at the Middle-Upper Palaeolithic boundary. Even if such sites as the Abri Pataud imply specialized hunting, they have their counterparts in the Middle Palaeolithic. Moreover, the hunting of large herds of migratory game animals appears to have been a common feature in the Middle Palaeolithic of Europe, so that the biological and cultural phenomena associated with the transition cannot be explained in that way. The evidence from Europe implies (contrary to the hypotheses formulated by L. R. Binford following his study of the South African data) that Middle Palaeolithic peoples were not only competent but also purposeful hunters. While the study of Middle Palaeolithic hunting techniques has not advanced to the point of giving us more than glimpses of this facet of the subsistence system, all the evidence indicates that whatever techniques were used were highly effective. Moreover, the faunal data imply that sharing was a regular feature of Middle Palaeolithic economic life. As a whole, Middle Palaeolithic subsistence appears to have been remarkably similar to that of the early Upper Palaeolithic, indicating that we must look elsewhere to find an explanation of the Middle-Upper Palaeolithic transition.

SOCIAL AND INTELLECTUAL FACTORS

There is another level of organization that this paper cannot address. It is possible that the general intellectual and social context within which subsistence activities took place changed significantly from the Middle to the

Upper Palaeolithic (Chase and Dibble 1987). The question of whether or not Neanderthals and their predecessors used language to the same extent that it is used today is currently a subject of considerable debate, one that involves brain morphology, vocal tract morphology, and the implications of tool manufacture. The inferences discussed above, which were made on the basis of faunal data, do little to settle this question. Other predators hunt cooperatively and share food, at least to some extent. Other animals show some foresight in terms of the transportation and caching of food. Many predators other than humans are very efficient and are perfectly capable of making a living hunting large game. While Middle Palaeolithic hominids carried these traits beyond the levels known among other species, it remains to be demonstrated that such behaviour would be impossible in the absence of language.

What language does do is to permit modern humans to operate in the context of shared information. This information is not simply information about the present location of resources (something social insects can communicate to one another) but also about the likely location and value of future resources – for example about the probability that a fruit will be abundant at a given location and the time when it is likely to ripen. Modern hunter- gatherers therefore have, at the very least, the option of operating on the basis of a body of information unavailable to other animals. Moreover, the habitual use of symbols permits the extension of social ties and, consequently, of social cooperation, to strangers, thus greatly extending the possible size of an information-gathering and information-sharing network (cf. Gamble 1983). It is possible that, if Upper Palaeolithic hunters were indeed more efficient than their predecessors, it is due less to an increase in technological-manipulative ability or or to an increase in foresight and planning, and more to the existence of language and the transfer of intelligence.

Unfortunately, this is a topic which, while relevant to the subject of this study, is quite beyond its scope. What the faunal data do indicate is that, by the Middle Palaeolithic, hominids were competent and efficient hunters of large game, and that their exploitation of these animals involved a degree of foresight and probably of cooperation which, archaeologically, is indistinguishable from those involved in modern hunting systems. By the same token, there is no change, either in the degree of specialization or in the nature of the animals targeted by European hunters, that can be recognized at the Middle-Upper Palaeolithic boundary. If the adaptation of Upper Palaeolithic hominids differed from that of the Middle Palaeolithic hominids, the change must lie not in this realm, but rather in the intellectual and social contexts in which food was obtained.

NOTE

1. The faunal remains from Starosel'e almost certainly do not represent a single episode of hunting, since they come from a considerable depth of sediment. It is quite probable, however, that the same is true of assemblages such as those from the Abri Pataud. The case of Starosel'e illustrates the

point that large numbers of animals in a faunal assemblage do not necessarily imply massive kills.

REFERENCES

Adam, K. 1951. Der Waldelefant von Lehringen, eine Jagdbeutes diluvialen Menschen. *Quartär* 5: 75-92.

Arambourou, R. and Jude, P.-E. 1964. *Le Gisement de la Chèvre à Bourdeilles* (Dordogne). Périgueux: Magne.

Bahn, P. G. 1983. *Pyrenean Prehistory: a Palaeoeconomic Survey of the French Sites*. Warminster: Aris and Phillips.

Beyries, S. 1987. *Variabilité de l'Industrie Lithique au Mousterien: Approche Fonctionelle sur Quelques Gisements Français*. Oxford: British Archaeological Reports International Series S238.

Binford, L. R. 1978. *Nunamiut Ethnoarchaeology*. New York: Academic Press.

Binford, L. R. 1982. The archaeology of place. *Journal of Anthropological Archaeology* 1: 5-31.

Binford, L. R. 1984. *Faunal Remains From Klasies River Mouth*. Orlando: Academic Press.

Binford, L. R. 1985. Human ancestors: changing views of their behavior. *Journal of Anthropological Archaeology* 4: 292-327.

Binford, L. R. and Binford, S. R. 1966. A preliminary analysis of functional variability in the Mousterian of Levallois facies. *American Anthropologist* 68: 238-295.

Binford, S. R. 1968. Early Upper Pleistocene adaptations in the Levant. *American Anthropologist* 70: 707-717.

Boessneck, J. and Driesch, A. von den 1973. *Die Jungpleistozänen Tierknochenfunde aus der Brillenhöhle*. Stuttgart: Miller and Gräff.

Bordes, F. 1967. Considérations sur la typologie et les techniques dans le Paléolithique. *Quartär* 18: 25-55.

Bordes, F. 1973. On the chronology and contemporaneity of different Palaeolithic cultures in France. In C. Renfrew (ed.) *The Explanation of Culture Change*. London: Duckworth: 217-225.

Bordes, F. 1979. *Typologie du Paléolithique Ancien et Moyen*. Paris: Centre National de la Recherche Scientifique.

Bordes, F. 1980. Le débitage Levallois et ses variantes. *Bulletin de la Société Préhistorique Française* 77: 45-49.

Bordes, F., Rigaud, J.-P. and Sonneville-Bordes, D. de. 1972. Des buts, problèmes, et limites de l'archéologie paléolithique. *Quaternaria* 16: 15-34.

Bordes, F. and Sonneville-Bordes, D. de. 1970. The significance of variability in Palaeolithic assemblages. *World Archaeology* 2: 61-73.

Bosinski, G. 1967. Die mittelpaläolithischen Funde im Westlichen Mitteleuropa. *Fundamenta* Reihe A, Band 4. Köln: Böhlau.

Bouchud, J. 1959. *Essai sur le Renne et la Climatologie du Paléolithique Moyen et Supérieur*. Unpublished Doctoral Thesis, University of Paris.

Bouchud, J. 1961. Etude de la faune du gisement des Cottés (Haute-Vienne). *L'Anthropologie* 65: 258-270.

Bouchud, J. 1968. L'Abri du Facteur II: la faune et sa signification climatique. *Gallia Préhistoire* 11: 1-121.

Bouchud, J. 1975. Etude de la faune de l'Abri Pataud. In H. L. Movius (ed.) *Excavation of the Abri Pataud, Les Eyzies (Dordogne)*. American School of Prehistoric Research Bulletin 30. Cambridge (Mass): 69-153.

Chase, P. G. 1986a. *The Hunters of Combe Grenal: Approaches to Middle Paleolithic Subsistence in Europe*. Oxford: British Archaeological Reports International Series S286.

Chase, P. G. 1986b. Relationships between Mousterian lithic and faunal assemblages at Combe Grenal, France. *Current Anthropology* 27: 69-71.

Chase, P.G. 1987. Spécialisation de la chasse et transition vers le Paléolithique supérieur. *L'Anthropologie* 91: 175-188.

Chase, P. G. 1988. Scavenging and hunting in the Middle Paleolithic: the evidence from Europe. In H. Dibble and A. Monet-White (eds) *The Upper Pleistocene Prehistory of Western Eurasia*. Philadelphia: University of Pennsylvania Press: 225-232.

Chase, P. G. and Dibble, H. L. 1987. Middle Paleolithic symbolism: a review of current evidence and interpretations. *Journal of Anthropological Archaeology* 6: 263-293.

Delpech, F. 1975. *Les Faunes du Paléolithique Supérieur dans le Sud-Ouest de la France*. Unpublished Doctoral Thesis, University of Bordeaux I.

Dibble, H. L. 1984. Interpreting typological variation of Middle Paleolithic scrapers: function, style, or sequence of reduction? *Journal of Field Archaeology* 17: 431-436.

Dibble, H. L. 1986. Reduction sequences in the manufacture of Mousterian implements of France. In O. Soffer (ed.) *Regional Perspectives on Old World Prehistory*. New York: Plenum Press: 34-45.

Dibble, H. L. 1987a. The interpretation of Middle Paleolithic scraper morphology. *American Antiquity* 52: 109-117.

Dibble, H. L. 1987b. Typological aspects of reduction and intensity of utilization of lithic resources in the French Mousterian. In H. Dibble and A. Monet-White (eds) *The Upper Pleistocene Prehistory of Western Eurasia*. In Press.

Donner, J. J. 1975. Pollen composition of the Abri Pataud sediments. In H. L. Movius (ed.) *Excavation of the Abri Pataud, Les Eyzies (Dordogne)*. American School of Prehistoric Research Bulletin 30. Cambridge (Mass): 160-173.

Fish, P. R. and Dibble, H. L. 1982. Comment on R. White: 'Rethinking the Middle/Upper Paleolithic transition'. *Current Anthropology* 23: 182-183.

Gábori, M. 1976. *Les Civilisations du Paléolithique Moyen entre les Alpes et l'Oural*. Budapest: Akadémiai Kiadó.

Gábori, M. 1979. Type of industry and ecology. *Acta Archaeologica Academiae Scientiarum Hungaricae* 31: 239-248.

Gamble, C. 1979. Hunting strategies in the Central European Palaeolithic. *Proceedings of the Prehistoric Society* 45: 35-52.

Gamble, C. 1983. Culture and society in the Upper Palaeolithic of Europe. In G. N. Bailey (ed.) *Hunter-Gatherer Economy in Prehistory: a European Perspective*. Cambridge: Cambridge University Press: 201-211.

Girard, C. and David, F. 1982. A propos de la chasse spécialisée au paléolithique moyen: l'exemple de Mauran (Haute-Garonne). *Bulletin de la Société Préhistorique Française* 79: 11-12.

Jelinek, A. J., Debénath, A. and Dibble, H. L. 1986. A preliminary report on evidence related to the interpretation of economic and social activities of Neanderthals at the site of La Quina (Charente), France. Paper presented to Colloquium on *L'Homme de Néandertal*, Liège, 1986. In Press.

Klein, R. G. 1969. The Mousterian of European Russia. *Proceedings of the Prehistoric Society* 35: 77-111.

Klein, R. G. 1976. The mammalian fauna of the Klasies River Mouth sites, southern Cape Province, South Africa. *South African Archaeological Bulletin* 31: 75-98.

Klein, R. G. 1986. Review of L. R. Binford: *Faunal Remains From Klasies River Mouth*. *American Anthropologist* 88: 494-495.

Laville, H., Rigaud, J.-P. and Sackett, J. 1980. *Rock Shelters of the*

Périgord. New York: Academic Press.
Lehmann, U. 1954. Die Fauna des 'Vogelherds' bei Stetten ob Lontal (Würtemberg). *Neues Jahrbuch für Geologie und Paläontologie. Abhandlungen* 99: 33-146.
Levine, M. A. 1983. Mortality models and the interpretation of horse population structure. In G. N. Bailey (ed.) *Hunter-Gatherer Economy in Prehistory: a European Perspective*. Cambridge: Cambridge University Press: 23-46.
Lumley, H. de. 1972. *La Grotte de l'Hortus (Valflaunès, Hérault)*. Etudes Quaternaires 1. Marseilles: Laboratoire de Paléontologie Humaine et de Préhistoire, Université de Provence.
Mellars, P. A. 1970. Some comments on the notion of 'functional variability' in stone tool assemblages. *World Archaeology* 2: 74- 89.
Mellars, P. A. 1973. The character of the Middle-Upper Palaeolithic transition in south-west France. In C. Renfrew (ed.) *The Explanation of Culture Change*. London: Duckworth: 255-276.
Mommějean, E., Bordes, F. and Sonneville-Bordes, D. de. 1964. Le Périgordien supérieur à burins de Noailles du Roc-de-Gavaudun (Lot-et-Garonne). *L'Anthropologie* 68: 353-316.
Mottl, M. 1960. Gedänken über Probleme der jungpleistozänen Warmzeiten im Ostalpengebiet. *Mammalia Pleistocaena: Anthropos Supplement* 1: 127-136.
Movius H. L. 1950. A wooden spear of third interglacial age from lower Saxony. *Southwestern Journal of Anthropology* 6: 139-142.
Musil, R. 1980-81. *Ursus spelaeus: der Höhlenbar*. Weimarer Monographien für Ur- und Frühgeschichte 2.
Orquera, L. A. 1984. Specialization and the Middle/Upper Paleolithic transition. *Current Anthropology* 25: 73-98.
Rolland, N. 1981. The interpretation of Middle Palaeolithic variability. *Man* 16: 15-42.
Spiess, A. E. 1979. *Reindeer and Caribou Hunters*. New York: Academic Press.
Straus, L. G. 1977. Of deerslayers and mountain men: Paleolithic faunal exploitation in Cantabrian Spain. In L. R. Binford (ed.) *For Theory Building in Archaeology*. New York: Academic Press: 41- 76.
Tode, A., Preul, F., Richter, K. *et al.* 1953. Die Untersuchung der paläolithischen freilandstation von Salzgitter-Lebenstedt. *Eiszeitalter und Gegenwart* 3: 144-220.
Tozzi, C. 1970. La Grotta di San Agostino (Gaeta). *Revista di Scienze Preistoriche* 25: 3-87.
Trinkaus, E. 1983. *The Shanidar Neanderthals*. New York: Academic Press.
Vézian, J. and Vézian, J. 1966. Les gisements de la Grotte de Saint-Jean-de-Verges (Ariège). *Gallia Préhistoire* 9: 93-130.

20. Technological Changes across the Middle-Upper Palaeolithic Transition: Economic, Social and Cognitive Perspectives

PAUL MELLARS

INTRODUCTION

One topic which seems to have become inextricably bound up with recent discussions of the 'origins and dispersal of modern humans' is the character and significance of the so-called 'Middle to Upper Palaeolithic transition' (e.g. Clark 1981; White 1982; Gilman 1984; Orquera 1984; Klein 1985 etc.). Exactly how this transition relates to changes in the anatomical or biological character of human populations is, of course, still an open question. Certainly, the discoveries at Saint-Césaire in western France and Skhūl and Qafzeh in Israel have demonstrated that any simplistic notion of a direct, one-to-one correlation between 'Middle Palaeolithic' technology and 'archaic' hominids on the one hand, and between 'Upper Palaeolithic' technology and 'anatomically-modern' hominids on the other, must now be abandoned (Lévêque and Vandermeersch 1980; Vandermeersch 1981; Stringer *et al.* 1984; Trinkaus 1984 etc.). Nevertheless the fact remains that over large areas of Europe, the major changes in both the anatomy of the human populations, and the technology of the associated archaeological assemblages, can be shown to have occurred over at least broadly the same range of time – i.e. broadly between *c.* 40 000 and 30 000 BP. To many workers, this suggests that the parallel changes in the biological and archaeological records within these regions are unlikely to be entirely coincidental or unrelated (Clark 1981; Trinkaus 1983; Howell 1984; Klein 1985; Foley 1987 etc.).

As a result of research over the past 10-20 years, the general character of technological changes over the period of the Middle- Upper Palaeolithic transition has now been documented in reasonable detail in most regions of Europe and Western Asia. The advances have come partly from a range of field investigations focused specifically on this part of the Palaeolithic sequence, and partly from new and more sophisticated approaches to the analysis of technological changes involving, for example, detailed core-reconstruction techniques, micro-wear analyses, and the experimental replication of flaking techniques (e.g. Anderson 1980; Beyries 1987; Marks and Volkman 1983 etc.). Surprisingly, relatively few studies have so far attempted to raise the more fundamental question of exactly what these well documented changes in technology might signify in more general behavioural and cultural terms. In other words, what factors might have been important in initiating basic changes in lithic and bone technology? How far might changes in certain specific aspects of technology (such as the adoption of large-scale blade production) have stimulated changes in other aspects of tool production (e.g. changes in the detailed forms of the tools produced)? How far are changes in tool technology related to more general shifts in the economic or subsistence strategies of the human groups? Or (most difficult, but perhaps most significant of all) how might changes in these and other aspects of technology be related to even more fundamental changes in the social organization of Palaeolithic groups, or the patterns of 'cognitive' or 'perceptual' development?

The assumption underlying this discussion, of course, is that there was indeed some kind of complex of closely inter-related technological changes which can be identified in a fairly consistent, repeated way, over large areas of Europe and Western Asia. A full analysis of these changes would require a separate study in its own right. In the present Symposium, useful analyses of the evidence from particular regions have been provided in the papers by Clark and Lindly, Marks, and Ohnuma and Bergman for the Middle East; Harrold and Farizy for Western Europe; Allsworth-Jones, Kozlowski and Soffer for Central and Eastern Europe; and Vermeersch and Van Peer for the Nile Valley and Northeast Africa. The details of this transition varied in subtle, and potentially highly significant, ways in different regions. In the Middle East, for example, there is evidence for a relatively gradual, cumulative process of technological change, which would seem to have commenced as early as 45-50 000 BP, and to have continued until at least 30-35 000 BP (Copeland 1975; Marks 1983). In Western Europe, by contrast, virtually all of the changes which are used to define the conventional Middle-Upper Palaeolithic transition would appear to be concentrated within a much shorter time-span, between *c.* 32 000 and 38 000 BP (Mellars 1973; Harrold, this volume). Nevertheless I would suggest that there was sufficient similarity in the *broad* patterns of these technological changes within the different regions to demand some discussion of these changes in more general and wide-ranging terms. How far similar changes can be identified in more distant regions such as Southern Africa, and Central and Eastern Asia is, of course, an entirely separate question (see for example the discussions by J. D. Clark, R. G. Klein, H. Deacon and R. Jones, this volume). The whole concept of a 'Middle-Upper Palaeolithic transition' has been confined, historically, to Europe and the immediately adjacent areas of Southwest Asia and Northeast Africa, and the present discussion will be focused specifically on the evidence from these regions.

THE CHARACTER OF THE TECHNOLOGICAL TRANSITION

The complex of technological changes which has conventionally been used to define the Middle-Upper Palaeolithic transition can perhaps be summarized under six main headings. Briefly, these may be summarized as follows:

1. A general shift from flake to blade producing technologies. Although still the most striking general technological feature of the Middle-Upper Palaeolithic transition, this generalization requires some qualifications. Relatively high frequencies of blade production have now been documented from a substantial number of Middle Palaeolithic contexts in Eurasia – in some cases extending back to the earliest stages of the last glaciation, or the last interglacial (as for example at Seclin and Crayford in Northern Europe, or in the 'Amūdian' and related industries in the Middle East (Tuffreau *et al.* 1985; Cook 1986; Otte, Jelinek, this Symposium). Usually, though not invariably, these were manufactured by some specialized variants of 'Levallois' or 'Levallois point' techniques (Boëda 1988). In the Middle Palaeolithic as a whole, however, these remain rare and relatively isolated occurrences. In virtually all regions of Europe and Western Asia there is evidence for a significant expansion in the scale of blade-producing technology, coinciding closely with the conventional transition from the Middle to the Upper Palaeolithic. Perhaps more significantly, there are frequently indications of a major change in the specific core-preparation and flaking techniques employed for blade production over this interval. A particularly well documented demonstration of this change has been provided by the detailed core reconstructions carried out by Marks and Volkman (1983) for the four stratified levels at Boker Tachtit in southern Israel – illustrating a basic shift from an opposed-platform 'Levallois point' technology in the lower levels, to an exclusively single-platform blade technique in the upper levels. Similar shifts can be documented in the detailed character of blade techniques in many early Upper Palaeolithic industries in Europe – as for example in the Châtelperronian industries of France (Farizy, this Symposium; Pélegrin 1988), in the early Aurignacian industries of Central Europe (Kozlowski 1988, and this Symposium), and in the Spitsynskaya industries of southern Russia (Soffer 1985a, and this volume). The generalization that there *were* significant shifts towards more extensive, more specialized, and more 'efficient' techniques of blade production closely associated with the conventional Middle-Upper Palaeolithic transition can now be documented from at least the majority, if not all, regions of Western Eurasia (Note 1). The only areas in which this generalization has been specifically questioned (as for example in northwest Spain) would appear to represent regions in which in which the poor quality of local raw materials made the adoption of sophisticated blade technology either difficult or effectively impossible to achieve (see Clark and Lindly, this volume).

2. The appearance of well defined (and generally relatively abundant) forms of both end-scrapers and burins. This again is a widely recognised feature of the Middle-Upper Palaeolithic transition, but is not entirely without exceptions. Relatively abundant 'burin' forms (defined in a strictly technological sense) have now been documented from a number of Middle Palaeolithic sites in the Middle East – although usually in forms which are both simpler and technologically less varied than those in later Upper Palaeolithic contexts (see for example Crew 1976: 99-105). End-scrapers

are more difficult to identify in typical form in Middle Palaeolithic contexts. The occasional 'end-scrapers' reported from Middle Palaeolithic sites are usually either extremely scarce or present in such 'atypical' form that their status as discrete artifact groups is open to serious doubt (see, for example, the comments of Copeland (1983: 230) on the occasional 'end-scrapers' recovered from the Amūdian levels at Adlun in Lebanon). In my own experience, fully convincing examples of end-scrapers are virtually absent from well-excavated Mousterian sites in France, and the occurrence of convincing burins is at best a rare and exceptional occurrence.

3. The appearance of a range of morphologically 'new' artifact types, which are qualitatively different from those encountered in earlier, Middle Palaeolithic contexts. As I have pointed out elsewhere (Mellars 1973: 257; 1982: 238), this is in many ways the most obvious and immediately recognizable contrast between Middle and Upper Palaeolithic technology. The point could be illustrated from early Upper Palaeolithic contexts in almost all regions of Europe and Western Asia – as for example in the appearance of typical 'Châtelperron points' and obliquely retouched blades in the French Châtelperronian industries; distinctive crescents in the Uluzzian industries of Italy; triangular, bifacially-worked points in the Streletskaya industries of south Russia; Emireh points, chamfered blades and a variety of distinctive burin forms in the 'transitional' industries of the Middle East, and so on (Broglio and Kozlowski 1986; Soffer 1985a; Copeland 1975; Marks 1983 etc.) (see Figure 20.1). The point to be emphasized is that these are qualitatively new, and visually highly distinctive forms (i.e. distinctive 'type fossils' in the traditional terminology) which make their first appearance specifically over the period of the Middle-Upper Palaeolithic transition. As discussed elsewhere (Mellars 1973: 257-258), it is virtually impossible to identify distinctive type fossils of this kind within earlier, Middle Palaeolithic contexts in Eurasia. The few 'type fossils' which have occasionally been claimed from Middle Palaeolithic industries (such as symmetrical, double-pointed *limaces* forms, bifacially-worked *blattspitzen*, typical backed knives, or small cordiform or triangular hand-axes) can all be shown to have an ancestry extending back (in certain contexts) well into the time range of the penultimate glaciation (i.e. >120 000 BP). The generation of qualitatively 'new' artifact forms therefore seems to be specifically associated with the earliest phases of the Upper Palaeolithic succession throughout all regions of Western Eurasia.

4. Closely related to the preceding point is the remarkable speed with which additional new artifact types appeared – and successively replaced each other – at many different points within the Upper Palaeolithic succession. This point is perhaps too widely recognized to require detailed documentation here (see for example Sonneville-Bordes 1960, 1973 and Mellars 1973 for a discussion of the French industries; Broglio and Kozlowski 1986, Klein 1969 and Soffer 1985a for Central and Eastern Europe; Marks 1983 and Gilead 1981 for the Middle East). It is of course this 'dynamic' or innovative aspect of technology which makes the division of the Upper Palaeolithic industries into a series of relatively sharply defined (and in

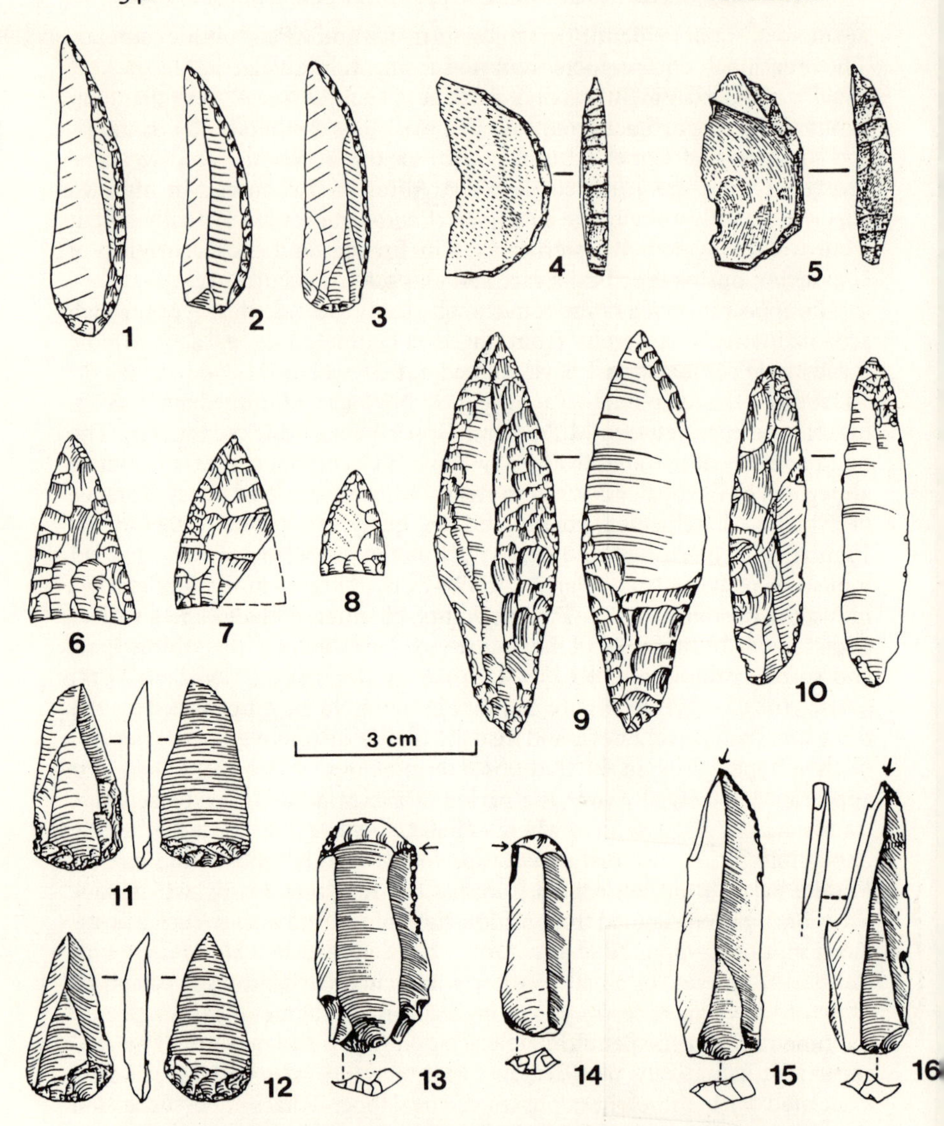

Figure 20.1. Some 'type fossils' of the earliest stages of the Upper Palaeolithic in Europe and Western Asia: 1-3 Châtelperron points (France); 4-5 Uluzzian crescents (Italy); 6-8 hollow-based Streletskaya points (south Russia); 9-10 Jerzmanovice points (Poland/Russia); 11-12 Emireh points (Israel); 13-14 chamfered blades (Lebanon); 15-16 Ksar Akil burins (Lebanon).

many cases very short-lived) chronological phases so easy. Again, the contrast with the Middle Palaeolithic must be emphasized. This is not to suggest that Middle Palaeolithic industries were in any way static or uniform in a *quantitative* sense – i.e. in terms of the relative frequencies of different tool forms, in the relative dimensions of the tools, or in the kinds of technology employed for flake production. As noted above, the point is simply that Middle Palaeolithic industries are (with a few rare exceptions) strikingly deficient in recognizable 'type fossils' which appear and disappear at specific points within the overall Middle Palaeolithic succession. And of course it is this feature – in sharp contrast to the Upper Palaeolithic – which makes the recognition of clearly defined chronological or technological 'phases' in the Middle Palaeolithic so elusive, and a topic of seemingly endless debate.

5. Almost exactly the same features as those discussed above can be documented in the production of bone, antler and ivory artifacts over the Middle-Upper Palaeolithic succession. Following a good deal of discussion in the literature (e.g. White 1982, and associated comments) it is now generally agreed that extensively *shaped* bone and antler artifacts are effectively lacking from Middle and Lower Palaeolithic contexts in Eurasia. Middle Palaeolithic groups certainly *used* bone and antler materials in various contexts, but with the exception of a few minimally-shaped awls, and occasional flaking of dense bone into simple *racloir* or denticulated forms (apparently copying the usual lithic forms) there is virtually no well documented evidence for the systematic shaping of these materials (Camps-Fabrer 1976). By contrast, there is now evidence for extensive and highly complex working of bone and antler reaching back to the earliest phases of the Upper Palaeolithic. The best examples, perhaps, are the various awls, needles, perforated teeth and bone 'rings' documented from the Châtelperronian levels at Arcy-sur-Cure (*c.* 33-34 000 BP) (Leroi-Gourhan and Leroi-Gourhan 1964), and the broadly contemporaneous forms of split-base bone points, perforated *batons,* elaborately shaped bead forms etc. recorded from early Aurignacian contexts in both Western and Central Europe (see White, this volume; also Leroy-Prost 1975, 1979; Hahn 1977). Outside Europe, this technology is rather less well represented. From both Ksar Akil (Lebanon) and Hayonim (Israel), however, there are examples of a variety of extensively shaped bone points from the early Aurignacian levels which must be at least broadly comparable in age to those in Central and Western Europe (Belfer-Cohen and Bar-Yosef 1981; Newcomer and Watson 1984).

The complexity of Upper Palaeolithic bone technology is of course striking in both the wide variety of different tool forms produced, and the highly complex and varied techniques employed in shaping the tools (cutting, sawing, grinding, polishing, perforating, 'groove and splinter' technology etc.: Camps-Fabrer 1976; Leroy-Prost 1975, 1979). As in the case of stone tools, the specific forms of bone and antler tools change repeatedly throughout the Upper Palaeolithic succession, and present a broadly similar succession of distinctive 'type fossils' to that documented in the sequence of lithic

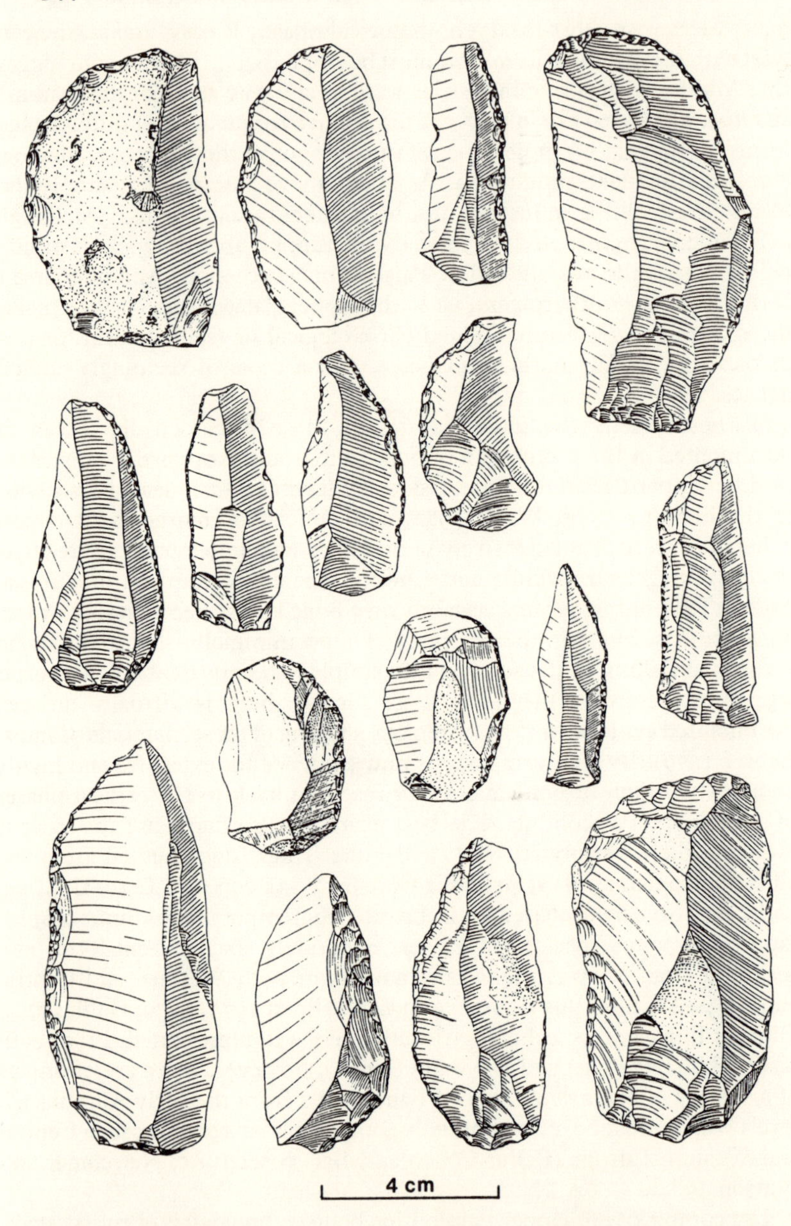

Figure 20.2. Backed knives (*'Couteaux à dos abattus'*) from French Mousterian of Acheulian tradition industries. Reproduced from Bordes 1981, Figs 35-37.

types (Sonneville-Bordes 1960, 1973; Klein 1969; Broglio and Kozlowski 1986).

6. Lastly, and perhaps most significantly, there would appear to be evidence for a much more tightly defined pattern of 'standardization' and deliberately 'imposed form' in the shaping of many Upper Palaeolithic tools than can at present be documented for the majority of Middle Palaeolithic artifacts. This is no doubt one of the most difficult and elusive features to document in lithic analysis, and one which would benefit most from careful scrutiny in future technological studies across the Middle-Upper Palaeolithic transition. The basic proposition being advanced here is that at least the majority of lithic tool forms encountered in Upper Palaeolithic assemblages are characterized by three features which are, at best, represented to a much less conspicuous degree in Middle Palaeolithic assemblages: (1) a relatively high degree of 'standardization' or 'redundancy' in the detailed forms of lithic artifacts; (2) a much clearer pattern of morphological separation between discrete artifact categories; and (3) a more obvious degree of 'imposed form' in the various stages of production and shaping of the tools.

There is no doubt that all these features require much more systematic analysis and documentation than they have received hitherto (but see recent studies by Dibble 1987, and this volume; Bricker and David 1984; David 1985 etc. for important steps in this direction). In the absence of systematic analyses, one has to rely primarily on intuitive impressions. In this context, there is little doubt that anyone approaching the analysis of Middle Palaeolithic assemblages for the first time is struck by the difficulty of dividing up the tools into clearly differentiated, neatly defined 'types'. In attempting to apply Bordes' typology, for example, there are usually numerous examples of tools which seem to grade almost imperceptibly between 'single' and 'double' edged *racloirs*; between 'lateral' and 'transverse' *racloirs*; between 'convergent' or '*déjeté*' *racloirs* and 'Mousterian points', and so on (see Dibble 1987, and this volume). The dilemma will be immediately familiar to anyone who has attempted to apply the Bordes system to a wide range of Mousterian assemblages or (even more acutely) to anyone involved in teaching lithic analysis to students. The classification of most Upper Palaeolithic industries, by contrast, usually poses far fewer problems. There may occasionally be times when a 'Gravette point' appears to grade into a 'Vachons point'; a 'Font-Yves point' into a '*lamelle Dufour*', or a 'carinate scraper' into a 'nosed scraper'. But these situations are (by comparison with the Mousterian) relatively rare, and often reflect some fairly obvious over-splitting in the taxonomy being applied. Distinctions between the major categories of end-scrapers, burins, perforators, backed blades, *raclettes*, tanged points, *fléchettes* etc. usually present far fewer problems (see Brézillon 1968). The types in most cases (and in sharp contrast to the Mousterian) usually appear to be relatively well defined, comparatively 'standardized', and clearly separated in morphological terms. This again is essentially a pragmatic observation which will no doubt be endorsed by anyone with first-hand experience of analysing both Upper and Middle Palaeolithic assemblages.

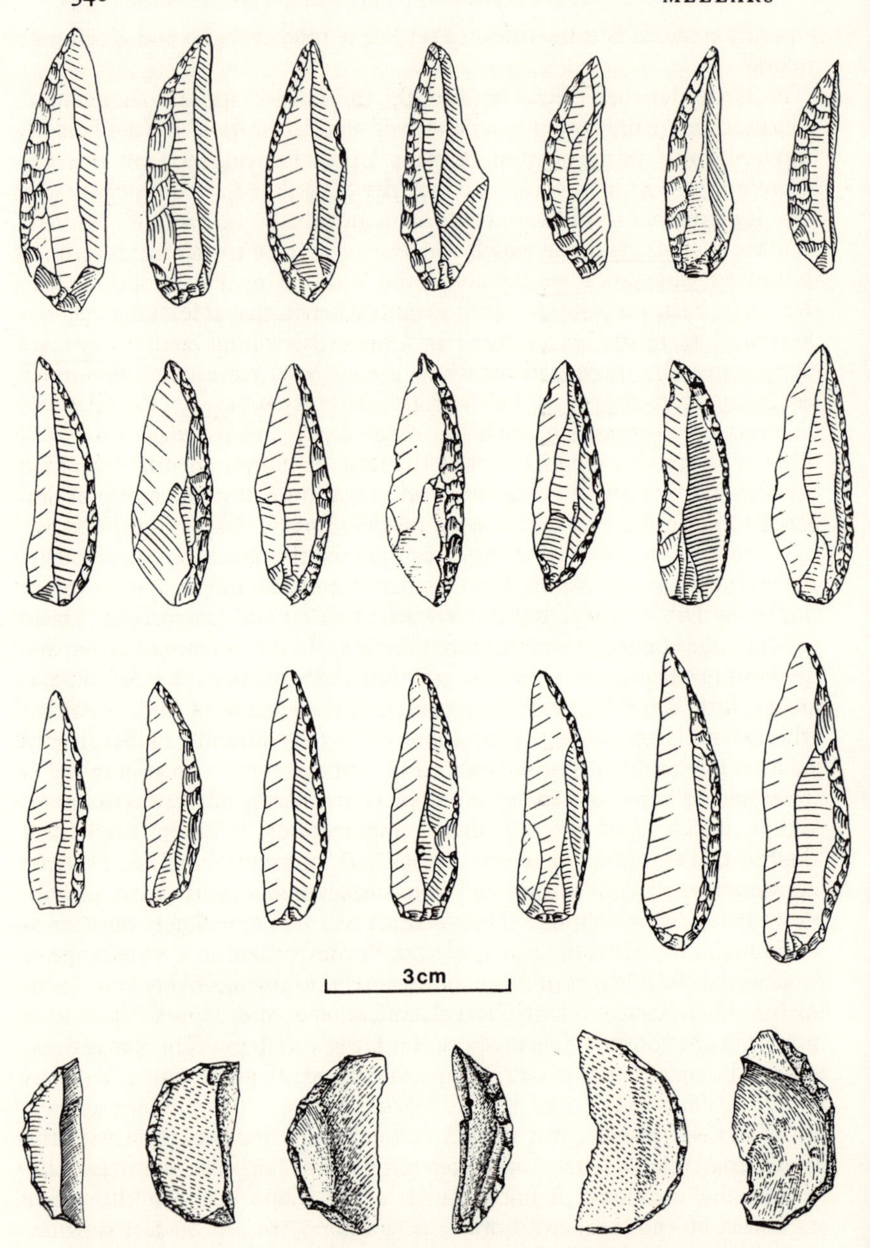

Figure 20.3. Upper three rows: 'Châtelperron points' from Châtelperronian/Early Périgordian sites in western France; lower row: 'Uluzzian crescents', from early Upper Palaeolithic sites in Italy. Reproduced from Sonneville-Bordes 1960, Figs 91-99, and Palma di Cesnola 1965, Figs 6-7.

The question of 'imposed form' in tool manufacture is closely related to the features discussed above, but perhaps requires further explanation. The suggestion, in essence, is that the majority (though by no means all) Upper Palaeolithic tools appear to reflect a much more obvious attempt to modify the original shapes of the flake or blade blanks in order to achieve some specific, sharply defined form. In other words, shaping of the tools usually involves the removal of large areas of the original flake or blade blanks, so that the final form of the tool bears little if any direct relationship to the shape of the original blank chosen. Numerous examples of this could be cited from Upper Palaeolithic contexts, ranging from such forms as Upper Périgordian Font Robert points and truncated elements, through the various laurel-leaf, willow-leaf or shouldered points of the Solutrean, to the wide range of triangular, trapezoidal, crescentic and other microlithic forms of the Magdalenian (Brézillon 1968). Perhaps one of the most graphic illustrations of this can be provided by comparing the shapes of early Upper Palaeolithic 'backed blade' forms, with those of the technologically-related 'backed knife' forms found in many Middle Palaeolithic industries (especially those of the Mousterian of Acheulian tradition group). As illustrated in Figure 20.2, the Mousterian backed knife forms reveal a good deal of extensive and careful retouch intended to blunt one edge of the original flake, but this retouch is almost invariably confined to the immediate margins of the original flake, and has relatively little effect on the overall shape of the resulting tool. Early Upper Palaeolithic 'backed blade' or 'backed point' forms, by contrast (as represented, for example, by the French Châtelperron points or the Italian Uluzzian crescents) typically reveal far more extensive reduction of the original blank to produce a much more tightly standardized and repetitive shape in the form of the finished tool. As will be seen from Figure 20.3, both Châtelperron points and Uluzzian crescents are in fact remarkably standardized products (in shape, and usually in size) which contrast sharply with the highly variable, fluid forms reflected in the Mousterian backed knife group. How far these changes in tool morphology may relate to changes in the intended functions of the tools is of course a separate question. But in at least the case of the Châtelperron points, there are reasons for believing that these fulfilled a broadly similar range of functions to those of the Mousterian 'backed knife' group (cf. Bordes 1972).

As already emphasized, far more research is needed to explore and document this kind of patterning in other groups of Middle and Upper Palaeolithic tools, but my impression is that these studies would reveal similar contrasts over a wide range of tool categories. As discussed further below, this aspect of change in tool morphology may perhaps emerge as one of the most culturally significant of all the technological changes currently documented across the Middle-Upper Palaeolithic transition (Note 2).

Possible explanatory models to account for these fundamental changes in technology across the Middle-Upper Palaeolithic transition will be dichotomized here into two broad categories: on the one hand those which emphasize the basic 'functional' or 'technological' aspects of the assemblages; and on the other hand those which focus on the potential role of more general and wide-ranging changes in the specific 'social', 'symbolic' or 'cognitive' contexts within which the artifact assemblages were employed. The relevant considerations may be outlined, rather briefly, as follows:

FUNCTIONAL/TECHNOLOGICAL MODELS

Blade technology

The first and perhaps most obvious question is how far the increase in the 'complexity' of stone-tool production which characterizes the Middle-Upper Palaeolithic transition may have been largely if not entirely dependent on one central technological development – that of the replacement of predominantly flake technologies by more specialized and standardized forms of blade production. This point is clearly crucial to the whole assessment of the significance of Upper Palaeolithic technology. If it could be argued that characteristically Upper Palaeolithic tool forms could not be produced effectively from non-laminar flake blanks, and if it could be argued that the large-scale production of carefully controlled blade forms would in some way 'stimulate' the appearance of a new range of tool forms (i.e. various forms of burins, end-scrapers, extensively shaped and standardized backed forms etc.), then one might have a potentially almost unicausal explanation for at least the majority of the documented changes in stone-tool morphology which characterize the Middle-Upper Palaeolithic transition in different regions of the world. The only challenge then would be to explain why this general shift from flake to blade technology had occurred within broadly the same time range over a wide range of geographical areas – whether by a simple process of technological diffusion, or by some kind of independent, convergent pressures towards the adoption of blade production, perhaps stimulated by increasing scarcity of raw material supplies or some similar phenomenon (as argued recently, for example, by Marks and Volman (1983) for the Middle-Upper Palaeolithic transition in the Middle East and by Ambrose (this Symposium) for the Howiesons Poort industries of Southern Africa).

This technological scenario should not be dismissed too readily. On the one hand it could be regarded as almost axiomatic that many of the most distinctive forms of Middle Palaeolithic tools (i.e most forms of side-scrapers, points and various related or intermediate forms) are in essence flake tools, and could not be produced effectively from typical blade blanks. A systematic shift from flake to blade technology might therefore *demand* a corresponding shift in the form of tools produced – requiring, for example, a general shift from side-scraper to end-scraper forms for skin-working tools, or corresponding shifts in the morphology of spear heads or other forms of weaponry. Conversely, it could be argued that the large-scale production of regular and standardized blade blanks would automatically open up a wide range of new 'opportunities' for the manufacture of novel tool forms which could not have been produced effectively within the context of a purely flake-dominated technology. These might include, for example, not only a variety of distinctive 'backed blade' forms, but also forms made from snapped or truncated blades (including perhaps several of the most distinctive burin forms), or tools made from small blade segments (for example, most of the microlithic forms such as trapezes, triangles or crescents). This is not to suggest that any of these forms are in any way inevitable or 'automatic' products of blade technology. The point is simply that a general and broad-ranging shift from predominantly flake to predominantly blade production would present flint workers with an essentially new range of opportunities to experiment with morphologically new varieties of stone tools. Effective, new forms of tools produced in this way might well diffuse from one area from another, and perhaps stimulate further experimentation in the production of further new tool forms.

One obvious attraction of this scenario is that it would help to explain the apparently 'premature' appearance of certain morphologically Upper Palaeolithic tool forms in a number of contexts which are demonstrably much earlier than the conventional Middle-Upper Palaeolithic transition. The obvious examples are the Amūdian industries of the Middle East, and the Howiesons Poort and related industries of Southern Africa. In both of these variants, the occurrence of a number of morphologically Upper Palaeolithic forms (i.e. simple end-scrapers and burins, backed blades and even – in the Howiesons Poort industries – a range of microlithic forms) are associated with technologies which are heavily biased towards blade as opposed to flake production (Singer and Wymer 1982; Copeland 1983; Jelinek, Ambrose, this Symposium). This would not of course explain why large-scale blade production emerged in these industries – still less, why this technology, once developed, would seem to have been largely superseded (in both the Middle East and Southern Africa) by a reversion to predominantly flake production. But the kind of mechanisms discussed above might perhaps help to explain why the rapid expansion of blade manufacture documented within these particular contexts (whatever its ultimate causes) was associated with the simultaneous production of certain tool forms which in other contexts are generally seen as distinctive hallmarks of the conventional Middle-Upper Palaeolithic transition.

Despite the simplicity and obvious attractions of this model, there are nevertheless a number of basic difficulties in attempting to see this as an adequate explanation for the total pattern of changes in stone-tool technology documented across the Middle-Upper Palaeolithic transition. At least four features are relevant in this context:

(a) First, it is now clear that blade technology in a broad, morphological sense is by no means confined entirely to conventionally Upper Palaeolithic industries, nor even to a few isolated and idiosyncratic Middle Palaeolithic variants such as the Amūdian and Howiesons Poort industries discussed above. A significant component of deliberate blade production has now been documented in a wide range of Middle (and even some Lower) Palaeolithic industries, in areas ranging from Europe, through Western Asia, to Southern Africa (see Boëda 1988; Kozlowski, Otte, Clark, this

Symposium). In all these industries, the essential 'opportunities' to experiment with various forms of blade-tool production clearly existed, but were evidently not adopted – or systematically developed – to any very substantial degree.

(b) Second, it should be emphasized that many of the distinctive 'new' tool forms which characterize Upper Palaeolithic industries are demonstrably *not* dependent on sophisticated blade technology. This applies not only to such forms as Aurignacian carinate and nosed scrapers, but also to several varieties of end-scrapers, most forms of borers and *raclettes*, and to a wide range of distinctive burin forms (e.g. *busqué* and parrot-beak burins, *burins sur encoche* etc.) (Brézillon 1968). All of these forms are manufactured predominantly if not entirely from flake blanks, and could therefore have been produced just as effectively within the context of a Middle Palaeolithic flake technology as in an Upper Palaeolithic blade tradition.

(c) Third (and most significant) recent research on Middle Eastern sites seems to have demonstrated fairly conclusively that the sequence of technological changes over the period of the Middle-Upper Palaeolithic transition did not necessarily or invariably involve an *initial* shift to blade production, followed by a subsequent development of characteristically Upper Palaeolithic tool forms. As Marks (1983: 89) has emphasized, the sequence documented at Boker Tachtit in southern Israel would seem to show that the basic shift from typically Middle Palaeolithic tool forms (side-scrapers, points, denticulates etc.) to Upper Palaeolithic forms (end-scrapers, burins etc.) clearly *preceded* the associated shift from flake to blade technology. Similar patterns would appear to be reflected in some of the earliest Upper Palaeolithic industries in Central Europe – most notably in the 'Bohunician' industries of Czechoslovakia and (more tentatively) in some of the formative stages of the Châtelperronian and Uluzzian industries in Western Europe (Kozlowski, this Symposium; Svoboda and Svoboda 1985; Pèlegrin 1988; Palma di Cesnola 1982). From these occurrences, one might well reverse the potential 'causative' relationships between blade technology and Upper Palaeolithic tools forms, and suggest that in at least some contexts it was the emergence of morphologically new tool forms which stimulated the more general shifts from predominantly flake to predominantly blade producing technologies.

(d) Finally, no hypothesis based purely on blade production seems adequate to explain the remarkable speed with which morphologically new varieties of stone tools appeared – and successively replaced each other – not only at the start of the Upper Palaeolithic sequence (i.e. coinciding with the initial adoption of dominant blade technology) but repeatedly at many points throughout the 20-30 000-year duration of the Upper Palaeolithic succession. This point will be discussed further at a later point in the discussion, but it may be noted here as a further objection to any simple hypothesis linking the dramatic proliferation of Upper Palaeolithic tool forms to the initial emergence of blade-using technologies.

Hafting technology

A second interesting possibility recently discussed by Desmond Clark (1983) and others is that much of the increased complexity apparent in Upper Palaeolithic technology might reflect the introduction and widespread application of hafting of stone tools as a technological device. As Clark points out, this could potentially have radical implications for almost all aspects of tool production. The adoption of hafting – and therefore the introduction of more complex, composite tools – would open up a potentially much wider range for the use of both stone and bone/antler tools, and would almost inevitably demand major changes in the form of the stone or bone components of the tools to accommodate different kinds of hafting procedures. With this development, there could be an almost quantum leap in the potential complexity of stone tools, with the production of a variety of specially shaped inserts for knives, spear-heads, skin-working tools etc., potentially hafted by a variety of different hafting techniques. Tools manufactured from blade blanks might well be more appropriate in many contexts for hafting than those manufactured from flakes, and this might have acted at least as a partial stimulant to the widespread adoption of blade-producing techniques.

At present, this possibility remains one of the most difficult to test from the available technological data. Despite the proliferation of lithic use-wear studies over the last 10-15 years, we are still remarkably ignorant as to both the specific functions of many Middle and Upper Palaeolithic tool forms, and the degree to which particular forms may or may not have been hafted. Almost certainly, many Upper Palaeolithic tools *were* hafted, and it is reasonable to speculate that this was applied to a wide range of forms including skin, wood and bone-working tools, as well as to a variety of missile heads. On the other hand, recent research points strongly to the conclusion that at least certain forms of Middle Palaeolithic tools were also hafted, possibly by a variety of different hafting techniques (Beyries 1987; Anderson 1980; Anderson-Gerfaud, this Symposium). Even allowing for the latter point, however, it remains possible and perhaps likely that the hafting of tools in Upper Palaeolithic contexts was applied on a much larger scale than in the Middle Palaeolithic, and most probably to a wider range of functional forms (particularly those involving complex, multi-component hunting missiles). As the evidence stands at present, it seems unlikely that this can be adopted as a *total* explanation for the overall increase in tool complexity between the Middle and Upper Palaeolithic – for reasons which will be discussed in the following section. But it would be surprising if the increased use of hafting procedures and the associated range of composite tool forms was not at least one of the factors underlying the generally increased complexity of Upper Palaeolithic tool forms. Obviously, this is an area in which much more research is urgently required.

General increase in economic and technological complexity between the Middle and Upper Palaeolithic

The preceding discussion on the role of hafting leads on to the more general question of how far the increased range and complexity of Upper Palaeolithic tool forms may reflect simply a *general* broadening in the total range and complexity of economic and technological activities engaged in by human groups over the period of the Middle-Upper Palaeolithic transition. Binford, for example, has recently been arguing (1985, 1986 etc.) that one of the critical developments which may have underlain the whole of the behavioural transition from the Middle to the Upper Palaeolithic may have been a sharp increase in the scale and intensity of hunting strategies – as opposed to a pattern of essentially 'opportunistic' scavenging throughout at least the greater part of the Lower and Middle Palaeolithic. This scenario of course remains highly controversial, but if it were substantiated it might have major implications for the range and complexity of material equipment produced across the Middle-Upper Palaeolithic transition. The most obvious demands would presumably be for new forms of weaponry, involving no doubt new forms of projectile points, as well as several forms of barbs or other components of composite weapon heads. But the demands of large-scale, intensive hunting might well extend much further – requiring, for example, more specialized tools for the effective and rapid butchery of large numbers of animal carcasses, or for the processing and storage of large quantities of meat (cf. Orquera 1984).

Similar arguments could no doubt be extended to several other aspects of potential economic and technological change over the Middle-Upper Palaeolithic transition. Either with or without the demands imposed by more intensive and specialized hunting strategies, it is not difficult to imagine that simultaneous changes were occurring in many other fields of technology – as for example in the working of animal skins to produce more effective clothing or tent coverings, the shaping of wooden artifacts to produce a range of new utensils, containers or living structures, or more extensive use of plant resources in the human food supply. Viewed in these terms, then, the increasing complexity apparent in the form and diversity of Upper Palaeolithic stone tools (as well, no doubt, as bone and antler artifacts) might well be seen as a direct reflection of this general broadening in the total range of human economic and technological activities across the Middle-Upper Palaeolithic transition.

In many ways, this might well be seen as perhaps the most simple and economical explanation for the overall pattern of increasing technological complexity between the Middle and Upper Palaeolithic. There are, however, at least three fundamental respects in which neither this explanation nor the earlier hypothesis of increased hafting complexity can provide an adequate explanation for the total pattern of technological change as this is reflected in the available archaeological record:

1. First, it should be emphasized that the increased 'complexity' reflected in Upper Palaeolithic stone-tool morphology is to a large extent a composite

phenomenon, created by juxtaposing the total range of distinctive Upper Palaeolithic tool forms – documented across the whole of the chronological and geographical range of the Upper Palaeolithic – against the vastly greater simplicity apparent in the corresponding range of Lower and Middle Palaeolithic tools. If attention were focused on any *individual* Upper Palaeolithic assemblage, however, it is doubtful whether one would be able to document a very much wider range of morphologically or functionally discrete 'types' than that represented in the majority of individual Middle Palaeolithic assemblages (as, for example, in a typical Mousterian of Acheulian Tradition or Quina Mousterian assemblage). It is doubtful, in other words, whether the extent of morphological or functional complexity displayed in the majority of *individual* Upper Palaeolithic industries is very much greater than that displayed in the majority of Middle Palaeolithic assemblages (see Sackett 1988 for a fuller discussion of this point).

2. Secondly, it is difficult to see how any appeal to simple 'functional' mechanisms of the kind discussed above (i.e. either increased use of hafting procedures, or increasing diversity and specialization of economic and technological activities across the Middle-Upper Palaeolithic transition) can account for the rapid and abrupt *changes* in the forms of stone tools recorded at many different points within the Upper Palaeolithic sequence – or for that matter in the similar degree of variation documented in geographical/spatial terms. As discussed in Section 1 above, the majority of the most distinctive and idiosyncratic forms of Upper Palaeolithic tools have a strikingly limited temporal and spatial distribution, which can be used to divide up the whole of the Upper Palaeolithic universe into a series of discrete and relatively tightly defined units. From our current understanding of the specific functions of these tools, it seems inconceivable that *all* this patterning can be attributed to simple variations in the basic economic or technological functions of the tools (see Sackett 1988). This is not of course to deny that some significant economic or technological shifts may have occurred at certain stages of the Upper Palaeolithic sequence, nor that some of the documented changes in the tool-type repertoires may be explicable in these terms. But it seems unlikely that any model will ever succeed in accounting for more than a limited part of the total range of spatial and chronological variability documented within Upper Palaeolithic tool forms in any simple economic or 'functional' terms.

3. Thirdly, there would seem to be equal difficulties in accounting for the sharply increased degree of 'imposed form' and 'morphological standardization' reflected in the majority of Upper Palaeolithic tool types in terms of any simple economic or functional models. The use of hafting techniques might conceivably go some way towards accounting for these features – since it could no doubt be argued that certain types of hafting procedures do impose tighter constraints on the shapes and standardization of the individual components employed in composite tools than for tools held simply in the hand. Even so, it is difficult to see this as more than a partial explanation for the sharply accentuated degree of morphological patterning and standardization reflected in the majority of Upper Palaeolithic tools,

or to account for these features in terms of any other simple shifts in the economic or 'functional' aspects of the tools.

SOCIAL, SYMBOLIC AND COGNITIVE FACTORS IN UPPER PALAEOLITHIC TECHNOLOGY

The preceding discussion leads to the conclusion that whilst a variety of essentially 'functional' factors may well account for at least certain features of the increasing patterns of technological complexity documented across the conventional Middle-Upper Palaeolithic transition, these can hardly account for the totality of the changes reflected in the available archaeological record. To recapitulate, there would seem to be at least three aspects of technological change which largely defy explanation in simple functional terms:

1. The *rapid* changes in tool morphology reflected at frequent and closely spaced intervals throughout the Upper Palaeolithic sequence;
2. The equally complex patterning of Upper Palaeolithic tool forms in spatial and geographical terms;
3. The much more conspicuous element of 'imposed form' and 'morphological standardization' apparent in the majority of Upper Palaeolithic tool forms, in contrast to the much more variable, relatively unstandardized forms characteristic of the majority of Lower and Middle Palaeolithic tools.

It is these features in particular which suggest that any attempt to account for the overall pattern of technological change across the Middle-Upper Palaeolithic transition must go beyond the level of the purely 'functional' or 'technological' features discussed in the preceding sections. Two features in particular deserve attention in this context: on the one hand the potential role of various 'social' factors in generating new patterns of complexity in artifact production; and on the other hand, the role of more fundamental changes in the levels of perception, or 'cognitive' structure, which lay behind the production of the tools. Briefly, the relevant scenarios may be outlined as follows:

Social factors

A good deal of discussion has been devoted over the past few years to the role of various 'social' factors in shaping the patterning and distribution of Upper Palaeolithic art forms (e.g. Hahn 1972, 1977; Jochim 1983; Gamble 1983; Conkey 1978), but rather less attention has been given to how this might relate to patterning in artifact assemblages. Any discussion of these questions of course runs the risk of becoming embroiled in the recent debates over the meaning and cultural significance of 'style' in artifact production, which has generated an impressive literature over the past decade. The brief discussion given here relies mainly on the general discussions of style provided by Sackett (1977, 1982, 1985, 1986 etc.), Wobst (1976, 1977) and others, and more specifically, on the detailed studies carried out by Polly Wiessner (1983, 1984) on the character of stylistic patterning in various categories of San (i.e. Bushman) artifacts. Wiessner's work is based on a detailed analysis of the ethnoarchaeological context of the San artifacts, and provides, perhaps, the most thorough basis at present available for assessing the specific role of 'stylistic' variation within the particular context of hunting and gathering communities.

In essence, Wiessner has defined two major forms of stylistic patterning within the San artifacts, which she refers to respectively as 'emblemic' and 'assertive' style. According to this distinction, 'emblemic' style is used to identify the individual with some specific, socially defined group. In Wiessner's studies, the clearest examples of this were found in the forms of the metal-tipped hunting arrows, which would appear to correlate closely with the major territorial and linguistic divisions within the present-day San communities (Wiessner 1983). In other contexts, however, Wiessner emphasises that 'emblemic' style might well operate at a much more restricted, individual level, either to link particular individuals with specific kinship groups within the local community, or to reflect the special economic or social status of the individual within the group. The category of 'assertive' style is reserved for variations which are entirely personal in character, and serve, in effect, to express the individuality or personality of the individual within the community as a whole. The best examples of this were found in the elaborate ornamental beadwork manufactured and worn by San women (Wiessner 1984). Needless to say, this is a highly over-simplified account of Wiessner's results, which hardly does justice to her discussions of the much more complex ways in which stylistic patterning may operate in a range of other social and personal contexts. Nevertheless, the central implication of Wiessner's work is that the element of 'style' in artifact manufacture can operate on at least two clearly differentiated levels within hunting and gathering communities: on the one hand at the very broad level of the individual 'ethnic' or territorial group; and on the other hand at the much narrower level of the personal role or status of the individual within the society.

To varying degrees, both of these forms of 'stylistic' or 'social' patterning are likely to have contributed to the sharply increased complexity reflected in Upper Palaeolithic tool production. The notion of highly complex 'social geography' has now become almost a commonplace of recent characterizations of the Eurasian Upper Palaeolithic (e.g. Isaac 1972; Clark 1975; Conkey 1978; Gamble 1983; Jochim 1983; Klein 1985 etc.). Recent studies by Jochim (1983), Hahn (1977) and others have used this concept in analysing the complex regional patterning of Upper Palaeolithic art forms, while the same notion has been applied more explicitly to variations in Upper Palaeolithic artifact assemblages by both Smith (1966, 1973) and David (1973, 1985). Smith, for example (1973), has postulated a number of distinct regional groupings within the Middle and Later Solutrean populations of Western Europe, based on clear regional variations in the forms of the bifacially-flaked missile heads. David (1973, 1985) has postulated similar patterns for the various groups of 'Upper Périgordian' and 'Noaillian' assemblages of France and adjacent areas, based on variations in both the character of the lithic assemblages and the associated bone and antler implements (see also Bricker and David 1987). Closely similar variations could no doubt be documented for many of the contemporaneous and later Upper Palaeolithic industries over wide areas of Central and Eastern Europe (see for example Broglio and Kozlowski 1986; Gamble 1986; Clark 1975 etc.).

As Jochim (1983) and others have pointed out, this kind of sharply increased 'regionalization' in material culture might well be seen as an almost inevitable outcome of more general changes in demographic and residential patterns associated (at least broadly) with the transition from the Middle to the Upper Palaeolithic (see also Whallon, this volume). There is now strong if not conclusive evidence from several areas of Europe that Upper Palaeolithic groups were living in much higher population densities than those of the Lower and Middle Palaeolithic, leading, apparently, to a much tighter 'packing' of individual group territories in some of the more ecologically favourable areas, such as southwestern France, or the Russian plain (Mellars 1973, 1985; Soffer 1985b, this volume; Jochim 1983). In addition, there is increasing evidence that Upper Palaeolithic groups may have occupied certain key locations on a more stable, semi-permanent basis, which would almost inevitably act as a further incentive to the definition of more sharply defined social territories, and to a more formalized pattern of reciprocal relationships between the occupants of adjacent territories (Jochim 1983; Mellars 1985; Soffer 1985b; Dyson-Hudson and Smith 1978). If we follow the arguments advanced by Wiessner (1983) and others, these are precisely the conditions which would be likely to foster a high degree of stylistic patterning in artifact manufacture, as a means of both symbolising and reinforcing the formalized social relationships between neighbouring territorial groups (Wobst 1976, 1977). Direct analogies with the Bushman data might suggest that these stylised variations in tool forms would be likely to emerge most clearly in the forms of hunting missiles since these (as Wiessner and others have pointed out) are the tools most likely to be involved directly any form of 'boundary maintenance' or disputes between adjacent groups. The available ethnographic record however provides ample illustrations of how more prosaic 'domestic' artifacts can provide an equally sensitive reflection of ethnic relationships in many historically documented contexts (e.g. Hodder 1978), and there seems no reason to doubt that similar mechanisms could have operated on an equally wide range of material equipment in the context of Palaeolithic groups.

The recognition of stylistic patterning at a more personal, individual level no doubt raises more controversial issues – especially when applied to essentially utilitarian (and undecorated) items such as stone tools. In the case of the San arrowheads, Wiessner could find no evidence of clear-cut personal variations in the forms of the arrows manufactured by different hunters, and little evidence that any other forms of stylistic patterning were used as a formal means of symbolising the particular economic or social role of the individual within the local residential group. On the other hand, it should be kept in mind that the San groups studied by Wiessner are examples of relatively small scale, highly egalitarian societies, in which the small size of the local residential groups provides little scope for any complex or highly structured separation of individual social or economic roles within the local communities. Other ethnographically documented groups of hunter-gatherers are known to have lived in much larger residential groups, and to have incorporated much more clearly defined divisions in the economic or social roles of the individuals within these groups. The best documented examples, of course, are the northwest coast Indian groups of North America, the Japanese Ainu, and some of the coastal eskimo communities of the Canadian arctic. Since most of these more 'complex' hunter-gatherer groups are now effectively extinct, we are poorly equipped with information on how far these individual roles may have been reflected in the associated material equipment. From the limited records available however, there are some well documented instances of this kind – most notably perhaps in the clear distinctions between the forms of 'men' and 'womens' knives in many Eskimo groups, and in other forms of equipment associated with specialized ceremonial roles such as those of Shamen or sorcerers (e.g. McGhee 1977; Birket-Smith 1945). How far these and other kinds of social distinctions might be reflected at the level of stone-tool technology is still a largely open question. But in any event it is now becoming clear that many of the Upper Palaeolithic communities in the periglacial zones of Eurasia are more likely to have resembled the modern groups of 'complex' hunters than the much simpler and more egalitarian groups represented by the present-day Bushmen (cf. Jochim 1983; Soffer 1985b; Mellars 1985). If this is so, then we should at least allow for the possibility that some of the morphological patterning recorded in Upper Palaeolithic tool forms could reflect essentially personal stylistic variation in the tools used by individuals occupying specialized economic or social roles within the societies.

Cognitive factors

The last and potentially most interesting possibility is that the increased 'complexity' apparent in Upper Palaeolithic technology reflects – at least in part – some kind of fundamental change in the basic structure of human thinking or cognition associated (at least broadly) with the transition from 'archaic' to 'modern' human populations. Speculations of this kind have of course been pursued in a number of more general studies of the development of hominid mental and cultural patterns (e.g. Wynn 1979, 1985; Gowlett 1984) but have not, so far as I am aware, been applied specifically to the developments in basic technology or tool production across the Middle-Upper Palaeolithic transition (but see Wynn 1986; Chase and Dibble 1987). The central notion is essentially that Upper Palaeolithic technology reflects a much more highly formalized or 'structured' approach to the production of artifacts (in stone as well as bone, antler and no doubt other materials), and that this may have fundamental implications for the ways in which human perception or thought processes were organized.

Before attempting to spell out these speculations in more detail, it may be useful to recapitulate some of the essential contrasts between Middle and Upper Palaeolithic tool morphology which have been discussed in earlier sections and which are most relevant in the present context. The central

propositions are as follows:
1. Upper Palaeolithic assemblages – taken as a group – display a far wider range of tool forms than that documented in the totality of Middle and Lower Palaeolithic industries – reflected partly in stone-tool morphology, and partly in the associated range of bone and antler artifacts.
2. The majority of Upper Palaeolithic tool forms display a much tighter degree of 'standardization' than those of the Middle Palaeolithic. This is reflected partly in the overall shapes of the tools, and partly in the choice of specific blank forms for tool manufacture, and the types and positioning of different kinds of retouch applied in shaping the tools.
3. The relatively tight standardization in the forms of most Upper Palaeolithic tools (coupled in many cases with the complexity of the forms themselves) leads to a much more sharply defined *separation* between discrete morphological categories.
4. Lastly, many if not most Upper Palaeolithic tool forms display a significant degree of 'imposed form' during the process of shaping the tools, which is largely if not entirely lacking in at least the majority of Lower and Middle Palaeolithic tools. In other words, the shapes of the tools are not only more sharply defined in a morphological sense, but appear to reflect more clearly defined 'mental templates' in the technological processes which lay behind the the production of the tools (see Figs. 20.2 and 20.3).

All of these points would benefit from much more specific documentation than I have provided here, and no doubt any of the individual generalizations could be countered by reference to specific tool forms in specific Middle or Upper Palaeolithic assemblages. As a generalization of the *overall* changes in tool morphology across the Middle-Upper Palaeolithic transition, however, I believe that the features are sufficiently clear and self-evident to call for explanations which go beyond the discussions in the preceding sections of this paper.

If we accept, therefore, that these are real points of contrast between the general patterns of Upper as opposed to Middle Palaeolithic tool morphology, and if we can further accept that there is no simple 'functional' or 'technological' explanation for this patterning, then we would seem to have evidence for a change in the ways in which the forms and procedures of artifact production were somehow *perceived* by Upper Palaeolithic artisans. The whole character of tool production in the Upper Palaeolithic suggests a much more sharply defined conceptual taxonomy or mental categorization of tool forms, which is reflected (at best) to a much more limited degree in the production of Lower and Middle Palaeolithic tools. As noted above, this is reflected not only in the degree of standardization and morphological separation in the overall *shapes* of the tools, but also in the various technological procedures used in producing the tools – the choice of specific blank forms for distinct artifact categories, the choice of different types of retouch for shaping the tools, the positioning of this retouch at specific points around the margins of the tools, and so on. All of these stages of manufacture would have required separate decisions on the part of the artisan, and all of these decisions would have contributed to the overall forms of the tools produced. The whole sequence of tool manufacture would therefore seem to have involved a series of formalized, structured decisions, which led to an equally formalized structure in the total morphological repertoire of the tools produced. It is perhaps not unrealistic to suggest that this clear-cut morphological and technological structure was not only *perceived* by the Upper Palaeolithic artisans, but was actively *imposed* on the tools in response to the mental and conceptual processes which lay behind the production of the tools. The minimal interpretation would be to see this as evidence of more tightly structured and formalized patterns of thought among Upper Palaeolithic groups, which would be difficult to conceive in the absence of relatively complex, structured language. At a more speculative level, it might well be seen as a form of explicit symbolism, relating in some way to the particular ways in which the different tool forms functioned within the overall cultural and cognitive structure of Upper Palaeolithic society.

Further speculation on the kinds of symbolic 'meaning' which might have been attached to particular tool forms in this context may or may not be useful. In the case of stone and bone artifacts, however, it is not difficult to imagine how specific symbolic 'meaning' could have been attached either to the specific economic functions of the tools, or to the particular social or economic contexts in which the tools were used. Eskimo groups, for example, make sharp conceptual distinctions between the tools used for processing different forms of plant *versus* animal products, between tools used on different social occasions or at different seasons of the year, or (as noted earlier) between tools used by males and females (e.g. Birket-Smith 1945; McGhee 1977). Potentially, therefore, formalized, perceptually-defined differences in the forms of stone and bone artifacts could have been tied into a much wider framework of symbolism and symbolically defined behaviour embracing many different aspects of the social and economic organization of Upper Palaeolithic groups.

In any event, it is now clear that the notion of a strong 'symbolic' or 'cognitive' component in artifact manufacture would be entirely in keeping with our understanding of other aspects of Upper Palaeolithic culture and society. The existence of highly developed symbolic systems is of course thoroughly documented from studies of both Upper Palaeolithic cave and mobiliary art, and from studies of the abstract signs and symbols on bone artifacts and ornaments (e.g. Marshack 1972, 1985; Leroi-Gourhan 1968). The chronology of these explicit symbolic manifestations is now known to extend back to a very early stage of the Upper Palaeolithic, coinciding (at least in Europe) with the earlier stages of the Aurignacian, around 32-34 000 BP (Marshack 1972; Delluc and Delluc 1978; Hahn 1972; White, this volume). Similar symbolic manifestations in Southern Africa may go back much earlier than this, associated, interestingly, with a similarly complex range of morphologically 'Upper Palaeolithic' tool forms (Singer and Wymer 1982). Few people would question the inference that these highly developed symboling systems must reflect the existence of relatively complicated and highly structured forms of language throughout the greater part if not the whole of the Upper Palaeolithic sequence. It is tempting to see the sharply increased morphological 'complexity' and 'structure' of Upper Palaeolithic tool forms as one further manifestation of this 'symbolic explosion' in the Upper Palaeolithic, paralleling – and no doubt closely associated with – the simultaneous development of language and art.

NOTES

1. This is of course not to suggest that all Upper Palaeolithic industries were based on high levels of blade production. There are a number of well documented exception to this – as for example in some of the earliest Magdalenian ('Badegoulian') industries in western France (Sonneville-Bordes 1960: 387) and apparently in some of the 'Levantine Aurignacian' industries in the Middle East (Marks 1983). In the Upper Palaeolithic as a whole, however, these are relatively rare occurrences. The point is simply that in almost all regions of Europe and Western Asia it is possible to document a substantial expansion in the *scale* of blade production (and often in the *techniques* of production) coinciding closely with the period of the conventional Middle-Upper Palaeolithic transition. As noted in the text, the apparent exception to this pattern in northwestern Spain is almost certainly due to the poor quality of local raw materials (principally poor-quality quartzite) which made the adoption of sophisticated blade production difficult if not possible to achieve (see Clark and Lindly, this volume).

2. I am not suggesting that *all* Upper Palaeolithic tool types exhibit a high degree of 'imposed form' in the sense described here. Some of the simpler types, such as end-scrapers and several of the simpler burin forms, exhibit this to a much less obvious degree. Also, as Sackett (1988) has recently pointed out, it may well be that many of the Aurignacian industries of Western Europe show less evidence for this kind of patterning than many of the later Upper Palaeolithic industries, such as those of the Upper Périgordian/Gravettian, or the Solutrean and Magdalenian. Nevertheless it can hardly be disputed that some of the most distinctive type fossils of the earliest stages of the Upper Palaeolithic do exhibit a strong element of 'imposed form' – as for example in the French Châtelperron points, the Italian Uluzzian crescents, or the hollow-based triangular points of the Russian Streletskayan industries (see Figures 20.1, 20.3). In the case of the Aurignacian, of course, similar 'imposed form' is clearly reflected in the shaping of several varieties of bone points (split-base, lozangic, biconical etc.) as well as in the production of bead forms, carved animal statuettes, and other art forms (see White, this volume; Hahn 1972, 1977 etc.).

If one were looking for evidence of comparable 'imposed form' in Middle Palaeolithic artifacts, the best case could perhaps be made out for some of the distinctive biface forms – such as certain varieties of triangular and cordiform hand-axes in the French Mousterian of Acheulian Tradition, and in the *blattspitzen* forms of Central Europe (Bordes 1981). Whether this kind of patterning can be documented convincingly in any of the major categories of Middle Palaeolithic flake tools (side-scrapers, points, denticulates etc.) is much more debatable (see Dibble 1987, and this volume).

REFERENCES

Anderson, P. 1980. A testimony of prehistoric tasks: diagnostic residues on stone tool working edges. *World Archaeology* 12: 181-194.

Bahn, P. 1982. Inter-site and inter-regional links during the Upper Palaeolithic: the Pyrenean evidence. *Oxford Journal of Archaeology* 1: 247-268.

Belfer-Cohen, A. and Bar-Yosef, O. 1981. The Aurignacian at Hayonim Cave. *Paléorient* 7: 19-42.

Beyries, S. 1987. *Variabilité de l'Industrie Lithique au Moustérien: Approche Fonctionelle sur Quelques Gisements Français*. Oxford: British Archaeological Reports International Series S238.

Binford, L. R. 1985. Human ancestors: changing views of their behavior. *Journal of Anthropological Archaeology* 4: 292-327.

Binford, L. R. 1986. Isolating the transition to cultural adaptations: an organizational approach. Paper presented to the Advanced Seminar on *The Origins of Modern Human Adaptations*, Santa-Fé, April 1986. In Press.

Birket-Smith, K. 1945. *Ethnographical Collections from the Northwest Passage*. Copenhagen: Report of the Fifth Thule Expedition 1921-1924: Vol. 6 (No. 2).

Boëda, E. 1988. Le concept laminaire: rupture et filiation avec le concept Levallois. In M. Otte (ed.) *L'Homme de Néandertal*. Vol. 8: *La Mutation*. Liège: Etudes et Recherches Archéologiques de l'Université de Liège 35: 41-59.

Bordes, F. 1972. Du Paléolithique moyen au Paléolithique supérieur: continuité ou discontinuité? In F. Bordes (ed.) *The Origin of Homo Sapiens*. Paris: UNESCO: 211-218.

Bordes, F. 1981. *Typologie du Paléolithique Ancien et Moyen*. Paris: Centre National de la Recherche Scientifique: Cahiers du Quaternaire I.

Brézillon, M. 1968. *La Dénomination des Objêts de Pierre Taillé: Materiaux pour un Vocabulaire de Préhistoriens de Langue Française*. 4th Supplement to *Gallia Préhistoire*. Paris: Centre National de la Recherche Scientifique.

Bricker, H. M. and David, N. 1984. *The Excavation of the Abri Pataud, Les Eyzies (Dordogne): The Perigordian VI (Level 3) Assemblage*. Cambridge (Mass): Peabody Museum of Archaeology and Ethnology, Bulletin 34.

Broglio, A. and Kozlowski, J. 1986. *Il Paleolitico: Uomo, Ambiente e Culture*. Milan: Jaca Books.

Camps-Fabrer, H. 1976. Le travail de l'os. In H. de Lumley (ed.) *La Préhistoire Française*. Vol. 1: *Les Civilisations Paléolithiques et Mésolithiques*. Paris: Centre National de la Recherche Scientifique: 717-722.

Chase, P. G., and Dibble, H. L. 1987. Middle Paleolithic symbolism: a review of current evidence and interpretations. *Journal of Anthropological Archaeology* 6: 263-293.

Clark, J. D. 1981. 'New Men, strange faces, other minds': an archaeologist's perspective on recent discoveries relating to the origins and spread of Modern Man. *Proceedings of the British Academy* 67: 163-192.

Clark, J. D. 1983. The significance of culture change in the early Later Pleistocene in northern and southern Africa. In E. Trinkaus (ed.) *The Mousterian Legacy: Human Biocultural Change in the Upper Pleistocene*. Oxford: British Archaeological Reports International Series S164: 1-12.

Clark, J. G. D. 1975. *The Earlier Stone Age Settlement of Scandinavia*. Cambridge: Cambridge University Press.

Conkey, M. W. 1978. Style and information in cultural evolution: toward a predictive model for the Paleolithic. In C. L. Redman, M. J. Berman, E. V. Curtin, W. Y. Langhorne, N. M. Versaggi and J. C. Wanser (eds) *Social Archeology: Beyond Subsistence and Dating*. New York: Academic Press: 61-85.

Cook, J. 1986. A blade industry from Stoneham's Pit, Crayford. In S. N. Collcutt (ed.) *The Palaeolithic of Britain and its Nearest Neighbours*. Sheffield: Department of Archaeology and Prehistory, University of Sheffield: 16-19.

Copeland, L. 1975. The Middle and Upper Paleolithic of Lebanon and Syria in the light of recent research. In F. Wendorf and A. E. Marks (eds) *Problems in Prehistory: North Africa and the Levant*. Dallas: Southern Methodist University Press: 317-350.

Copeland, L. 1983. The Palaeolithic industries at Adlun. In D. A. Roe (ed.) *Adlun in the Stone Age: the Excavations of D. A. E. Garrod in the Lebanon, 1958-1963*. Oxford: British Archaeological Reports International Series S159: 89-260.

Crew, H. L. 1976. The Mousterian site of Rosh ein Mor. In A. E. Marks (ed.) *Prehistory and Paleoenvironments in the Central Negev, Israel*, Vol. 1. Dallas: Southern Methodist University Press: 75-112.

David, N. C. 1973. On Upper Palaeolithic society, ecology and technological change: the Noaillian case. In C. Renfrew (ed.) *The Explanation of Culture Change*. London: Duckworth: 277-303.

David, N. C. 1985. *The Excavation of the Abri Pataud, Les Eyzies (Dordogne): The Noaillian (Level 4) Assemblages and the Noaillian Culture in Western Europe*. Cambridge (Mass): Peabody Museum of Archaeology and Ethnology, Bulletin 37.

Delluc, B. and Delluc, G. 1978. Les manifestations graphiques aurignaciennes sur support rocheux des environs des Eyzies (Dordogne). *Gallia Préhistoire* 21: 213-438.

Dibble, H. L. 1987. The interpretation of Middle Paleolithic scraper morphology. *American Antiquity* 52: 109-117.

Dyson-Hudson, R. and Smith, A. E. 1978. Human territoriality: an ecological reassessment. *American Anthropologist* 80: 21-41.

Foley, R. A. 1987. Hominid species and stone-tool assemblages: how are they related? *Antiquity* 61: 380-392.

Gamble, C. 1983. Culture and society in the Upper Palaeolithic of Europe. In G. N. Bailey (ed.) *Hunter-Gatherer Economy in Prehistory: a European Perspective*. Cambridge: Cambridge University Press: 201-211.

Gamble, C. 1986. *The Palaeolithic Settlement of Europe*. Cambridge: Cambridge University Press.

Gilead, I. 1981. Upper Palaeolithic tool assemblages from the Negev and Sinai. In J. Cauvin and P. Sanlaville (eds) *Préhistoire du Levant*. Paris: Centre National de la Recherche Scientifique: 331-342.

Gilman A. 1984. Explaining the Upper Palaeolithic revolution. In M. Spriggs (ed.) *Marxist Perspectives in Archaeology*. Cambridge: Cambridge University Press: 115-126.

Gowlett, J. A. J. 1984. Mental abilities of early Man: a look at some hard evidence. In R. Foley (ed.) *Hominid Evolution and Community Ecology: Prehistoric Human Adaptation in Biological Perspective*. London: Academic Press: 167-192.

Hahn, J. 1972. Aurignacian signs, pendants and art objects in Central and Eastern Europe. *World Archaeology* 3: 252-266.

Hahn, J. 1977. *Aurignacien: das Altere Jungpaläolithikum in Mittel- und Osteuropa*. Köln: Fundamenta Reihe A9.

Hodder I. (ed.) 1978. *The Spatial Organisation of Culture*. London: Duckworth.

Howell, F. C. 1984. Introduction. In F. H. Smith and F. Spencer (eds) *The Origins of Modern Humans: a World Survey of the Fossil Evidence*. New York: Alan R. Liss: xiii-xxii.

Isaac, G. Ll. 1972. Chronology and tempo of cultural change during the Pleistocene. In W. W. Bishop and J. A. Miller (eds) *Calibration of Hominoid Evolution*. New York: Wenner Gren Foundation for Anthropological Research: 381-430.

Jochim, M. A. 1983. Palaeolithic cave art in ecological perspective. In G. N. Bailey (ed.) *Hunter-Gatherer Economy in Prehistory: a European Perspective*. Cambridge: Cambridge University Press: 212-219.

Klein, R. G. 1969. *Man and Culture in the Late Pleistocene: a Case Study*. San Francisco: Chandler.

Klein, R. G. 1985. Breaking away. *Natural History* 94 (1): 4-7.

Kozlowski, J. K. 1988. L'apparition du Paléolithique supérieur. In M. Otte (ed.) *L'Homme de Néandertal*. Vol. 8: *La Mutation*. Liège: Etudes et Recherches Archéologiques de l'Université de Liège 35: 1-21.

Leroi-Gourhan, A. 1968. *The Art of Prehistoric Man in Western Europe*. London: Thames and Hudson.

Leroi-Gourhan, A. and Leroi-Gourhan, Arl. 1964. Chronologie des grottes d'Arcy-sur-Cure (Yonne). *Gallia Préhistoire* 7: 1-64.

Leroy-Prost, C. 1975. L'industrie osseuse aurignacienne: essai régional de classification: Poitou, Charentes, Périgord. *Gallia Préhistoire* 18: 65-156.

Leroy-Prost, C. 1979. L'industrie osseuse aurignacienne: essai régional de classification: Poitou, Charentes, Périgord. *Gallia Préhistoire* 22: 205-370.

Lévêque, F. and Vandermeersch, B. 1980. Découverte de restes humains dans un niveau castelperronien à Saint-Césaire (Charente-Maritime). *Comptes-Rendus de l'Académie des Sciences de Paris (Série II)* 291: 187-189.

McGhee, R. 1977. Ivory for the Sea Woman: the symbolic attributes of a prehistoric technology. *Canadian Journal of Archaeology* 1: 141-149.

Marks, A. E. 1983. The Middle to Upper Paleolithic transition in the Levant. In F. Wendorf and A. E. Close (eds) *Advances in World Archaeology* Vol. 2. Orlando: Academic Press: 51-98.

Marks, A. E. and Volkman, P. W. 1983. Changing core reduction strategies: a technological shift from the Middle to the Upper Paleolithic in the southern Levant. In E. Trinkaus (ed.) *The Mousterian Legacy: Human Biocultural Change in the Upper Pleistocene*. Oxford: British Archaeological Reports International Series S164: 35-51.

Marshack, A. 1972. *The Roots of Civilization*. New York: McGraw-Hill.

Marshack, A. 1985. *Hierarchical evolution of the human capacity: the Paleolithic evidence*. New York: American Museum of Natural History.

Mellars, P. A. 1973. The character of the Middle-Upper Palaeolithic transition in southwest France. In C. Renfrew (ed.) *The Explanation of Culture Change*. London: Duckworth: 255-276.

Mellars, P. A. 1982. On the Middle-Upper Paleolithic transition: a reply to White. *Current Anthropology* 23: 238-240.

Mellars, P. A. 1985. The ecological basis of social complexity in the Upper Paleolithic of southwestern France. In T. D. Price and J. A. Brown (eds) *Prehistoric Hunter-Gatherers: the Emergence of Cultural Complexity*. Orlando: Academic Press: 271-297.

Newcomer, M. and Watson, J. 1984. Bone artifacts from Ksar 'Aqil (Lebanon). *Paléorient* 10: 143-147.

Orquera, L. A. 1984. Specialization and the Middle/Upper Paleolithic transition. *Current Anthropology* 25: 73-98.

Palma di Cesnola, A. 1965. Il Paleolitico superiore arcaico (facies

Uluzziana) della Grotta de Cavallo (Lecce). *Rivisti di Scienze Preistoriche* 20: 33-62.
Palma di Cesnola, A. 1982. L'Uluzzien et ses rapports avec le Protoaurignacien en Italie. In L. Banesz and J. K. Kozlowski (eds) *Aurignacien et Gravettien en Europe*. Liège: Etudes et Recherches Archéologiques de l'Université de Liège 13 (Vol. 2): 271-288.
Pèlegrin, J. 1988. Observations technologiques sur quelques séries du Châtelperronien et du MTA du sud-ouest de la France: une hypothèse d'évolution. Paper presented to the Colloquium *Le Fin du Paléolithique Moyen, Debut du Paléolithique Supérieur en Europe,* Nemours, 1988.
Sackett, J. R. 1977. The meaning of style in archaeology: a general model. *American Antiquity* 42: 369-380.
Sackett, J. R. 1982. Approaches to style in lithic analysis. *Journal of Anthropological Archaeology* 1: 59-112.
Sackett, J. R. 1985. Style and ethnicity in the Kalahari: a reply to Wiessner. *American Antiquity* 50: 154-159.
Sackett, J. R. 1986. Isochrestism and style: a clarification. *Journal of Anthropological Archaeology* 5: 266-277.
Sackett, J. R. 1988. The Mousterian and its aftermath: a view from the Upper Paleolithic. In H. L. Dibble and A. Montet-White (eds) *Upper Pleistocene Prehistory of Western Eurasia*. Philadelphia: University Museum of Pennsylvania Monograph 54: 413-426.
Singer, R. and Wymer, J. 1982. *The Middle Stone Age at Klasies River Mouth in South Africa*. Chicago: University of Chicago Press.
Smith, P. E. L. 1966. *Le Solutréen en France*. Bordeaux: Delmas.
Smith, P. E. L. 1973. Some thoughts on variations among certain Solutrean artifacts. In *Estudios Dedicados al Professor Dr. Louis Pericot*. Barcelona: Universidad de Barcelona Instituto de Arqueologia y Prehistoria: 67-75.
Soffer, O. 1985a. *The Upper Paleolithic of the Central Russian Plain*. Orlando: Academic Press.
Soffer, O. 1985b. Patterns of intensification as seen from the Upper Paleolithic of the Central Russian Plain. In T. D. Price and J. A. Brown (eds) *Prehistoric Hunter-Gatherers: the Emergence of Cultural Complexity*. Orlando: Academic Press: 235-270.
Sonneville-Bordes, D. de. 1960. *Le Paléolithique Supérieur en Périgord*. Bordeaux: Delmas.
Sonneville-Bordes, D. de. 1973. The Upper Palaeolithic. In S. Piggott, G. Daniel and C. McBurney (eds) *France Before the Romans*. London: Thames and Hudson: 30-60.
Stringer, C. B., Hublin, J.-J. and Vandermeersch, B. 1984. The origin of anatomically modern humans in Western Europe. In F. H. Smith and F. Spencer (eds) *The Origins of Modern Humans: a World Survey of the Fossil Evidence*. New York: Alan R. Liss: 51-136.
Svoboda, J. and Svoboda, H. 1985. Les industries de type Bohunice dans leur cadre stratigraphique et écologique. *L'Anthropologie* 89: 505-514.
Trinkaus, E. 1983. Neandertal postcrania and the adaptive shift to modern humans. In E. Trinkaus (ed.) *The Mousterian Legacy: Human Biocultural Change in the Upper Pleistocene*. Oxford: British Archaeological Reports International Series S164: 165- 200.
Trinkaus, E. 1984. Western Asia. In F. H. Smith and F. Spencer (eds) *The Origins of Modern Humans: a World Survey of the Fossil Evidence*. New York: Alan R. Liss: 251-293.
Tuffreau, A., Revillon, S., Sommé, J., Aitken, M. J., Huxtable, J. and Leroi-Gourhan, A. 1985. Le gisement Paléolithique moyen de Seclin (Nord-France). *Archäologisches Korrespondenzblatz* 15: 131-138.
Vandermeersch, B. 1981. *Les Hommes Fossiles de Qafzeh (Israël)*. Paris:

Centre National de la Recherche Scientifique.

White, R. 1982. Rethinking the Middle/Upper Paleolithic transition. *Current Anthropology* 23: 169-192.

Wiessner, P. 1983. Style and social information in Kalahari San projectile points. *American Antiquity* 48: 253-276.

Wiessner, P. 1984. Reconsidering the behavioral basis for style: a case study among the Kalahari San. *Journal of Anthropological Archaeology* 3: 190-234.

Wobst, H. M. 1976. Locational relationships in Paleolithic society. *Journal of Human Evolution* 5: 49-58.

Wobst, H. M. 1977. Stylistic behavior and information exchange. In C. Cleland (ed.) *For the Director: Essays in Honor of James B. Griffin*. Ann Arbor: University of Michigan Museum of Anthropology, Anthropology Papers 61: 317-342.

Wynn, T. 1979. The intelligence of later Acheulean hominids. *Man* 14: 371-391.

Wynn, T. 1985. Piaget, stone tools and the evolution of human intelligence. *World Archaeology* 17: 32-43.

Wynn, T. 1986. Archaeological evidence for the evolution of modern human intelligence. In M. Day, R. Foley and Wu Rukang (eds) *The Pleistocene Perspective*. Precirculated papers of the World Archaeological Congress, Southampton, 1986. London: Allen and Unwin.

21. Production Complexity and Standardization in Early Aurignacian Bead and Pendant Manufacture: Evolutionary Implications

RANDALL WHITE

INTRODUCTION

Recently there has been intense interest in the transition to culturally- and biologically-modern humans in Europe (Trinkaus 1983, 1988; Conkey 1983; Straus 1983; White 1982a; Mellars 1973, 1982). A key area of inquiry concerns the evolution of material representations that are generally referred to the domains of art and body ornamentation. Clear consensus exists that there was an explosion of art and ornamentation at the beginning of the European Upper Palaeolithic, even if isolated earlier examples have been proposed (Martin 1909; Wetzel and Bosinski 1969). Indeed, these early examples lend credence to my position (White 1982a) that a restructuring of social relations, rather than changes in neural structure or manual dexterity, selected for the dramatic increase in objects of personal adornment. As always, (White 1982a), I deny here any causal role for the biological replacement of Neanderthals by *Homo sapiens sapiens* in this increase in body ornamentation.

The sample upon which our consensus on an explosion of early body ornamentation is founded is heavily Franco-centric. Nevertheless, there is clear evidence from a number of European and other regions (see Belfer-Cohen and Bar-Yosef 1981; Hahn 1972, 1986; Bader 1970; Szombathy 1925) to suggest that the increase in body ornamentation was a widespread, although probably not universal, phenomenon. This is not to suggest that the same factors selected for an increase in body ornamentation activities everywhere that they occurred. Rather, each region must be examined in its own right for causal factors, including those mitigating against any evidence for body ornamentation whatsoever in the basal Upper Palaeolithic (see Straus, this Symposium).

There has been a tendency on the part of many authors, including myself (White 1982a), to focus on *late* Upper Palaeolithic objects of art and ornamentation when drawing comparisons between the Middle and Upper Palaeolithic. This approach results in contrasts between widely separated periods rather than an assessment of the nature of change between 35 000 and 30 000 years ago. There has been relatively little discussion of art and ornamentation closer to the transition itself and, with few exceptions (Freeman 1978; Hahn 1971, 1975, 1986; Hahn *et al.* 1977; Champagne and

Espitalié 1981; Mazière and Raynal 1983; Leroi-Gourhan 1961; Mouton and Joffroy 1958), additions to our inventory of late Mousterian/early Upper Palaeolithic art and ornamentation have not been striking.

The bias in favour of the late Upper Palaeolithic is especially important to overcome in the light of recent suggestions that somehow the early Upper Palaeolithic is qualitatively different from the period after 20 000 BP (Straus 1983). In fact, there are almost certainly those among us who would like to see the early Upper Palaeolithic as the real transitional grey area between the Middle and Upper Palaeolithic. In essence, we need to know just how dramatic the changes were between the late Mousterian and the early Aurignacian/Châtelperronian (see Note 1).

One of the sources of bias in favour of the late Upper Palaeolithic grows out of a concern with body ornamentation *solely* within the context of mortuary analyses (cf. S. Binford 1968). It has not generally been recognized that this focus on mortuary practices virtually excludes early Aurignacian body ornaments from consideration, since few if any of the abundant early Aurignacian body ornaments were recovered from burial contexts. Indeed, circumstances of excavation preclude attributing shell and ivory pendants to burial context even at the classic site of Cro-Magnon (Lartet 1869), which in any event seems to have been no earlier than late Aurignacian in age. Thus, it is not clear that body adornment and burials were in any way related during this period. In sum, the importance of body ornamentation in the early Upper Palaeolithic remains poorly understood and grossly underestimated.

The general goal of this paper is to address the issue of the explosion of items of body adornment at the beginning of the Upper Palaeolithic and to understand more about the cultural contexts in which they were operative. The author's theoretical orientation hinges on a concern with the processes of 'self' definition and social display, universal among modern humans, and their evolutionary causes and consequences (White 1989). To contribute to knowledge in this important problem area it is necessary to understand the technological, socioeconomic and ideational constituents of the earliest material evidence for bodily adornment. To this end, the present study proposes to examine:

1. The complexity of technology in early Aurignacian bead and ornament production (Note 2);
2. The relationship between early Aurignacian body ornaments and the procurement/exchange of exotic raw materials from which they were often made;
3. The degree of standardization of ornament production within and between early Aurignacian sites;
4. Inter-site variation in bead and ornament frequencies and manufacturing activities;
5. Similarities and differences in the designs used to decorate early Aurignacian body ornaments within a well-defined region;
6. The patterns according to which only the teeth of certain animal species were being appropriated for social display.

In sum, the thrust of this paper will be to nudge us away from (1) the prevalent emphasis on mortuary aspects of body ornamentation; and (2) the undue emphasis on the *later* Upper Palaeolithic in discussions of the Middle-Upper Palaeolithic transition. This will be accomplished by summarizing a set of observations and hypotheses emerging from ongoing research on a large sample of early Aurignacian beads and pendants. *The data base to be employed consists of 835 stone and ivory beads and pendants and 369 unfinished objects along the bead production sequence recovered exclusively from non-mortuary Aurignacian I contexts in southwest France* (Note 3).

BACKGROUND

Most overviews of the Middle-Upper Palaeolithic transition recognize bodily adornment as one of the most significant developments of the basal Upper Palaeolithic. Given the importance of this phenomenon, the *published sample* of beads and pendants from the first 5000 years of the European Upper Palaeolithic remains surprisingly small. The large quantities of early art and ornamentation from Vogelherd (Germany) and the Abri Blanchard (France) were recovered more than 75 and 50 years ago respectively (Didon 1912; Riek 1934) and the rich body of material from Blanchard was never adequately studied and published before it was dispersed (Delluc and Delluc 1981; White 1986a, 1986b). The large number of early Aurignacian beads recovered by Peyrony from Abri Castanet, and only summarily published (Peyrony 1935), have never been studied since. This leaves us with an important part of the Upper Palaeolithic record that has largely been forgotten in museum drawers.

While there have been well-documented discoveries subsequently (Barandiaran 1980; González Echegaray and Freeman 1971, 1973; Kozlowski 1982; Leroi-Gourhan 1961; Mazière and Raynal 1983) it is clear that the number of European basal Upper Palaeolithic sites (early Aurignacian and contemporaneous Châtelperronian) that contain abundant art and body ornaments is small. For example, only four of the 25 Aurignacian sites studied by Hahn (1972) contained 'art objects', which he distinguished from simple markings. This fact may be significant in itself and may pattern spatially or temporally. Nevertheless, if we are to say much of importance about the earliest art and ornamentation, our published sample, research methods and chronological control must improve considerably. Fortunately, there are abundant well-provenienced, unpublished materials that can contribute to such improvements in the immediate future.

Early in this century American museums acquired, by purchase or excavation, significant collections of the earliest stone, bone, antler and ivory objects modified by humans for artistic and social purposes. They came from Didon and Castanet's excavations of Aurignacian I levels at Abri Blanchard (Didon 1911), Collie's (1928) excavations of several Aurignacian levels at Abri Cellier, and Marcel Castanet's sieving of disturbed early Aurignacian and Magdalenian deposits at La Souquette, all in the Vézère Valley of southwest France. The detailed study of the American collections of body ornaments from the sites of Abri Blanchard and La Souquette,

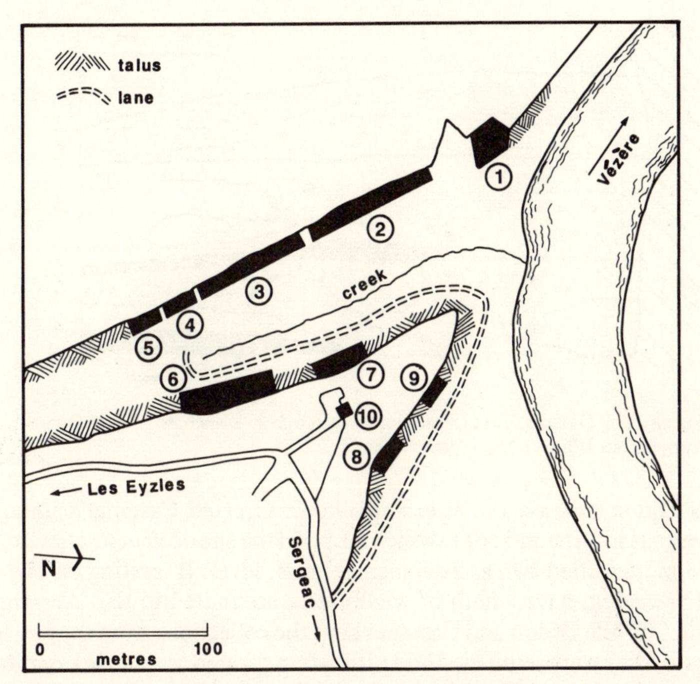

Figure 21.1. Map showing the relative locations of Palaeolithic sites in the Vallon de Castelmerle: (1) La Souquette; (2) Abri Labattut; (3) Roc de l'Acier; (4) Abri Reverdit; (5) unexcavated shelter; (6) Abri Castanet; (7) Abri Blanchard I; (8) Abri des Merveilles; (9) Abri Blanchard II; (10) Castanet farm and auberge.

neither of which has been previously studied or published in any detail, led the author to examine complementary collections remaining in France. These collections consist of additional materials from Abri Blanchard and La Souquette, but most importantly, from the nearby site of Abri Castanet (Peyrony 1935), excavated by Marcel Castanet under Peyrony's direction.

THE SITES STUDIED AND THEIR ARCHAEOLOGICAL CONTEXT

The sites of Abri Blanchard, Abri Castanet and La Souquette are located in the same karstic dry valley, known as the 'Vallon de Castelmerle' (Fig. 21.1), which houses numerous other sites (Reverdit, Roc de l'Acier, Labattut, Merveilles, Abri Blanchard II) spanning the whole of the Upper Palaeolithic sequence.

Abri Blanchard

The Abri Blanchard (also known as the 'Abri Blanchard-des-Roches' and 'Abri Didon'), the first of the three Aurignacian sites to be excavated, was superficially tested by Reverdit (1882) who abandoned the excavation after descending upon the massive rock fall overlying the lowermost Aurignacian layer. In 1909, Marcel Castanet, unaware of previous research at the site, found a bead on the surface and dug a test trench. He convinced

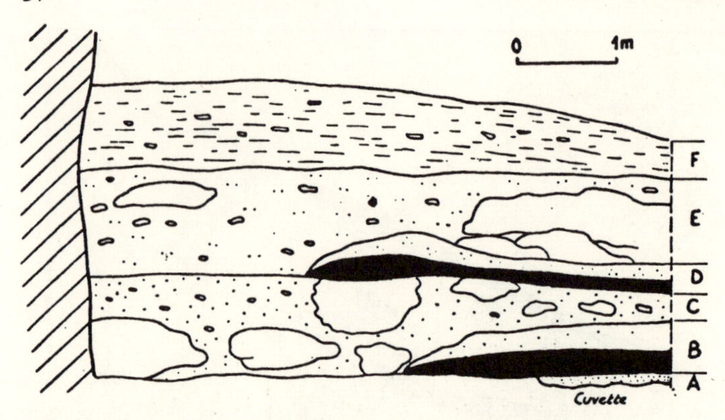

Figure 21.2. Didon's (1911) stratigraphy of the Abri Blanchard. Layer B = Aurignacian I. Layer D = Aurignacian II.

Louis Didon to lease the site, and Didon directed Castanet's important excavations in 1910 and 1911, which emptied the site of all of its contents.

Didon identified two archaeological units, levels B (resting on bedrock) and D (see Fig. 21.2), both of which were attributed to the 'Aurignacien typique'. While Didon and Castanet kept the collections from the two levels separate, they were published and ultimately curated as a single assemblage. Fortunately, Marcel Castanet maintained an active correspondence with Didon to report on his findings. The resulting archives serve, as Delluc and Delluc (1981) have pointed out, as field notes, providing stratigraphic provenience for all of the non-lithic artifacts. In thoroughly researching these archives, Delluc and Delluc (1981: 5) have determined that the upper archaeological level (Aurignacian II):

> ". . . contenait la majorité des bâtons percés, isolés les uns des autres et situés dans la partie externe de la fouille. Elle n'a livré aucune sagaie à base fendue, mais des sagaies 'pointues des deux cotés'. *La couche inférieure (Aurignacien I) a fourni toutes les sagaies à base fendue, les pendeloques, les perles et les coquillages percés.* En revanche, elle n'a livré qu'un seul baton percé" (emphasis added).

Sonneville-Bordes argues convincingly that the two levels recognized by Didon are referable to the Aurignacian I and II respectively of Denis Peyrony's classic sequence. Most importantly, it seems clear that the Abri Blanchard and the Abri Castanet, while bearing different site designations, are really contiguous areas of the same stratigraphic section (Sonneville-Bordes 1960: 100). Fortunately, the Abri Castanet was well excavated by Peyrony and objects from the two archaeological levels were curated separately. Combined with the letters from Castanet to Didon cited above, this fact, as will be clear below, allows a high degree of certainty with regard to the stratigraphic provenience of the beads from the Abri Blanchard. There is no doubt that, like those from Abri Castanet, they were excavated from the lowermost level, associated with large numbers of split-based antler points.

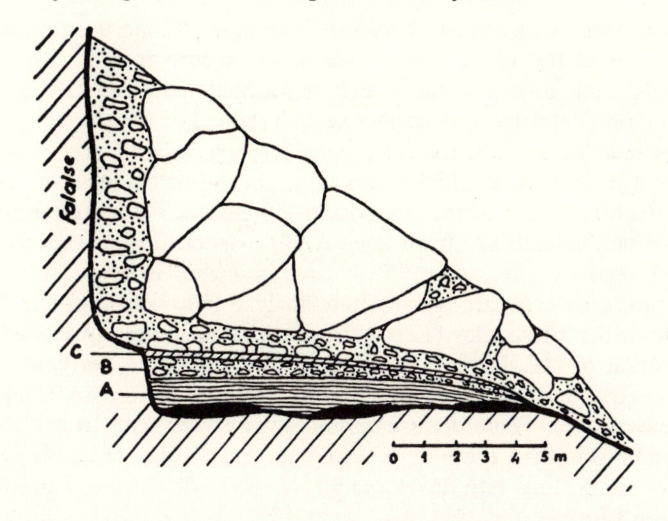

Figure 21.3. Peyrony's (1935) stratigraphy of the Abri Castanet. Layer A = Aurignacian I. Layer C = Aurignacian II.

Clearly, the Abri Blanchard stands out as one of the richest Aurignacian bone, antler and ivory assemblages ever excavated (Sonneville-Bordes 1960: 100). Moreover, 17 engraved/painted limestone blocks and 21 *pierres à anneaux* (holed stones) were recovered from this site, making it of capital importance for the origins of graphic representation (Delluc and Delluc 1978). In the mid 1920s, important segments of the Abri Blanchard assemblages, including many objects of art and body ornaments, were acquired from Louis Didon by representatives of American institutions (the American Museum of Natural History, Logan Museum of Anthropology, and the Field Museum of Natural History).

Abri Castanet

The Abri Castanet is situated to the south-south-east of the Abri Blanchard on the same shelter platform as the latter. During excavations at the Abri Blanchard, Marcel Castanet tried unsuccessfully to convince Didon to prolong his excavations to include the contiguous Abri Castanet deposits (Delluc and Delluc 1978). As at the Abri Blanchard, Peyrony (1935) recognized two archaeological levels (A and C) (see Fig. 21.3). The lowermost level, resting on bedrock, contained a rich assemblage of bone, antler and ivory objects indistinguishable from those described by Didon from the Abri Blanchard. Important for present purposes is the fact that all beads and production debris from the Abri Castanet are from the lowermost level A. The uppermost level C yielded virtually no organic artifacts: only "*quelques poinçons*" and "deux pointes losangiques à base non fendue" according to Sonneville-Bordes (1960: 104). My own examination of the collections fully confirms her observations.

Thus, the Abri Castanet provides a kind of stratigraphic anchor for the Castelmerle Aurignacian sites, and a confirmation of Marcel Castanet's excavation notes concerning the Abri Blanchard. While a stratigraphic analysis done in the 1930s may not be a very secure anchor, it must be emphasized that, unlike sections such as La Madeleine and La Ferrassie where Peyrony's stratigraphy has been much refined by recent excavations, Abri Castanet leaves little room for such revision since its archaeological levels are thin (maximum thicknesses of 30 cm and 20 cm for levels A and C respectively) and significant subdivision of them is difficult to envision.

The artifact assemblage from level A at Castanet has long been considered, by seriation (Sackett 1966), by its typological indices (Sonneville-Bordes 1960), by comparative attribute analysis (Brooks 1979) and by its bone and antler projectiles (Léroy-Prost 1975), to represent the earliest manifestation of the classic Aurignacian sequence in the Périgord. Only the proposed 'Aurignacian 0' at La Ferrassie is considered older (Delporte and Mazière 1977). Therefore in examining body ornaments from Castanet level A, and in linking these objects to other sites in the Vallon de Castelmerle, we are dealing with developments of the *earliest* Upper Palaeolithic in Western Europe.

La Souquette

Of all of the Castelmerle Aurignacian sites, the Abri de la Souquette has suffered most from historical processes of destruction. First, it was partly destroyed by a quarrying operation in historic times. Then in 1902-3, the Abbé Landesque undertook excavations, which were never published before the recovered material was dispersed. He was followed by two local workers, Costes and Letellier (Delage 1938), who destroyed much of the site. Then came the infamous Hauser in 1911, who instructed his workers to seek only long blades and knives (Delluc and Delluc 1981: 12), leaving the back dirt full of everything from ivory beads to split-based points. Subsequently, the back dirt from all of this destruction was carefully water-sieved by Marcel Castanet, using the same techniques he had applied at Abri Castanet and Abri Blanchard. The results of this sieving were published by Delage (1938), who was able to clearly identify an Aurignacian and a Magdalenian component. In 1929, a large proportion of this material was acquired by the Field Museum, Chicago. Accompanying this material was a significant amount of Magdalenian lithic and bone material, easily distinguishable by the dominance of backed bladelets, delicate burin and blade technology and at least one '*bec-de-perroquet*' burin (in the Institut de Paléontologie Humaine). Harpoons were apparently absent.

For present purposes, the important fact is that Castanet's water-sieving of the La Souquette deposits produced a total of 434 beads, almost all of ivory and steatite, which are indistinguishable in raw material, technique of manufacture (see Fig. 21.4) and in the distribution of their lengths (Fig. 21.10) from identical materials from the Aurignacian levels at Abri Blanchard and Abri Castanet. In addition, a large number of bead production stages (Fig. 21.12), identical to those from Castanet and Blanchard, were recovered from the La Souquette deposits. The beads recovered from La Souquette, Blanchard and Castanet, and the techniques used to produce them, are unknown in the Western European Magdalenian (see White 1988 for example). There is therefore no doubt, especially given the results of Roussot's recent excavations described below, that the beads can be attributed to the basal Aurignacian as found at Castanet and Blanchard.

No quantitative study has been carried out on the Aurignacian lithic material from La Souquette, but Sonneville-Bordes (1960: 106) sees close resemblances with the assemblages from Castanet and Blanchard. The organic artifacts from La Souquette are indistinguishable from those recovered from level A at Abri Castanet and include split-based points (exclusively), *pièces à languettes,* pierced batons (one with a 'threaded' hole as at Abri Blanchard and Castanet), a limestone block with an engraved vulva, and a *pierre à anneau* (in the Field Museum, Chicago).

Most recently, Alain Roussot (1982) undertook careful excavations at La Souquette and established a detailed stratigraphy in an area seemingly peripheral to the main zone of previous excavation. In two square metres he uncovered an intact Aurignacian I level (with 40 retouched tools) resting on bedrock (as at Blanchard and Castanet). While no beads were recovered from this limited area, two pierced shells and two fragments of mammoth-ivory were found in place. All sediments above this level were disturbed and/or recent in age. Since no beads were found we must remain mildly cautious concerning the exact provenience of the enormous sample of Aurignacian body ornaments from La Souquette (but see Fig. 21.11). However, no such caution is required for objects from Abri Castanet and Abri Blanchard which came exclusively from the lowermost (Aurignacian I) layer of a single stratigraphic section.

PRELIMINARY OBSERVATIONS ON EARLY AURIGNACIAN BODY ORNAMENTS

In this section I wish to outline a variety of observations on early Aurignacian body ornaments, drawing on both the published literature and the enormous sample from the Castelmerle sites described above.

1. *In contrast to popular conception, Aurignacian body ornaments are remarkably abundant.* Examination of secondary sources dealing with the Upper Palaeolithic gives one the impression that Aurignacian body ornaments are scarce, and that they are comprised almost exclusively of perforated shells and animal teeth. In fact, shells and teeth, while numerous, are far outnumbered by purposely-manufactured beads and pendants, almost always manufactured from mammoth-ivory or imported serpentinite. This reality is hinted at in a small number of old and obscure articles in the specialist literature (Didon 1912; Delage 1938; Peyrony 1935), but has been masked by the fact that many Aurignacian body ornaments from early research in the Vézère Valley were exported to the United States (White 1986a, 1986b) where they have been ignored by European and American scholars alike. Hundreds of others have remained in private hands. The sample of finished ivory and stone beads from the three Castelmerle sites, not including pierced teeth and shells, exceeds 835.

2. *Contrary to popular belief, body ornaments have seldom if ever been recovered*

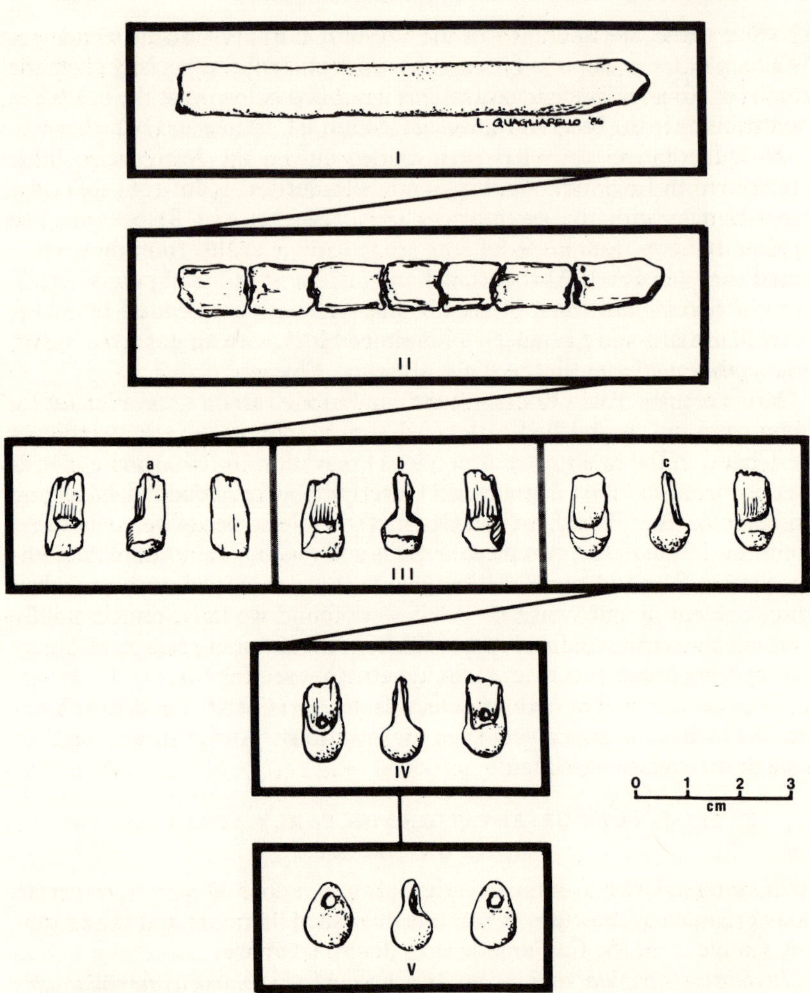

Figure 21.4. The sequence of production for the manufacture of early Aurignacian ivory beads as reconstructed from collections from Abri Blanchard, Abri Castanet and La Souquette. I: Pencil-like ivory baton. II: Bead-blanks created by circumincision and snapping of baton. III: Bilateral thinning of one end of blank by splitting along natural laminae of ivory: preliminary polishing (optional). IV: Perforation of thinned blank near 'bulb'. V: Polishing and grinding to remove most of thinned portion, and to round the bulbar end.

in quantity from Aurignacian burial contexts, but were excavated, in most cases, from living surfaces. Body ornaments and burial of the dead have often been treated as if they were coterminous phenomena. Sally Binford's (1968) now classic study of Middle and Upper Palaeolithic mortuary practices masks the fact that, despite a near absence of recovered burials, Aurignacian body ornaments are abundant and diverse. Indeed, the sites which have yielded the greatest numbers of body ornaments (Abri Blanchard, Abri Castanet,

La Souquette, Geissenklösterle and Vogelherd) have produced virtually no Aurignacian human skeletal material. As Oakley (1969) and Mussi (1985) have convincingly pointed out, the heavily decorated burials from Grimaldi, excavated before 1900 and often referred to as 'Aurignacian', are in fact late Würm in age. It can now reasonably be argued (Hahn 1983; White 1988) that the majority of Aurignacian body ornaments discovered to date represent decoration of everyday or special-occasion clothing, an hypothesis presently being examined by Scanning-Electron- Microscope analysis of the beads and pendants themselves.

3. *The vast majority of Aurignacian body ornaments are not pierced shells or teeth, but rather are formed ivory or serpentinite beads, which are the result of a complex production sequence that includes piercing, grinding and polishing.* The collections from the Castelmerle sites housed in the United States and France contain large numbers of unfinished beads and pendants, allowing a detailed reconstruction of the production sequence (see also Otte 1974) for the manufacture of these objects (Fig. 21.4). The result is a degree of production complexity, unanticipated by the present author, which does not conform with the presumption that production complexity was greater in the late Upper Palaeolithic than at the beginning. Moreover, a heavy use of grinding and polishing, used to mitigate severe limitations in drilling and perforating technology, does not conform to the idea that these processes appeared relatively late in the Upper Palaeolithic.

4. *The earliest evidence for long-distance procurement of raw materials (i.e. marine shells) is linked to body ornaments.* Mellars (1973) emphasized that one of the most striking shifts across the Middle-Upper Palaeolithic transition was the appearance of long-distance procurement of raw materials. Equally interesting however, is the fact that these exotic materials are most often transformed into objects of body ornamentation (Note 4). The relationship between exchange and items of social display is frequently encountered in the ethnographic literature (Strathern and Strathern 1971; Weiner 1987). The absence of both in the Middle Palaeolithic must be seen as evolutionarily significant, perhaps implying major differences in social organization. I have suggested elsewhere (White 1985) that the existence of a complex form of language-as-we-know-it, which was probably developed during the Middle Palaeolithic/Middle Stone Age, played a role in these trends.

A small number of Aurignacian sites accounts for most of the known body ornaments. These are the very sites that have yielded the greatest quantities of exotic materials, which were most frequently used in the production of body ornaments and objects of graphic representation. Taborin (1985) has described the Vallon de Castelmerle as a kind of 'market' due to the remarkable density of marine shells found in the three Aurignacian sites located there (Blanchard, Castanet, La Souquette). Because her analysis did not focus on beads and pendants from these sites, she did not describe the particularly strong coincidence that exists between these latter and exotic shells.

Nor, to my knowledge, has it been recognised that the ivory and steatite

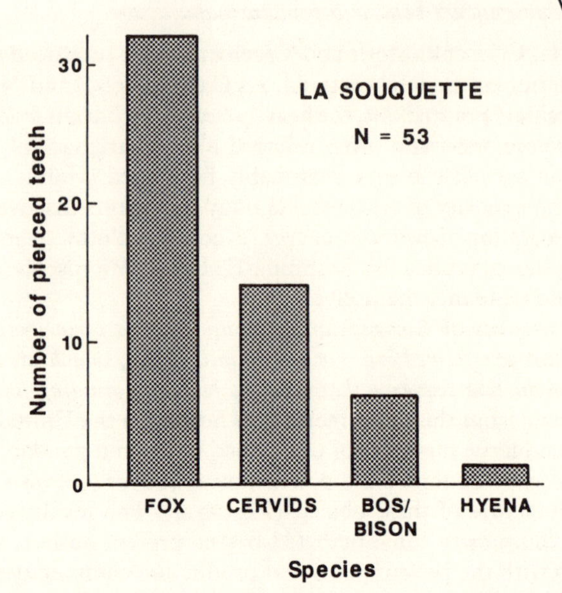

Figure 21.5. Rough breakdown by species of perforated teeth from the Aurignacian at La Souquette (based on collections in the Field Museum of Natural History, Chicago).

from which beads and pendants were manufactured were themselves obtained from some distance. No mammoth bones and only one intact section of tusk has been recovered from the Vallon de Castelmerle sites. Therefore, the large quantities of ivory used in bead manufacture seem to have been imported, either as tusk segments or in the form of the cylindrical rods found in the collections studied to date. These rods, seldom more than 10 cm long, and from 0.45 to 1.40 cm in diameter, represent the largest single units of *worked* ivory found in these rich sites. The absence of mammoth bones and the rarity of the early stages of tusk reduction are in direct contrast to Aurignacian sites in southern Germany, where mammoth bones are present and numerous ivory fragments allow the detailed reconstruction of the initial stages of tusk-working (Hahn 1986). Since tusks *and* mammoth bones were carried back to the German sites, it is hard to advance a credible argument that in southwest France mammoth bones and primary tusk-working are absent because the Aurignacians left mammoth carcasses at kill and processing localities outside of habitation sites.

With respect to the dozens of steatite beads and pendants, the case for exotic sources is much clearer. This stone is not native to the Périgord. Whether the material is derived from the Massif Central or the Pyrenees, the distance involved is in excess of 100 km.

There has always been a troublesome question as to what might have been exchanged in return for exotic shells. I suggest that the answer in the early Aurignacian may be serpentinite and ivory, given that mammoths seem to have been exceedingly rare in southwest France during the Aurignacian, and indeed during most of the Upper Palaeolithic (Delpech 1975).

As Taborin (1985) has suggested, the Vallon de Castelmerle may have been an agreed-upon point of exchange. If the above argument stands, this exchange would have been between groups with access to the coast (and/or fossil shell beds), and those with access to the ranges occupied by woolly mammoth.

5. *When animal teeth or shells were appropriated for ornamentation, there was very restricted selection of species.* In the early Aurignacian sample studied, carnivores dominate the sample of pierced teeth, with fox far outnumbering any other species (Fig. 21.5), even when their remains in midden deposits are rare. This pattern makes it possible to show significant quantitative differences in species representation between cave and portable art on the one hand and objects of social display on the other. The focus on fox canines has been demonstrated for Aurignacian sites in Germany (Hahn 1972, and pers. comm.) and Gravettian sites throughout Central and Eastern Europe (Soffer 1985). There seems little question, if the numerous ethnographic examples of social appropriation of especially meaningful species can be extended to the present example, that the fox had some special significance for early Upper Palaeolithic people across a broad region. It must be noted however, that Brooks (1979) has shown a shift through time away from fox toward cervids in the manufacture of tooth pendants within the Abri Pataud Aurignacian sequence.

6. *Aurignacian art and ornamentation in the Vézère Valley were characterized by a series of simple decorative motifs found in more-or-less contemporaneous levels at different sites.* Hahn (1972) found considerable redundancy in the simple decorations found on Aurignacian worked bone, antler and ivory in Central Europe. Although the motifs appear to be different from those of Central Europe, the same kind of inter-site redundancy in motif is evident in the Vézère Valley.

At Abri Blanchard, Abri Castanet and La Souquette meandering rows of punctuations were used to decorate several bone and ivory objects, including pendants. This pattern seems to mimic that found on Atlantic sea shells from the same levels. Both the pattern and the probability that it was inspired by the natural punctuations on exotic sea shells has seemingly been ignored by Marshack (1972) in arguing for lunar notation (see also White 1982b). This appropriation of a natural pattern is further verified by the fact that these same shells, complete with meandering punctate pattern, were replicated in ivory at La Souquette and worn as pendants (Fig. 21.6).

Another widespread motif in the Vézère Valley sites is the paired incision, which most frequently is arranged across the surface of an object (Fig. 21.7). On occasion, paired incisions and punctuations appear on the same object (Fig. 21.8). Several pieces of bone and ivory from the Vézère sites have lines etched across their surfaces, which inter-finger with marginal notches (Fig. 21.9).

7. *The degree of standardization in the manufacture of stone and ivory beads is striking.* Metric data have been recorded for beads from the three Castelmerle sites (Fig. 21.10). The predominant form of bead – that usually described as 'basket-shaped' – shows a remarkably homogeneous size dis-

Figure 21.6. Ivory replicas (upper) of sea shells (lower) recovered from the Aurignacian of La Souquette, with natural shell markings recreated by means of punctuations (Field Museum of Natural History, Chicago). Scales in mm.

tribution, probably related to conventional patterns of clothing decoration, suggesting well-defined production standards within and between sites. The inter-site similarities in dimensions of finished beads for the Castelmerle sites are noteworthy. There are important implications concerning craft specialization, quantitative production standards and sharing/reproduction of complex systems of information, that will be explored in future analyses.

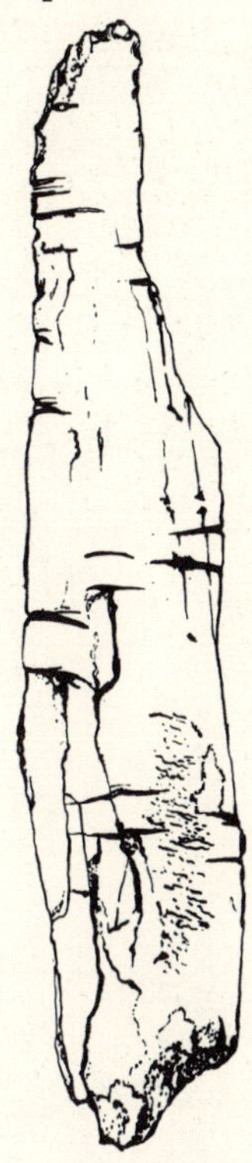

Figure 21.7. Long-bone fragment (‘*compresseur*’) with ‘paired incision’ arrangement, from the Aurignacian at Abri Cellier (Logan Museum, Beloit College). Length = 16.1 cm.

8. *Some severe limitations in drilling technology were making the bead-making process more labour-intensive than it need have been.* Pierced animal teeth and bead blanks both show severe gouging (Fig. 21.11) to thin an object, usually to a thickness of *c.* 0.2 cm, so that it could be pierced by pressure or occasionally by rotation of an acute tool edge. Very rough holes in bead blanks were disguised by subsequent labour-intensive techniques of grinding and polishing. This is linked to a virtual absence of appropriate drilling technology. Not surprisingly, perforators are exceedingly rare in Aurignacian assemblages (Sonneville-Bordes 1960; Brooks 1979), and those that do exist are blunt compared to those found in later periods of the Upper Palaeolithic.

If perforation technology in the early Aurignacian can be considered rudimentary, serious doubt is cast on the notion that Mousterian prototypes for Upper Palaeolithic beads and pendants were manufactured in non-preservable media such as wood. If this were true, one would expect to find already well-developed techniques of perforation by the time tooth and ivory were employed as media. This is simply not the case.

Two pierced objects from the Mousterian of La Quina (if they are not

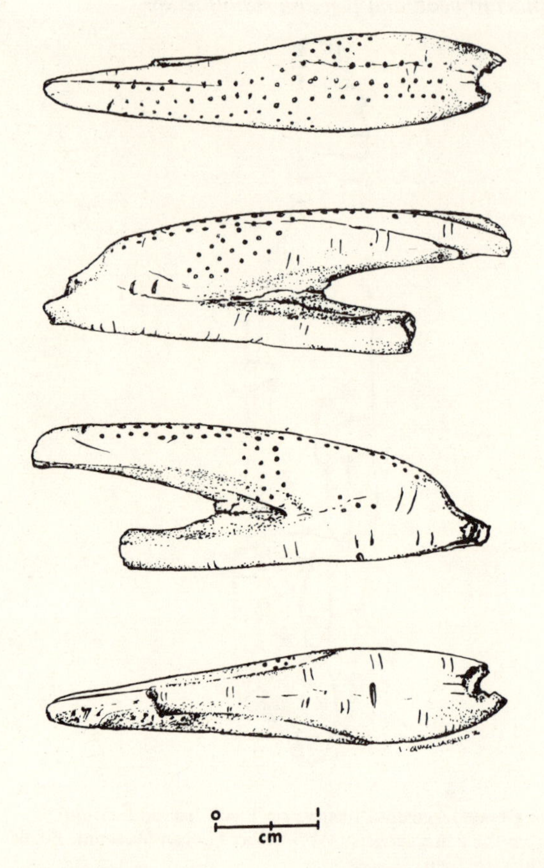

Figure 21.8. Ivory pendant with both 'paired incision' and 'meandering punctuation' motifs from the Aurignacian I at Abri Blanchard (Logan Museum, Beloit College).

intrusive from the overlying Aurignacian) (Note 5) and the two from the Central European Micoquian (= late Mousterian) at Bocksteinschmiede (Wetzel and Bosinski 1969) (if they are indeed the work of humans), suggest that prior to the Upper Palaeolithic the rarity of perforated objects results from the absence of the social, ideational, and technological context for use, rather than the inability to make or conceive of holes. In any case, there were more pierced/perforated objects in any given square metre of the Abri Blanchard than existed for all of Europe during preceding periods. This surely indicates some important qualitative cultural differences across the transition to the Upper Palaeolithic, at least in Western Europe.

9. *There are major differences between sites and levels in numbers of body ornaments found.* Striking quantitative differences between Périgord sites seem not to be attributable to differences in the quality of excavation techniques since, for example, Rigaud's (1982) meticulous excavations at Le Flageolet I over 15 seasons of excavation have produced a total of only

Figure 21.9. Bone objects from the Aurignacian at La Souquette showing characteristic pattern of marginal notches interfingering with surficial lines (Field Museum of Natural History, Chicago). Scales in mm.

three Aurignacian body ornaments. Moreover, Movius's careful excavations of numerous early Aurignacian levels at Abri Pataud (Brooks 1979) yielded a total of 12 body ornaments (one pierced shell and 11 pierced animal teeth). In contrast, Didon and Castanet's very early excavations at Abri Blanchard yielded at least 200 ivory beads and pendants, not to mention dozens more pierced teeth and shells. This raises interesting questions

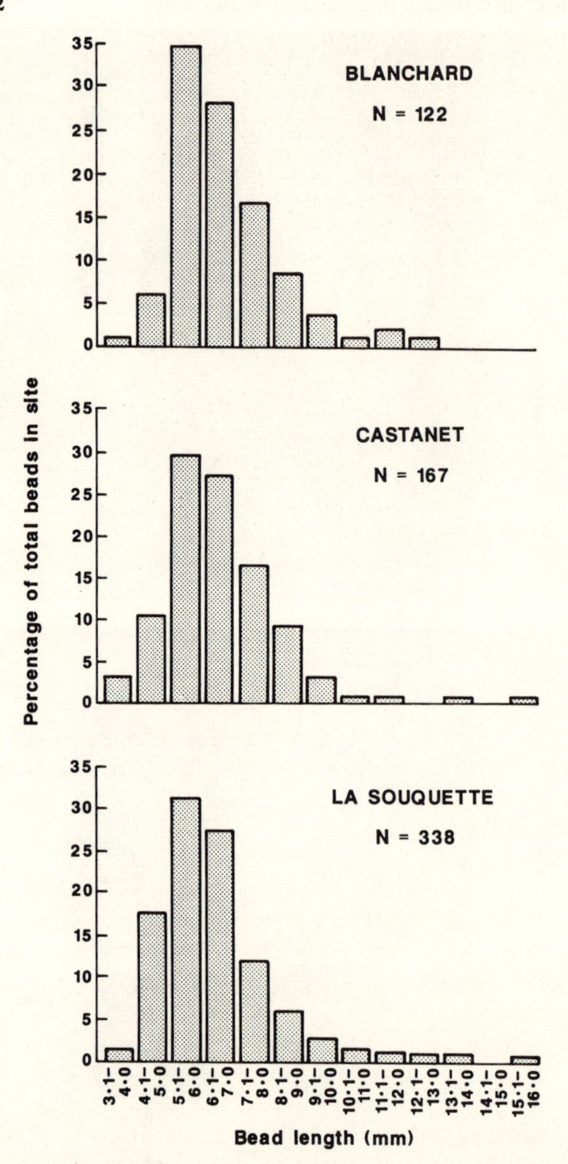

Figure 21.10. Frequency distribution for ivory, bone and steatite bead lengths from Castelmerle Aurignacian sites. Only unbroken beads are represented here.

about the varying contexts in which body ornaments were being produced, used and deposited in the archaeological record, which at 35-28 000 years ago (Mellars *et al.* 1987) is frustratingly coarse-grained. There is obviously considerable need to examine the differential distribution of body ornaments within a regional settlement/subsistence framework that also takes account of varying social contexts. There is also urgent need for greater chronological control to determine whether time is a significant axis of ornament fre-

Figure 21.11. An Aurignacian cervid tooth from La Souquette, illustrating the characteristic Aurignacian pattern of preparatory gouging to reduce thickness prior to piercing by pressure (Field Museum of Natural History, Chicago). Length = 2.7 cm.

quency variation. There is considerable reason to believe that the latter is the case. It seems that, while body ornaments are abundant in some Aurignacian I levels, they are rare or totally absent from levels representing succeeding stages of the Aurignacian.

10. *Despite overall similarities among sites in the representation of the different bead production stages, there are some interesting contrasts.* Within the Vallon de Castelmerle, where all beads and production stages were recovered by a single archaeologist, worries about the contribution of collector-bias to patterns of variation are alleviated. Inter-site comparison of the proportions of the various production stages (Fig. 21.12) produces interesting results. La Souquette stands out from Abri Blanchard and Abri Castanet in its abundance of stage IV (bifacially thinned and perforated ivory blanks). Stage IV was being produced almost exclusively at La Souquette (where there are three examples with unfinished holes in addition to 31 complete ones). It is possible that (1) La Souquette represents the primary locality at Castelmerle where stage IV's were being turned into finished beads; or (2) that a stockpile of stage IV's, ready for polishing and grinding, existed at La Souquette.

Beyond the confines of the Vallon de Castelmerle there is additional provocative evidence. In the lowermost Aurignacian I level (level B) at Les Rois (Charente) Mouton and Joffroy (1958: 87-88 and Figure 39) identified a cache of ivory objects composed of the following: one stage I baton, two stage II blanks, and two stage III thinned blanks, all indistinguishable both formally and metrically from the Castelmerle materials. These were found

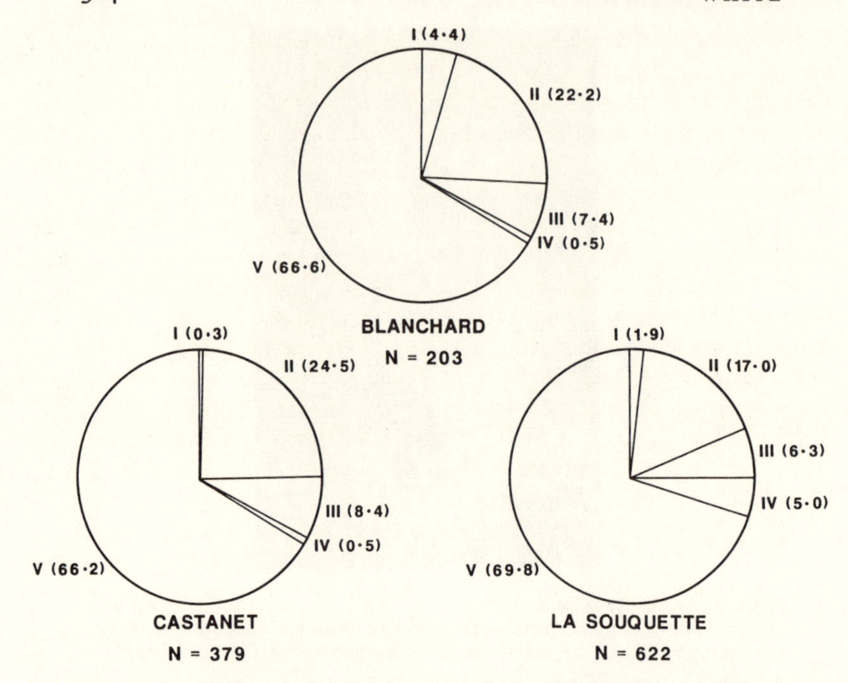

Figure 21.12. Comparison of the proportions of representation of the different bead production stages at Abri Blanchard, Abri Castanet and La Souquette. Finished bead category includes all beads, even if broken. Production stages (I-V) correspond with those illustrated in Figure 21.4.

together in an area of only a few square centimetres and were associated with two very fine quartzite cobbles. The stage III's had already been partially polished (perhaps using the quartzite cobbles as fine abrasives?) but had not yet been pierced. Although the sample is minute, the absence of stages IV and V is in contrast to the Castelmerle assemblages. Mouton and Joffroy believed that they had found the remains of a small sack containing a kit for manufacturing the stage III thinned blanks, which they mistakenly saw as the final goal of production and which they termed '*coins*' (wedges). Indeed, it seems plausible that this is a maintenance kit for the manufacture and replacement of clothing beads, which were produced as needed or as time allowed.

If the excavations in the Vallon de Castelmerle can be viewed as several locations within the same site, as Sonneville-Bordes (1960) has suggested, then we may be monitoring different activity areas. However, it is unclear how beads and production stages became similarly/differentially deposited in the archaeological record. Caching of unfinished and finished beads, storage or abandonment of articles of clothing, workshop loss and discard, and loss through breakage and attrition are all possibilities, perhaps acting in concert to produce the observed pattern. A lack of spatial provenience within sites prevents us from going further.

11. *Despite great similarities in lithic technology between the Aurignacian of*

Germany and that of southwest France, the differences in the technology and form of beads and pendants are striking. Inter-regional variation in the Aurignacian is only subtly expressed, if at all, in items of technology. For example, lithic and bone projectile technology is virtually identical between southwest France and southern Germany (see Sonneville- Bordes 1965). In contrast, there are critical differences in the domain of body ornamentation.

According to Joachim Hahn's (pers. comm.) recent analyses of body ornaments from the south German Aurignacian, basket-shaped ivory beads and the associated production sequence do not exist there. Rather, a very different production sequence for the manufacture of numerous small pendants or buttons predominates, as it does in Belgium (Otte 1974). Nevertheless, the pattern of appropriation of animal parts (e.g. canine teeth) and species is virtually identical between the two regions, suggesting a common ideational base for body ornamentation. In essence, expression of variation among regional groups seems especially pronounced in the form and arrangement of highly visible items of personal and social display. Such a pattern is clearly coherent with well established archaeological models of stylistic expression, such as that of Wobst (1977), which emphasize visibility as a major criterion for the choice of particular objects in which to invest stylistic information.

CONCLUSIONS

It is clear then, that focusing on body ornaments solely within a mortuary context has suppressed important questions and observations concerning early Upper Palaeolithic symbolic constructs, production techniques and standards, systems of social meaning and regional/inter-regional systems of exchange. I have attempted here to show the analytical potential of Aurignacian body ornaments, which will only be totally fulfilled with the discovery and careful excavation of additional ornament-bearing sites, where spatial provenience can be recorded.

It is also clear however, that Aurignacian body ornamentation explodes onto the scene in southwest France during the early Aurignacian (i.e. between 35 000 and 33 000 BP). It appears to have been complex conceptually, symbolically, technically and logistically right from the very beginning. Thus, it is totally unreasonable to see Aurignacian body ornamentation as somehow formative of or transitional to a Magdalenian florescence. This sudden, intrusive, and complex character of the earliest body ornamentation remains one of the greatest explanatory challenges in all of hominid evolution.

NOTES

1. Having been denied access to the unpublished bone and ivory objects from the Aurignacian and Châtelperronian at Arcy-sur- Cure, I am unable to draw comparisons between Aurignacian and Châtelperronian ornaments.
2. The term early Aurignacian as used here refers to assemblages with split-based points, virtually no burins (especially burins busqués) and few perforators. This Aurignacian I 'stage' has withstood the test of time, and

has been especially well described and dated at the Abri Pataud (Brooks 1979).

3. There may indeed be earlier body ornamentation in Central and Eastern Europe, where the Aurignacian appears to be older than in the west (Kozlowski 1982).

4. A cursory examination of the lithic assemblages from Abri Castanet and Abri Blanchard suggests extraordinarily high frequencies of exotic flint, especially *silex bergeracois*.

5. From the perspective of (1) the lack of delicacy in perforating technology and (2) the choice of fox, the attempt to pierce a fox canine, attributed to the Mousterian at La Quina (Martin 1909), looks perfectly Aurignacian and might well be derived from the overlaying Aurignacian layers.

ACKNOWLEDGEMENTS

Part of the research described here was supported by small grants from the National Endowment for the Humanities and the National Science Foundation. Helpful advice and comments for revision were provided by Heidi Knecht, Anne Pike Tay and Paul Mellars. Crucial support in the field was provided by Denis and Aguéda Vialou (in Paris), and Alain Roussot and Danielle Robin at Reignac. Alain Roussot also generously granted permission to use unpublished results of his La Souquette excavations. Necessary access to relevant collections was facilitated by Glen Cole (Field Museum, Chicago); Jane Troszak (Logan Museum, Beloit College, Beloit, Wisconsin); Jean Guichard, André Morala, Roger Rousset and Philippe Jugie, (Musée National de Préhistoire, Les Eyzies); Jean-Jacques Cleyet-Merle and Dominique Buisson (Musée des Antiquités Nationales, Saint Germain-en-Laye); and Henry de Lumley and Marie Pérpère (Musée de l'Homme). Monsieur René Castanet was most generous in giving me total access to the staggering collections from La Souquette made by his father Marcel, whose careful recovery of even the tiniest bead made this study possible. With full justification, this article is dedicated to the memory of Monsieur Marcel Castanet.

REFERENCES

Bader, O. 1970. The boys of Sungir. *Illustrated London News* (March) 7: 24-26.

Barandiaran, I. 1980. El Yacimiento de la Cueva de El Pendo. J. González Echegaray (ed.) *La Cueva del Pendo*. Madrid: Bibliotheca Praehistorica Hispana 18.

Belfer-Cohen, A. and Bar-Yosef, O. 1981. The Aurignacian at Hayonim Cave. *Paléorient* 7: 19-42.

Binford, S. 1968. A structural comparison of disposal of the dead in the Mousterian and the Upper Paleolithic. *Southwestern Journal of Anthropology* 24: 139-154.

Brooks, A. 1979. The Significance of Variability in Palaeolithic Assemblages: an Aurignacian Example from Southwestern France. Unpublished Ph.D. Thesis, Harvard University.

Capitan, L. and Bouyssonie, J. 1924. *Limeuil: Son Gisement à Gravures sur Pierres de l'Age du Renne*. Paris: Nourry.

Champagne, F. and Espitali;, R. 1981. *Le Piage, Site Préhistorique du Lot.* Paris: Mémoires de la Société Préhistorique Française 15.

Collie, G. 1928. *The Aurignacians and their Culture.* Beloit (Wis): Logan Museum Bulletin 1.

Conkey, M. 1978. Style and information in cultural evolution: towards a predictive model for the Paleolithic. In C. L. Redman, M. J. Berman, E. V. Curtin, W. T. Laughorne Jr., N. Versaggi and J. C. Wasner (eds) *Social Archeology: Beyond Subsistence and Dating.* New York: Academic Press: 61-85.

Conkey, M. 1983. On the origins of Paleolithic art: a review and some critical thoughts. In E. Trinkaus (ed.) *The Mousterian Legacy: Human Biocultural Change in the Upper Pleistocene.* Oxford: British Archaeological Reports International Series S164: 201-227.

Dauvois, M. 1977. Travail expérimental de l'ivoire: sculpture d'une statuette feminine. In H. Camps-Fabrer (ed.) *Méthodologie Appliquée à l'Industrie de l'Os Préhistorique.* Paris: Centre National de la Recherche Scientifique: 269-273.

Delage, F. 1938. L'Abri de la Souquette. *Bulletin de la Société Historique et Archéologique du Périgord* 65: 3-25.

Delluc, B. and Delluc, G. 1978. Les manifestations graphiques aurignaciennes sur support rocheux des environs des Eyzies (Dordogne). *Gallia Préhistoire* 21: 213-438.

Delluc, B. and Delluc, G. 1981. La dispersion des objets de l'abri Blanchard (Sergeac, Dordogne). *Bulletin de la Société d'Etudes et de Recherches Préhistoriques des Eyzies*, 30:1-19.

Delpech, F. 1975. *Les Faunes du Paléolithique Supérieur dans le Sud-Ouest de la France.* Unpublished Doctoral Thesis, University of Bordeaux.

Delporte, H. 1979. *L'Image de la Femme dans l'Art Préhistorique.* Paris: Picard.

Delporte, H. and Mazière, G. 1977. L'aurignacien de la Ferrassie: observations préliminaires à la suite des fouilles récentes. *Bulletin de la Société Préhistorique Française* 74: 343-361.

Didon, L. 1911. L'Abri Blanchard des Roches (commune de Sergeac). Gisement aurignacien moyen. *Bulletin de la Société Historique et Archéologique du Périgord* 87: 246-261, 321-345.

Didon, L. 1912. Faits nouveaux constatés dans une station aurignacienne, L'Abri Blanchard des Roches près de Sergeac. *L'Anthropologie* 23: 603.

Freeman, L.G. 1978. Mousterian worked bone from Cueva Morín (Santander, Spain): a preliminary description. In L. G. Freeman (ed.) *Views of the Past.* The Hague: Mouton: 29-51.

González Echegaray, J. and Freeman, L.G. 1971. *Cueva Morín: Excavaciones 1966-1968.* Santander: Patronato de las Cuevas Prehistóricas.

González Echegaray, J. and Freeman, L.G. 1973. *Cueva Morín: Excavaciones 1969.* Santander: Patronato de las Cuevas Prehistóricas.

Hahn, J. 1971. La statuette masculine de la grotte du Hohlensteinstadel (Wurtemberg). *L'Anthropologie* 75: 233-244.

Hahn, J. 1972. Aurignacian signs, pendants and art objects in Central and Eastern Europe. *World Archaeology* 3: 252-266.

Hahn, J. 1975. *Der Vogelherd, eine Wohnhohle der Altsteinzeit im Lonetal bei Stetten, Gemeinde Niederstotzingen Ldkr.* Heidenheim: Kulturdenkmale in Baden-Wurttemberg, kleine Fuhrer 15.

Hahn, J. 1983. Eiszeitliche jager zwischen 35000 und 15000 vor heute. In H.-J. Muller-Beck (ed.) *Urgeschichte in Baden-Wurttemberg.* Stuttgart: Theiss: 273-330.

Hahn, J. 1986. *Kraft und Aggression.* Verlag Archaeologica Venatoria 7. Tübingen: Institut für Urgeschichte.

Hahn, J., Koenigswald, W. von, Wagner, E. and Willie, W. 1977. Das Geissenklosterle bei Blaubeuren, Alb-Donau-Kreis, eine altsteinzeitliche Hohlenstation der Mittleren-Alb. *Fundberichte aus Baden-Wurttemberg* 3: 14-37.

Hauser, O. 1911. *Le Périgord Préhistorique*. Le Bugue: Réjou.

Julien, M. 1981. *Les Harpons Magdaléniennes*. 17th supplement to *Gallia Préhistoire*. Paris: Centre National de la Recherche Scientifique.

Kozlowski, J. 1982. *Excavation in the Bacho Kiro Cave (Bulgaria). Final Report*. Warsaw: Panstwowe Wydawnictwo Naukowe.

Lartet, L. 1869. Une sépulture des troglodytes du Périgord à Cro-Magnon. *Matériaux pour Servir à l'Histoire Primitive et Naturelle de l'Homme* 5: 97-105.

Leroi-Gourhan, A. 1961. Les fouilles d'Arcy-sur-Cure. *Gallia Préhistoire* 4: 3-16.

Leroi-Gourhan, A. 1965. *Préhistoire de l'Art Occidental*. Paris: Mazenod.

Léroy-Prost, C. 1975. L'industrie osseuse de l'Aurignacien. Essai regional de classification: Poitou, Charente, Périgord. *Gallia Préhistoire* 18: 65-156.

Marshack, A. 1972. *The Roots of Civilization*. London: Weidenfeld and Nicolson.

Martin, H. 1909. *Recherches sur l'Evolution du Mousterien dans le Gisement de la Quina (Charente):* Vol. 2. Paris: Schleicher.

Mazière, G. and Raynal, J.-P. 1983. La grotte du Loup (Cosnac, Corrèze), nouveau gisement stratifié à Castelperronien et Aurignacien. *Comptes-Rendus de l'Académie des Sciences de Paris* 296: 1611-1614.

Mellars, P. A. 1973. The character of the Middle-Upper Palaeolithic transition in southwest France. In C. Renfrew (ed.) *The Explanation of Culture Change*. London: Duckworth: 255-276.

Mellars, P. A. 1982. On the Middle/Upper Palaeolithic transition: a reply to White. *Current Anthropology* 23: 238-240.

Mellars, P. A., Bricker, H. M., Gowlett, J. A. J. and Hedges, R. E. M. 1987. Radiocarbon accelerator dating of French Upper Palaeolithic sites. *Current Anthropology* 128-133.

Mouton, P. and Joffroy, R. 1958. *Le Gisement Aurignacien des Rois à Mouthiers (Charente)*. 9th Supplement to *Gallia Préhistoire*. Paris: Centre National de la Recherche Scientifique.

Movius, H. L. 1975. *Excavation of the Abri Pataud, Les Eyzies (Dordogne)*. American School of Prehistoric Research Bulletin 30. Cambridge (Mass).

Movius, H. R. 1977. *Excavation of the Abri Pataud, Les Eyzies (Dordogne): Stratigraphy*. American School of Prehistoric Research Bulletin 31. Cambridge (Mass).

Mussi, M. 1985. On the chronology of the burials found in the Grimaldi Caves. *Antropologia Contemporanea* 8: 1-9.

Oakley, K. 1969. *Frameworks for Dating Fossil Man*. London: Weidenfeld and Nicolson.

Otte, M. 1974. Observations sur le débitage et le façonnage de l'ivoire dans l'Aurignacien en Belgique. In H. Camps-Fabrer (ed.) *Premier Colloque International sur l'Industrie de l'Os dans la Préhistoire*. Paris: Centre National de la Recherche Scientifique: 93-96.

Otte, M. 1979. *Le Paléolithique Supérieur Ancien en Belgique*. Monographies d'Archéologie Nationale 5. Brussels: Musées Royaux d'Art et d'Histoire.

Pales, L. 1969. *Les Gravures de La Marche I: Felins et Ours*. Bordeaux: Delmas.

Peyrony, D. 1935. Le gisement Castanet, Vallon de Castelmerle, Commune de Sergeac (Dordogne): Aurignacien I et II. *Bulletin de la*

Société Préhistorique Française 32: 418-443.
Peyrony, D. 1946. Le gisement préhistorique de l'Abri Cellier, au Ruth, commune de Tursac (Dordogne). *Gallia Préhistoire* 4: 294-301.
Reverdit, A. 1882. *Station des Roches, Commune de Sergeac*. Toulouse: Durand, Pillous and Lagarde.
Riek, G. 1934. *Die Eiszeitjugerstation am Vogelherd im Lonetal*. Tübingen: Die Kulturen.
Rigaud, J.-P. 1982. *Le Paléoithique en Périgord: les données du sud-ouest Sarladais et leurs implications*. Unpublished Doctoral Thesis, University of Bordeaux I.
Roussot, A. 1982. *Abri de la Souquette, commune de Sergeac (Dordogne): Rapport de Fouilles. Recapitulatif 1980, 1981, 1982*. Unpublished Report to the Direction des Antiquités Préhistoriques d'Aquitaine, Bordeaux.
Sackett, J. 1966. Quantitative analysis of Upper Paleolithic stone tools. In J. D. Clark and F. C. Howell (eds) *Recent Studies in Paleoanthropology. (American Anthropologist* Vol. 68 No. 2, Pt. 2): 356-394.
Soffer, O. 1985. *The Upper Paleolithic of the Central Russian Plain*. Orlando: Academic Press.
Sonneville-Bordes, D. de. 1960. *Le Paléolithique Supérieur en Périgord*. Bordeaux: Delmas.
Sonneville-Bordes, D. de. 1965. Observations statistiques sur l'Aurignacien du Vogelherd, Lonetal, Wurtemberg, fouilles G. Riek. *Fundberichte aus Schwaben* 17: 69-73.
Strathern, A. and Strathern, M. 1971. *Self-decoration in Mount Hagen*. London: Duckworth.
Straus, L. 1983. From Mousterian to Magdalenian: cultural evolution viewed from Vasco-Cantabrian Spain. In E. Trinkaus (ed.) *The Mousterian Legacy: Human Biocultural Change in the Upper Pleistocene*. Oxford: British Archaeological Reports International Series S164: 73-111.
Szombathy, J. 1925. Die diluvialen Menschenreste aus der Furst-Johanns-Hohle bei Lautsch in Mahren. *Die Eiszeit* 2: 1-34.
Taborin, Y. 1977. Quelques objets de parure. Etude technologique: les percements de incisives de bovinés et des canines de renards. In H. Camps-Fabrer (ed.) *Methodologie Appliquée à l'Industrie de l'Os Préhistorique*. Paris: Centre National de la Recherche Scientifique: 303-310.
Taborin, Y. 1985. Les origines des coquillages paléolithiques en France. In M. Otte (ed.) *La Signification Culturelle des Industries Lithiques*. Oxford: British Archaeological Reports International Series S239: 278-301.
Trinkaus, E. (ed.) 1983. *The Mousterian Legacy: Human Biocultural Change in the Upper Pleistocene*. Oxford: British Archaeological Reports International Series S164.
Trinkaus, E (ed.) 1989. *Patterns and Processes in Later Pleistocene Human Emergence*. Cambridge: Cambridge University Press. In Press.
Vezian, J. and Vezian, J. 1970. Les gisements de la grotte de Saint-Jean-de-Verges (Ariège). *Bulletin de la Société Préhistorique de l'Ariège* 25: 29-65.
Weiner, A. 1987. *The Trobrianders of Papua New Guinea*. New York: Holt, Rinehart and Winston.
Wetzel, R. and Bosinski, G. 1969. *Die Bocksteinschmied im Lonetal*. Veroffentlichungen des Staatlichen Amtes für Denkmalpflege Stuttgart, Reihe A.
White, R. 1982a. Rethinking the Middle/Upper Paleolithic transition.

Current Anthropology 23: 169-192.

White, R. 1982b. The manipulation of burins in incision and notation. *Canadian Journal of Anthropology* 2: 129-135.

White, R. 1985. Thoughts on social relationships and language in hominid evolution. *Journal of Social and Personal Relationships* 2: 95-115.

White, R. 1986a. Rediscovering French Ice Age art. *Nature* 320: 683-684.

White, R. 1986b. *Dark Caves, Bright Visions: Life in Ice Age Europe*. New York: American Museum of Natural History.

White, R. 1988. Objets magdaléniens provenant de l'Abri du Soucy (Dordogne): La collection H.-M. Ami au Royal Ontario Museum, Toronto, Canada. *L'Anthropologie*. In Press.

White, R. 1989. Toward a contextual understanding of the earliest body ornaments. In E. Trinkaus (ed.) *Patterns and Processes in Later Pleistocene Human Emergence*. Cambridge: Cambridge University Press. In Press.

Wobst, M. 1977. Stylistic behavior and information exchange. In C. Cleland (ed.) *For the Director: Essays in Honor of James B. Griffin*. Ann Arbor: Anthropology Papers of the University of Michigan 61: 317-342.

22. The Origins of Some Aspects of Human Language and Cognition

PHILIP LIEBERMAN

Human speech, which differentiates us from other animals, depends on the species-specific morphology of the basicranium and mandible. These structures have a functional value and may be better indices of the evolution of modern human beings than other aspects of the skull. Comparative studies of fossils and living primates show that the morphology of the basicranium and mandible are also indices for the presence of neural mechanisms that are necessary for the production and perception of human speech. Comparative studies of skeletal morphology allow us to determine whether particular fossil hominids had human speech. Insofar as rule-governed syntax and cognition may also derive from neural mechanisms that first evolved to facilitate the complex articulatory manoeuvres that underlie speech, we can infer their presence from this same skeletal morphology. Therefore, we can make inferences concerning the linguistic and cognitive abilities of particular fossil hominids by studying the skeletal structures of the basicranium and mandible that relate to the supralaryngeal vocal tract and some other aspects of speech production. The studies described here suggest that classic Neanderthal hominids appear to be deficient with respect to their linguistic and cognitive ability. At minimum their speech communications would be nasalized and more susceptible to perceptual errors: they probably communicated vocally at extremely slow rates and were unable to comprehend complex sentences. They may also have been deficient in cognitive tasks that involve rule-governed logic. In contrast, fossils like Petralona, Broken Hill and Skhūl V had some aspects of modern human speech.

INTRODUCTION

It has become evident that human language involves a number of innate and genetically transmitted anatomical and neural mechanisms. Some of these, including the ability to acquire and to use words, may involve the elaboration of a neural substrate that is present in non-human animals. Other aspects of human language, particularly those relating to speech and rule-governed syntax, are however found only in present-day humans. We can derive some inferences on the origins of modern humans by means of comparative studies of the speech-producing anatomy, brains and behaviour of living primates and the anatomy of fossil hominids. These studies are consistent with the hypothesis that the biological substrates for human speech, syntax and rule- governed cognitive ability are linked (Lieberman 1984), and evolved in some hominid lineages over the past 500 000 years, and probably reached their present state in the last 100 000 years. The objective of this paper is to suggest how we may be able to trace the evolution of human speech, language and cognition, and by doing so trace the evolution of modern humans.

HUMAN SPEECH

Until the 1960s it was not realised that human speech is itself an important component of human linguistic ability. Linguists thought that any set of arbitrary sounds would suffice to transmit words. Research that was initially directed at making a machine that would read books to blind people, demonstrated that the sounds of speech had a special status. Speech allows us to transmit vocally phonetic 'segments' (which are approximated by the letters of the alphabet) at an extremely rapid rate, up to 25 per second. It is, in contrast, impossible to identify non-speech sounds at rates that exceed 7-9 items per second (Miller 1956). A short sentence, such as this one, contains about 50 speech sounds. The 50 speech sounds or phonetic segments can be uttered in two seconds, and human listeners have no particular difficulty in understanding what has been spoken. If this sentence were transmitted at the non-speech rate, it would take so long that a listener might well forget the beginning of the sentence before hearing its end. The engineers working on the reading machine discovered that only speech sounds would allow people to understand the meaning of even moderately complex sentences.

The high transmission rate of human speech is thus an integral part of human linguistic ability, as it allows complex thoughts to be transmitted within the constraints of short-term memory. Although sign language can also achieve a high transmission rate, the signer's hands cannot be used for other tasks. Nor can viewers see the signer's hands except under restricted conditions. Visual hand signs still function as part of the linguistic code (McNeill 1985), but the primary linguistic channel is vocal. Vocal language represents the continuation of the evolutionary trend towards freeing the hands for carrying and tool use that started with upright bipedal hominid locomotion. Human speech also has some lesser selective advantages; the sounds of human speech are less susceptible to perceptual confusion than the sounds that other primates can make. These perceptual factors, which we will discuss below, may have had a primary adaptive role in the initial stages of the evolution of human speech.

The supralaryngeal vocal tract and larynx.

Research that started in the time of Johannes Müller (1849) shows that the key biological mechanisms that are necessary for human speech are the supralaryngeal vocal tract and 'matching' neural mechanisms that decode the acoustic cues for linguistic information in the speech signal, and govern the complex articulatory manoeuvres that underlie speech. The high transmission rate of human speech is achieved through the generation of 'formant-frequency' patterns and rapid temporal and spectral cues by the species-specific human supralaryngeal airway. Formant frequencies are simply the frequencies at which maximum acoustic energy will get through the supralaryngeal airway, which acts in a manner similar to the way a pipe organ lets maximum acoustic energy through at certain frequencies. Both the pipe organ and the supralaryngeal airway act as 'filters', letting relatively more acoustic energy through at these 'formant' frequencies. During the production of speech we continually change the shape and make small adjustments in the length of the supralaryngeal airway, thereby generating a changing formant frequency pattern. The sounds of human speech differ with respect to their formant frequency patterns, as well as with respect to other temporal factors and the acoustic source filtered by the supralaryngeal airway.

The larynx acts as a source of acoustic energy, either generating noise from turbulent air flow or periodic phonation as the vocal cords rapidly open and close. A sequence of puffs of air moves up from the larynx into the supralaryngeal airway as the vocal cords open and close. The fundamental frequency of phonation, which reflects the rate at which the puffs of air occur, determines the 'pitch' of a speaker's voice. The fundamental frequency pattern can convey linguistic information like the pitch 'tones' that differentiate words in languages like Chinese, independent of the formant frequency pattern. The rapid change of formant frequencies is, however, the key to the speed of human speech.

Figure 22.1 illustrates the filtering effect of the supralaryngeal vocal tract. The plot in the centre shows the spectrum of the acoustic signal produced by the larynx. This is roughly similar to the sound that you would hear (a raspy buzz) if you held the reed of a wood-wind instrument in your hand and blew through it. Acoustic energy is present at the fundamental frequency of phonation, the lowest line in the graph at 500 Hz, and its harmonics. As Figure 22.1 shows, the energy in the spectrum of the sound produced by the larynx generally falls off with frequency. The upper plot in Figure 22.1 shows the filter function of the supralaryngeal vocal tract for the vowel [i], the vowel of the word *bee*. The formant frequencies, F1, F2 and F3, are the frequencies at which maximum acoustic energy will get through the supralaryngeal airway. The lower plot shows the net effect of the filter on the glottal source. The frequencies of the formants are marked by the circled 'X's'.

Neural mechanisms for speech perception

Formant frequency 'extraction'. Note that there is no energy present in the output signal at the exact frequencies of the formants, the X's. Almost 200 years of research demonstrates that human beings are equipped with neural devices that, in effect, calculate these formant frequencies from the speech signal. We do this even when very little acoustic information is present, as

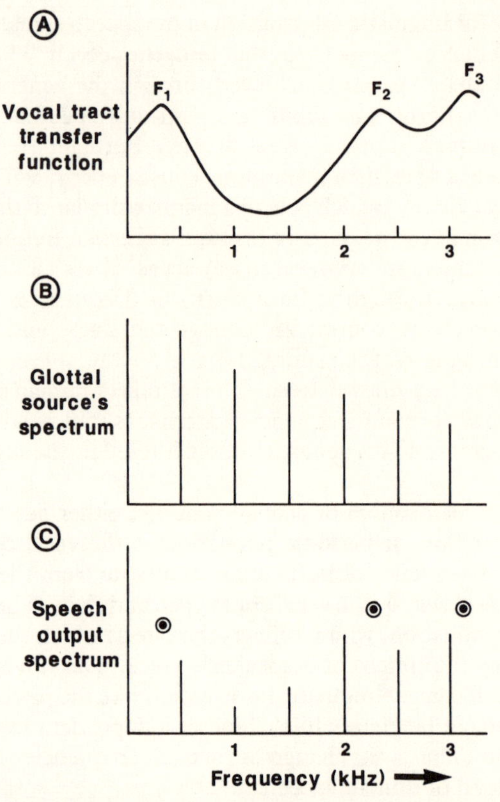

Figure 22.1. (A) The filter function of the supralaryngeal vocal tract for the vowel [i]. (B) The frequency spectrum of a possible glottal source generated by the larynx. (C) The net effect of the filter on the glottal source. The frequencies of the formants are marked by the circled crosses.

is the case on a telephone. We appear to be equipped with a neural computational device that 'knows' the filtering characteristics of the human supralaryngeal vocal tract. The almost perfect identification of the vowels of people who have high fundamental frequencies of phonation (Ryalls and Lieberman 1982) probably follows from the fact that human listeners have a complex neural formant frequency 'detector' that calculates the formant frequencies on the basis of an internal representation of the physiology of speech production. Computer algorithms that go through this process are able to calculate the formant frequencies of un-nasalized sounds with reasonable accuracy. One important point to note in connection with formant frequency detection is that it is very difficult to calculate the formant frequencies of vowels when they are nasalized. When the nose is connected to the rest of the supralaryngeal vocal tract, it introduces nasal formants and 'zeros' which obscure the formant frequency patterns that differentiate vowels. Human listeners have similar problems: nasalized vowels are misidentified 30 to 50 per cent more often than similar non-nasal vowels (Bond 1976). In other words, nasalized speech is inherently less perceptible.

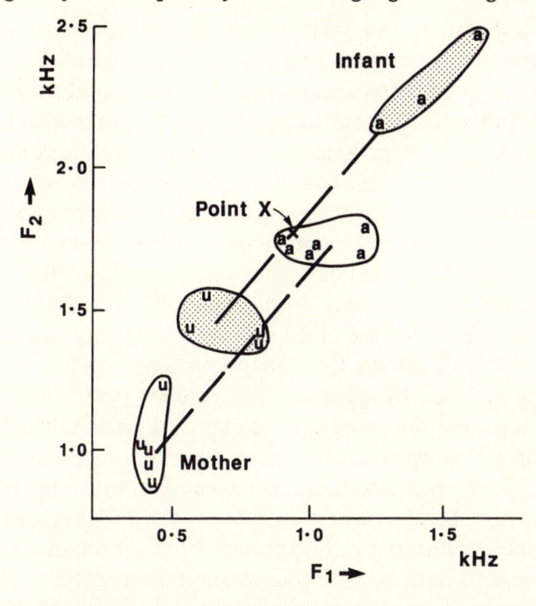

Figure 22.2. The first two formant frequencies (F1/F2) of the vowel sounds [u] and [a] that a mother and a 3 month-old infant made as they engaged in verbal play. (After Lieberman 1984).

Vocal-tract-normalization. Human listeners do some other remarkable feats as they interpret the linguistic significance of different formant frequency patterns. We have to estimate the probable length of a speaker's supralaryngeal airway in order to assign a particular formant frequency pattern to a particular speech sound. The length of the human supralaryngeal airway differs greatly. Those of young children are half the length of those of adults. There is overlap between the formant frequency patterns that convey different speech sounds because of this variation. For example, the word *bit* spoken by a large adult male speaker can have the same formant frequency pattern as the word *bet* produced by a smaller person. For a particular individual, the formant frequencies of *bit* are higher than those of *bet*. The longer supralaryngeal airway of the adult male, however, produces lower frequency formant frequencies for his *bit* than that of the smaller person, and his *bit* can match the smaller person's *bet*. When we listen to speech we are not aware of these differences; at some level of neural processing we 'normalize' and take these effects into account.

The imitative behaviour of infants by the age of 3 months demonstrates that this is the case. In Figure 22.2 we have plotted the first two formant frequencies of the vowel sounds that a mother and a 3 month-old child made as they engaged in verbal play. The mother starts with an [a] that changes to [u]. The child imitated her vowel quality, overlapping with his mother's vocalization. The first two formant frequencies of a vowel essentially determine its phonetic identity (Fant 1960) so these plots can be viewed as indicators of phonetic, i. e. linguistic, imitation by the child.

The child cannot imitate the absolute formant frequencies of his mother because his superalaryngeal airway is much shorter than hers. The principles of physical acoustics and the constraints of anatomy absolutely preclude the child's being able to imitate the actual formant frequencies of his mother. What the child does is to produce a frequency-normalized version of his mother's speech. He produces a sound whose formant frequencies are proportionate to the ratio between the length of his supralaryngeal airway and hers. Note that the child does not produce a vowel having the formant frequencies of point 'X' in response to his mother's [a]. This sound would be a closer match to her actual formant frequencies. The child instead produces the frequency-scaled child [a] when imitating the mother's [a]. Adult human listeners always frequency-normalize speech signals in this manner (Ladefoged and Broadbent 1957; Nearey 1978). We don't realise that we are doing this; the process is similar to the size normalization that occurs in vision where we will recognize someone's face independently of the size of the image projected on our retina. We will return to the issue of vocal-tract-normalization when we discuss the fossil record; the vowel [i] which can be produced by the species-specific human supralaryngeal vocal tract is an optimum cue for vocal-tract normalization. Other sounds can be used to estimate the length of a speaker's supralaryngeal vocal tract, but [i] is best. It is impossible to produce an [i] with a pongid or Classic Neanderthal supralaryngeal tract. Hence we can make some reasonable inferences about their speech. We will return to this topic later.

Phonetic feature detectors. Human beings when listening to speech also integrate an ensemble of acoustic cues and contextual constraints, such as rate, phonetic environment, etc., that are related by the physiology of speech production. They assign patterns of formant frequencies and short-term spectral cues into discrete phonetic categories in a manner that is consistent with the presence of neural 'detector' mechanisms. These neural detectors appear to be 'matched' or tuned to respond to the particular acoustic signals that the human speech-producing anatomy can produce. It is impossible to even survey the extensive research literature pertaining to the perception of human speech; at least one hundred independent studies are consistent with the premise that we are equipped with genetically transmitted neural devices that facilitate the perception of the particular sounds that occur in human speech. Many of these studies are noted in Lieberman (1984) where some of the biological bases of human speech are discussed in detail.

Neural mechanisms for speech production

The production of human speech likewise involves species-specific neural mechanisms. Broca first identified the area of the brain in which the motor programs that control the production of human speech may be stored or processed. The articulatory manoeuvres that underlie the production of human speech are among the most complex that human beings attain. Until the age of 10 years, normal children are not up to adult criteria for even basic manoeuvres like the lip positions that are necessary for different vowels (Watkins and Fromm 1984). Human speakers are able to execute complex voluntary articulatory manoeuvres involving the tongue, lips, velum, larynx and lungs that are directed towards linguistic goals, e.g. producing a particular formant frequency pattern. Lesions in Broca's area result in deficits in speech production. The victim is unable to produce particular speech sounds though he can move individual articulators, or use his tongue and lips to swallow food.

Pongids who lack Broca's area likewise cannot be taught to control their supralaryngeal airways to produce any human speech sounds. Chimpanzees are not able to produce any reasonable approximations to human speech. Although their non-human, pongid airway is inherently unable to produce all of the sounds of human speech, it could produce a subset of human speech sounds. Computer modelling studies show that chimpanzees have the anatomy that could produce nasalized vowels like [I], [ae], [e] and consonants like [t], [d], [b], [p]. However, all attempts over the past 300 years to teach them to produce any approximations to human speech have failed completely. It is apparent that non-human primates who lack Broca's area are unable to produce the voluntarily muscular manoeuvres that underlie human speech.

Pongids furthermore appear to have difficulty in the *intentional*, voluntary control of their vocal signals. Goodall (1986), for example, notes that chimpanzees are not able to suppress food-barks, even when this is in their best interest. Chimpanzee vocalizations seem to be tied to oro-facial gestural patterns that determine their acoustic quality. The regulation of their vocalizations appears to derive from the limbic system (MacLean 1985) rather than the cortical areas that control speech in modern human beings. The acoustic quality of the call, e.g., formant-frequency lowering because of lip rounding, derives from the limbically controlled oro-facial expression. It is probable that the elaboration of the precentral cortex in the course of hominid evolution (Deacon 1985) ultimately yields Broca's area, which appears to facilitate the intentional control of speech signals.

THE HUMAN SUPRALARYNGEAL VOCAL TRACT AND THE FOSSIL RECORD

Figure 22.3 shows the typical non-human airway in which the tongue is positioned entirely within the oral cavity, where it forms the lower margin. The midsaggital view shows the airway as it would appear if the animal were sectioned on its midline from front to back. The position of the long, relatively thin tongue reflects the high position of the larynx. The larynx moves up into the nasopharynx during respiration, providing a pathway for air from the nose to the lungs that is isolated from any liquid that may be in the animal's mouth. Non-human mammals and human infants, who have this same morphology until an age of 3 months, can simultaneously breathe and drink. The ingested fluid moves to either size of the raised larynx, which resembles a raised periscope protruding through the oral cavity, connecting the lungs with the nose. During ontogenetic development the human palate moves back with respect to the bottom, i. e., the base of the skull. The base of the human adult skull is restructured in a

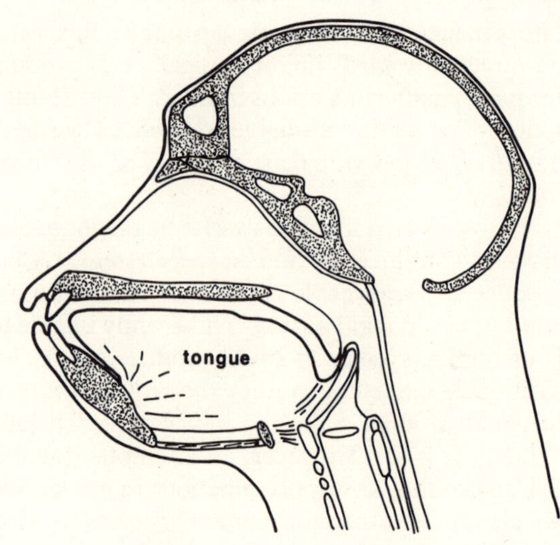

Figure 22.3. Midsagittal view of the typical non-human upper airway. The tongue is positioned entirely within the oral cavity. (Adapted from Laitman and Heimbuch 1982).

manner unlike that of all other mammals to achieve the adult human supralaryngeal airway (Laitman and Crelin 1976).

Figure 22.4 shows the adult human configuration. The larynx has lowered into the neck. The tongue's contour in the midsagittal plane is round, and it forms the anterior margin of the pharynx as well as the lower margin of the oral cavity. Air, liquids and solid food make use of the common pharyngeal pathway. Humans thus are more liable than other terrestrial animals to choke when they eat, because food can fall into the larynx, obstructing the pathway into the lungs. The peculiar deficiencies of the human supralaryngeal vocal tract for swallowing have long been noted. Darwin (1859: 191) for example noted 'the strange fact that every particle of food and drink which we swallow has to pass over the orifice of the trachea, with some risk of falling into the lungs'.

The adult human configuration is also less efficient for chewing because the length of the palate and of the mandible have been reduced compared to those of non-human primates and archaic hominids. The reduced length of the palate and mandible also crowd our teeth, presenting the possibility of infection due to impaction – a potentially fatal condition until the advent of modern medicine. These vegetative deficiencies are offset, however, by the increased phonetic range of the human supralaryngeal airway.

Phonetic advantages of human speech

1. Non-nasal sounds: The minimum selective advantage of this increased phonetic range first involves our ability to produce sounds that are *not* nasalized. The sharp bend in the human supralaryngeal vocal tract and the short span sealed by the velum make it possible to seal the nose off from the rest of the airway when we speak. The velum in the non-human airway

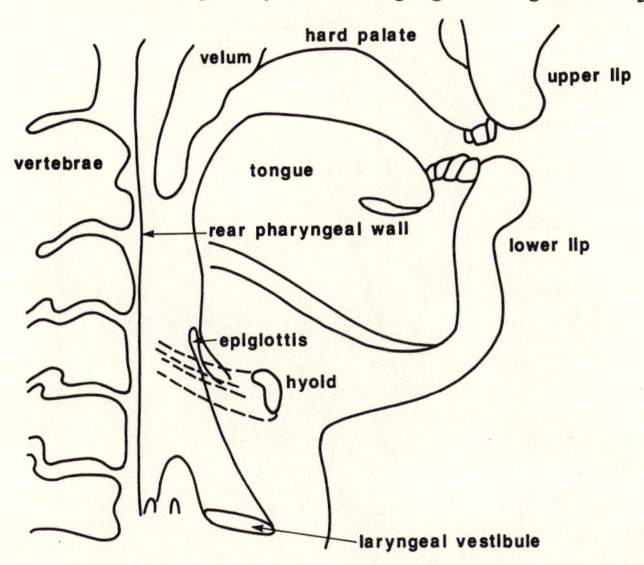

Figure 22.4. Midsagittal view of the adult human upper airway. The larynx has lowered into the neck. The tongue's contour in the midsagittal plane is round, and it forms the anterior margin of the pharynx as well as the lower margin of the oral cavity.

is adapted to forming a seal with the epiglottis, rather than to closing off the nasal cavity from the rest of the airway. Nasalized sounds occur when we don't seal off the nose; as noted earlier, it is harder to determine their formant frequency patterns. They are misidentified by human listeners 30 to 50 per cent more often than are non-nasalized sounds (Bond 1976). This obviously interferes with the effectiveness of vocal communication and human languages tend to avoid using nasal sounds (Greenberg 1963).

2. Quantal vowels: There are further phonetic advantages that derive from the morphology of the human supralaryngeal airway. The round human tongue moving in the right-angle space defined by the palate and spinal column can generate formant frequency patterns like those that define 'quantal' sounds (Stevens 1972) like the vowels [i], [u], and [a] (the vowels of the words *meet, boo,* and *mama*) and consonants like [k] and [g]. These sounds have formant frequency patterns that make them, and sounds like the consonants [b], [p], [d], [t], better suited for vocal communication than other sounds. They occur more often in different human languages (Greenberg 1963); moreover, children (Olmsted 1971) and adults (Peterson and Barney 1952) identify these sounds better than other speech sounds. The error rate for misidentification of the vowel [i] is particularly low and it can serve as an optimum cue in the perceptual process of vocal tract 'normalization' which we discussed above.

Is human speech a selective force for the human vocal tract?

The effects of malformations of the human supralaryngeal vocal tract have

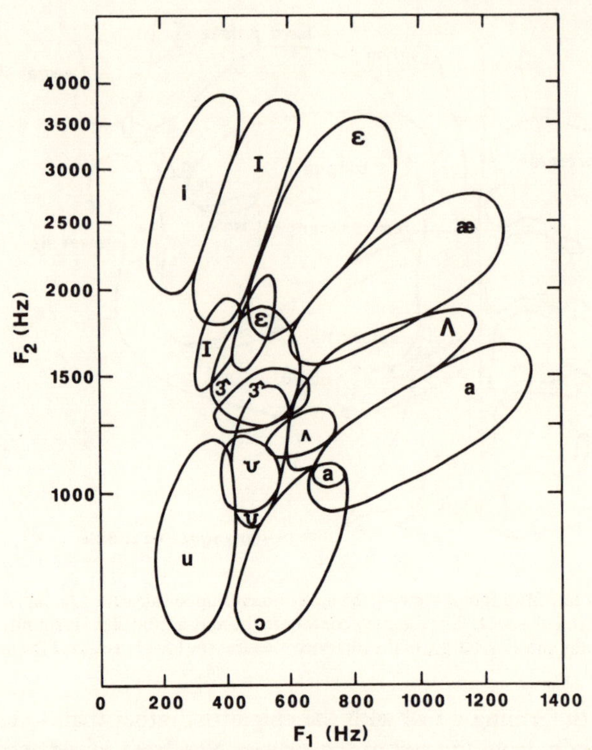

Figure 22.5. A computer modelling of an Apert's subject. The loops of the Peterson and Barney (1952) study of normal subjects are shown on the F1 versus F2 formant frequency plot. The vowel symbols show the formant frequencies of the subject's vowels. (After Landahl and Gould 1986).

long been noted. The effects of a cleft palate, which results in nasalization and unintelligible speech, have been studied in great detail (Folkins 1985). The studies of anomalies of the vocal tract can constitute 'experiments in nature' which allow us to assess the selective value of different aspects of vocal tract morphology. Apert and Cruzon's syndromes result in anomalous supralaryngeal vocal tracts. The palate in these pathologies is positioned in a posterior position relative to its normal position and the pharynx is constricted. In the course of ontogenetic development the palate continues to move back along the sphenoid bone on the base of the skull, past its 'normal' position. Landahl and Gould (1986) use acoustic analysis, psychoacoustic tests, and computer modelling based on radiographic data to show that the phonetic output of subjects having these syndromes is limited by their supralaryngeal vocal tracts. The Aperts and Cruzons syndrome subjects attempt to produce normal vowels but are unable to produce the normal range of formant frequency values. Psychoacoustic tests of their productions yield a 30 per cent error rate for the identification of their vowels. Acoustic analysis shows that they are unable to produce the vowels [i] and [u].

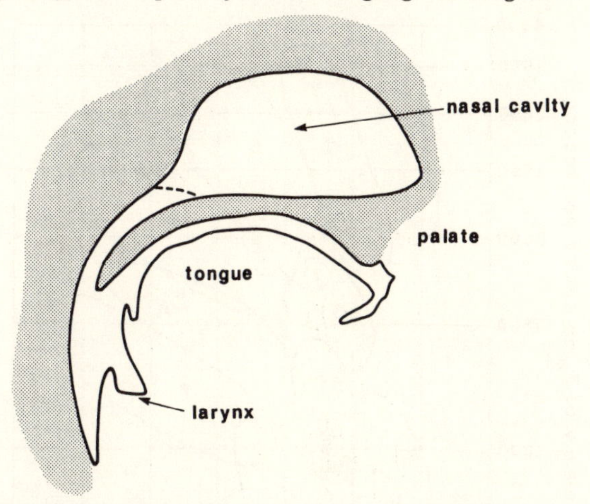

Figure 22.6. Crelin's reconstruction of the upper airway of the La Chapelle-aux-Saints fossil.

Figure 22.5 shows the results of a computer modelling of an Apert's subject. The loops of the Peterson and Barney (1952) study of normal subjects are shown on the F1 versus F2 formant frequency plot. The vowel symbols clustered towards the centre of these loops represent the formant frequency patterns that the Apert's supralaryngeal vocal tract can produce when the experimenters attempted to perturb it towards the best approximation of a normal vocal tract's configuration for each vowel. Note the absence of [i]'s or [u]'s. The computer-modelled plot in which the Apert's vocal tract was pushed to its phonetic limits is virtually identical to the formant frequencies that the Apert's subject actually produced. The ontogenetic restructuring of the human supralaryngeal vocal tract thus appears to be driven by its phonetic function.

These data are consistent with the hypothesis that in the course of human evolution, the final selective force on the restructuring of the human supralaryngeal vocal tract was its phonetic output. They show that the 'normal' configuration of the human supralaryngeal vocal tract yields the maximum formant frequency range. Configurations in which the palate is anterior (Lieberman *et al.* 1972) or posterior (Landahl and Gould 1986) yield a reduced range of formant frequency patterns. The anomalous Apert's and Cruzon's positions furthermore do not appear to be detrimental to upright bipedal posture and locomotion.

The evolution of the human supralaryngeal vocal tract

Classic Neanderthals. The first attempt to chart the evolution of the human supralaryngeal vocal tract must be credited to Victor Negus and Arthur Keith. Negus (1949), in *The Comparative Anatomy and Physiology of the Larynx*, demonstrated that the supralaryngeal vocal tracts of all other mammals differ from that of adult humans. Negus presents sketches of the reconstructed supralaryngeal airways of the Broken Hill fossil which he

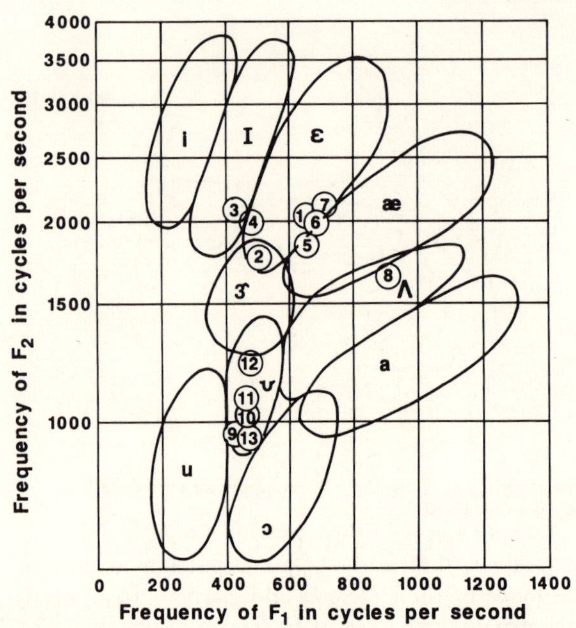

Figure 22.7. Computer modelling of the vowels that can be produced by the reconstruction of the upper airway of the La Chapelle-aux-Saints fossil. The loops of the Peterson and Barney (1952) study of normal human subjects are shown on the F1 versus F2 formant frequency plot. The vowel symbols (1-13) show the formant frequencies of the Neanderthal vowels.

and Keith concluded was similar to that of a modern human, and a Neanderthal reconstruction which lacks a human tongue and pharynx. The details of the reconstruction are unfortunately not specified. It is significant, however, that Negus and Keith, who were not aware of the functional significance of human speech or the role of the round human tongue and pharynx, nonetheless reached the same conclusion as that noted in Lieberman and Crelin (1971). Crelin's reconstruction, which was based on comparative studies of the basicranium and mandible of human newborns, chimpanzees, and the La Chapelle-aux-Saints fossil (Lieberman and Crelin 1971; Lieberman *et al.* 1972), yielded the non-human configuration shown in Figure 22.6.

Computer modelling of this airway showed that its phonetic output was quite similar to that of non-human primates and human newborns (Lieberman 1968; Lieberman *et al.* 1969, 1972). The output of the computer modelling is shown in Figure 22.7. Note that like the Apert's patient, the Neanderthal vocal tract inherently cannot produce vowels like [i] or [u]. Its output is also nasalized and its speech would thus inherently be subject to higher phonetic errors. If the Neanderthal hominid had the full perceptual ability of modern human beings, his speech would thus have at minimum a phonetic error rate that was 30 per cent higher than ours.

Detailed discussions of the skeletal features that are involved in recon-

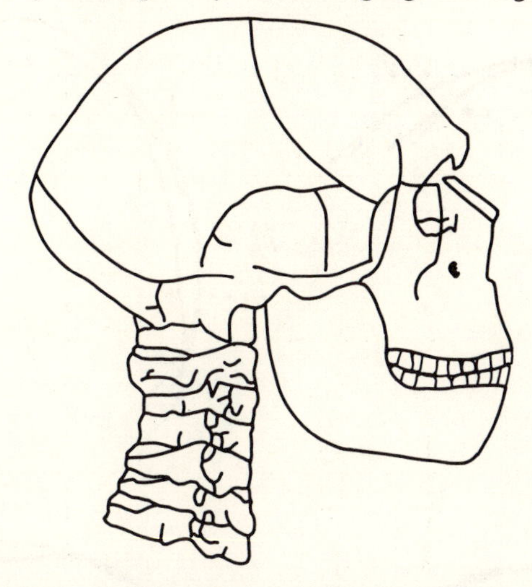

Figure 22.8. The La Chapelle-aux-Saints skull and mandible placed on a modern vertebral column.

structing the supralaryngeal airway have been published in several papers and books (see Lieberman 1984 for references) but some studies (Falk 1975; Dubrul 1977) claim that Neanderthal hominids had a human supralaryngeal airway and that they could produce the full range of speech-sounds. It is easy to demonstrate that this is impossible and in doing so point out the key factors that differentiate the human supralaryngeal vocal tract and basicranium from that of archaic hominids.

In Figure 22.8 the La Chapelle-aux-Saints skull and mandible have been placed on a modern vertebral column. What we have to 'give' to the Neanderthal fossil to allow him to produce the full range of human speech sounds is a supralaryngeal vocal tract that has a curved tongue body which forms both the floor of the oral cavity and the anterior wall of the pharynx. A number of independent studies (Russell 1928; Perkell 1969; Ladefoged *et al.* 1972; Nearey 1979) have demonstrated that the posterior contour of the human tongue is almost circular. Figure 22.9 shows tongue contours during the production of different vowels of English (Ladefoged *et al.* 1972). Note that the contours are almost circular and that the tongue moves as an undeformed body in the production of these vowels. The movements of the round undeformed human tongue in the right angle space defined by the vertebral column and basicranium are necessary to generate the full range of sounds of human speech.

This means that the span of the tongue within the oral cavity is equal to the vertical distance between the hard palate and epiglottis. The larynx consequently is positioned low, but it is positioned within the neck; the laryngeal opening at rest is between the fifth and sixth cervical vertebrae. In Figure 22.10 we have 'given' the Neanderthal tongue 'T' from Figure 22.9, making an [I] vowel. The modern human tongue must span the long Neanderthal oral cavity. Since the distance from the prosthion of the

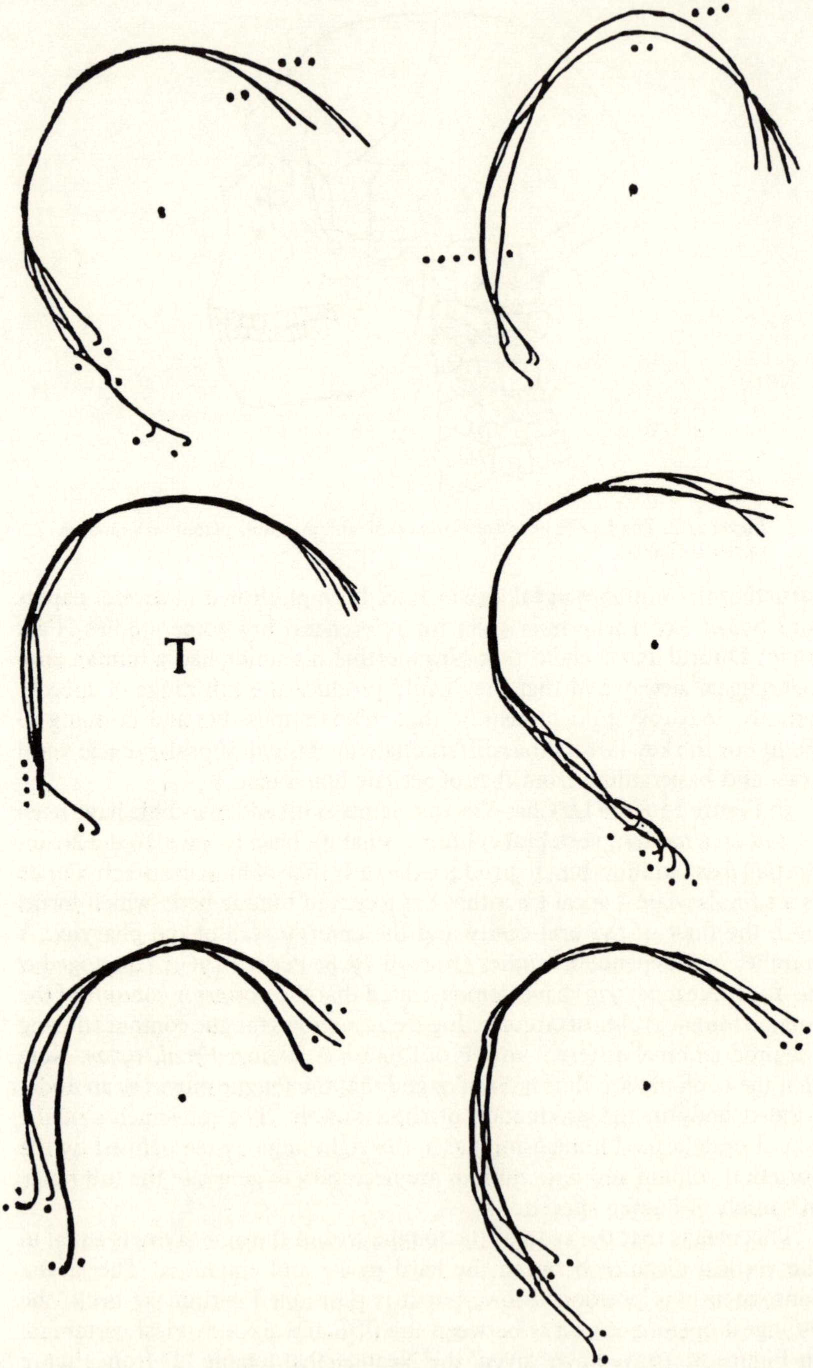

T

Figure 22.9. Midsagittal views of human tongue contours during the production of different vowels of English. (After Ladefoged et al. 1972).

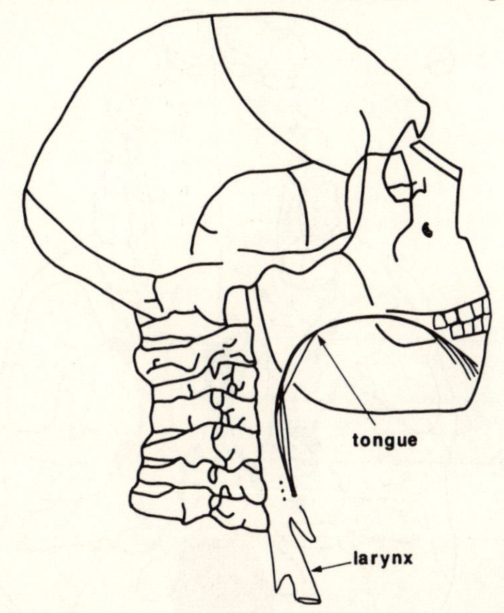

Figure 22.10. Tongue "T" from Figure 22.9, making an [I] vowel on the Neanderthal skull and mandible. This places the larynx below the cervical vertebrae in the Neanderthal chest because of the requirement for a round human tongue. The reconstruction yields an impossible creature.

endobasion is long, the radius of the human tongue that we have fitted to the Neanderthal skull must also be long. This places the larynx below the cervical vertebrae in the Neanderthal chest because of the requirement for a *round* human tongue. The reconstruction yields an impossible creature; no mammal has its larynx in its chest.

We actually have given the human-Neanderthal supralaryngeal vocal tract reconstruction the benefit of all possible doubts. The particular tongue, T, that we selected was that of an adult female. There is a slight degree of dimorphism in modern humans with regard to the tongue shape. Males tend to have a slightly longer pharynx (Fant 1960; Lieberman 1986). The larynx is closer to the basicranium in the production of [I] than some other vowels, and, finally, the Neanderthal vertebral column may be shorter than that of modern humans (Boule and Vallois 1957). All of these factors would place the larynx still lower in the chest if we insist, as was once proposed, that a classic Neanderthal hominid is no different from other passengers who may be riding the New York Subway. The correct reconstruction is one in which the long unflexed basicranium matches a non-human supralaryngeal vocal tract.

The general method and preliminary results. Figure 22.11 shows the significant basicranial landmarks for the reconstruction of the supralaryngeal vocal tract on a chimpanzee skull. The distance between hard palate and vertebral column, staphylion to endobasion, matches up with the high position of the non-human larynx and the pharynx which is positioned behind it. There

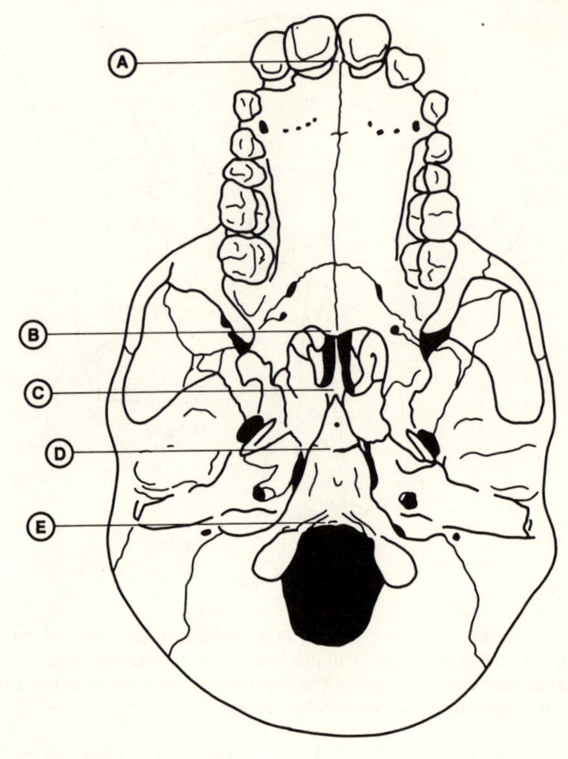

Figure 22.11. Basicranial landmarks for the reconstruction of the supralaryngeal vocal tract on a chimpanzee skull. (After Laitman, Heimbuch and Crelin 1978).

has to be room for a high larynx that can lock into the nasal cavity. The long unflexed pongid basicranium reflects the fact that a long thin tongue is positioned within the oral cavity with a high larynx. Figures 22.12 and 22.13 from the quantitative studies of Laitman and his colleagues show the change in basicranial angle in living primates and in fossil hominids.

Figure 22.14 shows one possible scenario for the evolution of the human supralaryngeal vocal tract. The process may, like other important transitions, involve preadaption. The flexure of the basicranium in *Homo erectus* grade fossils like KNM-ER 3733 is greater than that of living pongids or Australopithecines, possibly indicating a lowering of the larynx to facilitate mouth- breathing, as is the case for human infants at an age of 3 months. The skull base in fossils like Skhūl V (at the end of the right branch in Figure 22.14) is completely modern. Qafzeh likewise had a completely modern supralaryngeal vocal tract around 92 000±5000 BP. It could *not* support a non-human supralaryngeal airway; there is not enough room for a high larynx almost in line with the tongue and pharynx. The basicranial skeletal configuration and supralaryngeal vocal tract would have had the full range of vegetative deficits and phonetic advantages of present-day humans. These fossil hominids probably had modern speech and language.

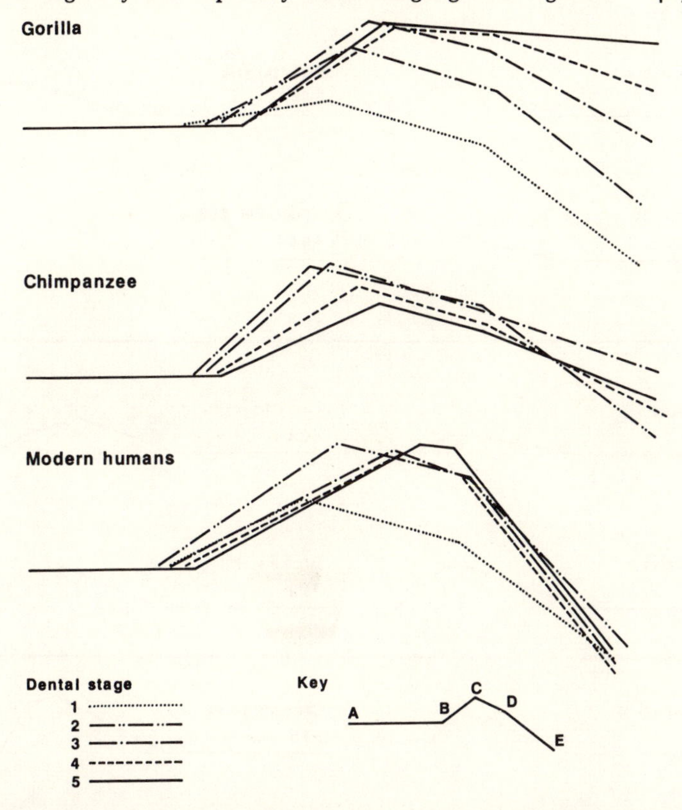

Figure 22.12. Ontogenetic development of the basicranial line in gorillas (upper), chimpanzees (middle), and modern human beings (lower). (Adapted from Laitman, Heimbuch and Crelin 1978).

The branch to the left groups fossils who appear to retain the non-human supralaryngeal airway. They would not have had modern human speech.

The course of evolution along the line leading to the modern human basicranium is, however, not even. Steinheim has a palate that is within the human range, but has less flexure. Broken Hill has modern basicranial flexure with a longer palate. Its vocal tract would appear to be functionally modern, though not optimum for the production of quantal vowels (Stevens 1972). The changes are not consistent with the pattern that we might expect if it were simply a case of some regulatory gene changing the pattern of ontogenetic development of the basicranium (a kind of inverse neoteny), shifting the palate backwards along the sphenoid with corresponding changes in flexure. The fossils noted in the central area are harder to interpret. Saccopastore has more cranial flexure than classic Neanderthals, but its long palate precludes human speech. Monte Circeo I has a shorter palate but has basicranial flexure outside the modern human range. The fossil record thus is not consistent with a sudden, coordinated restructuring of the basicranium and mandible to yield a modern human configuration. The

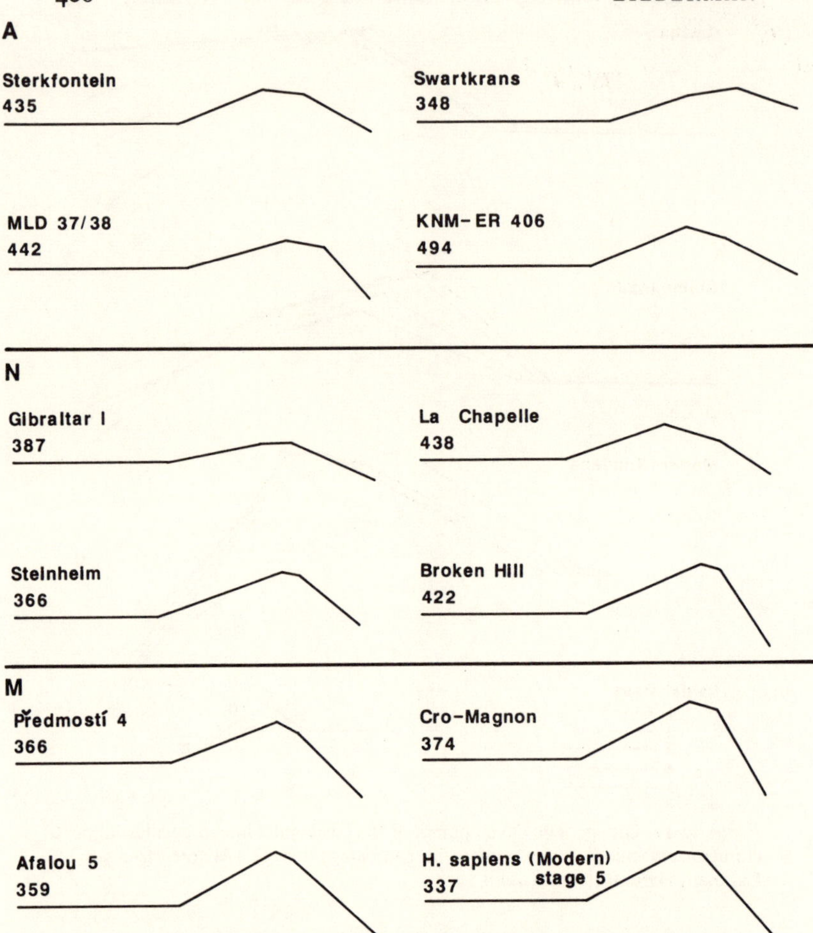

Figure 22.13. (A) Basicranial lines showing entire length of palate for the gracile australopithecine Sterkfontein 5 and MLD-37/38 fossils, and the robust australopithecine Swartkrans 47 and KNM-ER-406 fossils. (N) Basicranial lines for the Neanderthal Gibraltar I and La Chapelle-aux-Saints fossils. (M) Basicranial lines for the Stenheim and Broken Hill fossils and modern adult Homo sapiens. (After Lieberman 1984).

evolution of the human supralaryngeal vocal tract probably involved a number of preadaptions, e.g., mouth breathing and different patterns of chewing. These, in turn, may reflect changes in diet, social organization, etc. However, as we noted above, the 'final' restructuring of the supralaryngeal vocal tract appears to have been driven by phonetic considerations. We are in a position to assess the phonetic consequences of fossil skeletal morphology, drawing on comparative studies of human pathologic 'experiments in nature' noted above. Biological anthropologists working with computer modelling techniques and properly defined pathologic populations can solve these problems.

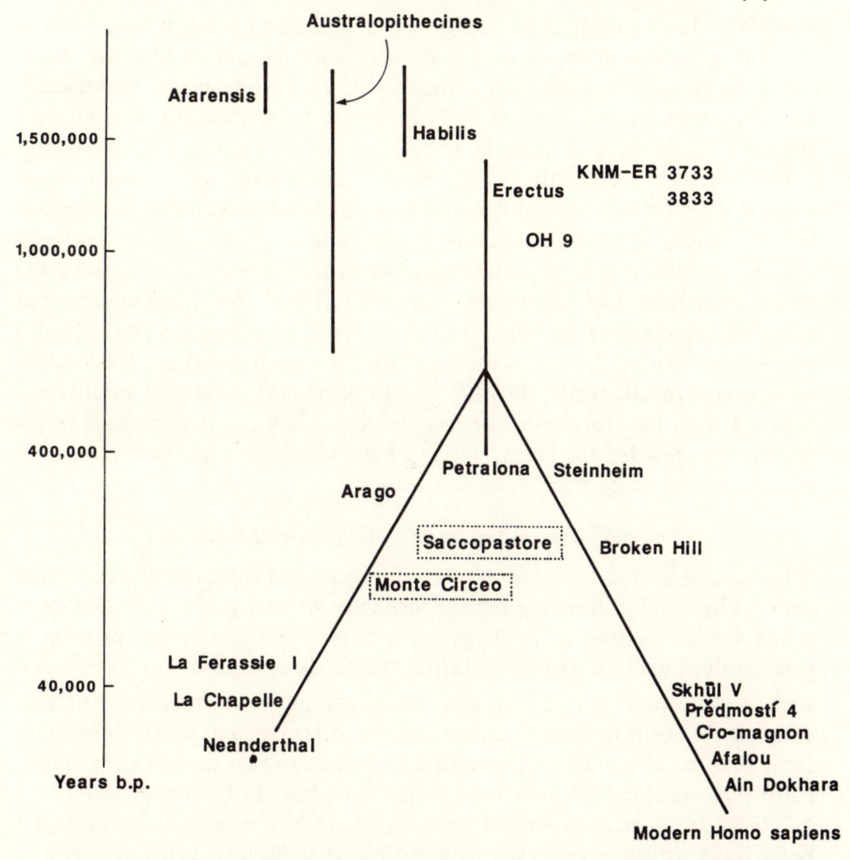

Figure 22.14. The possible evolution of the human supralaryngeal vocal tract. (After Lieberman 1984).

SYNTAX AND LOGIC – INTERPRETING THE FOSSIL RECORD

Though the reconstruction and modelling of the phonetic possibilities of a fossil vocal tract is itself interesting, it can lead to other insights if we take account of the following premises.

1. Vocal communication clearly exists in all primates. Negus (1949) demonstrated that the larynges of social mammals are adapted for phonation at the expense of respiration. Hence a stage in hominid evolution in which communication was entirely gestural is most unlikely. Early hominids may resemble present-day chimpanzees insofar as they may not have been able to produce vocalizations that were decontextualized from gestural displays. Therefore, it is possible that gestural communication for referential communication was at one time more important than is presently the case.

2. In this connection, the vocal anatomy of living non-human primates would be sufficient for complex vocal communication if they could achieve voluntary control of the complex motor controls necessary for human

sion. We also know that we make use of syntax to further facilitate the transmission of complex thoughts. We also make use of rule-governed logic in non-linguistic domains. These human abilities are probably biologically linked in modern humans and so we have to ask the question of when they appeared in the fossil hominids whom we think are among our ancestors.

It is clear that the only thing that the modern human supralaryngeal vocal tract is better adapted for is speech, and that speech depends on our having matching neural mechanisms that control its production. A fossil who had a human-like supralaryngeal vocal tract some 100 000 years ago would also have had to execute the rapid articulatory manoeuvres that encode human speech to fully take advantage of its adaptive value. He also would have to have the neural mechanisms for speech decoding. If he lacked these neural mechanisms, the human supralaryngeal vocal tract would have been a deficit that increased the likelihood of choking, impeded his respiration and crowded his teeth. The human supralaryngeal vocal tract thus is an index for these neural mechanisms.

The evolution of syntax and rule-governed behaviour

What can be said about the evolution of syntax and complex rule-governed logic? The evolution of the neural mechanisms that govern the rule-governed syntax of human language may involve the Darwinian process of preadaption. Broca's area is also implicated in the comprehension of syntax.
speech. Modulations of pitch, and some formant transitions and patterns, can and do seem to occur in chimpanzee calls (Goodall 1986). Thus it is clear that the evolution of the neural architecture that facilitates *voluntary* control of vocal communication is one of the keys to human speech.

3. The earliest stages of further specialization for human speech could have been built up on a general primate base *if* voluntary neural control of vocalization were in place. The human supralaryngeal airway would have conferred phonetic selective advantages, by allowing the production of non-nasal sounds that were less readily confused. But the selective advantages of more efficient vocal communication would only be realised if voluntary control of speech existed. Therefore it is apparent that a fossil hominid who has a modern human supralaryngeal vocal tract has, at minimum, the neural substrate that confers voluntary control of speech.

The initial increase in fitness derived from more efficient vocal communication might possibly be derived without additional neural modifications for speech perception beyond the non-human pongid-like base. Studies of the perception of human speech by chimpanzees (Savage-Rumbaugh *et al.* 1985) are addressing this question. Initial data show that chimpanzees can perceive human speech using formant transitions and fundamental frequency contours. It is, however, not clear whether chimpanzees perceive human speech as efficiently as human beings, or whether chimpanzees can make use of the full ensemble of acoustic cues that humans use.

4. The evolution of the human supralaryngeal vocal tract in fossil hominids could thus initially have been selected for by phonetic factors. However, we know that our speech, the end point, is encoded for rapid data transmis-

The speech production and syntactic deficits of chimpanzees, and the agrammatic victims of Broca's aphasia, thus appear to be functionally linked (Lieberman 1984, 1985). Recent data relating the acquisition and deterioration of complex speech-motor-control and syntactic comprehension are consistent with this theory, which claims that the neural substrate that evolved to facilitate speech motor control is the preadaptive basis for human rule-governed syntactic and cognitive ability (Lieberman 1984, 1985). In recent studies, Emery (1982) and Kynette and Kemper (1986) show that aged people who are otherwise cognitively 'normal' show deficits in the production and comprehension of written and spoken sentences. Subjects are not able, for example, to identify the subject in passive sentences like 'Susan was kissed by Tom', nor are they able to comprehend sentences that have embedded clauses. Subjects generally are able to use semantic and pragmatic information and have no difficulty with sentences like 'Bill ate the apple'. The deficits noted by Emery appeared to mirror ones that have been noted in young children. Emery's data also showed deficits in Piagetian cognitive tasks that also mirror the cognitive development of young children. These phenonema occurred in 'normal' aged people who were not demented.

We have replicated some aspects of Emery's study; we also have correlated syntactic deficits with the rate of speech production. Our pilot experiment involves residents of the Jewish Home for the Aged in Providence, Rhode Island. We assessed their syntactic comprehension and some aspects of speech production. Each person was tested using the 'Rhode Island Test of Syntactic Ability' (Engen and Engen 1982). This test was originally devised for hearing-impaired persons. It consists of 100 sentences which are each read to the subject. The person being tested is asked to point to one of three sketches. One sketch correctly illustrates the meaning of the sentence. The sentences differ in syntactic complexity. However they all make use of similar vocabulary. The test has been designed to avoid any meaning distinctions that are conveyed by morpheme differences; these distinctions are hard to perceive and would introduce errors for elderly subjects who may have hearing deficits. The scoring of the Rhode Island Test furthermore reveals 'clustered' error patterns which can differentiate syntactic deficits from difficulties that result from hearing loss. These clustered error patterns also can reveal difficulty with particular syntactic distinctions. Normative data are available, since the test has been administered to over 4000 individuals. In general, normal children over the age of 6 years make no errors. We also had each aged person speak the syllables [si] and [su], [di] and [du], [ti] and [tu], as well as a series of sustained vowels and sentences. We obtained at least 5 tokens of each syllable. The durations of these syllables were measured using a computer system.

Our data replicate those of Emery (1982) and Kynette and Kemper (1986); we find errors in syntactic comprehension that are age-related. We, however, find strong individual differences; some elderly subjects make no errors while others have a 35 per cent error rate. Many of the aged people who had high syntactic error rates had long vowel durations and/or speech production deficiencies. The subjects who had the highest syntactic error rates talked at a slower rate. The durations of their syllables was on average 331 msec compared to 277 msec for subjects who were error-free. The difference is significant $p < 0.02$ ($t = 2.79$, $df = 14$, two-tailed test). Our data, though preliminary, show a possible correlation between deficits in speech production and syntax; they are consistent with data derived from the study of aphasia. These studies are consistent with the theory discussed here concerning the neural substrate that underlies rule-governed behaviour in human beings. Data correlating deficits in speech production and mathematical operations (Deloche and Seron 1984), speech and logic (Emery 1982) and language and cognition (Gopnick and Meltzoff 1985; Wills 1973) in normal and mentally retarded humans are also consistent with this theory. Further experiments are in progress looking at possible links between the neural substrates for speech production, syntax and logic. Will we, for example, see links between particular levels of linguistic ability and the ability to produce certain types of tools? If so, the presence of certain types of tools in the archaeological record may, in turn, signify a particular level of cognitive ability.

To conclude, the supralaryngeal vocal tracts of fossil hominids can be reconstructed. This allows us to assess the presence of the anatomical prerequisites for human speech and neural mechanisms adapted for speech production. These neural mechanisms may also be implicated in syntax and logic. Therefore, the basicranium and mandible, which determine the morphology of the supralaryngeal vocal tract, are functional skeletal markers for the evolution of modern humans.

REFERENCES

Bond, Z. S. 1976. Identification of vowels excerpted from neutral nasal contexts. *Journal of the Acoustical Society of America* 59: 1229-1232.

Boule, M. and Vallois, H. V. 1957. *Fossil Men*. New York: Dryden Press.

Darwin, C. 1859. *On the Origin of Species*. (Facsimile ed. 1964). Cambridge (Mass): Harvard University Press.

Deloche, G. and Seron, X. 1984. Some linguistic components of acalculia. In F. C. Rose (ed.) *Advances in Neurology*. Vol. 452: *Progress in Aphasiology*. New York: Raven Press.

Deacon, T. W. 1985. *Connections of the Inferior Periarcuate Area in the Brain of Macaca fascicularis: an Experimental and Comparative Neuroanatomical Investigation of Language Circuitry and its Evolution*. Unpublished Ph.D. Dissertation, Harvard University.

Emery, O. B. 1982. *Linguistic Patterning in the Second Half of the Life Cycle*. Unpublished Ph.D. Dissertation, University of Chicago.

Engen, E. and Engen, T. 1983. *Rhode Island Test of Language Structure*. Baltimore: University Park Press.

Fant, G. 1960. *Acoustic Theory of Speech Production*. The Hague: Mouton.

Folkins, J. W. 1985. Issues in speech motor control and their relation to the speech of individuals with cleft palate. *Cleft Palate Journal* 22: 106-122.

Goodall, J. 1986. *The Chimpanzees of Gombe*. Cambridge (Mass): Harvard University Press.

Gopnick, A. and Meltzoff, A. 1985. From people, to plans, to objects: changes in the meaning of early words and their relation to cognitive

development. *Journal of Pragmatics* 9: 495-512.
Greenberg, J. 1963. *Universals of Language*. Cambridge (Mass): Massachusetts Institute of Technology Press.
Kynette, D. and Kemper, S. 1986. Ageing and the loss of grammatical forms: a cross-sectional study of language performance. *Language and Communication* 6: 65-72.
Ladefoged P. and Broadbent, D. E. 1957. Information conveyed by vowels. *Journal of the Acoustical Society of America* 29: 98-104.
Ladefoged, P., De Clerk, J., Lindau, M. and Papcun, G. 1972. An auditory-motor theory of speech production. *UCLA Working Papers in Phonetics* 22: 48-76.
Laitman, J. T. and Crelin, E. S. 1976. Postnatal development of the basicranium and vocal tract region in man. In J. Bosma (ed.) *Symposium on Development of the Basicranium*. Washington (DC): U.S. Government Printing Office: 206-219.
Laitman, J. T. and Heimbuch, R. C. 1982. The basicranium of Plio-Pleistocene hominids as an indicator of the upper respiratory systems. *American Journal of Physical Anthropology* 59: 323-344.
Laitman, J. T., Heimbuch, R. C. and Crelin, E. S. 1979. The basicranium of fossil hominids as an indicator of their upper respiratory systems. *American Journal of Physical Anthropology* 51: 15-34.
Landahl, K. L. and Gould, H. J. 1986. Congenital malformation of the speech tract in humans and its developmental consequences. In R. J. Ruben, T. R. Van de Water and E. W. Rubel (eds) *The Biology of Change in Otolaryngology*. Amsterdam: Elsevier: 131-149.
Lieberman, P. 1968. Primate vocalizations and human linguistic ability. *Journal of the Acoustical Society of America* 44: 1574- 1584.
Lieberman, P. 1973. On the evolution of language: a unified view. *Cognition* 2: 59-94.
Lieberman, P. 1975. *On the Origins of Language: an Introduction to the Evolution of Human Speech*. New York: Macmillan.
Lieberman, P. 1984. *The Biology and Evolution of Language*. Cambridge (Mass): Harvard University Press.
Lieberman, P. 1985. On the evolution of human syntactic ability: its preadaptive bases – motor control and speech. *Journal of Human Evolution* 14: 657-668.
Lieberman, P. 1986. Some aspects of dimorphism and human speech. *Human Evolution* 1: 67-75.
Lieberman, P. and Crelin, E. S. 1971. On the speech of Neanderthal Man. *Linguistic Inquiry* 2: 203-222.
Lieberman, P., Klatt, D. H. and Wilson, W. A. 1969. Vocal tract limitations on the vowel repertoires of rhesus monkey and other nonhuman primates. *Science* 164: 1185-1187.
Lieberman, P., Crelin, E. S. and Klatt, D. S. 1972a. Phonetic ability and related anatomy of the newborn, adult human, Neanderthal man and the chimpanzee. *American Anthropologist* 74: 287-307.
Lieberman, P. , Harris, K. S., Wolff, P. and Russell, L. H. 1972b. Newborn infant cry and nonhuman primate vocalizations. *Journal of Speech and Hearing Research* 14: 718-727.
MacLean, P. D. 1985. Evolutionary psychiatry and the triune brain. *Psychological Medicine* 15: 219-221.
McNeill, D. 1985. So you think gestures are nonverbal? *Psychological Review* 92: 350-371.
Miller, G. A. 1956. The magical number seven, plus or minus two: some limits on our capacity for processing information. *Psychological Review* 63: 81-97.
Müller, J. 1848. *The Physiology of the Senses, Voice and Muscular Motion*

with the Mental Faculties (Translated by W. Baly). London: Walton and Maberly.

Nearey, T. 1978. *Phonetic Features for Vowels.* Bloomington (Ind): Indiana University Linguistics Club.

Negus, V. E. 1949. *The Comparative Anatomy and Physiology of the Larynx.* New York: Hafner.

Olmsted, D. L. 1971. *Out of the Mouth of Babes.* The Hague: Mouton.

Perkell, J. S. 1969. *Physiology of Speech Production: Results and Implications of a Quantitative Cineradiographic Study.* Cambridge (Mass): Massachusetts Institute of Technology Press.

Peterson, G. E. and Barney, H. L. 1952. Control methods used in a study of the vowels. *Journal of the Acoustical Society of America* 24: 175-184.

Russell, G. O. 1927. *The Vowel.* Columbus: Ohio State University Press.

Ryalls, J. H. and Lieberman, P. 1982. Fundamental frequency and vowel perception. *Journal of the Acoustical Society of America* 72: 1631-1634.

Savage-Rumbaugh, S., Rumbaugh, D. and McDonald, K. 1985. Language learning in two species of apes. *Neuroscience and Biobehavioral Reviews* 9: 653-665.

Stevens, K. N. 1972. Quantal nature of speech. In E. E. David and P. B. Denes (eds) *Human Communication: a Unified View.* New York: McGraw Hill.

Watkin, K. and Fromm, D. 1984. Labial coordination in children: preliminary considerations. *Journal of the Acoustical Society of America* 75: 629-632.

Wills, R. H. 1973. *The Institutionalized Severely Retarded.* Springfield (Ill): Charles C. Thomas.

23. The Implications of Stone Tool Types for the Presence of Language During the Lower and Middle Palaeolithic

HAROLD L. DIBBLE

INTRODUCTION

Of central concern to all anthropologists is the discovery of when language and essentially modern cultural systems made their first appearance in the course of hominid evolution. Two sources of evidence are crucial in this regard: the fossil evidence itself, and the more direct evidence of behaviour that is reflected primarily in assemblages of stone tools, or lithics.

There are two basic approaches to the biological data that have been pursued. The first of these is to evaluate the ability of fossil hominids to produce modern speech sounds. The studies by Lieberman (Lieberman and Crelin 1971; Lieberman *et al.* 1972; Lieberman, this volume) were based on comparisons of a reconstruction of the vocal tract of the La Chapelle-aux-Saints Neanderthal with vocal tracts of chimpanzees, newborn and adult humans. While computer simulations showed that the range of possible vowel sounds produced by that individual were perhaps limited, their study has been criticized on the grounds that, among other things, their reconstruction of the soft tissues associated with Neanderthal vocal tracts may not have been accurate (see Nett 1973; Carlisle and Siegel 1974; Falk 1975). As a possible way around such reconstructions, Laitman *et al.* (1978, 1979) have shown that the amount of flexion in the basicranial lines from prosthion to endobasion is a good reflection of the position of the larynx and pharynx. Their work, involving comparison of basicranial morphology of non-human primates, modern humans (at varying stages of development) and various fossil hominds, supports the general conclusions of Lieberman *et al.*) for especially the La Chapelle fossil, and other Classic Neanderthal forms as well, in that they 'probably had a more restricted vocal range than that of modern adult or subadult humans' (Laitman *et al.* 1979: 31). However, the Neanderthals, unlike the gracile Australopithecine represented by Sterkfontein 5, showed considerable divergence from pongid basicranial lines.

But, as pointed out by many researchers (see Hewes 1973; Falk 1980a, 1980b), there is more to language than the production of specific sounds. A more important consideration, therefore, may be what changes are taking place in the brain during the span of human evolution that may indicate changes in cognition related to language, either in the form of speech or something else – for example, gestures. This second approach, based on studies of fossil cranial endocasts, has been pursued by Holloway (1981a, 1981b, 1981c, 1983a, 1983b; Holloway and de la Coste-Lareymondie 1982; see also Falk 1980b, 1985). As summarized by Holloway (1983b: 110), a number of aspects of fossil brains are relevant to the question of language origins. One critical class of such evidence concerns the development of brain asymmetries. This feature, which is 'suggestive of handedness and differentiation of symbolic and visuospatial integration cognitive tasks between hemispheres' (Holloway 1983b: 111), does seem to develop early in the hominid line, at least as early as the earliest *Homo* (Holloway and de la Coste-Lareymondie 1982). Equally early, i.e. with the early *Homo* remains KNM-ER 1470 and KNM-ER 1805 from Koobi Fora, Kenya, is the appearance of human-like cortical sulcal patterns (Falk 1983).

However, brain asymmetries do occur in other primates and even non-primate species (e.g. Japanese macaques: Seyfarth 1986: 447) and the exact and initial cause for such lateralization is not known. In fact, it has been argued by Frost (1980; see also Steklis 1985: 167) that initial lateralization may have occurred in conjunction with handedness related to repeated tool manufacture. Evidence for deliberate and repeated lithic tool manufacture extends back to beyond two million years (Isaac 1984) and there is some evidence in these early lithic industries for handedness (Toth 1985a) and more evidence later (Cornford 1986). The development of handedness as a response to tool manufacture may then have prompted the original neurological changes seen in the early hominids. This neural reorganization may then have served as a preadaptation for the later development of symbolic language and related structures in the left side of the brain (see Falk 1908b).

Thus, the biological evidence is equivocal in its demonstration for or against language or linguistic behaviour before the advent of modern *Homo sapiens*. The other major class of evidence that may be relevant to this question are the stone tools. Arguments that relate lithic attributes and defined types to the presence of language have been made by most of those cited above who have concerned themselves primarily with the biological evidence (Lieberman 1975, and this volume; Lieberman *et al.* 1972: 302; Falk 1980a: 102; 1980b: 76; see especially Holloway 1969, 1981b). Most of these arguments are based on the assumption that typological variation in these early stone tools reflects arbitrary and non-iconic symbolic variation, or styles, that could be linked to linguistic categories. For example, Holloway (1981b: 295) says that 'since many (but not all) of the various tools are highly standardized (as reflected in the archaeologists' recognition and naming of them), they suggest that social behavioral adaptations [involved] symbolic communication and learning of arbitrary standards.' For Holloway, then, stone tools represent arbitrary impositions of form on the natural environment by early hominids, and this is his major criterion for the recognition of modern cultural systems.

As archaeologists learn more about prehistoric lithic technologies, it has become clear that similarities in form of lithic artifacts can be the result of many factors, including raw material and technological constraints. Variability that is a result of these constraints is probably not profitably viewed as being truly arbitrary and therefore not representative of cultural symbolic variation. It is the purpose of this paper to review data on some lithic tools from Lower and Middle Palaeolithic contexts which suggest that much of the recognized morphological patterning in them is, in fact, due to basic technological factors. Three basic tool classes will be covered: flake scrapers, bifaces and Levallois flakes. The implications of these results are that the lithic evidence in at least these kinds of stone tools cannot be used to demonstrate the early presence of linguistically-structured thought or behaviour.

SCRAPER REDUCTION AND TYPOLOGY

The typology of Lower and Middle Palaeolithic stone tools which is widely used in Europe and the Near East recognizes 63 discrete types (Bordes 1950, 1961a; Bordes and Bourgon 1951). Of these, more than one-third are various types of *racloirs,* or scrapers. Among the scrapers, four major classes can be discerned: (1) simple laterally-retouched single-edged side-scrapers (types 9-11 in Bordes' typology); (2) double scrapers with two non-joining retouched edges (types 12-17); (3) convergent scrapers, which have two adjacent retouched edges that usually form a point on the distal end (types 8 and 18-21); and (4) transverse scrapers with retouch on the edge opposite the striking platform (types 22-24). These classes represent the most common types of scrapers in all Lower and Middle Palaeolithic assemblages, while scrapers in general represent – along with bifaces, denticulates and notches – one of the primary diagnostic features of those periods (Bordes 1953; Dibble 1988; Rolland 1977, 1981; Jelinek, 1984; Geneste 1985). Thus, variability among scrapers represents a significant portion of Palaeolithic assemblage variability, the interpretation of which is a major question for Old World prehistorians (Binford 1973; Binford and Binford 1966; Bordes 1961b; Bordes and de Sonneville-Bordes 1970; Mellars 1969, 1970).

The aspect of typological variation to be addressed here is whether the various types of scrapers are standardized end- products that reflect clearly-defined mental images held by the hominids who made them. That is to say, whether each type was made to have a particular form related to its ultimate function or whether the types represent arbitrary forms that reflect past cultural norms. If so, then this patterning may reflect the presence of more-or-less modern cognitive, linguistic and/or cultural symbolic structures. However, in a recent series of papers (Dibble 1984a, 1987a), it was proposed that typological variation among these tools instead reflects continuous reduction of the pieces through resharpening and remodification until their eventual discard. In this case, the presence of such types would not be relevant to the recognition of symbolic thought.

In the studies of Middle Palaeolithic scrapers by the present author, two distinct reduction sequences have been noted. The first involves a sequence from single-edged side-scrapers through double-edged side-scrapers to con-

Table 23.1 Summary statistics on observatons (in mm) of artifacts representing different typological classes of complete scrapers and unretouched flakes from Tabūn and La Quina, reproduced from Dibble (1987b). Data from La Quina are taken from Dibble (1986a). T-tests of differences between means of Tabūn and La Quina material are not significant at the 0.05 level, except where noted (* = P<.05, † = P<.01).

Scraper Types		Length		Width		Thickness		Surface Area		Platform Area		Ratio of Surface/Platform Area		Median Retouch Intensity	
		Tabūn	La Quina	Tabūn	La Quina	Tabūn	La Quina	Tabūn	La Quina	Tabūn	La Quina	Tabūn	La Quina	Tabūn	La Quina
SINGLE	mean	59.37	60.86	36.46	35.85	12.63	10.17†	2213	2229	281.19	246.61	14.26	19.48	2.50	2.59
	std	12.00	13.48	9.38	8.75	4.35	4.28	922	925	211.69	200.41	18.99	24.35	1.14	1.02
	N	152	117	153	117	156	117	152	117	156	117	152	117	153	120
DOUBLE	mean	59.63	61.51	38.71	34.99	13.31	9.88*	2358	2228	319.60	173.49*	11.48	19.73	3.37	2.71*
	std	7.96	13.58	12.94	9.62	4.00	4.69	1067	1181	218.44	152.15	8.01	9.74	0.88	1.10
	N	16	35	16	35	16	35	16	35	16	35	16	35	10	35
CONVERGENT	mean	54.43	52.81	34.04	35.99	13.73	13.07	1893	1923	368.03	328.89	7.87	9.69	3.60	3.22
	std	13.15	10.93	8.24	8.59	4.04	4.38	823	724	208.61	276.44	7.09	8.73	0.70	0.88
	N	33	71	33	71	33	71	33	71	33	71	33	33	71	74
TRANSVERSE	mean	51.55	47.31*	33.71	38.08†	13.43	13.27	1759	1816	506.17	570.80	5.38	7.30	3.11	3.31
	std	11.36	12.58	8.84	11.28	5.16	13.27	690	780	329.04	380.29	4.89	13.28	0.98	0.88
	N	69	137	69	137	69	137	69	137	68	137	68	137	69	140
UNRETOUCHED FLAKES	mean	51.25	N/A	33.93	N/A	10.03	N/A	1835	N/A	251.45	N/A	15.44	N/A	N/A	N/A
	std	14.75		11.35		4.15		1190		243.26		22.46			
	N	160		160		160		160		160		160			

vergent scrapers. The second sequence involves the continuing reduction of a single edge. Typologically, this sequence is represented first by the single- edged types which, as the reduction continues, can be transformed into transverse scrapers. Why one sequence or another is followed probably depends on the initial shape of the flake blank. But in either case, single-edged scrapers would represent the least reduced pieces while convergent and transverse scrapers represent those most reduced. In both of these collections it appeared that reduction continued along this continuum until a particular minimum width was attained and the piece discarded. Thus, flake blanks that were originally larger could undergo more reduction until this point was reached.

These models of scraper reduction have been tested with material from several Mousterian assemblages from the Near East (Tabūn, Bisitun, Warwasi and Kunji) and France (La Quina, Combe-Grenal, Combe-Capelle) (see Dibble 1984a, 1987a, 1987b, and unpublished data). In all cases, observed morphology of the scrapers exhibits patterns that are completely consistent with these reduction models. For example, data from two sites – La Quina and Tabūn – are reproduced in Table 23.1. In this table, it can be seen that as reduction proceeds there is an overall decrease in length across the tool categories, except between single and double scrapers, which would not be expected to lose length during reduction. In a similar fashion, surface area of the tools also decreases across the typological classes; as does the ratio of surface-area to platform-area (which should more accurately reflect the amount of reduction of tools relative to the original size of the flake blanks on which they are made). Also as expected, the median retouch intensity is lightest for the single scrapers, suggesting that more material was removed during the manufacture of double, convergent and transverse forms. But the fact that width is not significantly different among the four typological classes – a finding that has been true for each of the assemblages studied – suggests that this dimension is a primary factor in determining at what point in the reduction sequence a particular piece was discarded. In other words, it would appear that tools are discarded when they reach a certain minimum width.

The data on unretouched flakes complement the tool data. It is apparent that flakes not chosen for retouching are those that were already too small in terms of average width, length or surface area to allow for further reduction. Other indications that the originally larger flakes were selected for retouching are that the retouched flakes are thicker, and that they have larger platform areas than the unretouched flakes.

Thus, in every respect these data from Tabūn and La Quina are consistent with a general reduction model in accounting for variability among these types of Middle Palaeolithic scrapers. But another surprising fact is that the average values of most of the variables measured for one of these tool series are virtually identical to the values for the other. In other words, the average lengths and widths of single, double and convergent scrapers are not significantly different for the tools of these two sites. Likewise, the ratios of flake surface-area to platform-area for each stage of reduction is

Figure 23.1. Graph of two indices of biface shape. The vertical axis is the ratio of the total length to the distance from the butt end to the widest part of the biface. The horizontal axis is the ratio of width at midpoint to the maximum width of the piece. The lines represent boundaries for different handaxe types in Bordes' typology: I = triangular; II = subtriangular; III = cordiforms; IV = ovals, discs and limandes. (Reproduced from Bordes 1961, Figure 7).

the same. The fact that there is such a close correspondence between the French and Israeli material in itself suggests that the tool morphology is primarily affected by very basic technological considerations.

Thus, it would appear on the basis of the above discussion that the different scraper types do not represent desired end-products, but rather end-results of continuous remodifications which were finally discarded. If Mousterian scraper morphology reflects a continuum of reduction then the types recognized today cannot be used as evidence of native classification systems.

Even the point at which the pieces were discarded was probably only related to prehension of the artifact (as reflected by the same average widths) and not to cultural rules of any sort. The scraper assemblages from La Quina and Tabūn, for example, are quite distinct in terms of space and, probably, time. It would be unlikely, therefore, that they were made by culturally-related people. Yet in terms of the processes of tool reduction and the absolute forms of the discarded pieces, they are virtually identical. This suggests that during this time, technological constraints exerted a greater effect on morphology than any cultural pattern.

BIFACE VARIABILITY

In addition to the 63 flake tools, Bordes' typology also recognizes 21 different types of handaxes, or bifaces. Since these are fairly complex tools it is reasonable to suggest that they may also reflect different symbolic or cognitive categories that reflect linguistic categories.

In fact, the existence of natural categories of handaxe shape has been difficult to demonstrate and, again, Bordes typology may be partitioning a continuum of shape variation. This is suggested on the basis of a graph used by Bordes (1961: Figure 7) to show how two indices of biface shape are used to distinguish triangular, subtriangular, cordiform and other types. Reproduced here as Figure 23.1, this graph does not show any evidence of natural categories that could be used as evidence of linguistic categories. In a similar vein, Alimen and Vignal (1952) present frequency distributions of several dimensions and dimensional indices of a series of bifaces from the Atelier Commont at Saint-Acheul, which also do not show evidence of polymodality that might be reflecting native categories.

Recently, Gowlett (1984: 185) has argued that samples of early Middle Pleistocene Acheulian handaxes from Kilombe, Kenya, show 'a high degree of standardization, and must imply a well-defined mental image of the desired end-product'. These standards are shown by high correlations between, for example, length and width of the specimens. These correlations, computed on five different samples from the site, vary between 0.80 and 0.93. Again, what is being at least implied in this argument is that similarity of form of the lithic tools reflects not technological constraints but rather cultural rules. The study by Alimen and Vignal (1952) also gives results that are remarkably similar to those of Gowlett. They also obtain an *r* value of 0.90 and, as can be seen in Figure 23.2, the slopes of the regression lines are virtually identical (though it is not possible to test the

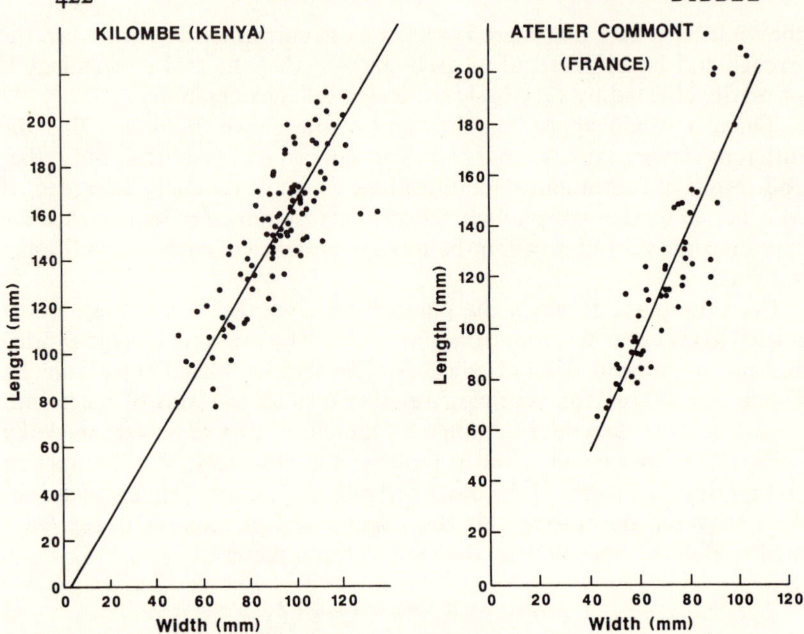

Figure 23.2. Left: graph of length versus width of sample of bifaces from Kilombe, Kenya (after Gowlett 1984, Figure 7.4). Right: graph of relationship of same dimensions of sample from the Atelier Commont, Saint-Acheul. (After Alimen and Vignal 1952, Figure 10).

similarity without access to the original data).

These data suggest that handaxe shape was not random, but a lack of randomness does not in itself necessitate forethought or conscious standards. The real question is whether the morphological patterns are arbitrary. While arbitrariness is difficult to demonstrate, the concept itself suggests that we should see such patterns differ in different parts of the world and at different times, assuming that the cultural rules would vary from place to place. But we see in these two studies of handaxes that the results are virtually identical. As was the case for the similarity between the scrapers of La Quina and Tabūn, the correspondence between the French and African bifaces is also surprising if the shape relationships were truly arbitrary.

It is also important to bear in mind that some of the standardization of these data may be an artifact of classification. According to these data, bifaces at both sites tend to between one and a half to two times as long as they are wide. But what causes the high correlation is that there are no pieces as wide or wider than they are long. Many of the rounder and thicker artifacts would most probably be called cores instead of bifaces, and, by definition, a biface is always longer than it is wide. Thus, we are imposing limits to variability in the process of categorization and these limits can lead to effects similar to those reported by Gowlett, Alimen, and Vignal.

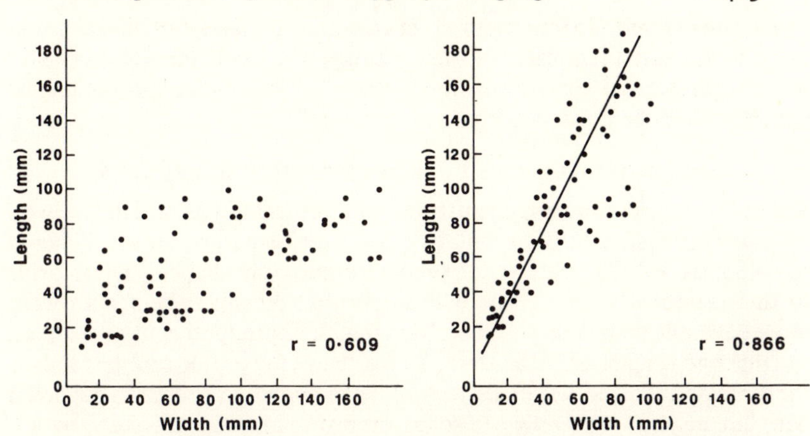

Figure 23.3. Left: plot of 100 random lengths and widths, where one dimension does not exceed three times the other. Right: plot of the same dimensions, but in this case lengths and widths have been transposed if the original width was longer than the length (see text).

The effect of such limits on the relationships of length and width can be shown with a very simple computer simulation. Suppose we program the computer to generate a series of bifaces that vary randomly in length. Then, for each of those bifaces, let us calculate a width that is a random function of that length. For purposes of this simulation, limits have been set so that one dimension cannot exceed three times that of the other. Such a constraint would seem to be realistic given the level of most Lower Palaeolithic biface technology. So, in other words, our simulated bifaces vary completely randomly in length and width within given technological limits.

We would expect that, given two random variables, any relationship between them should also be random. In fact, such is not the case, as can be seen in Figure 23.3a. Because of only the single technological limit that we imposed, there is not complete randomness but rather a computed *r* value of 0.609. But in this graph, we are allowing bifaces to be wider than they are long which is, according to the typology, not possible. Therefore, Figure 23.3b takes the same random lengths and widths as in Figure 23.3a, but transposes the two dimensions where necessary to ensure that length is always greater than the width. In other words, if the width of one of those simulated bifaces in Figure 23.3a was greater than the length, then the length becomes the width and the width becomes the shorter length. Now there is a significant reduction in allowable variability and our *r* value is 0.866, which is well within the range of *r* values given by Gowlett (1984: Table 7.1) for different loci of Kilombe (in fact, repeated simulations with different random numbers give a similar range of *r* values). Moreover, the slopes of the regression appear to be quite similar to those seen in the two different sets of Acheulian data.

So, to some extent, and perhaps to a very significant one, a reduction in variation (i.e. a high correlation) may simply be due to basic technological constraints as well as to our methods of classification, and thus have nothing to do with mental templates of the hominids who made them. Obviously, more sophisticated quantitative analyses of actual biface morphological variability will be needed to confirm this.

TECHNOLOGICAL PATTERNS IN LEVALLOIS FLAKES

Another question concerns the use of Levallois technique, which has been defined (Bordes 1947; Bourgon 1957; see Brézillon 1977: 79-80; Tixier *et al.* 1980: 44-50) as a method for predetermining the shape of flakes prior to their removal. If true, this may imply that certain flake shapes were more desirable than others, and perhaps that their production may be linked to language categories. Moreover, Lieberman (1975: 168-170) has linked the process involved with this technique to cognitive processes involved with language. It is therefore of some interest to investigate flake variation related to this technique.

The data presented here (originally presented in Dibble 1984b) were taken from samples of complete, unretouched flakes drawn from collections of five Lower and Middle Palaeolithic sites located in southern France – namely Combe-Grenal, Orgnac III and the three excavated localities of Pech de l'Azé (sites I, II and IV). In all there are a total of 27 discrete stratified assemblages represented by over 8700 artifacts. The observations of this material were made by this author and Arthur Jelinek of the University of Arizona. Three technological classes of flakes are presented here: Levallois, biface retouch, and 'normal'. The last category consists of flakes for which no special technique of production was recognized. Flakes produced through other techniques, such as blade or disc core techniques, were excluded because there were so few.

If by 'predetermination' it is meant that there is an element of standardization within Levallois flakes, then we would expect to find less variability in size and/or shape of Levallois flakes than exists for 'normal' and biface retouch flakes. To investigate this, coefficients of variation were computed for flake dimensions and various dimension ratios within each assemblage and technological class. If Levallois flakes were standardized with regard to size or shape, then they would be expected to have lower coefficients of variation than either normal or bifacial retouch flakes.

Table 23.2 presents average (across assemblages) coefficients of variation for several variables which reflect both size and shape of the flakes. There are no differences in variability among the three technological classes in length, width and surface area. Thus, for these measures the flakes produced by all three techniques are equally variable. This would suggest that Levallois flakes are not standardized, or at least not recognizably so, in terms of these measures. However, in terms of other variables (thickness, and the ratios of length to width, width to thickness and surface area to thickness), normal flakes are significantly more variable than both Levallois flakes and biface retouch flakes.

The important point about these data is, however, that Levallois flakes are not less variable than biface retouch flakes in terms of any measure

Table 23.2. Mean coefficients of variation for different technological classes of flakes. T-tests of differences between columns are not significant except where noted (* = P<.05). There are no significant differences in coefficients of variation between Levallois and Biface retouch flakes.

	Average Coefficients of Variation				
	Levallois Flakes		Normal Flakes		Biface Retouch Flakes
Length	27.1		29.7		27.1
Width	30.4		30.9		30.1
Thickness	40.6	*	47.3	*	42.1
Surface Area	54.1		59.6		57.6
Area/Thickness	44.5	*	50.8		47.0
Length/Width	31.6	*	35.5	*	27.7
Width/Thickness	36.6	*	46.6	*	40.1

tested. Now, it is doubtful that the size or shape of biface retouch flakes was consciously predetermined. But, if it is concluded that the reduction of variability in Levallois flakes *vis-à-vis* the normal flakes reflects predetermination of form, then we would have to conclude the same for the biface retouch flakes, which is a conclusion that not many would accept.

Actually, these data suggest an alternative perspective on Levallois techniques. Instead of defining Levallois as a method of core preparation leading to the production of a single flake with a predetermined size and shape, it may be more accurate instead to see it as a specific method or technique for core reduction that leads to the production of many flakes from a single core – a reductive strategy (see Boëda 1986, 1988). Similarly, biface retouch flakes are the result of a consistent technology and are, therefore, more consistent in shape and size. Likewise, it is undoubtedly true that consistent application of *any* technology, such as blade or disc core techniques, will also produce consistent results. Our category of 'normal' flakes, on the other hand, probably includes several different technologies, and for this reason the flakes express more variability.

Even within the technological class of Levallois flakes, it has been shown (Dibble 1985) that variability in size is more a function of the particular site – and presumably, therefore, of local raw material sources – than of time or assemblage type (see Fig. 23.4). Thus, again, standardization, or a reduction of variability can be explained on the basis of raw material and technology, without having to suppose the presence of linguistic rules, structures, or categories.

DISCUSSION

Our focus in this paper has been on the relevance of various lithic types to the question of whether or not they reflect linguistic categories. The evidence of the scrapers and of even more complex types, such as handaxes,

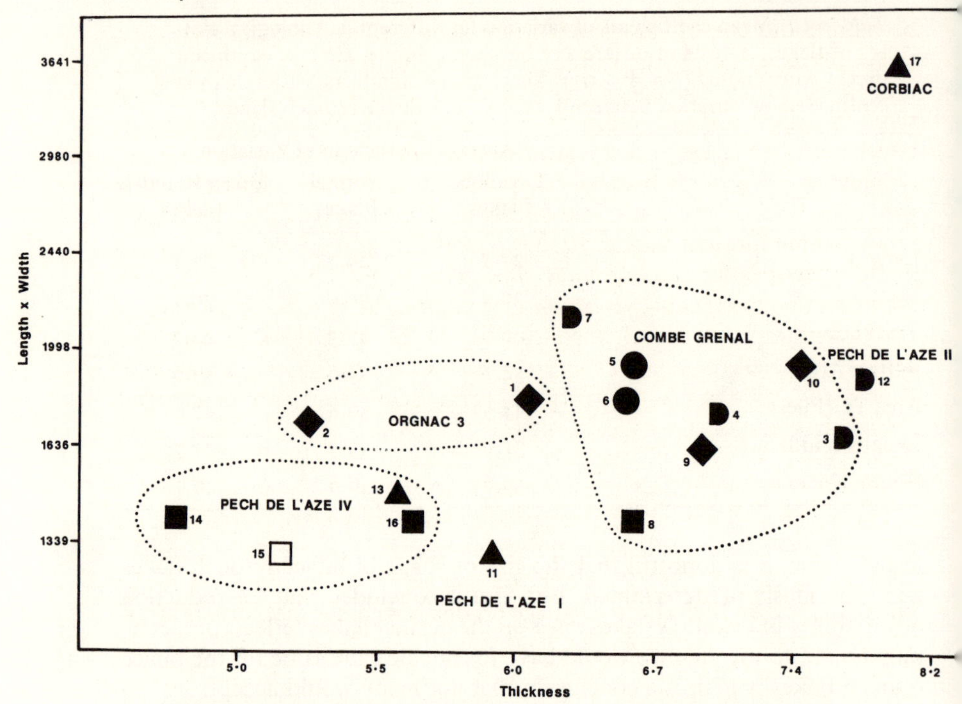

Figure 23.4. Plot of average flake surface area (length x width) against average flake thickness for assemblages of complete Levallois flakes. Symbols represent different industrial variants: Diamonds = Acheulian; Squares = Typical Mousterian; Triangles = Mousterian of Acheulian Tradition; D = Denticulate Mousterian; Circles = Charentian Mousterian; Open square = Asinipodian. (Reproduced from Dibble 1985).

suggests that basic technological, raw material, and even our own classificatory methods are primarily responsible for the aspects of shape variability presented here. For the scrapers, morphological variation was shown to be the result of continuous reduction, with virtually identical patterns being seen in industries separated by time and space. If linguistic categories do underlie these scraper forms, then we must conclude that the hominids of La Quina and Tabūn were speaking very similar languages. Other similarities are seen between aspects of French and African bifaces, but in this case the suggested standardization may in part be only an artifact of the classification method. The possibility does remain that other aspects of biface shape variation relate to linguistic categories, but this needs to be demonstrated on the basis of more comprehensive and detailed studies of overall shape variation instead of the examination of a limited number of indices. Other kinds of standardizations, for example in Levallois flakes, biface retouch flakes, and blades, are due only to the application of consistent technologies.

There are, of course, many other tool types recognized for the Lower and Middle Palaeolithic and it is certainly possible that some of them may be more indicative of linguistic categorization or iconological style than those discussed here. But it does remain to be demonstrated that those types are evidence of language, and to do this it must be shown that those types represent distinct, arbitrary, patterns not related to technology or raw material. In this regard, it is always important to remember that Bordes' typology is, like all archaeological typologies, a modern construct. If it partitions what is really continuous morphological variability, then those types do not represent standardized products that necessarily reflect meaningful categories for the hominids who made them. This, however, does not mean that the typology is bad or inaccurate. The function of this typology is for description and categorization of the industries. In fact, the demonstration of the relationship between the scraper types and reduction clearly supports the use of this typology for interpretation of Lower and Middle Palaeolithic assemblage variability (see Dibble 1988). But, while these types reflect *our* linguistic or symbolic categories, there is no reason to assume that they reflect prehistoric ones.

The question is not whether the production of stone tools represents an imposition of form on the natural environment. Clearly all tool manufacture, including that done by other non- human primates, imposes form. The real question relevant to language origins is whether that form is arbitrary or symbolic. The presence during the Middle Palaeolithic of this sort of variability in the lithics – what Sackett (1982) calls 'iconological style' – remains to be demonstrated. On the other hand, it could be argued that inter- and intra-regional variations in technology or the choice of raw material – the factors that have been shown here to be fundamental – represent different 'isochrestic' styles that reflect different ethnic groups. But such technological traditions reflect the fact that tool making is a learned behaviour and one that was transmitted within the social group. Now, the tools of the Lower and Middle Palaeolithic are clearly much more complicated than anything produced by non-human primates, and Palaeolithic traditions spanned much longer periods of time. But like many other learned behaviours in primates, there is nothing in these kinds of technologies that necessarily forces us to assume a linguistic mode of transmission.

Other recent studies of Lower and Middle Palaeolithic industries also suggest that many aspects of lithic assemblage variability are linked only to aspects of technology and raw materials (Isaac 1986; Toth 1985b; Villa 1983). The simplicity of these industries, and the constraints imposed by technological and raw material factors, indicates that extreme caution should be exercised in the degree to which lithic assemblage variation is attributed to stylistic or cultural factors in general, and to linguistic categories in particular. Thus, our ability to recognize discrete populations of toolmakers during this period of human development may be limited simply because variability in lithic morphology is reflecting such basic technological factors. The same caveats are relevant to reconstructions of intelligence based on lithic categories (Wynn 1979, 1981, 1985; see also Atran 1982).

The purpose of this paper was to argue against the use of lithic types as evidence for early symbolic behaviour. The primary conclusion is not that pre-modern hominids did not talk, but rather that the kinds of lithic evidence discussed here do not demonstrate that they did. Of course, the lack of direct evidence for such behaviour in the easily preserved stone tools does not necessarily mean that such behaviour was not manifested in other ways which are not present in the archaeological record. In this regard, comparative anatomy and primate ethology may be crucial. There is, for example, evidence that non-human primate calls have some semantic component, which may mean that early hominids probably had from the beginning some level of vocal communication, and it may have developed in complexity throughout the Pleistocene (Cheney *et al.* 1986; Steklis 1985; Falk 1980b). But until more concrete evidence is advanced from the archaeological and fossil record, archaeologists and physical anthropologists must not assume the presence of fully symbolic language too early in the hominid line. It may also be that the simplicity of these industries and the general lack of clear symbolic or stylistic variability within and among them reflects a lack of language and other forms of symbolic behaviour before the advent of the Upper Palaeolithic (Butzer 1981; Chase and Dibble 1987; Klein 1973; White 1982). Thus, the question of when language did appear is still very much open to debate.

ACKNOWLEDGEMENTS

The author expresses his thanks to A. Jelinek for permission to analyse the Tabūn collections and to publish the flake measurements from other French sites collected by him, and to M. Baumler, W. Goodenough, M. Voigt, B. Wailes, and especially P. Chase for helpful comments.

REFERENCES

Alimen, H. and Vignal, A. 1952. Etude statistique de bifaces acheuléens: essai d'archéometrie. *Bulletin de la Société Préhistorique Française* 49: 56-72.

Atran, S. 1982. Constraints on a theory of hominid tool-making behavior. *L'Homme* 22: 35-68.

Bar-Yosef, O. 1980. Prehistory of the Levant. *Annual Review of Anthropology* 9: 101-133.

Beyries, S. 1987. *Variabilité de l'Industrie Lithique au Moustérien: Approche Fonctionelle sur Quelques Gisements Français*. Oxford: British Archaeological Reports International Series S238.

Binford, L. R. 1973. Interassemblage variability: the Mousterian and the functional argument. In C. Renfrew (ed.) *The Explanation of Culture Change*. London: Duckworth: 227-254.

Binford, L. R. and Binford, S. R. 1966. A preliminary analysis of functional variability in the Mousterian of Levallois Facies. *American Anthropologist* 68: 236-295.

Boëda, E. 1986. *Approche Technologique du Concept Levallois et Evaluation de son Champ d'Application*. Unpublished Ph.D. Thesis, University of Paris.

Boëda, E. 1988. Le concept Levallois et evaluation de son champ d'application. In M. Otte (ed.) *L'Homme de Néandertal*. Vol. 8: *La Mutation*. Liège: Etudes et Recherches Archéologiques de l'Université de Liège 35: 13-26.

Bordes, F. 1947. Etude comparative des différentes techniques de taille du silex et des autres roches dures. *L'Anthropologie* 51: 1-29.
Bordes, F. 1950. Principes d'une méthode d'étude des techniques de débitage et de la typologie du Paléolithique Ancien et Moyen. *L'Anthropologie* 54: 19-34.
Bordes, F. 1953. Essai de classification des industries 'Moustériennes'. *Bulletin de la Société Préhistorique Française* 50: 457-66.
Bordes, F. 1961a. Mousterian cultures in France. *Science* 134: 803-810.
Bordes, F. 1961b *Typologie du Paléolithique Ancien et Moyen*. Bordeaux: Publications de l'Institut de Préhistoire de l'Université de Bordeaux.
Bordes, F. and Bourgon, M. 1951. Le complexe Moustérien: Moustérien, Levalloisien et Tayacien. *L'Anthropologie* 55: 1-23.
Bordes, F. and Sonneville-Bordes, D. de. 1970. The significance of variability in Palaeolithic assemblages. *World Archaeology* 2: 61-73.
Bourgon, M. 1957. *Les Industries Moustériennes et Pré-Moustériennes du Périgord*. Archives de l'Institut de Paléontologie Humaine Mémoire 27. Paris: Masson.
Brézillon, M. 1977. La Dénomination des Objects de Pierre Taillé. 4th Supplement to *Gallia Préhistoire*. Paris: Centre National de la Recherche Scientifique.
Butzer, K. 1981. Cave sediments, Upper Pleistocene stratigraphy and Mousterian facies in Cantabrian Spain. *Journal of Archaeological Science* 8: 133-183.
Carlisle, R. and Siegel, M. 1974. Some problems in the interpretation of Neanderthal speech capabilities: a reply to Lieberman. *American Anthropologist* 76: 319-322.
Chase, P. G. and Dibble, H. L. 1987. Middle Paleolithic symbolism: a review of current evidence and interpretations. *Journal of Anthropological Archaeology* 6: 263-296.
Cheney, D., Seyfarth, R. and Smuts, B. 1986. Social relationships and social cognition in nonhuman primates. *Science* 234: 1361-66.
Copeland, L. and Hours, F. 1983. Le Yabroudien d'el Kowm (Syrie) et sa place dans le Paléolithique du Levant. *Paléorient* 9: 21-37.
Cornford, J. 1986. Specialized reshaping techniques and evidence of handedness. In P. Callow and J. Cornford (eds) *La Cotte de St. Brelade 1961-1978: Excavations by C. B. M. McBurney*. Norwich: Geo Books: 337-351.
Dibble, H. L. 1984a. Interpreting typological variation of Middle Paleolithic scrapers: function, style, or sequence of reduction? *Journal of Field Archaeology* 11: 431-436.
Dibble, H. L. 1984b. Technological variation in French Lower and Middle Paleolithic flake assemblages. Paper presented at the 49th Annual Meeting of the Society for American Archaeology, Portland (Ore) 1895.
Dibble, H. L. 1985. Raw material variability in Levallois flake manufacture. *Current Anthropology* 26: 391-393.
Dibble, H. L. 1987a. The interpretation of Middle Paleolithic scraper morphology. *American Antiquity* 52: 109-117.
Dibble, H. 1987b. Comparisons des séquences de réduction des outils Moustériens de la France et du Proche-Orient. *L'Anthropologie* 91: 189-196.
Dibble, H. L. 1988. Typological aspects of reduction and intensity of utilization of lithic resources in the French Mousterian. In H. L. Dibble and A. Montet-White (eds) *The Upper Pleistocene Prehistory of Western Eurasia*. Philadelphia: University of Pennsylvania Press: 181-197.
Falk, D. 1975. Comparative anatomy of the larynx in Man and the

chimpanzee: implications for language in Neanderthals. *American Journal of Physical Anthropology* 43: 123-132.

Falk, D. 1980a. Hominid brain evolution: the approach from Paleoneurology. *Yearbook of Physical Anthropology* 23: 93-107.

Falk, D. 1980b. Language, handedness, and primate brains: did Australopithecus sign? *American Anthropologist* 82: 72-78.

Falk, D. 1983. Cerebral cortices of East African early hominids. *Science* 221: 1072-1074.

Frost, G. 1980. Tool behavior and the origins of laterality. *Journal of Human Evolution* 9: 447-459.

Geneste, J.-M. 1985. *Analyse Lithique d'Industries Moustériennes du Périgord: Une Approche Technologique du Comportement des Groupes Humains au Paléolithique Moyen*. Unpublished Doctoral Thesis, University of Bordeaux.

Gowlett, J. A. 1982. Procedure and form in a Lower Palaeolithic industry: stoneworking at Kilombe, Kenya. *Studia Praehistorica Belgica* 2: 101-09.

Gowlett, J. A. 1984. Mental abilities of Early Man: a look at some hard evidence. In R. Foley (ed.) *Hominid Evolution and Community Ecology: Prehistoric Human Adaptation in Biological Perspective*. London: Academic Press: 167-192.

Holloway, R. 1966. Cranial capacity, neural reorganization, and hominid evolution: a search for more suitable parameters. *American Anthropologist* 68: 103-121.

Holloway, R. 1981a. Volumetric and asymmetry determinations on recent hominid endocasts: Spy I and II, Djebel Ihroud I, and the Sale *Homo erectus* specimens, with some notes on Neandertal brain size. *American Journal of Physical Anthropology* 55: 385-393.

Holloway, R. 1981b. Culture, symbols and human brain evolution. *Dialectical Anthropology* 5: 287-303.

Holloway, R. 1983a. Human brain evolution: a search for units, models and synthesis. *Canadian Journal of Anthropology* 3: 215- 230.

Holloway, R. 1983b. Human paleontological evidence relevant to language behavior. *Human Neurobiology* 2: 105-114.

Holloway, R. and Coste-Lareymondie, C. de la. 1982. Brain endocast asymmetry in Pongids and Hominids: some preliminary findings on the paleontology of cerebral dominance. *American Journal of Physical Anthropology* 58: 101-110.

Hours, F., Copeland, L. and Aurenche, O. 1973. Les industries Paléolithiques du Proche-Orient: essai de correlation. *L'Anthropologie* 77: 229-280, 437-496.

Isaac, G. Ll. 1984. The archaeology of human origins: studies of the Lower Pleistocene in East Africa, 1971-1981. In F. Wendorf and A. E. Close (eds) *Advances in World Archaeology* Vol. 1. New York: Academic Press: 1-87.

Isaac, G. Ll. 1986. Foundation stones: early artefacts as indicators and abilities. In G. N. Bailey and P. Callow (eds) *Stone Age Prehistory: Studies in Memory of Charles McBurney*. Cambridge: Cambridge University Press: 221-242.

Jelinek, A. J. 1977. The Lower Paleolithic: current evidence and interpretations. *Annual Review of Anthropology* 6: 11-32.

Jelinek, A. J. 1981. The Middle Paleolithic in the Southern Levant from the perspective of the Tabūn Cave. In J. Cauvin and P. Sanlaville (eds) *Préhistoire du Levant*. Paris: Centre National de la Recherche Scientifique: 265-280.

Jelinek, A. J. 1982. The Tabūn Cave and Paleolithic Man in the Levant. *Science* 216: 1369-1375.

Jelinek, A. J. 1984. Mousterian variability and reduction intensity: a comparison of Levantine and Perigordian industries. Paper presented at the 49th Annual Meeting of the Society for American Archaeology, Portland (Ore) 1984.

Jelinek, A. J., Farrand, W. R., Haas, G., Horowitz, A. and Goldberg, P. 1973. New excavations at the Tabūn Cave, Mount Carmel, Israel, 1967-1972: a preliminary report. *Paléorient* 1: 151-183.

Klein, R. G. 1973. *Ice-Age Hunters of the Ukraine*. Chicago: University of Chicago Press.

Laitman, J. T., Heimbuch R. C. and Crelin, E. S. 1978. Development change in a basicranial line and its relationship to the upper respiratory system in living primates. *American Journal of Anatomy* 152: 467-482.

Laitman, J. T., Heimbuch, R. C. and Crelin, E. S. 1979. The basicranium of fossil hominids as an indicator of their upper respiratory systems. *American Journal of Physical Anthropology* 51: 15-34.

LeMay, M. 1975. The language capability of Neanderthal man. *American Journal of Physical Anthropology* 42: 9-14.

Lieberman, P. 1975. *On the Origins of Language*. New York: Macmillan.

Lieberman, P. 1986. Some aspects of dimorphism and human speech. *Human Evolution* 1: 67-75.

Lieberman, P., Crelin, E. S. and Klatt, D.H. 1972. Phonetic ability and related anatomy of the newborn, and adult human, Neanderthal Man, and the chimpanzee. *American Anthropologist* 74: 287-307.

Mellars, P. A. 1965. Sequence and development of Mousterian traditions in southwestern France. *Nature* 205: 626-627.

Mellars, P. A. 1969. The chronology of Mousterian industries in the Perigord region of South-west France. *Proceedings of the Prehistoric Society* 35: 134-171.

Nett, E. G. 1973. A note on phonetic ability. *American Anthropologist* 75: 1717-1719.

Rolland, N. 1977. New aspects of Middle Palaeolithic variability in Western Europe. *Nature* 266: 251-252.

Rolland, N. 1981. The interpretation of Middle Palaeolithic variability. *Man* 16: 15-42.

Seyfarth, R. 1986. Vocal communication and its relation to language. In B. Smuts, D. Cheney, R. Seyfarth, R. Wrangham and T. Struhsaker (eds) *Primate Societies*. Chicago: University of Chicago Press: 440-451.

Steklis, H. 1985. Primate communication, comparative neurology, and the origin of language re-examined. *Journal of Human Evolution* 14: 157-173.

Tixier, J., Inizan, M. and Roche, H. 1980. *Préhistoire de la Pierre Taillé:* Vol. 1: *Terminologie et Technologie*. Valbonne: Cercle de Recherches et d'Etudes Préhistoriques.

Toth, N. 1985a. Archeological evidence for preferential right-handedness in the Lower and Middle Pleistocene, and its possible implications. *Journal of Human Evolution* 14: 607-614.

Toth, N. 1985b. The Oldowan reassessed: a close look at early stone artefacts. *Journal of Archaeological Science* 12: 101-121.

Villa, P. 1983. *Terra Amata and the Middle Pleistocene Archaeological Record of Southern France*. University of California Publications in Anthropology 13.

White, R. 1982. Rethinking the Middle/Upper Paleolithic transition. *Current Anthropology* 23: 169-192.

Wynn, T. 1979. The intelligence of later Acheulian hominids. *Man* 14: 371-391.

Wynn, T. 1981. The intelligence of Oldowan hominids. *Journal of Human Evolution* 10: 529-541.

Wynn, T. 1985. Piaget, stone tools and the evolution of human intelligence. *World Archaeology* 17: 32-43.

24. Elements of Cultural Change in the Later Palaeolithic

ROBERT WHALLON

INTRODUCTION: UPPER PALAEOLITHIC EXPANSIONS

Two major events of human demographic history occurred in the earlier part of the Upper Palaeolithic: the expansion of human populations into Australia and a similar expansion into Siberia, leading ultimately to the colonization of the New World. The explanation of these two events, occurring late, suddenly, and at closely the same time, relative to the long, preceding periods of human presence on the earth, must be a serious goal for archaeologists. We will argue here that both these events demand in large measure similar explanations. These explanations postulate not only the emergence of new socio-cultural structures as others also have proposed (Gamble 1983), but that these structures, in turn, would have required the development at this time of even more fundamental human capacities for conceptualization and communication. We will try to specify what some of those capacities were and why they were essential to the development of these socio-cultural structures. The implications of the existence or lack of such capacities, however, go further, and include such considerations as whether or not early human groups were capable of carrying out certain economic strategies and what some of the characteristics of the immediately preceding cultural systems may have been.

The suggestion that it was the emergence of new socio-cultural structures, rather than the development of new technologies alone, that allowed the expansion of human populations into Australia, Siberia, and eventually the New World is based on the fact that both the Australian desert and the Siberian arctic tundra appear to share the general features of low resource density, diversity, and predictability in comparison to previously-inhabited environments, yet the specific characteristics of these areas in terms of climate and the particular plants and, especially, animals available are quite different. The arguments may be outlined essentially as follows:

RESOURCES AND HUMAN POPULATIONS

Regional and local densities

Human population density in any region is closely tied to available resource density (e.g. Birdsell 1953, 1968: 230). This is not only true when technology is held relatively constant but appears to hold generally within a given mode of production, regardless of the particular technologies available. For example, certain cases in which more effective technology for resource procurement was incorporated in a hunter-gatherer economy have shown dramatically that the environment was unable continuously to support a much higher rate of hunter-gatherer exploitation, resulting in a negative impact on the human population (Campbell 1978). Further, recent theoretical modelling of hunter-gatherer resource use has shown that there is an optimal level of exploitation, dependent upon the characteristic rates of replacement or recovery of the resources involved, at which productivity is greatest. Increasing exploitation rates beyond this level results over time in an absolute decrease in resource procurement. Hunter-gatherer economic returns in any given environmental context cannot be increased on a sustained basis simply by investing more time and effort in resource exploitation, i.e. by attempting simply to harvest more of the same resources (Winterhalder *et al.* 1988). Therefore, given resources with similar rates of recovery from exploitation, human population densities will be lower in regions of lower resource densities.

Local group size, in the same way, largely will be related to local resource density or availability. Thus, it is possible to have local groups of equal size in environments of different overall, regional resource density, and, conversely, local groups of varying size in environments of equivalent resource density. However, small patch size may be rather common in low-resource environments, and it is apparent that even with much clumping of resources, one must reach a point as regional resource densities decline at which local group size of the human populations dependent on those resources must decline also.

Contact and movement among local groups

Given certain biological constraints intrinsic to the species and its reproductive capacities, there must be a point, therefore, at which local group sizes will fall below the minimum size at which the population can maintain itself. Below this point, local groups must have contact for procreation among themselves adequate to encompass a total breeding population of at least minimum equilibrium size, whether this be an open or a closed mating network (cf. Wobst 1974, 1976). If adequate between-group contacts and mating are impossible, and local groups therefore cannot become smaller and still remain reproductively viable as the capacity of local resource patches to support them decreases, then these groups either must migrate to an area of higher resource density or die out.

As both local group size and regional population density decline, there will be fewer alternatives available for finding and obtaining access to possible mates, either locally or regionally. Therefore, the extent of such contact, in terms of the number of groups and thus of the geographical extent of the territory involved, will vary inversely with the size of the local groups.

In addition, no environment is completely stable from year to year, and those with low resource densities seem particularly prone to significant fluctuations. This means, to the degree that these fluctuations are unpredictable, that local groups will not be able to subsist continuously in one place. They will have to move and use resources in other areas at least occasionally. The number of such other areas and the abundance of resources available at them will be a function of overall environmental productivity and predictability, and also a function of resource diversity. As either or both decline, the number of alternatives open to a group in a situation of local resource failure will decrease.

Also, the degree to which resource fluctuations are correlated over space conditions significantly the importance of different scales of geographical movement and contact among hunter-gatherer groups. To the extent that fluctuations in resource abundance are independent and occur at an areal scale smaller than the range of local human groups, these groups will tend to be capable of economic (subsistence) independence. However, as the scale of areal synchrony of resource fluctuation increases, local groups will begin to experience resource failures (and abundance) that encompass their entire ranges and, eventually, larger areas. Under these circumstances inter-group contact and movement will become increasingly advantageous or necessary. At some point, of course, the regional scale of such correlated resource fluctuations will extend beyond the limits of the contact and communicational abilities of the groups involved, and the entire regional human population will be affected more or less uniformly by resource abundance or failure. If such resource failures are severe enough, sustained human habitation of the region will not be possible. Occupation of environments that exhibit relatively large-scale geographical homogeneity thus may require certain levels of human abilities for communication and movement among separate, local groups. Desert and tundra environments seem likely to be of this type.

In short, as resource density, diversity, and predictability become less in any environment, regional human population densities can be expected to decline and local group size can be expected to decrease. As a consequence, therefore, contacts beyond the local group to find appropriate mates, as well as moves to escape shortages and find adequate resources elsewhere will necessarily increase in frequency. Further, the range of alternatives available to a local group in both cases will decrease.

Assurance of access to mates and resources

In such conditions of growing necessity and decreasing alternatives, fruitless contacts or wrong moves would have increasingly serious negative consequences: declining rates of reproduction, partial or total starvation of the group. Therefore, as conditions become more severe, local groups need greater assurance that any given contact or move will be successful. Success means not only that a mating partner will exist or that adequate resources will be available at any particular location, but equally importantly that access to these partners or resources will be possible. In such conditions, reproduction and subsistence require both more, and more reliable, information from beyond the social and geographical bounds of the local group itself, and ever surer means of access to the necessary mates or resources thus known to the group.

LIMITS TO HUMAN OCCUPATION

Sources of limitation

There must be, of course, some ultimate limit to possible human occupation imposed by a level of resources in an environment (taking both densities and recovery rates into account) inadequate to support a biologically viable hunter-gatherer population, no matter how effectively that environment may be exploited. Above this ultimate level, limits to sustained human occupation of any area must be variable and depend on the cultural factors of human technological and organizational abilities. These limits are set in the first instance by the technological means available to procure and utilize the resources present. Purely technical means alone, however, cannot determine entirely the ability of a group to exploit a given environment, since their effectiveness depends in large measure on the capacities of the human group to organize appropriate strategies within which the available technical means are put to use with varying effect.

Within a given, broad level of technological competence, for example, environmental limits to human occupation will be set to a large degree, as indicated above, by the abilities of local groups to establish and utilize contacts among themselves for the transmission of information and to allow movement of individuals and groups to cope with differential availability of potential mating partners and with unpredictable fluctuations of subsistence resources. Inability to obtain adequate information, either in terms of reliability and accuracy or of the area covered, on the existence and location of needed mating partners or subsistence resources, subjects individuals and groups to risks of failure to reproduce at a sufficient rate or of nutritional stress and eventually starvation. As outlined above, such risks and the penalties for failure grow with decreasing resource density and predictability. Inability to obtain access to mates or subsistence, even if located, leads to similar risks of failure and to high stress on individuals and groups in the form of competition and conflict. As individuals or groups increasingly are unable to obtain necessary information or, alternatively, are able to obtain this information but unable to gain necessary access, the result is decreasing ability to reproduce or survive, and ultimately extinction. Consistent failure of this sort in any environment obviously renders it uninhabitable on a sustained basis.

Informational limits

As population density decreases and local group size declines in response to lessening environmental productivity and predictability, a biologically viable human population will encompass an increasingly large geographical area and a greater number of individual local groups. Under these conditions, appropriate mates will be more widely scattered over a larger number of separate groups and a wider geographical extent, while the variation in availability of resources and the geographical dispersion of resource patches likewise will tend to increase. Therefore, the proportion both of the total number of available mates in the population and of the overall configuration of resource presence and absence in the environment that can be known to any local group through direct knowledge and immediate, face-to-face contact inevitably must decline. Thus, it will become increasingly risky to rely only on directly perceived and reported knowledge, and there will be a growing advantage to the ability to obtain knowledge indirectly – that is from beyond the limits of direct perception. Ethnographically known hunter-gatherers in desert and arctic tundra environments appear to represent recent extremes in this direction, and such groups are well known for their extensive travelling, visiting, or ceremonies, which serve in various ways to facilitate and maintain the flow of such indirect information over large distances.

Clearly, therefore, at a certain threshold of population density and local-group size it must become impossible to obtain adequate information for continuous survival through direct perception and immediate, face-to-face communication. Such inability would constitute an effective block to colonization and sustained occupation of such environments by any human group not capable of indirect communication and circulation of information.

Note that the existence of this threshold is not due to a lack of ability to represent symbolically information on the perceived existence of mates or resources and to communicate that information to another individual verbally, in some form of 'language'. The ability to do this of course is important, and as Reynolds and Ziegler (1979) have demonstrated, the area that a local group is able to exploit optimally is highly limited without it. However, basic symbolic reference to concepts, including perceived phenomena, and the capacity to communicate about them on a simple topic-comment basis (cf. Bickerton 1981: 268-269) appear to be adequate for this.

We will argue instead that such a threshold would exist because of limitations in language capacities of a more complex sort. The abilities to form clauses and to use verbs of reporting and perceiving are needed to obtain and communicate information indirectly, since these are the means by which 'displacement' – that is reference to things beyond the here-and-now – is accomplished in language. Not only are these abilities more complex than symbolic reference to concepts and topic-comment communication, but it has been proposed on linguistic grounds that they most probably developed later in the evolution of language (Bickerton 1981: 268-278).

Social (access) limits

Possessing adequate information on the existence and location of mates and resources is not alone sufficient to assure survival, however. As already stressed, access to these must be assured. Such assurance becomes increasingly important as alternatives decrease and the penalties for failure in any given contact or move consequently grow greater. Here, too, there must be a threshold below which the risks of failure become too great in the absence of assured access to permit sustained human presence in an environment. Lack of a reliable means to guarantee access to known mates and resources thus becomes a block to the colonization and long-term occupation of such environments. Reliable granting of access to either mates or resources requires a mechanism whereby the integration of individuals or groups coming together is relatively easy, where mutual rights and obligations may be established immediately or comparatively rapidly among all the actors involved. The contrast with the integration of individuals or groups into primate (particularly ape) societies is striking and instructive here. Such moves among primates typically take time (which may not readily be available in environments with low resource density and few alternatives), usually involve display that often or almost invariably leads to fighting and sometimes injury (an unadaptive result in situations of low population density: Wrangham 1987: 66-68), and above all are uncertain in their outcome (e.g. de Waal 1982: 56-58).

CULTURAL RESPONSES TO LIMITS

Alliance networks

In human societies, Gamble (1983) has suggested that it is generally 'alliance networks' which serve the purpose of assuring such access, as well as of circulating information. Alliances of many sorts creating such networks are established among ethnographically known hunter-gatherers by face-to-face contact and negotiation (cf. Gamble 1986: 54). It is conceivable that simple face-to-face relationships could serve as one level of assurance for access to mates and resources. The prerequisite of simple symbolic representation and communication necessary for such relationships undoubtedly would have been available in the most rudimentary, and presumably earliest, languages (cf. Bickerton 1981: 268). Such alliances, however, could not extend far, probably not beyond adjacent local groups, if indeed that far, and in particular could not be used to establish relations among individuals who had had little or no previous face-to-face contact. A mechanism that would allow the extension of relations beyond adjacent local groups through the establishment of rights and obligations among individuals without prior face- to-face negotiation of such relations would require more than simple symbolic representation of concepts and topic-comment communicative abilities. It would require the ability to discourse beyond the here-and-now (i.e. 'displacement'), and the development of full symbolic systems within which individuals are identified not as unique persons but in terms of symbolic categories, among which mutual rights and obligations are defined by the system, whatever the history of actual interpersonal encounters among the members of the group sharing that system.

Kinship systems

Kinship systems are of course the universal and most fundamental systems of this sort in human society. They appear to be basic to the establishment of many of the other sorts of alliance systems seen in hunter-gatherer societies (e.g. Wiessner 1982: 66), especially as they constitute the only reliable mechanism for the extension of relationships beyond the range of regular face-to-face contacts. Examples are not uncommon of the intentional manipulation of kinship systems to allow such extensions for the integration of individuals and groups coming together, often for the first time, and the network of kin relations provides the primary foundation on which most other alliance systems are built and most movement of hunter-gatherers, either individually or in groups, is based, whether on visits or in need.

Simple systems of linguistic representation of face-to-face relationships, especially those which are biologically (and perhaps socially) most directly perceivable, could well have existed at even the hypothetically most elementary level of language development, in which not only primary concepts presumably existed, but also secondary concepts formed from combinations of the primary ones. However, ethnographically studied kinship systems regularly include representation of relationships that are not directly perceivable in biological terms, including some that are not biologically based at all (e.g. affinal relations). These systems also allow reference to relationships that may or may not exist in any particular social group at a given moment, simply as a matter of chance. The expansion of simple 'kinship' systems in these directions requires more than just the symbolic representation and combination of concepts. It requires minimally an ability to refer beyond the here and now – i.e. displacement – and it requires the construction of symbolic systems that in major part, if not totally, are created from the manipulation of symbols and lack a concrete perceptual basis.

CULTURAL RESPONSES IN THE UPPER PALAEOLITHIC

Thus, it would seem that two of the important blocks to human expansion at any given level of technology into environments of low resource density and predictability are defined by communicational and social abilities: the ability to know about mates and resources within the physically attainable region, and the ability to assure access to these once known. Both the arctic tundra (as well as the adjacent boreal forest) and the Australian desert appear in general to be such environments. The climatic, vegetational, and faunal differences between them make it highly unlikely that their colonization and the subsequent maintenance and expansion of human populations in both areas in the earlier Upper Palaeolithic was due simply to the invention in this period of the requisite technical means for their exploitation.

On the other hand, the development of the communicational abilities necessary for: (1) the adequate transfer of information, and (2) the creation of the social networks necessary for such information flow, as well as for assurance of access to mates and resources over an adequately large area and sufficient number of local groups, both seem to repose on the emergence of a roughly equivalent level of language competence. An explanation of the major demographic events of the Upper Palaeolithic in these terms would complement and refine in some details Gamble's (1983, 1986: 344, 377-378) suggestions that it was primarily social rather than technological developments that allowed human occupation of such environments. This picture may be tempting, particularly if we tie it into the numerous other observations that there was a sudden 'explosion' of symbolic capacity and behaviour with the advent of the Upper Palaeolithic. However, there are some other factors still to consider which suggest that it is not so clear that the relatively sudden 'familiarity' of culturally organized behaviour in the Upper Palaeolithic is due only to the development at this time of human symbolic capacities, with the consequent emergence of alliance networks, including kinship systems, and information exchange among local groups.

IMPLICATIONS OF UPPER PALAEOLITHIC RESPONSES

Kinship and the size of biologically stable populations

In the first place, we must remember that kinship systems strongly influence the minimum equilibrium size of a human (breeding) population, so much so that within a range of reasonable biological parameters for a hunting-gathering society one can say that the minimum equilibrium size is largely a function of the extent of the 'incest taboo' as defined by marriage restrictions inherent in the kinship system of the population (Wobst 1974). Thus, as kinship comes to replace presumably instinctive and relatively invariant patterns of incest avoidance, the minimum equilibrium size moves from a relative constant for any given set of birth and death rates to a variable, dependent as much on cultural factors as on the biology of the species. It is already well known that in certain instances these cultural factors (i.e. kinship systems) vary systematically in relation to environmental productivity. However, in Australia the effect has been observed to be one of extending marriage prohibitions through increasing complication of the kinship system, thereby increasing the minimum size of a reproductively viable population, as environmental productivity declines (Yengoyan 1968). A consequence of this relationship between environmental productivity and complexity of the kinship system is an extension of the network of necessary mating relations over an ever-larger area (the more so since population densities are also declining with decreasing environmental productivity) as resources become scarcer. One of the significant effects of this is the incorporation of an ever-greater range of environmental variability within the extent of a mating network in less productive areas. The incorporation of such increasing ranges of territory and potential environmental variation may be advantageous to the exploitation of such sparse, unpredictable environments, perhaps even to long-term survival in them.

It seems, therefore, that instead of simply *allowing* contact among local groups over a large area (i.e. beyond the bounds of face-to-face contact among adjacent groups) – and thus permitting regional alliance networks, the flow of information over a wide area, and access to mates and resources over this area – kinship systems may in fact sometimes *require* these things in a population. Kinship systems, and alliance networks predicated on their existence, therefore do not create just a permissive potential to cultural systems. Rather, they may vary in such a way as to constitute a significant mechanism that determines a number of population and cultural parameters in situations of adaptation to certain kinds of environments. In the case of Australia, this mechanism extended 'incest' prohibitions eventually to extremes unheard of in other species, and the example is particularly interesting with respect to the Upper Palaeolithic colonization of this area.

However, among the Eskimo in arctic tundra environments we do not find anything like the elaboration of kinship seen among the Australians. Although kinship systems are fundamental to inter-group relations in Eskimo society and are used as networks for communication of information and help in times of resource failure, these systems are of comparative simplicity.

Logistical economic organisation

The Eskimo, on the other hand, do show another aspect of adaptation to low resource density in one of its most highly developed forms as we know it ethnographically, and that is logistically organized exploitation of their environment – what has been called a 'collector' strategy, as opposed to a 'forager', or non-logistically organized strategy, typical of Australian hunter-gatherers (Binford 1980). We can expect the use of logistical strategies to be increasingly advantageous as the differential spatial distribution of resources within the environment becomes more marked and as the diversity of adequate alternative resources declines – in other words as resource distributions tend to become more 'patchy' and these patches tend to consist of a smaller variety of resources. Under such environmental conditions, resources tend to become more strongly differentially distributed in time as well as space, and storage tends to become more important as a part of logistical strategies of exploitation. Binford (1980) further has related all three of these trends to large-scale geographical variation in length of growing season and/or differential distribution of rainfall throughout the year. From all points of view, the arctic tundra represents an extreme in these aspects of environmental conditions and resource distributions, while Australia and other desert environments represent significantly less extreme situations.

Although the use of either 'logistical' or 'foraging' strategies is not (as it is often is portrayed in the current archaeological literature) a strictly 'either/or' proposition, in terms of the overall organization of any hunter-gatherer economy, it seems likely that here, too, there may be a threshold in the ability of a human group to occupy certain environments without adopting such strategies for at least part of their subsistence activities. As resources become increasingly differentiated in space (and time), there must be a point at which logistical strategies of resource exploitation ultimately become indispensable for assured survival and long-term occupation. The arctic tundra seems to represent one such case, where these strategies are necessary rather than simply adaptively advantageous. If human groups were unable to organize themselves effectively to carry out these strategies, they would be blocked from colonization and sustained occupation of such environments. Yet, the organization of logistical exploitation strategies requires the existence of certain capacities that, although taken for granted in fully modern man and cultural systems, cannot be assumed to have been always present, and must have come into being sometime during the course of Palaeolithic cultural evolution.

Division of labour

An immediately obvious organizational prerequisite is an effective division of labour within the group, creating clearly defined rights and obligations with respect to subsistence activities for each category of actors within the system. The level of organizational capacities required for at least a rudimentary division of labour may not be very high, however. At its simplest level, it demands an ability to recognize stable, categorical distinctions among members of the group and to associate differentially a set of obligatory behaviours and expected rights with each category. Such a basic organizational division of labour in subsistence pursuits seems perhaps possible even prior to the development of any language capabilities. Still, a division of labour does not seem to exist in contemporary primate societies.

At least basic symbolic representations may be prerequisite for the systematic definition of the categorical divisions involved and the differential association of rights and obligations with these divisions, which in all likelihood must have been perceived in the first instance as the biologically and behaviourally obvious ones of (adult) male and (adult) female. If so, even the most fundamental level hypothesized as the first stage in the emergence of language, in which primary and secondary concepts are given symbolic expression and manipulated verbally with reference to here-and-now relationships (Bickerton 1981: 268- 269), should be adequate for a basic division of labour. (As discussed briefly below, the extension of division of labour to the more complex 'separation of labour', which can convey significant adaptive advantages to hunter-gatherer groups, perhaps requires much more evolved language systems, including the more abstract and 'displaced' symbolic systems of reference typical of developed kinship systems.) At least a rough division of labour therefore ought to have been possible very early in the course of cultural evolution. However, with respect to the Upper Palaeolithic colonization of the arctic tundra and the role of logistical exploitation strategies in the occupation of this environment, there is another prerequisite to such strategies that seems to be of more relevance and critical importance.

Future planning. Logistical organization and decisions require planning ahead, as has been stressed strongly by others (e.g. Gamble 1986: 15, 38, 47 *et passim*). The capacity to plan ahead certainly exists on an individual basis even among primates. The capacity to do this on a group basis, however, requires an ability to communicate among the members of the group, and in so doing to refer to and discuss anticipated events in the future. This ability is not a component of the hypothesized earliest stages in the development of language competence, and in fact the 'tense-modality-aspect systems', which in language permit the incorporation of the future into discourse, are, for reasons quite unrelated to our considerations here, hypothesized to have been the last major elements added to the developing human capacity for language (Bickerton 1981: 278). If this is so, it may well be significant in terms of a relatively late development of logistically organized adaptation to environments characterized by increasing degrees of differential distribution, abundance, and diversity of resources.

If future planning is essential to the organization of logistical economic strategies, the emergence of the capacity to refer to the future in language may be expected to have played a decisive role in the appearance of these adaptive strategies. Groups lacking such language capabilities would have been incapable of regularly and reliably organizing themselves for logistically exploiting their environments, and therefore would have operated with less and less efficiency in environments of increasing differential resource distribution and decreasing resource variety. Ultimately, they would have been unable to colonize and permanently occupy certain environments that exhibited extremes of these characteristics, including very probably the arctic tundra.

It may be worthwhile here to consider some of the other consequences for human cultural systems of developing language capabilities. As we just have outlined, the ability to refer to the future opens up important, perhaps in some environments critical, new possibilities for the organization of resource procurement activities. The implications of the addition of a past tense seem equally important. In Bickerton's (1981: 280- 286) hypothetical reconstruction, past tense distinctions were added as the last element in the development of tense-modality- aspect systems, immediately after the addition of a future tense, and were the final step in the evolution of basic language competence.

Time depth of adaptation to resource fluctuations

Reference to the past is critical in order to make predictions about the future. It is on the basis of generalizations drawn from past experience that future events may be anticipated (cf. Bickerton 1981: 225-226, 270). Moving from a situation in which knowledge of the past is restricted to individual memories to the sharing of such memories through an ability to refer to them and talk about them creates something new: a group memory. Group memory provides entirely new opportunities, and results in the development of a number of cultural mechanisms for information storage, retrieval, and use, whose roles within cultural systems are complex but highly important (cf. Minc 1986). Without going into the details of any of these, we only will consider briefly here the fact that the fundamental effect of any and all of these mechanisms, beginning with simple verbally-shared individual memories, the basic and possibly earliest form of group memory, is that they increase the time depth of information available to the members of any group significantly beyond the life-span of an individual.

The most obvious effect of this expansion of the time depth of information available to a hunter-gatherer group is that it allows adaptation to environmental fluctuations with a periodicity greater than the average individual life span. An initial extension of the time span over which human populations can adapt to environmental variability will be available immediately to a group within which individual memories are shared and discussed to form a 'group memory'. One can expect that this immediate expansion, and the subsequently observable period over which information on environmental fluctuations and potential adaptive responses are maintained in simple hunting- gathering societies by direct oral transmission of information, would be on the order of three to five generations, some 60-125 years. The development of other cultural mechanisms such as ritual, myth, etc., with important functions for longer-term storage of information and adaptive responses, would increase this time depth significantly (cf. Minc 1986), although for any cultural system there will be environmental fluctuations whose impact cannot be avoided or mitigated, because of their severity and low frequency. Regardless of the exact depths of time involved, the important consequence would be that human populations would shift from being limited by critical resource cycles one generation in length to cycles several times that length, in effect shifting the focus of adaptive responses from shorter-term, smaller resource fluctuations toward increasingly longer-term, larger fluctuations.

Carrying capacity and human population fluctuations

One result of these shifts, universally observable today among hunter-gatherers and the source of frequent comment and speculation, is that human population densities would appear more and more of the time to be below the supposed 'carrying capacity' of their environment. The resource 'lows' to which the existence of a shared, group memory allows a population to adapt would be less frequent, with the consequence that periods of resource stress due to fluctuations that reach or exceed a group's abilities to respond effectively would occur at wider intervals. The phenomenon of adaptive responses being tuned to longer-term, larger resource fluctuations, so that field studies and observation (including ethnographic expeditions) normally find populations under non-stressed conditions in which many important adaptive responses are latent and cannot be seen in operation, is well known also in the ecological study of many species (cf. Wiens 1977). (It may be well to bear this phenomenon in mind in archaeology, also, when thinking about our ability to use archaeologically reconstructed 'moments in time', be they individual occupations, settlement systems, or whatever, as data in the study of processes of prehistoric adaptation and cultural evolution.) The important point with respect to human populations is that evolving cultural mechanisms allow, at certain points, the significant expansion of the time span over which adaptation to environmental fluctuation is possible.

Human population fluctuations accordingly could be expected to be damped with adaptation to longer-term resource cycles. If, as seems likely, populations tend to grow or decline in response to changes in environmental productivity whose periodicity is just beyond their ability to perceive, such fluctuations should decrease in frequency as this ability extends over longer periods of time, eventually reaching a point at which population size and density will be seen at most times to be stable, although 'well below carrying capacity'.

We might suggest further that the damping of population fluctuations that are due to variations in resource availability may be a significant part of the process of development and maintenance of cultural systems as a basis for the organization of human behaviour. This suggestion is predicated on the fact that even the simplest cultural systems do not exist complete within any individual's repertoire of behaviour and knowledge. Cultural systems, as some (e.g. Binford 1965: 205) have stressed, are participated in differentially by the members of any society, and the diverse elements of these systems are not shared equally among individuals. Thus, if continual and unpredictable fluctuations in size are a characteristic of local populations of any species, and if such fluctuations continually result in the periodic random (non-selective, accidental) removal of elements (behaviours, knowledge) composing the cultural system of that population, it would be virtually impossible for such a system either to stabilize or to evolve.

Therefore, the appearance of ways to extend the range of temporal perception (time depth) of the environment must significantly enhance the ability of cultural systems to evolve and stabilize as human populations become less buffeted by unpredictable variability in their environments. In human societies, in fact, one probably reaches a point in this way at which parameters of human population biology and cultural organization of behaviour become as, or more, important than variations in environmental productivity in their impact on demography.

Nonetheless, a population in which behaviour largely is learned rather than genetically programmed cannot develop and maintain responses to environmental variation that is beyond its range of (temporal) perception. One means to mitigate the negative impact of random removal of elements from cultural systems exposed to such fluctuations is maintenance of as great uniformity as possible in the sharing of these elements among individuals – in other words, increasing redundancy among individuals as participants in the system. It is perhaps worth considering also that this may play a role in the observed maintenance of strongly egalitarian organization in small-scale human societies in marginal environments.

EGALITARIAN SOCIAL ORGANISATION

Group size and hierarchical organization

One of the major cultural factors affecting demographic arrangements in human populations is the strong tendency for social differentiation and hierarchicalization of group organization as the number of effective decision-making units in a group reaches and begins to exceed six. It has been shown several times that, although decision-making ability is improved by increasing the number of units involved up to six, the quality of decision-making decreases rapidly beyond that point (Johnson 1978, 1982, 1983; Reynolds 1984). In larger groups there are increasing pressures for the introduction of a hierarchy – either 'vertical' or 'serial' in Johnson's terms – to facilitate or allow effective group organization and action.

As we have suggested above, however, there may be equally strong counteractive forces in certain situations, in which the introduction of social differentiation and hierarchy into a human society might be significantly disadvantageous and selected against. In such circumstances, the organizational optimum of six decision-making units to an effective social group would create a 'limit' to group size. It is perhaps in this light that the universally observed number of some 25-30 members (6 nuclear families) in recent hunter-gatherer minimal bands may be seen to have an organizational basis. Both the increasing quality of decisions with the involvement of more decision-makers (Johnson 1978) and the exponentially decreasing risk of group extinction (Wobst 1974: 171-172) would confer advantages to larger groups up to the level of six family units, beyond which it becomes increasingly difficult to maintain both effective decision-making and an egalitarian organization.

This systematic change in decision-making quality with changing numbers of participating decision-makers is an empirically observed relation. We do not know the exact basis for this relationship other than that it must somehow be linked to human information handling capacities. These in turn must somehow be related further to individual mental capacities. But, since language is perhaps the prime means by which humans represent concepts (not only to others but even more basically to themselves) and manipulate them in thought (cf. Bickerton 1981: 218 *et passim*), it seems rather likely that the development of human information processing, individually as well as in groups, is tied closely to the development of language. The conclusion therefore must be that the constraints on human decision-making abilities evolved along with the evolution of language, and that the relation between quality of decisions and number of decision-makers must also be a product of this evolution and not have existed from the beginning of hominid cultural evolution.

Evolutionary position of egalitarian cultural systems

We do not know what relationship may have existed between group size (number of decision-makers) and quality of decisions prior to that which is now empirically observable. If we look quickly at something that may represent the opposite extreme, and take the 'voting' model for decision-making in groups without language or other symbolic communicational abilities (cf. Reynolds and Ziegler 1979), one might suggest that adaptive advantage would accrue with increasing group size as greater numbers allowed more thorough territorial coverage and resulted in more accurate representation of the resource distributions thus perceived in 'voting' movements. In this case, the long-term evolutionary trend in egalitarian group size as the emergence of language enhanced communicational abilities might be expected to be from larger to smaller minimal local groups. In any case, the contemporary observation of a relatively constant size of 25-30 individuals for minimal bands among hunter-gatherers ought not to be projected too rigidly or too far into the past. It is better to consider minimal local group size as a variable, whose observed value at different times and in different environments may inform us (if we know how to interpret it in relation to other factors) on aspects of the total organizational system characteristic of the population involved.

Among contemporary hunter-gatherers, the decline in decision-making ability beyond the 'limit' of six decision-making units in a group becomes an impediment to increasing group size only if there is an impediment to hierarchicalization – i.e. an adaptive or organizational disadvantage to social differentiation or to the institutionalization of status differences within the society. What, however, would constitute adaptive or organizational disadvantage in these cases? Given the decision-making advantages inherent in either or both 'serial' or 'vertical' hierarchicalization (cf. Reynolds 1984), and given that it has been argued that in situations of potential differential distribution of status, wealth, or authority, even random processes might be expected to produce a non-egalitarian organization significantly more often than not (Mayhew and Schollaert 1980), it seems to us that the question of how 'complex' societies arose (which has been a central question behind much anthropological and archaeological research) might profitably be reversed. It is perhaps more difficult to explain why certain societies remain egalitarian and why we observe in them so many cultural mechanisms to ensure the maintenance of egalitarian organization (cf. Wiessner 1982), than it is now to explain the development of status differentiation and hierarchical organization in many situations.

Distribution of resources

Probably many of the factors that significantly select for egalitarian organization in human societies are related to conditions of relatively low resource density and hence relatively low human population density – the conditions in fact under which the major Upper Palaeolithic demographic expansions occurred. In such environments, the factor that comes immediately to mind is the economic advantage, or even necessity, of at least relatively equal access to resources, which most frequently implies their more or less equable distribution. This is so because success in resource procurement is typically unequal in situations of low resource density, and especially so as these resources become less predictable in their spatial or temporal availability. Distribution thus is necessary in order to assure access by all group members. As at least roughly equal access to resources becomes of greater importance in the adaptation and survival of the population, egalitarian organization concomitantly becomes more advantageous, particularly as minimum-equilibrium-size populations become spread over geographical ranges that approach the maximum that physically can be covered by available means of transportation.

Under these conditions, stress on, and eventual loss to the population of any sub-group (perhaps eventually even individuals) capable or potentially capable of reproduction will have increasingly negative consequences for survival of the population as a whole. Further, if there are geographical and social limits to the ability of local groups in this population to obtain information or to have access to mates and resources over an area and from other local groups, such consequences would appear at significantly higher population densities, as minimum equilibrium population sizes are reached in smaller geographical areas.

'Separation of labour'

Egalitarian organization is conducive to another aspect of economic organization that may confer significant adaptive advantages to groups in environments characterized by unpredictable differential availability of resources in space and/or time, and that is *separation* of labour. Separation of labour should be distinguished from *division* of labour, discussed above. In a division of labour, individuals or sub-groups carry out different tasks, the results or fruits of which are combined in various ways to make up an integrated range of economic activities in a society. In separation of labour, by contrast, the individuals or sub-groups involved carry out substantially or exactly the *same* activities, but separately, in different places or at different times. The individuals or sub-groups thus are not accomplishing different aspects of a single, integrated economic activity. They rather are all performing the same economic activity, but this performance is now dispersed in space and time. The result when applied to the search for unpredictably scattered resources is to increase substantially the probability that such resources will be found by one or another of the individuals or sub-groups. Such separation of labour can exist of course within a system of division of labour. Good examples among hunter-gatherers would be male hunters individually setting out on different mornings or in different directions to look for game, and females dividing up into a series of small parties that move off in different directions from camp to search for plant foods.

For such separation of labour to provide an advantage to the economic system of a group, however, the results of any individual's or sub-group's activities, both substantively and in terms of information gathered, must be made available to the group as a whole (or at least the major part of it). If such activities are not carried out in close proximity, so that verbal calls suffice to bring the entire group together for joint exploitation of the resources located by an individual or sub-group, cultural means to ensure the equable distribution of resources found and brought back to the group are essential. If these means exist, it is clear that separation of labour, where possible in any aspect of a group's exploitation of uncertain resources, can increase substantially the probabilities of success in that part of the group's economic activities. Winterhalder (1986) has demonstrated the significant reduction in risk in hunter-gatherer procurement systems that such separation of labour provides. An egalitarian organization of access to the fruits of procurement activities is thus essential to the separation of labour as a means to exploit environments of scarce and unpredictable resources with less risk.

Adaptive significance of egalitarian organization

It is important to stress once again here that egalitarian organization is not something that 'just happens' in human society. Rather, it is something that is created and actively maintained by cultural mechanisms. This is not true only in contrast to the emergence of socially differentiated, non-egalitarian systems at one end of the evolutionary spectrum. If we adopt a basically primate model (more specifically ape and in particular chimpanzee) for the hypothetical 'other end' of this spectrum, we can see, among other things, that the typical pattern of dominance relations among individuals (especially if it involved highly differential access to certain resources: e.g. de Waal 1982: 185-186, 200) would not be particularly adaptive in environments of low resource density and predictability. It would prevent an effective internal distribution of resources in the group, which therefore could not take advantage of the risk reduction available with division of labour and particularly with separation of labour.

Given such a primate model, we can see that local populations would have had to be at or above minimum equilibrium size (even if that were significantly smaller than for modern man, it seems unlikely that it could have been as low as the 25-30 individuals typical of minimum hunter-gatherer bands in contemporary low-resource environments), or they would have been restricted as are primates today to environments where individuals could have survived alone for sufficient time to establish the direct, face-to-face contacts and dominance relations necessary for between-group moves. In either case, the conclusion must be that a primate model for such early hominid societies implies an adaptive restriction to more resource-rich and predictable environments. The developments that would have led to a breaking of that restriction would have had to include, importantly, the replacement of ape-like systems of interpersonal dominance established through relatively constant display, combat, and trial and error, by systems of at least relatively egalitarian, stable, and reliable relations of rights and obligations among individuals both within and between local groups.

CONCLUSION

A general model for the Upper Palaeolithic

Thus, it is clear that the major demographic expansions of the Upper Palaeolithic into two quite different but resource-limited and locally variable environments are probably not quite so simply explained by the equation: emergence of alliance networks = possibilities of information exchange and movement among local groups = capacity to colonize marginal and unpredictable environments, even if we add to this equation the postulate of emerging linguistic underpinnings for alliances and information exchange. Certainly, much may lie in this equation, but it seems more than likely that the major population expansions of the Upper Palaeolithic into Australia and into the Siberian tundra – ultimately leading on into the New World – were made possible also by a series of other, newly developed, organizational properties of human cultural systems, including at least logistical exploitative strategies, significant extension of the time depth of adaptive response to environmental variation, separation in addition to division of labour, extended kinship systems, and mechanisms to create and maintain egalitarian relations within social groups, as well as the emergent language capabilities that are prerequisite to these organizational characteristics. We therefore must take into account the biological evolution of human capacities for language (cf. Laitman 1983; Bickerton 1981; Lieberman, this volume) as well as cultural evolution in the Palaeolithic.

Emergence of Upper Palaeolithic cultural systems

It seems fairly clear that all the basic elements of the full capacity for language and of the capacity for cultural organization existed in the Upper Palaeolithic. The relatively sudden expansion at this time into new and difficult environments, whose exploitation on a sustained basis demands all these capacities, argues further for their emergence at this time, breaking the communicational and organizational 'blocks' to such colonization that must have existed previously (cf. Gamble 1983). All this does not mean, though, that fully developed language and cultural systems sprang up suddenly in the Upper Palaeolithic out of nothing. In fact, it seems most unlikely that fully developed language capacities could have emerged from anything other that an already evolving system of symbolic communication (cf. Bickerton 1981: 261, *et passim*). It seems equally unlikely that kinship, for example, as an organizing principle in cultural systems, could have emerged from an organization entirely lacking in the definition of social roles and positions independent of the qualities and characters of the specific individuals involved. Just as any evolutionary process involves the transformation of structures and populations that have the potential for development in certain directions, so also with the evolution of language and culture we can expect the process to have led through a series of intermediate stages from presumed 'animal' or 'primate-like', pre-human beginnings to their present, ethnographically familiar forms.

Thus, although the earlier Upper Palaeolithic may be seen as a period of rapid evolutionary change, there is no reason to expect that the evolving cultural and communicational systems (when examined in detail) will show an even and uniform progression of development in all elements or capacities simultaneously. For example, Bickerton's (1981: 268-290) hypothesized sequence of additions to human language capabilities consists of a number of steps, in which the capacities for 'displacement' that are perhaps necessary to the elaboration of kinship systems are proposed to have begun before the full development of the ability to use tenses and to refer to future and past events – i.e. the features which we have seen here as prerequisite to the organization of logistical economic strategies and stable adaptation to environments with important resource fluctuations on a time scale significantly longer than the average human life span. Would such a sequence of increments to human language, and thus organizational, abilities explain the presently apparent difference of some thousands of years in the earlier colonization of Australia followed by a later expansion into the Siberian tundra (cf. Gamble 1986: 382)? We do not know, and such suggestions are for the moment highly speculative. Yet we do know that the organizational and communicational (i.e. cultural) requirements are different in the two cases, and we do not imagine early human populations just waiting on the fringes of these areas until the fancy strikes to colonize them.

Nature of earlier cultural systems

Similarly, on a larger scale, the Middle Palaeolithic cannot be characterized simply as 'pre-cultural', without some more accurate specification of what the communicational and organizational systems of this period might be expected to have consisted of, and what elements they lacked in comparison to the Upper Palaeolithic and ethnographically known systems. This is a more difficult problem, yet at least some critical guidance may be available from our discussions above. In the first place, it seems unlikely that immediately pre-Upper Palaeolithic populations could not 'speak'. The proposition is simply that the language (or perhaps rather 'proto-language') that was spoken probably lacked certain elements typical of fully developed human languages today. We might propose that such proto-languages lacked at least tense-modality-aspect systems. Probably they also lacked the ability to form clauses, and hence displacement, in discourse. These are the elements that were proposed above as essential to such features as the elaboration of extended kinship systems, communication beyond face-to-face encounters and exchange of information beyond the here-and-now, the organization of logistical economic strategies, and the extension of the time depth of adaptation to environmental fluctuations. Therefore, all of these organizational features may be proposed to have been absent as well, leading to the 'blocking' of human populations in this period from colonization and long-term occupation of both tundra and desert environments.

If so, what did such systems look like in positive terms? Here it is much more difficult to be precise, even on a speculative basis. Local groups would probably have been capable, at least on a face-to-face, here-and-now basis, of communicating information on perceived conditions in their immediate territories, and thus of foraging optimally within them. If such communication were possible, and if even restricted 'kinship' systems of relationships, rights and obligations among individuals existed to allow access to mates and resources in adjacent groups and their territories, this would have consequences in the extent of contact and use of resources available to any local group within a region. Cultural systems of this period should differ in this respect from more complex, subsequent systems, as well as less complex, earlier ones. Such differences may be monitored to some degree in some archaeological situations through the determination of the patterns of raw material distributions. This has been documented in support of the proposed emergence of regional alliance networks in the Upper Palaeolithic (Gamble 1986) and could have interesting results if systematically applied also to earlier periods. (Such data as are currently readily available do suggest the existence of shifts of appropriate relative magnitude in the scale of areal distribution of raw material, but their timing implies that Lower and Middle Palaeolithic cultural evolution is more complex than the brief comments offered here can do justice to.)

At the same time, a relative limit on the geographical extent of possible contacts and access to mates would imply (unless significant biological and ecological differences affected birth and death rates so as to lower the minimum equilibrium sizes of these populations) that minimal local groups would have had to be larger than in the Upper Palaeolithic and later periods. Such larger local group sizes may have been more feasible in these 'proto-human' societies because they may well have been under less organizational pressure to differentiate internally and develop more complex, hierarchical decision-making structures, if these pressures are (as presumed) linked to the complexities generated in language-based information processing. On the other hand, the shorter time depth of information on environmental variability available to these groups, if they lacked past tense reference and consequently a usable 'group memory', would imply greater and more frequent fluctuations of the human population, with observable population densities at any moment being closer to the calculated carrying capacity of their environment than is the case among ethnographically observed hunter-gatherers, or, we assume, in the Upper Palaeolithic.

Models: variables versus analogies

The challenge now is to develop further, more specifically and in more detail, such hypothetical models of Upper Palaeolithic and earlier adaptive systems and demographic arrangements. From these we must then proceed to draw concrete implications for the archaeological record to allow confrontation of these models with data from the past. Undoubtedly, many of these implications will demand new sorts of observations on the Palaeolithic archaeological record in our effort to determine just what happened and when in the course of human biological and cultural evolution. In this process, we have so far managed to identify certain of the surely relevant variables and to understand at least partially the relations among them, some of which have been outlined and discussed in this paper.

One thing, however, is abundantly clear: we have to stop thinking about all of the Palaeolithic archaeological record as simply the remains of variously organized hunter-gatherer systems, somewhat on the model of simple, or simplified, ethnographically-documented hunter-gatherers. In the course of hominid evolution, we observe a progression from more or less 'no culture' to 'full culture'. As should be evident from the considerations touched on in this paper, not only are ethnographically known hunter-gatherers no analogy for the early and intermediate stages of this sequence, but neither are chimpanzees or any other primate species. The adaptive systems and demographic patterns of these stages may have been very different from any such contemporary models, just as human biology manifestly was different. Archaeologically, the only way to understand these now-vanished stages is to build fully hypothetical models, not based directly on any modern analogy but constructed instead from a knowledge of the major relevant variables involved and their interrelationships. We hope that some of the points made or touched on above will prove useful in this process.

ACKNOWLEDGEMENTS

The author would like to thank Anick Coudart, Clive Gamble, John Speth, and Bruce Winterhalder for extensive comments and criticisms, as well as for several helpful discussions. None are responsible, however, for my perversity in not following their advice and for the problems that therefore remain in this paper. Many thanks also must go to Clive Gamble for providing his extremely useful publications in this area.

REFERENCES

Bickerton, D. 1981. *Roots of Language*. Ann Arbor: Karoma.

Binford, L. R. 1965. Archaeological systematics and the study of culture process. *American Antiquity* 31: 203-210.

Binford, L. R. 1980. Willow smoke and dogs tails: hunter-gatherer settlement systems and archaeological site formation. *American Antiquity* 45: 4-20.

Birdsell, J. B. 1953. Some environmental and cultural factors influencing the structuring of Australian Aboriginal populations. *American Naturalist* 87: 171-207.

Birdsell, J. B. 1968. Some predictions for the Pleistocene based on equilibrium systems among recent hunter-gatherers. In R. B. Lee and I. Devore (eds) *Man the Hunter*. New York: Aldine: 229-240.

Campbell, J. M. 1978. Aboriginal human overkill of game populations: examples from interior North Alaska. In R. C. Dunnell and E. S. Hall, Jr. (eds) *Archaeological Essays in Honor of Irving B. Rouse*. The Hague: Mouton: 179-208.

Gamble, C. 1983. Culture and society in the Upper Palaeolithic of Europe. In G. Bailey (ed.) *Hunter-Gatherer Economy in Prehistory*. Cambridge: Cambridge University Press: 201-211.

Gamble, C. 1986. *The Palaeolithic Settlement of Europe*. Cambridge: Cambridge University Press.

Johnson, G. A. 1978. Information sources and the development of decision-making organizations. In C. L. Redman *et al.* (eds) *Social Archeology: Beyond Subsistence and Dating*. New York: Academic Press: 87-112.

Johnson, G. A. 1982. Organizational structure and scalar stress. In A. C. Renfrew, M. J. Rowlands and B. A. Segraves (eds) *Theory and Explanation in Archaeology: the Southampton Conference*. New York: Academic Press: 389-421.

Johnson, G. A. 1983. Decision-making organization and pastoral nomad camp size. *Human Ecology* 11: 175-200.

Laitman, J. T. 1983. The evolution of the hominid upper respiratory system and implications for the origins of speech. In E. de Grolier (ed.) *Glossogenetics*. New York: Harwood Academic Publishers: 63-90.

Mayhew, B. H. and Schollaert, P. T. 1980. The concentration of wealth: a sociological model. *Sociological Forces* 13: 1-35.

Minc, L. 1986. Scarcity and survival: the role of oral tradition in mediating subsistence crises. *Journal of Anthropological Archaeology* 5: 39-113.

Reynolds, R. G. 1984. A computational model of hierarchical decision systems. *Journal of Anthropological Archaeology* 3: 159- 189.

Reynolds, R. G. and Ziegler, B. D. 1979. A formal mathematical model for the operation of consensus-based hunting-gathering bands. In A. C. Renfrew and K. Cooke (eds) *Transformations: Mathematical Approaches to Culture Change*. New York: Academic Press: 405-418.

Wiens, J. A. 1977. On competition and variable environments. *American*

Scientist 65: 590-597.

Wiessner, P. 1982. Risk, reciprocity and social influences on !Kung San economics. In E. Leacock and R. Lee (eds) *Politics and History in Band Societies*. Cambridge: Cambridge University Press: 61-84.

Winterhalder, B. 1986. Diet choice, risk and food sharing in a stochastic environment. *Journal of Anthropological Archaeology* 5: 369-392.

Winterhalder, B., Baillargeon, W., Cappelletto, F., Daniel, R. and Prescott, C. 1988. The population ecology of hunter-gatherers and their prey. *Journal of Anthropological Archaeology* 7: 289-328.

Wobst, H. M. 1974. Boundary conditions for Paleolithic social systems: a simulation approach. *American Antiquity* 39:

Wobst, H. M. 1976. Locational relationships in Paleolithic society. *Journal of Human Evolution* 5: 49-58.

Wrangham, R. W. 1987. The significance of African apes for reconstructing human social evolution. In W. G. Kinzey (ed.) *The Evolution of Human Behavior: Primate Models*. Albany: State University of New York Press: 51-71.

Yengoyan, A. A. 1968. Demographic and ecological influences on Aboriginal Australian marriage sections. In R. B. Lee and I. Devore (eds) *Man the Hunter*. New York: Aldine: 185-199.

25. Evolution of the Human Psyche

RICHARD D. ALEXANDER

> The gap (between us and our nearest living relatives, the apes. . .) is largest, and most difficult to comprehend, in terms of mind. . . As human beings are distinguished so much by their minds, . . . those minds must be a legitimate object of evolutionary studies (Gowlett 1984: 167 and 188).

INTRODUCTION

The purpose of this essay is to develop and test hypotheses about the process and pattern by which the human psyche evolved, and to seek to understand why humans, and humans alone, differ strikingly in mentality from their closest relatives – and evidently from all other organisms. Understanding the human psyche is a key to understanding human sociality (1) as it relates to the behaviour of individuals in different circumstances and after different kinds of learning experiences or developmental events and (2) as it yields variations in cultural patterns in different environments, and following different histories, including extreme and complex phenomena such as the rise of nations.

By the human 'psyche' I mean the entire collection of activities and tendencies that make up human mentality. I include concepts such as (1) *consciousness* and all of its correlatives or components, such as subconsciousness, self-awareness, conscience, foresight, intent, will, planning, purpose, scenario-building, memory, thought, reflection, imagination, ability to deceive and self-deceive, and representational ability; (2) *cognition* (i.e. learning, logic, reasoning, intelligence, problem-solving ability); (3) *linguistic ability;* (4) *the emotions* (grief, depression, elation, excitement, enthusiasm, anger, fear, indignation, embarrassment, despair, guilt, uncertainty, etc.); and (5) *personality traits* (stubbornness, pliancy, subservience, timidity, persistence, arrogance, audacity, etc.).

One can analyse human mentality by (a) morphological and physiological studies of the brain and its functions; (b) psychological and psychoanalytical investigation of behaviour and its underlying motivations and other correlates; (c) inquiries into artificial intelligence, including modelling with machines or mathematics; (d) archaeological and anthropological analysis of fossils and artifacts (focusing primarily on the most direct possible evidence and the *pattern* of evolutionary change); or (e) comparative study of humans and other animals, especially close relatives, combined with adaptive modelling (utilizing primarily predictiveness from knowledge of the *process* of evolutionary change). The last method is the one principally employed here.

First, a unique selective situation is postulated to account for humans departing as far as they have, psychically and in other regards, from their closest living relatives, and it is compared to alternative hypotheses. The human psyche is then characterized in terms of the probable reproductive significance of its different aspects, thereby generating additional hypotheses about its selective background. Finally, an effort is made to test the hypotheses generated by the first and second parts of the discussion.

THE POSTULATED SELECTIVE SITUATION

Background of the Hypothesis

There is probably general agreement that explaining human evolution is to a large extent a question of understanding how human mental attributes evolved. The problem is not only why brains evolved to be bigger and intellects to be more complicated, but also why they became so dramatically different from those of our closest living relatives. Humphrey (1976) suggested that the selective situation was primarily a social one, with evolving humans providing their own selective challenge; as with others who have made suggestions in this direction, however (see references below), he did not explain what forces caused humans to continue to live under the social conditions responsible for the expenses of intense competition and the resultant manipulations, deception, and favouring of social cleverness that he and others postulate. Thus, he did not account for the fact that humans alone have followed an evolutionary pathway leading to what he called a 'runaway intellect'.

Humans are not just another unique species, rather they are unique in many and profound ways – that is, in many attributes, and also in ways that unexpectedly set them apart from all primates, all mammals, or even all life (e.g. Alexander and Noonan 1979; Tooby and DeVore 1987; Wilson 1975; Wrangham 1987). For example, rapid evolution usually means more speciation, but humans, whose brains are regarded as having evolved according to an 'autocatalytic' model – increasingly rapidly – during at least the past two million years (Godfrey and Jacobs 1981; Stringer 1984), have no living close relatives. By this I mean that there are no closely similar or sister species, no congeneric species, no interfertile species, no species, even, with the same number of chromosomes. Why? Why do we have to go back two, five (or more) million years to find the most recent phylogenetic juncture with our nearest *living* relatives (Ciochon 1987)?

Human social groups are also unique (currently) in being huge and socially complex, while also having all individuals both genetically unique (excepting monozygotic twins) *and expecting to reproduce;* and only humans (apparently) play competitively, group-against-group (currently on a large scale). Although we became a virtually world-wide, highly polytypic species with numerous geographic variants, until recently those different variants have not easily mixed or lived in sympatry; and there seems to be no universally accepted evidence that multiple species of hominids ever lived together. With increases in the world population of humans, moreover, we did not come to live in a single, huge, dense, amorphous, universally beneficent population; rather, we have always lived with tense national boundaries, patriotism, xenophobia, and almost continual and destructive intergroup competition and conflict. Today we have a terrible international arms race as a central horror in our lives. We have at least been primed by evolution so as to allow these things to happen. How were we so primed?

Components of the Hypothesis

A. The most unpredictable and demanding aspects of the environment of evolving humans have always been its *social* aspects, not the physical climate or food shortages, as is often implied. The human psyche was designed primarily to solve *social* problems within its own species, not physical and mathematical puzzles, as educational tests and some concerns of philosophers might cause us to believe. Darwin (1859, 1871), Keith (1949), Bigelow (1969), Wilson (1973), Hamilton (1975), Alexander and Noonan (1979), and Alexander (1967-1988) have all suggested some parts or versions of this model, but Humphrey (1976) probably described it most clearly (independent hints toward it are also numerous – e.g. Trivers 1971; Fox 1980; Kurland and Becker 1985; Box and Fragaszy 1986; Burling 1986). This hypothesis implies that even the solving of mathematical, physical, and nonhuman biotic problems had its central significance (in the broadest sense, its reproductive rewards) in social contexts. (For example, Lenneberg (1971) argues that 'mathematical ability may. . . be regarded as a special case of the more general ability that also generates language. . .' and Burling (1986) that 'the. . . evolution of the. . . capacity to learn and use highly complex language is unlikely to be explained primarily by any subsistence or technological advantages that language offers. Rather, language probably served social purposes'.) In other words, this hypothesis rejects the notion that complex intellects evolved because they saved early humans from starvation, predation, climate, weather, or some combination of such challenges. All other organisms have solved these kinds of problems, in a variety of ways, without complex human-like intellects. If humans solve such ordinary problems in extraordinary ways, I am suggesting, it is because they are using an intellect evolved in a different context (For comparative purposes, Isaac (1979) lists six hypotheses bearing on human evolution: Dart's (1949, 1954) 'hunting' hypothesis; Jolly's (1970) 'seed-eating' hypothesis; Tanner and Zilman's (1976) 'gathering' hypothesis; Isaac's (1978) 'food-sharing' hypothesis; Parker and Gibson's (1979) 'developmental' hypothesis, based primarily on food skills; and Lovejoy's (1981) 'shortened birth interval' hypothesis). Dart's is the closest to that espoused here; the others all depend on non-human biotic or physical threats as primary forces. As I am restricting it, my hypothesis also requires that human proficiency in tool construction and use is also a secondary or incidental effect of the evolution of an intellect designed to be effective in social contexts. Wynn (1979) and Gowlett (1984) discuss the relationship between the manufacture of tools, and especially the transport of materials involved in their construction, to the evolution of planning and foresight. In this connection it is relevant that chimpanzees show evidence of planning, and perhaps scenario-building, in such tool-using behaviour as the selection, preparation and carrying of termiting sticks (Goodall 1986; Ghiglieri 1988). It is obviously important to my hypothesis that they also seem to show considerable foresight in social activities (Goodall 1986; Ghiglieri 1988).

B. Human mental abilities evolved as a result of *runaway social competition*, an unending within-species process dependent upon interminable (and intense) conflicts of interest, compared (below) to Fisher's (1958) concept of runaway sexual selection (see Alexander 1987).

C. *Balance* (or *imbalance of power races*) between social groups, either within or between (very similar) species, facilitated runaway social competition by favouring complex social living, and abilities to behave cooperatively and competitively within (and between) social groups. Such races 'trapped' humans into social interdependence, led to within-group amity and between-group enmities, and in part created the selective situation that gave rise to our creative intellects. Humans may not be the only species to engage in social reciprocity and cooperation-to-compete (with conspecifics), but they are probably the only one in which this combination of activities is a central aspect of social life.

D. These processes became paramount partly because the *ecological dominance* of evolving humans diminished the effects of 'extrinsic' forces of natural selection such that within-species competition became the principal 'hostile force of nature' guiding the long-term evolution of behavioural capacities, traits, and tendencies, perhaps more than in any other species. The evidence for this having happened is the current ecological dominance of humans; the only problem is when and how it came about. One might ask if (1) the ecological dominance of humans allowed the evolution of complex intelligence or (2) complex intelligence enabled humans to become ecologically dominant. I would argue, rather, that the two went hand in hand, reinforcing one another at every stage, and I suggest (below) that, aside from the human presence, chimpanzees may already have attained the required dominance.

The reference to 'extrinsic forces' above is to Darwin's Hostile Forces of Nature – parasites, predators, diseases, food shortages, climate, and weather – as an exhaustive list of the features of natural selection that determine the reproductive success and failure of different genotypes. Darwin (1859, 1871) distinguished between natural selection and sexual selection, so he did not include in the list, as I do, mate 'shortages' (meaning, ultimately, variations in mating success, including quality as well as quantity of mates). Darwin's emphasis on sexual selection to account for the evolution of many human traits is in accord with the idea presented here, if the context is expanded to include other kinds of social competition, and

I believe that the current idea supports his general suggestion.
E. The combination of (1) *balance-of-power races* between human social groups, (2) *runaway social competition* and the emphasis on creative and manipulative intellects, allowed or facilitated by (3) *human ecological dominance,* can also be used to help explain the changes in social structure that occurred as human social groups expanded toward their present sizes and took the forms (bands, tribes, and nation-states –'egalitarian', despotic, totalitarian, or democratic societies) represented across human history.
F. The central evolved function of the human psyche, then, is to yield an ability to anticipate or predict the future – explicitly the social future – and to manipulate it in the (evolutionary, reproductive) interests of self's genetic success. In the hypothesis developed here, all other effects or properties of the psyche are secondary to this strategic function. This general situation came about because evolving humans (a) came to live in highly cooperative social groups and (b) became ecologically dominant, these two conditions together (i) reducing the significance of hostile forces of nature other than conspecifics and (ii) leading to cooperation to compete against conspecifics who were doing precisely the same thing. In this fashion the combination of an unending runaway social competition and an unending balance-of-power race was set in motion, which continues within and among human populations today. This general situation allowed and caused the radical departure of humans from their closest relatives, in psychical and other attributes.

A central feature of the human psyche is the construction of alternative scenarios as plans, proposals, or contingencies in a manner or form perhaps appropriately termed *social-intellectual* practice for social interactions and competitions (practice which lacks a prominent physical component). This hypothesis of *scenario-building* sheds light simultaneously on a collection of human enterprises that have seemed virtually impossible to connect to evolution – such as humour, art, music, myth, religion, drama, literature, and theatre – because they are involved in *surrogate scenario-building,* a form of division of labour (or specialization of occupation) that may be unique to humans (partly because language is required for communication of mental scenarios between individuals). The centrality of scenario-building in human sociality (which will be related to the concept of *play*) is connected with the appearance of *rules* (hence, moral and legal systems) through (in part) the value of limiting the extent to which the elaborate and expensive scenarios (plans) of others can be thwarted by selfish acts (Alexander 1987, see also below). Finally, part of the game of human social competition involves concealing how it is played, and some of such concealment involves concealing it from one's self (*self-deception*). This in turn compounds the problem of understanding ourselves because of the difficulty of bringing into the conscious items that have been kept out of it by natural selection, most particularly items involving social motivations.

The hypothesis assumes that some version of these social functions initially drove the evolution of consciousness and other aspects of the human psyche, and that other uses of the psyche, such as in predicting or dealing with aspects of the physical universe, are (or were initially) incidental effects (in the evolutionary sense). This scenario does not preclude adaptive functions of the psyche in dealing with nonsocial phenomena throughout human evolution, only that such functions could not have caused the evolution of consciousness, cognition, linguistic traits, and the emotions as a set of human attributes. The emphasis on manipulation and deception is because the hypothesis holds that the human psyche would not have evolved in a world dominated by truth-telling, so that its complexity is tied to its use in deception. Once efforts at deception are widespread, successful, and complicated, truth-telling also becomes difficult to identify or prove. The argument is that truth is approached only when necessary – that is, cost-effective.

General Comments on Natural Selection

Ultimately, there must be compatibility between our view of the functions of the human psyche and our understanding of the selective background that gave rise to it. I am going to develop the argument from the beginning, because there can be no agreement, or adequate evaluation of arguments, unless common ground has been established from the outset.

If we accept the view of modern biology that natural selection is the principal guiding force of evolution, this means, first, that to understand traits we must concentrate on their reproductive significance and discard most of the old notions about adaptive function, such as survival of the individual (at all cost – i.e. even when survival is opposed to reproductive success), benefit to the population or species (again, when there is conflict with benefit to the individual's reproduction), progress, or any kind of goal-oriented or orthogenetic trend. We are not free to assume that genetic drift or other random events can account for elaborate attributes, just because they seem to give an unprejudiced, amoral, or value-free aspect to evolution or because they can account for minor differences between populations (Alexander 1979, 1987).

Rules for Applying Selective Thinking

Continuing from this initial assumption, I assume five general rules in applying natural selection to the attributes of organisms (for the first two, see Williams 1985):

First, we must consider the question of *adaptation,* not according to some notion of optimality or ends to be achieved, but rather according to the now widely accepted usage, from Williams (1966), of simply *better versus worse in the immediate situation.* This view implies that long-term trends occur because particular selective forces remain in place for long times, so that step-by- step small changes sometimes give a false retrospective appearance of goal-oriented or orthogenetic trends. As Williams (1966) emphasized, we must also distinguish between incidental effects of traits and their evolved functions or evolutionary 'design'.

Second, natural selection must always work from 'last year's model' – a fact often referred to by modern biologists under concepts like phylogenetic and ontogenetic inertia, or structural laws of development and evolution. This particular rule implies that phenomena like allometry or neoteny are in general maintained as a result of selection and not in spite of it; that when such phenomena cause some kinds and degrees of evolutionary 'inertia', they must be presumed to have developed the potential for such effects as a result of past selection. To invoke physiological, developmental, or phylogenetic constraints to explain evolved phenomena is thus an argument of last resort.

Third, random events such as mutations and drift introduce noise into the adaptive process but do not guide long-term directional change.

Fourth, selection is more potent at lower levels in the hierarchy of organization of life (Williams 1966, 1985; Hamilton 1964, 1975; Lewontin 1970; Dawkins 1976-1986; Alexander and Borgia 1978; Alexander 1979, 1987), so that, as Williams (1966) first argued convincingly, 'most of the characteristics of organisms, including social behaviour, must be the result of differential fitness at the level of individual genotypes' (Lewontin 1966).

Fifth, to understand traits, it is effective, and parsimonious, to seek or hypothesize singular selective causes (or contexts or changes) in evolution, as opposed to accepting multiple ones too readily. This is so because (1) it is difficult to falsify individual causes when multiple contributing factors are accepted uncritically; (2) single causes can be sufficient, even when multiple contributory factors are known; and (3) once a particular event, such as group-living, has occurred, then secondary effects will appear that (especially without attention to the possibility of single sufficient causes) can be confused with the primary cause (in other words, single *different causes* may occur *in sequence* without violating these arguments).

Hypothesizing singular causes, I believe, is a way of making one's ideas maximally subject to falsification, if they are incorrect, and therefore of advancing knowledge most effectively. It is a way of going most forcefully after the actual driving forces in evolutionary change, and of unravelling most quickly and completely the actual patterns of change. This 'rule' for applying selection is the most controversial one, and the controversy arises primarily because some see it as a way of over-simplifying causation in human social affairs. This criticism, in turn, is prevalent among those who believe that knowledge (or supposed knowledge) of history yields ideology for the future. This problem arises in large part out of ignorance about the relationship between genes and phenotype, heredity and development, rigidity and plasticity in expression of behaviour. It cannot be erased, however, by emasculating the procedures of science. Rather, it must be removed from importance by explaining why the argument that history justifies ideology cannot be sustained.

I have given no rules regarding the mechanisms by which the phenotype is acquired, such as learning or maturation. There are two reasons: First, the operation of natural selection can be understood without knowing about, implying, or eliminating any particular ontogenetic or physiological mechanisms. Second, predictions about expected proximate mechanisms in particular evolutionary situations are extremely difficult and complicated. Although ignorance about design mechanisms makes it more difficult to be confident about interpreting function, so long as no particular restrictions are assumed (e.g. that a behaviour is 'innate', learned in some particular fashion, etc.), analysis of such mechanisms can usually be postponed without necessarily causing error.

Group Selection

Group selection (*versus* selection at lower levels in the hierarchy of organization of life: Alexander and Borgia 1978) is an issue that is central, yet seems to remain complex and confusing. Two different situations may be implied by the term 'group selection'. The first appears to be relatively unimportant, for the reasons given. The second is the one to which I referred (Alexander 1974) when I suggested that humans are an excellent model for group selection (see also Alexander and Borgia 1978).

1. *Group Selection when there are Conflicts of Interest between Individual and Group Levels*. In this kind of selection, the spread or maintenance of alleles is determined at group levels, regardless of conflicts of interest within groups, because selection at the group level is simply more powerful than that at lower levels. In other words, the maintenance and spread of alleles is determined primarily by the differential extinction or reproduction of groups, regardless of what is happening at individual or other levels. This is the kind of selection that Wade (1976, 1978), D. S. Wilson (1975, 1980), and others seek to validate theoretically (as feasible or likely in natural populations) and demonstrate in laboratory experiments. Their work, however, actually indicates that group effects are weak in the face of the strength of selection at lower levels (Williams 1985; Dawkins 1986; Alexander 1987). Thus, to create potent group selection they have been forced to postulate populations with attributes much like those of individuals. They invoke groups that are founded by one or a few individuals (thus as near as possible to being single broods of offspring), and last about one generation (thus have the same generation time as individuals). In the laboratory they create populations (sometimes highly artificial) with minimal within-population genetic variance and maximal between-population genetic variance. The effort is to maximize genetic variance and minimize generation time at population levels, because the intensity of selection depends on these attributes (Fisher 1930; Lewontin 1970), which are virtually always more favourable to selection at lower levels. There are multiple indications, in the behaviour and life histories of organisms, that this kind of group selection does not often prevail (Williams 1966; Alexander 1979, 1987).

Group selection that is weaker than individual selection may often affect the *rate* of selective change as a result of the differential reproduction of individuals, but group selection probably is only rarely potent enough to affect the *direction* of selection within species in natural situations when it runs counter to selection at lower levels. Selection that results in one of two competing species becoming extinct is this kind of group selection, but such interspecies selection occurs under conditions when the differences between groups cannot be compromised as a result of interbreeding. Interbreeding and gene flow between adjacent groups reduce the potency of group selection within species because they reduce genetic differences between adjacent populations (Hamilton 1964, 1975; Alexander and Borgia 1978).

Group selection of the sort just posited ultimately will result in individuals that sacrifice their genetic reproduction for the good of the group because differential extinction and/or reproduction of alleles at the group level exceeds in importance that at the individual level. This is an effect postulated by Wynne-Edwards (1962, 1986) to explain what many ecologists saw a few decades ago as 'intrinsic' population regulation that adaptively avoids over-use of the environment. It has been invoked in ways implying that there are no special difficulties in the postulated sacrifices. Many people believe (erroneously) that this kind of 'genetic altruism' (Alexander 1974; 1979; 1987) would necessarily typify a species, human or otherwise, that had evolved through a process significantly involving group selection. Only a predominance of this kind of selection, it would seem, could prime the individuals of a species to participate readily in the kind of 'greatest good to the greatest number' utilitarianism proposed by many social philosophers. The required kind of sacrifice of one's reproduction, however, is not expected to have evolved, and no natural cases of this kind of selection have been documented or are widely suspected.

2. *Selection with Confluences of Interest at Individual and Group Levels.* In this kind of situation the individual who gives his life for the group (all or any part of the population) gains genetically from the act because his individual interests are identical with those of the group. Identity of evolutionary interests will be temporary in sexually recombining organisms because individuals in such species will for the most part be genetically unique. Identity of interests, on the other hand, is permanent in clones (barring mutation). Altruism between genetically non-identical individuals can thus evolve when neighbours (interactants) are alike genetically – compared to more distant conspecifics (as in Hamilton's (1975) 'viscous population' – cf. Nunney (1985)) – to a degree that over-compensates the increased competitiveness for resources likely to result from proximity.

Even when individual and group interests are identical, presence and prevalence of alleles are likely to be more affected by lower levels of selection because of the greater potency of selection there (Williams 1966; Lewontin 1970; Dawkins 1976; Alexander 1979). Such selection would, however, be enhanced by selection in the same direction at group levels. In sexually reproducing organisms, such as humans, confluences of interest within groups are likely to occur when different groups are in more or less direct competition. As a result, the kind of selection alluded to here would be expected to produce individuals that would cooperate intensively and complexly *within* groups but show strong and even extreme aggressiveness *between* groups. Such tendencies characterize modern humans and chimpanzees, and there are no convincing reasons for believing that they did not characterize humans throughout their evolution. The kinds of intergroup interactions that occur among modern humans and chimpanzees probably occurred throughout the evolution of the great apes and hominids. In humans, especially, the impression of group selection can be given because of costs imposed socially for failures to be altruistic.

One reason why humans have been described as a good model for within-species evolution by group selection is that opposing groups often behave toward one another as if they were different species. Social rules, morality, law, and a great deal of culture represent the imposition of costs and benefits on the actions of individuals and subgroups designed to force a convergence of their interests with those of the social group as a whole, thus tending to create the second kind of situation described above.

It seems to me that, after more than 50 years of discussion and analysis, the appropriate conclusion from all the arguments on the topic of self-sacrifice, heroism, and altruism toward others is still that suggested by Fisher (1930: 265): 'The mere fact that the prosperity of the group is at stake makes the sacrifice of individual lives occasionally advantageous, though this, I believe, is a minor consideration compared with the enormous advantage conferred by the prestige of the hero upon all his kinsmen'. Even the extreme cases of Japanese kamikaze pilots in World War II, in which all of the men in sizeable units volunteered willingly, and even insistently and competitively (Morris 1975) – if they are to be queried in evolutionary adaptive terms – must be analysed in the light of ceremonies, honours, and other described effects having to do with relatives of the volunteers, and connected to the costs of cowardice and rewards for heroism, during the long-term cultural history of Japan.

A Note on Competition

Humans live in groups, and individual interests are expressed in cooperation and competition at all levels of social organization. I interpret human social organization of virtually all kinds to be cooperation, either for the explicit purpose of direct competition with other humans also living in groups or as a part of the indirect competition of non-intentional or non-interactive differential reproduction. With respect to the general process of evolution, the concept of competition must be taken in this broad sense, in which it stands alone, with no real or existing opposite. Thus, cooperation, and all parallel activities, cannot be regarded as *alternatives* to competition; as such they could not evolve. Cooperation must exist because it has aided reproductive competition, however indirectly. One individual or group can be said either to compete or cooperate with another; but the cooperation, if it depends on evolved tendencies, also represents either direct or indirect competition with still other such units. In evolution, only genetic altruism (Alexander 1974, 1979, 1987) could be regarded as opposing competition, and for this reason such altruism will not evolve and will not be maintained if it appears incidentally.

Distinctive Aspects of Selection on Humans

The nature and all-inclusiveness of organic evolution requires us to assume that the human psyche evolved as a vehicle serving the genetic or reproductive interests of its individual possessors. These interests are expressed in individual humans via success in (1) survival and social integration across juvenile (and adult) life; (2) mate-seeking and -holding as an adult; (3) offspring-production and -tending; (4) beneficence (and the seeking of beneficence) with respect to collateral kin; and (5) various forms of (direct and indirect) social reciprocity to both kin and non-kin (or close and distant kin), which means beneficence dispensed in situations in which returns with interest are expected (Trivers 1971, 1985; Axelrod and Hamilton 1981; Alexander 1979, 1987).

Why Live in Groups?

To understand any group-living species, such as humans, we must first ask why organisms live in groups and then ask, in turn, why humans live in groups. I have previously discussed these two questions in detail (Alexander 1974, 1979: 58-65; 1987: 79-81). Partly because the above view of natural selection is fairly new in biology, the answers are not the same as those that have been prevalent in anthropology and the other social sciences.

Sociality by definition can exist only in organisms that live in groups. Efforts to understand sociality for a long time rested upon the intuitive view that groups exist for the good of the species, and individuals for the good of the group. If selection is more potent at individual than at group levels, however (see above), we should expect organisms to behave as if their own reproductive success is what matters (and they do – see examples in Alexander 1979, 1987).

Except in clones, the life interests of individuals within groups are rarely identical. If behaviour evolves because it helps its individual possessors, then group-living inevitably entails expenses to individuals, such as increased competition for all resources, including mates, and increased likelihood of disease and parasite transmission. Why, then, should animals live in groups? Why be social, beyond the minimum required to mate and raise a family (indeed, why even keep one's offspring near for a time)? If the answer is that individuals living in groups reproduce more than individuals not living in groups, what are the reasons?

Theoretically, the causes of group-living include enhanced access to some resource, or enhanced ability to exploit a resource, which more than offsets, for individuals, the automatic detriments. I have previously argued (Alexander 1974) that reasons for group-living are few in number:
(1) lowering of susceptibility to predation either because of aggressive group defence or because of what Hamilton (1971) called *selfish herd* effects (e.g. a more effective predator detection system or the opportunity to place another individual between one's self and a predator); (2) cooperative securing of fast, elusive, aggressive, or hard-to-locate prey; and (3) localization of resources (food, safe sleeping sites, etc.) that simply forces otherwise solely competitive individuals to remain in close proximity (see also Alexander 1975, 1979).

It is obviously useful to distinguish, when possible, between the primary causes of group-living and its secondary results. Postulated causes of group-living other than those listed above are probably secondary. For example, it seems unlikely that cooperative defence of clumped resources could ever be a primary cause of group living, since clumping of resources would at first yield the third kind of group-living just listed (Alexander 1974). My own opinion is that groups, such as in primates, that cooperate now, whether to defend food, females, or territory, probably evolved originally because of predator influence (even if what was at first involved was only one or a small group of females protecting their young) and secondarily (after having evolved cooperative tendencies or abilities in other contexts) began to defend food resources in those groups. Similarly, I would suppose that defence of large prey items in cooperatively hunting canines and felines is also secondary, evolving after group hunting. Again, group hunting most likely took the initial form of one or two parents hunting with their own offspring. It is obviously easier to understand cooperative interactions within groups of relatives, especially parents and offspring, than among nonrelatives.

Wrangham (1980), Cheney and Wrangham (1987), and others have downplayed the role of predation in causing group-living in primates, mainly because of the paucity of observations of predation. Nevertheless, predation is probably responsible for herding in ungulates and colonial life in a great many vertebrates that live in groups similar to the groups of related females discussed in various primates by Wrangham (1980), and that have little possibility of being explained as defenders of food bonanzas or any other resources. Moreover, Cheney and Wrangham (1987) list estimated predation rates on 30 primate species as averaging about 6.5% per annum.

Humans have affected predators negatively by their own actions more than prey species, and as well may deter predators completely by their presence as observers, so the above figures can scarcely be regarded as insignificant. Indeed, the only other significant source of primate mortality discussed in Smuts *et al.* (1987), is infanticide by conspecifics. Predation is also rarely observed in many species where the observers do not doubt that it has been responsible for chemicals, powerful senses, mimicry, cryptic coloration, or the patterning of social life or group living (e.g. Alcock 1984). If one wishes to answer the question of why groups form and persist, moreover, observed rates of predation on normally structured social groups are far less significant than observations of the fates of individuals outside their social groups, or in groups that, for example, lack large males or are abnormally small.

Causes of Human Grouping

Early groups of humans are widely believed to have been group hunters. This idea is not incompatible with evidence that gathered foods provided most of the diet of early humans (e.g. Lee 1968; Tooby and DeVore 1987); but gathering seems less likely to have led to complex cooperative tendencies (for perhaps the best case for the alternative possibility, see Kurland and

Beckerman 1985). If arguments given earlier are correct, the predecessors of the earliest hominid group hunters almost certainly lived in groups because, as with probably all modern group-living nonhuman primates, they were the hunted rather than the hunters. By all indications, humans are the only primate that became to some significant extent a group-hunter – the only group-living primate who, at least for a time, has escaped having its social organization essentially determined by large predators (chimpanzees are nearest to being an exception in both regards, but they obtain far less meat from hunting, and they are obviously subject to human and other kinds of predation: see King 1976; Goodall 1986; Cheney and Wrangham 1987). For this reason it is not surprising that we empathize to some degree with canine and feline social groups. The human brand of sociality appears to have been approached by various other primates because they are our closest genetic relatives, but by canines and felines (lions) because they most nearly (among nonprimates) do, socially, what we did for some long time.

But the organization and maintenance of recent and large human social groups cannot be explained by a group-hunting (or gathering) hypothesis (Alexander 1974, 1979). The reason is that the upper size of a group in which each individual gained because of the group's ability to locate and secure large game or other food bonanzas would be rather small. Indeed, according to such a hypothesis, as weapons, skills, and cooperative strategies improved, group sizes should have gone down, owing to the expenses (to individuals) of group-living, which tend to be exacerbated as group sizes increase (acceptance of this argument depends on the assumption of powerful selection below the group level). Cooperative group hunters among nonhumans tend to live in small groups (canines, felines, cetaceans, some fish, and pelicans); and most large groups (e.g. herds of ungulates) are probably what Hamilton (1971) called 'selfish herds', whose evolutionary *raison d'être* is security from predation. (The relevant security is to individuals, but not necessarily to the species as a whole; as Hamilton pointed out, the population can actually suffer higher overall predation, but because existence of groups causes predation on lone individuals to become even more severe, group-living in selfish herds can nevertheless continue to evolve.) Even groups evidently evolved to cooperate against predators are typically small (chimpanzees, baboons, musk ox). But maximum human group sizes, beginning at times and places not easily ascertainable, went up – right up, eventually, to nations of hundreds of millions.

Modern human groups are unique, suggesting that explaining them will call for selective situations that are in some sense unique. Thus, other complexly cooperative groups either tend to be small, as with cooperatively hunting groups of canines or felines, or else they are structurally unlike human groups. In human groups, for example, coalitions exist at many levels of stratification, and in many functional contexts. Clones are often very large, as are modern human groups, but clones are composed of individuals with continuously identical evolutionary interests, while human groups are not. Eusocial insect colonies, such as those of ants and termites, are often both huge (up to 15-20 million: Wilson 1971) and complexly social, as with humans. But they represent variations on a nuclear family theme. In them, a single female (usually) and one or a few males produce all the offspring, and all but these few reproductive individuals tend to be full or half siblings to one another. Even in a eusocial colony of millions, every individual is closely related to every other one. Completely unlike this, large human groups are composed of close and distant relatives, and nonrelatives; and every individual expects to reproduce unless special circumstances intervene. Moreover, in the huge modern eusocial insect colonies, such as with ants and termites, the interests of the different colony members, whether queen or workers, may be virtually identical. The reason is that all will realise reproductive success only through the rather small group of reproductively mature individuals that will emigrate and found new colonies. Not only are the workers and the queen likely to be similarly related to these reproductives (especially in forms with diplo-diploid sex determination, as with termites), but they are likely to have no other opportunity to reproduce. A good comparison in familiar terms would be represented by a species in which the male and female tend to be obligately monogamous, bonded for life. If opportunities for differential assistance to nondescendant relatives, and for philandering, are rare or non-existent, then, even though the male and female may be completely unrelated, their reproductive interests are identical. Each will reproduce only via the offspring they produce together; and in most cases the two parents will be equally related to the offspring. In such cases the male and female are expected to behave as though their evolutionary interests are identical, as with members of clones, and queen and workers in some eusocial colonies (Alexander 1987).

In this light, conflicts of interest take on special significance in the huge modern social groups of humans: such conflicts are more or less continual, and they can become exceedingly complicated. They wax and wane in response to competitive and cooperative interactions of groups with one another, as well as in response to the interactions of individuals within groups. As we shall see, chimpanzees live in multi-male groups that in some ways parallel those of humans, and this is significant for efforts to understand human evolution and the nature of the human psyche.

In trying to explain how modern humans developed such huge and unified political groups, and became involved in the current international arms race, it is possible to argue that the early benefits of group-living – whatever they might have been – were so powerful that they produced humans with such strong tendencies to be socially cooperative that the huge groups of recent history developed as more or less incidental effects, despite widespread deleterious effects on the reproduction of the individuals that comprised them. Such a view may at first seem correct, in the sense that living in small, highly cooperative groups may have produced humans who readily adopt competitive or adversarial attitudes toward members of other groups, and continually compare the relative strengths of groups and strive to produce or maintain strength (through extension and intensification of cooperation) in their own group (Hamilton 1975). But this view allows continual readjustments that return positive effects on the reproduction of individuals living in groups. In any other sense, the argument implies a degree of rigidity, or a kind of genetic control, of behaviour that I would like to regard as an argument of last resort. Moreover, if this latter kind of rigidity were the correct view, then we should not be able to identify widespread advantages from living in large groups currently, and alternative hypotheses to explain modern human groups should be difficult to apply. Neither is the case.

Imbalances of Power and Runaway Social Competition

The general hypothesis that I support to account for the maintenance and elaboration of group-living and complex sociality in humans, described earlier, derives from a theme attributable to Darwin (1871) and Keith (1949), and developed by a succession of more recent authors (Bigelow 1969; Carneiro 1970; Wilson 1973; Pitt 1978; Strate 1982; Betzig 1986; Alexander 1967-1988; Alexander and Noonan 1979; Alexander and Tinkle 1968). It includes group-against-group, within-species competition as a central driving force, leading to balance-of-power races with a positive feedback upon cooperative abilities and social complexity. It implies that the only plausible way to account for the striking departure of humans from their predecessors and all other species with respect to mental and social attributes is to assume that humans uniquely became their own principal hostile force of nature.

This proposition is immediately satisfying, for it (perhaps alone) can explain any size or complexity of group (as parts of balance-of-power races). It accords with all of recorded human history. It is consistent with the fact that humans alone play competitively *group-against-group* (and, indeed, they do this on a large and complex scale) – if play is seen, broadly, as practice. And it accords with the ecological dominance of the human species and the disappearance of all its close relatives.

No other sexual organisms compete in groups as extensively, fluidly, and inexorably as do humans. In no other species, so far as we know, do social groups have as their main jeopardy other social groups in the same species – hence, the unending selective race toward greater social complexity, intelligence, and cleverness in dealing with one another at every social level. No other species deals in war so as to make it a centre-piece of social cooperation and competition. I am not aware of hypotheses other than that given above which can deal with all of these issues.

Most of the evolution of human social life, I am hypothesizing, and the evolution of the human psyche, has occurred in the context of within- and between-group competition, within-group competition shaped by between-group competition, and the centrality of social competition resulting from the ecological dominance of the human species. Once cooperation among individuals (and subgroups) became the *central means of within-species competition*, the race toward intellectual complexity was on. Without the pressure of between-group competition, within- group competition would have been relatively mild, or at least dramatically different, because groups would have remained small and would have required less unity and different kinds of cooperativeness. There would have been no selective pressure that could produce the modern human intellect.

The situation I am postulating is not simply that described by Darwin's observation that, because of their similarity to one another, the members of a species are their own worst competitors for food, shelter, mates, and so forth. Rather, this view calls for a species that has so dominated its environment that all other hostile forces have been manipulated and modified into relative trivialities, *compared to the effects of competitive and cooperative conspecific neighbours*. If there are forces that remain potent (for humans, parasites are the most obvious example), then my argument would suggest that they could not be neutralized effectively by human effort in ways that led to major long-term trends in behaviour that could account for the evolution of the human intellect (the reasons might be erratic or infrequent appearance, rapid evolution, invisibility, or other causation that has somehow been outside human knowledge or capabilities of thwarting). In present circumstances the AIDS virus, a 'social' disease, might be seen as a counterexample to my argument. It seems beyond doubt that this disease is at least temporarily modifying human sexual and social behaviour in a significant fashion. Particularly interesting is the extent to which people who previously regarded as immoral actions that increase the likelihood of contracting AIDS use this jeopardy to promote their particular views of morality. To place AIDS-like diseases in an appropriate perspective with regards to human evolution, however, one has to consider what the reaction to them would have been without modern technology and knowledge from it. It seems unlikely that connections would be easily established between sexual interactions and physical deterioration many years later, or that sexual behaviour would have been severely modified prior to modern medical knowledge.

Runaway social competition can be understood by considering three features: (1) interminable conflicts of interest that cause social competition to be unending; (2) runaway aspects that can come into play most powerfully when the competition is within species; and (3) minimizing of brakes or direction changes because of the ecological dominance of the human species.

Interminable conflicts of interest cause unending evolutionary races. Such races occur, for example, between predators and their prey, so long as two species remain in this relationship to one another. They also occur within species, as between males and females, when conflicts of interest exist between the sexes in regard to the social interactions that sexual reproduction requires them to undertake. For example, males in many insect species use their genitalia in ways contrary to the interests of females, such as by holding the female longer than is to her advantage during copulation, so as to decrease the likelihood of another copulation and competition from the sperm of another male. Females may be expected to evolve to extricate themselves sooner, males to evolve to hold them more effectively; and the race is potentially unending. Similarly, in many mammals, males can maximize their reproduction only by mating with multiple females and showing little or no paternal care, while females in the same species can only maximize their reproduction by securing more paternal care than is advantageous for males to give. Such conflicts can lead to rapid evolutionary change, and sometimes unending races, in which each party evolves in response to the particular changes that occur in the other: what is beneficial for a male will depend on what countering changes occur in females, and *vice versa* (although extrinsic environmental changes may also be crucial in both of these examples).

One of the relevant facts from the 'balance (or imbalance) of power' argument described above for humans is that in social- intellectual-physical competition (such as physical competitions in which intelligence and the ability to gain and use support from others are important), conspecifics are likely to be – as no other competitors or hostile forces can be – inevitably no more than a step behind or ahead in any evolving system of strategies and capabilities. (The exception is when geography restricts contact, hence prevents more or less continual transfers of information via either aggression or cooperation and allows either cultural or genetic divergence or both.) Evolutionary unending races are thus set in motion that, because of the presumed paucity or absence of hostile influences extrinsic to the human species, have a severity and centrality as in no other circumstance. In other words, human social competition may be expected to involve a 'runaway' aspect, comparable to Fisher's runaway sexual selection, that is not likely in evolutionary races between, say, predators and prey. Indeed, the postulated process could be more extreme than runaway sexual selection.

Fisher (1930: 58) used the term 'runaway' sexual selection for situations in which females (usually) begin to favour extremeness of traits in males, leading to greater mating success by males that possess extremes of traits that are deleterious in every other respect (Trivers 1972). Within-species social competition is likely to take on 'runaway' aspects for three reasons: (1) the interdependence of the adversarial parties causes the significance of change in one to depend on the traits of the other; (2) the traits involved in the competition are likely to be arbitrary (and deleterious) in all other contexts; and (3) within-species groups of an ecologically dominant species such as humans are relatively immune to effects from other selective agents. When one's adversary continually remains similar or identical to one's self in all but the particular trait that is at the moment changing, when changes in one party depend solely upon changes in the other, and when other hostile forces are insignificant, then there are few or no brakes on change in the traits used in the competition, and little extrinsic guidance (cf. West Eberhard 1979, 1983). I believe that the current human arms race is the prime example of such a process, and as well a logical outcome of the history that such a process suggests for the human species (Alexander 1987).

Runaway social competition would (perhaps alone) account for the fact, stressed earlier, that human evolution has resulted in a single species, with all the intermediate forms having become extinct along the way. (I also speculate that the evolving human line has for a long time been a severe predator and competitor of apes, and is at least partly responsible for the low number of surviving Pongidae.) Indeed, unlike any other hypothesis so far advanced, it appears to *require* this outcome. It could also account not only for an acceleration of the relevant changes in the psyche, in social organization, and in culture in general, at certain stages of human social evolution, but as well for deceleration or even reversal of the direction of evolution of the psyche at other stages (Alexander 1971; Pitt 1978). Stringer (1984) summarizes the evidence that '. . . the autocatalytic model of endocranial volume increase seems most appropriate since there is an increasing rate of change until the late Pleistocene, when endocranial capacity values stabilize or even decline'. All that is required for the presumed stabilization or decline is that (1) social change eventually creates large societies in which the kinds of abilities and actions that preserve the entire group are possessed and used appropriately by smaller proportions of the society's members (in their own interests); and (2) group success and the social structure somehow lessen the reproductive disadvantages previously suffered within (and between) societies by those who lack the qualities of such leaders or governors. As numbers of leaders diminish in relation to numbers of followers – with increases in the sizes of social and political groups – the probability of producing a sufficient number of individuals with the necessary qualities to lead or govern effectively (whatever these qualities may be) would not necessarily diminish, owing partly to effects of genetic recombination. This condition (absence of advantage for increased complexity of mental activity) may exist in all large societies today (e.g. Vining 1986; compare with Alexander 1988). Whether or not it accounts for what Stringer describes (which could also reflect changes in brain structure consistent with the previous trend towards greater brain size, body size changes, or other forces) is another question.

In human intergroup competition and aggression, there are two prominent facilitators that unbalance the power of competitive groups, leading to more dramatic outcomes of confrontations and a greater likelihood of significant group selection in the form of unilateral extinction or one group taking over another's women and resources (Alexander 1971). These are (1) social and cooperative abilities that allow or cause larger (hence, variable) group sizes, and more concerted and effective group actions; and (2) culture and technology, which can provide one side or the other with superior competitive ability through means as diverse as language, weapons, and patriotic or religious fervour or perseverance. Changes in these regards can repeatedly adjust balances of power and fuel the kind of runaway social competition here postulated.

This argument may be compared to Gowlett's (1984) comment that 'It has become widely accepted that. . . biological evolution and cultural evolution affect one another in a positive feedback relationship, thus providing both change and its cause'.

'Initial Kicks' or What Started the Process?

As Wilson (1975: 568) notes: 'Although internally consistent, the autocatalysis model contains a curious omission – the triggering device.' Gowlett (1984) also argues that any such view as the above '. . . accounts for a process that is going on, rather than for the start of it. It does not satisfy the desire for a single perceptible factor, an "initial kick" which starts off human evolution. Consequently hypothesis after hypothesis has emerged in which such a kick has been found, and in some of them its momentum dominates the whole story.' In other words, even if the above scenario were acceptable, it is still essential to know how, when, and why the ancestors of modern humans became involved in an evolutionary, balance-of-power, runaway social competition.

In chimpanzees and a few other primates that live in multi-male groups – as in humans – females rather than males move between groups (Wrangham 1979; Pusey and Packer 1987; Cheney 1987). This change allowed males to bond, in connection with defending the home area as bands of relatives. Cheney (1987) and Manson and Wrangham (n.d.) believe that the initial resource involved in intergroup aggression was females; and they cite chimpanzees as most like us in these regards. King (1976) regards the pivotal resource as territory, but the point is the same: male cooperativeness and territoriality, coupled with female transfer, lead to intergroup aggression. Following King, Manson and Wrangham also lay great stress on variability in sizes of attacking and attacked groups, which they say is likely in human hunter-gatherers and chimpanzees partly because they live in 'communities', in which (as Reynolds 1965, pointed out) 'travelling parties are almost always *subgroups* of the politically autonomous unit'. Wilson (1975) speaks of 'the fluidity of chimpanzee social organization' being 'truly exceptional'. King refers to this fluidity as temporary fragmenting of otherwise stable societies, while Manson and Wrangham use Kummer's (1971) phrase 'fission and fusion'. In other words, through temporary alliances with individuals and subgroups with whom relationships have been maintained in the larger 'community', chimpanzees – as with humans on an immensely larger scale – meet threats, some of which are posed by other cooperating groups of conspecifics (Goodall 1986; Manson and Wrangham, n.d.).

Chimpanzees are evidently also most like humans psychically, using as criteria their performances at linguistic and other tasks in the laboratory, their use of tools, their tendencies to hunt cooperatively and to cooperate against 'enemies', and the evidence that they possess a self-awareness in some sense paralleling our own (which, however, at least orangutans share – Gallup 1970; Suarez and Gallup 1981; see also Premack and Woodruff 1978). Manson and Wrangham (pers. comm.) are sceptical that the cognitive abilities shared by humans and chimpanzees are relevant to the conduct of intergroup aggression. On the other hand, cooperation, coordination of emotions by displays, bluffing, and anticipation (if not planning) all seem to occur in connection with chimpanzee attacks on members of other groups (references in Goodall 1986; Smuts *et al.* 1987). My view is that the evidence for (1) cooperation to compete against conspecifics and (2) social reciprocity, foresight, and deception in both chimpanzees and humans implies that this combination of attributes has functional significance.

In general, the above arguments are compatible with the scenario developed here and earlier for the driving forces in human evolution (Alexander 1967-1988; Ghiglieri 1987, 1988). They help clarify the similarity between humans and chimpanzees, and that similarity implies that the processes suggested here were initiated quite early, before the primates involved would have been termed 'human' by modern investigators (see also Wrangham 1987).

These arguments tend to support an affirmative answer to Gowlett's (1984) question whether or not 'the earliest known men were hunters' (see also Tooby and DeVore 1987). The question remains whether chimpanzee (and pre-human) males initially cooperated to defend against predators, to hunt, to defend territory (or females or both), or to 'export' aggression to other males (Manson and Wrangham, n.d.). My scenario has these behaviours occurring in the order just given (Alexander 1979: 223), and suggests that, *in a primate physically, socially, and psychically like chimpanzees or ourselves, cooperative hunting or defence of territory or females are all adequate 'initial kicks' for intergroup balance-of-power races if extrinsic hostile forces are sufficiently insignificant.*

All of these comparisons imply to me that if by some chance the human species should be extinguished while chimpanzees were not, there is a fair chance that chimpanzees would embark upon an evolutionary path paralleling in some important regards that taken by human ancestors across the past million years or so. Indeed, they also imply that chimpanzees have been kept in their current status by the predatory and competitive actions of humans (Alexander 1974: 335), and that if they were even more like humans than they are, they would have long ago suffered the same fate that I believe had to befall the closer relatives of modern humans: extinction by their closest relatives, the evolving human line. Cheney and Wrangham (1987) refer to humans as 'particularly dangerous predators', of baboons, so classified with lions, and remark that '. . . human predation no doubt accounts for the greatest number of primate deaths, even in areas where hunting methods are still primitive. The regular hunting of primates by hunter-gatherers suggests that humans were important predators of nonhuman primates long before the advent of firearms and that human hunting may have exerted an influence on the evolution of antipredator behaviour and even social structure. . .' (see also Tenaza and Tilson 1985). I emphasize that the scenario I am developing does not require that the relevant competition be restricted to members of the same species; rather, it would be expected that any species similar to a species in which intergroup competition had become regular and intense would also be in jeopardy.

Female dispersal, and male relatives defending territory cooperatively in both chimpanzees and humans, cause even more intrigue to be attached to the relationship between male competition for females – both within and between groups – and the perplexing problem of (1) the rise of despotism (including extreme polygyny for despots) in societies of intermediate sizes and social complexity (tribes, chiefdoms,), yet (2) the institution of socially-imposed monogamy, reverence for the nuclear family, and suppression of extended families and kin networks in still-larger human societies (nation-states) (Alexander 1979, 1987, in press; Betzig 1986; Harpending 1986). Various authors (Levi-Strauss 1949; Irons 1981; Flinn and Low 1986) have pointed out that, in humans alone, males treat females as commodities, to be bargained with and for, in connection with their movement between groups. (Two major differences between chimpanzee and human societies are that human males show paternal care, while chimpanzee males do not; and human females conceal ovulation, thus affording males some confidence of paternity, while chimpanzee females advertize their ovulation and mate promiscuously – Goodall 1986; Alexander and Noonan 1979). Whether or not women were the resource that led to the initiation of intergroup aggression, and even if sexual selection has remained prominent in the activities of war and the admonitions given to young men of fighting age (Alexander 1987; Manson and Wrangham, n.d.), it is most unlikely that women are still a central resource at issue in the international arms races that baffle us all (see also below).

> The nettlesome question, of course, is why are [chimpanzees] territorial? Where is the survival advantage in risking one's life for land? The answer appears to be that winning more habitat enhances a group's mating success. Because ecological resources limit the number of females who can live in any region, the success of males in expanding, or at least holding, their territory determines the upper limit of their reproductive potential. No wonder they are territorial; if they were pacifists, or even individualists, their more coordinated neighbours would carve their territory into parcels and annex them. Thus armies are introduced into the natural arms race. Once this happens, solidarity between a community's males becomes essential (Ghiglieri 1987: 70).

Difficulties in Advancing Evolutionary Understanding of Ourselves

Evolution proposes to explain the explainers themselves. The difficulty of this proposition lies not only in the fact that some of the traits to be explained must be used in their own explanation, but also that one trait of the explainers is that they do not always wish to be explained – at least not too completely to anyone else – and that humans, more than any other species, are evolved to be exceedingly clever at deceiving other humans. Moreover, there may be no task of learning or teaching that is imaginably more difficult than that of bringing into the conscious items that have been kept out, not incidentally but by natural selection, and most particularly by selection that has disfavoured conscious knowledge of *motivation* as a social strategy.

There are probably two other main reasons for the slow progress of evolutionary understanding, especially with respect to ourselves:

1. Applying evolution to the understanding of organisms is not easy, even

if the process itself is deceptively simple. Such understanding is difficult because organisms are exceedingly complex. It calls for understanding the relationships between gene action and development as well as physiological, morphological, and behavioural outcomes and their variations. It is helped greatly by a repeated or continual necessity of dealing with these problems, a difficulty that many biologists face on a daily basis while most non-biologists do not.

2. The preceding difficulty is exacerbated by two facts: first, most people don't care about evolution, believing that it has little effect on their personal everyday lives; and, second, some people care too much. Evolutionary arguments seem to many to threaten cherished beliefs about humans and their history. Evolutionary arguments about humans also come from biology – a field distinct from the social sciences and the humanities, and one traditionally preoccupied with nonhuman species. Moreover, evolution has had a notoriously poor record in explaining humans in the past: during the decades when the social sciences were developing, biologists simply did not know how to apply selection to understand behaviour. For the most part they didn't even try, and so the social sciences developed more or less independently of biology and evolutionary theory. Finally, science and the humanities – disciplines preoccupied, respectively, with searches for undeniable facts and meaning or values – clash when biologically oriented scientists begin to analyse human actions in terms of their functions or effects, because such analyses seem to infringe on questions of meaning and value, hence to represent ideologies (Alexander 1988).

Cooperation also seems always to be a matter of congeniality and pleasantness. The concept of competition, and the idea of evolution by natural selection, on the other hand, imply nastiness. As already noted, however, in evolutionary terms cooperation and competition are not opposites, but rather, cooperation is *necessarily* a form of competition, which can be either indirect and remote, or quite direct. We have evidently evolved tendencies to develop proximate feelings that make it convenient to ignore the competitive effects of cooperative activities that give pleasure to us and our associates. It is also widely believed that group selection implies peaceful and non-competitive existence, as compared to selection at lower levels. As the definitions and descriptions of group selection given earlier indicate, this is not necessarily so: group selection and its mimics, as with some of the most pleasurable and intimate forms of cooperation, can also imply the most heinous and destructive kinds of between-group competitive interactions. The vilest of discriminatory jokes can be told in an atmosphere of warmth and conviviality; and, as Bigelow (1969) noted, 'A hydrogen bomb is an example of mankind's enormous capacity for friendly cooperation'. He suggested, with irony, that its successful construction might seem an occasion for us to 'pause and savor the glow of self-congratulation we deserve for belonging to such an intelligent and sociable species'.

Only with such considerations in mind, I think, is there likelihood of a thorough understanding of the human psyche, or other distinctive human features. Even if it seems inappropriate to emphasize the competitive and not-so-honourable sides of human action, motivation, and history, neither is it helpful to dismiss summarily the possibility of distasteful kinds or intensities of reproductive competition from human history as over-simplifications or reductionisms not deserving of consideration as causative forces in determining our present psychological makeup and socio-cultural forms.

THE NATURE OF THE HUMAN PSYCHE

There have been few efforts to characterize the human psyche in terms useful to those who would understand and reconstruct its functional aspects from a modern evolutionary viewpoint (but see Premack and Woodruff 1978; Griffin 1978; Savage-Rumbaugh *et al.* 1978, and the accompanying commentaries). I think the key argument (Humphrey 1976, 1978, 1983; Alexander 1979, 1987) is that consciousness represents a system of (1) building scenarios or constructing possible (imagined) alternatives; (2) testing and adjusting them according to different projected circumstances; and (3) eventually using them according to whatever circumstances actually arise. Earlier, I referred to such abilities as the capacity to over-ride immediate rewards and punishments in the interests of securing greater rewards visualized in the future (Alexander 1987). In this view, consciousness, cognition, and related attributes – which probably represent the core of the problem in understanding the human psyche – have their value in social matters, and the operation of consciousness can be compared to the planning that takes place in a game in which the moves of the other players cannot be known with certainty ahead of time. In other words, by this hypothesis, the function of consciousness is to provide a uniquely effective foresight, originally functional (*sensu* Williams 1966) in social matters, but obviously useful, eventually, in all manner of life circumstances. I will argue (below) that the emotions, linguistic ability, and personality traits are primarily communicative devices, hence, also social in their function.

The above view of the psyche is compatible with that of cognitive psychologists, such as Neisser (1976). Cognitive psychologists, however, concentrate more on mechanisms than on function, and so the idea that the use of cognition might have evolved explicitly in the context of social competition seems not to have emerged in their arguments. Nevertheless, Neisser's insistence on use of the concept of 'schemata' as plans, representing what is here called scenario-building, is a close parallel to Humphrey's arguments and my own. It is clear that a merging of ideas is likely to be easy, and profitable.

Functions of The Psyche

Learning would appear to include two forms: (1) accumulating memory banks; and (2) modifying memory banks, when 'memory bank' means a store of information that influences abilities and tendencies to act. In some sense all phenotypes are memory banks, in which some (genetic) information carried over from the previous generation (and, to a decreasing extent, from increasingly distant ancestors) has been 'interpreted' ('read out') by the environment of the phenotype (organism), including its associates (e.g.

parents) as a result of what is commonly called epigenesis, or ontogenetic or experiential plasticity.

Humour and Play. One can learn (1) by trial and error or successive approximation of the actual performance that is useful or desired, or of surrogates of it (practice? play?); (2) by observing and then imitating or avoiding; or (3) by being told about (taught) or by thinking about (and imitating or avoiding). The last two methods, at least, imply 'observing in the mind'. Learning by observing in the mind parallels the concept of play as practice. Play can be solitary-physical (as with a cat practising predation by playing with a twig or a bunch of dry grass); social-physical (as in practice-fighting or play-fighting); or social-intellectual (i.e., without a prominent physical component, as with building of social scenarios through thinking, dreaming, planning, humour, art, or theatre). Presumably, there are also intellectual (or mental) components to both solitary-physical and social-physical play (e.g. for the latter, in team sports involving complex strategies, bluff, and deception or trickery).

I agree with Fagen (1981) in regarding the concept of *practice* (including low-cost testing) as representing the best general theory of play, and I so use the concept of play throughout this paper (for a best-case dissenting argument, see Martin and Caro (1985) who note that 'at present, there is no direct evidence that play has any important benefits, with the possible exception of some immediate effects on children's behavior'). Fagen concludes (p. 388) that 'Current understanding of the functions of animal play suggests that individuals play in order to obtain physical training, to train cognitive strategies, and to develop social relationships'. He also reviews an extensive literature attempting to connect play behaviour to human deception, self-deception, dance, music, literature, painting, and sculpture (pp. 467 ff.; see also Wilson 1975). He describes 'hints at essential relationships between play and creative thought' in the words of Einstein and the thoughts of some other scientists, noting that 'these unsatisfactory metaphors are the best currently available links between play and human creation'. Klopfer's (1970) brief comment probably comes closest to the discussion of social- intellectual play developed here. Describing aesthetics as 'the pleasure resulting from biologically appropriate activity' and play as 'the tentative explorations by which the organism "tests" different proprioceptive patterns for their goodness of fit', Klopfer suggested that 'thought and abstraction in man is but a form of play' and 'Abstractions may be the play through which we learn how to think well' (Klopfer 1970: 402-403).

To Fagen's conclusions (above), I would add that play sometimes represents low cost repetitions and out-of-context or pretend 'run-throughs' in the interests of (1) practising for predictable situations that cannot actually be experienced beforehand; (2) preparing for different preconceived alternatives in unpredictable situations; and (3) assessing skills and abilities of one's self and others. As Humphrey (1986) says, '. . . play is a way of experimenting with possible feelings and possible identities without risking the real biological or social consequences'. It is also obvious that playing individuals can learn about one another and establish (accept) dominance relationships in low cost situations which may persist into high- cost situations; conversely, they may also learn how to reverse such relations in their own interests. Symons (1978b; pers. comm.) argues that 'dominance rankings are very unlikely to be established during play'. But I know from personal experience that, at least in humans, they can be either established or altered during play; and that play may be entered into with such goals explicitly in mind. I have done both, and I suspect that few humans do not share this experience.

Loizos (1967) exemplifies the authors who present objections to the general theory that play is practice (see also Martin and Caro 1985; for arguments very similar to mine, see Fagen 1982; Symons 1978a). One of Loizos' objections distinguishes play from practice: '. . . t is not necessary to play in order to practise – there is no reason why the animal should not just practise". But I regard play as a form of practice, and so believe that the mistake is precisely the other way around; a playing animal is 'just practising'. Second, Loizos, and Martin and Caro, note that not just juveniles but also adults play. But adults also practise extensively, and there is no reason to expect that this particular kind of practice should be absent in adults, especially long-lived adults with complex sociality who may be subjected to new social situations almost endlessly. Third, Loizos believes that '. . . it is simply not necessary to play in order to learn about the environment'. I would say, however, that it is often *useful* to play to learn about the *social* environment. Loizos notes that '. . . it is inevitable that during play, or during any activity, an animal will be gaining additional knowledge about what or who it is playing with; but if this is the major function of play, one must wonder why the animal does not use a more economical way of getting hold of this information'. I suggest that, with regard to the *social* environment, there often is no more effective and inexpensive way of securing information (again, Humphrey 1983: 76-79, comes closest to saying the same thing).

Martin and Caro (1985) argue that because play 'has only minor time and energy costs', is 'highly variable and labile', and 'is curtailed or absent under many naturally occurring conditions, it seems unlikely that it is essential for normal development'. Leaving aside the conservatism of the phrase '*essential for normal development*', however, the low costs of play can be cited as reasons for its use in developing social capabilities and increasing predictability of social outcomes. Moreover, feeding is curtailed in the presence of predators and sexually receptive mates, and planning is curtailed when immediate circumstances demand attention; but this does not mean that either feeding or planning is functionless. Any activity having its significance primarily in social behaviour is expected to be variable. Their estimates that play uses 4-9% of a kitten's calories (from Martin 1984) and 1-10% of total time in most species (from Fagen 1981) do not seem convincing for the purpose for which they use them. Thus, one might ask what per cent of calories and time are spent by various species in, say, the act of copulation.

Martin and Caro also question whether play should be suspected, as is commonly the case, of having its primary benefits later in life. They seem to disparage the notion that juvenile life has evolved as a preparation for success in adulthood; but there is no other *raison d'être* for juvenile stages (Alexander, in press, n.d.). Moreover, benefits that occur a long time in the future are those most likely to be difficult to identify and evaluate.

Expanding primarily from the arguments of Humphrey (1976-1986), I would relate the evolution of the psyche, and the representational (scenario-building) capacity of the human mind, to social play, as practice. I suggest that, during their 'runaway', group-against-group, social-intellectual evolution, humans went from social-physical play (typical of all social species) eventually to social-intellectual play (as scenario-building and practice), which probably occurs in at least rudimentary forms in all complexly social mammals, and team competitions (evidently unique, as play, in humans). Social-intellectual play I hypothesize to be practice for later, more consequential social-intellectual (and physical) competitions (that is, direct competition for mates or resources), just as solitary- and social-physical play represents practice for later, more consequential solitary- and social-physical activities or competitions (cf. Smith 1982).

I suggest that social-intellectual play led to an expanding ability and tendency to elaborate and internalize social-intellectual-physical scenarios. Along with the increasing elaborateness of internal scenario-building came an increasing elaborateness of social communication, including language and the evolution of linguistic ability. Every trait and tendency that represents or typifies the human psyche – every mental, emotional, cognitive, communicative, or manipulative capability of humans – I regard as a part of, derived from, or influenced by the elaboration of social-intellectual-physical scenario-building, and of the use of such scenarios – and of the emotions, language, and personality – to anticipate and manipulate cause-effect relations in social cooperation and competition. This would happen ultimately in the context of winning or losing both as an individual within a social group and as a member of a social group, the survival of which depends, in the end, on success in group-against-group competitions within the species.

Just as I believe that the evident radical departure of the human psyche from the mentalities of the closest relatives of humans can only be explained by assuming that humans themselves kept driving the selection in a peculiar way, I also believe that only other humans represent a sufficiently complex and unpredictable force to drive the evolution of the psyche in regard to its special ability to make and test social predictions. In other words, we became progressively better at practising for our social competitions through internal scenario-building because our adversaries and competitors were doing precisely the same thing. And because we all belonged to the same species, so that those with differently useful expressions of the psyche were parts of the same interbreeding population, there has for a very long time been a positive feedback involved in the evolution of increased human mental capacities, with the 'losers' or 'followers' never more than a step or two behind the 'winners' or 'leaders'.

To the extent that social-intellectual play can be carried out by observing in the mind, it can also be (1) accomplished (secondarily) in solitary (e.g. we laugh at jokes when alone); and (2) concerned with not only social-intellectual striving or competition but also social-physical or even solitary-physical striving (again, secondarily: note Neisser's (1976) relating of his 'schema' to locomotion). Moreover, effects of what previously was 'pure' play, as practice, can begin to influence actual contests over resources; a simple example would be carry-overs, into the resource competition of adults, of dominance rankings established during play among juveniles. Such secondary effects ought not to confuse our identification of primary causes.

Humans are probably not the only organisms capable of social- intellectual practice or play that does not have prominent physical concomitants. Perhaps all organisms that give evidence of dreaming utilize scenario-building of some sort in their social activities. It is easy to suspect, as Darwin (1871) did, that dogs, as well as apes and some other primates (e.g. Humphrey 1983: 90), do these things. But it is possible that humans alone engage in what I see as the next stage of evolution of the intellect in respect to scenario-building, and that is to reward or compensate others for building surrogate scenarios that are even more condensed (less time-consuming), more elaborate (hence, more effective), and more risk-free than one's own efforts. Once scenario-building has become widely useful, status and livelihoods can be secured by intellectual-social as well as other forms of occupational specialization – not merely by taking on intellectually demanding or specialized tasks, but by using unusual abilities and experiences to develop and conduct scenarios for others – hence, actors, artists, musicians, writers, comedians, orators, shamans, chiefs, generals, scientists, priests, preachers, teachers, and even professional players in sports. In this fashion, a number of human activities, which have until now seemed inaccessible from an approach stressing evolution or reproductive success, can be understood as a part of explaining the reproductive significance of the human psyche.

Humour and Play. Expanding from previous arguments (Alexander 1986, 1987), I explicitly identify humour as a form of social- intellectual *play*, unusual because of its emphasis among adults, which influences resource competition directly through *status shifts*. To illustrate my arguments here about the social use of intellect in regard to scenario-building, humour can be seen as operating in several different ways:

1. It can represent social practice for later competitions that will be more direct or more expensive because they will involve the actual resources of reproduction (jobs, money, mates, etc.). Such practice, as noted above, can be accomplished secondarily even in solitary, just as one can practise the moves of chess either while alone (even within one's mind) or while playing with others (i.e., one can laugh at a joke, and gain from the practice afforded, even if alone).

2. It can sometimes represent the actual competition for the resources, in

the sense that the people engaging in the humour may be those with whom one will actually compete later for significant resources; the competition may involve reputation or status that can be demonstrated so convincingly beforehand, using humour, as to turn aside expensive interactions that would otherwise have occurred.

3. It can involve surrogate scenario-building, in which one solely or primarily learns through observing scenarios built by another, such as a clown, comedian, or writer. Such professional humourists are compensated for building scenarios for others, more elaborately, more rapidly, or less expensively than these others can do it for themselves.

4. The vicarious aspect of humour can be carried further, in the sense that observers can alter their status among friends and associates (competitors and cooperators) by the kinds of humour they exert effort to observe (or use), and by how they respond to surrogate scenario-building via particular forms of humour.

All of the above four uses of humour involve only its directly competitive effects within groups. In the context of indirect competition, through within-group affiliation, humour can also operate in testing, promoting, or ensuring compatibility, and willingness to cooperate, and simultaneously in establishing group limits and thereby identifying competitors outside the group (Alexander 1986, 1987).

5. Humour can be directed against one's self, in a version of Zahavi's (1975) Handicap Principle, in which the humourist demonstrates that he can denigrate himself, or reveal embarrassing information that causes humour in others, and still maintain superior status. As with the superior racehorse handicapped with extra weight or the golfer handicapped with extra strokes, both of which may still manage to win the contest, the ultimate effect can be an enormous rise in status, worth far more than the prize for the particular contest being waged. In these examples – and particularly in the case of self-directed humour – even if the handicapped individual loses the immediate contest, it can win (because of the rewards for status in human societies) in the long run because of how well it did in spite of the handicap.

Elsewhere (Alexander 1988, n.d.) I have argued that the physical incompetence of the human baby (its *physical* helplessness or *altriciality*), as well as that of certain other organisms, is a correlate of *precociality* in respect to attributes that will improve its performance as an adult; and for humans this precociality is largely social-intellectual. I speculate that the early and astonishing acquisition of complex language ability in the juvenile human is related to its freedom (from the necessity of protecting itself) to devote itself to acquiring the necessary skills and knowledge of social communication, including practice and the analysis and acquisition of strategies, in the interests of becoming a socially and intellectually more capable adult.

Observing in the mind implies *consciousness* and scenario-building. It also implies being able to view and modify the memory bank with the option of saving changes or not, as if two copies existed of the memory bank during the scenario-building process; the question might be raised whether consciousness is somewhat like a viewing screen (relating it, perhaps, to the concepts of short-term and long-term memory). To observe (involve) one's self in scenarios in the mind is, I think, what is called *self-awareness*. To practise by observing one's self in the mind must in some sense be a description of the source of *foresight, purpose, planning, intent, and deliberateness*. Such practice gives rise to the concept of *free will* as freedom to choose among alternatives visualized in the future. This view of free will contrasts with the more widely discussed alternative implying questions about the presence, absence, or nature of physical causation (Alexander 1979, 1987).

A parallel view, expressed in different terms, is that of Neisser (1976, especially p. 20):

> In my view, the cognitive structures crucial for vision are the anticipatory schemata that prepare the perceiver to accept certain kinds of information rather than others and thus control the activity of looking. Because we can see only what we know how to look for, it is these schemata (together with the information actually available) that determine what will be perceived. Perception is indeed a constructive process, but what is constructed is not a mental image appearing in consciousness where it is admired by an inner man. At each moment the perceiver is constructing anticipations of certain kinds of information, that enable him to accept it as it becomes available. Often he must actively explore the optic array to make it available, by moving his eyes or his head or his body. These explorations are directed by the anticipatory schemata, which are plans for perceptual action as well as readinesses for particular kinds of optical structures. The outcome of the explorations – the information picked up – modifies the original schema. Thus modified, it directs further exploration and becomes ready for more information.

Once planning, anticipating, 'expecting' organisms are interacting without complete over-lap (confluence) of interests, then each individual may be expected to include in its repertoire of social actions special efforts to thwart the expectations of others, explicitly in ways designed to be beneficial to himself and, either incidentally or not, costly to the others (not necessarily consciously in either case). The expense of investing in one's scenarios, or expectations, and of having such scenarios thwarted, are involved in the invention of rules (see also Rawls 1971: 6; Alexander 1987: 96). Rules are aspects of indirect reciprocity (Alexander 1979, 1987) beneficial to those who propose and perpetuate them, not only because they force others to behave in ways explicitly beneficial to the proposers and perpetuators but because they also make the future more predictable so that plans can be carried out. One of their effects, especially as the rule-makers and -enforcers come to represent larger proportions of the group (e.g. through democratic processes), is to converge the interests of individuals and group.

Cognition, or problem-solving ability, can, I think, easily be related to the above arguments about the function of consciousness. Logic rationality, and cognition – as ability to perceive cause-effect relations correctly – can be viewed in the contexts of dealing with either (1) social possibilities (which entails assessing probable responses of living actors); or (2) nonsocial puzzles (some of which involve only the somewhat more predictable logic of physical laws). The process of selecting the most profitable (self-beneficial) among possible social alternatives involves *conscience,* as ability to recognize and evaluate consequences (ultimately, reproductive costs and benefits), especially as a result of the existence of rules. But in the sense or to the extent that conscience is linked to being good or bad (moral or immoral) – and to a failure to be conscious that one's motivation is to serve one's own reproduction – either ignorance or *self-deception* (or both) is an obligate concomitant. Trivers (1971, 1985) and Alexander (1979, 1987) have argued that self-deception, via the subconscious, is a social phenomenon, evolved as a system for deceiving others, most generally through denial of pursuit of self-interests, in turn through denial of any broad or precise knowledge of the nature of self-interests.

I regard *the emotions and their expression,* as well as self- deception and *personality traits,* as, in the main, an extraordinarily complex system evolved in the interests of deceiving or manipulating competitors. Deception is a crucial aspect of competition, because only through deception can the predictable outcomes of contests between competitors of unequal strength or resource-holding-power be altered (Parker 1974). The possibility of deception, moreover, and the difficulty of determining its effectiveness, can almost unimaginably complicate predictiveness about the outcomes of contests.

Because humans are, like most other organisms, sexual reproducers, they have evolved to behave, as individuals and families and collections of related families, as if their life interests (which translate as genetic or reproductive interests) are unique – different from those of other such units. Differences of interest between genetically unique individuals may be small (as between close relatives or between spouses in monogamy), but they do not disappear except under special circumstances, and then only temporarily. Understanding such considerations, and the long history of human interactions, provides the only way, I believe, for comprehending why individuals, families, social groups, and nations compete today – fiercely, continuously, and unendingly – even when no seemingly valid or sufficient reasons are evident, or can be given by the participants. (These arguments are expanded in Alexander 1987.)

To summarize, I have suggested that social-intellectual play, as scenario-building without extensive physical concomitants, is restricted to a small number of intensely or complexly cooperative mammals, such as group hunters, and may often be indicated by evidence of dreaming; in humans it is demonstrated by the communication of representational ability. Surrogate scenario-building, or the rewarding of others to build some of our scenarios for us, is probably restricted to humans, as is evidently also true of rules. Morality, I have suggested, represents the placing of more or less agreed-upon restrictions on actions that interfere too severely with the social-intellectual scenarios and plans of other societal members, and leads to convergence of individual and group interests.

The idea of fantasizing as play, and as problem-solving, is by no means original here. Piaget (1945: 131) saw all imaginative thought as "interiorized play". Symonds (1949), Singer (1966), and Klinger (1971) all saw fantasy as related to play and to later problem-solving. Novel here are (1) the association of scenario-building with social problems and deception; (2) the primacy of scenario-building as social-intellectual practice, leading to the prominence of surrogate scenario-building in human sociality; and (3) the argument connecting these activities to a history of intergroup competition.

The Ultimate Mystery: Discrepancies between the Functions of the Psyche and Our Knowledge of its Functions

Part of the difficulty in understanding ourselves arises out of the fact that if the human psyche is evolved to promote inclusive fitness maximizing (i.e., genetic reproduction via both descendant and nondescendant relatives: see Hamilton 1964; below), it clearly is *not* evolved to tell us precisely that this is its function and ours. This discrepancy makes it difficult to understand what the psyche is evolved to do, and difficult to construct a statement about what humans are evolved to do and not to do, that makes any sense to humans themselves, in terms of their conscious knowledge. I want to approach this problem indirectly. I am interested first in constructing the most general and explicit statement possible about how organisms – eventually, and in particular, humans – are expected, from evolutionary theory, to behave. Specifically, I wish to describe, in the most general terms, that sense in which behaviour is evolutionarily determined, or to describe what I expect organisms, because of their evolutionary history, *are not able to avoid doing* (Alexander 1979; 1987). In this fashion I propose to get at the question of what the human psyche *is* evolved to do. The reason for interest in what organisms are not able to avoid doing is roughly as follows: It is obvious that genes contribute to the behaviour of organisms. They determine how particular environments affect the developing phenotype. Equally obviously, it is not accurate to say that any particular behaviour of any particular organism is 'genetically determined'. The reason is that, unless it refers explicitly to the differences between variant behaviours being genetically determined, any such statement leaves out the effects of the environment. Thus, if such a statement were made, outside the context of causes of behavioural *variations,* it is quite probable that some one could eventually identify a change in the environment that would alter the behaviour, thus proving, in some sense, that the statement was wrong and the behaviour was in fact not 'genetically determined'.

Alternatively, one might say that a particular behaviour – say, how to recognize or behave differentially toward kin – is learned, if he knows that particular social experiences are necessary to cause the behaviour. But one could then ask: Was the tendency to accept or use the learning situation in that particular way also learned? Such questions then continue, like the turtles under the turtles in a storied Eastern philosopher's conception of the universe which had it 'after that, turtles all the way down'. But we know very well that, without some very special definitions, it cannot be '*learning* all the way down' because, even if they are only potentials to action, there are genes down there, in the form, sometimes, of alternative alleles giving rise to potentials for different actions.

In What Sense is Behaviour Evolutionarily Determined?

Human zygotes give rise to human organisms, honeybee zygotes to honeybees, etc. This means that, regardless of environmental variations, particular sets of genes produce particular phenotypes that do not vary beyond certain limits on any axis. It is fair to say that no one expects a human zygote to give rise to anything but a human phenotype. What is it fair to say, generally, about the limits of variation in behaviour, given a primacy for an evolutionary process guided principally by natural selection and effective primarily at the genic or individual level, or some other low level in the hierarchy of organization of life?

The most general statement is this: *No organism is expected to act in a way contrary to its genetic interests, except through error or miscalculation.*

To understand the significance of this statement, we must first establish precisely what are an organism's genetic interests, so that we can recognize whether it is doing as we expect. For example, we must understand that genetic survival is ultimately all that counts in evolution, and that genetic survival results from reproduction. Success in reproduction typically involves producing more lasting copies of one's genetic materials than are produced by competitors and potential competitors, leading to numerical preponderance, and among other things reducing the likelihood of accidental extinction. Moreover, copies of one's alleles appear in collateral relatives as well as descendant relatives, and not only in offspring but in subsequent generations as well. As a result we expect social organisms to measure abilities of relatives available for assistance to translate assistance into increases in reproductive success. We expect them to judge alternative ways of using life effort, comparing their reproductive costs and benefits accurately. In short, we expect the organism continually to be evolving to behave so as to maximize its *inclusive fitness* (Hamilton 1964). Even more explicitly, we expect organisms to behave so as to maximize, on average, the likelihood of survival of their genes – even, that is, if some copies of their genes are in other genomes, and even if some individuals die in the attempt because they are taking risks that are perfectly appropriate. I mean by this that, even if a behaviour results in some copies of alleles being lost, the behaviour may nevertheless maximize the likelihood that not all copies will be lost (cf. Dawkins' 1982 concept of the "extended phenotype").

Second, we must also recognize all of the possible ways that an organism can miscalculate, and the antecedent events that adjust its likelihood of miscalculation. For example, we must understand the concept of evolutionary novelty, and realise that organisms may make reproductive mistakes because they have been subjected to learning experiences or other events which alter their phenotypes yet were not encountered by their ancestors during all the time that the traits and tendencies expressed today were being moulded by selection. We must realise that competitors and predators will evolve to *cause* miscalculations in their adversaries and their prey. We must understand that unpredictable events may catch organisms in unprepared states. We must recognize that selection against 'errors', especially as side effects of adaptive behaviour, can only be effective if the cost of the mistake is greater than the value of the adaptive extreme that leads incidentally to it. We must understand that because organisms are (evidently) selected to maximize inclusive fitness only via the accomplishment of a wide array of more proximate ends or goals (such as avoiding pain, ingesting sufficient food of the right kinds, favouring one mate over another, or risking survival to save a brood of offspring – see below), there are innumerable ways for inclusive-fitness-maximizing to be sidetracked. Finally, we must recognize that we, as observers, will sometimes be able to identify actions more reproductive than those taken by organisms, but for historical or other reasons not available to the organism.

Applying the 'Evolutionary Determinism' Question to Humans

When the problem of making a general statement about evolutionary determinism is applied to humans, perplexing complications arise. Humans have what we call an 'awareness' of at least some of their intentions and their purposes in life: they can anticipate and reflect upon their activities and their goals.

If this conscious understanding about personal behaviour were tuned precisely in the interests of inclusive-fitness-maximizing through direct and conscious seeking of explicitly that goal, then our task would probably not be greatly complicated. In such event, to make the initial statement above most meaningful for humans, we might modify it to say: *No human is expected knowingly to act in a way contrary to its own genetic interests, except as a result of error or miscalculation.*

Unfortunately, we know immediately that this is not the correct prediction from evolutionary theory, because we know that whatever it is that humans have brought into their consciousness, it is not a maximizing of understanding, or even a steadily increasing understanding, of the process of inclusive-fitness-maximising. Humans are not only unaware of this process until it is explained to them, they are instantly reluctant to believe that they are engaging in it. Even if they see some aspects of their behaviour as according with probable predictions from evolutionary theory, the most enthusiastic among them are unlikely to believe for a moment either that they always behave so as to maximize their inclusive fitness or that this is what they are evolved to use their consciousness to achieve. Most humans would assert immediately that they frequently act in ways contrary to their own interests, even though it might be possible to show that some of the acts for which they believed this to be true were in fact precisely according to their *genetic* interests, others were more or less predictable results of evolutionary novelty in their environment, and still others were side effects of adaptive behaviour or simple errors as a result of deficient information.

Paradoxically, we cannot say both 'knowingly' and 'genetic interests' in the above statement, because humans in general do not know what their genetic interests are or how to maximize them. But they do *think that* they know what their interests are, just as they may think, without careful reflection, that everything important about their behaviour must be conscious or readily available to consciousness. Because consciousness is the only way of considering behaviour, there is something that seems illogical to a conscious being about behavioural knowledge being inaccessible to conscious consideration. This disparity is the source of our greatest problem in understanding ourselves. It causes us to wonder what our brains were designed to accomplish, and to suppose that there are no challenges in everyday life that are sufficient to explain them. Jaynes (1977: 23) created an apt analogy: "It is like asking a flashlight in a dark room to search around for something that does not have light on it. The flashlight, since there is light in every direction it turns, would have to conclude that there is light everywhere. And so consciousness can seem to pervade all mentality when actually it does not".

Together with the reasons for the physical altriciality or helplessness of the human juvenile, and our response to them (Alexander 1988, n.d.), the above difficulty may have helped cause two prominent evolutionary theorists (Hutchinson 1965; Williams 1966) to advance the notion that the intellect of humans evolved solely or primarily to assist the juvenile in social interactions, with effects on adults mere incidental 'overshoots'.

The picture, then, does not fall into place in the way we would expect it to if human consciousness had evolved as a steady improvement of personal understanding of one's own behaviour in the light of the goal of inclusive fitness maximizing. We can be certain that the reason we do not yet know all about inclusive fitness maximizing, and how the human psyche works, is not simply that the psyche has not had time to evolve far enough in that direction. In fact, it has evidently been evolving in some different direction.

Consciousness and Evolutionary Determination of Behaviour

Consciousness is the part of the human psyche that enables us to know what we know – or so it might seem. In actuality, it may be designed to enable us to know *certain* things but not others, and to keep from us some of the things that we nevertheless do 'know' in the more general sense of being able to act on possessed information (see Chomsky's 1980: 69 concept of 'cognizing'). The psyche may be designed specifically to keep us from knowing precisely (in the conscious sense) what we know and what is the evolutionary significance of our existence. That kind of information we may have to learn from the evolutionarily novel approach of science and technology.

If consciousness is indeed evolved, then it must be evolved to enable its bearer to maximize inclusive fitness. If it is not evolved to bring the realisation of its own purpose into the conscious understanding of its bearer (it is not necessary for humans to understand Darwinian theory for some version of such understanding to be present), then it has to be evolved to bring something different into the conscious understanding of its bearer.

To identify this something else is surely a first step in understanding the evolution of the human psyche.

One procedure for determining the evolutionary function of the human psyche would be to construct a model of a psyche evolved to deliver into conscious understanding the direct goal of maximizing inclusive fitness and then seek to describe the ways in which the human psyche actually deviates from this model.

Kin Recognition (i.e. measuring r in Hamilton's, (1964) formula: $k>1/r$, suggesting the situations in which beneficence can profitably – in terms of reproduction – be given to a relative; r refers to relatedness, k to the environmental costs and benefits of the situation).

First, we might expect that a psyche evolved to render the human individual acutely conscious of the goal of inclusive fitness maximizing would develop the ability to measure the relative genetic over-lap between itself and its various relatives. A growing body of evidence suggests that the human psyche is indeed evolved to accomplish this end, although not in an explicitly conscious way (i.e., the psyche does not automatically deliver to the bearer the conscious realisation of the purpose of the ability, or even, necessarily, the existence of the ability). In other words, any human in a normal social situation can usually identify which of any two of its relatives is more closely related to it. In part, at least, this accomplishment appears to be carried out by some kind of counting of genealogical links. Such counting generally works perfectly well, since each additional link halves (on average) the likelihood of any genes possessed by one of the two relatives also being possessed by the other as a result of their relatedness through immediate descent.

Similarly, the fact does not seem explicitly revealed to our conscious selves that our closest relatives are either approximately or precisely 50% likely to carry any particular gene in our own genomes as a result of relatedness through immediate descent (meaning, at least when an allele first appears in the population – Alexander 1979: 129), and that each link reduces this percentage by one half. It is likely that we are somehow programmed, developed, or instructed to treat relatives as if these things were true; but we are not consciously aware of any such instructions. It is difficult to think of a way in which we could gain by being conscious of such details (again, it is not necessary to be aware of the facts of meiosis or the particulate nature of inheritance to approach this kind of realisation, or to possess a ready acceptance of the significance of such facts when they do become available to us). This realisation highlights the facts that (1) there must be a great deal of knowledge that will do us no more good if conscious than if not; and (2) conscious time may be restricted and valuable, so that different potentially conscious items may compete for the available circuits. These possible kinds of limitations on consciousness, however, are not the ones that most concern us here. We are primarily interested in whether or not items or connections have been excluded from consciousness specifically in the interests of preventing the conscious picture from being complete and accurate, not simply because they are no less effective outside conscious circuits. Said differently, we are interested in the extent to which self-deception is a social phenomenon – a system of deceiving others through restriction of self-understanding and corresponding adjustments of social signals.

Environmental Costs and Benefits (Measuring k). Continuing our description of the hypothetical (but unreal) human psyche, designed to understand inclusive fitness maximizing and to know about it, we might also expect such a psyche to be evolved to develop into a superb and acutely conscious evaluator of the costs and benefits involved in helping relatives, spouses, and friends. Again, although it would appear that we are capable of such judgments, there is every evidence that, when we do it, the operation is not typically brought into or kept precisely in our consciousness. Sometimes we do indeed seem to make conscious judgements – especially if the contemplated act is quite expensive, the returns are not anticipated soon, or the potential recipient of substantial beneficence is a casual interactant or a distant relative (i.e., there is considerable risk involved). Even then it does not seem likely that all aspects of the judgement are manipulated on conscious circuits, or that the eventual reasons for decisions are fully conscious.

Again, it could be argued that no advantage is to be gained by bringing such details into presumably expensive conscious circuits. At some point, however, we must begin to wonder if there are kinds of information that can be made conscious only at a (reproductive) cost so high that selection works to exclude them even when there may be available conscious time that is not very expensive.

On the other hand, although we have gone through the central items in Hamilton's (1964) formula for inclusive fitness maximizing, we seem not to have identified any items yet for which the evolution of consciousness would be particularly advantageous or required. Nonhuman as well as human organisms maximize inclusive fitness, and neither human nor nonhuman forms appear to have evolved a conscious realisation of the fact. So we are still equally intrigued by the items that supposedly make consciousness an advantage in inclusive fitness maximizing and other items that would be disadvantageous if conscious.

Let us, then, consider the question from a different direction. Rather than continue trying to identify the kinds of items that we might expect to have been placed into the consciousness of humans, let us see if we can characterize those that have indeed been placed there, particularly in light of the manner in which inclusive fitness is maximized and how we think about the operation of natural selection as a result of information from the modern science of evolutionary biology.

With What, Then, is the Human Psyche Preoccupied?

Do we use our consciousness (psyche) primarily to learn how to do the things that incidentally maximize inclusive fitness, explicitly when other humans are the main competitors and adversaries, and social skills are the kind we need (that is, when other humans are the main hostile forces of nature)? Do we learn by extremely sophisticated and complicated kinds of social-intellectual play (as practice) how to do the calculating necessary to fulfil Hamilton's formula, and then perhaps move many of the actual calculations into the nonconscious, as with the actions of our fingers in playing musical instruments (Lieberman 1984 calls this "automatization")? Do we also practise continually at a kind of sincere hypocrisy (Campbell 1975) (self deception) in which we strive to present some picture other than one of seeking to serve our own interests as individuals? If so, then how do we answer the original question about the evolved limits of behaviour? Perhaps as follows: *No human is expected to behave in ways contrary to his own genetic interests, but all humans are expected to believe that they do so (if asked) because, in human sociality, a continual effort to serve one's own interests in a conscious deliberate way is not the best way to accomplish the purpose.* It is sometimes *reproductively* disadvantageous (but, I stress, *not necessarily disadvantageous or undesirable in any other terms*) not only to know that one is serving one's own interests, but even to know what those interests are. Because social interactions are typically long-term and repetitive, because multiple potential (alternative) partners in reciprocity are typically available, and because motivations are often used to judge suitability of individuals for later interactions, the most effective ways to deal with human competitors are: (1) sincerity achieved through self- as well as other-deception; and (2) the ability to see ourselves as others see us, so as to cause them to see us as we would like them to rather than as they would like to. The human psyche is evidently evolved to excel at such practices. I hypothesize that the human psyche achieves and maintains this excellence through continual social-intellectual play – and other practice – in such forms as humour, partying, oratory, theatre, soap operas, planning, purpose, and other kinds of social exchanges and scenario-building. Examples are outdoing a competitor in a game or in banter, being first in any competition that yields the prestige of being thought best at that particular game, etc. Such play or practice involves payoffs and expenses that are typically trivial compared to those for which successful practice can eventually reward the player more handsomely and directly – such as the reality of getting the job, wife, husband, friend, contract, commission, tenure, grant, award, raise, business, farm, inheritance, etc. – thus, actually winning a disproportionate share of the resources of reproduction and the freedom to use them in your own interests or as you see fit. The social functions of the psyche are thus realised in (1) partial-cost play or practice episodes; and (2) full-cost or 'real-life' episodes. I would venture that recognizing the social function of the intellect, and the centrality of self-interest, is a largely unexploited opportunity for those who would analyse human mentality in terms of operations paralleling those of machines (i.e., via 'artificial intelligence'); at the least, the actual nature and complexity of the activities of the psyche seem most likely to be revealed through analysis of social manipulations motivated by self-interest (in the form, of course, of being ultimately genetic and reproductive).

Deception and the Backgrounds of Information in the Subconscious

I assume, then, that when information is pressed (or kept) out of the conscious (by selection), this happens either because it is likely to be more useful in the subconscious or because it is not useful enough to warrant saving.

Evidently, some information moves into the subconscious as a result of repetition because it no longer requires conscious effort (e.g. playing a musical instrument, typing, or language). Some such information may be lost from retrieval to the conscious (even from the memory bank entirely) as a result of disuse (e.g. a little-used language). Still other information may never have been conscious, but may remain in the subconscious (and be available to the conscious under particular circumstances). It may have gotten into the subconscious by a process evolved to deal with useful information that was never conscious either (1) in the individual or (2) in the species. Or it may have been kept there by selection, in which case the analogy by Jaynes (1977) – mentioned earlier – between consciousness and a flashlight, would have to be modified to include that the flashlight cannot see into every corner but does not know it.

Deception, as with any life goal, can be deliberate or conscious or not. The conscious motive can be: (1) correct; (2) wrong because the real motive is inaccessible to the senses (e.g. the furthering of genic survival); or (3) wrong because the real motive is concealed from consciousness – either it was pressed out of the conscious or prevented from getting there.

The last must involve deception by self-deception. But why? Perhaps because conscious intent is not needed or conscious intent would interfere. Are we, then, ignorant of the self (genetic)-serving nature of our behaviour because (1) the true nature of our striving is not accessible to our senses; (2) there is no value in using the conscious; or (3) keeping our goals nonconscious aids us in serving them in social circumstances? In each case self-deception that effects deception of others is the aspect that I think is by far the most important in understanding the human psyche, and the one that I will develop here. There would be no premium on self-deception in nonsocial circumstances if this is the case, and this suggests the beginnings of a test. The concealment of ovulation in human females is an excellent case to analyse in terms of conflicts and confluences of interests in the parties involved (see Alexander and Noonan (1979) and Daniels (1983) – the latter especially for a review of published discussions following Alexander and Noonan). Mitchell (1986) criticizes Alexander and Noonan's discussion of deception and self-deception in connection with concealment of ovulation as "a confusion of a lack of information with deception". He does not, however, seem to grasp our argument that, if an event as central to reproduction and as physiologically profound as ovulation – and as available to both sexes as it is in perhaps all other mammals – is not conscious (and cannot be made conscious), there is a strong implication that it has been kept out of the conscious by selection. Alexander and Noonan noted that imperfect concealment – from either self or others – does not necessarily deny that selection has favoured concealment: we simply argued that less information about ovulation is available to the consciousness of either women or men than would be expected if selection had not favoured its exclusion from consciousness. If this conclusion is incorrect, then in this age of strong desires to control pregnancy there would surely be considerably less of a market for contraceptives.

Bernard Crespi (pers. comm.) has noted that males may react negatively to indications that their mates are aware of ovulation (implying control of fertilization and the possibility of cuckoldry), so that human females may be evolved primarily to avoid giving any such indication (for example, by maintaining a more or less unchanging interest in copulation during the ovulatory cycle) rather than to be entirely ignorant (unconsciously as well as consciously) of ovulation. This argument actually seems to predict the precise condition that exists in modern women – some ability to predict ovulation, but a general failure of such abilities to be acutely conscious, or even possible, sometimes, without modern medical information.

Self-Deception, Deception, and Intergroup Conflict

Arguments that the complex cooperativeness of human social life has been driven by intergroup competition and conflict call for all of what has just been said to be cast in terms of intergroup interactions. I have already argued that self-deception is a social phenomenon related to deception of others. Now I will argue further that self-deception explicitly plays a role in fostering and maintaining group unity, and that this role is intricated with the practice and prominence of familial, tribal, ethnic, racial, or regional myths, including organized religion. Indeed, I speculate that self-deception is a central factor in the group-unifying effects of patriotism, organized religion, and similar phenomena because it leads to acceptance of dogmas and myths that impart, at least temporarily, unity of purpose, interests, and striving. Myths need not represent the truth if their only significance is group unity: they need only be accepted. Acceptance, in turn, is expected to depend not on plausibility *per se* but on conviction, or acceptance, of a myth's value in group unification; scientific arguments about humans, for example, may be rejected because they do not have unifying effects, yet are seen as myths (as world views, ideologies, or even religious). Similarly, social, political, or religious leaders may find even otherwise highly laudable goals rejected by the populace if the effect of exhortations concerning them is divisive, restrictive, or self-deprecatory (for example, emphasizing misuse or over-use of environmental resources or projecting guilt rather than pride). Acceptance of unifying myths or information or goals depends on the individual's acceptance of the value of group unity, including the position or status of himself that will result, or other effects on himself and his intimates (children, spouse, relatives, reciprocants). Even myths widely regarded as counterfactual may be accepted, repeated, and elaborated if their effect is seen as unifying. The extent to which directly group-unifying effects of self-deception followed or paved the way for self-deception as a means of deceiving others within one's group seems moot (even self-deception with respect to one's own likelihood of recovering from pain or illness – in, say, a terminal illness – probably has as its primary function deception of others).

What about self-deception in cases of maladaptively extreme risk-taking, as in compulsive gambling or extreme heroism? In part the question seems always to be which is (1) extreme risk-taking as a result of self-deception either (a) about the risks directly or (b) as a result of coercion or gullible acceptance of exhortations from others; and which is (2) inaccurate assessment of risks owing to (a) incomplete information (and failure to assess properly the likelihood of its incompleteness), (b) inaccurate information, or (c) imperfect internal cost-benefit assessment machinery (including developmental misinformation and pathologies such as obsessiveness or addiction). Careful analysis of risk-taking dissected into some such categories seems necessary to resolve this question.

Evolution of the Emotions

The emotions can be defined as various complex reactions with both psychical and physical manifestations, as love, hate, anger, fear, grief, etc. (Webster's Unabridged Dictionary 1977; see also Panskepp 1982, and its following commentaries). Students of human behaviour are apt to regard the emotions as one of the principal features of the human psyche, along with consciousness, cognition, linguistic ability, and personality traits.

From either logic or the above definition, one can consider the emotions as comprising three more or less separate aspects:

1. The *expression* of the emotions (e.g. blushing, smiling, crying, frowning, screaming);
2. The *feelings* we associate with their expression (also sometimes used in definition, without mention of expression, as with "strong, generalized feeling; psychical excitement" or "any specific feeling");
3. The *underlying physiological activities or changes*.

Both the *expressions* of the emotions and the *feelings* associated with them can be significant to us either when they occur in ourselves or when they occur in others. Presumably, some of the underlying physiological activities or changes can occur without extrinsic expression or even feelings that we might term emotional (I emphasize the assumption that the first two aspects of the emotions above – at least as we know and experience them – would not be necessary for appropriate actions outside social contexts – i.e., in more or less completely solitary-living organisms). Also, presumably, the underlying physiological activities or changes have been modified (elaborated, altered) by the evolution of the expressions of the emotions and of the feelings we associate with the term.

How did emotions evolve to assume their current form and degree of expression in humans?

Stage 1: It seems reasonable to assume that there was a time, in the ancestry of humans, when the emotions were still incidental effects of physiological events that cause appropriate behaviour in specific circumstances, unnoticed by other individuals; this condition probably exists now in many or most nonsocial organisms. Such physiological activities or changes, which prime or adjust the organism to respond in ways favourable to its own survival or reproduction, could have (and must have) sometimes yielded, *strictly incidentally,* both extrinsic effects and internal feelings in the organism experiencing the physiological changes.

Stage 2: Incidental extrinsic changes reflecting physiological changes must have been noticed and eventually used by other organisms. Presumably, such observers would use the evidence of emotions in other organisms to their own advantage rather than to the advantage of the organism showing emotions, when there were differences in their interests. Such uses could include fleeing if a stronger individual showed evidence of anger or likelihood of attacking; taking advantage when another individual showed evidence of indecision or fear; searching for evidence of danger suggested by the emotional state of another so as to place one's self in a more advantageous position or condition, perhaps with respect to the individual showing the emotion; etc. External evidence of changes in the emotions could also be used by other individuals to assist them in inducing changes in the emotions of another, presumably in directions that benefited the individual inducing the changes.

Stage 3: Once individuals became capable of recognizing emotional states in other individuals, then it seems virtually certain that selection would alter both this ability and the emotional states themselves, or the external evidence of them, in ways that would be called communicative. In other words, as Darwin (1898) knew, at this point, the external expression of the emotions would surely be poised to become a major source of communication, especially in social species, and most especially in species with complex sociality in which the flow of social interactions tended to involve multiple and rapid emotional changes among many different states. This is the point (in evolution) at which it would become important for us to know about and assess our own 'feelings' or emotions – because we could then manipulate them to affect use by others of evidence about them.

Presumably, any organism that altered its emotional expressions under the influence of natural selection would do so in a way that affected its own interests *positively*. If the selection occurred because other organisms were already evolving to use the incidental expressions of the emotions to *their* advantage, then we can see that the organisms would tend to evolve to alter external expressions of their own emotions in such fashions as to thwart their use by others, at least when the others were using them to serve interests that differed from those of the individual showing them. This means that, at least most of the time, organisms would evolve to change their emotions in one or more of at least four ways: (a) to conceal some emotion being experienced; (b) to suggest an emotion not felt; (c) to indicate one emotion when actually experiencing a different one; or (d) to suggest either more or less intensity of emotion than felt.

All of these changes imply deception or manipulation of others. But scarcely anyone is likely to believe that all communication involves solely manipulation and deception. One wishes to explore the question whether or not expressions of the emotions have ever been altered during evolution in such ways as to convey *true* feelings – to tell the truth, so to speak, about one's emotions. Presumably, this could happen if social partners or companions were using the expressions of each other's emotions to help themselves because their interests were (at least temporarily) coincident. I presume that if two individuals sharing (at least temporarily) the *same* interests were to detect changes in one another's emotional states, in each case the detecting individual would use the information in its own interests, although such use should be imagined to include assisting either itself *(directly)* or the other individual (hence, in this case, itself *indirectly*). Such uses might include calming the other individual if its emotional state were placing it (or both individuals) in danger; trying to determine what was responsible for the emotional state of the other individual so as to respond to that environmental factor appropriately too; making some effort directly to attain an emotional state similar to that of the other individual if it seemed likely that the situation would call for cooperative effort; etc. On the other hand, if the interests of two individuals differed even slightly, we should expect each individual showing emotions to alter their expression so as to cause the responding individual to give a slightly different response than it might if following strictly its own interest. Situations may be rare in which two individuals share the same interests in such ways or to such degrees that neither can gain by deceiving the other into a little more assistance than it would give if it were acting according to complete and truthful information about the situation or the other individual's motivations.

The above arguments imply that expressions of the emotions are either incidental effects or else communicative, largely in the context of manipulation and deception. They also imply that virtually any extrinsic expression of the emotions, in an organism as complexly and continuously social as humans, is likely to have been noticed and used enough that some evolutionary modification has occurred in the context of communication (even non-noticeable, non-extrinsic expressions of the emotions are so used now, in polygraphs, or so-called 'lie detectors'). It seems likely that a significant effect has been caused on how we feel about what we call our emotions. In other words, some, much, or perhaps virtually all of the ways that we feel, consciously, when we experience what we think of as changes in our emotional states, are results of emotions having evolved to be communicative. In all likelihood, selection on the expression of the emotions has modified not only the way we feel about our emotions, but the actual physiological events that underlie the emotions as well. There must have been considerable feedback among these three aspects of the emotions all during human evolution. Paradoxically, because so much of communication, especially that involving expressions of the emotions, may be non-conscious or even self-deceptive (as use of polygraphs suggests), it is difficult for us to accept that physiological changes resulting in appropriate behaviours can occur without the feelings that we associate with expressions of the emotions. This is so because the emotions have actually evolved to be communicative and presumably would not be experienced by us in the way they are if they had not evolved such a function. Again, the potential for confusing primary and secondary effects is evident.

In turn, it is difficult to argue that *because* we blush or smile or laugh or frown or grieve when *alone (as well as when with others)*, this means that the emotions, as we experience them now, are often simply ways of changing ourselves physiologically to meet *nonsocial* eventualities, and may not be social or communicative at all. Presumably, however, if expressions of the emotions have evolved to be communicative, they may occur (secondarily) when we are alone either (1) because we cannot easily eliminate such non-social expressions, owing to the insignificance of their expense and the expense of eliminating them while retaining appropriate expressions in social situations; or (2) because we have evolved to use them in social scenario-building or planning when we are alone. It may be noticed that, to the extent that the latter is true, expressions of the emotions when one is alone should be honest and true reflections of at least the emotions that would be felt in the real situation that is being modelled in a mental scenario. In other words, truth in emotions may sometimes be expected when we are communicating with ourselves, if in no other situation.

As noted earlier, it is significant that the communicative function of expression of the emotions has not become entirely conscious and deliberate. Humans obviously do not have complete control of their emotions, including sexual excitement. If the function of expression of the emotions is, as I have suggested above, communicative, then why this should be true becomes a significant question. I believe that the answer is that, first, evidence of complete control of the emotions would indicate to others that there is no reliable way of assessing the effects of social events, or their own or others' presence and actions, on others. Accordingly, it seems likely to be a disadvantage in social matters to give the impression of such control over one's own emotions. We are in awe of actors and actresses who can produce emotions at will, but we are suspicious and negative toward individuals who do the same thing in our social interactions with them (consider such derogatory remarks as "I think she is just turning on the tears!"). Anyone engaged in establishing an important social interaction (such as seeking a long-term or lifetime mate) is bound to respond negatively to actions making it appear that the prospective partner can control at will its reactions to social, emotional, or sexual intimacy with us. We expect social interactants sometimes to behave in a certain fashion despite any conscious intentions. We look for evidence of such effects, we try with increasing effort to cause them to occur, and we are likely to regard their absence as evidence that the other party is less interested in us than we would like or may require.

Accordingly, in the degree of consciousness of expression of the emotions, we humans tread a fine line that can exemplify the aspects of consciousness that are most difficult to understand, and that cause consciousness – which represents the means by which we examine ourselves in the first place – be to quite poorly suited to self-analysis, even if, paradoxically, it is the only analytical device available to us. For it seems apparent that none of the attributes we need most to understand if we are to comprehend our psychical nature is likely to be more completely available to the conscious than are the emotions. The reason is evidently that humans have evolved to be so adept at identifying falseness in deliberate (or conscious) actions and motivations that they have also evolved to deceive by keeping many aspects of motivation out of the conscious. Even more paradoxically, this complexity would not have arisen without the evolution of consciousness in the first place. To use Humphrey's (1986) analogy of the inner eye, there is no inner eye eyeing the inner eye (of consciousness), so that we are left to analyse these problems (create such an eye) by using the procedures of science. This is, to some extent, a procedure advanced by Sigmund Freud. It would appear, however, that to continue the process – to understand motivations ever more deeply so as to understand the human psyche and all our mental activities and tendencies more deeply as well – we will be required to refer continually to the best available understanding of natural selection, because that is the ultimate designer of motivation.

Group-Coordination and the Emotions

Scenarios constructed earlier in this essay for the early evolution of hominids included three major stages (see also Alexander 1979):

1. Group-living because of predation (probably most group-living primates and early hominids);
2. Group-living that includes cooperative group-hunting (chimpanzees, humans);
3. Group-living that includes direct intergroup competition or aggression (chimpanzees, humans).

Beginning with this sequence, a major question is what kinds of mechanisms enabled group-cooperative humans to conduct intergroup aggression cooperatively. How did humans manage the coordination necessary to carry out raids efficiently, especially against enemies belonging to their own species and possessing the same general abilities and tendencies? What kinds of evolutionary change elaborated and perfected the ability to coordinate cooperative efforts of individuals in complex fashions? Although group hunting may have been the initial circumstance in which mechanisms of cooperation evolved, the greatest challenges would obviously have been in connection with intergroup conflicts and raids involving conspecifics.

Coordination of the emotions almost certainly plays a central role in group cooperation during intergroup aggression, as it does in group hunting. Demagogues can coordinate group emotions, and recognition of the value of leaders in such contexts could lead to acceptance of despotism as group sizes increase. Ritual, myth, religion, patriotism, xenophobia, ceremonies, cheerleading, and pep rallies can all be seen as related to the coordination (and testing) of emotions in connection with specific cooperative tasks. One can hardly fail to see parallels between the elaborate and ceremony-like expressions of excitement among diverse organisms such as African wild dogs about to depart on a hunt (Lawick and Lawick-Goodall 1971), and humans engaged in stirring their fellows to participation in risky activities.

Robert Hinde has suggested (in a lecture at the University of Michigan, April 1987) that emotions in modern wars (as opposed to the raids of bands or tribal groups on neighbours) are tuned not to developing and showing anger and aggressive tendencies and passions but to the support of an institution (or institutions); that patriotism, and economic, political, and religious responsibility, are called upon by the orators and demagogues and leaders; that modern soldiers fight out of responses to these kinds of exhortations; and that we must understand the genesis of iinstitutions to understand modern war. This argument is probably slightly over-simplified. Thus, when I was in the US Army, we were exhorted by being told, first, that we should hate 'gooks' (the enemy); second, that we were, ultimately, defending our sweethearts, sisters, wives, mothers, and children; third, that we were, again, ultimately, defending our homes and land. (These last two exhortations were especially clear during World War II, when there were also emotional songs about home and family, such as "This is worth fighting for!" But both were also used when the US was fighting in Korea: "If we don't fight them there, we'll be fighting them here!"); and, fourth, that we were defending democracy and all the good institutions that are America (I also entered the Army with these final, electrifying words from my own mother: "Although I did not raise you to be a soldier, I *know you will be a good one!*"). But the idea is worth considering that – together with the emotions *per se* – Hinde's proposition can be related to the difficult problem of why despotism rises then wanes as social systems change so as to allow or cause larger and larger groups to be unified (Alexander 1979; Betzig 1986). Presumably, with very small groups, the emotions of the moment determine the efficacy of a raid. Perhaps the exhortations of leaders and others, through shows and exaggerations of their own emotions, cause everyone else to become aroused enough to carry out a raid and do it well. Maybe increasing extremes of despotism work similarly as group sizes enlarge – up to a point, but problems arise in managing very large groups through despotism (although multiple episodes in recent history show that in specific situations individual demagogues can be appallingly effective). Perhaps such problems provide part of the explanation for the rise of democracies and what I have previously called reproductive opportunity levelling (Alexander 1987).

Surely one of the most consequential uses of linguistic ability must have been in coordinating group efforts. And as it became significant, language would have become a vehicle for expression of the emotions, and as well would surely have altered their expression as a communicative device (Burling 1986).

Language and Scenario-Building

Laura Betzig (pers. comm.) has reminded me of the relationship between (1) what Hockett (1960) called "displacement" in human linguistic communication, and (2) scenario-building as a modelling and testing activity with respect to possible later events. Displacement refers to the human capability of communicating linguistically about events removed in time and space from the act of communication – the use of past and future tenses, and the discussion of events involving some different spatial location. Although many species may have evolved some capability of building and testing mental scenarios that involve displacement, communication between individuals about such displaced events would be exceedingly difficult without language, and, especially, the use of past and future tenses (functionally, consideration of past events would seem always to be in the interest of learning more about possible future events). Displacement, so defined, is in some sense not limited to human language, as Hockett knew: honeybee "dance language", in which distance and direction of sources of food or hive sites are communicated with precision and detail, is one of the phenomena of animal communication that has most intrigued and baffled students of human behaviour (Gould 1975). Leaving aside for the moment the problem of comparing adequately the physiological mechanisms of honeybee and human communication, the relationship between the rise of scenario-building in human mental activities and the value of evolving abilities to communicate about them is obviously worth the attention of those wishing to understand linguistic ability and the evolution of human mentality. Lieberman (1984: 248) suggests, from the work of Fouts, that both temporal and spatial displacement occur in the communication of chimpanzees, through signing, with humans.

TESTING THE GENERAL HYPOTHESIS

What Things Seem Right with the Hypothesis?

1. It has the potential to account for any and all sizes of socially complex groups. Critics of hypotheses giving 'war' a central role in human evolution sometimes have asserted that war cannot account for the rise of nations because different social or political groups have been at war more or less continually without having turned their social systems into nation-states. But this criticism misses the significance of particular kinds of expenses in increasing group sizes in some localities – such as uncrossable mountain ranges or rivers. An intergroup competition hypothesis includes the condition that suitable adversaries must exist to account for continual increases in group size and complexity (see Carneiro 1970; Alexander 1979, and references cited therein).
2. It accords with recorded history with respect to prevalence of intergroup competition. This fact seems to me at least to shift the burden of proof to those who would claim that prehistoric humans did not live in situations that would have caused them to evolve tendencies and abilities to be aggressive when circumstances demanded.
3. It accords with the unique human attribute of group-against-group competition in play, and with the centrality of such play in human sociality. I repeat my acceptance of the general theory that play represents practice and low-cost testing.
4. It accords with the ecological dominance of the human species. This, of course, is just a modification of the widespread anthropological description of humans as the species that, more than any other, creates its own environment. As remarked earlier, I am not referring to ecological dominance of a sort that could only postdate agriculture, but rather a kind that, except for the presence of humans, is probably possessed even by chimpanzees.

5. It accords with the disappearance of all close human relatives despite our rapid evolution (and probably *requires* it). This requirement thus approaches becoming a falsifying proposition (that fails: see also, comments below on apes and dolphins).
6. It accords with the 'autocatalytic' model of brain size increase (Stringer 1984). Of course, internal changes in the brain that increased intellectual capacity were surely occurring simultaneously with increases in size of the brain itself, or of the brain cavity; and, as already noted, it is possible for social structures to be achieved in which strong selection for increases in brain size might taper off and disappear, as the fossil record suggests.

What Things Seem Wrong with the Hypothesis?

1. Early humans and pre-humans are generally assumed to have very low population densities – too low for intergroup competition to have been significant (e.g. Martin 1981). We must, however, wonder how much of this assumption is due to inadequate information, and to the tendency to assume *a priori* that weather, climate, and food were early humans' greatest problems. To maintain my argument I have to conclude that densities *per se* were not critical, or else that estimates of densities were wrong. It is probably more important to know what kinds of social groups people lived in, and why, than to know densities *per se*. We must consider the possibility, I think, that social groups may have been in intense competition with one another for scarce or localized resources even if overall population density was low (e.g. Ember 1978). Hamilton (1975) has stressed that life in small kin groups – such as presumably would occur under low population densities – could well exacerbate the tendency to be ethnocentric and xenophobic.

The question of densities also seems to bear on hypotheses about rates of movement between geographic regions, and thus would appear to bear on the alternatives of (1) a multi-regional hypothesis involving a single species in which genetic (or cultural) changes appearing in a few or many separate localities are repeatedly spread throughout the species; or (2) a single- locality hypothesis involving appearance of either a separate species or a strikingly different form in one locality, spreading without much (or any) interbreeding to cover eventually the entire range of modern humans and replacing the forms previously living there (e.g. Wolpoff, this volume). Either of these alternative hypotheses is compatible with the hypothesis advanced here. Intergroup competition could have been (and probably was) between both conspecific social groups and similar species.

That humans reached Australia and New Guinea about 40 000 years ago – apparently across a sizeable stretch of water – and penetrated to southern South America after crossing the Bering Strait 10 000 or 15 000 years later, and also the broad overall distribution of evolving hominids for several million years, implies considerable ability of hominids to move great distances – hence, to interbreed, and to interact with strange groups. It also implies a strong likelihood of evolution according to the amity-enmity polarity of intergroup competition as outlined here, regardless of actual densities or general way of life. Only a close intragroup cooperativeness, leading almost automatically to intergroup hostility, is necessary. The entire recorded history of humanity, and our direct knowledge of ethnocentricity and xenophobia (see Reynolds *et al.* 1987), is also consistent with these implications, hence, in this sense, with the general hypothesis developed here.

2. Early humans and pre-humans are generally assumed to have been under great food stress. There is, however, more than one reason for food stress. As others have pointed out, a dominant male in a highly polygynous species, at the height of his breeding performance, is almost certain to be under 'food stress'. So may be a subordinant male, ostracized from his social group by a dominant male or for any other reason, or a subordinate *group*.

3. War and large groups are both recent; as discussed below, however, there is no unambiguous evidence of either early aggression or its absence.

4. Great apes have some of our mental attributes, including a kind of self-awareness (Suarez and Gallup 1981), but perhaps only chimpanzees currently have an appropriate social structure. It might be possible to argue that orangutans and gorillas do not have intellects that are sufficiently like those of humans to require (by my hypothesis) that they have lived in social groups that would have caused the kind of runaway intellectual evolution I have been describing. But I cannot easily reject the notion that the intellects of the great apes may in fact demand explanation in terms of my hypothesis. I am led, then, to wonder – entirely without empirical evidence – whether or not orangutans and gorillas once lived in social groups more like those of chimpanzees and humans than is presently the case: in other words, multi-male groups, and perhaps multi-male groups in which the males were cooperative in hunting, or even in intergroup aggression (see Wilson 1975: 568, and Wrangham 1987: 60, for comparisons of some relevant attributes). Wilson (1975: 36) suggests that orangutans may have once been more social than they are now, and that conditions leading to extreme sexual dimorphism (exaggerated in both orangutans and gorillas) may have reduced sociality (Ciochon (1987) summarizes information on ape and hominid phylogeny, and Robert Smuts has suggested to me that juvenile orangutans may be more social than expected from the current social life of the species.

5. The problem of explaining the size and complexity of dolphin brains, and their apparently remarkable learning abilities, parallels that of understanding the great apes (Connor and Norris 1982; Schusterman *et al.* 1986; Herman 1980; Connor, pers. comm.). Dolphins may be as unusual in these respects compared to other inhabitants of the sea as humans (or humans and apes) are compared to other inhabitants of terrestrial habitats (e.g. Worthy and Hickie 1986). Dolphins do not construct tools or other complex artifacts. They are evidently subject to severe predation from sharks (Krushinskaya 1987), and are remarkable navigators (Kellogg 1958; Norris et al. 1961; Klinowska 1986). If these features of their environment are not responsible for their large brains and apparently complex mental abilities we are left with the complexity of their social life by default. Unfortunately,

because of rudimentary knowledge of the details of everyday dolphin sociality (Norris and Dohl 1980; Krushinskaya 1986), we can make little further comment on this topic. This ignorance, coupled with the unusual brain and learning abilities of dolphins, has caused a great deal of public interest and considerable speculation among scientists (for example, the highly publicized speculations about dolphin communication, and hypotheses that dolphins may engage in reciprocity (Connor and Norris 1982) and that odontocetes may stun their prey with high- intensity sounds (Norris and Mohl 1983; Morris 1986)). The importance of examining both dolphin and ape sociality more intensively seems apparent.

How Can the Hypothesis Be Falsified (beyond the above difficulties)?

1. One obvious possibility of falsifying the general argument that humans have evolved largely through a process of runaway social competition and imbalances of power between competing groups would be to show that such activities are too recent in human history to account for evolution of humans from nonhumans or for evolution of modern humans from archaic humans. I think that most accounts of human social evolution (e.g. see Mann 1986, and references cited therein) imply that this is in fact the case. If, however, chimpanzees have already embarked upon a path involving significant intergroup aggression – a proposition developed independently of the argument generated here and earlier (compare Alexander 1967-1988, with King 1976; Goodall 1986; Wrangham 1987; and Manson and Wrangham, n.d.), then this potential falsifier, it seems to me, is itself falsified. Nevertheless, it seems necessary to understand how it might be that current accounts of the evolution of civilization do not always seem to support the ideas I am espousing. Part of the reason may result from views about hunter-gatherers that may be untenable (e.g. see Ember 1978; Alexander 1979). A second part involves the nature of evidence about intergroup aggression.
2. Absence of indicators of significant intergroup aggression is a second possible falsifier of my arguments. My previous comments on this question (Alexander 1979: 227-228) are here paraphrased and supplemented:

Two kinds of evidence bear on the question whether or not intergroup competition and aggression have played a central role in human evolution. One kind is physical evidence of aggression, including fossils. Little or none of this evidence is unequivocal: spear points, arrowheads, and stone axes all have been called 'tools' or 'weapons', depending on one's bias, and they could have been either or both; skulls could have been crushed by predators or damaged after death; evidence of cannibalism could have been interpreted differently if it came from ceremonial affairs within groups rather than from wars; etc. For example, as suggested by Darwin (1871), Pilbeam (1966), Wolpoff (1971), Lovejoy (1981), and others, in the light of the evidence that humans are willing and adept at complex inter- and intragroup competition, stone 'tools' (weapons) could have lowered the usefulness of teeth *as weapons* (of defence or offence, and against predators as well as conspecifics) as much as (or rather than) "removed the need for use of the anterior teeth in aiding the hands to hold various utensils or materials such as skin or wood" (Howells 1976).

Even continuous intergroup hostility and aggression do not necessarily leave a record for archaeologists to trace. If there were no written records, what evidence would there be to tell us what happened to the Tasmanians and the Tierra del Fuegians? Without written records could we have been unequivocal thousands of years later about what the invading Europeans did to the Native Americans on both continents of the New World? Consider the most monstrous cases of genocide in recorded history: can we even be sure, again without written words, that what happened in the twentieth century at Buchenwald and Auschwitz, and in Nigeria and Cambodia, would be properly interpreted, say, a million years from now? Yet more people may have been killed in these places than existed in all of the time before recorded history. Such questions, it seems to me, cast doubt on the interpretation that equivocal evidence of human aggression, not to say the milder yet potentially continual and crucial forms of intergroup competition, must automatically be discarded.

The second kind of evidence comes from interpreting recent history and the behaviour of modern humans, and then asking about the legitimacy of extrapolating backward in time, both to postulate what happened and to interpret the otherwise equivocal evidence from archaeology and palaeontology. We know that intergroup competition and aggression have been continuous across nearly the whole face of the earth throughout recorded history. We know that cooperativeness on the grandest scale, and the greatest of all the alliances of history, were in response to upsets in balances of power and the aggression of one nation against another. We know that competition is continuous among the various kinds of political groups, large and small, that exist across the whole earth. We know that atomic fission, space travel, and probably most of the remarkable modern advances in science and technology occurred or were accelerated as a consequence of intergroup competition or outright war.

Not only are there two kinds of evidence with respect to intergroup aggression, but the nature or effects of intergroup aggression on human evolution may be better understood if it is considered in at least two major stages. The stage that is more understandable to us, and better represented by evidence, is the later stage, involving organized military interactions, extensive weaponry and strategizing, armies, and sometimes very large scale operations. This is the kind of intergroup aggression that is typically called 'war'. It extends from the beginnings of recorded history to the present, virtually continuously, is often highly organized and complex, and was evidently instrumental in the development and maintenance of the kinds of social systems that have prevailed across recorded history. As already mentioned, these facts place a certain burden of proof on those who would have intergroup aggression disappear as one moves back into prehistory.

Intergroup aggression prior to recorded history is more difficult to substantiate directly. As already suggested, tools may have been weapons, and most evidence is equivocal. It seems legitimate to consider chimpanzees as approximating a model of some early stage of such aggression. The time between such a stage and the appearance of full-fledged 'war' could have involved several hundred thousands of generations and, at times, more than a single species. It would have been during this period that the transition from the pre-human to the human condition would have occurred.

3. Show that, historically, the *topics, timing,* or *sequences* of representational ability (evidenced by tools, weapons, art, ceremony) are wrong; or, by other means, that the complexity of the psyche is not tied primarily to success in social matters.

Any interpretation of the function or operations of the modern human psyche, and of its manner of evolution, must be compatible with the fragments of evidence that exist in fossil or other forms (tools, sculptures, paintings, evidence of planning and social structure, items associated with burials, and what is known about the behaviour of nonhuman primates). Meanings appropriate to the rest of the arguments must eventually be derivable from the nature of the artifacts, their ages, and their sequence. One wishes to ask whether or not evidence of appropriate representational ability appears at the right times and in the right sequences during human history to support the arguments here generated. What are the (1) real and (2) expected sequences of changes in the fossil and other evidence of scenario-building (as well as tool- and weapon-use and social structure) across the relevant periods of human evolution? When and how did scenario-building ability become a means of acquiring status – as in sexual selection or leadership? What kinds of findings would support or falsify the ideas expressed here? Two kinds of evidence are available to us on this question: (1) indications of changes in sizes, compositions, and interactions of human social groupings leading toward the ranges of variation found in modern humans; and (2) artifacts, fossils, and remains relevant to psychical abilities of the sort found in modern humans. How can they be used to test the arguments advanced here? The seminal contributions to this projected test appear to be those of Wynn (1979, 1981) and Gowlett (1984). Using the artifactual record, these authors begin tracing the gradual appearance of consciousness, self-awareness, foresight, and the internal representational abilities of modern humans. Although the data are sparse, they are at least able to suggest that *Homo erectus,* as we would also imagine from its phylogenetic position, possessed an elaborated kind of 'great apes' mentality appropriate to the predecessor of modern humans. So far as I can tell, the meagre evidence from this kind of analysis as yet casts no doubt on the hypotheses advanced here.

4. Show that the greatest increases in intellectual or mental capacities (brain size?) occurred when nonhuman or nonbiotic hostile forces were most severe *rather than vice versa*. To me this test seems the most unequivocal, and the most likely to be useful (see Stringer 1984). It requires evidence as to whether or not the greatest changes in intellect occurred during maximum extent of glaciation (and near the glaciers or in otherwise severe – perhaps xeric – climates), and when population densities were lowest as a result of

such extrinsic hostile forces; or on the other hand under mild climatic conditions when food was relatively abundant and population densities were highest.

Additional tests may be possible from psychological studies that bear on the use of the human intellect in social matters *versus* other circumstances. No such test is obvious to me now, since it would appear that usefulness of the brain in other circumstances may have evolved concomitantly in such fashion as to render inextricable the two aspects of brain function. I suspect, however, that continued confirmation of the significance of the emotions and personality traits in social and communicative matters, and especially in manipulating and deceiving others, would provide strong support for the arguments advanced here. This will be especially true if such features of the psyche are also related to linguistic ability and the various aspects of consciousness and cognition in ways that reinforce the argument for social significance.

When I described to a friend the problem of titling this essay about the human psyche, he suggested with a sly smile that I might call it "Psychology". Having completed the essay, I discover that, indeed, I have argued, in agreement with Humphrey (1976-1986) and using his phrase, that the function of the human psyche is to "do psychology" – that is, to study itself as a phenomenon, in ourselves and other conspecific individuals, and to manipulate, in particular, the versions of itself found in those other individuals. When I read this statement back to the same friend, he nodded and added, "Unconsciously".

ACKNOWLEDGEMENTS

I thank Theodore H. Hubbell, Richard C. Connor, Robert W. Smuts, Donald Symons, Pat Overby, Lars Rodseth, Laura Betzig, Martin Daly, Margo Wilson, George C. Williams, William D. Hamilton, Daniel Otte, Robert Foley, Katharine M. Noonan, Mark V. Flinn, Paul Turke, Cynthia K. Sherman, Paul W. Sherman, John Speth, Bernie Crespi, Kyle Summers, Randolph M. Nesse, Milford Wolpoff, and Frank Livingstone for stimulating discussions on this and related topics, and for help with the manuscript. Joseph Manson and Richard Wrangham allowed me to discuss their unpublished manuscript. Bernie Crespi and Aina Bernier helped immensely with the literature search. Richard C. Connor, in particular, assisted in developing ideas about reciprocity and the stratification of coalitions. Financial support is acknowledged from the Frank Ammerman Fund of the Insect Division of the University of Michigan Museum of Zoology, and the Evolution and Human Behavior Program of The University of Michigan College of Literature, Science, and the Arts.

REFERENCES

Alexander, R. D. 1967. Comparative animal behavior and systematics. In (Anonymous) *Systematic Biology*. National Academy of Science Publication 1692: 494-517.

Alexander, R. D. 1971. The search for an evolutionary philosophy of man. *Proceedings of the Royal Society of Victoria* (Melbourne) 84: 99-120.

Alexander, R. D. 1974. The evolution of social behavior. *Annual Review of Ecology and Systematics* 5: 325-383.

Alexander, R. D. 1975. Natural selection and specialized chorusing behavior in acoustical insects. In D. Pimentel (ed.) *Insects, Science and Society*. New York: Academic Press: 35-77.

Alexander, R. D. 1979. *Darwinism and Human Affairs*. Seattle: University of Washington Press.

Alexander, R. D. 1986. Ostracism and indirect reciprocity: the reproductive significance of humor. *Ethology and Sociobiology* 7: 253-270.

Alexander, R. D. 1987. *The Biology of Moral Systems*. Hawthorne (NY): Aldine.

Alexander, R. D. 1988. The evolutionary approach to human behavior: what does the future hold? In L. L. Betzig, M. Borgerhoff Mulder and P. W. Turke (eds) *Human Reproductive Behavior: a Darwinian Perspective*. Cambridge: Cambridge University Press: 317-341.

Alexander, R. D. (in press). Über die Interessen der Menschen und die Evolution von Lebensabläufen. In H. Meir (ed.) *Die Herausforderung der Evolutionsbiologie*. München: Piper: 129-171.

Alexander, R. D. n.d. Why human babies are helpless: a general theory of altriciality. Unpublished manuscript.

Alexander, R. D. and Borgia, G. 1978. Group selection, altruism and the levels of organization of life. *Annual Review of Ecology and Systematics* 9: 449-474.

Alexander, R. D. and Noonan, K. M. 1979. Concealment of ovulation, parental care and human social evolution. In N. A. Chagnon and W. G. Irons (eds) *Evolutionary Biology and Human Social Organization: an Anthropological Perspective*. North Scituate (Mass): Duxbury: 436-453.

Alexander, R. D. and Tinkle, D. W. 1968. Review of K. Lorenz: *On Aggression* and R. Ardrey: *The Territorial Imperative*. *Bioscience* 18: 245-248.

Alcock, J. 1984. *Animal Behavior: an Evolutionary Approach*. Sunderland (Mass): Sinauer. Third Edition.

Axelrod, R. and Hamilton, W.D. 1981. The evolution of cooperation. *Science* 211: 1390-1396.

Betzig, L. L. 1986. *Despotism and Differential Reproduction: a Darwinian View of History*. Hawthorne (NY): Aldine.

Bigelow, R. S. 1969. *The Dawn Warriors: Man's Evolution toward Peace*. Boston (Mass): Little, Brown.

Box, H. O. and Fragaszy, D. M. 1986. The development of social behaviour and cognitive abilities. In J. G. Else and P. C. Lee (eds) *Primate Ontogeny, Cognition and Social Behaviour*. Cambridge: Cambridge University Press: 119-128.

Burling, R. 1986. The selective advantage of complex language. *Ethology and Sociobiology* 7: 1-16.

Campbell, D. T. 1975. Conflicts between biological and social evolution and between psychology and moral tradition. *American Psychologist* 30: 1103-1126.

Carneiro, R. L. 1970. A theory of the origin of the state. *Science* 169: 733-738.

Cheney, D. 1987. Interactions and relationships between groups. In B. B. Smuts, D. Cheney, R. M. Seyfarth, R. W. Wrangham and T. T. Struhsaker (eds) *Primate Societies*. Chicago: University of Chicago Press: 267-281.

Cheney, D. and Wrangham, R. W. 1987. Predation. In B. B. Smuts, D. L. Cheney, R. M. Seyfarth, R. W. Wrangham and T. T. Struhsaker (eds) *Primate Societies*. Chicago: University of Chicago Press: 227-239.

Chomsky, N. 1980. Rules and representations. *Behavioral and Brain Sciences* 3: 1-61.

Ciochon, R. L. 1987. Hominid cladistics and the ancestry of modern apes and humans. In R. L. Ciochon and J. G. Fleagle (eds). *Primate Evolution and Human Origins*. Hawthorne (NY): Aldine.

Connor, R. C. and Norris, K. S. 1982. Are dolphins reciprocal altruists? *American Naturalist* 119: 358-374.

Daniels, D. 1983. The evolution of concealed ovulation and self-deception. *Ethology and Sociobiology* 4: 69-87.

Dart, R. 1949. The predatory implemental technique of *Australopithecus*. *American Journal of Physical Anthropology* 7: 11-38.

Dart, R. 1954. The predatory transition from ape to man. *International Anthropological and Linguistic Review* 1: 201-213.

Darwin, C. R. 1859. *On the Origin of Species*. Facsimile of the first edition with an Introduction by Ernst Mayr. Cambridge (Mass): Harvard University Press, 1967.

Darwin, C. R. 1871. *The Descent of Man and Selection in Relation to Sex* (2 Vols). New York: Appleton.

Darwin, C. R. 1898. *The Expression of the Emotions in Man and Animals*. New York: Appleton.

Dawkins, R. 1976. *The Selfish Gene*. Oxford: Oxford University Press.

Dawkins, R. 1982. *The Extended Phenotype: the Gene as the Unit of Selection*. San Francisco: Freeman.

Dawkins, R. 1986. *The Blind Watchmaker*. New York: Norton.

Ember, C.R. 1978. Myths about hunter-gatherers. *Ethnology* 17: 439-448.

Fagen, R. 1981. *Animal Play Behaviour*. New York: Oxford University Press.

Fagen, 1982. Evolutionary issues in development of behavioral flexibility. In P. P. G. Bateson and P. H. Klopfer (eds) *Perspectives in Ethology*. New York: Plenum Press: 365-383.

Fisher, R. A. 1958. *The Genetical Theory of Natural Selection*. New York: Dover. Second Edition.

Flinn, M. and Low, B. S. 1986. Resource distribution, social competition, and mating patterns in human societies. In D. I. Rubenstein and R. W. Wrangham (eds) *Ecological Aspects of Social Evolution*. Princeton (NJ): Princeton University Press: 217-243.

Fox, R. 1980. *The Red Lamp of Incest*. New York: Dutton.

Gallup, G. G. 1970. Chimpanzees: self-recognition. *Science* 167: 86-87.

Ghiglieri, M. P. 1987. Toward a strategic model of hominid social evolution. In *Understanding Chimpanzees*. Chicago: Chicago Academy of Sciences.

Ghiglieri, M. P. 1988. *East of the Mountains of the Moon: Chimpanzee Society in the African Rain Forest*. New York: Free Press.

Godfrey, L. and Jacobs, K. H. 1981. Gradual, autocatalytic and punctuational models of hominid brain evolution: a cautionary tale. *Journal of Human Evolution* 10: 255-272.

Goodall, J. 1986. *The Chimpanzees of Gombe: Patterns of Behavior*. Cambridge (Mass): Belknap Press.

Gould, J. 1975. Honeybee communication: the dance-language controversy. *Science* 189: 685-693.

Gowlett, J. A. J. 1984. Mental abilities of early man: a look at some hard evidence. In R. Foley (ed.) *Hominid Evolution and Community Ecology: Prehistoric Human Adaptation in Biological Perspective*. New York: Academic Press: 167-192.

Griffin, D. R. 1978. Prospects for a cognitive ethology. *Behavioural and Brain Sciences* 1: 527-538.

Hamilton, W. D. 1964. The genetical evolution of social behaviour I, II. *Journal of Theoretical Biology* 7: 1-52.

Hamilton, W. D. 1971. Geometry for the selfish herd. *Journal of Theoretical Biology* 31: 295-311.

Hamilton, W. D. 1975. Innate social aptitudes of man: an approach from evolutionary genetics. In R. Fox (ed.) *Biosocial Anthropology*. New York: Wiley: 133-155.

Harpending, H. 1986. Review of L. Betzig: *Despotism and Differential Reproduction: a Darwinian View of History*. *American Scientist* 75: 87.

Herman, L. M. (ed.) 1980. *Cetacean Behavior: Mechanisms and Functions*. New York: Wiley.

Hockett, C. F. 1960. Logical considerations in the study of animal communication. In W. E. Lanyon and W. N. Tavolga (eds) *Animal Communication*. Washington (DC): American Institute of Biological Sciences: 392-430.

Howells, W. W. 1976. Explaining modern man: evolutionists *versus* migrationists. *Journal of Human Evolution* 5:477-495.

Humphrey, N. K. 1976. The social function of intellect. In P.P.G.Bateson and R. A. Hinde (eds) *Growing Points in Ethology*. Cambridge: Cambridge University Press: 303-317.

Humphrey, N. K. 1978. Nature's psychologists. *New Scientist* 78: 900-903.

Humphrey, N. K. 1979. Nature's psychologists. In B. Josephson and B. S. Ramchandra (eds) *Consciousness and the Physical World*. New York: Pergamon: 57-75.

Humphrey, N. K. 1983. *Consciousness Regained: Chapters in the Development of Mind*. Oxford: Oxford University Press.

Humphrey, N. K. 1986. *The Inner Eye*. London: Faber and Faber.

Hutchinson, G. E. 1965. *The Ecological Theatre and the Evolutionary Play*. New Haven: Yale University Press.

Irons, W. 1981. Why lineage exogamy? In R. D. Alexander and D. W. Tinkle (eds) *Natural Selection and Social Behavior: Recent Research and New Theory*. New York: Chiron Press: 476-489.

Isaac, G. Ll. 1978. Food-sharing and human evolution: archaeological evidence from the Plio-Pleistocene of East Africa. *Journal of Anthropological Research* 34: 311-325.

Isaac, G. Ll. 1979. Evolutionary hypotheses. *Behavioral and Brain Sciences* 2: 388.

Jaynes, J. 1977. *The Origin of Consciousness in the Breakdown of the Bicameral Mind*. Boston: Houghton Mifflin.

Jolly, C. 1970. The seedeaters: a new model of hominid differentiation based on a baboon analogy. *Man* 5: 5-26.

Keith, A. 1949. *A New Theory of Human Evolution*. New York: Philosophy Library.

Kellogg, W. N. 1958. Echo ranging in the porpoise. *Science* 128: 982-988.

King, G. E. 1976. Society and territory in human evolution. *Journal of Human Evolution* 5: 323-332.

Klinger, E. 1971. *Structure and Functions of Fantasy*. New York: Wiley-Interscience.

Klinowska, M. 1986. The cetacean magnetic sense: evidence from strandings. In M. M. Bryden and R. Harrison (eds) *Research on*

Dolphins. Oxford: Clarendon Press: 401-432.

Klopfer, P. H. 1970. Sensory physiology and esthetics. *American Scientist* 58: 399-403.

Krushinskaya, 1986. The behaviour of cetaceans. In G. Pilleri (ed.) *Investigations of Cetacea* 19: 115-273. Berne (Switzerland): Brain Anatomy Institute.

Kummer, H. 1971. *Primate Societies: Group Techniques of Ecological Adaptation*. Chicago: Aldine-Atherton.

Kurland, J. A. and Beckerman, S. J. 1985. Optimal foraging and hominid evolution: labor and reciprocity. *American Anthropologist* 87: 73-93.

Lawick, H. van and Lawick-Goodall, J. van. 1971. *Innocent Killers*. Boston: Houghton-Mifflin.

Lee, R. B. 1968. What hunters do for a living, or how to make out on scarce resources. In R. B. Lee and I. DeVore (eds) *Man the Hunter*. Chicago: Aldine: 30-48.

Lenneberg, E. 1971. Of language knowledge, apes and brains. *Journal of Psycholinguistic Research* 1: 1-29.

Levi-Strauss, C. 1949. *Les Structures Elémentaires de la Parenté*. Paris: Plon.

Lewontin, R. C. 1966. Review of G. C. Williams: Adaptation and Natural Selection. *Science 52:* 338-339.

Lewontin, R. C. 1970. The units of selection. *Annual Review of Ecology and Systematics* 52: 1-18.

Lieberman, P. 1984. *The Biology and Evolution of Language*. Cambridge (Mass): Harvard University Press.

Loizos, C. 1967. Play behaviour in higher primates: a review. In D. Morris (ed.) *Primate Ethology*. Chicago: Aldine: 226-282.

Lovejoy, C. O. 1981. The origin of man. *Science* 211: 341-350.

Mann, M. 1986. *The Sources of Social Power*. Vol. 1: *A History of Power from the Beginning to AD 1760*. Cambridge: Cambridge University Press.

Manson, J. and Wrangham, R. W. n.d. The evolution of hominoid intergroup aggression. Unpublished manuscript.

Martin, P. 1984. The time and energy costs of play behaviour in the cat. *Zeitschrift für Tierpsychologie*. 64: 298-312.

Martin, P. and Caro, T. 1985. On the functions of play and its role in behavioral development. *Advances in the Studies of Behavior* 15: 59-103.

Martin, R. A. 1981. On extinct hominid population densities. *Journal of Human Evolution* 10: 427-428.

Mitchell, R. W. 1986. A framework for discussing deception. In R. W. Mitchell and N. S. Thompson (eds) *Deception: Perspectives on Human and Nonhuman Deceit*. Albany: State University of New York Press: 3-40.

Morris, I. 1975. *The Nobility of Failure: Tragic Heroes in the History of Japan*. New York: Holt, Rinehart and Winston.

Morris, R.J. 1986. The acoustic faculty of dolphins. In M. M. Bryden and R. Harrison (eds) *Research on Dolphins*. Oxford: Clarendon Press: 369-399.

Neisser, U. 1976. *Cognition and Reality: Principles and Implications of Cognitive Psychology*. New York: Freeman.

Norris, K. S. and Dohl, T. P. 1980. The structure and functions of cetacean schools. In M.M. Bryden and R. Harrison (eds) *Research on Dolphins*. Oxford: Clarendon Press: 369-399.

Norris, K. S. and Mohl, B. 1983. Can odontocetes debilitate prey with sound? *American Naturalist* 122: 85-104.

Norris, K. Prescott, J. H., Asa-Dorian, P. V. and Perkins, P. 1961. An experimental demonstration of echolocation behavior in the porpoise,

Tursiops truncatus (Montagu). *Biological Bulletin* 20: 163-176.
Nunney, L. 1985. Group selection, altruism and structured-deme models. *American Naturalist* 126: 212-230.
Panskepp, J. 1982. Toward a general psychobiological theory of emotions. *Behavioral and Brain Sciences* 5: 407-467.
Parker, G. A. 1974. Assessment strategy and the evolution of fighting behavior. *Journal of Theoretical Biology* 47: 223-243.
Parker, S. T. 1984. Playing for keeps: an evolutionary perspective on human games. In P. K. Smith (ed.) *Play in Animals and Humans*. Oxford: Blackwell: 271-293.
Parker, S. T. and Gibson, K. R. 1979. A developmental model for the evolution of language and intelligence in early hominids. *Behavioral and Brain Sciences* 2: 367-408.
Piaget, J. 1962. *Play, Dreams and Imitation in Childhood*. New York: Norton. Revised Edition.
Pilbeam, D. R. 1966. Notes on *Ramapithecus*, the earliest known hominid, and Dryopithecus. *American Journal of Physical Anthropology* 25: 1-6.
Pitt, R. 1978. Warfare and hominid brain evolution. *Journal of Theoretical Biology* 72: 551-575.
Premack, D. and Woodruff, G. 1978. Does the chimpanzee have a theory of mind? *Behavioral and Brain Sciences* 1: 515-526.
Pusey, A. E. and Packer, C. 1987. Dispersal and philopatry. In B. B. Smuts, D. Cheney, R. M. Seyfarth, R. W. Wrangham and T. T. Struhsaker (eds) *Primate Societies*. Chicago: University of Chicago Press: 250-266.
Rawls, J. 1971. *A Theory of Justice*. Cambridge (Mass): Harvard University Press.
Reynolds, V. 1965. Some behavioural comparisons between the chimpanzee and the mountain gorilla in the wild. *American Anthropologist* 67: 691-706.
Reynolds, V., Falger, V. S. E. and Vine, I. (eds) 1987. *The Sociobiology of Ethnocentrism*. London: Croom Helm.
Savage-Rumbaugh, E. S., Rumbaugh, D. M. and Boysen, S. 1978. Linguistically mediated tool use and exchange by chimpanzees *(Pan troglodytes)*. *Behavioral and Brain Sciences* 1: 539-554.
Schusterman, R. J., Thomas, J. A. and Wood, F. G. (eds) 1986. *Dolphin Cognition and Behavior: a Comparative Approach*. New York: Erlbaum.
Singer, J. L. 1966. *Daydreaming: an Introduction to the Experimental Study of Inner Experience*. New York: Wiley.
Smith, P. K. 1982. Does play matter? Functional and evolutionary aspects of animal and human play. *Behavioral and Brain Sciences* 5: 139-184.
Smuts, B. B., Cheney, D. L., Seyfarth, R. M., Wrangham, R. W. and Struhsaker, T. T. (eds) 1987. *Primate Societies*. Chicago: University of Chicago Press.
Speth, J. Early hominid subsistence strategies in seasonal habitats. *Journal of Archaeological Science* 14: 13-29.
Strate, J. M. 1982. *An Evolutionary View of Political Culture*. Unpublished Ph.D. Thesis, University of Michigan, Ann Arbor.
Stringer, C. 1984. Human evolution and biological adaptation in the Pleistocene. In R. Foley (ed.) *Hominid Evolution and Community Ecology: Prehistoric Human Adaptation in Biological Perspective*. New York: Academic Press: 55-83.
Suarez, S. D. and Gallup, G. G. 1981. Self-recognition in chimpanzees and orangutans, but not in gorillas. *Journal of Human Evolution* 10: 175-188.
Symonds, P. M. 1949. *Adolescent Fantasy*. New York: Columbia

University Press.
Symons, D. 1978a. *Play and Aggression: a Study of Rhesus Monkeys*. New York: Columbia University Press.
Symons, D. 1978b. The question of function: dominance and play. In E.O. Smith (ed.) *Social Play in Primates*. New York: Academic Press: 193-230.
Tanner, N. and Zihlman, A. 1976. Women in evolution. Part 1: innovation and selection in human origins. *Signs: Journal of Women in Culture and Society* 1: 585-608.
Tenaza, R. and Tilson, R. 1985. Human predation and Kloss's gibbon (*Hylobates klossii*) sleeping trees in Siberut Island, Indonesia. *American Journal of Primatology* 8: 299-308.
Tooby, J. and Devore, I. 1987. The reconstruction of hominid behavioral evolution through strategic modelling. In W.G. Kinzey (ed.) *The Evolution of Human Behavior: Primate Models*. Albany (NY): State University of New York Press: 183-237.
Trivers, R. L. 1971. The evolution of reciprocal altruism. *Quarterly Review of Biology* 46: 35-57.
Trivers, R. L. 1972. Parental investment and sexual selection. In B. Campbell (ed.) *Sexual Selection and the Descent of Man*. Chicago: Aldine: 136-179.
Trivers, R. L. 1985. *Social Behavior*. Menlo Park: Benjamin/Cummins.
Vining, D. R. 1986. Social *versus* reproductive success: the central theoretical problem of human sociobiology. *Behavioral and Brain Sciences* 9: 167-187.
Wade, M. J. 1976. Group selection among laboratory populations of *Tribolium*. *Proceedings of the National Academy of Science (USA)* 173: 4604-4607.
Wade, M. J. 1978. A critical review of the models of group selection. *Quarterly Review of Biology* 53: 101-114.
West Eberhard, M. J. 1979. Sexual selection, social competition and evolution. *Proceedings of the American Philosophical Society* 123: 222-234.
West Eberhard, M. J. 1983. Sexual selection, social competition and speciation. *Quarterly Review of Biology* 58: 155-183.
Williams, G. C. 1966. *Adaptation and Natural Selection*. Princeton (NJ): Princeton University Press.
Williams, G. C. 1985. In defense of reductionism in evolution. *Oxford Surveys in Biology* 2: 1-27.
Wilson, D. S. 1975. New model for group selection. *Science* 189: 8701.
Wilson, D. S. 1980. *The Natural Selection of Populations and Communities*. Menlo Park: Benjamin/Cummins.
Wilson, E. O. 1971. *The Insect Societies*. Cambridge (Mass): Harvard University Press.
Wilson, E. O. 1973. The queerness of social evolution. *Bulletin of the Entomological Society of America* 19: 20-22.
Wilson, E. O. 1975. *Sociobiology: the New Synthesis*. Cambridge (Mass): Belknap Press.
Wolpoff, M. H. 1971. Competitive exclusion among Lower Pleistocene hominids: the single species hypothesis. *Man* 6: 601-614.
Worthy, G. A. J. and Hickie, J. P. 1986. Relative brain size in marine mammals. *American Naturalist* 128: 445-459.
Wrangham, R. W. 1979. On the evolution of ape social systems. *Social Science Information* 18: 334-368.
Wrangham, R. W. 1980. An ecological model of female-bonded primate groups. *Behaviour* 75: 262-299.
Wrangham, R. W. 1987. The significance of African apes for

reconstructing human social evolution. In W. G. Kinzey (ed.) *The Evolution of Human Behavior: Primate Models*. Albany (NY): State University of New York Press: 51-71.
Wynn, T. 1979. The intelligence of later Achulean hominids. *Man* 124: 371-391.
Wynn, T. 1981. The intelligence of Oldowan hominids. *Journal of Human Evolution* 10: 529-541.
Wynne-Edwards, V.C. 1962. *Animal Dispersion in Relation to Social Behaviour*. Edinburgh: Oliver and Boyd.
Zahavi, A. 1975. Mate selection – a selection for a handicap. *Journal of Theoretical Biology* 53: 205-214.

26. Culture, Constraint and Community: Semantic and Coercive Compensations for the Genetic Under-Determination of Homo Sapiens Sapiens

ERNEST GELLNER

The existence of culture amongst species other than man is debatable. There are writers who have gone as far as to affirm the absence of culture amongst other species (Note 1). The claim that culture is incomparably more important amongst men than amongst members of any other species is far less contentious. This, much less disputable premise, is all we really require for the present argument.

What is meant by *Culture?* Basically, a non-genetic mode of transmission, located in an on-going community. A community is a population which shares a culture. One might say that *culture* refers to whatever is transmitted non-genetically. The two notions, culture and community, are intimately linked.

Culture and community are defined in terms of each other: culture is what a population shares and what turns it into a community. A community is a sub-population of a species, which shares its genetically transmitted traits with the species, but which is distinguished from that wider population by some additional characteristics: these in some way or other depend on what the members of that community or sub-population *do,* rather than on their genetic equipment. It shares a series of traits which are transmitted semantically: what is reproduced is *behaviour,* but the limits imposed on that behaviour depend on markers carried by the society and not by the genes of its members. Cultural behaviour is not dictated genetically, and cannot be reproduced either by some genetic inner *Diktat,* or even by a mere conjunction of genetic programming with external non-social stimuli. Hence its boundaries or limits must be defined by something or other in possession of the community within which this reproduction of behaviour takes place. Such non-genetic delimitation of boundaries of conduct or of perception, in the keeping of a community, is about as good a definition of *meaning* as we possess. Meaning, culture, community – these notions interlock with each other. The circularity of their definitions, their interdependence, does not matter.

What the human species does share genetically is an unbelievable degree of behavioural plasticity or volatility. The diversity in actual conduct found amongst members of this single species is quite incredible: they do very very different things, they speak remarkably different languages, observe different codes, and so forth. It is reasonably obvious that these differences are not transmitted genetically: infants drawn from one population pool can be socially and linguistically incorporated and successfully reared in totally different communities. If the new community possesses a self-image incompatible with some traits which are genetically transmitted, some trouble does ensue. If, for instance, a community which thinks of itself as belonging to one pigmentational category, socializes an infant adopted from another such category, the child in question may later have some difficulty in securing full moral incorporation. This however is not in conflict with the claim that the child in question normally experiences no difficulty in successfully assimilating all the socially or semantically transmitted traits.

Racialists are defined by the conviction that some socially important traits are specially common, or even exclusively present, in certain *genetically* defined populations: that some genetically defined populations possess strong concentrations of either desirable or undesirable characteristics. It is doubtful whether this is true to any significant extent: the traits considered morally significant seem to depend far more on socially than on genetically transmitted elements. Morally exciting or repellent features have often been transformed historically with a speed which makes it most implausible, or impossible, to attribute the metamorphosis in question to genetic change.

What *does* seem genetically based in humankind is the plasticity, the volatility itself. All members of the species are endowed with it, and no other species possesses it. But possibly the most important single sociological fact about mankind is that this plasticity is very seldom much in evidence *within* single on-going communities. On the contrary: members of the same community in the main resemble each other to a marked degree. Generally they speak the same language, both in a literal and in a broader sense: they use similar linguistic tokens; and the general way in which they use them, their *culture*, is also fairly similar.

This point may need some modification with respect to complex civilizations, which contain a wide diversification of roles within their social structure, and where the occupants of diverse roles often do markedly different things, wear different clothes, are obliged to speak in a different manner, are dominated by distinct values, and so on. Though such diversity indisputably prevails amongst the members of complex societies, this does not really militate against our main point: men are, all things considered, astonishingly well-disciplined and restrained in their conduct and their thought. Members of simpler societies may, as Durkheim stressed, all do and think and feel the same things, whereas members of more complex societies may complement each other by their diversity; but neither the former nor the latter normally stray very much from what is culturally expected of them. Where diversity of roles is expected or imposed by culture, behaviour is nontheless constrained within rather narrow bounds. Genuinely chaotic and unpredictable conduct, crossing the bounds of culturally recognized alternatives, is astonishingly rare.

To sum up the argument so far: if we look at mankind inter- culturally, we find an amazing diversity. If we look at mankind within any one community/culture, we find an equally amazing discipline and restraint. Question: how can a species, genetically granted by Nature such remarkable freedom and licence, nevertheless observe such restraint, such narrowly defined limits, in its actual conduct? Man is born genetically free but is everywhere in cultural chains. How is this possible?

What are the necessary conditions of this cultural enslavement? What are its sufficient conditions? And what are its functions, or, to give functionality its old Aristotelian name, what is its Final Cause? In a world in which natural – and social – selection can be assumed to operate in some measure, Final Causes or Functions are of great interest: they provide significant clues to efficient causes. That which serves a purpose, which in turn is a precondition of survival, legitimately constitutes an element in the explanation of that survival.

To begin with the function: it is reasonable to suppose that without *local* homogenization, standardization, discipline, human communities would be unviable. Highly unpredictable, and wildly diversified members of a local group or herd, simply would not be capable of cooperating, and would cease to exist as any kind of group, and, presumably, as individuals. A plausible theory is available which claims that *Homo sapiens sapiens* acquired his intelligence in the course of working out devices for predicting, and managing, the waywardness, deviousness and cunning of his fellows: we were propelled into intelligence, so to speak, by being obliged to keep up with each other's smartness (Note 2).

There may be a great deal in this account of the Gadarene or competitive rush into cleverness, but it could only have worked if the waywardness and unpredictability stayed within certain well defined limits. I can stretch my intelligence to its limits in my attempts to outsmart a clever chess player or poker player or intriguer. But if he is so volatile that that he is in effect playing a series of quite unconnected games (as happens in Tom Stoppard's play *The Real Inspector Hound*), then no effective strategy is possible. In fact, he ceases to be any kind of a player, and there is no game.

What is the necessary condition of this essential, indispensable intra-cultural or intra-communal discipline? A species which is genetically capable of such diversity, but at the same time requires great homogeneity and discipline within arbitrarily selected sub-populations, simply *must* be capable of operating and recognizing an extremely rich system of markers, which set bounds to what is and what is not done in any one community. Given the astonishing internal complexity and external diversity of cultures, and also their well demonstrated capacity for rapid transformation, it seems to me obvious that this system of markers must also possess the features which, as a result of the central ideas and insights of Chomsky, have now come to be associated with *language:* the system of markers must be able to achieve infinite results by finite means.

A language is a so to speak modular system of markers, within which the same elements can be combined and re-combined in a wide variety of ways, engendering the delineation of quite distinct and alternative boundaries of conduct. For instance, a language containing the notions of seven days of the week, and the opposition 'work' and 'non-work', can command the members of a given culture not to work on Sundays; and within the same system of ideas, the notion of prohibition of work on any other day is also easily intelligible, as is the idea of the obligation to work on this or that day.

The semantic/cultural constraints on conduct are exceedingly rich twice over: in any one language, prohibitions other than those actually imposed, are conceivable and understood. In other words, members of the culture live in a *world*, such that the actual state of affairs is but one possibility amongst others. Its possibilities, so to speak, greatly exceed its actuality, but they are understood and, in a sense, present. A world is a system of intelligible possibilities, only a small fraction of which is actually realised. Without a language and its modular combination of diverse elements, it is not clear whether it makes much sense to talk of a sentient being living in a world. A world only comes into being for conceptualizers who can specify the options, the unrealised possibilities. A set of stimuli do not add up to a world. A modular system of markers, generating alternative possibilities, all of them conceptualizable, but only one actually realised – *that* is a 'world'. Without language, there is no *world*. Language-less animals, living only within realised possibilities plus perhaps a small range of fears and anticipations, in a way do not inhabit a *world* at all.

The second and more abstract sense in which mankind is rich in possibilities is that, over and above the alternative possibilities present in any one language, we are evidently capable of acquiring and internalizing more than one set of elements ready for combination. Languages and cultures differ radically in the ways in which they construct a world. Men evidently have the capacity to internalize and submit to quite distinct visions. Languages are rich, but mankind is also rich in alternative possible kinds of language.

A culture is a system of constraints, limiting an endlessly labile set of possibilities, within bounds which are themselves also very very complex, and which apply to a very wide range of situations. It seems to me quite inconceivable that the principles making such a complex system possible should be invented and erected *ad hoc* in the case of each culture; still less that each prohibition, each proscription within it should all be invented and instilled separately. Chomsky has popularized the argument that the data available to a language-learning child simply are not rich enough to enable it to catch on to the structure of the language which it is learning to acquire. It could not learn to speak, were it not for the fact that the child is already highly and, so to speak, generically language-prone. I am arguing that, similarly, we should not be able to live in a culture were we not already very prohibition-prone. We are culture-prone as well as language-prone, and these two intimately linked susceptibilities must be rooted in some well-structured general predispositions, without which our volatility would be our undoing.

The *specific* prohibitions to which we are subject are culturally idiosyncratic: but our strong tendency towards observing some reasonably coherent set of prohibitions, and our sheer capacity to perceive them with precision and to comply with their requirements, is species-generic. Moreover, The linguistic and the cultural-discipline predispositions in great measure overlap. Language is, initially and basically, a system of prohibitions. *Am Anfang war das Verbot.* In the beginning was the prohibition. Language consists of markers indicating what thou shalt not do. The rules, governing the combination of the elements, are deployed to form markers, and are themselves obeyed by a linguistically, culturally most biddable humanity.

The referential aspect of language has of course been greatly exaggerated by empiricist theoreticians, who project our own empirical fastidiousness onto other users of language. Referentiality is not wholly absent amongst earlier users, in as far as the markers need to be triggered off by *something:* but that something is in the main social, and only in much smaller measure natural. Social stimuli dominate and trump natural ones. Sophisticated scientific languages, whose terms are *operationalized,* as we now say, in terms of natural conditions, and at the same time fairly independent of social ones, constitute an exception, a late and unusual development. A referential system, sensitive to nature and blind to society, is a rare achievement, a part of what we mean by 'science'.

Diversity is the clue to the history of mankind and, one may add, to the success mankind has had in dominating the planet. This diversity has two aspects, intra-social and inter-social. The fact that societies differ so markedly allows diverse societies to explore diverse options or strategies: many cultures have adopted ways which were not markedly successful. What brought mankind to its present condition is that so very many options could be tried out. It must be assumed that the successful ones were always in a minority, perhaps a tiny minority. It did not matter from an evolutionary viewpoint, in as far as so many alternatives were tried, and the successful ones could be emulated by the others, or eliminate them.

Intra-social diversity (compatible with finely-tuned conformity and discipline) is just as important, though for another reason. It and it alone permits great complexity of social organization, which in turn allows the exploration of cultural alternatives of great power. Options could be tried which simply would not be available to societies whose members resembled each other closely. This was one of Emile Durkheim's central insights. It was he who proposed the distinction between Mechanical and Organic Solidarity, where the former engenders social cohesion based on the similarity of the cohering elements, be they individuals or groups, whilst the latter achieved cohesion based on complementarity and on mutual interdependence. Whether or not Durkheim was right about the mechanisms of social cohesion, there can hardly be any doubt concerning the point that great diversity of roles permits an incomparably wider range of possible social organizations, much broader than that which would have been possible on the basis of a mere accumulation of elements which fundamentally resemble each other.

This point had already been made by Adam Smith, *nur mit ein bisschen anderen Worten.* It was he who most eloquently and persuasively proposed the idea that the key to progress was the Division of Labour. The system of differentiated and minutely proscribed activities which make up a given society, is not predetermined, but can be elaborated and developed to a point of refinement where it greatly enhances our well being and prosperity. That was Smith's point.

Durkheim later suggested that what really mattered were the socio-political rather than the economic consequences of the division of labour: its impact on human sociability was more important than its impact on the production of pins. But Adam Smith was not unaware of the political importance of the phenomenon either. What was wrong with Smith's theory was that, perhaps because he lived so long before Darwin, his basic anthropology was so naive: it took far too much for granted. It simply assumed, as *given*, certain basic features of human nature, which in reality are most mysterious, and must not be taken as self-evident: our capacity to conceptualize very specific aims, very specific procedures, often stretching over long periods of time – in other, words, our capacity to enact and persist in highly distinct and specific roles. This presupposes a conceptually organized world within which alternatives are easily grasped. This capacity may seem obvious to people addicted to a kind of pan-human ethnocentrism, to seeing our shared human condition as natural, obvious, a given and unquestioned birthright. But in reality, it is nothing of the kind. Other species lack it. After Darwin, we need to explain our distinctive endowments.

The division of labour requires and presupposes this ability to conceptualize. Not all specializations, role-specificities,are economic, and not all of them are beneficial: but before we can specialize economically, and specialize in a manner which enhances our efficiency and productivity, we must first of all be capable of conceptualization which is both precise and persistent and independent of immediate stimuli. We must master and respect a conceptual, modular system of behaviour-delimiting markers. This is a gift which is not shared, or only shared in incomparably smaller measure, by our primate cousins.

So we return, via Adam Smith, to our initial point. Man is defined by his plasticity, which enables mankind to display its amazing diversity, both within individual societies and between them. But this potential for diversity would be useless, were it not also restrained by some compensatory mechanism. This compensating mechanism restores behaviour to relatively narrow limits in any one cultural milieu, but not to the *same* limits in all milieux. Assume that man is indeed as volatile as manifestly he is, but also that learning and knowledge operate in the manner in which simple-minded empiricist philosophers had supposed they did work: by the interaction of individual minds with 'experience'. On this model, each individual would build up his own system of associations, in the light of an inevitably diversified and idiosyncratic experience. Divergence between individual minds would be enormous, and ever increasing at a tremendous rate. Our minds would suffer from semantic cancer, because meanings would expand through idiosyncratic associations. Would such volatile minds ever be capable of either communicating or cooperating? It is unthinkable. Without plasticity, no diversity; and without diversity, none of that rapid exploration of alternative strategies which has made mankind what it is. But without cultural restraint, the plasticity would become to speak malignant and excessive, and move much too fast. It would also be unable, through its very volatility, to *retain* any advantages gained. It would simply, as it were, skate over them and move on much too quickly. Any gain would rapidly be lost. So the plasticity needs to be counterbalanced by restraint and constraint. Language constitutes the major part of the system of markers which indicate the tolerated boundaries imposed in any one culture at any one time.

Note that two famous definitions of man – man is a rational animal, and man is a social animal – are not really basic or elementary, but derivative. They are, each of them, corollaries of our more fundamental point – mankind is the plastic or volatile species. This being given, rationality and sociability become necessary consequences: the volatile animal cannot survive unless he is both rational and sociable. Or rather, he must be rational, and in order to be rational, he must also be social. The idea of rationality used here is that of Emile Durkheim: it simply means being susceptible to and restrained by shared, socially imposed concepts. A volatile being devoid of this capacity would be *far too* volatile.

But man must also be social: it is difficult to see how a very small community could either perpetuate a really rich system of markers, or inhibit an over-rapid development of such systems. Every new deviation from the norm could and would, much too easily, become a new norm. It takes a population of a certain size – not a very large size, but at least a band – to impose a sensitivity to the distinction between what is accepted usage and what constitutes a deviation. So, rationality and sociability are corollaries of volatility; without them, the plastic animal would be too unconstrained to be viable.

So speech and plasticity came together: neither is possible without the other, twice over. Each presupposes the other; each renders the other necessary. It is hard to conceive full language existing prior to volatility of conduct: what use would this highly adjustable, rich system of markers be? A rich, alternative-engendering system of markers, accompanying a rigid, invariant, genetically dictated form of behaviour, would be a totally useless or noxious luxury. A chained being has no use for the capacity to conceptualize alternative paths to freedom. But it is equally impossible to imagine the absence of language *after* the arrival of volatility. So it is reasonable to suppose that the marks of volatility of conduct are indices of the origins of language.

But if speech is the necessary condition of restraining *excessive* plasticity, is it also sufficient? The doctrine that indeed it is, is one form of sociological idealism. This is the view that it is our systems of meaning which constrain us, and *suffices* to make a society, or to define it. Another way of putting this would be to say – give me the system of meanings, in other words the culture, and I will tell you what kind of society you are dealing with. No system of constraints over and above culture need be considered: the specification of the culture will suffice. I have very grave doubts about the adequacy of this widely held view.

Now it does indeed seem to me true that, culturally and linguistically, we are astonishingly biddable, docile and well-behaved; and it may also be true that for pre-agrarian mankind, such a culturally oriented anthropology may indeed be appropriate, and perhaps very nearly sufficient. I am not sure that this is so, but it is at least conceivable, and it may constitute an approximation of the truth. Pre-agrarian humanity does not have very much to fight about, and the mere description of a shared culture may constitute something approaching a near-complete account of the maintenance of social order in the society in question. Perhaps it lacks elaborate coercive systems. There are also some grounds for doubting this, but at least it is a *prima facie* possibility, and deserves investigation.

It does however lose all plausibility when we come to food-producing, resource-storing societies. As no principle of resource-distribution is either self-evident or self-enforcing, but many are *conceivable*, and every single one is unfair to *some* participants, these societies are inherently conflict-prone, and have to be endowed with coercive systems, which in turn go beyond the cultural system. One and the same cultural system of shared meanings and markers is compatible not merely with diverse occupancies of key positions of the society, but also with quite diverse authority structures and methods of enforcement. Structure and culture are often independent of each other to a marked degree. One and the same culture or marker-generating semantic system may be compatible with quite diverse organizational, coercive systems.

My point about the difference between foraging and agrarian humanity could be put as follows: mankind is susceptible to two kinds of constraint, cultural/linguistic, and coercive. All in all, men obey the grammatical and cultural rules of their community without needing much in the way of sanctions, though it could be said that the ridicule which follows phonetic or sartorial solecisms is a sanction of a kind, and a very powerful one at that. But when it comes to the occupancy of positions in the economic and political hierarchy, sanctions rather more potent that the mere 'grammar of the culture' are generally required.

My claim is that the balance between these two kinds of constraint changed profoundly with the discovery of agriculture and the presence of a stored surplus. Amongst foragers, semantic systems were no doubt reinforced by some violence, which was presumably also important in the allocation of roles to individuals; but amongst agrarian societies, elaborate coercive organization reinforce and maintain the society and its organization to a far greater extent. Perhaps this is a mere matter of stress: physical violence was no doubt also present amongst foragers, and placid submission to mere custom, without threat of force, is not unknown amongst agriculturalists. But even if it is a matter of stress rather than of a radical discontinuity, a very very significant shift in the balance between these two elements must have taken place.

Our general point was that man is basically plastic, but that in any one society his comportment is restrained within remarkable narrow limits.

The genetically under-determined species is constrained by new, semantically transmitted bonds. What mankind *is* given genetically is, precisely, that leeway, *and* the capacity to construct an open-ended system of markers, which will compensate for the genetic flexibility or under-determination. Man is also endowed with a remarkable docility in the face of these systems of markers, and a capacity to recognize the limits which they impose and to comply with them. In terms of this overall scheme, our point concerning the Neolithic Revolution and the implications of food-production and storage for the size and complexity of societies, is this: very complex systems of roles arise with food production and storage, and these can no longer be sustained by culture alone, but also need systematic duress. To put all this in an old-fashioned way: production and storage bring about the state, or institutional coercion. We can speak of the state proper when the agents of this institutional coercion are concentrated in one part of society.

Culture, or the constraint of conduct by conceptual means, is of course not absent or unimportant in the agrarian age. Agrarian societies generally aspire towards stability, and their complex organization tends to be underwritten and supported by a correspondingly elaborate cultural machinery. Where simpler societies had the dance and the story, these societies possess doctrines as well as myths, in the keeping of a corps of ideological specialists. The basic formula for these doctrines is that what is, *must be*, and rightly so. The proofs offered in support of this sacralization of social reality cannot possibly be genuine and independently valid. Hence the agrarian age must appear, to those no longer enslaved to it, as an age of superstition and coercion. This is precisely how the Enlightenment saw it.

A general formula which may be applicable to agrarian societies is that they need *both* cultural/semantic and elaborate coercive systems of social control. Their complexity and demandingness and, one might add, their inequitableness are all so extreme, that without coercion and fear, social order would crumble. But coercion alone would never have been able to sustain these elaborate societies either: what they need is, so to speak, coercion at *a distance*. A herd of concept-less animals cannot be constrained to do something highly specific and to do it *now*, by the threat of dire punishments at a later date. Animals will only obey rather crude instructions, enforced by crude and currently perceptible sanctions. They can admittedly be drilled into performances, or abstentions, but the range and refinement of such internalized instructions is rather small. They could never engender the intricate and sensitive patterns of conduct presupposed by the complex social order which we take for granted. Complex human civilizations require men to do very minutely circumscribed things, and to do them in the light of sanctions that are not immediately present and operative. Complex civilizations would not be possible unless men were capable both of *fear* and of abstract, alternative-conceptualizing *thought*. They need coercive action-at-a-distance, fear-at-a-distance, which is finely tuned, precise, discriminating. We need to be both clever and frightened.

Or one might put it all as follows: semantic or cultural systems alone do not suffice as compensations for our genetic under- determination. They need to be supplemented by coercive systems. This need becomes conspicuous and overwhelming at the point at which the institutionalization of food production and storage opens up the possibility of populous, complex and diversified societies. Coercion without meaning is blind, meaning without coercion is feeble. Meaning on its own enforces cultural, but not political conformity. Only jointly can force and signification construct those masterpieces of social organization which emerge in the agrarian age. De Maistre observed that the executioner is the foundation of social order. But he is not sufficient. Civilized humanity needs to be under the sway of both executioners *and* generative grammar.

The co-presence of semantic and coercive techniques for imposing order on behaviour is reflected in the prominence, and in the insulation, of the specialists of legitimation and the specialists of violence. These two supremely important and dominant social categories generally preside, in some kind of often uneasy cooperation, over complex societies. Their pre- eminence constitutes eloquent testimony to the fact that the maintenance of social discipline is highly problematic, and that it is seldom attainable without them. Society could not manage without their help, and once endowed with them, cannot easily resist their demands. So they generally secure great privileges for themselves. The markers delimiting the boundaries of sub- groups and of required behaviour need to be serviced and maintained, and this confers power on the priests. The tools of coercion need likewise be deployed and kept in readiness, and that bestows power on their frequently monopolistic possessors. Why are these two categories so frequently distinct and separate? Presumably the nature of the two sets of skills is such that their deployment, and/or the training which leads to the acquisition of virtuoso status, are not easily combined. The correct question seems to me not why institutions such as divine kingship or priestly rule are occasionally encountered, but rather, why they are so often absent. The separation of, and rivalry between, these two categories of dominators, may well constitute one of the important clues to the question of how we managed to escape from the agrarian order. Priests helped us to restrain thugs, and then abolished themselves in an excess of zeal, by universalizing priesthood. First Canossa, then the Reformation.

We now know that it is indeed possible to escape from the agrarian age of Fear and Faith. We know it because we have indeed escaped from it, though the romantics amongst us would prefer to say that we were expelled from it. We know it happened, though we do not fully understand how this escape or expulsion came about. What we do know is that the rules of the game have changed radically once again. Both coercion and culture are still with us, but in a wholly new form. Coercion has diminished in degree, at least in liberal societies. Agrarian society was inescapably Malthusian, with population constantly pressing on resources; the distribution of those resources could not but be invidious, and hence required a good deal of coercive enforcement, often very brutal. Post-agrarian society is affluent and can afford, or at any rate has frequently allowed itself, the luxury of a marked relaxation of coercion. Coercion is by no means absent, but is greatly softened, at any rate in some societies.

When we come to the new role of culture, we find not just a change in degree, but in kind. In its cognitive life, this new kind of society respects *science*. This is made up of a curious system of markers largely disconnected from any social triggers, and related in a systematic manner to something extra-social ('nature', 'experience'). As a set of imperatives, it only commands hypothetically, and relates to the choice of means, not of ends. It is not expected to be stable.

In its productive life, this society is similarly unusual: stability is not expected here either, and there is an unconstrained free choice of methods and personnel. In other words, both cognition and production are liberated from the restrictive constraints on role and method innovation which had dominated them, in the interests of social stability, throughout the agrarian age. It is as if that genetic volatility, which had characterized man for nearly two hundred thousand years, had at long last, and for the first time, been allowed to manifest itself to the full, not merely in the form of variety *between* cultures or societies, but *within* one of them. A kind of society has arisen which could tolerate a relative liberation of human plasticity, even within the bounds of a single social order. The explosion of cognitive and productive innovation is of course linked to a rather special new and inwardly imposed restraint, which can for convenience be referred to as distinctively modern rationality. This mysteriously allows men to be orderly and social, even whilst they freely re-combine elements of production and cognition. An inwardly sanctioned constraint leads producers and researchers to observe and respect formal rules, even in the course of substantive innovation. Discipline has moved on to a higher plane, permitting an astonishing amount of innovation at the ground level, whilst social order is preserved.

So constraints are not absent in this society either, but they take a new form. The compulsive sacralization of important concepts, which had so preoccupied Durkheim, is replaced by the second-order sacralization of procedural propriety, of the rule of treating like cases alike, of conceptual tidiness, of the unification of referential concepts in an ideally unified system, and of their separation, to a remarkable extent, from the markers delimiting social conduct. The crucial imperatives are formal, not substantive. No individual concepts are heavily underwritten and rendered immovable by awesome ritual. Our literal rituals are playful and semi-serious. A unified and orderly and open system of referential concepts, now lives in relative isolation from the markers guiding social conduct.

The requirement of orderly symmetry is at the heart both of the social and the cognitive ethic of modern society. Absence of ritual has become the most potent ritual, absence of graven images the most pervasive fetish. Both inculcate orderly and experimental treatment of nature and help engender the technology which is the new basis of society. Bribery by economic growth in some measure replaces fear as the corner-stone of the social edifice.

The importance of the sociology of Max Weber lies in the fact that it offers a theory of how this new kind of restraint has come into being. As so often, the merit of the theory resides more in its highlighting of a previously unperceived and important problem, than in the solution it offers for that problem. Weber has made us sensitive to the difference between a world in which formal order is sacred but all specific objects are equal, and the older world in which some substantive objects were much more sacred than others – the world analysed by Durkheim.

This new kind of society, and associated vision also, quite transforms the relationship between High Culture and Low Culture (Note 3), which had been in mutual tension throughout the later part of the agrarian age. High Culture had been perpetuated by script and formal education, Low Culture by informal socialization within the local community. The new nature of work, which consists of the context-free manipulation of meanings and people, not of things, requires the overwhelming majority of the population to be in possession of a High, literate culture. Its acquisition is the most valued qualification for most individuals, and a precondition of effective membership: men passionately identify with it, and this is known as 'nationalism'. Social control now operates in part through such nationalism, or the ardent identification with either an existing or a desired nation-state. It is also enforced by a centralized and pervasive state, and made palatable by a shared stake in affluence and the promise of its continuous enhancement (Note 4).

My argument has been that genetic under-programming must have been linked to the presence of a compensating system of cultural/linguistic restriction. These cultural systems, and systems of coercion, have complemented each other in diverse ways at different stages. The volatility must obviously have had its own genetic preconditions, so that our volatility, and our endowment with compensating talents and propensities, must have arrived jointly. The consequence has been the emergence of a species in whose life both social or semantic transmission and institutionalized coercion have become far more important than genetic mutation, making it possible for change to be astonishingly rapid.

NOTES

1. See Robert A. Hinde, *Individuals, Relationships and Culture: Links between Ethology and the Social Sciences*. Cambridge: Cambridge University Press, 1987, p.3: ". . . we can thus speak of the possession of 'culture' as being a uniquely human attribute".
2. See N. Humphrey, *Inner Eye*. London: Faber, 1986.
3. See S. N. Eisenstadt, *The Origins and Diversity of Axial Age Civilizations*. Albany (N.Y.): State University of New York Press.
4. See E. Gellner, *Nations and Nationalism*. Oxford: Blackwell, 1983.

SECTION III

Regional Case Studies

27. Biological and Behavioural Perspectives on Modern Human Origins in Southern Africa

RICHARD G. KLEIN

INTRODUCTION

Long ignored in discussions of modern human origins, Southern Africa now figures prominently, because it contains what may be the oldest known anatomically-modern human fossils. The possibility that these indicate an African origin for modern humans is clearly supported by the occurrence of penecontemporaneous or slightly younger, modern or near-modern fossils in Eastern and Northern Africa (references below) and especially by genetic evidence that all living humans may derive from an African population that existed within the last 200 000 years (Wainscoat *et al.* 1986; Jones and Rouhani 1986; Cann *et al.* 1987; Wainscoat 1987: see also papers by Rouhani, Wainscoat *et al.*, and Stoneking and Cann, this volume). My purpose here is to outline briefly the Southern African data for modern human origins, with reference to their broader African context and with special emphasis on whether the biological modernity inferred from the fossils was accompanied by behavioural modernity as inferred from associated artifacts and animal bones.

THE FOSSILS

The Southern African fossils that bear on modern human origins come from five sites (Fig. 27.1): Florisbad (a partial cranium comprising facial, frontal, and parietal fragments) (Bräuer 1984; Rightmire 1984; Clarke 1985); Border Cave (an infant's skeleton, an adult skull, two partial adult mandibles, and some postcranial bones) (Beaumont *et al.* 1978; Beaumont 1980; Rightmire 1979, 1984); Die Kelders Cave (isolated teeth) (Klein unpublished); Klasies River Mouth (five partial mandibles, a maxilla, various small cranial fragments, isolated teeth, and postcranial bones) (Wymer and Singer 1982; Rightmire 1984, 1987; Deacon, in press); and Equus Cave (a partial mandible and isolated teeth) (Beaumont *et al.* 1984; Grine and Klein 1985). A conclusive assessment of the fossils is complicated by two fundamental problems – most are very fragmentary and their geologic age is uncertain. A special irony is that the most diagnostic specimen – the adult skull from Border Cave – remains especially problematic from a dating perspective, because, like other Border Cave specimens, it was not found *in situ* during tightly controlled excavations. As a group, the Border Cave

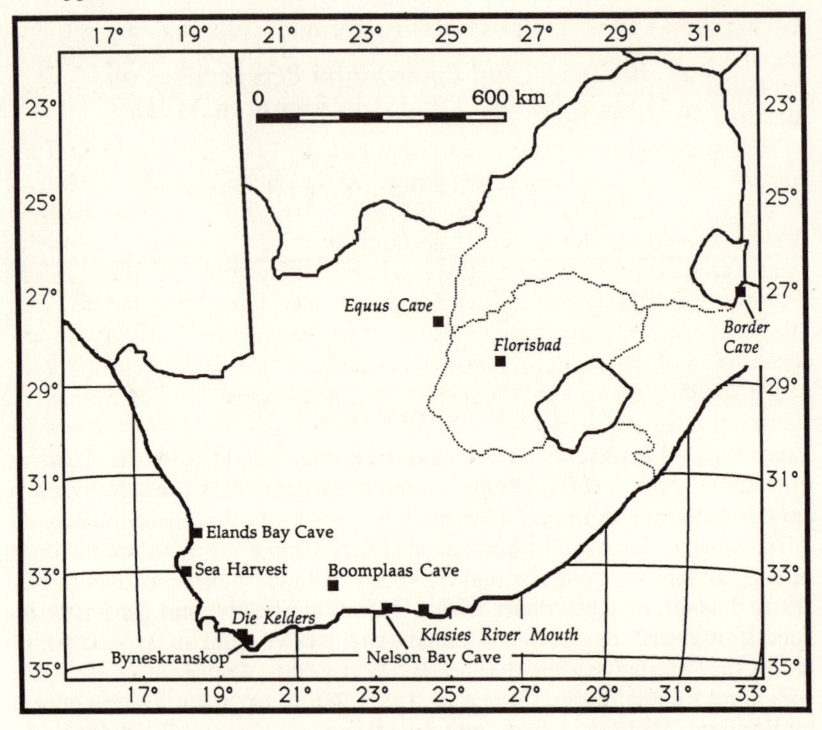

Figure 27.1. The approximate locations of the sites mentioned in the text (italics indicate sites with modern or near-modern human fossils).

fossils are significantly better preserved and more complete than animal bones from the same horizons, and they could represent intrusive burials from much later, even protohistoric times. This issue may be partly resolved if fresh excavations recover additional human fossils in well-documented context, or if it proves possible to obtain reliable radiometric dates on the existing specimens.

With regard to the remaining sites, the basic antiquity of the fossils is not in question. Minimal or infinite radiocarbon dates indicate that the fossils at both Florisbad and Klasies River Mouth lie well beyond the 30-40 000 year limit of conventional radiocarbon dating. At Klasies, Die Kelders, and Equus Cave, the fossils are all associated with Middle Stone Age artifacts, and numerous dates from a wide range of sites now show that the Middle Stone Age also terminated at or beyond the range of conventional radiocarbon technology (Vogel and Beaumont 1972; Volman 1984). Thus, all the fossils in question almost certainly antedate 30-40 000 BP, but more precise estimates are problematic. They depend partly on presumed correlations between the climatic events recorded in site profiles and those inferred from the global oxygen-isotope stratigraphy, and partly on absolute dates provided by controversial or still-developing methods like amino-acid racemization, Uranium-series disequilibrium, and electron-spin resonance.

Among these methods, Uranium-series disequilibrium and electron-spin resonance (or its cousin, thermoluminescence) probably have the greatest potential, but for the moment, most results should be regarded as only tentative or provisional.

The stratigraphic context of the Florisbad skull and the associated fauna suggest a late Middle Pleistocene age, somewhat before 130 000 BP (Butzer 1984a; Kuman and Clarke 1986). The skull is probably the oldest Southern African fossil considered here, and interestingly, it is the only one with clearly archaic features, including a supraorbital torus whose shape recalls that of the famous Broken Hill ('Rhodesian Man') skull (Rightmire 1984, 1987). In other features, including the relative thinness of the torus and the doming of the frontal, Florisbad is clearly advanced over Broken Hill, and it could represent a population linking local archaic, mid-Middle Pleistocene *Homo sapiens* (Broken Hill and allies) to Upper Pleistocene, fully modern people (other fossils discussed here) (Rightmire 1984, 1987). However, fresh, more complete, late Middle Pleistocene specimens will be necessary to demonstrate this.

At Klasies River Mouth, faunal remains, palaeoclimatic interpretation of the profile, and both Uranium disequilibrium and electron-spin resonance determinations indicate that most of the human fossils date from the Last Interglaciation *sensu lato* (oxygen-isotope stage 5) (Deacon, this volume), between perhaps 115 000 and 80 000 BP. A smaller number probably date from the earlier part of the Last Glaciation, between roughly 80 000 and 60 000 BP. At Die Kelders, geologic context suggests an early last glacial age (Tankard and Schweitzer 1876), probably sometime between 75 000 and 60 000 years. More tentatively, geology and faunal associations imply a similar, early last glacial age for the Equus Cave fossils (Grine and Klein 1985). The age of the Border Cave specimens is obviously indeterminate because of their uncertain stratigraphic provenience, but the Middle Stone Age artifacts with which they may have been associated are broadly similar to ones dating from the Last Interglaciation at Klasies River Mouth (Beaumont 1980). A last interglacial age is consistent with a revised palaeoclimatic interpretation of the profile (Butzer 1984b *contra* Butzer *et al.* 1978).

In terms of sample size, parts represented, and the probable reliability of the dating, the most important fossils are clearly those from Klasies River Mouth. Especially noteworthy are a fronto-nasal fragment, lacking any sign of a strongly developed supraorbital torus, and a relatively complete mandible, whose morphology is totally modern in all observable respects, including the presence a strongly developed chin. Other pieces are less diagnostic, but none argue for the presence of non-modern people. There are significant differences in size or robusticity, but these may only reflect substantial sexual dimorphism, assuming that only one population is represented.

Likewise, neither the size nor the morphology of the Die Kelders and Equus Cave teeth differentiate them meaningfully from those of living Africans, and it is tempting to conclude that the early Upper Pleistocene inhabitants of Southern Africa were totally modern. However, only the insecurely provenienced Border Cave sample contains a sufficient range of parts to argue for a fully modern morphology, and for the moment, perhaps the most accurate conclusion is that early Upper Pleistocene people in Southern Africa were clearly more modern than their Neanderthal contemporaries in Europe. The full contrast is apparent when the Southern African fossils are considered together with ones of possible or probable earlier Upper Pleistocene age (between >40 000 BP and *c.* 130 000 BP) in Northern and Eastern Africa. These include a maxillary fragment and isolated tooth from the Mugharet el 'Aliya, a nearly complete skull, a skull cap, and a partial mandible from Jebel Irhoud Cave, a partial adult skull, a child's skull, two partial mandibles and isolated teeth from Dar es Soltan Cave 2, a mandible and isolated canine from Zouhrah Cave, and a mandible and occipito-parietal fragment from Smugglers Cave (Témara), all in Morocco (Debénath 1980; Debénath *et al.* 1982, 1986); two partial mandibles from the Haua Fteah in Libya (McBurney 1967; Tobias 1967); a skull from Singa in the Sudan (Stringer 1979), a partial mandible from Diré-Dawa (Porc-Epic) (Vallois 1951; Clark *et al.* 1984) and two relatively complete skulls and fragments of a third from the Kibish Formation in the lower Omo Valley (Day and Stringer 1982), both in Ethiopia; a skull from Eliye Springs, West Turkana in Kenya (Bräuer and Leakey 1986); and a skull from the Ngaloba Beds at Laetoli (Day *et al.* 1980; Magori and Day 1983) and partial dentitions from Mumba Cave, both in Tanzania (Bräuer 1984).

Within the total African sample, the more complete skulls and jaws range from clearly archaic (especially Jebel Irhoud, Omo- Kibish 2, Eliye Springs, Haua Fteah, and Smugglers' Cave) to only marginally archaic (Singa and Ngaloba) to essentially modern (Zouhrah Cave, Dar es Soltan, Omo-Kibish 1, Border Cave, and Klasies River Mouth). The variability is certainly greater than within earlier Upper Pleistocene European (Neanderthal) populations and may partly reflect the much larger geographic area being sampled as well as possible misdatings, especially the inclusion of some late Upper Pleistocene or even Holocene fossils. However, even putting aside specimens like Border Cave and Omo 1 whose ages are debatable, there are no African fossils with indisputably Neanderthal features (as described, for example, by Trinkaus 1986). Some of the mandibles are large and rugged, but where the appropriate parts are preserved, they never have retromolar spaces and they usually have undoubted chins. Together with other facial bones, the mandibles indicate that early Upper Pleistocene Africans tended to have relatively short, flat faces. Similarly, some of the skulls (Jebel Irhoud, Omo- Kibish 2, Singa, Eliye Springs, and Ngaloba) are ruggedly built, with large brow ridges and in some cases relatively prominent transverse occipital tori (Jebel Irhoud and Omo-Kibish 2), as well as pronounced bony crests or mounds in the occipital-mastoid region. However, in general, in overall vault form, they resemble modern skulls much more than they do Neanderthal skulls.

Trinkaus (*in litt.* 24 April 1987) points out that on present knowledge, early Upper Pleistocene Africans were more modern than the Neanderthals only in their craniofacial anatomy and especially in their smaller, flatter faces. Since the Neanderthal face and much associated cranial anatomy (particularly in the occipito-mastoid region) may reflect habitual use of the anterior teeth as tools (Trinkaus 1986), the more modern craniofacial structure of the early Upper Pleistocene Africans suggests an important behavioural difference. The Neanderthals also possessed several postcranial specializations or peculiarities that were probably behaviourally significant (Trinkaus 1986), suggesting overall that they accomplished physically tasks that their fully modern successors accomplished technologically. Unfortunately, the postcranial anatomy of early Upper Pleistocene Africans is still too poorly known to determine whether they differed from the Neanderthals postcranially as well as cranially.

A denser fossil record and better dating methods may one day show that the early Upper Pleistocene African sample is highly variable mostly because it represents an evolving lineage including parent, near-modern populations and derived, fully modern ones. As it stands, the record is certainly adequate to argue that modern or near-modern populations appeared earlier in Africa than in Europe, and to highlight the possibility that the Neanderthals represent a specialized offshoot of *Homo sapiens* that contributed few if any genes to modern populations. From a fossil perspective, perhaps the main obstacles to demonstrating that Africa was the principal, if not sole cradle of modern humanity are the possibility that modern people were present just as early in the Near East (Bar-Yosef and Vandermeersch 1981; Vandermeersch 1982; Bar-Yosef 1989, and this volume) and the murkiness (some would say contrariness) of the record in the Far East (Wolpoff *et al.* 1984; Wolpoff 1985). The issue is resolvable, but the well-dated, relatively complete fossils necessary for resolution may be a long time in coming.

ARTIFACTS

If the fossils from Klasies River Mouth and other early Upper Pleistocene sites in Southern Africa represent anatomically modern humans, then it is reasonable to ask if the associated artifacts suggest fully modern behaviour. Here, I think there would be widespread agreement that the answer is 'not clearly'. Since the pioneering work of Goodwin (1928, 1929), corrected mainly by advances in dating, the relevant artifact assemblages have been assigned to the Middle Stone Age (MSA). They are generally dominated by flakes or flake-blades, often from well- prepared cores, and the principal retouched pieces are scrapers, points, and denticulates similar to those found in the contemporaneous Mousterian (Middle Palaeolithic) of Europe and the Near East (Sampson 1974; Volman 1984).

Middle Stone Age (MSA) assemblages resemble Mousterian ones in two further fundamental respects: (1) they lack formal bone, ivory, or shell tools and art objects, with some very rare exceptions that could represent undetected intrusions from later horizons; and (2) they exhibit relatively little variability in time and space that cannot be attributed to differences or changes in the availability of lithic raw materials. In fact, typologically and technically, there is little to distinguish the MSA from the Mousterian, and at present the basic difference is geographic: assemblages that would be called 'Mousterian' in Europe, the Near East, or North Africa are called 'Middle Stone Age' in sub-Saharan Africa. Further, in such fundamental features as the absence of formal bone artifacts and art and the relatively limited degree of spatial and temporal variability, the MSA differs from the succeeding Later Stone Age (LSA) in much the same way that the Mousterian differs from the succeeding Upper Palaeolithic.

In Europe, of course, the Mousterian is closely linked to the Neanderthals, and if human fossils were absent from MSA sites in Southern Africa, we might guess that the people were Neanderthals, or at least that they differed from modern people to the same extent that Neanderthals did. Clearly, we would be wrong, and the well-established association in the Near East between Mousterian artifacts on the one hand and both Neanderthals and early moderns on the other has long shown that there is no necessary correlation between physical type and artifacts. Equally germane, it suggests that a major change in physical type may precede a major change in artifacts.

The end of the MSA/beginning of the LSA in sub-Saharan Africa is even less well defined and more poorly dated than the end of the Mousterian/beginning of the Upper Palaeolithic to the north. The interface or transition may prove particularly difficult to study in Southern Africa, because many sites were abandoned in the 60- 30 000 BP interval when it occurred, probably because of widespread hyperaridity in the middle of the Last Glaciation. However, problems with identifying and dating the earliest LSA aside, modern or near-modern human fossils clearly antedate it.

They also apparently antedate the puzzling phenomenon known as the 'Howiesons Poort' Industry or Variation, which deserves mention here, if only as the one sure exception to the generalization that MSA industries did not vary greatly through time. The Howiesons Poort is defined by a dramatic increase in the abundance of well-made backed pieces, mainly segments ('crescents') and trapezoids, in otherwise basically MSA assemblages at numerous sites south of the Limpopo River (Volman 1984). At one time, the Howiesons Poort appeared to postdate the MSA *sensu stricto* and was regarded as a possible transitional industry to the LSA. Above all, the segments resemble LSA pieces by the same name, though on average they are much larger. However, it is now clear that typical MSA industries lacking numerous well-made backed pieces succeeded the Howiesons Poort, and that it antedates the LSA by a substantial interval.

Dating the Howiesons Poort is a problem (Parkington, this Symposium, Volume 2) and not all occurrences need be the same age. However, wherever it occurs in long MSA sequences, it is always at or near the top, suggesting broad contemporaneity among occurrences. Where dates seem most secure, as at Boomplaas Cave and Klasies River Mouth in the southern Cape Province of South Africa, the Howiesons Poort appears to date from the transition between the Last Interglaciation and the Last Glaciation, and an age centered on 70 000 BP seems reasonable (Deacon, this volume).

In Southern Africa, the Howiesons Poort represents basically the same kind of puzzle that the broadly coeval Pre-Aurignacian/Amūdian and

Aterian industries do in the Near East and North Africa respectively. If the Howiesons Poort genuinely belongs to the last interglacial/last glacial transition, then perhaps it is most fruitfully seen as a response to the climatic stress that local MSA populations experienced (Deacon, this volume). Future finds may also show that it actually relates in some way to modern human origins, but at Klasies River Mouth and perhaps Border Cave, most of the modern or near modern fossils almost certainly antedate it (Wymer and Singer 1982; Deacon, this volume; Beaumont 1980).

SUBSISTENCE

Because of unfavourable bedrock lithology, most known MSA sites in Southern Africa do not preserve animal bones or shells. However, animal remains do occur at some sites, obviously including those with human fossils. At one of these sites, Equus Cave, the rarity of artifacts, the abundance of hyena coprolites, and the frequency of carnivore-altered bones indicate that hyenas were the main bone accumulators. At the remaining sites, however, abundant artifacts and other evidence for human presence (especially hearths), frequent cut marks on bones, and the overall rarity of evidence for carnivore activity, implicate people as the main, if not the only bone collectors. At these sites, the animal remains presumably represent MSA food debris, and it is reasonable to ask if they suggest fully modern subsistence behaviour. I argue here that they do not.

A full answer to the question would require comparisons between MSA and preceding, Acheulian (Early Stone Age) faunal assemblages. However, like most other parts of the world, Southern Africa contains no Acheulian sites where it is clear that people were the main bone collectors (Klein 1988). Known Acheulian bone/artifact associations occur mostly at open-air sites near ancient water sources where it is possible the bones represent mainly natural deaths or carnivore kills, possibly, but not necessarily even scavenged by the people who left the artifacts. What is needed is a cave where context and associations point clearly to Acheulians as the main bone collectors, but Southern Africa, like most other parts of the world, contains few caves or rock-shelters with deposits of the right (Lower or Middle Pleistocene) age, and the small number of Southern African caves with abundant Acheulian artifacts either do not preserve bones or have not provided large enough samples for meaningful analysis. It should be stressed that the problem does not appear to be that Acheulian people avoided caves. Where cave deposits of the right age exist, they generally do contain Acheulian occupation debris. The problem is that deposits of the right age do not exist in most caves, either because the caves did not exist in Acheulian times or because they were flushed in the post-Acheulian period.

The rarity of Acheulian cave sites and the eye-catching, immediately diagnostic character of Acheulian artifacts in open- air (especially surface) contexts may give the impression that Acheulian settlement patterns differed greatly from MSA and LSA ones, but this cannot really be determined from the available data. Likewise, the absence of Acheulian caves with abundant humanly accumulated bones precludes any attempt to determine

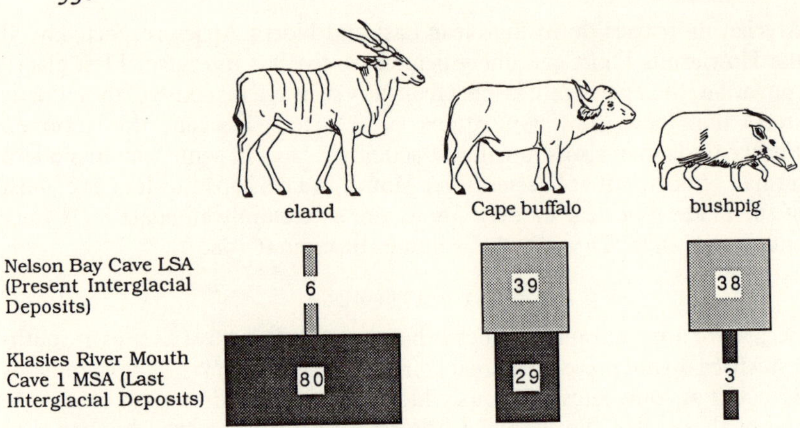

Figure 27.2. The minimum number of eland, Cape buffalo, and bushpig in the Present Interglacial deposits of Nelson Bay Cave and the Last Interglacial deposits of Klasies River Mouth Cave 1 (from Klein 1983b).

if Acheulian people were less effective at obtaining animals than their successors. The only meaningful comparisons possible are between MSA and LSA sites, especially caves where context and associations indicate that people were the main bone collectors. Almost certainly, all LSA people were anatomically-modern, and this is firmly established for the more recent (Holocene) ones who figure most prominently in comparisons I have discussed before (Klein 1974, 1977, 1979, 1983a) and summarize below. The comparisons thus have the potential to tell us whether MSA people were behaviourally modern, at least in the sense that LSA people were.

Besides differences in human behaviour, differences between faunal assemblages may also reflect differences in past environments. Thus, if the goal is to isolate behavioural differences, comparisons must be limited to faunal assemblages that accumulated under essentially the same environmental conditions. Additionally, the assemblages must be reasonably large, to limit the possibility that observable differences were caused by chance. By good fortune, the southern Cape Province of South Africa has provided MSA and LSA assemblages that meet these palaeoenvironmental and sample-size criteria to an extent that is unparalleled elsewhere in Africa and perhaps also among like-aged assemblages in Eurasia. Comparisons of the southern Cape samples suggest two basic contrasts:

1. Bones of fish and flying sea birds (cormorants, gulls, etc.) are very abundant in coastal LSA sites occupied during the Present Interglaciation (Holocene), but are rare or absent in coastal MSA sites occupied during the Last Interglaciation.

2. Present interglacial LSA sites contain ungulate species in rough proportion to their historical abundance near the sites, while last interglacial MSA sites tend to contain ungulate species in rough proportion to the danger involved in hunting them. In particular, compared to LSA faunas, MSA ones are much richer in the relatively docile eland and much poorer in the

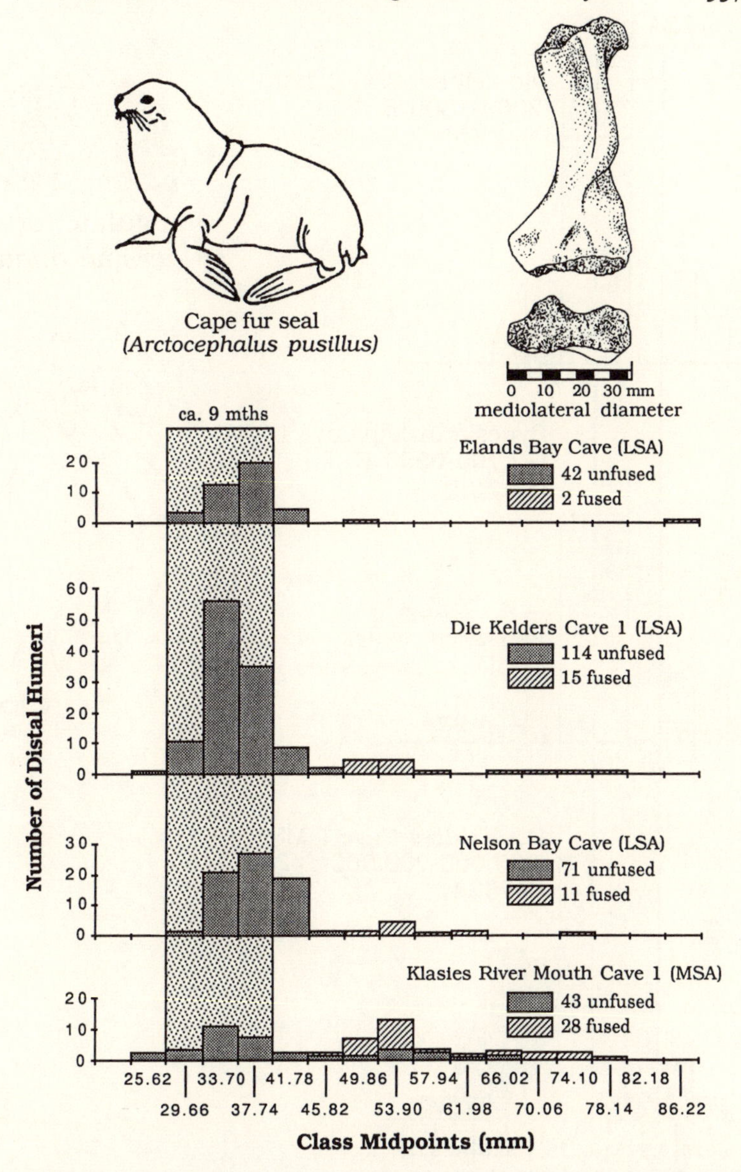

Figure 27.3. The mediolateral diameters of Cape fur seal distal humeri in the Later Stone Age layers of Elands Bay Cave, Die Kelders Cave 1, and Nelson Bay Cave, and in the Middle Stone Age layers of Klasies River Mouth Cave 1. The Later Stone Age samples differ from the Middle Stone Age one in two related ways: they contain significantly fewer shafts with fused epiphyses, and they show a more distinct peak in unfused shafts from individuals that were probably recently weaned (that is, about 9 months old) based on their size similarity to known-age individuals obtained by Graham Avery of the South African Museum. The relatively small differences among the Later Stone Age samples may reflect slightly different seasons of site occupation.

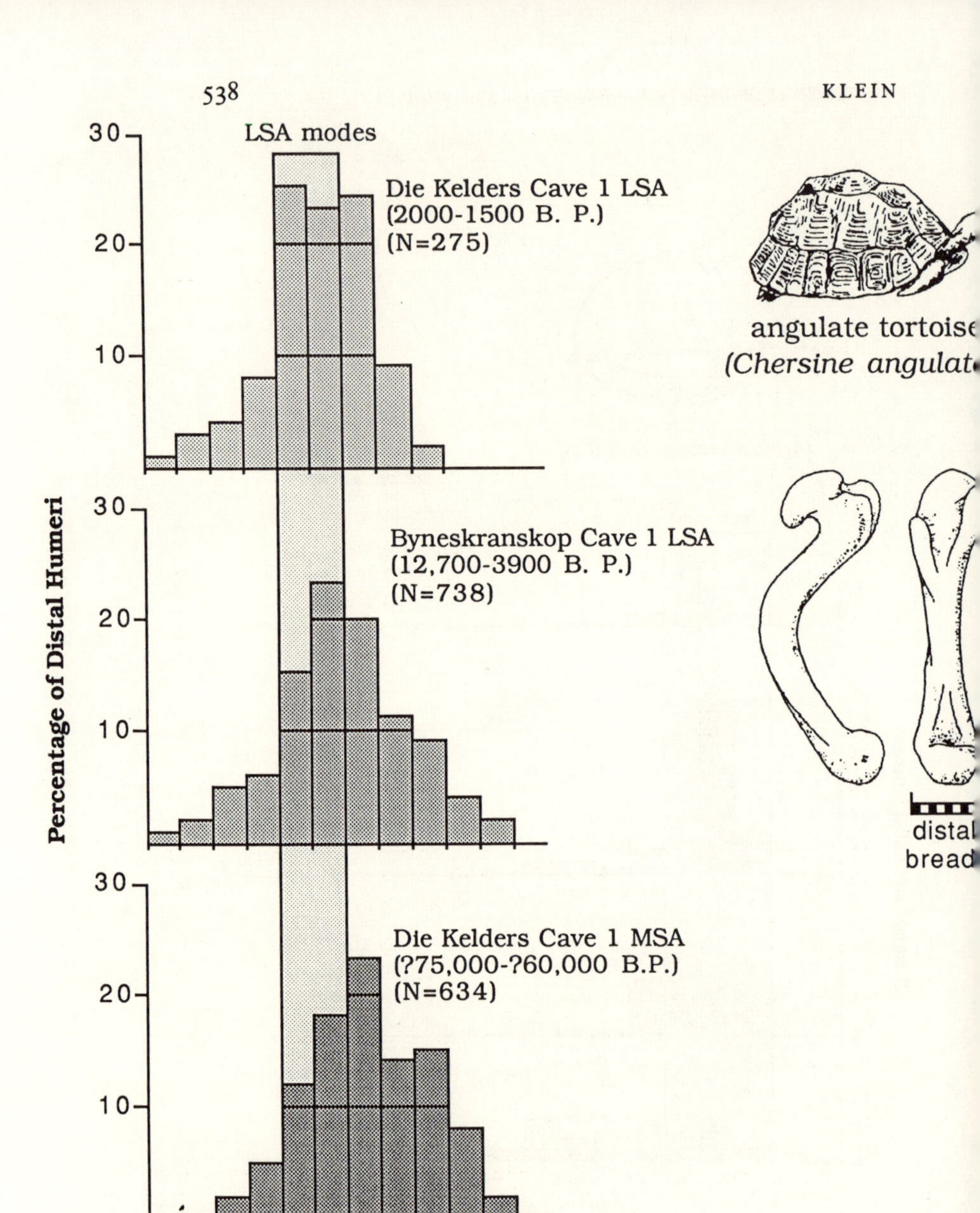

Figure 27.4. The mediolateral diameters ('breadths') of angulate tortoise distal humeri from the Middle and Later Stone Age deposits of Die Kelders Cave 1 and the Later Stone Age deposits of Byneskranskop Cave 1. The histograms lump humeri from long time periods, and subdivision of the samples by layer shows that there was size variation within these periods, particularly at Byneskranskop. However, the layer- by-layer samples reveal the same broad pattern as the histograms: on average, the Middle Stone Age humeri are significantly larger than the Later Stone Age ones. (For additional discussion, see Klein and Cruz-Uribe 1983.)

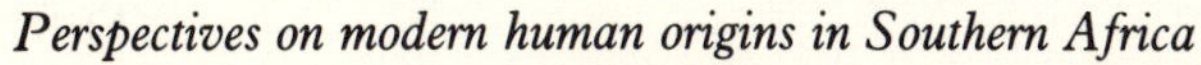

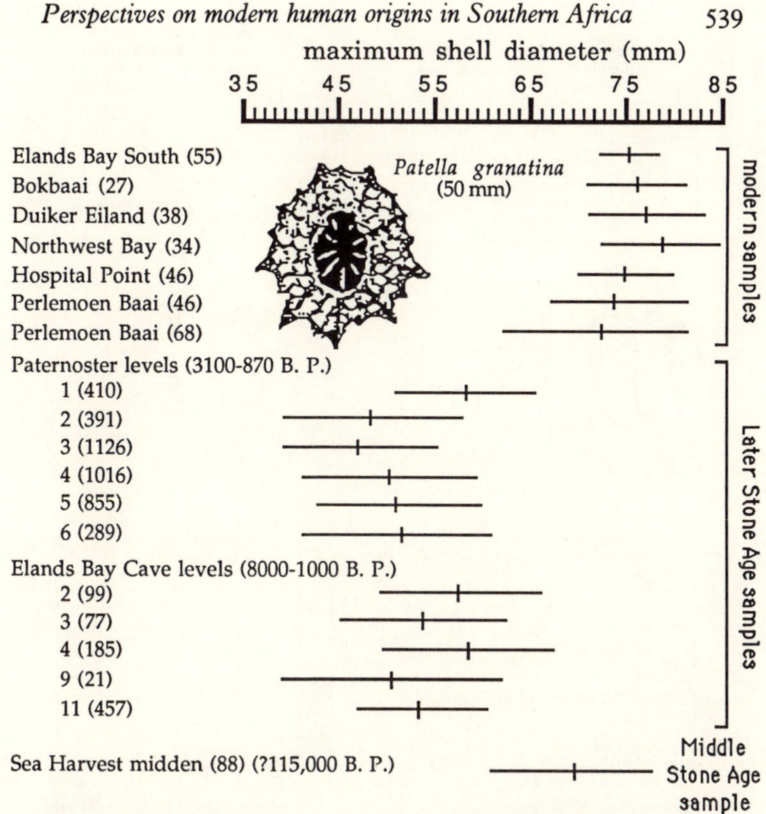

Figure 27.5. The maximum diameter of limpet *(Patella granatina)* shells in modern samples, Later Stone Age samples from the Paternoster Midden and Elands Bay Cave, and the Middle Stone Age sample from the Sea Harvest Midden site. In each case, the vertical line indicates the mean, the horizontal line one standard deviation from the mean, and the figure in parentheses after the sample name is the number of measured specimens. The modern samples were collected by students in 10-minute intervals from intertidal rocks that are not being exploited today. (The modern and Later Stone Age data are from Buchanan et al. 1978; the Sea Harvest data are from Parkington, pers. comm.)

more dangerous Cape buffalo and bushpig (Fig. 27.2). The significantly greater abundance of buffalo and bushpig *versus* eland in LSA faunas is reminiscent of their relative abundance in the historic environment.

I have interpreted these contrasts to indicate that, unlike LSA people, MSA people did not actively fish and fowl and that they lacked the ability to hunt truly dangerous animals on a regular basis. There is support for this hypothesis in the contrast between MSA and LSA artifact assemblages. Only LSA assemblages contain artifacts that are reasonably interpreted as fishing and fowling gear, together with ones that probably or certainly indicate the presence of the bow and arrow. With the bow and arrow, LSA people could have attacked dangerous species from a distance at reduced personal risk, and even if their success rate was low, the higher proportion of initial trials would lead to a higher proportion of dangerous animals in

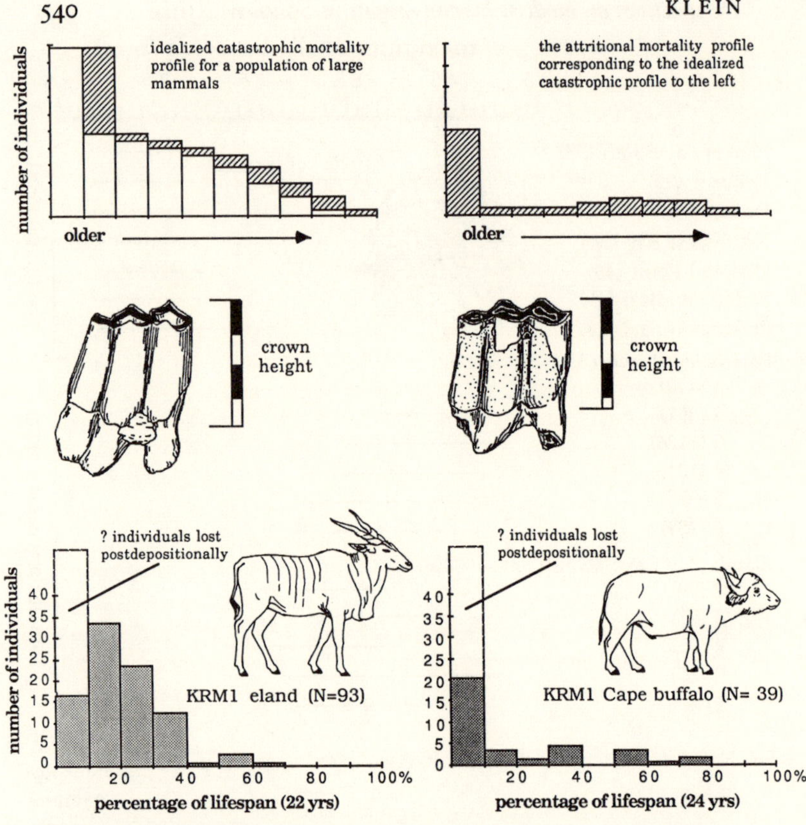

Figure 27.6. Top left: A schematic catastrophic age (mortality) profile for a population of large mammals that is basically stable in size and age structure. The blank bars represent the number of individuals that survive in each successive age cohort, the hatched bars the number that die between successive cohorts. Top right: a separate plot of the hatched bars, showing the corresponding, schematic attritional age profile. The basic form of corresponding catastrophic and attritional profiles is the same for all large mammals, but the precise form will differ from population to population depending on species biology and specific mortality factors. Bottom: Mortality profiles based on dentitions for eland (left) and Cape buffalo (right) from the Middle Stone Age layers of Klasies River Mouth Cave 1. It is probable that postdepositional leaching, profile compaction, and other destructive factors have selectively destroyed teeth of the youngest eland and buffalo. Taking this into account, the eland profile is clearly catastrophic and the buffalo profile is attritional. (For additional discussion, see Klein 1982.)

their food debris.

There are other faunal contrasts that support the idea of a fundamental difference in subsistence ecology between MSA and LSA peoples. One that I have not reported before concerns the age composition of the fur seals from the last interglacial levels of Klasies River Mouth and from a variety of present interglacial LSA horizons, including those at Nelson Bay Cave, Die Kelders Cave, and Elands Bay Cave. The LSA samples comprise mainly seals near the age when they were probably weaned, while the Klasies MSA sample contains a much wider range of individuals, including significantly more full adults (Fig. 27.3). The most likely reason for the age clustering in the LSA sites is that the people timed their visits to coincide with the weaning season, when newly weaned individuals often become exhausted at sea and come ashore, where they may be easily killed or scavenged. The age distribution of the Klasies seals may indicate that MSA people were unaware of the seasonally increased availability of seals and that they simply killed or scavenged occasional individuals they encountered during less seasonally focused visits.

The larger average size of limpets and angulate tortoises in MSA versus LSA sites further suggests an important difference (Figs. 27.4 and 27.5). The most straightforward interpretation is that MSA people preyed less intensively on both shellfish and tortoises, because MSA populations were less dense. Smaller MSA population density could be both a cause and consequence of more limited ability to acquire various animals.

To the listed contrasts could be added Binford's (1984) argument, based mainly on the pattern of skeletal part representation at Klasies River Mouth, that MSA people, unlike their successors, primarily scavenged rather than hunted large bovids. However, I see two basic problems with Binford's interpretation. While it could partly account for the 'attritional' mortality (age) profiles that characterize buffalo and most other large bovid species at Klasies, it is totally inconsistent with the 'catastrophic' mortality profile that characterizes eland, the most common large bovid in the site (Fig. 27.6). Second, and equally important, the same pattern of skeletal part representation found at Klasies characterizes a wide range of other sites, including both LSA and Iron Age sites in Southern Africa (Klein and Cruz-Uribe 1987; Voigt 1983). I have no compelling explanation for why the pattern is so widespread, but it might partly reflect the notorious 'schlepp effect' of Perkins and Daly (1968) and partly the widespread occurrence of selective, pre- and post-depositional destructive factors that are totally independent of how people obtained animals or bones. Whatever the case, at least so far, there is no evidence for a systematic difference between the MSA and LSA in patterns of skeletal part representation. Nor have I found any systematic difference in ungulate mortality profiles, though this might be due to the relatively small size of most ageable LSA samples.

Alternative explanations for the faunal contrasts I recognize are obvious. MSA people may have avoided fish, flying sea birds, and pigs because of food taboos or tastes, not because they were unable to catch them, and the abundance of flying birds in LSA sites may partly reflect the people's interest in bird bones as raw material for artifact manufacture (Deacon, this volume). It will be recalled that MSA people did not make formal bone tools. I appreciate these alternatives, but I feel that the occurrence of widespread food taboos or preclusive tastes in the MSA and their disappearance in the LSA itself requires an explanation, and LSA interest in flying birds partly as sources of artifact raw material does not diminish the potential evolutionary significance of the contrast between MSA and LSA sites in bird abundance.

If there is a problem with my interpretations, I think it is with the limited data on which they are based. There are numerous LSA sites with abundant bones of fish and flying sea birds, but coastal MSA sites with samples adequate for comparison are limited to Klasies River Mouth and Die Kelders. The contrast in the proportions of dangerous species has been developed almost exclusively from a comparison between the large Klasies River Mouth last interglacial MSA sample and the equally large one from the present interglacial LSA deposits at nearby Nelson Bay Cave. Among MSA sites, only Klasies has provided enough seal bones for an age analysis, while Klasies cannot be used to check the contrast in shellfish size, since shells were not systematically collected during the large-scale excavations of 1966-68. The shellfish size contrast is based on data from the Sea Harvest last interglacial MSA midden and from nearby present interglacial LSA sites such as Elands Bay Cave. The tortoise size contrast also cannot be confirmed with Klasies data, because tortoises are rare in the MSA deposit, just as they were in the historic environment. It is based mainly on measured bones from the MSA levels of Die Kelders Cave and the LSA horizons of nearby Byneskranskop Cave and Die Kelders itself, where the historic abundance of tortoises is mirrored in the deposits.

The argument that LSA people were better at obtaining animals is thus based on a patchwork data base, which nonetheless reflects a great deal of archaeological time and effort. It will not be improved quickly, and it may be many years before the interpretations I have offered here can be tested with new data. It is further important to stress that, even if the interpretations are correct, it does not follow that MSA people completely lacked special subsistence skills. For example, the 'catastrophic' mortality profile of the Klasies MSA eland suggests the people knew that eland are relatively easy to drive over cliffs or into other traps where whole groups can be dispatched. Additionally, it is possible that in common with local LSA people, they deliberately fired the veld to encourage the growth of geophytes whose underground storage organs (corms) are a potentially rich source of carbohydrates. Deacon (this volume; Deacon *et al.* 1986) suggests that the carbonized plant residues associated with hearths at the Klasies River Mouth sites derive from geophytes and that the abundance of hearths at Klasies, particularly near the top of the sequence, implies the people could make fire at will.

Finally, it is important to emphasize that the contrast between MSA and LSA faunas involves mainly last interglacial and present interglacial occurrences. The problem with expanding it to the intervening Last Glaciation is the rarity of known last glacial occupations, especially ones with large faunal samples. The rarity may always be difficult to redress, if, as I suggested above, it reflects adverse climatic conditions (hyperaridity) during much of the last glacial. Moreover, even if more last glacial sites are found, we may never discover ones showing the origins of coastal fishing and fowling, which probably occurred on the now-drowned continental shelf.

The end result is that it is not yet possible to say that the contrasts that

I have reported truly characterize the entire MSA *versus* the entire LSA, or to determine if LSA advances evolved slowly or abruptly. In the context of the present paper, however, the data are adequate to contend that early Upper Pleistocene faunal remains, like the accompanying artifacts, reflect less than fully modern behaviour. If the people were indeed anatomically-modern, then, with the artifacts, the faunal remains may be used to argue that the evolution of the modern physical form preceded the evolution of fully modern behaviour.

SUMMARY

The number of Southern African fossils bearing on modern human origins is small, and many specimens are relatively undiagnostic fragments. In addition, some specimens come from questionable stratigraphic contexts, and even those whose provenience is well documented are dated only imprecisely or provisionally. Nonetheless, considered together with fossils from other parts of Africa, the Southern African ones suggest that anatomically-modern people were present before 60 000 years ago, certainly earlier than in Europe and perhaps earlier than in the Near East. In contrast, the artifacts and animal remains associated with early modern or near modern fossils in Southern Africa suggest non-modern behaviour, perhaps broadly on a par with that of the contemporaneous European Neanderthals. The implication is that the appearance of the modern physical form preceded the appearance of fully modern behaviour. The fossil and archaeological data on which this conclusion is based are still limited, and it is thus only a working hypothesis to be tested against future discoveries.

ACKNOWLEDGEMENTS

I thank the National Science Foundation for supporting my own research reported here and J. Deacon, P. Mellars, J. E. Parkington, G. P. Rightmire, C. B. Stringer, E. Trinkaus, T. P. Volman, and C. A. Wolf for critical comments on a preliminary draft.

REFERENCES

Bar-Yosef, O. 1989. Upper Pleistocene human adaptations in Southwest Asia. In E. Trinkaus (ed.) *Patterns and Processes in Later Pleistocene Human Emergence*. Cambridge: Cambridge University Press. In Press.

Bar-Yosef, O. and Vandermeersch, B. 1981. Notes concerning the possible age of the Mousterian layers in Qafzeh Cave. In J. Cauvin and P. Sanlaville (eds) *Préhistoire du Levant*. Paris: Centre National de la Recherche Scientifique: 281-285.

Beaumont, P. B. 1980. On the age of Border Cave hominids 1-5. *Palaeontologia Africana* 23: 21-33.

Beaumont, P. B., Villiers, H. de and Vogel, J.C. 1978. Modern Man in sub-Saharan Africa prior to 49 000 BP: a review and evaluation with particular reference to Border Cave. *South African Journal of Science* 74: 409-419.

Beaumont, P. B., van Zinderen Bakker, E. M. and Vogel, J. C. 1984. Environmental changes since 32 KYRS BP at Kathu Pan, northern Cape, South Africa. In J.C. Vogel (ed.) *Late Cainozoic Palaeoclimates of the Southern Hemisphere:* Rotterdam: Balkema: 324-338.

Binford, L. R. 1984. *Faunal Remains From Klasies River Mouth*. Orlando: Academic Press.

Bräuer, G. 1984. A craniological approach to the origin of anatomically modern *Homo sapiens* in Africa and implications for the appearance of modern Europeans. In F. H. Smith and F. Spencer (eds) *The Origin of Modern Humans: a World Survey of the Fossil Evidence:* New York: Alan R. Liss: 327-410.

Bräuer, G. and Leakey, R. E. 1986. The ES-1693 cranium from Eliye Springs, West Turkana, Kenya. *Journal of Human Evolution* 15: 289-312.

Buchanan, W. F., Hall, S. L., Henderson, J., Olivier, A., Pettigrew, J. M., Parkington, J. E. and Robertshaw, P. T. 1978. Coastal shell middens in the Paternoster area, southwestern Cape. *South African Archaeological Bulletin* 33: 89-93.

Butzer, K. W. 1984a. Archeogeology and Quaternary environment in the interior of Southern Africa. In R. G. Klein (ed.) *Southern African Prehistory and Paleoenvironments*. Rotterdam: Balkema: 1-64.

Butzer, K. W. 1984b. Late Quaternary environments in South Africa. In J. C. Vogel (ed.) *Late Cainozoic Palaeoclimates of the Southern Hemisphere*. Rotterdam: Balkema: 235-264.

Butzer, K. W., Beaumont, P. B. and Vogel, J. C. 1978. Lithostratigraphy of Border Cave, KwaZulu, South Africa: a Middle Stone Age sequence beginning *c* 195 000 BP. *Journal of Archaeological Science* 5: 317-341.

Cann, R. L., Stoneking, M. and Wilson, A.C. 1987. Mitochondrial DNA and human evolution. *Nature* 325: 31-36.

Clark, J. D., Williamson, K. D., Michels, J. W. and Marean, C. A. 1984. A Middle Stone Age occupation site at Porc-Epic cave, Diré-Dawa. *African Archaeological Review* 2: 37-71.

Clarke, R. J. 1985. A new reconstruction of the Florisbad cranium with notes on the site. In E. Delson (ed.) *Ancestors: the Hard Evidence*. New York: Alan R. Liss: 301-305.

Day, M. H., Leakey, M. D. and Magori, C. 1980. A new hominid fossil skull (L.H. 18) from the Ngaloba Beds, Laetoli, northern Tanzania. *Nature* 284: 55-56.

Day, M. H. and Stringer, C. B. 1982. A reconsideration of the Omo Kibish remains and the *erectus-sapiens* transition. In H. de Lumley (ed.) *L'Homo erectus et la Place de l'Homme de Tautavel Parmi les Hominidés Fossiles*. Nice: Centre National de la Recherche Scientifique / Louis-Jean Scientific and Literary Publications: 814-846.

Deacon, H. J., Geleijnse, V., Thackeray, A. I., Thackeray, J. G. and Tusenius, M. L. 1986. Late Pleistocene cave deposits in the southern Cape: current research at Klasies River. *Palaeoecology of Africa* 17: 31-37.

Debénath, A. 1980. Nouveaux restes humains atériens du Maroc. *Comptes-Rendus de l'Académie des Sciences de Paris (Série D)* 290: 851-852.

Debénath, A., Raynal, J.-P. and Texier, J.-P. 1982. Position stratigraphique des restes humains paléolithiques marocains sur la base des travaux récents. *Comptes-Rendus de l'Académie des Sciences de Paris (Série D)* 294: 972-976.

Debénath, A., Raynal, J.-P., Roche, J., Texier, J.-P. and Ferembach, D. 1986. Stratigraphie, habitat, typologie et devenir de l'Atérien Marocain: données récentes. *L'Anthropologie* 90: 233-246.

Goodwin, A. J. H. 1928. An introduction to the Middle Stone Age in South Africa. *South African Journal of Science* 25: 410-418.

Goodwin, A. J. H. 1929. The Middle Stone Age. *Annals of the South African Museum* 29: 95-145.

Grine, F. E. and Klein, R. G. 1985 Pleistocene and Holocene human remains from Equus Cave, South Africa. *Anthropology* 8: 55-98.
Jones, J. S. and Rouhani, S. 1986. How small was the bottleneck? *Nature* 319: 449-450.
Klein, R. G. 1974. Environment and subsistence of prehistoric man in the southern Cape Province, South Africa. *World Archaeology* 5: 249-289.
Klein, R. G. 1977. The ecology of early man in Southern Africa. *Science* 197: 115-127.
Klein, R.G. 1979. Stone age exploitation of animals in Southern Africa. *American Scientist* 67: 151-160.
Klein, R.G. 1982. Patterns of ungulate mortality and ungulate mortality profiles from Langebaanweg (early Pliocene) and Elandsfontein (Middle Pleistocene), south-western Cape Province, South Africa. *Annals of the South African Museum* 90: 49-94.
Klein, R.G. 1983a. The stone age prehistory of Southern Africa. *Annual Review of Anthropology* 12: 25-48.
Klein, R. G. 1983b. Paleoenvironmental implications of Quaternary large mammals in the fynbos region. *South African National Scientific Programmes Report* 75: 116-138.
Klein, R. G. 1988. The archaeological significance of animal bones from Acheulean sites in Southern Africa. *African Archaeological Review*. In Press.
Klein, R. G. and Cruz-Uribe, K. 1983. Stone age population numbers and average tortoise size at Byneskranskop Cave 1 and Die Kelders Cave 1, southern Cape Province, South Africa. *South African Archaeological Bulletin* 38: 26-30.
Klein, R. G. and Cruz-Uribe, K. 1987. Large mammal and tortoise bones from Elands Bay Cave and nearby sites, western Cape Province, South Africa. In J. E. Parkington and M. Hall (eds) *Papers in the Prehistory of the Western Cape, South Africa*. British Archaeological Reports International Series S332: 132-163.
Kuman, K. and Clarke, R.J. 1986. Florisbad: new investigations at a Middle Stone Age hominid site in South Africa. *Geoarchaeology* 1: 103-125.
Magori, C. C. and Day, M. H. 1983. Laetoli Hominid 18: an early *Homo sapiens* skull. *Journal of Human Evolution* 12: 747-753.
McBurney, C. B. M. 1967. *The Haua Fteah (Cyrenaica) and the Stone Age of the South-East Mediterranean*. Cambridge: Cambridge University Press.
Perkins, D. and Daly, P. 1968. A hunter's village in Neolithic Turkey. *Scientific American* 219: 96-106.
Rightmire, G. P. 1979. Implications of Border Cave skeletal remains for later Pleistocene human evolution. *Current Anthropology* 20: 23-35.
Rightmire, G. P. 1984. *Homo sapiens* in Sub-Saharan Africa. In F. H. Smith and F. Spencer (eds) *The Origin of Modern Humans: a World Survey of the Fossil Evidence*. New York: Alan R. Liss: 295-325.
Rightmire, G.P. 1987. Africa and the origin of modern humans. In R. Singer and J.K. Lundy (eds) *Variation, Culture and Evolution in African Populations: Papers in Honour of Dr. Hertha de Villiers*. Johannesburg: Witwatersrand University Press: 209-220.
Sampson, C.G. 1974. *The Stone Age Archaeology of Southern Africa*. New York: Academic Press.
Stringer, C. B. 1979. A re-evaluation of the fossil human calvaria from Singa, Sudan. *Bulletin of the British Museum of Natural History (Geology)* 32: 77-83.
Tankard, A. J. and Schweitzer, F. R. 1976. Textural analysis of cave

sediments: Die Kelders, Cape Province, South Africa. In D. A. Davidson and M. L. Shackley (eds) *Geoarchaeology:* London: Duckworth: 289-316.

Tobias, P. V. 1967. The hominid skeletal remains of Haua Fteah. In C. B. M. McBurney (ed.) *The Haua Fteah (Cyrenaica) and the Stone Age of the South-East Mediterranean*. Cambridge: Cambridge University Press: 338-352.

Trinkaus, E. 1986. The Neanderthals and modern human origins. *Annual Review of Anthropology* 15: 193-218.

Vallois, H. V. 1951. La mandibule humaine fossile de la grotte du Porc Epic près Diré Daoua (Abyssinie). *L'Anthropologie 55:* 231- 238.

Vandermeersch, B. 1982. The first *Homo sapiens sapiens* in the Near East. In A. Ronen (ed.) *The Transition from Lower to Middle Palaeolithic and the Origin of Modern Man*. Oxford: British Archaeological Reports International Series S151: 297-299.

Vogel, J. C. and Beaumont, P. B. 1972. Revised radiocarbon chronology for the Stone Age in South Africa. *Nature* 237: 50-51.

Voigt, E. A. 1983. Mapungubwe: an archaeozoological interpretation of an Iron Age community. *Transvaal Museum Monograph* 1: 1-203.

Volman, T. P. 1984. Early prehistory of Southern Africa. In R. G. Klein (ed.) *Southern African Prehistory and Paleoenvironments*. Rotterdam: Balkema: 169-220.

Wainscoat, J. 1987. Out of the garden of Eden. *Nature* 325: 13.

Wainscoat, J. S., Hill, A. V. S., Boyce, A. L., Flint, J., Hernandez, M., Thein, S. L., Old, J. M., Lynch, J. R., Falusi, A. G., Weatherall, D. J. and Clegg, J. B. 1986. Evolutionary relationships of human populations from analysis of nuclear DNA polymorphisms. *Nature* 319: 491-493.

Wolpoff, M. H. 1985. Human evolution at the peripheries: the pattern at the eastern edge. In P. V. Tobias (ed.) *Hominid Evolution: Past, Present and Future*. New York: Alan R. Liss: 355-365.

Wolpoff, M. H., Wu Xinzhi and Thorne, A. G. 1984. Modern *Homo sapiens* origins: a general theory of hominid evolution involving the fossil evidence from East Asia. In F. H. Smith and F. Spencer (eds) *The Origin of Modern Humans: a World Survey of the Fossil Evidence*. New York: Alan R. Liss: 411-483.

28. Late Pleistocene Palaeoecology and Archaeology in the Southern Cape, South Africa

H. J. DEACON

INTRODUCTION

There is a growing consensus that human skeletal remains representative of anatomically-modern people are of higher antiquity in Africa than elsewhere in the Old World. In some sense this seems to be a vindication of claims made in the 1970s, particularly in respect of the Border Cave finds (Beaumont *et al.* 1978; Rightmire 1979). The available fossil evidence has been reviewed recently by Bräuer (1984) and Stringer (1984, 1985) among others. It is the appreciation that the Eurasian Neanderthal populations may have had no place in the ancestry of modern peoples as much as the claimed high antiquity of any African finds that has encouraged the suggestion that the centre of origin and dispersal of modern people lay in Africa.

Recent studies of mitochondrial and nuclear DNA polymorphism in modern populations (Johnson *et al.* 1983; Wainscoat *et al.* 1986) show a basic genetic division between African and non-African populations, with the Africans exhibiting greater genetic diversity and thus age. The implications are that the dispersal of anatomically-modern people out of Africa resulted in replacement of archaic Eurasian populations and explosive radiation and range expansion. Although the molecular biological data are a source of information on human ancestry that is independent of the fossil record, the scope of the present studies involving only a few genetic loci is too limited for a very precise estimate of the temporal scaling of genetic distances between populations.

The plausibility of the interpretation of an African centre of origin for anatomically-modern peoples from the fossil or molecular evidence rests on the precision with which the archaeological evidence can be dated and with which the context of the finds can be established. In sub-Saharan Africa, claims for the high antiquity of the relevant fossils have been made for occurrences in Eastern and Southern Africa. As the age of the deposits lies beyond the range of conventional radiocarbon dating, this is not easily established on a chronometric scale, and potential alternative dating methods do not have the same level of routine application. The context and associations of the human skeletal finds are not uniformly well known. Some, like the mandible from Dire Dawa (Clark *et al.* 1984) are from old

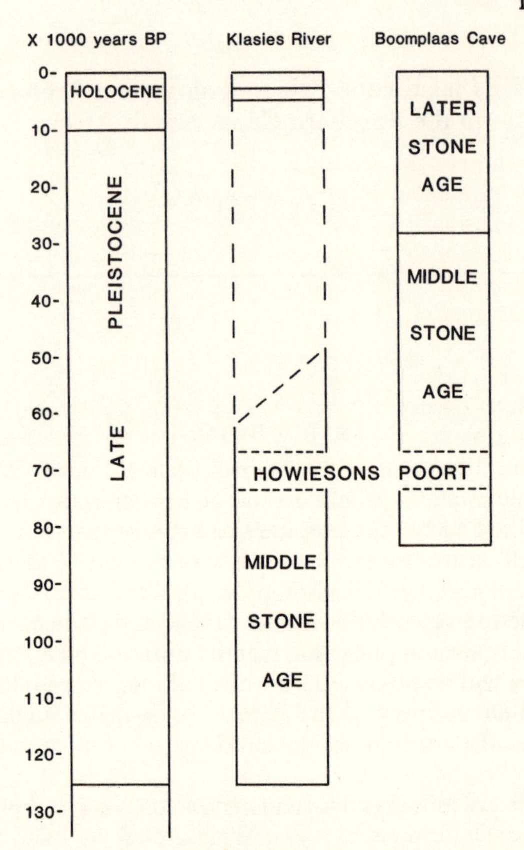

Figure 28.1. Temporal relationship of the Klasies River and Boomplaas sequences.

excavations that lack modern standards of stratigraphic control and others like those from Border Cave (Beaumont *et al.* 1978; Beaumont 1980) have not been recovered *in situ*. This does not preclude any or all of the human fossils from being acceptable evidence for the early representation of anatomically-modern people in Africa, but it highlights the need for this to be demonstrated beyond all reasonable doubt.

The Klasies River main site (southern Cape Province, South Africa) has developed an apparently unique importance not only as the find site of the putatively oldest remains of an anatomically-modern person (Singer and Wymer 1982), but also because it provides a culture-stratigraphic sequence for correlation with other occurrences. Estimates of the age of the Border Cave remains, for example, depend on such correlations, as do those from Equus Cave (Grine and Klein 1986). Klasies River main site has offered more independent lines of dating evidence, but the interpretation of the chronology and stratigraphy of the deposits has been the subject of debate (Binford 1984). Main site is not an isolated occurrence and the Late Quaternary chronometric, biostratigraphic and culture-stratigraphic sequence in the southern Cape is relatively well known. In particular, the Boomplaas

Cave deposit (Deacon 1979; Deacon *et al.* 1984) provides a very well studied record of the last 80 000 years that complements that from Klasies River (Fig. 28.1).

This paper discusses the implications of research carried out at Boomplaas between 1973 and 1983 and at Klasies River between 1984 and the present, for the origins of anatomically-modern people and the history of human behaviour for the southernmost part of the African continent. Research at these sites has developed as an investigation of the effects of climatic forcing on human habitats and their productivity, and consequently on the maintenance of populations.

THE KLASIES RIVER MAIN SITE

The Klasies River occurrences on the Tsitsikama coast between Druipkelder Point and the mouth of the Klasies River have been adequately described in the literature (Singer and Wymer 1982; Deacon *et al.* 1986). It is only the deposits somewhat confusingly labelled separately as caves or shelters 1, 1A, 1B, 1C and 2 which together comprise the main site, and, with the exception of 1B, represent a single set of deposits, that is of concern here. The deposits were extensively exposed and sampled during the 1967-68 excavations, and the 1984-86 field work has been directed primarily at lA and in extending the sampling and analysis of the contents of the deposit for comparison with Boomplaas. This has led to a better understanding of main site and has resolved some problems of possible ambiguity in the interpretation of the 1967-68 investigation.

The sedimentary sequence at main site has built up to a thickness of some 16 metres against an overhanging cliff face (lA) that does not appear to have been roofed. In the initial stages of this accumulation, openings into the rock face like caves 1 and lC were loci for primary archaeological deposition, but as the pile grew in thickness and the centre of the cone progressed upslope from the immediate coast, caves 1 and lC occupied a foot- slope position and became blocked and largely inaccessible to human occupation. This is particularly relevant to the interpretation of the stratigraphy of cave 1 which functioned as a drainage sump and was only occupied by carnivores in the later stages. Appreciation of this, for example, contradicts inferences drawn from the cave 1 sequence evidencing progression in human hunting behaviour (Binford 1984), or again inferences drawn on past sea-level stands (Butzer 1978).

Apart from the sampling of the Holocene midden by Binneman, no further archaeological excavations have been carried out in cave 1 since 1967-68. Cuttings have been made in the floor below the cultural horizon for stratigraphic purposes and the standing sections have been logged in detail.

A source of confusion has been the description of the sediments on bedrock as 'shingle' (Singer and Wymer 1982). Bedrock is overlain by angular clastic materials with a derived sand component. In the rear of the depository there is a remnant beach gravel preserved that is overlain by clastic materials and is similarly non-fossiliferous. These basal deposits are capped by a flowstone sheet, informally termed a 'crust', which is associated with the oldest phase of speleothem development. The development of large, free-standing stalagmites has been suggested to represent a considerable time lapse (Hendey and Volman 1986) but the possibilities of dating this phase of primarily carbonate deposition represented in caves 1 and 1C by the Uranium disequilibrium method are still being explored. The point that needs to be made is that the oldest Middle Stone Age occupation deposits that overlie the calcareous crust in cave 1 have no direct relationship to the remnant beach deposit in cave 1. Sands derived from a back-of-a-beach deposit, however, are associated with the basal Middle Stone Age occupation at cave 5 to the east, as well as cave 1 (Deacon and Geleijnse 1988) and this suggests that occupation of the caves dates from the fall in sea level from the high stand of the beginning of the Last Interglacial. The dating of the oldest Middle Stone Age occupation also rests on the determination of the oxygen-isotope values for shell from the basal middens (Shackleton 1982) and the more recent and as yet unpublished Uranium disequilibrium dating results obtained by Vogel (pers. comm.) for samples from a second and younger phase of speleothem formation that is stratigraphically younger than the oldest Middle Stone Age occupation (Deacon *et al.* 1986).

Further support for an earliest Late Pleistocene age is provided by the Aspartic Acid dates (Bada and Deems 1975) and Electron Spin Resonance determinations (Goede and Hitchman 1987) although the latter appear to be anomalously old and are explicable on current thinking only in terms of groundwater enrichment in uranium. The convergence of various lines of evidence provides strong support for an early Last Interglacial age for the beginning of the Middle Stone Age occupation at Klasies River, and this may have followed shortly after the maximum transgression correlated with the deep-sea oxygen-isotope stage 5e. The only other chronometric dating for a Last Interglacial Middle Stone Age occupation is that at Herolds Bay (Brink and Deacon 1982) where Vogel has determined a Uranium disequilibrium age equivalent to the Last Interglacial for a flowstone. At Herolds Bay the Middle Stone Age occupation lies directly on beach sediments and these are overlain by thin aeolian sand lenses representing a fall in sea level, and a thickness of coarse angular material, probably derived by haloclastic weathering of the wall rock in which a hyaena/porcupine-accumulated fauna is preserved. The flowstone caps the whole sequence. There seems no reason to question the association of the Middle Stone Age with the Last Interglacial, a correlation that was established by Rogers (1905) some 80 years ago and became evident when the Last Interglacial high sea level could be dated (see Imbrie and Imbrie 1979).

In cave 1, Wymer (Singer and Wymer 1982) recognized layers 38 and 37 as the initial (MSA I) occupation. Layer 38 is equivalent to the LBS member of the present investigation and Layer 37 was provisionally included in a RBS member. The LBS member (38) consists of sand intercalated with artifact-bearing shell-rich horizons and hearths, and the RBS consists of more carbonized horizons and is extensively drip eroded, exhibits load and flame structures, and includes pool clays. The RBS (37) wedges out towards the rear of cave 1 and so cannot be directly related to the second phase of stalagmite formation. It was initially presumed that the RBS represents a discrete series of deposits conformable with the LBS member and dating to more than 100 000 years BP, but the interpretation now offered is that it is the lower part of the overlying SAS member and separated in time from the LBS by a hiatus. The hiatus in deposition is indicated by the growth of significant-sized stalagmites in free air during the second phase of speleothem formation in cave 1 and represents an important time-break. The deposits overlying the LBS, including layer 7, are probably all younger than 100 000 years.

In the 1A and 1B sequences a discontinuity correlated with this hiatus has been mapped and separates the deposits into an older LBS series and a younger SAS series. In cave 1B there are deposits as old as those in the base of cave 1, but the human mandible accessioned as '41815' comes from just above the discontinuity. This explains why the oxygen-isotope values for shell from 1B (layer 10), related stratigraphically to the 41815 mandible, differ significantly from the values for shell from the oldest Middle Stone Age horizon (38) in cave 1. Singer and Wymer (1982) suggested the 1B mandible was more than 100 000 years old on the basis of association with the MSA 1 substage. The new observations indicate a younger age, but this may not be significantly younger. Electron Spin Resonance data (Goede and Hitchman 1987) do not support the placement of the 1B deposits above the disconformity, late in time relative to the 1A culture stratigraphic sequence, as suggested by Binford (1984). They are appreciably older rather than younger than the Howiesons Poort levels in 1A and of equivalent age to the layer 37-17 interface, the base of the SAS member in caves 1 and 1A. Thus while oxygen-isotope correlation with stage 5c or 5b rather than 5e may be indicated, and an age closer to 90 000 years rather than >100 000 years is more acceptable, this best estimate does not materially affect the contention that the mandible is of high antiquity. Any debate rests on the phylogenetic significance of this specimen rather than its age.

The deposits that overlie layer 37 in cave 1, associated with the MSA II, were designated layers 17-14 by Wymer (Singer and Wymer 1982). In the current study, these have been mapped in section as the SASB, SASU, SASW and SASR sub-members. A persistent ashy marker horizon was used to define the interface between SASN and SASU, and together these would be equivalent to layers 16 and 17. Towards the entrance they include shell lenses but the lenses lose their distinction towards the rear as the rubble component increases. Although there was primary occupation in cave 1 in this time, the SASW equivalent to Singer and Wymer's layer 15 is a steeply dipping talus of sands with stone and shell lines that accumulated rapidly and largely blocked the entrance. The overlying SASR equivalent to Singer and Wymer's layer 14, is composed of rubble and cobbles and occupies a channel-like feature. No artifacts or fauna in SASW (15) and SASR (14) are in archaeological context, although the dank recesses of the blocked-off cave 1 were probably still accessible to carnivores as the leopard remains reported by Singer and Wymer (1982) indicate. These are the footslope deposits of the 1A accumulation, and cave 1 functioned as a drainage sump. The prominent cobble and pebble component in the SASR (14) sub-member has no relevance to any Late Pleistocene sea levels, as has been suggested in the literature, and a now-buried Plio-Pleistocene raised beach is a more probable source. The WS member (13) is a white sand with lenses of pool clays and marks the end of the Late Pleistocene deposition in cave 1. It truncates the SASR (14) and is in part a slope-wash deposit. The grading characteristics are similar to sands in the top of the 1A sequence and in the Pliocene cliff top (Geelhoutboom) dune. The context of materials included in these sands is geological and not archaeological. The Bada and Deems (1975) Aspartic Acid date for the WS member may date the source deposit, an accumulation of sands in the top of the 1A sequence now largely removed by erosion. It may prove possible to date these sands by thermoluminescence, but an age of the order of 60 000 years is a reasonable estimate. The strata of 1A represent an expanded and more complete sequence accumulated in the same depository. The deposits are a series of sands with a variable coarser clastic component of wall rock, interbedded with human occupation deposits, notably shell lenses, carbonized partings and ash features. The whole sequence is richly fossiliferous, with marine invertebrates, large mammal bones and artifacts representing the main trace fossils of human activities, as well as small vertebrates, including reptiles, amphibians, fish and small mammals and terrestrial invertebrates, comprising the main fossils in the non-human occupation horizons. The frequency of occurrence of *Tropidophora*, a land snail that is a chance inclusion, has been used as an indicator of relative rates of deposition, and this shows a higher rate of accumulation for the deposits in the SAS member, with deposition slowing towards the top of the sequence.

The 1A sequence has been sampled by a series of small cuttings (one square metre or smaller) that expose the major stratigraphic units (Fig. 28.2). From these sampling points there has been the systematic collection of all remains, cultural and non-cultural, from charcoals to sediments, and these have been analysed in the laboratory at Stellenbosch. The deposits up to 20 m above sea level tend to be shell-rich, and those in the top of the sequence are more carbonized and shell-poor.

The basal LBS sands have a grain-size distribution resembling the modern beach, and in part this is their source. They suggest a high sea-level stand which has been correlated with oxygen-isotope stage 5e by Shackleton (1982), a correlation supported in more recent work in this project. An unconformity separates the LBS sands from the overlying finer SAS sands derived from the cliff-top fossil dune. The massive middens in grid squares T50 upwards to the sequence in grid square K48, equated with the SAS member, suggest a coastline not too distant from the site. These shell-rich deposits are the equivalent of the SASB (17) to SASR (14) members in cave 1 and are associated with the 'MSA II' of Wymer. A dating to between 100 000 and >80 000 BP would be consistent with the present evidence.

A *Donax* midden horizon below the Rockfall member (layer 22) indicates

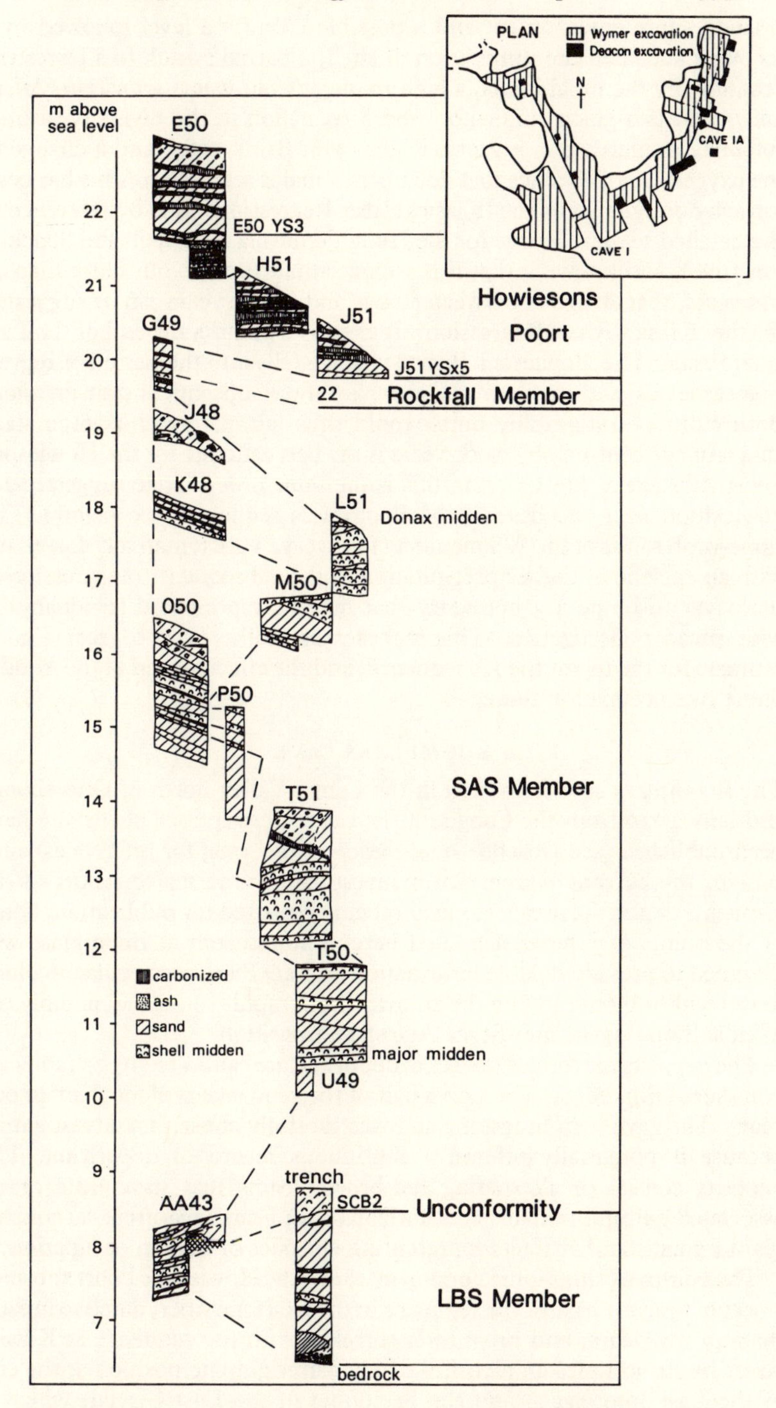

Figure 28.2. Section through the cave 1A sequence at Klasies River.

a sandy shore environment and a possible fall in sea level followed by a recovery and then the diminution of shell, a partial switch to a terrestrial economy by the inhabitants, a high frequency of *Otomys saundersii* (Avery 1987) that is a glacial indicator, and a reduction in the diversity of microfauna, an increase in grazers (Klein 1976; Brink 1987) and a change in the oxygen-isotope values that document a major regression. This has been formally designated as the 'Klasies River Regression'. It is by reference to the detailed sea-level curve for the Huon Peninsula (Chappell and Shackleton 1986), and knowing that it is younger than 100 000 but older than 49 000 years, that an age of between 80 000 and 60 000 years can be suggested for the 'Klasies River Regression'. It effectively marks the end of the Last Interglacial. The Howiesons Poort substage falls into the period of regressing sea levels, and the 'MSA III' into part of the subsequent transgression. Both culture-stratigraphic units would thus fall in oxygen-isotope stage 5a-4. An age centred on 70 000 years is the best estimate for the Howiesons Poort substage at present, and this is the same order of age suggested for its position near the base of the Boomplaas sequence (see below). The slope-wash sands of the WS member (13) in cave 1 are tentatively correlated with an episode of higher precipitation in the subsequent transgression at the oxygen-isotope 4-3 boundary that may be represented regionally by widespread podsolization. This makes a date of *c.* 60 000 years a good estimate for the top of the 1A sequence, and the effective end of the Middle Stone Age occupation there.

BOOMPLAAS CAVE

The Boomplaas cave is located in the Cango Valley north of Oudtshoorn and only 4 km from the Cango tourist cave. Descriptions of the site have been published (see Deacon 1979; Deacon *et al.* 1984 for references) and, as with the current Klasies River investigation, the major report to the Human Sciences Research Council is being prepared for publication. Some of the points of interest are cited here. The research at Boomplaas was designed to provide reliable information on Late Pleistocene palaeoecology in the fynbos biome and on the culture-stratigraphic sequence, notably the Middle Stone Age–Later Stone Age stage transition.

The deposits are some 5 metres in depth and are calibrated by 22 radiocarbon dates (Fig. 28.3). The lower half of the sequence is older than 40 000 years. The cave is in limestone and was carefully chosen for investigation because it potentially offered a continuous record of deposition. The deposits consist of alternating red-brown loams that have little or no associated cultural remains and horizons with hearth features, carbonized plant remains and artifacts representing episodes of human occupation.

The culture stratigraphic horizon marker – the Howiesons Poort substage – occurs near the base of the sequence in the OCH member, dated to greater than 49 000 years, and provides a correlate with the sequence at Klasies River lA. It occupies an identical climate-stratigraphic position at the end of the Last Interglacial and the beginning of the Last Glacial, which is perhaps best shown in a trend towards lower species diversity in the small

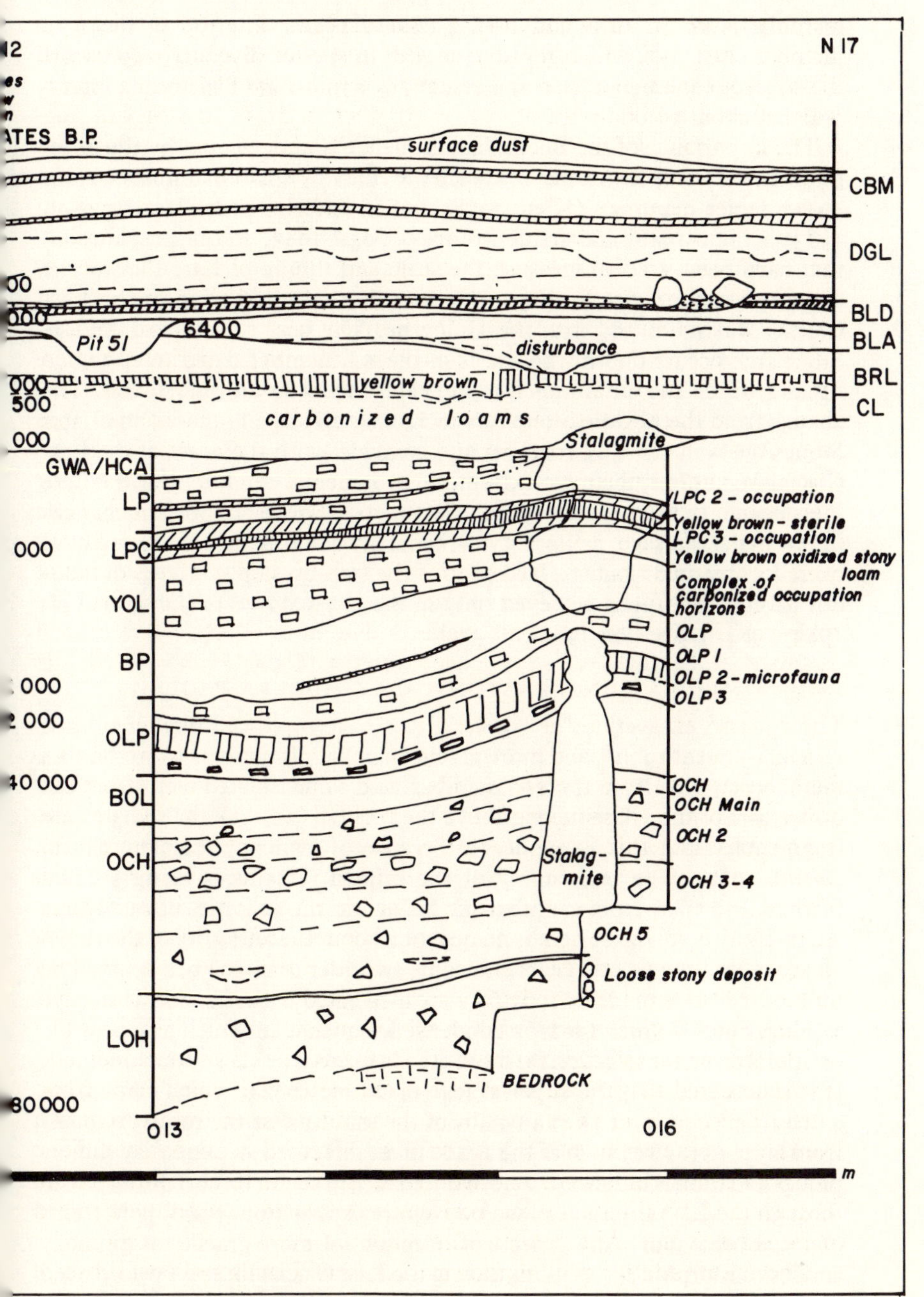

Figure 28.3. Section through the Boomplaas Cave sequence.

mammal fauna in OCH and BOL 4 (Avery 1982). The top of the BOL member (BOL 1-3) which registers a peak in species diversity, represents the subsequent amelioration and essentially a mid-Late Pleistocene interstadial at about 60 000 years.

The importance of the Boomplaas sequence is underscored by the different lines of complementary evidence provided by small mammals (Avery 1982), larger mammals (Klein 1978), pollen and charcoals (Deacon *et al.* 1983; Scholtz 1986) and stable isotopes (Vogel 1983, and in preparation) that have been used to measure the scale and timing of Late Pleistocene environmental change leading up to and through the period of the Last Glacial Maximum and into the Holocene. The best represented Middle Stone Age occupation in the site is in the BP member dated to *c.* 32 000 BP and which the environmental indicators mark as an interstadial. The change from the Middle Stone Age technological stage to that of the Later Stone Age is in the YOL member and coincides with the onset of the Last Glacial Maximum, the most extreme environmental conditions of the Late Pleistocene. In this sequence for the last *c.* 20 000 years, important changes in raw material usage, artifact styles, economy and function of the site have been documented against a backdrop of the step-by-step synthesis of modern habitat conditions achieved only in the last 5000 years (Deacon *et al.* 1983, 1984; J. Deacon 1984).

IMPLICATIONS: ANATOMICALLY-MODERN PEOPLE

The 1984-86 excavations at Klasies River have produced further human remains – burnt robust and more gracile maxillary fragments from the LBS member, an ulna from the SAS member, and some isolated teeth from the upper part of the 1A sequence above the rockfall (22). All the remains are fragmentary and this is explicable because of evidence for cannibalism (White 1987). The taphonomy of the human remains is being studied further, and apart from cannibalistic behaviour the activities of carnivores are possibly involved. There is no question about the antiquity of the finds: all are older than 60 000 years and some are older than 100 000 years. The bulk of the finds made in 1967-68 were associated with the MSA II in cave 1 (Singer and Wymer 1982), a condensed sequence in which any possible temporal trend for selection for more gracile forms is less easily documented. If it is accepted that the super-gracile mandible 16424, which came from a disturbance in layer 15 (SASW), is of the same age as the robust remains from layer 14 (SASR) such as the 13400 mandible, then an extremely dimorphic population is indicated. The maintenance of such a level of dimorphism through the Last Interglacial would require explanation, possibly in terms of social behaviour. Any process of selection for more gracile morphology and less dimorphic populations during the Last Glacial likewise would need explanation. Environmental conditions may have been an important factor in this selection. The fertility of modern San populations (Howell 1979) appears to be finely tuned to environmental conditions and to the development of mechanisms to maintain populations under conditions of periodic scarce and variable resources.

One of the maxillary fragments from the base of the 1A sequence derives from a young but very robust individual. The palate is not as broad as that of Kabwe, but superficially the morphology as a whole is very similar. It could represent the same lineage. The reassessment of all the Klasies River finds is planned, but they could represent a Southern African endemic type ancestral to the modern San. Long isolation of populations in Southern Africa is also supported on cultural grounds and, as discussed below, the distinct Howiesons Poort substage of the Middle Stone Age, for example, is confined to areas south of the Zambezi.

IMPLICATIONS: THE EVIDENCE FOR BEHAVIOUR

The distribution of Middle Stone Age sites in the southern Cape is the same as that for the Later Stone Age and departs markedly from that of the Acheulian distribution. Acheulian sites are tied to valleys (as around the type locality of Stellenbosch) and to water sources on the coastal platform. By contrast, Middle and Later Stone Age sites are found high up in the Cape mountains as well as on the coast, and frequent use was made of rock shelters. The differences in site distribution may reflect the use or otherwise of water containers, but it is more likely that it represents a change in the perception of the potentials of the environment. The Cape mountains are a very special habitat, and the nutrient poor substrates are associated with a species-rich 'fynbos' vegetation (Deacon *et al.* 1983). The 'fynbos' is a sclerophyllous evergreen vegetation characteristically found in mediterranean-type climate regions. One of the features of the 'fynbos' is the diverse geophyte flora associated with it, in which a large part of the biomass is concentrated below ground. Natural fields of geophytes form resource-rich patches along the mountains and in acid soils along the coastal margin. It is the distribution of these resources that explains the wide range of Middle and Later Stone Age locations. The conclusion is that Middle Stone Age people did not differ from the Later Stone Age people in their basic subsistence ecology, and that subsistence behaviour was essentially modern.

The same point has been argued specifically for the Klasies River main site (Deacon 1985) where a shift from an essentially coastal economy to a terrestrial one can be seen in the 1A sequence. The Howiesons Poort levels at the main site, with their conspicuous lenses of carbonized materials, represent the same kind of accumulations of the inedible residues of geophytic plant food as have been documented from Melkhoutboom Cave in the Holocene (Deacon 1976) and, by contrast, are absent from the coastal cave at East Guanogat on Robberg Peninsula (Deacon 1972, 1976). One of the arguments relating to the use of geophytes is that they are a slowly renewing resource, and to be exploited in any sustained way would require controlled burning. There are abundant hearths in the Middle Stone Age levels at Klasies River main site, and at Boomplaas the hearth features are identical to those in the Later Stone Age. This provides evidence for the ability to make fire at will and to manage geophyte patches in the Middle and Later Stone Ages in a way that was not available to the Acheulian population.

A subsistence base that is carbohydrate-rich and low in protein requires episodic ingestion of higher quality food – which is how hunting and scavenging for meat relates to gathering in this system. Shellfish are a necessary source of minerals in this environment of pure groundwaters of high pH, rather than a staple resource. At Klasies River main site there is no archaeological evidence for fishing. The fish remains are most abundant in the non-occupation horizons, and derive from very small fishes, which would have been accumulated by animals like birds roosting in the cliff above the site, rather than by people. The use of shellfish was essential for nutrients, but the investment of time and energy in fishing is likely to have had more to do with the availability of other potentially higher-ranked protein sources than the perception of fish as food and the ability to catch them. Klein (1980) has noted the absence of flying birds in the avifauna from the Middle as opposed to Later Stone Age sites, suggesting that this also reflects a lower perception of potential foods; however it can be argued that in Later Stone Age contexts it was the value of bird *bone* as a raw material, rather than their carcasses as food, that accounts for their presence.

Another argument that has been put forward with regard to Middle Stone Age subsistence evidence at Klasies River (Binford 1984) has stressed the importance of scavenging *versus* hunting, and has linked a postulated small size of the food packages to an absence of food sharing and an absence of base-camp living in the Middle Stone Age. In a 'catch as catch can' opportunistic approach to obtaining a meat supplement to a basic plant food diet, distinctions between hunted and scavenged foods become blurred and the size of the individual meat packages provides no reliable criterion for the identification of sharing. Holocene hunter- gatherers in the same habitats are known to have used the small nocturnal and territorial bovids, and even ground game like hyrax and tortoise as meat sources. These are foodstuffs which occur in small packages, but there is no question that such groups practised food sharing. It seems unwarranted to assume that the Middle Stone Age groups using the Klasies River main site were incapable of, or quantumly less effective in, hunting selected prey species than their Holocene counterparts, or that with the control of fire they would have been less effective in scavenging from carnivore kills before the kills had been reduced to the bare bones. There is at present a dearth of good information on the taphonomy of Middle Stone Age archaeological faunas because few adequate samples are available, but one recently completed study (Brink 1987) shows that the Middle Stone Age groups at Florisbad were selectively hunting small- to medium-sized bovids and scavenging hippopotamus. To postulate a scenario of the Middle Stone Age groups at Klasies River main site eking out a living by cracking bones collected during safe daylight hours (Binford 1984: Fig. 6.1) from nearby waterholes is a somewhat extreme interpretation of the faunal evidence.

The view offered here is that the Middle Stone Age groups in the southern Cape had essentially the same perception of their environment as their Holocene successors. In their subsistence behaviour they show the same reliance on carbohydrate-rich plant foods, supplemented by animal protein and the use of shellfish as a source of nutrients, as found in the Later Stone Age. The implication of this review of Middle Stone Age subsistence is that in this respect behaviour and the ability to solve problems relating to resources was modern.

Another source of information on the behaviour of Late Pleistocene human populations in the southern Cape is provided by the cultural remains found at the sites they occupied. The present stage of research into the Middle Stone Age has concentrated on defining the chronology and contents of sequences, and the detailed study of habitation areas in the Late Pleistocene is only just beginning. Thus while there is little information on the intra-site or inter-site patterning in the Middle Stone Age in the southern Cape, there are good data on sequential cultural changes. The sequences at Klasies River and Boomplaas are central to any discussion because they provide a dated framework. Singer and Wymer (1982) in their study of the Klasies River sites were able to show some significant changes in artifact typology through the sequence, and labelled these as the 'MSA I', 'MSA II', 'Howiesons Poort', 'MSA III' and 'MSA IV' substages. This sequence gave the first clear demonstration of the position of the Howiesons Poort substage, distinctive in typology and raw material usage, within the Middle Stone Age sequence, and some indication of its age. They were also able to show the presence of ochre pencils but the absence of personal ornaments in shell or other materials and, with the exception of a bone point possibly associated with the Howiesons Poort levels, an absence of bone tools.

There is a technological continuity throughout this Middle Stone Age sequence in the production of standardized flake-blade blanks from prepared cores. Shaping through secondary retouch is infrequent but there are good examples (Singer and Wymer 1982: Figs 5.10, 5.11) of the reduction of the base and unifacial or bifacial retouching of points and, in the Howiesons Poort levels, of the use of blunting or backing. The substages defined by Singer and Wymer (1982) are gross culture stratigraphic-divisions and have a real basis, but any periodization belies the technological continuity within this sequence. Their initial study was designed to determine the major typological changes rather than the finer details of shifting norms in artifact production; follow-up studies to shed more light on these questions are currently in progress.

With a better knowledge of technological systems like the Wilton (J. Deacon 1972, 1984), the Middle Stone Age can be seen as evidencing subtle change in stylistic attributes over time. The horizon-marker of the MSA II recognized by Singer and Wymer (1982), a short 'stubby' (Levallois-type) point, is one stylistic element, and is replaced by segments and trapezes in the overlying Howiesons Poort levels. The latter artifacts in the Klasies River sequence have a prescribed stratigraphic position that falls into a period of regressing sea level that equates with a glacial event, and cooler and drier climates are indicated in the associated biological and isotopic data. The termination of the Howiesons Poort horizon and the 'transition' to the MSA III approximates to the maximum of the regression. A feature of a glacial event in the latitudes of the southern Cape, well documented for the Last Glacial Maximum in the Boomplaas sequence, is the rapid subsequent amelioration of climates which followed the full glacial conditions. The MSA III thus falls in the same glacial event as the Howiesons Poort, but during the period of recovering sea levels and ameliorating climate which followed this regression event.

The coincidence of the Howiesons Poort horizon with a period of variable and deteriorating environmental conditions suggests that this archaeological phenomenon may be linked to the problems of coping with increasing environmentally-related stress. Such stress may have more to do with the maintenance of populations and their social structures under conditions of lowered habitat productivity, than simply coping with colder and drier climates. The novelty of the Howiesons Poort horizon lies in the introduction of standardized backed tools and in preferential (but not exclusive) selection of a high-cost raw material – silcrete – over quartzite, for making formal tools. Standardization of artifacts such as segments and trapezes is enforced by hafting, and functionally equivalent backed tools in Later Stone Age contexts are considered to be projectile points related to hunting. There is ethnographic evidence for the hafting of backed tools for use as arrowheads, but in a Middle Stone Age context (given the size of the pieces) they are more likely to have functioned as spearheads and barbs. Among the modern San, projectiles are male artifacts, and a recent study by Wiessner (1983) shows that the style of hafting can define high levels of social or linguistic identity, and arrows feature strongly in gift exchange within groups. The marking of boundaries and the intensification of social networks would explain the novelty of the Howiesons Poort industry in structuralist terms, rather better than functionalist arguments relating to the needs of new tools for new environments. The release of stress through amelioration of conditions removed cultural selection for the same level of symbolic behaviour, and the technology of the subsequent MSA III industries reverts to type.

The behavioural implications are that the Middle Stone Age people in the southern Cape and elsewhere in Southern Africa were using social mechanisms – reflected in stylistic shifts in artifacts analogous to those documented in the Later Stone Age and the ethnographic present – to cope with considerations of stress. This is cogent evidence for modern social behaviour in the Middle Stone Age.

As a cultural marker the Howiesons Poort is useful in delimiting a distinct technological province south of the Zambezi. The northern limit corresponds to the major ecological boundary with the Miombo woodland of Central Africa. The same boundary separates scraper-dominated Southern African Later Stone Age Wilton-like artifact assemblages from segment-dominated assemblages of the Nachikufan and Makwe-like artifact industries of Zambia in the Holocene (Deacon and Deacon 1980). The naturalistic rock art region of Southern Africa has the same northern limit (Phillipson 1985). That this distinctive Southern African region has had an integrity in evincing high levels of similarity in material culture throughout the last 100 000 years or more is taken to indicate containment of populations, if not complete isolation, and indicates a level of genetic and behavioural continuity between the Middle and Later Stone Ages.

The Boomplaas sequence provides evidence that the Middle Stone Age-Later Stone Age transition corresponds to the initiation of a glacial event similar to that documented for the MSA II-Howiesons Poort transition, but 50 000 years later. The final phase of the Middle Stone Age in the BP member dated to *c.* 32 000 years is characterized by long flake-blades and the use of prepared cores. In the succeeding YOL member there is a very low density of artifact finds, but they include novelties like bone points. The use of the prepared core technique for the production of flake- blade blanks falls away and miniaturization of at least the blade component becomes pronounced from 21 000 and particularly from 18 000 years ago. The climatic amelioration after the Last Glacial Maximum corresponds to the development of the classic Later Stone Age sequence for the southern Cape which has been detailed elsewhere (J. Deacon 1984).

The crucial question is whether the Middle Stone Age-Later Stone Age transition *per se* carries any implication for a major change in the quality of human behaviour. The answer must be negative because the level of technology practised is not correlated with behaviour potentials. A pertinent illustration that the use of a prepared core technology does not signify any distinctive level of behaviour and is epiphenomenal in character is provided by the evidence from the Orde River region in Western Australia, where this technology was adopted in the last 2000 years among modern people and replaced a microlithic technology (Dortch 1977: 117). On the same grounds it can be argued that the use of bone tools, shell ornaments, leather clothing and the practice of rock painting, as can be seen in the Later Stone Age at Boomplaas, are not markers of a different quality of human behaviour and are in a sense epiphenomena. The absence of these attributes does not preclude the Middle Stone Age populations from being behaviourally modern.

CONCLUSION

The fragmentary human remains from Klasies River main site associated with the Middle Stone Age and assessed to be anatomically-modern (Singer and Wymer 1982; Bräuer 1984), are of the same order of age as has been previously claimed. This provides clear support for the 'Out of Africa' hypothesis for the origins of anatomically-modern people. These remains relate to local populations in the southern Cape that showed a perception of their environment, and the use of artifacts as symbols to cope with stress, that indicates a modern quality of behaviour. The conclusion is that not only were people in the southern Cape some 100 000 years ago anatomically-modern, but they were also behaviourally modern.

ACKNOWLEDGEMENTS

Archaeological research at Klasies River and Boomplaas has been supported by the Human Sciences Research Council, Pretoria, and the University of Stellenbosch, and their assistance is gratefully acknowledged. I should also like to thank the Council for Scientific and Industrial Research, through the Fynbos Biome Programme, and the National Programme for Atmosphere, Weather and Climate, and the L.S.B. Leakey Foundation for funding that has contributed to the interpretation of the Klasies River and Boomplaas sequences.

REFERENCES

Avery, D. M. 1982. Micromammals as palaeoenvironmental indicators and an interpretation of the late Quaternary in the southern Cape Province, South Africa. *Annals of the South African Museum* 85: 183-374.

Avery, D. M. 1987. Upper Pleistocene coastal environment of the southern Cape Province of South Africa: micromammals from Klasies River Mouth. *Journal of Archaeological Science* 14: 405-421.

Bada, J. L. and Deems, L. 1975. Accuracy of dates beyond the C-14 dating limit using the aspartic acid racemization reaction. *Nature* 255: 218-219.

Beaumont, P. B. 1980. On the age of Border Cave hominids 1-5. *Palaeontologia Africana* 23: 21-33.

Beaumont, P. B., Villiers, H. de and Vogel, J. C. 1978. Modern man in sub-Saharan Africa prior to 49 000 BP: a review and evaluation with particular reference to Border Cave. *South African Journal of Science* 74: 409-419.

Binford, L. R. 1984. *Faunal Remains From Klasies River Mouth*. Orlando: Academic Press.

Bräuer, G. 1984. A craniological approach to the origin of anatomically-modern *Homo sapiens* in Africa and its implications for the appearance of modern Europeans. In F. H. Smith and F. Spencer (eds) *The Origins of Modern Humans: a World Survey of the Fossil Evidence*. New York: Alan R. Liss: 327-407.

Brink, J. S. 1987. *The Archaeozoology of Florisbad,* Orange Free State. Unpublished M.A. Thesis, University of Stellenbosch.

Brink, J. S. and Deacon, H. J. 1982. A study of a Last Interglacial shell midden and bone accumulation at Herolds Bay, Cape Province, South Africa. *Palaeoecology of Africa* 15: 31-39.

Butzer, K. W. 1978. Sediment stratigraphy of the Middle Stone Age sequence at Klasies River Mouth, Tsitsikamma coast, South Africa. *South African Archaeological Bulletin* 33: 141-151.

Chappell, J. and Shackleton, N. J. 1986. Oxygen isotopes and sea level. *Nature* 324: 137-140.

Clark, J. D., Williamson, K. D., Michels, J. W. and Marean, C. A. 1984. A Middle Stone Age occupation site at Porc Epic cave, Dire Dawa. *African Archaeological Review* 2: 37-71.

Deacon, H. J. 1972. A review of the post-Pleistocene in South Africa. *South African Archaeological Society Goodwin Series* 1: 26-45.

Deacon, H. J. 1976. *Where Hunters Gathered: a Study of Holocene Stone Age People in the Eastern Cape*. Claremont: South African Archaeological Society Monograph 1.

Deacon, H. J. 1979. Excavations at Boomplaas Cave: a sequence through the Upper Pleistocene and Holocene in South Africa. *World Archaeology* 10: 241-257.

Deacon, H. J. 1985. Review of L. R. Binford: *Faunal Remains From Klasies River Mouth. South African Archaeological Bulletin* 40: 59-60.

Deacon, H. J. and Deacon, J. 1980. The hafting, function and distribution of small convex scrapers with an example from Boomplaas Cave. *South*

African Archaeological Bulletin 35: 31-37.

Deacon, H. J., Deacon, J., Scholtz, A., Thackeray, J. F., Brink, J. S. and Vogel, J. C. 1984. Correlation of palaeoenvironmental data from the Late Pleistocene and Holocene deposits at Boomplaas Cave, southern Cape. In J.C. Vogel (ed.) *Late Cainozoic Palaeoclimates of the Southern Hemisphere*. Rotterdam: Balkema: 339-352.

Deacon, H. J., Geleijnse, V. B., Thackeray, A. I., Thackeray, J. F. and Tusenius, M. L. 1986. Late Pleistocene cave deposits in the southern Cape: current research at Klasies River. *Palaeoecology of Africa* 17: 31-38.

Deacon, H. J., Scholtz, A. and Daitz, L. D. 1983. Fossil charcoals as a source of palaeoecological information in the fynbos region. In H. J. Deacon, Q. B. Hendey and J. J. N. Lambrechts (eds) *Fynbos Palaeoecology: a Preliminary Synthesis*. Pretoria: South African National Scientific Progress Report 75: 174-182.

Deacon, H. J. and Geleijnse, V. B. 1988. The stratigraphy and sedimentology of the main site sequence, Klasies River, South Africa. *South African Archaeological Bulletin* 43: 5-14.

Deacon, J. 1972. Wilton: an assessment after 50 years. *South African Archaeological Bulletin* 27: 10-45.

Deacon, J. 1984. *The Later Stone Age of Southernmost Africa*. Oxford: British Archaeological Reports International Series S213.

Dortch, C. 1977. Early and late stone industrial phases in Western Australia. In R. V. S. Wright (ed.) *Stone Tools as Cultural Markers*. Canberra: Australian Institute of Aboriginal Studies: 102-132.

Goede, A. and Hitchman, M. A. 1987. Electron spin resonance (ESR) analysis of marine gastropods from coastal archaeological sites in Southern Africa. *Archaeometry 29:* 163-174.

Grine, F. E. and Klein, R. G. 1985. Pleistocene and Holocene human remains from Equus Cave, South Africa. *Anthropology* 8: 55-98.

Hendey, Q. B. and Volman, T. P. 1986. Last Interglacial sea levels and coastal caves in the Cape Province, South Africa. *Quaternary Research* 25: 189-198.

Howell, N. 1979. *Demography of the Dobe !Kung*. New York: Academic Press.

Imbrie, J. and Imbrie, K. P. 1979. *Ice Ages: Solving the Mystery*. London: Macmillan.

Johnson, M. J., Wallace, D. C., Ferris, S. D., Rattazi, M. C. and Cavalli-Sforza, L. L. 1983. Radiation of human mitochondrial DNA types analysed by restriction endonuclase cleavage pattern. *Journal of Molecular Evolution* 19: 255-271.

Klein, R. G. 1976. The mammalian fauna of the Klasies River Mouth sites, southern Cape Province, South Africa. *South African Archaeological Bulletin* 31: 74-98.

Klein, R. G. 1978. A preliminary report on the larger mammals from the Boomplaas Stone Age cave site, Cango Valley, Oudtshoorn district, South Africa. *South African Archaeological Bulletin* 33: 66-75.

Klein, R. G. 1980. Environmental and ecological implications of large mammals from Upper Pleistocene and Holocene sites in Southern Africa. *Annals of the South African Museum* 81: 223-283.

Phillipson D. W. 1985. *African Archaeology*. Cambridge: Cambridge University Press.

Rightmire P. 1979. Implications of Border Cave skeletal remains for later Pleistocene human evolution. *Current Anthropology* 20: 23-35.

Rogers, A. W. 1905. A raised beach deposit near Klein Brak River. *Annual Report of the Geological Commission of the Cape of Good Hope* 10: 293-296.

Scholtz A. 1986. *Palynological and Palaeobotanical Studies in the Southern Cape*. Unpublished M.A. Thesis, University of Stellenbosch.

Shackleton, N. J. 1982. Stratigraphy and chronology of the Klasies River Mouth deposits: oxygen isotope evidence. In R. Singer and J. Wymer (eds) *The Middle Stone Age at Klasies River Mouth in South Africa*. Chicago: Chicago University Press: 194-199.

Singer, R. and Wymer J. 1982. *The Middle Stone Age at Klasies River Mouth in South Africa*. Chicago: Chicago University Press.

Stringer, C. B. 1984. Human evolution and biological adaptation in the Pleistocene. In R. Foley (ed.) *Hominid Evolution and Community Ecology: Prehistoric Human Adaptation in Biological Perspective*. London: Academic Press: 55-83.

Stringer, C. B. 1985. Middle Pleistocene hominid variability and the origin of Late Pleistocene humans. In E. Delson (ed.) *Ancestors: the Hard Evidence*. New York: Alan R. Liss: 289-295.

Vogel, J. C. 1983. Isotopic evidence for past climates and vegetation of South Africa. *Bothalia* 14: 391-394.

Wainscoat, J., Hill, A., Boyce, A., Flint, J., Hernandez, M., Thein, S. L., Old, J. M., Lynch, J. R., Falusi, Y., Weatherall, D. J. and Clegg, J.B . 1986. Evolutionary relationships of human populations from an analysis of nuclear DNA polymorphisms. *Nature* 319: 491-493.

Wiessner P. 1983. Style and social information in Kalahari San projectile points. *American Antiquity* 48: 253-276.

White, T. D. 1987. Cannibalism at Klasies? *Sagittarius* 2: 6-9.

29. The Origins and Spread of Modern Humans: a Broad Perspective on the African Evidence

J. DESMOND CLARK

INTRODUCTION

Recently interest has focused on Africa as the possible continent where anatomically-modern humans made their first appearance. This is because the molecular data from both nuclear and mitochondrial DNA appear to point to an early, exclusively African population as having been ancestral to all modern humans (Wainscoat *et al.* 1986, and this volume; Cann *et al.* 1987; Stoneking and Cann, this volume). Cann *et al.* go on to calculate, on the basis of 2-4% divergence in mitochondrial DNAA over a million years, that the first female Modern lived between 140 000 and 290 000 years ago. There are problems with this time-scale (Wainscoat 1987) and it seems likely that a group, rather than a single individual is involved, but the ball is now in the court of human palaeontologists and archaeologists to re-examine their own data to see what sort of a bearing they have and how they might contribute to a more precise assessment of where, when, why and how this transformation and distribution took place.

DATES AND FOSSILS

If the DNA findings stand up (and they are still controversial) they not only support the hypothesis that Modern humans originated in one nuclear region of the Old World and spread very rapidly, but they indicate also that the archaic *Homo erectus* forms in Eurasia did not contribute anything to the gene-pool of modern populations which are derived solely from an African source. Crucial to the problem, therefore, is the fossil evidence from the African continent which has recently been critically reviewed by Rightmire (1984, 1985), Bräuer (1984a, 1984b), Hublin (1985) and Stringer (1985). A number of human fossils of Middle and early Upper Pleistocene age, some relatively complete, some fragmentary, show a variable degree of mixing of *Homo erectus* and Modern characteristics (see Fig. 29.1) These are, among others, Bodo, Broken Hill, Ndutu, Saldanha and Salé, and they appear to date to between 400 000 and 200 000 years ago on faunal evidence and various other dating methods (Fig. 29.2). Between >100 000 and ?75 000 years ago, forms are present which, on the one hand, still retain archaic features to a significant degree, (e.g. Florisbad, Ngaloba and Omo 2) and, on the other hand are contemporary with essentially Modern forms such

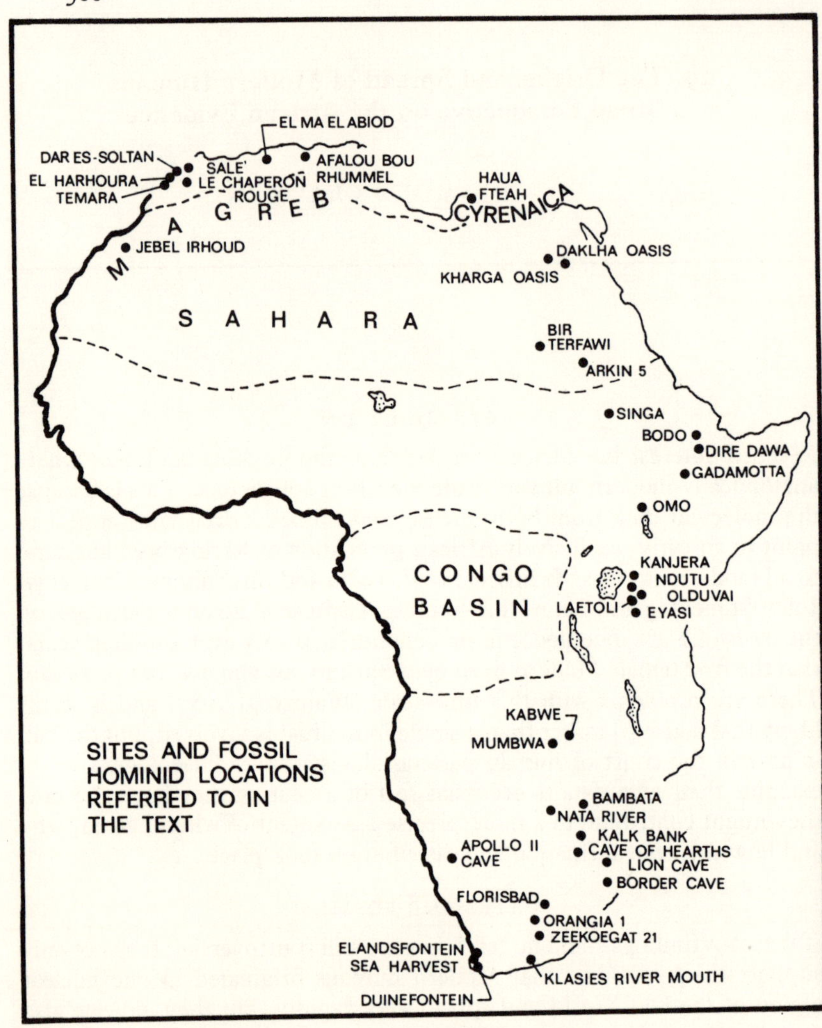

Figure 29.1. Map to show sites and fossil hominid localities referred to in the text.

as Omo 1 and those from Klasies River Mouth, Border Cave, Singa, Dar es Soltan 2 and Témara. These anatomically-Modern fossils appear to be older than the earliest Moderns in Eurasia, so that Modern traits would be recognisable earlier in Africa than they are elsewhere. But is this really so? Or does it only *seem* to be so? Absolutely crucial to any attempt to use the fossil or the archaeological evidence to try to track down the place of origin of the Modern genotype and the beginnings of complex behaviour which is the hallmark of all Modern populations, is a reliable chronology in terms of years before the present. As yet, we do not possess this. Radiocarbon is of little use beyond 40 000 years; Uranium-series disequilibrium, Thermoluminescence, Electron Spin Resonance and the

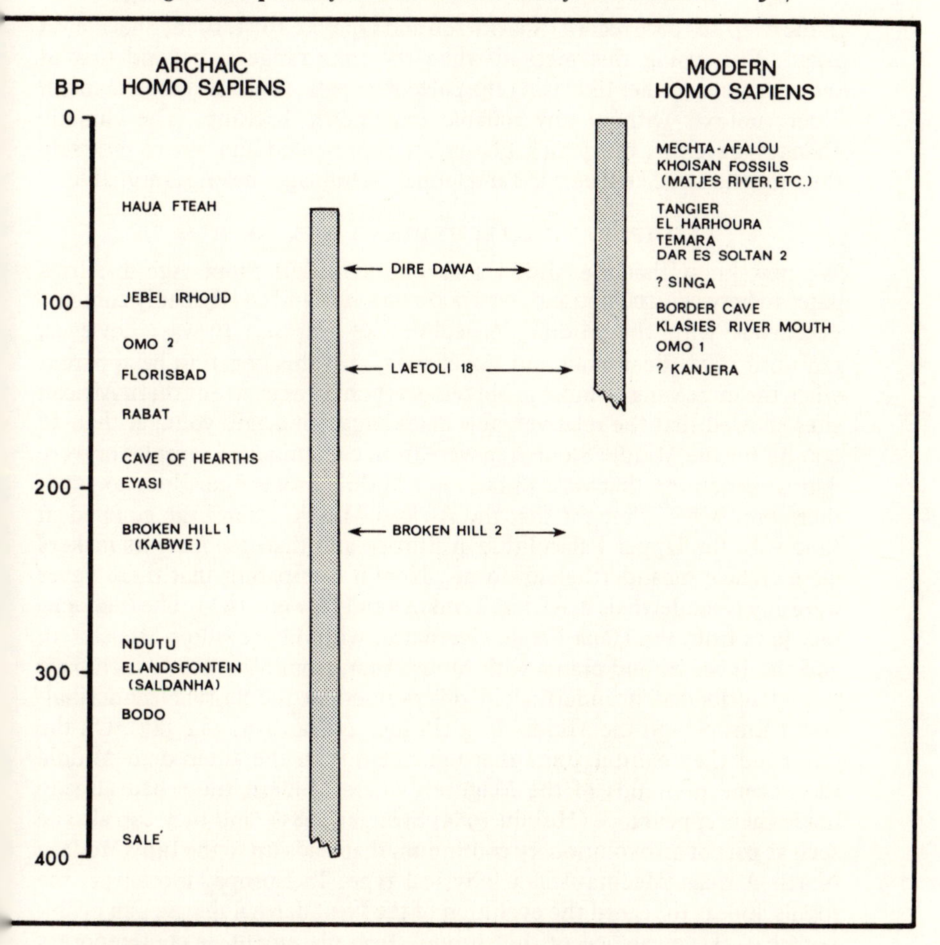

Figure 29.2. Probable chronological positions of African Archaic and Modern human fossils.

Aspartic Acid methods of dating are not yet as precise or as reliable as they need to be, and Potassium-argon, even the laser method, is not yet proven to be able to provide reliable absolute dates in the 100 000 year time-range. This is the major problem that palaeoanthropologists and prehistorians throughout the world are facing today, and until it is possible to make correlations in which we can have confidence between widely separated parts of the Old World, any attempted reconstruction of the series of events leading to the appearance of Modern humans must remain unproven.

This does not, of course, mean that we should not examine the evidence we have to see to what extent it may help to eliminate some possibilities while emphasizing the likelihood of others. Much can be and has already been done to order into a time sequence the archaeological and fossil assemblages using the evidence for changing climates and environments in the later Pleistocene seen in the sequence of oxygen-isotope stages present in the deep-sea core record (Shackleton and Opdyke 1973, 1976; Shackleton 1977). But, using this method when the time range is beyond that of radiocarbon is rather like using the palaeomagnetic reversal record in much older contexts without any reliable radiometric back-up. The climatic changes are clear, but which phases are represented and where precisely they and the associated cultural and faunal assemblages belong is arguable.

COMPARATIVE TECHNOLOGY AND FOSSILS

We now know that the Middle Palaeolithic/Middle Stone Age in Africa dates to between 100 000 and ≥ 30 000 years ago, and so is broadly contemporaneous with the Middle Palaeolithic of Eurasia. It was, however, not until 1972 (Beaumont and Vogel 1972) that this began to be apparent when the increasing number of old radiocarbon dates from Southern African sites showed that the relatively few dates suggesting ages younger than 30 000 BP for the Middle Stone Age were from contaminated samples or were dating something that was, in fact, not Middle Stone Age. Prior to 1972, therefore, it was thought that the African Middle Stone Age equated in time with the Upper Palaeolithic in Europe and that *ipso facto* its makers were archaic 'neanderthaloid' forms. Now it is apparent that there never were any Neanderthals in Africa (Trinkaus and Howells 1979) The fragmentary jaws from the Haua Fteah, Cyrenaica, with the Levallois-Mousterian and the Jebel Irhoud crania with Mousterian, from Morocco, which have been described as 'neanderthaloid' do not resemble the classic Neanderthalers of Europe and the Middle East (Bräuer 1984a: 359, 384-387). On the one hand they exhibit traits that link them with the antecedent Middle Pleistocene hominids of the Maghreb where Modern traits had already made their appearance (Hublin 1985; Stringer 1985), and they can also be seen as part of an evolutionary continuum that leads up to the fully Modern North African Mechta-Afalou physical type. In Europe, moreover, the fossils appear to record the evolution of the Neanderthal lineage, an evolution that, if we can accept the chronological placement, is contemporary with the anatomically-Modern *Homo sapiens* in Africa. If the Southern African Border Cave fossils are not as tightly tied into the stratigraphy and archaeology as could be desired, this does not seem to be the case for the Klasies River Mouth fossils that are clearly associated with the early Middle Stone Age (Singer and Wymer 1982: 139-149). The artifacts associated with the Laetoli 18 (Ngaloba Beds) cranium (Day *et al.* 1980) belong to the Middle Stone Age but are, as yet, undescribed (Harris and Harris 1981). Middle Stone Age artifacts are present in the sediments that produced the Kanjera crania (D. D. Davis, pers. comm.); the Dire Dawa mandible fragment clearly belongs with the Middle Stone Age occupation of Porc Epic Cave (Clark and Williamson 1984). The cranial remains of three individuals from Dar es Soltan Cave 2 (Debénath 1975), together with the remains from the Témara (Roche and Texier 1976) and El Harhoura (Debénath *et al.* 1982) Caves, belong to the robust Mechta-Afalou type of Modern hominid (Ferembach 1976a, 1976b; Debénath, in press) and are directly associated with Middle and Upper Aterian. The Aterian in Morocco is dated by radiocarbon to between 23 700 and >40 000 BP (Debénath *et al.* 1986). The oldest epi-Palaeolithic assemblages in the Maghreb date to 21,900±400 BP at Taforalt (Debénath *et. al.* 1986) and 20 600±500 BP at Tamar Hat (Saxon *et al.* 1974). The Aterian in the Western Desert of Egypt, from Bir Tarfawi, has now been dated by the Uranium-series disequilibrium and Thermoluminescence methods to >100 000 BP (Wendorf, pers. comm.). The Aterian appears to be a direct derivative in Northwest Africa and the Sahara out of the Mousterian. In some regions the new Aterian elements (e.g. tangs and Upper Palaeolithic forms) make their appearance relatively early, as at El Guettar in Tunisia (Gruet 1954), or Bir Tarfawi/Bir Sahara (Wendorf and Schild 1980: 55-80). In others, such as Cyrenaica (McBurney 1967), the characteristic Aterian forms of tool are either very rare or absent (Chmielewski 1968). The Middle Palaeolithic in the Middle East appears to have the same time range (Jelinek 1981, 1982) and an early Middle Palaeolithic from the Rajasthan Desert in India is now dated by Thermoluminescence to *c.* 163 000 BP (Misra 1987: 12).

If, therefore, the overall transition from Lower to Middle Palaeolithic took place at approximately the same time in the more southerly parts of the Old World, why was it that the Modern genotype evolved in Africa rather than elsewhere – if, indeed, it did originate there? What is it possible to see in the African cultural record that is not also present in that from Eurasia? Since the original Modern group must have been small, some time must have elapsed for the improved technology attendant upon the behaviour of Modern humans to become recognizable in the archaeological record, and the chances of finding occurrences that lie near the threshold are likely to be small. If, however, technological changes were sufficiently innovative and revolutionary, then they can be expected to have spread as rapidly as did the migratory Modern population. A major problem to begin with in recognizing technical innovation is that much, or most, of this probably found expression in material objects made of perishable substances – wood, bark, basketry, etc. – that have not survived. Essentially what we have are stone artifacts, and these may well have been only a minor part of the tool-kits of early hominids. Bone does survive but, with the exception of the bone point with the Howiesons Poort at Klasies River Mouth (Singer and Wymer 1982: 115-116), the seven, apparently shaped 'daggers' of split and ground pig tusk from Border Cave (Beaumont *et al.* 1978: 412), the bone point and utilized ivory pieces from Broken Hill (Kabwe) (Clark *et al.* 1950) and the notched, minimally modified, opportunistic bone pieces from Klasies (Singer and Wymer 1982: 114-115) and the Apollo 11 Shelter, Namibia (Volman 1984: 215), bone tools with the Middle Stone Age are non-existent. This does not necessarily mean that the Middle Stone Age hominids had not reached a stage of perceiving the advantages of bone and antler technology. It may mean nothing more than that, where hardwoods are readily available, as in the African tropics and sub-tropics, they preferred to use wood which was also easier to work than bone, while antler was virtually absent from the continent. I still believe that more detailed analysis of stone artifacts made in relation to their vertical and horizontal stratigraphic contexts, examined in conjunction with controlled experiment to reconstruct the technological reduction process together with refitting and use-wear studies, will enable us to identify more exactly the times and nature of these changes.

At present, all that is available is the broad outline. In Africa, as in Europe, the bifaces of the Acheulian techno-complex seem to disappear between 200 000 and 100 000 years ago and the small flake tools, scrapers, etc., long present in Acheulian tool-kits, take over but are based now on refined techniques, e.g. the true Levallois cores in place of the proto-Levallois forms, and true blade cores. In the final Evolved Acheulian (e.g. stages VII and VIII of the Moroccan sequence: Biberson 1961: 335- 398), at El Ma el Abiod (Algeria) (Balout 1955: 224-231), the Fauresmith (South Africa) (Goodwin 1929: 71-94; Sampson 1972: 52- 59), as also the Mousterian of Acheulian Tradition in Europe (Bordes 1968: 98-120); or the Jabrudian of the Middle East (Rust 1950: 11-38), the handaxes become more diminutive, cordiform or triangular bifaces with carefully retouched butts suggesting perhaps use as adzes. Other bifaces emphasize fine points for piercing, while the bifacial backed-knife form suggests more certainly use for cutting and sawing. Some of the most informative assemblages are those from the mound springs in the eastern desert (Caton Thompson 1952: 54-73; Schild and Wendorf 1977: 17-100; 1981: 72-85). Might it be possible to see in these changes a greater emphasis on the use of wood and its by-products – bark and resin? Incidentally, the Sangoan assemblages from East and Central Africa – as yet undated but certainly all >40 000 years BP (Van Zinderen Bakker and Clark 1962; Clark 1974: 79) – and the Sangoan Njarasa Industry from the Eyasi Beds associated with Archaic *Homo sapiens* fossils in the Eyasi section of the Rift in northern Tanzania, estimated to be >130 000 years BP (Mehlman 1987) suggest an equal emphasis on the working of wood (Clark 1964). Use-wear studies of primary context artifacts should help to resolve on what material these different tools were used but, until it becomes possible to correlate *precisely* primary context assemblages and to examine their micro-typology and technology, it is unlikely that the stone artifacts alone will be able to contribute more.

What distinguishes the Middle Palaeolithic/Middle Stone Age of Africa from the Lower Palaeolithic/Earlier Stone Age? The chief difference in the artifact assemblages lies in the basic similarity exhibited by all those of the Earlier Stone Age. It is true that they are infinitely variable within the dual traditions (Acheulian/Developed Oldowan) that interdigitate and intermingle on the same activity areas at a multiple context site, but they are basically the same throughout the whole of the Old World where these great techno-complexes are found (Clark 1975). There are not many ecosystems in Africa in which they do not occur but they do not vary significantly, except in proportions, whether their makers were exploiting the grassland or woodland savanna, the steppe, montane grasslands or the sea coast. With the Middle Palaeolithic/Middle Stone Age, this changes, and we begin to see broad, regional specialization starting with the beginning of the later Pleistocene or Last Interglacial. This was a highly favourable time *c.* 125

000-80 000 BP, as has been well demonstrated (see Deacon, this volume), and not unlike the present inter-glacial period. The evolved Acheulian spread into the Sahara and quickly developed into regional forms of Middle Palaeolithic; it moved also into the empty area of the Congo Basin from which the forests had begun to retreat during the late Middle Pleistocene (Hamilton 1979). As more and more of the forest was replaced by woodland and grassland, so the human populations developed their very specialized stone tool-kits: some of the end-products of which – the long lanceolates – are among the finest examples of Palaeolithic stonework. Dates for the Lupemban lie between 38 000 and >30 000 BP (Van Noten 1982: 46-50), but are minimal ages.

A phenomenon that was apparent throughout the continent 100 000 or more years ago is the appearance of blade technology. Blades were sometimes present with the Acheulian and form small but significant components of the Evolved Acheulian of Morocco. In some regions, however, with the Middle Palaeolithic/Middle Stone Age, blades, together with long, triangular flakes struck from prepared cores, become a major primary form (e.g. Wendorf and Schild 1974; Sampson 1974: 151-175). To a great extent I believe the raw material has an important bearing on whether blades are the primary form of blank or not. Some materials lend themselves to the manufacture of blades – quartzite, for example, on the South African south coast sites and in the Sahara; hornfels (indurated shale) on the interior plateau of South Africa; obsidian in Ethiopia and flint/chert in Cyrenaica. The choice and the technical ability to produce blades as the most versatile standard form of blank stem from the perception of this advantage and the learned skills that lay within the brain of the producers. Once demonstrated, however, it is not difficult to acquire the skills to manufacture artifacts. This learning process needs little verbal instruction but can be acquired by observation and practice. In fact, in Africa, this is the way that young potters, for example, learn to become experts with minimal verbal instruction involved (I. Herbich, pers. comm.).

These blade and flake-blade blanks were, no doubt, used in a variety of ways but one of the characteristics of the early assemblages is the absence or rarity of retouch. Edge-damage and retouch are certainly present on the Ethiopian obsidian tools but this is probably more often the result of damage of easily fractured edges than intentional resharpening by retouch. Retouch to modify significantly the primary shape of a piece is more especially associated with the later Middle Palaeolithic/Middle Stone Age assemblages (e.g. Cave of Hearths (Mason 1962: 262-268) or Porc Epic Cave, Diré Dawa (Clark and Williamson 1984: 50-59)). It is possible that, at the beginning of the Middle Palaeolithic/Middle Stone Age, flakes and blades were not hafted. Certainly, the large flake-blades with the MSA I from Klasies River Mouth or with the Lower Pietersburg from the Cave of Hearths, would have been more effective unmounted. In the Middle Palaeolithic of North Africa and the Middle East, removal of striking platforms and reduction of the proximal ends of flake- blades and Levallois points is usually, and probably correctly, taken as evidence of hafting (Balout 1965). The earliest irrefutable evidence, however, is to be seen in the Aterian tang (Tixier 1967: 786-797). Aterian points, with the exception of the *pointe marocaine,* are usually thick and would have been relatively clumsy equipment if intended for use as a projectile point; it is more likely they were hafted for use as a hand-held cutting and scraping tool. Use-wear analysis of Later Mousterian flint artifacts from Kebara Cave, Mount Carmel by Shea (this volume) showed use for light duty activities such as working wood, bone and skin but also, where Levallois points are concerned, they were used for butchery and as projectile points.

Kebara is late, between 50 000 and 60 000 BP (Valladas *et al.* 1987), but the Qafzeh hominids are thought, on evidence from micro-mammals and TL dating, to be *c.* 90 000-100 000 years BP (Bar-Yosef, this volume). The Gademotta site in the Ethiopian Rift is, on Potassium-argon dating evidence, between 140 000 and 180 000 years old (Wendorf *et al.* 1975). Obsidian was the raw material used, and characteristic are Mousterian-type side- scrapers and unifacial and bifacial points. The unifacial points more usually retain the striking platforms showing removal from Levallois cores. Side-scrapers and points co-vary in percentages through time, and Wendorf and Schild suggest that they are cutting tools, mainly knives. Blades are a significant form and increase through time (Wendorf and Schild 1974: 158). In the Congo Basin, the long, early Lupemban, bifacial lanceolates are yet another expression of variability in Middle Stone Age Africa, and they are not produced by the Levallois method (Clark 1966). While we are unable to order all these various regional assemblages in a precise chronology, we still need to ask what does this significant regional variability mean? In part, it would seem to be related to the kind of raw material used – the size, shape, texture and flaking properties. Quartzite is hard and resistant and holds an edge well; flint is much the same but less hard and is likely to give a more regular edge microscopically; obsidian produces the sharpest edge that can be obtained in any material but it is easily broken and so, to produce a resistant edge in obsidian that will stand up to moderate usage, it is necessary to retouch it. Again, some materials are best suited for producing blades, some flakes, and so on. Possibly, therefore, subtle changes in form and style through time can best be seen in regions where a single raw material – such as flint, chert, obsidian or hornfels – was employed.

But raw material can hardly be the whole answer as to why there is so much variability in Africa during the Middle Palaeolithic/Middle Stone Age. The main reason must, I feel, be sought in the different ways the populations adapted their way of life to make the best use of the resources of their environment, which would have required different ways of organizing social and economic behaviour, different kinds of activity places and patterns, different sizes of territories, and different emphasis on what technological equipment was used. As skills and strategies for improved food-getting evolved, so the number of alternative choices of how to use resources more efficiently, and what new ones to bring into use, must also have grown. Moreover, with the changes in world climate during the early late Pleistocene, readjustments were essential, especially in times of harsher conditions as during a glacial. Such long-time adaptation must have contributed to, if not been largely responsible for, the two widely dispersed Modern populations of the late Pleistocene – the Mechta-Afalou physical type in Northern Africa and the Khoi-San physical type in Southern Africa. The origins of both are now clearly seen as being associated with the Middle Palaeolithic/Middle Stone Age (Debénath et al. 1986: 236; Rightmire 1984: 321). The Mechta-Afalou physical type was at one time thought to be restricted to the Maghreb but can now be seen to have been much more widely spread and to have been present in Nubia and Upper Egypt in the late Pleistocene, *c.* 20 000 years BP (Anderson 1968; Greene and Armelagos 1972; Wendorf and Schild 1986). Pertinent to the origins of the early Upper Palaeolithic 'Cro-Magnon' population of Western Europe, is the close resemblance this physical type bears to the skeletal remains of the Mechta-Afalou people (Arambourg *et al.* 1934). Does this suggest, perhaps, that the Cro-Magnon population is descended from an ancestral group that was already present in the Maghreb by the later Middle Palaeolithic?

PRE-AURIGNACIAN AND HOWIESONS POORT

Although blades have been an integral part and in some cases the most significant forms found with Middle Palaeolithic assemblages from the beginning, the most striking examples of technological change in the Middle Palaeolithic/Middle Stone Age in Africa are to be seen in the Pre-Aurignacian/Amudian of the Southeast Mediterranean and the Howiesons Poort Complex in Southern Africa south of the Zambezi. They appear to be more or less contemporary and are assigned either to a late oxygen-isotope stage 5 or to early stage 4, *c.* 80 000-75 000 years BP. Both are described by other participants in this Symposium (Deacon, Klein, Parkington) but it is necessary to stress that, while blade technology is rightly emphasized as the most significant characteristic of both, the usual Middle Palaeolithic prepared-core methods are *also* present. Indeed, at the Haua Fteah, there are, as well, the tips of what appear to be bifaces, suggesting that an 'Evolved Acheulian' tradition not unlike that of stage VIII in Morocco might be immediately ancestral (McBurney 1967: 89). The Levallois and disc core methods were also present in both North and South Africa prior to the Pre-Aurignacian and Howiesons Poort occurrences and they persist during this time and become dominant again after the blade complexes are replaced by conventional Middle Palaeolithic/Middle Stone Age technology.

The earliest Upper Palaeolithic-type blade industries appear in Africa in Cyrenaica and the Nile Valley between 40 000 and 30 000 years BP (McBurney 1967: 135-184; Wendorf and Schild 1976: 243- 250; Vermeersch *et al.* 1982). Some, like myself, tend to see a functional cause for the earlier, interpolated, blade-dominated occurrences within the Middle Palaeolithic and to look to palaeoenvironmental factors as a possible cause of economic and technological change. Climatic fluctuation in the form of a major cold period and regression of sea level is certainly demonstrated at Klasies River Mouth (Deacon, this volume) and a similar change is testified to at Haua Fteah when the Pre-Aurignacian is replaced by the Levallois-Mousterian in oxygen-isotope stages 4 and 3. Faunal changes are apparent and shellfish and marine animals (for example at Klasies) appear to be replaced as major resources by geophytes and grassland herbivores (Deacon, this volume). At Haua Fteah, changes in the terrestrial mammalian fauna between the Pre-Aurignacian and the Levallois-Mousterian appear more like those of degree. The significant features of the Pre-Aurignacian fauna are the marine molluscs, collected presumably for food, which are absent from the Mousterian and early Upper Palaeolithic levels (McBurney 1967: 16-71; Klein and Scott 1986). Similar shellfish are also recorded from Aterian coastal sites in Morocco and Algeria and with the Mousterian at Devil's Cave, Gibraltar. While the molluscan remains from the South African coastal sites represent the oldest known sea foods in a Pleistocene context (Klein and Scott 1986: 520-21; Volman 1978), no simple correlation of this early blade technology and artifacts for the exploitation of sea foods is possible since the technology is also found inland in the Levant and Southern Africa where other resources must have been used (Rust 1950).

Another explanation favoured by Deacon (this volume) is that of stylistic change, and he points to such forms as stubby, triangular points and snapped blade segments as examples of stylistic preference earlier than the Howiesons Poort backed segments. The selection of silcrete and chert for making these involved transport of raw material over several kilometres, however, to the Klasies River Mouth site, so that style alone may not be the sole answer, but rather that some behavioural advantages are involved such as new ways of mounting the segments as knives or the cutting parts of spears. Traces of mastic are present on a utilized blade from the MSA 3 levels at Apollo 11 Cave, Namibia (Volman 1984: 215). A spear with cutting parts that leaves a blood spoor and weakens the animal in addition would have a considerable advantage over a wooden pointed spear that, while it penetrates, can only be fatal if it strikes a vital organ. There may also be a change here from short stabbing spears (for example, of the Clacton type) to lighter, throwing spears. If, however, blade technology is so superior to flake technology, as the Upper Palaeolithic tool-kit suggests, why did these early manifestations of it disappear *c.* 80 000-75 000 years ago, only to reappear some 35 000 years later? Until a considerably more refined timetable of events and more 'in-depth' analysis of occurrences become available, there appears to be no satisfactory answer. One possible explanation might be that there were at that time no efficient methods of hafting and that blades, struck by direct percussion only, and hand-held, were no more efficient than flake forms, which, indeed, replaced them, until efficient methods of mounting were introduced ⩾ 50 000 BP. Some of the blades appear to have been punch-struck, resulting in a lighter and more standard pre-form from which numerous types of retouched tool were produced for mounting as the working parts of composite tools.

DISCUSSION: BEHAVIOURAL TRAITS

Anatomically-modern humans are associated with Middle Palaeolithic/Middle Stone Age occurrences in Africa well before 40 000 years BP. It is, therefore, necessary to try to recognize which, if any, of the traits that are well seen later in the products of the complex behaviour of Eurasian Upper Palaeolithic societies can also be seen in any antecedent of the Middle Palaeolithic/Middle Stone Age traditions. Most of them appear to be present to some rudimentary or less evolved degree, if the interpretations usually placed on the evidence are close to being the right ones. Alternative explanations surely exist, however, and it is necessary to take a leaf out of the book of the prehistorian investigating the activity places of the earliest tool-making hominids and, after identifying possible alternatives, to narrow these down by a process of actualistic testing and elimination.

Some traits distinguish Middle Palaeolithic occupation sites from those of the Acheulian and show a pattern similar to that of the Upper Palaeolithic/ Later Stone Age. The most significant is the general 'sameness' to be seen in the Lower Palaeolithic artifact assemblages, and the absence of chronological and spatial specialization in the Earlier Stone Age. For example, there is nothing to compare with the regional specialization seen in the Howiesons Poort, the Lupemban and the Aterian. While Lower Palaeolithic groups *did* make use of caves, this was a late development and they do not seem to have done so to the extent that Middle Palaeolithic or Upper Palaeolithic groups did. As to settlement size, I personally feel that the evidence is insufficient to make distinctions. There are large and small Lower, Middle and Upper Palaeolithic activity sites, but can we say that there is an increasing number of larger settlements during the Upper Palaeolithic/Later Stone Age? Continual use of some sites (e.g. Haua Fteah and Klasies River Mouth Caves) over long periods is clear from the accumulations of occupation debris they contain. Cave sites such as these must regularly have been used over long periods of time, presumably on a seasonal basis. The nature of cave and rock-shelter sites is to confine and preserve superimposed occupational debris within relatively limited areas. At open sites, on the other hand, there is no such constraint, and if the locality is one that was adjacent to important resources and was strategically situated for their exploitation and the pursuit of other activities, it is likely to have been used as long as these resources were present. Activity areas at open sites are generally randomly distributed *en echelon* and not superimposed, so that if the site was used over a long period it is likely to cover an extensive area. Open sites, therefore, offer a greater opportunity of being able to identify individual occupation assemblages relating to a single episode of activity. In the case of caves this is more difficult (if not impossible) except over small areas, because of the compaction of the superimposed deposits. The homogeneous assemblage is the clue to the activity it represents.

Such 'large' sites have come to be called 'base camps', but this term should not necessarily be taken to imply any long, uninterrupted period of use. More probably during the Middle Palaeolithic/Middle Stone Age, as indeed in much later times also, occupation would have been seasonal, each time of no long duration and dictated by game movements and times when plant and water resources were available and abundant. Seasonality is difficult to prove but can be inferred from the associated faunal and botanical remains. It is essential, therefore, to try to identify what resources were being exploited – terrestrial, marine, freshwater, plant or animal – for food or as raw material. In addition to regularly reoccupied sites are others that, by reason of the paucity of cultural remains, are interpreted as relating to more ephemeral, brief episodes of activity by small groups. Some of these have been identified as transitory hunting camps, such as that at Nata River, Botswana (Bond and Summers 1954), or Orangia I, Orange Free State (Sampson 1968: 25-26); or, again, as butchery places like Duinefontein, Western Cape (Klein 1976) and Kalkbank, Transvaal (Mason *et al.* 1958). Other special-purpose contexts are the shell middens at Sea Harvest, Western Cape (Volman 1978) or quarries for raw material for tools (e.g. Arkin 5, Nubia: Chmielewski 1968: 134-147) or for pigment (Swaziland: Beaumont 1973). Where plant remains are absent, as at most Middle Palaeolithic/Middle Stone Age sites, the problem of resource identification is compounded but can be attempted by the use of models based on comparative studies of the palaeogeographical evidence with modern analogues of plant communities in similar habitats (Sept 1984) and ethnographic studies of plant use by hunter/gatherers and others (Vincent 1985).

Lack of evidence for storage, the uncertainty of certain seasonal resources, the reliability of others, as well as competition for these at times of adversity and reduced availability, strongly suggest that Middle Palaeolithic/Middle Stone Age groups most probably had definite territories over which they moved in a regular routine and had priority rights to exploit from traditionally selected localities. It behooves us, therefore, to try to identify such localities using ecological models based upon detailed knowledge of contemporary biomes and other palaeohabitats.

Each site with activity debris represents, therefore, one part only of the total record of the annual use by a group of its territory, so that every site needs to be studied in relation to all the others, not as if each were an entity to itself. This is the approach advocated by Glynn Isaac (1980) and, with the archaeological data as the starting point, such an expanded strategy can be expected to open up new opportunities for collaboration with natural scientists and others, so leading to a better understanding of the motives and pressures that underlie the selective choices and adaptations that were made by the different regional populations.

In terms of subsistence, one significant change that distinguishes the Middle Palaeolithic/Middle Stone Age from the earlier periods is the use of sea foods, proven at cave and open sites in South Africa and the Maghreb and at Haua Fteah in Cyrenaica (Klein and Scott 1986). Middle Palaeolithic/Middle Stone Age populations are considered to have hunted medium to small animals and scavenged remains of larger ones (Binford 1984), but a case can be made for the hunting of large game also (Klein 1976) – and probably even by late Acheulian times or before (e.g. at Elandsfontein: Klein 1978). Deacon (this volume) believes that the quantity of carbonized plant remains within the Howiesons Poort levels at Klasies River Mouth shows the beginnings of reliance on geophytes (i.e. buried plant foods). These plant remains are not present in the earlier and later occupation deposits.

Middle Palaeolithic/Middle Stone Age, like Later Stone Age groups, made regular use of fire, and hearths are generally the same, except for the elaborate stone-built ones of the Upper Palaeolithic that appear to be absent from earlier contexts. Such regular use of fire from Morocco to the Cape, carries a strong implication for the ability also to make it.

There is some evidence for the use of habitation structures. With the Aterian in Morocco, post-holes and stone concentrations that supported posts are present at both open and cave sites, and strongly suggest the use of some kind of structure (Debénath *et al.* 1986: 236-238). The best evidence for Middle Stone Age structures comes from the Orange Free State where at Zeekoegat 21 evidence is present for a large semi-circular structure of stone blocks (Sampson 1968).

Storage of food surplus is difficult to identify at archaeological sites unless storage pits are present, since above-ground containers decay without trace, except in late contexts. Two ways of conserving meat can be mentioned, however. The first is preserving marrowbone by 'storing' the whole bone. The Ovatchimba hunters in northern Namibia, still using stone tools, conserved the bones of zebra they had killed by placing them on ledges on a cliff face with a view to returning later to retrieve the marrow (MacCalman and Grobbelaar 1965: 2). Marrow, so Louis Leakey told me, remains good to eat for as long as one month. Another possible storage method is to dry the meat, when it will remain highly nutritious for a long time. The general way of making 'biltong' is to hang strips of meat on bushes or racks and use the sun to dry it but, during the rains in Malawi, drying is done in rock shelters. The characteristic, long, ashy hearths with much broken bone found in Later Stone Age contexts there are physically the same as those used today in Central Africa. The disarticulated sections of an animal are jugged (i.e. pounded to break the bones with the meat still on them) for easier handling on the drying racks (Clark 1973). It might be suggested that the significant amount of ash, often in lenses, carbonized material and burned, comminuted bone that has been reported from Middle Stone Age levels at both Klasies River Mouth (Singer and Wymer 1982: 115) and Border Cave (Beaumont 1980: 24-26) might be evidence for meat drying, in particular by Howiesons Poort groups whose presence coincided with a time of cold and arid climate and, no doubt, fog banks along the coast.

Turning to traits symbolic of complex behaviour, these all seem to cluster in the Upper Palaeolithic, with only slender evidence in the Middle Palaeolithic/Middle Stone Age. This is particularly so with personal ornamentation, and the few examples of such with the Middle Stone Age are the *Conus* shell with the infant burial and the notched rib fragment from Border Cave (Beaumont 1973), and the notched bone splinters and incised ostrich eggshell from the Apollo 11 Shelter (Wendt 1976: 8). Border Cave is thought to preserve evidence of intentional burial (Beaumont *et al.* 1978: 414-415), and the apparent caching of human remains within piled stones against the cave wall in the Middle Stone Age levels at Mumbwa Cave (Dart and del Grande 1930-31: 389-390, 426; Jones 1940) seems to imply a care for the dead. However, no burials are known, as yet, with the Aterian. Pigment was unquestionably used in Middle Stone Age times. It was mined at Lion Cave, Swaziland (Beaumont 1973) and there are facetted crayons of haematite and ochre from many Middle Palaeolithic/Middle Stone Age sites such as Klasies River Mouth (Singer and Wymer 1982: 117) and Bambata Cave in Zimbabwe (Armstrong 1931). At the Porc Epic Cave in Ethiopia, similar facetted pieces of pigment include an ammonite which had undergone alteration to haematite (Clark and Williamson 1984: 50). As yet there is no evidence of graphic art, and presumably the colour was used for painting the body, skins or implements; nevertheless, clearly, the production of powder for paint implies that a good deal of experimentation with colour must have been taking place during the Middle Palaeolithic/Middle Stone Age. A claim has even been made for music in the form of the perforated bone tube with the Pre-Aurignacian at Haua Fteah, which is seen as a flute (McBurney 1967: 90).

Turning to the stone artifact assemblages themselves, Middle Palaeolithic/Middle Stone Age technology and typology does not seem to show the rapid change and variability through time that the Upper Palaeolithic/Later Stone Age does. Transitions and changes from one mode to another *appear* to take longer. Nevertheless, there seems to be care in the selection of raw materials for certain purposes, for example in the use of flint for certain tools at the Aterian Chaperon-Rouge sites near Rabat, which necessitated transporting it from elsewhere since there were no local sources (Debénath *et al.* 1986: 238). The same can be said for the silcretes and cherts preferred by the Howiesons Poort groups at Klasies River Mouth (see Deacon, this volume). More can be learned, it is sure, from a more precise analysis of assemblages in sealed context, including identification of sources of raw materials and their transportation in and out of a site, all of which can provide clues to the thought processes underlying the technology. Bone, except for opportunistic bone 'tools' seldom seems to have been important in Africa before Later Stone Age times, and its place is likely to have been taken by hardwoods.

In summary, therefore, the behavioural traits of the Middle Palaeolithic/Middle Stone Age in Africa can be seen as the prototype for those of the Epi-Palaeolithic/Later Stone Age there.

All the kinds of archaeological evidence we have been considering can sometimes be seen much better in Europe and Western Asia. Probably there is not quite the same degree of regional variation in the Eurasian Mousterian industries as there is in the African Middle Stone Age, other than the interdigitating or succeeding traditions in Western Europe, but the Szeletian of Central Europe is certainly sufficiently distinctive (Allsworth-Jones, this Symposium). In my opinion, the African evidence does not show any highly innovative, technological breakthrough such as might be expected from the handiwork of Modern Man. The same evidences as are present in Africa we find also in Eurasia. This brings us back to the original position: without a reliable, precise chronology and more in-depth analysis of all aspects of occupation sites and their contents, there can be

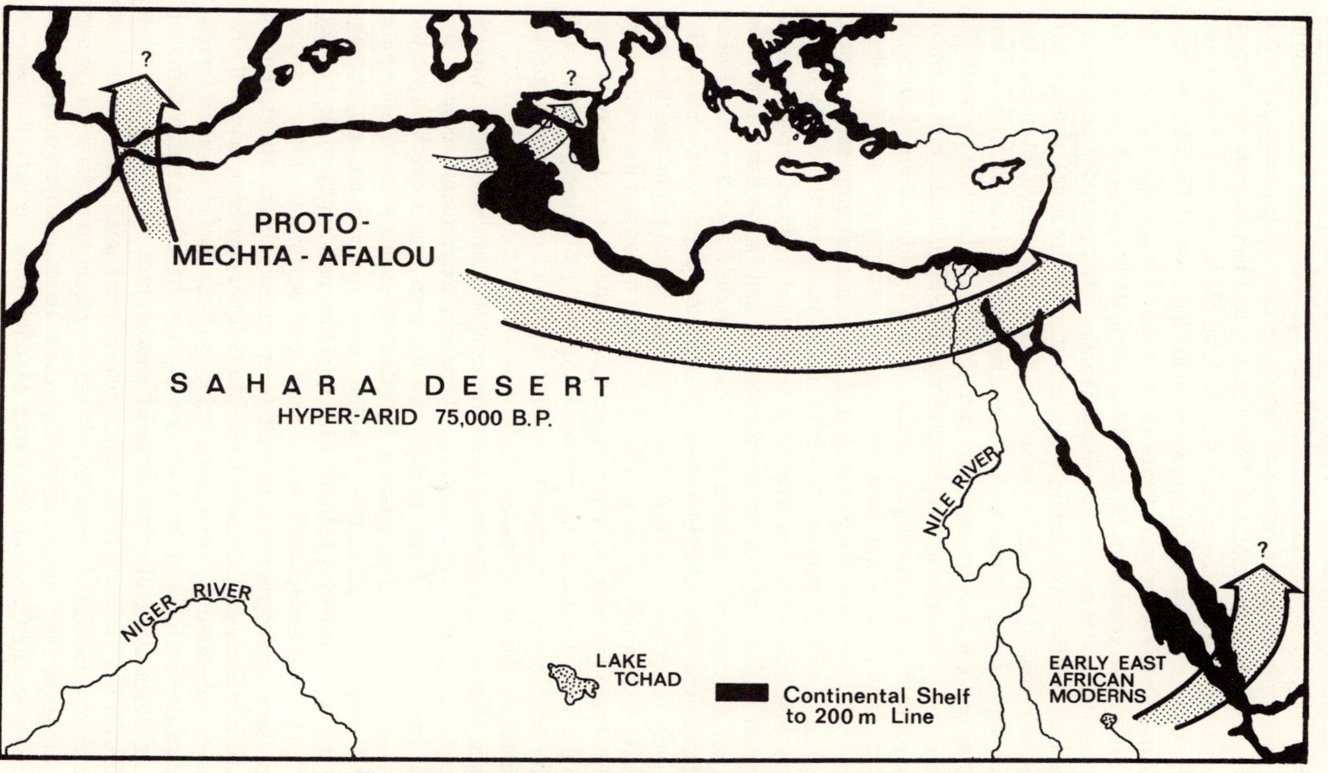

Figure 29.3. Possible migration routes out of Africa during the early Last Glacial and earlier periods of lowered sea level and of aridity in the Sahara.

no – and perhaps never will be *any* – formal answer. The fossil evidence appears to point to the emergence of the Modern genotype in both South, East and probably North Africa at a time anterior to its presence in Europe and probably also in the Middle East. On the other hand, might the African evidence be contemporaneous with the first Moderns from Qafzeh and Skūhl? Only more precise dating will provide the answer. The environmental changes coincident with a time of regressive sea-level, aridity and the onset of glacial conditions at the end of the Last Interglacial can be seen as favouring movements out of inhospitable regions, such, for example, as the Sahara, which was a region of hyperaridity during times of significant global lowering of temperatures and glaciation in the high latitudes. The Sahara was a favoured region during the time of the Last Interglacial, but with the onset of the very cold early part of the Last Glacial (oxygen-isotope stage 4: *c.* 75 000 BP), severe desertification resulted in the movement of game animals and human populations out of the desert into less harsh habitats. The effect would have been similar at earlier periods of severely lowered temperature. Such considerable lowering of temperature would have produced ice caps on the Atlas mountains, and the plains of the Maghrebian plateau would have been windy and cold and certainly not a preferred habitat for human populations. It can be postulated that the Saharan and northern plateau populations would, therefore, have been displaced, and one such movement would have been northward to the relatively restricted area of the Mediterranean coastal plains which would not have been much more extended even at the maximum lowering of sea levels. These movements (e.g. oxygen-isotope stages 4 and 5) could have brought about disequilibrium among the hunting/gathering groups already established there, resulting in competition for resources and a greater emphasis on the use of marine foods.

Such a situation could have been the catalyst that caused some groups to move out of Africa into Eurasia. There are three possible routes which could have been used into Europe and Western Asia – overland via the Isthmus of Suez, and by crossing two short stretches of open water, one from Morocco via the Straits of Gibraltar into southern Spain and the other over the Strait of Bab el Mandeb from the Horn into Southern Arabia (see Fig. 29.3). A possible fourth route can also be postulated on the evidence of its use in Holocene times, namely from Tunisia via the Mediterranean islands into Sicily and southern Italy. The Gibraltar crossing may well have involved only some 10 km of open water, and the Bab el Mandeb Straits, which today are some 40 km wide, are likely to have been much narrower during times of significantly lowered temperature, as is indicated by a confirmed low sea level of 90 m at Zeila on the Somali coast. In view of the fact that the first human populations to reach Australia some 40 000 years ago must have crossed successfully some 100 km of open water (Jones, this volume) the Gibraltar and Bab el Mandeb crossings would appear to have presented no particular problem to human groups adapted to the use of swimming logs and rafts – though even if the first Australoid population was so adapted, it does not necessarily mean of course that those in Africa were also. Moreover, strong currents at the Straits could have effectively prevented a successful crossing of even the smallest number of individuals. The apparent absence in Southern Arabia of Middle Stone Age industries comparable to those of Africa (see Note 1), and the fact that tanged artifacts do not occur in Spain until well on in the Upper Palaeolithic, suggest that these routes may not have been viable for the early Middle Palaeolithic/Middle Stone Age groups in Africa. If this was so, then movement out of the continent would have been confined to the land bridge at the Isthmus of Suez. How much interaction there was between the human populations in North Africa and those in Western Asia during Middle to early Upper Palaeolithic times still remains to be shown. Connections with Cyrenaica are likely on the evidence from Haua Fteah and other sites there, and the interchange of African and Asiatic faunas during the later Middle Pleistocene is established (Tchernov 1981; Jaeger 1975). Might not this interchange also have included human groups?

Material and behavioural traits that foreshadow the complex patterns of the Upper Palaeolithic and other late Pleistocene societies throughout the Old World had already begun to manifest themselves in the Middle Palaeolithic/Middle Stone Age in both Africa and Eurasia but, limited by the broad spectrum approach which is all we can use today, no one region appears more favourable than another *on the basis of the archaeology,* as the nuclear area where the Modern genotype emerged. My personal bias is that what made the Modern genotype so successful was the possession of a full language system, similar to our own; a system that made it possible to convey precise information and abstract ideas, while the unique stimulation to intellectual thought processes enabled the late Pleistocene populations to develop the complex behavioural systems that are apparent from their archaeological remains. Language does not, unfortunately, fossilize, so that the archaeological evidence (in chronological context) is all the more crucial to helping answer the questions of when, where, and how Modern populations replaced all other forms of humankind. It is time to begin a more systematic and rigorously designed approach to the problem; one that is based, moreover, on interdisciplinary and international collaboration.

Models that can be tested against the hard evidence from the archaeological data themselves are an important guide to the ways in which Modern humans might have evolved and spread to the ultimate exclusion of all more archaic forms of hominid. Such models may also help to show to what extent, if at all, these archaic populations might have contributed to the Modern genotype. To have any value, however, models *must* be based on the archaeological evidence in context, and this can come only from rigorous field investigations. We simply need more data, and it is clear that appreciably more *could* be obtained from survey and excavation than is actually the case. I believe it is time to leave the drawing board and go back to the field for, until we have more factual, contextual data, understanding of social and economic behaviour will remain insubstantial. More studies of distribution patterning at individual sites and interrelationships across the landscape are needed. It is necessary to identify contemporary, homogeneous assemblages on activity surfaces in order to understand their behavioural implications, and to see what meaning these may have in terms of social organization we must be able to look beyond changes in technology and style. Distribution patterns on contemporary activity surfaces are much more meaningful if they are excavated as an entity that can be visually studied in the field, and not simply reconstructed from coordinates afterwards in the laboratory. In my experience, it should also be emphasized that open sites provide the best means of identifying contemporaneous assemblages on sealed, uncontaminated horizons. The parameters of the activity area can be located, and those parts of greater density and concentrated activity are more readily identified if a large enough area is uncovered. With small excavations there exists the danger of recovering only an incomplete range of the activities that took place there. One only has to look at the plots of existing hunter/gatherers' camps to see how different activities take place at different locations (e.g. Yellen 1977).

Again, more attention is needed to understand the context of the assemblages, to determine the extent to which the original pattern may have been modified both before and after burial and what agencies – fluvial, animal, geological – contributed to this modification. To this end, more systematic, actualistic studies of, for example, surface modification of bone by carnivores and other animals need to be carried out, to distinguish this from human modification, and the ways in which water, wind or gravitation can redistribute material need to be clearly understood.

Where flint or chert is almost exclusively used, as in a large part of Europe and Western Asia, the problems of understanding the textural qualities of the raw material are appreciably less. Where a variety of other rocks in a range of sizes was used, however (as for example in much of Africa, the Far East and Southeast Asia) the flaking properties of these need to be understood, since they will have an important effect on the morphology of the primary products as well as on the nature of the retouch. Identification of sources of raw materials will show something of group movements and perhaps exchange patterns, and assemblage composition will vary depending on the distance of the site from the source. Refitting studies combined with systematic flaking experiments can provide a wealth of new understanding of the whole reduction process and also show the extent to which artifacts have been moved or transported in or out of an activity area. When such refitting studies are combined with use-wear analysis there exists a unique opportunity of reconstructing the activities of a group, where these tasks were carried out, and the sequence in which they took place. The very fine study of the late Upper Palaeolithic site at Meer, Belgium (Van Noten *et al.* 1978) shows what would be possible also with Middle Palaeolithic/Middle Stone Age occupation sites. There are still problems with some identifications, but these will be overcome and it is likely that micro-wear studies will become a specialized field for a small group of recognized experts. Use- wear studies, in particular on flint implements, provide one of the best indications of function: how and on what material the piece was used (see papers by Shea and Anderson-Gerfaud, this Symposium). Linked with studies of use-polish and micro-chipping of utilized edges is the still experimental but very exciting research on identifying organic trace elements on the edges of artifacts (Loy 1983) that can recognize blood protein on artifacts as much as 100 000 years old (Loy, pers. comm.). With these advances go also studies permitting the identification of fire and burning on sediments, bone and stone (Shipman *et al.* 1984; Clark and Harris 1985). Clearly, these and other techniques need further refinement but it is, I believe, by an increasing liaison with science and by refinement of our own recovery methods from the sites themselves, that appreciably more evidence of activities, and so of behaviour, can be extracted from the archaeological record. When this expanded data-base is tied into a chronological framework, in which we can have confidence, for the time-range 40 000-200 000 BP, the means should exist to monitor the appearance and spread of the Modern genotype with much greater certainty than is possible at the present time. A resurgence of fieldwork – of survey and excavation – leading to increased input of information from the sites themselves could be the basis for more meaningfully specific rather than general modelling and, in due course, could be the main source of information on social and economic patterning in the Middle Palaeolithic/Middle Stone Age, and on the events and processes that brought about the appearance and rapid dispersal of Modern humans.

Note

1. Levallois cores (radially prepared and point cores) together with bifaces have recently been discovered at Wadi Muqqah and Hayd al'Ghalib near Shabwa in the Hadramawt in South Yemen (Inizan and Ortlieb 1987). These appear to show resemblances to the cultural assemblages from the lower member of the Later Pleistocene *tug* (wadi) sediments in northern Somalia. It is not impossible, therefore, that a connection between South-west Arabia and the Horn of Africa could have existed at that time.

References

Anderson, J. E. 1968. Late Palaeolithic skeletal remains from Nubia. In F. Wendorf (ed.) *The Prehistory of Nubia*. Dallas: Southern Methodist University Press: 996-1040.

Arambourg, C., Boule, M., Vallois, H. and Verneau, R. 1934. *Les Grottes paléolithiques des Beni Ségoual (Algérie)*. Paris: Archives de l'Institut de Paléontologie Humaine 13.

Armstrong, A. L. 1931. Rhodesian archaeological expedition (1929): excavations at Bambata Cave and researches on prehistoric sites in southern Rhodesia. *Journal of the Royal Anthropological Institute* 61: 239-276.

Balout, L. 1955. *Préhistoire de l'Afrique du Nord*. Paris: Arts et Métiers Graphiques.

Balout, L. 1965. Données nouvelles sur le problème du Moustérien en Afrique du Nord. *Actas del V Congresso Panafricano de Prehistoria y de Estudio del Cuaternario* 1. Santa Cruz de Tenerife: 137-145.

Beaumont, P. B. 1973. The ancient pigment mines of Southern Africa. *South African Journal of Science* 69: 140-146.

Beaumont, P. B. 1980. On the age of Border Cave hominids 1-5.

Palaeontologia Africana 23: 21-33.
Beaumont, P. B., Villiers, H. de and Vogel, J. C. 1978. Modern Man in sub-Saharan Africa prior to 49 000 years BP: a review and evaluation with particular reference to Border Cave. *South African Journal of Science* 74: 409-419.
Beaumont, P. B. and Vogel, J. C. 1972. On a new radiocarbon chronology for Africa south of the Equator. *African Studies* 31: 66-89, 155-182.
Biberson, P. 1961. *Le Paléolithique Inférieur du Maroc Atlantique*. Rabat: Service des Antiquités du Maroc.
Binford, L. R. 1984. *Faunal Remains From Klasies River Mouth*. Orlando: Academic Press.
Bond, G. and Summers, R. 1954. A late Stillbay hunting-camp on the Nata River, Bechuanaland. *South African Archaeological Bulletin* 9: 89-95.
Bordes, F. 1968. *The Old Stone Age*. New York: McGraw-Hill.
Bräuer, G. 1984a. A craniological approach to the origin of anatomically modern *Homo sapiens* in Africa and implications for the appearance of modern Europeans. In F. H. Smith and F. Spencer (eds) *The Origins of Modern Humans: a World Survey of the Fossil Evidence*. New York: Alan R. Liss: 327-410.
Bräuer, G. 1984b. The Afro-European sapiens-hypothesis and hominid evolution in East Asia during the late Middle and Upper Pleistocene. In P. Andrews and J. L. Franzen (eds) *The Early Evolution of Man with Special Emphasis on Southeast Asia and Africa*. Senckenberg: Courier Forschungsinstitut 69: 145-165.
Cann, R. L., Stoneking, M. and Wilson, A. C. 1987. Mitochondrial DNA and human evolution. *Nature* 325: 31-36.
Caton Thompson, G. 1952. *Kharga Oasis in Prehistory*. London: Athlone Press.
Chmielewski, W. 1968. Early and Middle Palaeolithic sites near Arkin, Sudan. In F. Wendorf (ed.) *The Prehistory of Nubia*. Dallas: Southern Methodist University Press: 110-147.
Clark, J. D. 1964. The Sangoan culture of Equatoria: the implications of its stone equipment. In E. Ripoll Perelló (ed.) *Miscelánea en homenaje al Abate Henri Breuil:* Vol. 1. Barcelona: Casa Provincial de Caridid: 309-325.
Clark, J. D. 1966. *The Distribution of Prehistoric Culture in Angola*. Lisbon: Diamang: Museu do Dundo Publiçacões Culturais 73.
Clark, J. D. 1973. Archaeological investigation of a painted rockshelter at Mwana wa Chencherere, north of Dedza, central Malawi. *The Society of Malawi Journal* 26: 28-46.
Clark, J. D. 1974. *Kalambo Falls Prehistoric Site:* Vol. 2. Cambridge: Cambridge University Press.
Clark, J. D. 1975. A comparison of the Late Acheulian industries of Africa and the Middle East. In K. W. Butzer and G. Ll. Isaac (eds) *After the Australopithecines*. The Hague: Mouton: 605-659.
Clark, J. D. and Harris, J. W. K. 1985. Fire and its roles in early hominid life ways. *African Archaeological Review* 3: 3-27.
Clark, J. D., Oakley, K. P., Wells, L. H. and McClelland, J. A. C. 1950. New studies on Rhodesian Man. *Journal of the Royal Anthropological Institute* 77: 7-32.
Clark, J. D. and Williamson, K. D. 1984. A Middle Stone Age occupation site at Porc Epic Cave, Diré Dawa (east-central Ethiopia): Part 1. *African Archaeological Review* 2: 37-64.
Dart, R. A. and Grande, N. del. 1930-1931. The ancient iron- smelting cavern at Mumbwa. *Transactions of the Royal Society of South Africa* 19: 379-427.

Day, M. H., Leakey, M. D. and Magori, C. 1980. A new hominid fossil skull (L.H. 18) from the Ngaloba Beds, Laetoli, northern Tanzania. *Nature* 284: 55-56.

Debénath, A. 1975. Découverte de restes humains probablement atériens. *Comptes-Rendus de l'Académie des Sciences de Paris* 281: 875-876.

Debénath, A. (in press). Moroccan Aterian: Man and his tools. *Journal of Human Evolution.* In Press.

Debénath, A., Raynal, J.-P. and Texier, J.-P. 1982. Position stratigraphique des restes humains paléolithiques marocains sur la base des travaux récents. *Comptes-Rendus de l'Académie des Sciences de Paris* 294: 1247-1250.

Debénath, A., Raynal, J.-P., Roche, J., Texier, J.-P. and Ferembach, D. 1986. Stratigraphie, habitat, typologie et devenir de l'Atérien marocain: données récentes. *L'Anthropologie* 90: 233-246.

Ferembach, D. 1976a. Les restes humains de la Grotte de Dar-es- Soltane 2 (Maroc): campagne 1975. *Bulletins et Mémoires de la Société d'Anthropologie de Paris (série 3)* 13: 183-193.

Ferembach, D. 1976b. Les restes humains atériens de Témara (campagne 1975). *Bulletins et Mémoires de la Société d'Anthropologie de Paris (série 3)* 13: 175-180.

Goodwin, A. J. H. 1929. The Fauresmith Industry. In A. J. H. Goodwin and C. van Riet Lowe (eds) *The Stone Age Cultures of South Africa.* Annals of the South African Museum 37: 71-94.

Greene, D. L. and Armelagos, G. L. 1972. *The Wadi Halfa Mesolithic Population.* Amhurst: University of Massachusetts Department of Anthropology Research Reports 11.

Gruet, M. 1954. El Guettar. *Karthago* 5: 1-87.

Hamilton, A. C. 1979. The significance of patterns of distribution shown by forest plants and animals in tropical Africa for the reconstruction of Upper Pleistocene palaeo-environments: a review. *Palaeoecology of Africa* 9: 63-97.

Harris J. W. K. and Harris, K. 1981. A note on the archaeology of Laetoli. *Nyame Akuma* 18: 18-21.

Hublin, J. J. 1985. Human fossils from the North African Middle Pleistocene and the origin of Homo sapiens. In E. Delson (ed.) *Ancestors: the Hard Evidence.* New York: Alan R. Liss: 283-288.

Inizan, M. L. and Ortlieb, L. 1987. Préhistoire dans la région de Shabwa au Yemen du sud (R. D. P. Yemen). *Paléorient* 13: 5-22.

Isaac, G. Ll. 1980. Casting the net wide: a review of archaeological evidence for early hominid land-use and ecological relations. In L.-K. Königsson (ed.) *Current Argument on Early Man.* Oxford: Royal Swedish Academy of Sciences: 226-251.

Jaeger, J.-J. 1975. The mammalian faunas and hominid fossils of the Middle Pleistocene of the Maghreb. In K. W. Butzer and G. Ll. Isaac (eds) *After the Australopithecines.* The Hague: Mouton: 399-418.

Jelinek, A. J. 1981. The Middle Palaeolithic of the Levant. In J. Cauvin and P. Sanlaville (eds) *Préhistoire du Levant.* Paris: Centre National de la Recherche Scientifique: 299-302.

Jelinek, A. J. 1982. The Middle Palaeolithic in the Southern Levant with comments on the appearance of modern *Homo sapiens.* In A. Ronen (ed.) *The Transition from Lower to Middle Palaeolithic and the Origin of Modern Man.* Oxford: British Archaeological Reports International Series S151: 57-104.

Jones, T. R. 1940. Human skeletal remains from the Mumbwa Cave, Northern Rhodesia. *South African Journal of Science* 37: 313-319.

Klein, R. G. 1976. A preliminary report on the 'Middle Stone Age' open-air site of Duinefontein 2 (Melkbosstrand, south-western Cape

Province, South Africa). *South African Archaeological Bulletin* 31: 12-20.
Klein, R. G. 1978. The fauna and overall interpretation of the 'Cutting 10' Acheulian site at Elandsfontein (Hopefield) south-western Cape Province, South Africa. *Quaternary Research* 10: 69-83.
Klein, R. G. and Scott, K. 1986. Re-evaluation of faunal assemblages from the Haua Fteah and other late Quaternary sites in Cyrenaican Libya. *Journal of Archaeological Science* 13: 515-542.
Loy, T. H. 1983. Prehistoric blood residues: detection on tool surfaces and identification of species of origin. *Science* 220: 1269-1271.
MacCalman, H. R. and Grobbelaar, B. J. 1965. Preliminary report of two stone-working OvaTjimba groups in the northern Kaokoveld of South West Africa. *Cimbebasia* 13: 1-39.
Mason, R. J. 1962. *Prehistory of the Transvaal*. Johannesburg: Witwatersrand University Press.
Mason, R. J., Dart, R. A. and Kitching, J. W. 1958. Bone tools at the Kalkbank Middle Stone Age site and the Makapansgat australopithecine locality, central Transvaal. *South African Archaeological Bulletin* 13: 85-116.
McBurney, C. B. M. 1967. *The Haua Fteah (Cyrenaica) and the Stone Age of the South-East Mediterranean*. Cambridge: Cambridge University Press.
Mehlman, M. J. 1987. Provenience, age and associations of archaic *Homo sapiens* crania from Lake Eyasi, Tanzania. *Journal of Archaeological Science* 14: 133-162.
Misra, V. N. 1987. Evolution of the landscape and human adaptations in the Thar Desert. Presidential Address, Section of Anthropology and Archaeology. *Proceedings of the 74th Indian Science Congress (Bangalal)*. Calcutta: 1-24.
Rightmire, G. P. 1984. *Homo sapiens* in sub-Saharan Africa. In F. H. Smith and F. Spencer (eds) *The Origins of Modern Humans: a World Survey of the Fossil Evidence*. New York: Alan R. Liss: 295-325.
Rightmire, G. P. 1985. The tempo of change in the evolution of mid-Pleistocene Homo. In E. Delson (ed.) *Ancestors: the Hard Evidence*. New York: Alan R. Liss: 255-264.
Roche, J. and Texier, J.-P. 1976. Découverte de restes humains dans un niveau atérien supérieur de la Grotte des Contrabandiers à Témara (Maroc). *Comptes-Rendus de l'Académie des Sciences de Paris* 2821 45-47.
Rust, A. 1950. Die Hühlenfunde von Jabrud (Syrien). Neumünster: Karl Wachholtz.
Sampson, C. G. 1968. The Middle Stone Age industries of the Orange River Scheme area. *Bloemfontein: Memoirs of the National Museum* 4: 1-111.
Sampson, C. G. 1972. *The Stone Age Industries of the Orange River Scheme and South Africa*. Bloemfontein: National Museum Memoir 6.
Sampson, C. G. 1974. *The Stone Age Archaeology of Southern Africa*. New York: Academic Press.
Saxon, E. C., Close, A., Cluzel, C., Morse, V. and Shackleton, N. J. 1974. Results of recent investigations at Tamar Hat. *Libyca* 22: 49-82.
Schild, R. and Wendorf, F. 1977. *The Prehistory of Dakhla Oasis and Adjacent Desert*. Warsaw: Polska Akademia Nauk, Instytut Historii Kultury Materialnej.
Schild, R. and Wendorf, F. 1981. *The Prehistory of an Egyptian Oasis: a Report of the Combined Prehistoric Expedition to Bir Sahara, Western Desert, Egypt*. Warsaw: Polska Akademia Nauk, Instytut Historii Kultury Materialnej.
Sept, J. M. 1984. *Plants and Early Hominids in East Africa: a Study of*

Vegetation in Situations Comparable to early Archaeological Site Locations. Unpublished Ph.D. Dissertation, Department of Anthropology, University of California, Berkeley.

Shackleton, N. J. 1977. The oxygen isotope stratigraphic record of the Late Pleistocene. *Philosophical Transactions of the Royal Society of London*, B, 280: 169-182.

Shackleton N. J. and Opdyke, N. D. 1973. Oxygen isotope and palaeo-magnetic stratigraphy of Equatorial Pacific Core V28-238: oxygen isotope temperatures and ice volumes on a 10^5 year and 10^6 year scale. *Quaternary Research* 3: 39-55.

Shackleton, N. J. and Opdyke, N. D. 1976. Oxygen isotope and palaeo-magnetic stratigraphy of Pacific Core V28-239: late Pliocene to latest Pleistocene. *Geological Society of America Memoir* 145: 449-464.

Shipman, P., Foster, G. and Schoeninger, M. 1984. Burnt bones and teeth: an experimental study of color, morphology, crystal structure and shrinkage. *Journal of Archaeological Science* 11: 307-325.

Singer, R. and Wymer, J. 1982. *The Middle Stone Age at Klasies River Mouth in South Africa*. Chicago: Chicago University Press.

Stringer, C. B. 1985. Middle Pleistocene hominid variability and the origin of late Pleistocene humans. In E. Delson (ed.) *Ancestors: the Hard Evidence*. New York: Alan R. Liss: 289-295.

Tchernov, E. 1981. The biostratigraphy of the Middle East. In J. Cauvin and P. Sanlaville (eds) *Préhistoire du Levant*. Paris: Centre National de la Recherche Scientifique: 67-97.

Tixier, J. 1967. Procédés d'analyse et questions de terminologie concernant l'étude des ensembles industriels du Paléolithique récent et de l'Epipaléolithique dans l'Afrique du Nordouest. In W. W. Bishop and J. D. Clark (eds) *Background to Evolution in Africa*. Chicago: Chicago University Press: 771-812.

Trinkaus, E. and Howells, W. W. 1979. The Neanderthals. *Scientific American* 24: 118-133.

Valladas, H., Joron, J. L., Valladas, G., Arensburg, B., Bar-Yosef, O., Belfer-Cohen, A., Goldberg, P., Laville, H., Meignen, L., Rak, Y., Tchernov, E., Tillier, A.-M. and Vandermeersch, B. 1987. Thermoluminescence dates for the Neanderthal burial site at Kebara in Israel. *Nature* 330: 159-160.

Valladas, H., Reyss, J. L., Joron, J. L., Valladas, G., Bar-Yosef, O. and Vandermeersch, B. 1988. Thermoluminescence dating of Mousterian 'Proto-Cro-Magnon' remains from Israel and the origin of modern man. *Nature* 331: 614-616.

Van Noten, F. 1982. *The Archaeology of Central Africa*. Graz: Akademische Druk.

Van Noten, F., Cahen, D., Keeley, L. H. and Moeyersons, J. 1978. *Les Chasseurs de Meer*. Brugge: Dissertationes Archaeologicae Gandenses 18.

Van Zinderen Bakker, E. M. and Clark, J. D. 1962. Pleistocene climates and cultures in northeastern Angola. *Nature* 196: 639-642.

Vermeersch, P. M., Otte, M., Gilot, E., Paulissen, E., Gijselings, G. and Drappier, D. 1982. Blade technology in the Egyptian Nile Valley: some new evidences. *Science* 216: 626-628.

Vincent, A. S. 1985. *Wild Tubers as a Harvestable Resource in the East African Savannas: Ecological and Ethnographic Studies*. Unpublished Ph.D. Dissertation, Department of Anthropology, University of California, Berkeley.

Volman, T. P. 1978. Early archaeological evidence for shellfish collecting. *Science* 201: 911-913.

Volman, T. P. 1984. Early prehistory of Southern Africa. In R. G. Klein (ed.) *Southern African Prehistory and Palaeoenvironments*. Rotterdam: Balkema: 169-220.

Wainscoat, J. 1987. Out of the garden of Eden. *Nature* 325: 13.

Wainscoat, J. S., Hill, A. V. S., Boyce, A. L., Flint, J., Hernandez, M., Thein, S. L., Old, J. M., Lynch, J. R., Falusi, A. G., Weatherall, D. J. and Clegg, J. B. 1986. Evolutionary relationships of human populations from an analysis of nuclear DNA polymorphisms. *Nature* 319: 491-493.

Wendorf, F., Laury, R. L., Albritton, C. C., Schild, R., Haynes, C. V., Damon, P. E., Shafiqullah, M. and Scarborough, R. 1975. Dates for the Middle Palaeolithic of East Africa. *Science* 187: 740-742.

Wendorf, F. and Schild, R. 1974. *A Middle Stone Age Sequence From the Central Rift Valley; Ethiopia*. Warsaw: Polska Akademia Nauk Instytut Historii Kultury Materialnej.

Wendorf, F. and Schild, R. 1976. *Prehistory of the Nile Valley*. New York: Academic Press.

Wendorf, F. and Schild, R. 1980. *Prehistory of the Eastern Sahara*. New York: Academic Press.

Wendorf, F. and Schild, R. 1986. *The Wadi Kubbaniya Skeleton: a Late Palaeolithic Burial from Southern Egypt*. Dallas: Southern Methodist University Press.

Wendt, W. E. 1976. 'Art Mobilier' from the Apollo 11 Cave, South West Africa: Africa's oldest dated works of art. *South African Archaeological Bulletin* 31: 5-11.

Yellen, J. E. 1977. *Archaeological Approaches to the Present: Models for Reconstructing the Past*. New York: Academic Press.

30. Geochronology of the Levantine Middle Palaeolithic

O. BAR-YOSEF

INTRODUCTION

The dating of the Southwest Asian hominids, located as they are at an inter-continental crossroads, is of crucial importance in the current discussion of the origins of modern humans and their phylogenetic relationship to Neanderthals. The aim of this paper is to explore the various proposals for dating the archaeological deposits in which these Levantine hominids were found. Unfortunately, we have few radiometric dates for the late Middle Pleistocene and Upper Pleistocene in Southwest Asia, and those we do have are open to question (see Table 30.1). The methods currently used (such as Uranium-series and thermoluminescence) are relatively new, and are sometimes regarded as of dubious value by scholars in a situation reminiscent of the early days of the now generally accepted radiocarbon method. However, as every summary is bound to be out of date soon after its publication, I can only hope that this one will serve to stimulate further re-examination both of available Upper Pleistocene deposits and of the dating of archaeological layers, and that this in turn will contribute to the construction of a more accurate chronological framework.

The archaeological sequence of the Levantine Middle Palaeolithic is currently defined as including two major complexes – the Mugharan Tradition (formerly known as the 'Acheulo-Yabrudian', 'Yabrudian' and 'Amūdian') and the Mousterian. In earlier publications, the assemblages now included within the Mugharan Tradition were incorporated within the Lower Palaeolithic (Garrod and Bate 1937; Howell 1959; Gilead 1970; Bar-Yosef 1975). Only in recent conferences on Levantine prehistory has the 'Acheulo-Yabrudian' been classed (as suggested by Garrod 1962) as part of the Middle Palaeolithic, because of its transitional typological traits (Copeland 1975; Jelinek 1981). This terminological shift is quite important, since most Quaternary geologists and prehistorians who have worked in Southwest Asia have commonly used the European chronological scheme. In my opinion, their equation of the 'Middle Palaeolithic' with the 'Würm glaciation', and of the 'pre-Mousterian' and 'Acheulian' with the 'Riss glaciation' (oxygen-isotope stages 6-10) has distorted the geo- chronological interpretations of Levantine cave sequences.

Before examining the chronological issues, I will present a brief description of the archaeological entities involved. The earliest entity of the Middle Palaeolithic is that recently defined as the 'Mugharan Tradition' (Jelinek 1981, 1982a, 1982b; Copeland and Hours 1983). This industrial complex includes: (1) a facies rich in bifaces and in flake scrapers (which are often thick with steep or scalar retouch – the 'Acheulo-Yabrudian'); (2) a facies rich in scrapers but with few or no bifaces ('Yabrudian') and (3) a facies rich in blades, end-scrapers and burins ('Amūdian' or 'Pre-Aurignacian': e.g. Garrod 1956; Bordes 1977; Copeland and Hours 1983; Jelinek 1981, 1982a, 1982b).

The Mugharan Tradition is limited to the Northern and Central Levant. In the Southern Levant, where intensive and extensive surveys have been carried out in recent years, no assemblages related to this morphologically distinctive tradition have been reported, and only Upper or Late Acheulian sites or find spots are recorded (e.g. Rollefson 1984).

Acheulian sites with or without Levallois technique are now known from Egypt (Caton-Thompson 1952; Wendorf and Schild 1980; Wendorf *et al.* 1987). Since several Mousterian assemblages in Eastern Sahara have been dated to the pluvial conditions of the pre-Last Interglacial, it seems plausible that the earliest Mousterian in Egypt was earlier than the Northern and Central Levantine Mousterian (Wendorf *et al.* 1987).

The Early Levantine Mousterian is considered to have emerged relatively abruptly (Jelinek 1982a; Rust 1950). It is heavily dominated by blanks produced by the Levallois method, which is very rare in the Mugharan Tradition (Jelinek 1982a; Copeland and Hours 1983). The systematic use of Levallois technique began during the Late Levantine Acheulian (e.g. Gilead and Ronen 1977; Goren-Inbar 1985a, 1985b). The non-Levallois lithic tradition, which characterizes most of the facies of the Mugharan Tradition, may have come from Anatolia or the Zagros region (e.g. Bar-Yosef 1987) where such industries are predominant (Dibble 1984; Smith 1986). The same lithic tradition may be the basis of the early blade/point industry from El-Kowm (Northeastern Syria), known as the 'Hummalian'. It should be noted that the area of El-Kowm, in contrast with the Zagros and the neighbouring Taurus region, is rich in flint of good quality, which is found in large nodules (Besançon *et al.* 1982), so the morphological variability of the El-Kowm industries cannot be attributed simply to variations in the availability and accessibility of the local raw material.

Stratigraphically, the Hummalian precedes the Mousterian, which is dominated by the Levallois technique in the El-Kowm oasis. It succeeds the Yabrudian there, and is generally considered to be contemporary with the Early Levantine Mousterian. The Hummalian is characterized by a proliferation of blades and points (Copeland 1985), and thus resembles the industry from Tabūn layer D.

The most important hominid remains from Southwest Asia include the child burial at Teshik-Tash, Uzbekistan (Movius 1953), the two groups of adults in the Shanidar Cave in the Zagros mountains (Trinkaus 1983), and the skeletal remains uncovered in several sites in the Levant. These include the caves of Mount Carmel – Skhūl, Tabūn and Kebara – as well as Qafzeh

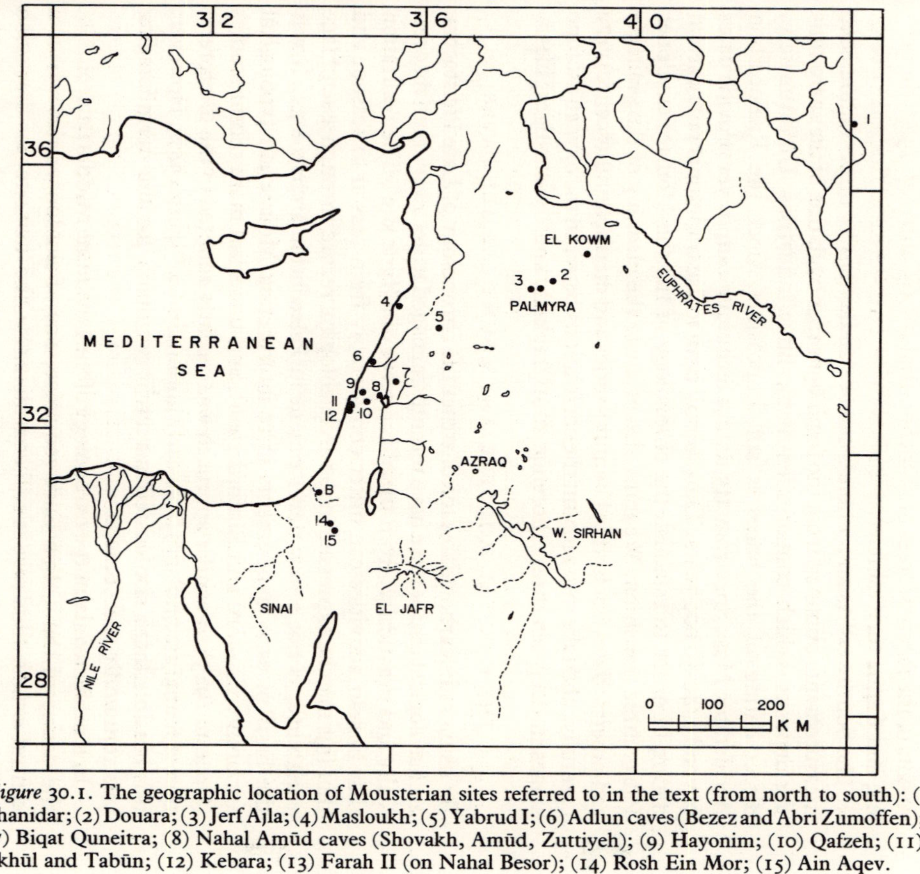

Figure 30.1. The geographic location of Mousterian sites referred to in the text (from north to south): (1) Shanidar; (2) Douara; (3) Jerf Ajla; (4) Masloukh; (5) Yabrud I; (6) Adlun caves (Bezez and Abri Zumoffen); (7) Biqat Quneitra; (8) Nahal Amūd caves (Shovakh, Amūd, Zuttiyeh); (9) Hayonim; (10) Qafzeh; (11) Skhūl and Tabūn; (12) Kebara; (13) Farah II (on Nahal Besor); (14) Rosh Ein Mor; (15) Ain Aqev.

and Amūd Caves, located about 35 km and 50 km to the east (McCown and Keith 1939; Vandermeersch 1981; Suzuki and Takai 1970). Most scholars agree that these human relics can be classified into two groups: (1) Neanderthals (Tabūn, Amūd, Kebara), and (2) 'Proto-Cro-Magnons' or early modern humans (Skhūl and Qafzeh). There is less agreement about whether the Neanderthals preceded the modern human forms, or whether the two populations were partially or fully contemporary. The dating of these fossils would provide an independent measure with which to test the results of the recent studies of mitochondrial DNA (Cann *et al.* 1987; Stoneking and Cann, this volume).

The Southwest Asian hominids can be dated either by direct radiometric measurements obtained from the fossil bones, or by dating of the surrounding deposits, which contain identifiable lithic industries. Unfortunately, direct dating of the bones is still impossible, since the Radiocarbon Accelerator Mass Spectrometry dating technique cannot yet provide dates beyond *c.* 40-50 000 years. Thus we still have to use traditional radiocarbon dating in order to establish the chronology of the the Mousterian-Upper Palaeolithic transition. We must also evaluate the meaning of the available radiometric dates (including Uranium-series and thermoluminescence) very carefully, and take into account the correlations between the oxygen-isotope stages in deep-sea cores, shoreline stratigraphies, and bio-stratigraphies.

THE COASTAL MOUSTERIAN: MOUNT CARMEL CAVES

By using radiocarbon dates to construct the chronology of Late Pleistocene archaeological entities, we have recently managed to get away, both theoretically and practically, from the old concept of a 'type site'. The realization that human activities can differ considerably from season to season, and that this may be expressed in the archaeological record through assemblage variability, is not a new development in Palaeolithic archaeology. It was the basis of early experiments in re-interpreting Mousterian typological variability (Binford and Binford 1966), and underlies more recent studies (Geneste 1985). A similar approach was recently adopted in the interpretation of faunal assemblages from early hominid sites (Speth 1987). However, it is possible that a site located in a territory suitable for long-term foraging was repeatedly reoccupied.

In the debates about the chronology of the Levantine Middle Palaeolithic, the dating of the Tabūn Cave sequence, one of the longest known, is a central issue as this cave is still considered, as the 'key site' for the region in spite of the recent trend away from using this concept. Even though the accumulating evidence indicates that, as in Europe, different archaeological entities in Southwest Asia may have been partly or completely contemporaneous, it is still expected that most of the sites within the limited area of the Levant (about 300 by 100 km), will provide the same cultural sequence. It is thus essential to begin this discussion by evaluating the dating of the Tabūn sequence (see Table 30.2).

The geological studies carried out during the last series of excavations in the Tabūn Cave (directed by A. Jelinek), which lies at 45 m above sea

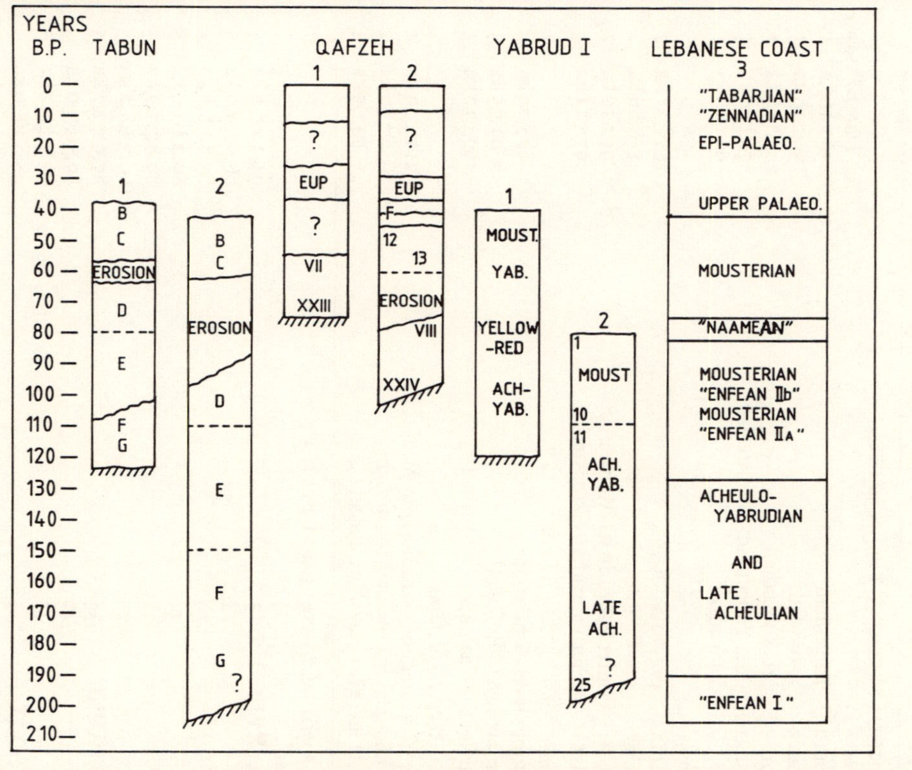

Figure 30.2. Chronological table comparing the stratigraphies of three Middle Palaeolithic sites with the sequence of the Lebanese shorelines (based on Sanlaville 1977, 1981). In each site column 1 is after Farrand (1979) and column 2 represents the version proposed in this paper.

Table 30.1. Uranium-series, thermoluminescence and radiocarbon dates for Levantine Middle Palaeolithic sites. The dates are taken from the following sources : (a) Henning and Hours 1982; (2) Schwarcz *et al.* 1980; (3) Leroi-Gourhan 1980; (4) Schwarcz *et al.* 1979; (5) Valladas *et al.* 1988; (6) Valladas *et al.* 1987; (7) Weinstein 1984; (8) Henry and Servello 1974; (9) Akazawa 1979; (10) Marks 1983. Sample materials for radiocarbon-dates were charcoal ('Ch'), bone ('B'), charred bone ('chB') and ostrich eggshell ('Ost'). (N.B. All Mousterian dates younger than 38,000 BP are omitted, on the assumption that this date already marks the onset of the Upper Palaeolithic, following the 'Transitional' Industry.)

URANIUM-SERIES (^{230}TH/^{234}U) DATES			
Kowm		*Date BP*	*Source*
Humm 2	Yabrudian	156±16 ka	1
Oumm 3	Yabrudian	139±16 ka	1
Oumm 5	pre-Yabrudian	245±16 ka	1
Oumm 4	inter Yabrudian – Mousterian	76±16 ka	1
Tell 6	Yabrudian	99±16 ka	1
Zuttiyeh			
76ZU 1a	pre-Yabrudian	164±21 ka	2
76ZU 4	pre-Yabrudian	148±6 ka	2
76ZU 1	inter Yabrudian – Mousterian	95±10 ka	2
76ZU 6	inter Yabrudian – Mousterian	97±13 ka	2
Naame			
Strombus level – Enfean II		90±20 ka	3
Strombus level – Enfean II		93±5 ka	3
vermet level – Naamean		90±10 ka	3
Nahal Aqev, fossil spring			
Layer B (average of 3 samples) pre-Mousterian		211±19 ka	4
76NZ6d-4 layer D below Mousterian		85.2±10 ka	4
76NZ1 layer D below Mousterian		74±5 ka	4

THERMOLUMINESCENCE DATES			
Qafzeh			
Layer XVII	13	94.3±8.8 ka	5
	14	106.0±9.6 ka	5
	29	107.2±8.8 ka	5
	33	89.2±8.4 ka	5
	34	87.8±7.2 ka	5
	36	100.7±8.2 ka	5
Layer XVIII	38	87.9±7.2 ka	5
	40	89.5±7.0 ka	5
	42	93.4±8.2 ka	5
Layer XIX	45	98.8±8.9 ka	5
	47	82.4±7.7 ka	5
	49	84.9±7.3 ka	5
	77	95.8±8.1 ka	5
Layer XXI	1	109.9±9.9 ka	5
	2	89.2±8.9 ka	5
	61	90.9±8.7 ka	5
Layer XXI	65	86.6±7.4 ka	5
	66	91.2±8.7 ka	5
	67	85.4±6.9 ka	5
Layer XXIII	76	95.0±7.7 ka	5
Weighted mean for the above:		92±5.0 ka	5

Kebara		
Unit VI	48.3±3.5 ka	6
Unit VII	51.9±3.5 ka	6
Unit VIII	57.3±4.0 ka	6
Unit IX	58.4±4.0 ka	6
Unit X	61.6±3.6 ka	6
Unit XI	60.0±3.5 ka	6
Unit XII	59.9±3.5 ka	6

RADIOCARBON DATES

Site	*Material*	*Sample Ref.*	*Date BP*	*Source*
Kebara	chB	Grn-2561	41,000±1000	7
Kebara (same sample)	'rest fraction'	Grn-2551	35,300±500	7
Tabūn B	Ch	GrN-2534	39,700±800	7
Tabūn C	Ch	GrN-2729	40,900±1000	7
Tabūn Unit I		GrN-7408	>47,000	7
Tabūn Unit I		GrN-7409	$51{,}000^{+4{,}800}_{-3{,}800}$	7
Tabūn Unit I		GrN-7410	$45{,}800^{+2100}_{-1600}$	7
Tabūn Bed 42	very black soil	LJ-2084	38,800±2,400	7
Geula cave B1	ChB	GrN-4121	42,000±1,700	7
Ras el Kelb	chB	GrN-2556	>52,000	7
Ksar Akil	dark clay band	GrN-2579	42,750±1,500	8
Jerf Ajla	Ch	NZ-76	42,00±2,000	8
Douarah layer E	Ch	TK-111	>43,900	9
IIIB	Ost	GrN-8638	$46{,}700^{+2{,}200}_{-1{,}700}$	9
IIIB	Ost	GrN-8058	>53.800	9
IVB	Ch	TK-165	38,900±1,700	9
IVB	Ch	TK-166	>43,200	9
IVB	Ch	TK-167	>43,000	9
IVB	Ch	TK-168	>43,200	9
IVB	Ch	GrN-7599	>52,000	9
(Fission-track IVB	Barite	Kyoto	75,000)	9
(Racemization IIIB			*ca.* 85,000)	9
(Racemization IIIB			*ca.* 110,000)	9
Rosh Ein Mor	Ost	Tx-1119	>37,000	7
Rosh Ein Mor	Ost	Pta-543	>44,000	7
Rosh Ein Mor	Ost	Pta-546	>50,000	7
Shanidar cave D	Ch	GrN-2527	46,800±1,500	8
Shanidar cave D	Ch	GrN-1495	50,600±3,000	8
Kunji cave 135cm	Ch	SI-247	>40,000	8
Kunji cave 145cm	Ch	SI-248	>40,000	8
Boker Tachtit (Earliest Upper Palaeolithic Transitional Industry)				
Level 1	Ch	GX-3642	>35,000	10
Level 1	Ch	SMU-580	47,280±9050	10
Level 1	Ch	SMU-259	44,930±2420	10
Level 1	Ch	SMU-184	>45,490	10
Level 4	Ch	SMU-579	35,055±4,100	10

level, were intended to clarify the ages of its various layers (Jelinek *et al.* 1973; Jelinek 1982a, 1982b; Goldberg 1973; Farrand 1979, 1982). The published reports make it clear that the lower part of the sequence (layers G, F and the lower half of E) 'consists predominantly of fine-grained, well-sorted sand, identical with modern dune sand' (Farrand 1979: 373). The upper part of layer E becomes increasingly silty, while the layer D deposit resembles loess. These features are taken to indicate that the 'sea level was quite low at the time, so that the local source of aeolian sand in the littoral area was stabilized: thus only long distance transport of dust, presumably from the Sinai and Egyptian deserts, contributed to the sedimentation in Tabūn' (Farrand 1979: 376).

This lower sandy unit in Tabūn (Garrod's layers G, F and part of E) contains evidence for two slumping events, apparently caused by the lowering of the water table. It was assigned to the Last Interglacial for several reasons:

1. The 39 metre beach near Tabūn identified by Michelson (1968) was assumed to be the source of the sand and could have been uplifted to its present position during the general diastrophic movements which characterize the Pleistocene (Kafri 1970).
2. There are only minor changes in the lower sedimentary cycle within the cave, except for the two minor slumping events, which would clearly indicate the scale of climatic fluctuations which could be expected during a glacial-interglacial regime.
3. Pollen evidence (Horowitz 1979) indicates a warmer climate for layers G and E, thus corroborating the dry interglacial age suggested for this unit.
4. The stratigraphy of the *kurkar* ridge, about 2 km west of Tabūn Cave (Farrand 1979: Fig. 2), has a lower cycle of sand accumulation, which ended in a long process of pedogenesis. This process was responsible for the formation of the 'red loam' soil (also known as the 'Mousterian hamra'), because of the morphological classification of artifacts uncovered in this deposit (Farrand and Ronen 1974; Ronen 1977).

The alternative chronological interpretation for the Tabūn sequence proposed here (see Table 30.2) is based on the following considerations:

1. During each of the pre-Mousterian glacials and interglacials, the shore of the Mediterranean reached the steep escarpment of Mount Carmel. The evidence for major tectonic uplifting during the Upper Pleistocene is highly debatable (Michelson 1971) and only small-scale tectonic movements along the western escarpment are discernible (Michelson 1968, 1971). Similar phenomena were observed by Sanlaville in the shorelines on the Ras Beirut promontory (Sanlaville 1977). The 35-45 m beach on Mount Carmel is well represented by numerous outcrops along the foot of Mount Carmel, including the 39 m exposure near Tabūn (Michelson 1968). Given the proximity of the cave mouth to the source of the dune sand (45 m above sea level), it is not surprising that the sand periodically accumulated in the cave over a very long period, perhaps during both interglacial and glacial conditions. The repeated retreat of the sea (isotope stages 8-7-6-5e?) could have caused the slumping phenomena observed in the deposits.

The sequence of shorelines on Mount Carmel is essentially similar to that of Lebanon (Sanlaville 1977; Slatkine and Rohrlich 1964, 1966; Michelson 1968, 1971; Horowitz 1979) and there is no definite evidence for considerable uplifting during the Upper Pleistocene on the scale of that observed in Haiti, Bermuda or New Guinea (e.g. Dodge *et al.* 1983; Chappell 1974; Aharon 1983).

2. The accumulation of the more silty upper part of layer E and the 'loessial' deposit of layer D are taken to indicate that the source of the sand (i.e. the shore) was further away from the cave at this time (as noted by Goldberg and Farrand). These conditions could be correlated with the initial deposition of the inland *kurkar* ridge, 2 km west of the cave. This was followed by soil formation, indicating vegetational cover as shown in the pollen samples from layer D. Such wetter conditions (Horowitz 1979) can be correlated with the formation phase of several inland lakes, such as Lake Lisan in the Levant (isotope stages 5e, 5d-5a?).

3. Humid conditions inside the cave may have been due to an active small spring, which would in turn indicate an even wetter climate. These conditions caused the erosion which truncated the top of layer D. Subsequently, the shape of the cave changed drastically when the 'chimney' opened and the front of the ceiling collapsed (isotope stages 5d-5a and early stage 4).

4. Layers C and B accumulated rapidly, incorporating both anthropogenic materials as well as large masses of washed-in *terra rossa* (isotope stages 4 and early 3).

To summarize: there are many ambiguities in the chronological interpretations of the Tabūn sequence. Acceptance of the currently advocated chronological scheme (Jelinek *et al.* 1973; Farrand 1979) parallels other options. The dating of the earliest industries of Tabūn (Late Acheulian and Mugharan Tradition) to the Last Interglacial (e.g. Jelinek 1982a, 1982b; Farrand 1979) resembles an early suggestion made by Howell (1959), at a time when the Western European school of prehistory viewed the Mousterian as of exclusively Würmian age.

The alternative interpretation proposed above is based on newly acquired data, including the biostratigraphic evidence, to be discussed below (Tchernov 1981, 1984, and in press), and the Uranium-Thorium dates from El-Kowm (Henning and Hours 1982) and Zuttiyeh (Schwarcz *et al.* 1979: see Table 30.1). In spite of this additional evidence it was felt that the chronological issue had not been satisfactorily resolved, and that many behavioural aspects of the Levantine Mousterian still needed investigation. These considerations led to a major excavation project in Kebara Cave, which began in 1982.

Kebara Cave is located on the western escarpment of Mount Carmel, as are the other cave sites of Wadi Mughara. During the recent excavations (Bar-Yosef *et al.* 1986), bedrock was reached in the interior chamber. The earliest layer (Unit XIV) is a sandy-silt deposit modified by standing water which subsequently eroded, possibly because of a drop in the water table (caused by a marine regression?). It was followed by the rather rapid accumulation of Mousterian layers approximately 5 metres in thickness, and rich in hearths, bones, charcoal and lithic artifacts. In Unit XII an almost complete burial of an adult male, classified as a Neanderthal, was uncovered in 1983 (Bar-Yosef *et al.* 1986). The lithic analysis (Meignen and Bar-Yosef 1988) indicates that the assemblages of Kebara are similar to the Tabūn B industry (Copeland 1975), which Ronen (1979) and Jelinek (1982a) suggested grouping with the Tabūn C industry under the taxon of Mousterian 'Phase 2-3'.

A series of thermoluminescence dates (Valladas *et al.* 1987: see Table 30.1) indicates that most of the Mousterian layers in Kebara span the range from 50 000 to 60 000 BP. This probably means that the first slumping event in the cave occurred at some time around 60 000-65 000 BP, at the transition from isotope stage 4 to 3, or perhaps even earlier at the beginning of stage 4.

Identifying isotope stage 4 in the palaeoclimatic records of the Levant and its environmental effects in both the cave and pollen sequences is of crucial importance. In the deep-sea cores, stage 4 reflects somewhat similar climatic conditions to those of stage 2. In Europe, the beginning of stage 4 is considered to represent the beginning of harsh glacial conditions. Unfortunately, in the Levant there are no well-dated palynological or deep-sea core data bearing on the climate of isotope stage 4. However, if we use the environmental data available for stage 2 in order to characterize stage 4, this would indicate that cold and dry conditions prevailed in the region (Bottema and Van Zeist 1981; Van Zeist and Bottema 1982; Goldberg and Bar-Yosef 1982; Bottema 1987). In this case the basal unit in Kebara may perhaps be correlated with the wet phase of Tabūn layer D (isotope stages 5d-5c) and the erosional phase which followed its deposition (stage 5b-5a?).

THE COASTAL LEBANESE MOUSTERIAN: MARINE TERRACES AND CAVE-SITES

The Lebanese coast provides a series of sites which are directly linked to the Pleistocene marine stratigraphy. The mountains of Lebanon drop directly into the sea, leaving either a very small coastal plain or none at all. Coastal abrasion with episodal and sporadic depositional events left a series of clear-cut marine terraces, dunal accumulations and beach-rocks. Sanlaville's detailed studies (1977) have provided a wealth of information which (when considered apart from European Alpine glacial chronology) can be used to resolve some of the current debates concerning the chronology of the Mount Carmel sites.

For the later part of the Quaternary period, Sanlaville defined a sequence of marine terraces of both transgressive and regressive character ranging in altitude between 20 m and 8 m above present sea level. He subdivided these into three transgressions termed respectively 'Enfean I and II' and 'Naamean', and tentatively dated them to the 'Riss-Würm' and 'Early Würm', in accordance with the common French sub-divisions of the time (Sanlaville 1977, 1981; Bonifay and Mars 1959; Gigout 1966).

Since there is hardly space to repeat the details of this scheme here, I shall restrict my observations to the fact that only the Enfean II shoreline contains the West African molluscan species which usually designate the Tyrrhenian faunas in the Mediterranean basin (Sanlaville 1977). *Strombus bubonius* (Lmk) is the 'guide fossil', and is accompanied in Lebanon by the following additional African species : *Arca afra* (Gm), *Tapes dura* (Gm), *Natica (Polynices) lactea* (Guild), *Natica turtoni* (Smith), *Cymatium costatum* (Born), *Bursa pustolosa* (Reeve), *Cantharus viverratus* (Kien) and *Connus testudinarius* (Wass) (Fleisch *et al.* 1971). In terms of 'archaeological time' the penetration of the Senegalian fauna into the Mediterranean was very slow. It apparently never reached the Southern Levant because of the dominant sandy environment of its shoreline, except for a rare occurrence reported from a terrace at the foot of Rosh Hanikra ridge (on the Lebanese border) and one isolated surface item on Mount Carmel (Issar and Picard 1969). Instead of molluscs, the Tyrrhenian fauna is represented by *Marginopora,* a foraminifera that inhabits warm sea water probably not more than 30 m deep (Reiss and Issar 1961). Sandy deposits which contain this species (in Mount Carmel between the 6 m and the 45 m beaches) were tentatively correlated with Tyrrhenian shorelines identified in the Western and Central Mediterranean Sea by the Senegalese shells (Issar 1968; Horowitz 1979). This chronological correlation was heavily criticized (Sanlaville 1981).

The penetration and distribution of the West African molluscs could have happened only during the warmest interglacials (Hearty 1986), which is probably why they are not found on the pre-Enfean II shorelines that offered the same rocky habitats suitable for this faunal association. Thus the suggestion made recently (Gvirtzman 1985) that the Enfean II shoreline should be correlated with isotope stage 5e is tenable. The oscillations designated by Sanlaville as 'Enfean IIb' and 'Naamean' could account for the rising sea levels in stages 5c and 5a.

Sanlaville's research (1977) clearly demonstrated that the *Strombus* beaches occur at various heights, from 14 m to 6 m above sea level. In several instances, Mousterian artifacts were found in the deposit above the *Strombus* beach conglomerate. Given the distinctive features of 'Yabrudian' scrapers, it is surprising that they did not appear above any of the *Strombus* beaches, unless they were mostly older and belong to the post-Enfean I phase (about 150 000-190 000 BP). The finds from several well-known caves, described below, illustrate this situation.

Enfean I and Enfean II shorelines were uncovered inside and in front of the caves of Ras el-Kelb as well as Bezez and Abri Zumoffen, two caves near Adlun (Garrod and Henri-Martin 1961; Roe 1983; Sanlaville 1977). A transgressive shoreline deposit without *Strombus* shells was identified at Abri Zumoffen. This 12 m beach is overlain by lithic assemblages of the Mugharan Tradition (Amūdian and Yabrudian). It was assigned to the Enfean II on the sole basis of altimetric considerations (shown by many examples along the Lebanese coast to be misleading when used as a dating method). This interpretation is undoubtedly one of the major sources of ambiguity. If the Mugharan Tradition in Abri Zumoffen is dated to post-Enfean I or to isotope stage 6 instead of to Enfean II, it would in general correspond to the time of most of Tabūn layer E (according to the interpretation advocated in this paper). In the other cave, Ras el-Kelb, the indoor beach conglomerate covered by thick Mousterian layers was traced by Sanlaville outside the cave to a *Strombus* beach.

Reliable dating of the *Strombus* shells from the Lebanese shorelines would be an important contribution to solving these problems. In this context it is worth noting that dated *Strombus* shells by Uranium-series in the Western Mediterranean were interpreted as indicating several penetrations from West Africa during isotope stage 7 until stage 5a (Hillaire-Marcel *et al.* 1986). However, both amino-stratigraphy and reservations concerning the validity of the Uranium-series dates, suggest a one-time event (Hearty 1986). Thus on the French and Italian coasts, the main *Strombus* presence is related to the Last Interglacial, and there are no definite indications that they reached the Eastern Mediterranean before this time.

To summarize: the dating of the Mousterian to the Last Interglacial or immediately after (isotope stage 5d) is supported by the site of Naame, which overlies a *Strombus* beach; by the deposits in Ras el-Kelb; and by the 'Bergy shoreline' in Beirut (Fleisch 1956, 1970; Garrod and Henri-Martin 1960; Kirkbride *et al.* 1983). This chronological interpretation simplifies the correlation between a large number of sites located in similar environments within a radius of 150 km. It implies that some archaeological entities are older than commonly thought, but this is in keeping with the dating of earlier archaeological entities. For example, the typologically Late Acheulian site of Brekhat Ram, now dated to >233 000 BP (Goren-Inbar 1985), fits in well with an estimated age of 0.5 million years for the Upper Acheulian of Gesher Benot Ya'acov (Bar-Yosef and Goren 1980). The spreading of the Upper Acheulian sequence over a much longer period than previously thought (Gilead 1970; Bar-Yosef 1975) agrees with the African chronology (Clark 1982). We may therefore cautiously conclude that the Levantine Mousterian sequence began either at the end of the Last Interglacial, during isotope stage 5e, or slightly later (stage 5d), possibly around 110-120 000 BP.

THE INLAND MIDDLE PALAEOLITHIC: CAVE SITES

Inland Middle Palaeolithic sites include both caves (Jerf Ajla, Douarah, Qafzeh, Shovakh, Amūd, Zuttiyeh, etc.), rock-shelters (Yabrud I, Abu Zif, Tor Sabiha, etc.) and open-air sites (Rosh Ein Mor, Ein Aqev, Fara II, Wadi Hasa 634 etc.). Good preservation in caves has provided rich faunal collections; bones are usually poorly preserved in rock-shelters, and are entirely absent from most of the open-air sites. Chronological correlations between these sites and their possible contemporaries in the coastal hilly range are based on mixed criteria including faunal, lithic and geomorphological considerations.

The dating of Yabrud I, (containing both Mugharan Tradition and Mousterian) and the dating of the lower layers in Qafzeh cave, which contained the hominid burials, are particularly relevant to the subject of this paper.

In the excavations of Yabrud shelter I (Rust 1950; Solecki and Solecki 1986), an 11-metre sequence of deposits was exposed. The entire sequence is built of stony *eboulis* mixed with sand (Goldberg 1971; Brunnacker 1970), and its lower portion is interrupted by a layer of well-sorted aeolian sand (*Flugsand*). The date of this dry (and apparently short) event, placed within the Mugharan Tradition (layers 25-11 in Yabrud) is important but unfortunately could not be accurately ascertained. Archaeological correlations served as the basis for two alternative interpretations (Farrand 1979), both based on the correlation of the *Flugsand* with isotope stage 5e. The generally arid conditions prevailing at Yabrud (in the rainshadow on the eastern slopes of the Anti-Lebanon mountains) suggest that the Middle Palaeolithic occupations were rather ephemeral; a longer chronology would also be in accordance with the coastal sequence. On the basis of similar evidence from the Last Glacial, the aeolian activity designates arid conditions during the maximum of the glacial; thus a correlation of the *Flugsand* event with the maximum of isotope stage 6 could provide a basis for a new geochronological interpretation, as follows:

layers 25-20 = early stage 6 (Mugharan Tradition)
layer 21 (the *Flugsand*) = maximum cold and dry of stage 6
layers 19-11 = late stage 6 (Mugharan Tradition)
layers 10-1 = stage 5 (mainly Mousterian of Tabūn D type).

The dating of the Mousterian lower layers of Qafzeh was considered problematic (Bar-Yosef and Vandermeersch 1981; Jelinek 1982b; Trinkaus 1984; Bar-Yosef 1988; Tchernov 1981, 1984, and in press). The palaeontological arguments for an early date for Qafzeh have been presented in detail in the papers cited above. They are based on the evidence collected by various zoologists about the behaviour patterns of the barn owl (*Tyto alba*), the main agent which introduces rodent bones into Mediterranean caves (Tchernov 1968; Bunn *et al.* 1982). The accumulation of rodent bones over time represents the immediate environment of the sampled site. When assemblages of rodent bones are seriated according to the principle that the older assemblages are those which contain archaic forms, while the younger ones include new arrivals, the following picture emerges:

1. The rodent assemblages retrieved from the lower layers in Qafzeh (where the human burials were uncovered) cluster with Tabūn layers F and E, despite the paucity of remains at the latter site.
2. The disappearance of two types of African rat, *Mastomys batei* and *Arvicanthis ectos* after these early layers in Qafzeh, together with the previous extinction of *Allocricetus jesreellicus* (a cold steppe hamster), is taken to indicate a faunal break.
3. The presence of a grey hamster, *Cricetulus migratorius*, in Mousterian deposits such as Tabūn C and B, Hayonim E etc. (Tchernov, in press) is considered as marking the arrival of a new species, thus indicating a later age for the relevant layers in the Mousterian sequence.
4. According to the local evolution of the Euro-Asian dormouse (*Myomimus roachi roachi*) (Haas 1973; Daams 1981), the type common in Qafzeh is earlier than that common in later Mousterian assemblages (Haas 1972,

1973; Tchernov 1981, 1984, 1986).
5. The latter contention is now supported by the thermoluminescence dates from Qafzeh (Valladas *et al.*; see Table 30.1) with a weighted mean age of 92 000±5000 BP.

Finally, two cave sites, Jerf Ajla and Douarah, located in the desertic belt of the Levant, have provided some intriguing information. Jerf Ajla, a cave in the Palmyra basin, contains two types of assemblages (Schroeder 1969). The uppermost was tentatively classified as a transitional Middle-Upper Palaeolithic industry. The rest of the sequence, about 5.5 metres in depth, is classified as Mousterian with high frequencies of blades and points, removed from uni- and bi-directional blade/point cores. It resembles the Tabūn D industry. It has been suggested that this could have been a Late Mousterian evolving into a 'transitional' industry (Schroeder 1969); this is based on the acceptance of a radiocarbon date of *c.* 43 000 BP from the Mousterian level one metre below the Upper Palaeolithic material. Goldberg (1971) suggested a rate of accumulation of one metre per 10 000 years, but if the transition to the Upper Palaeolithic occurred around 40 000 BP, a rate of 0.3 m per 1000 years gives an estimate of *c.* 60 000 BP for the beginning of the infilling of the shelter. On the other hand, if the morphological traits of the industry can also be considered as chronological markers, then most of the Jerf Ajla Mousterian should be contemporary with Tabūn layer D and within the age-range of isotope stages 5d-5c.

The second site, Douarah Cave, is mentioned here because the initial geomorphological evidence did not seem to fit in with the faunal interpretation (Payne 1983). Further fieldwork in the area (Sakaguchi 1987) concluded that the Palaeo-Palmyra Lake A was first filled with water during isotope stage 5. In spite of somewhat greater rainfall and the better vegetational cover, the area remained within the Irano-Turanian vegetational belt. The lithic industries from the cave indicate an early occupation with a Tabūn D-type assemblage, and a later Tabūn C/B type assemblage (Akazawa 1987).

The Inland Middle Palaeolithic Open-Air Sites

In situ open-air sites have been reported from various areas including the El-Kowm basin, the basalt plateau of the Golan, the desert around the Azraq oasis, the hilly region of southern Jordan and the terraces of various wadis in the Negev (e.g. Besançon *et al.* 1981; Goren-Inbar 1985b; Henry 1986; Clark *et al.* in press; Marks 1977; Goldberg 1984, 1986). Only the Negev sites have so far provided direct dating evidence.

The chronostratigraphy suggested by Goldberg indicates that Mousterian sites existed in the desert before the deposition of the 'massive, well rounded gravels. . . (which). . . point to higher and more sustained discharges under a climatic regime wetter than today's' – i.e., in the current interpretation, isotope stage 5d-5a. This reconstruction is supported by the Uranium-series dates of fossil travertines from the Ain Aqev area (Schwarcz *et al.* 1979, 1980: see Table 30.1) and by the few pollen spectra from adjacent Mousterian sites (Horowitz 1979).

The erosional phase which followed the accumulation of the gravelly unit is tentatively dated by Goldberg (1984) to 45 000-70 000 BP, and indicates arid conditions. However, in Nahal Besor the site of Fara II, which is embedded in silts, suggests that the return to somewhat wetter (and possibly colder) conditions took place before the beginning of the Upper Palaeolithic. This is also indicated by the deposits which underlie Boker Tachtit (Goldberg 1983). Perhaps the truly arid erosional phase was limited to isotope stage 4 (64-74 000 BP).

This would imply that the industries from Rosh Ein Mor and Ein Aqev which resemble Tabūn D and are regarded as 'Early Levantine Mousterian' (Crew 1975; Marks 1977; Munday 1979), date from pre- stage 4, while Fara II could be as old as 60 000 BP. Direct cultural continuation of the semi-desertic Early Mousterian into the Upper Palaeolithic Transitional industry is suggested to have occurred in the Trans-Jordanian high plateau (Marks 1983; Henry 1986).

If we consider the example of the distribution of sites in the hyper-arid region of Eastern Sahara (Wendorf and Schild 1980; Wendorf *et al.* 1987), we may conclude that human groups developed the technology which enabled them to survive in truly arid zones only during the later stages of the Upper Palaeolithic. It therefore seems unlikely that Mousterian groups would have occupied Levantine arid environments during dry periods (Marks 1983). The palaeoenvironment of Douarah Cave (Palmyra), if taken as reconstructed by Payne (1983), may perhaps serve as an example for Late Mousterian adaptation to the Irano-Turanian semi-arid steppic belt. In most other cases, the limited environmental evidence available indicates that Mousterian groups exploited the desertic region only under favourable climatic conditions, which would have been wetter than those of today.

ARCHAEOLOGICAL IMPLICATIONS

The bio-anthropological and archaeological implications of the preceding discussions can be briefly summarized in the following statements:

1. The Mugharan Tradition (or 'Acheulo-Yabrudian') is dated, on the basis of correlations between the Lebanese marine terraces and the isotope stages, to stage 6 and possibly to stage 5e (*c.* 120-190 000 BP), which fits in with the available Uranium-series readings (Table 30.1). This places the Zuttiyeh fragmentary skull in an earlier time span than previously thought (Garrod 1962; Gisis and Bar-Yosef 1974). The geographic distribution of the Acheulo-Yabrudian is limited to the Northern and Central Levant.
2. The Northern and Central Levantine Mousterian sequence began during stage 5e and/or 5d, i.e. about 110-128 000 BP. This is probably later than the dates suggested for the Middle Stone Age (South African Mousterian), the Saharan Mousterian (Wendorf *et al.* 1987) or even the early manifestations of the European Mousterian.
3. If these two statements are correct, then we must expect to find a Levantine Mousterian assemblage of a pre-Mugharan Tradition age or a contemporary one. New dates may indicate that the Acheulo-Yabrudian is of an even older age. Otherwise we will have to explain how the early Mousterian

evolved separately in Europe and Africa, avoiding the obvious crossroads of Southwest Asia (Bar-Yosef 1987) which made long distance inter-group communications a feasible option (Wobst 1976).

4. The linear, developmental sequence within the Levantine Mousterian has been used to justify the morphological ordering of the Levantine hominid fossils (e.g. Wolpoff 1983; Trinkaus 1984). This sequence however is not supported by the biostratigraphy of the rodents or by the results of recent thermoluminescence dating. Earlier suggestions to date the Qafzeh hominids, classified as 'Proto-Cro-Magnons' on morphological grounds to around 40 000 BP should be replaced by dates around 92 000 BP (Valladas *et al.* 1988).

5. It should be stressed that no simple relationship between the lithic industries and the human types can be satisfactorily demonstrated. Both Neanderthals and 'Proto-Cro-Magnons' are found with essentially similar Mousterian assemblages. Moreover, the occurrence of hearth structures in Kebara and Qafzeh, the presence of organized burials in various sites, and the similar use of artifacts as shown by edge damage (Shea, this volume) show that we cannot yet trace differences in behavioural patterns from the archaeological data alone.

6. Dating the Qafzeh hominids earlier than the Neanderthals supports the contention that the latter group penetrated Southwest Asia as the result of the rapid expansion of glacial conditions with the beginning of oxygen-isotope stage 4, at about 74 000 BP (Bar-Yosef 1988, 1989). Moreover, the dating of the 'Out of Africa' flow of modern humans, as suggested by the recent mtDNA studies (Cann *et al.* 1987), finds further support in the archaeological record.

ACKNOWLEDGEMENTS

I am grateful to my colleagues L. Copeland (Aston Rowant, England), P. Goldberg and N. Goren-Inbar (Institute of Archaeology, Hebrew University), L. Meignen (CRA, URA 28, CNRS), Y. Nir (Geological Survey of Israel, Jerusalem), W. Farrand, and J. Speth (Museum of Anthropology, University of Michigan) for their helpful comments on an earlier draft of this paper, and to Lindsey Taylor for editing it. However, any remaining ambiguities and inaccuracies are mine.

REFERENCES

Aharon F. 1983. 140 000-yr isotope climatic record from raised coral reefs in New Guinea. *Nature* 304: 720-723.

Akazawa, T. 1987. The ecology of the Middle Palaeolithic occupation at Douara Cave, Syria. *University Museum, University of Tokyo Bulletin* 29: 155-166.

Bar-Yosef, O. 1975. Archaeological occurrences in the Middle Pleistocene of Israel. In K. W. Butzer and G. Ll. Isaac (eds) *After the Australopithecines*. Chicago: Aldine: 571-604.

Bar-Yosef, O. 1987. Pleistocene connexions between Africa and Southwest Asia: an archaeological perspective. *African Archaeological Review* 5: 29-38.

Bar-Yosef, O. 1988. The date of the South-West Asian Neanderthals. In M. Otte (ed.) *L'Homme de Néandertal*. Vol. 3: *L'Anatomie*. Liège: Etudes et Recherches Archéologiques de l'Université de Liège 30: 31-38.
Bar-Yosef, O. 1989. Upper Pleistocene human adaptations in Southwest Asia. In E. Trinkaus (ed.) *Patterns and Processes in Later Pleistocene Human Emergence*. Cambridge: Cambridge University Press. In Press.
Bar-Yosef, O. and Goren, N. 1980. Notes on the chronology of the Lower Palaeolithic in the Southern Levant. In J. D. Clark and G. Ll. Isaac (eds) *Las Industrias mas Antiquas, Pre-Acheulense y Acheulense*. Proceedings of the UISPP Congress, Mexico City, 1981: 28-42.
Bar-Yosef, O. and Vandermeersch, B. 1981. Notes concerning the possible age of the Mousterian layers in Qafzeh Cave. In J. Cauvin and P. Sanlaville (eds) *Préhistoire du Levant*. Paris: Centre National de la Recherche Scientifique: 281-286.
Bar-Yosef, O., Vandermeersch, B., Arensburg, B., Goldberg, P., Laville, H., Meignen, L., Rak, Y., Tchernov, E. and Tillier, A.- M. 1986. New data concerning the origins of Modern Man in the Levant. *Current Anthropology* 27: 63-64.
Besançon, J., Copeland, L., Hours, F., Muhesen, S. and Sanlaville, P. 1981. Le Paléolithique d'El Kowm: rapport préliminaire. *Paléorient* 7: 33-55.
Besançon, J., Copeland, L., Hours, F., Muhesen, S. and Sanlaville, P. 1982. Prospection géographique et préhistorique dans le basin d'El Kowm (Syrie): rapport préliminaire. *Cahiers de l'Euphrate* 3: 9-26.
Binford, L. R. and Binford, S. R. 1966. A preliminary analysis of functional variability in the Mousterian of Levallois facies. *American Anthropologist* 68: 238-295.
Bonifay, E. and Mars, P. 1959. Le Tyrrhénien dans le cadre de la chronologie quaternaire méditerranéene. *Bulletin de la Société Géologique Française* 1: 62-78.
Bordes, F. 1977. Que sont le Pré-Aurignacien et le Iabroudien? *Eretz-Israel* 13: 49-55.
Bottema, S. 1987. Chronology and climatic phase in the Near East from 16 000 to 10 000 BP. In O. Aurenche, J. Evin and F. Hours (eds) *Chronologies Relatives et Chronologies Absolues dans le Proche Orient*. Oxford: British Archaeological Reports International Series S379: 295-310.
Bottema, S. and Van Zeist, W. 1981. Palynological evidence for the climatic history of the Near East 50 000-6 000 BP. In J. Cauvin and P. Sanlaville (eds) *Préhistoire du Levant*. Paris: Centre National de la Recherche Scientifique: 111-132.
Brunnacker, K. 1970. Die Sedimente des Schutzdaches I von Jabrud. *Fundamenta* A2: 189-198.
Bunn, D. S., Warburton, A. B. and Wilson, R. D. S. 1980. *The Barn Owl*. Calton: Poyser.
Cann, R. L., Stoneking, M. and Wilson, A. C. 1987. Mitochondrial DNA and human evolution. *Nature* 325: 31-36.
Caton-Thompson, G. 1952. *Kharga Oasis in Prehistory*. London: University of London, Athlone Press.
Chappell, J. 1974. Geology of coral terraces, Huon Peninsula, New Guinea: a study of Quaternary tectonic movements and sea-level changes. *Geological Society of America Bulletin* 85: 553-570.
Clark, J. D. 1982. The transition from Lower to Middle Palaeolithic in the African continent. In A. Ronen (ed.) *The Transition from Lower to Middle Palaeolithic and the Origin of Modern Man*. Oxford: British Archaeological Reports International Series S151: 235-255.
Copeland, L. 1975. The Middle and Upper Palaeolithic of Lebanon and

Syria in the light of recent research. In F. Wendorf and A. E. Marks (eds) *Problems in Prehistory: North Africa and the Levant.* Dallas: Southern Methodist University Press: 317-350.

Copeland, L. 1985. The pointed tools of Hummal Ia (El-Kowm, Syria). *Cahiers de l'Euphrate* 4: 177-190.

Copeland, L. and Hours, F. 1983. Le Yabroudien d'El-Kowm (Syrie) et sa place dans le Paléolithique du Levant. *Paléorient* 9: 21-38.

Crew, H. 1975. An evaluation of the relationship between the Mousterian complexes of the Eastern Mediterranean: a technological perspective. In F. Wendorf and A. E. Marks (eds) *Problems in Prehistory: North Africa and the Levant.* Dallas: Southern Methodist University Press: 427-438.

Daams, R. 1981. The dental pattern of the dormice *Dryomys, Myomimus, Microdryomys and Peridryomys. Utrecht Micropaleontology Bulletin. Special Publications* 3: 1-115.

Dibble, H. L. 1984. The Mousterian industry from Bisitun Cave (Iran). *Paléorient* 10: 23-34.

Dodge, R. E., Fairbanks, R. G., Benninger, L. K. and Maurasse, F. 1983. Pleistocene sea levels from raised coral reefs in Haiti. *Science* 219: 1423-1425.

Farrand, W. R. 1979. Chronology and palaeoenvironment of Levantine prehistoric sites as seen from sediment studies. *Journal of Archaeological Science* 6: 369-392.

Farrand, W. R. 1982. Environmental conditions during the Lower/Middle Palaeolithic transition in the Near East and the Balkans. In A. Ronen (ed.) *The Transition from Lower to Middle Palaeolithic and the Origin of Modern Man.* Oxford: British Archaeological Reports International Series S151: 105-108.

Farrand, W. R. and Ronen, A. 1974. Observations on the Kurkar-Hamra succession on the Carmel coastal plain. *Tel Aviv* 1: 45-54.

Fleisch, H. 1956. Depôts préhistoriques de la côte libanaise et leur place dans la chronologie basée sur le quaternaire marin. *Quaternaria* 3: 101-132

Fleisch, H. 1970. Les habitats du Paléolithique Moyen à Naame (Liban). *Bulletin du Musée de Beyrouth* 23: 25-93.

Fleisch, H., Comati J., Reynard P. and Elouard,P. 1971. Gisement à *Strombus bubonius* Lmk (Tyrrhénien) à Naame (Liban). *Quaternaria* 15: 217-237.

Garrod, D. A. E. 1956. Acheuléo-Jabroudien et 'Pré-Aurignacien' de la grotte du Tabūn (Mont Carmel). *Quaternaria* 3: 39-59.

Garrod, D. A. E. 1962. The Middle Palaeolithic of the Near East and the problem of Mount Carmel Man. *Journal of the Royal Anthropological Institute* 92: 232-259.

Garrod, D. A. E. and Bate, D. M. A. 1937. *The Stone Age of Mount Carmel*: Vol. 1. Oxford: Clarendon Press.

Garrod, D. A. E. and Henri-Martin, G. 1961. Rapport préliminaire sur la fouille d'une grotte au Ras el-Kelb. *Bulletin du Musée de Beyrouth* 16: 61-67.

Geneste, J.-M. 1985. *Analyse Lithique D'Industries Moustériennes du Périgord: une Approche Technologique du Comportement des Groupes Humains au Paléolithique Moyen.* Unpublished Ph.D. Thesis, University of Bordeaux I.

Gigout, M. 1966. Le quaternaire de la côte libanaise comparé à celui du Maroc atlantique. *Bulletin de la Société Géologique Française* 8: 17-20.

Gilead, D. 1970. Handaxe industries in Israel and the Near East. *World Archaeology* 2: 1-9.

Gilead, D. and Ronen, A. 1977. Acheulian industries from 'Evron on the

western Galilee coastal plain. *Eretz-Israel* 13: 56-86.
Goldberg, P. 1971. Analyses of sediments of Jerf 'Ajla and Yabrud rockshelters, Syria. In M. Ters (ed.) *Etudes sur le Quaternaire dans le Monde*. Proceedings of the 8th INQUA Congress, Paris, 1969: 747-754.
Goldberg, P. 1973. *Sedimentology, Stratigraphy and Paleoclimatology of El-Tabūn cave, Mount Carmel, Israel*. Unpublished Ph.D. Dissertation, University of Michigan, Ann Arbor.
Goldberg P. 1983. The geology of Boker Tachtit, Boker and their surroundings. In A. E. Marks (ed.) *Prehistory and Paleoenvironments in the Central Negev, Israel*, Vol. 3. Dallas: Southern Methodist University Press: 39-62.
Goldberg, P. 1984. Late Quaternary history of Qadesh Barnea, northeastern Sinai. *Zeitschrift für Geomorphologie* N.F. 28: 193- 217.
Goldberg, P. 1986. Late Quaternary environmental history of the Southern Levant. *Geoarchaeology* 1: 225-244.
Goldberg, P. and Bar-Yosef, O. 1982. Environmental and archaeological evidence for climatic changes in the Southern Levant. In J. L. Bintliff and W. Van Zeist (eds) *Palaeoclimates, Palaeoenvironments and Human Communities in the Eastern Mediterranean Region in Later Prehistory*. Oxford: British Archaeological Reports International Series S133: 399-414.
Goren-Inbar, N. 1985a. The lithic assemblage of the Berekhar Ram Acheulian site, Golan Heights. *Paléorient* 11: 7-28.
Goren-Inbar, N. 1985b. Un site Moustérien de plein air à Biqat Quneitra. *L'Anthropologie* 89: 251-254.
Gvirtzman, G. 1983. Coastal terraces and prehistoric sites in Lebanon: correlation with Quaternary sedimentary cycles in Israel and a chronological interpretation. *Abstracts of the Annual Meeting of the Israel Society for Quaternary Research, Jerusalem:* 14-20.
Haas, G. 1972. The microfauna of Djebel Qafzeh cave. *Palaeovertebrata* 5: 261-270.
Haas, G. 1973. The Pleistocene glirids of Israel. *Verhandlungen der Naturforschenden Gesellschaft* (Basel) 83: 76-110.
Hearty, P. 1986. An inventory of Last Glacial (*sensu lato*) age deposits from the Mediterranean Basin: a study of Isoleucine Epimerization and U-series dating. *Zeitschrift für Geomorphologie* N.F. Suppl. Bd. 62: 51-69.
Henning, G. J. and Hours, F. 1982. Dates pour le passage entre l'Acheuléen et le Paléolithique moyen à El Kowm (Syrie). *Paléorient* 8: 81-84.
Henry, D. O. 1986. The prehistory and palaeoenvironments of Jordan: an overview. *Paléorient* 12: 5-26.
Henry, D. O. and Servello, F. 1974. Compendium of C-14 determinations derived from Near Eastern prehistoric sites. *Paléorient* 2: 19-44.
Hillaire-Marcel, C., Carro, O., Causse, C., Goy, J. L. and Zazo, C. 1986. Th/U dating of *Strombus bubonius*-bearing marine terraces in southeastern Spain. *Geology* 14: 613-616.
Horowitz, A. 1979. *The Quaternary of Israel*. New York: Academic Press.
Howell, F. C. 1959. Upper Pleistocene stratigraphy and Early Man in the Levant. *Proceedings of the American Philosophical Society* 103: 1-65.
Issar, A. 1968. Geology of the central coastal plain of Israel. *Israel Journal of Earth Sciences* 17: 16-29.
Issar, A. and Picard, L. 1969. Sur le Tyrrhénien des côtes d'Israël et du Liban. *Bulletin de l'Association Française Pour l'Etude du Quaternaire* 6: 35-41.
Jelinek A. 1981. The Middle Paleolithic in the Southern Levant from the perspective of the Tabūn Cave. In J. Cauvin and P. Sanlaville (eds)

Préhistoire du Levant. Paris: Centre National de la Recherche Scientifique: 265-280.

Jelinek, A. 1982a. The Middle Palaeolithic in the Southern Levant with comments on the appearance of modern *Homo sapiens*. In A. Ronen (ed.) *The Transition from Lower to Middle Palaeolithic and the Origin of Modern Man*. Oxford: British Archaeological Reports International Series S151: 57-104.

Jelinek, A. 1982b. The Tabūn Cave and Paleolithic Man in the Levant. *Science* 216: 1369-1375.

Jelinek, A., Farrand, W. R., Haas, G., Horowitz, A. and Goldberg, P. 1973. New excavations at the Tabūn Cave, Mount Carmel, Israel 1967-1972: a preliminary report. *Paléorient* 1: 151-183.

Kafri, U. 1970. Pleistocene tectonic movements in the coastal plain of Israel emphasizing the Mount Carmel area. *Israel Journal of Earth Sciences* 19: 147-152.

Kirkbride, D., Saint-Mathurin, S. de and Copeland, L. 1983. Results, tentative interpretation and suggested chronology. In D. Roe (ed.) *Adlun in the Stone Age*. Oxford: British Archaeological Reports International Series S159: 415-432.

Leroi-Gourhan, A. 1980. Les analyses polliniques au Moyen Orient. *Paléorient* 6: 79-92.

Marks, A. E. 1977. *Prehistory and Paleoenvironments in the Central Negev, Israel*, Vol. 2. Dallas: Southern Methodist University Press.

Marks, A. E. 1983. The Middle to Upper Paleolithic transition in the Levant. In F. Wendorf and A. E. Close (eds) *Advances in World Archaeology* Vol. 2. New York: Academic Press: 51-98.

McCown, T. and Keith, A. 1939. *The Stone Age of Mount Carmel*, Vol. 2. Oxford: Clarendon Press.

Meignen, L. and Bar-Yosef, O. 1988. Variabilité technologique au Proche Orient: l'exemple de Kebara. In M. Otte (ed.) *L'Homme de Néandertal*, Vol. 4: *La Technique*. Liège: Etudes et Recherches Archéologiques de Université de Liège 31: 81-95.

Michelson, H. 1968. *The Geology of the Carmel Coast*. Unpublished M.Sc. thesis, Hebrew University of Jerusalem. (in Hebrew).

Michelson, H. 1971. Discussion: Pleistocene tectonic movements in the coastal plain of Israel emphasizing the Mount Carmel area (Kafri 1970). *Israel Journal of Earth Sciences* 20: 129-132.

Movius, H. L. 1953. The Mousterian cave of Teshik-Tash, southeastern Uzbekistan, Central Asia. *Bulletin of the American School of Prehistoric Research* 17: 11-71.

Munday, F. 1979. Levantine Mousterian technological variability: A perspective from the Negev. *Paléorient* 5: 87-104.

Payne, S. 1983. The animal bones from the 1974 excavations at Douara Cave. In K. Hanihara and T. Akazawa (eds) Palaeolithic Site of Douara Cave and Palaeogeography of Palmyra Basin in Syria. *University Museum, University of Tokyo Bulletin* 21: 1-108.

Reiss, Z. and Issar, A. 1961. Contribution to the study of the Pleistocene in the coastal plain of Israel. *Bulletin of the Geological Survey of Israel* 32.

Roe, D. A. 1983. *Adlun in the Stone Age* (2 Vols). Oxford: British Archaeological Reports International series S159.

Rollefson, G. O. 1984. A Middle Acheulian surface site from Wadi Uweinid, eastern Jordan. *Paléorient* 10: 127-133.

Ronen, A. 1977. Mousterian sites in Red Loam in the coastal plain of Mount Carmel. *Eretz-Israel* 13: 183-190.

Ronen, A. 1979. Palaeolithic industries. In A. Horowitz (ed.) *The Quaternary of Israel*. New York: Academic Press: 296-330.

Rust, A. 1950. *Die Höhlenfunde von Jabrud (Syrien)*. Neumunster: K.

Wachholtz.

Sakaguchi, Y. 1987. Paleoenvironments in Palmyra District during the Late Quaternary. *University Museum, University of Tokyo Bulletin* 29: 1-28.

Sanlaville, P. 1977. *Etude Géomorphologique de la Région Littorale du Liban* (2 Vols). Beyrouth: Université Libanaise.

Sanlaville, P. 1981. Stratigraphie et chronologie du Quaternaire marin du Levant. In J. Cauvin and P. Sanlaville (eds) *Prĥistoire du Levant*. Paris: Centre National de la Recherche Scientifique: 21-32.

Schroeder, B. 1969. *The Lithic Industries from Jerf Ajla and their Bearing on the Problem of a Middle to Upper Paleolithic Transition*. Ph.D. Dissertation, Columbia University. University Microfilms International.

Schwarcz, H., Blackwell, B., Goldberg, P. and Marks, A.E. 1979. Uranium series dating of travertine from archaeological sites, Nahal Zin, Israel. *Nature* 277: 558-560.

Schwarcz, H., Goldberg, P. and Blackwell, B. 1980. Uranium series dating of archaeological sites in Israel. *Journal of Earth Sciences* 29: 157-165.

Slatkine, A. and Rohrlich, V. 1964. Sur quelques niveaux marins quaternaires du Mont Carmel. *Israel Journal of Earth Sciences* 13: 125-132.

Slatkine, A. and Rohrlich, V. 1966. Données nouvelles sur les niveaux marins Quaternaires du Mont Carmel. *Israel Journal of Earth Sciences*. 15: 57-63.

Smith, P. E. L. 1986. *Palaeolithic Archaeology in Iran*. The American Institute of Iranian Studies Monograph 1. Philadelphia: The University Museum.

Solecki, R. S. and Solecki, R. L. 1986. A reappraisal of Rust's cultural stratigraphy of Yabrud Shelter I. *Paléorient* 12: 53-60.

Speth, J. D. 1986. Early hominid subsistence strategies in seasonal habitats. *Journal of Archaeological Science* 14: 13-29.

Suzuki, H. and Takai, F. 1970. *The Amud Man and his Cave Site*. Tokyo: University of Tokyo.

Tchernov, E. 1968. *Succession of Rodent Faunas during the Upper Pleistocene of Israel*. Hamburg: Parey.

Tchernov, E. 1981. The biostratigraphy of the Middle East. In J. Cauvin and P. Sanlaville (eds) *Préhistoire du Levant*. Paris: Centre National de la Recherche Scientifique: 67-98.

Tchernov, E. 1984. Faunal turnover and extinction rate in the Levant. In P. S. Martin and R. G. Klein (eds) *Quaternary Extinctions*. Tucson: University of Arizona Press: 528-552.

Tchernov, E. 1986. *Les Mammifères du Pleistocene Ancien de la Moyen Vallée du Jourdain à Obeidiyeh (Israël)*. Paris: Association Paléorient.

Tchernov, E. (in press). The succession of Mousterian faunas in the Southern Levant. In O. Bar-Yosef and B. Vandermeersch (eds) *A Mousterian Burial from Kebara Cave, Mount Carmel*. Paris: Centre National de la Recherche Scientifique.

Trinkaus, E. 1983. *The Shanidar Neandertals*. New York: Academic Press.

Trinkaus, E. 1984. Western Asia. In F. H. Smith and F. Spencer (eds) *The Origins of Modern Humans: a World Survey of the Fossil Evidence*. New York: Alan R. Liss: 251-293.

Valladas, H., Joron, J. L., Valladas, B., Arensburg, P., Bar-Yosef, O., Belfer-Cohen, A., Goldberg, P., Laville, H., Meignen, L., Rak, Y., Tchernov, E., Tillier, A. M., and Vandermeersch, B. 1987. Thermoluminescence dates for the Neanderthal burial site at Kebara (Mount Carmel), Israel. *Nature* 330: 159-160.

Valladas, H., Reyss, J. L., Joron, J. L., Valladas, G., Bar-Yosef, O. and Vandermeersch, B. 1988. Thermoluminescence dating of Mousterian 'Proto-Cro-Magnon' remains from Israel and the origin of modern man. *Nature* 331: 614-616.

Van Zeist, W. and Bottema, S. 1982. Vegetational history of the Eastern Mediterranean and the Near East during the last 20 000 years. In J. L. Bintliff and W. Van Zeist (eds) *Palaeoclimates, Palaeoenvironments and Human Communities in the Eastern Mediterranean Region in Later Prehistory*. Oxford: British Archaeological Reports International Series S133: 277-231.

Vandermeersch, B. 1981. *Les Hommes Fossiles de Qafzeh (Israël)*. Paris: Centre National de la Recherche Scientifique.

Weinstein, J. M. 1984. Radiocarbon dating in the Southern Levant. *Radiocarbon* 26: 297-366.

Wendorf, F., Close, A .E. and Schild, R. 1987. Recent work on the Middle Palaeolithic of Eastern Sahara. *African Archaeological Review* 5: 49-64.

Wendorf, F. and Schild, R. 1980. *Prehistory of the Eastern Sahara*. New York: Academic Press.

Wobst, M. 1976. Locational relationships in Paleolithic society. *Journal of Human Evolution* 5: 49-58.

31. A Functional Study of the Lithic Industries Associated with Hominid Fossils in the Kebara and Qafzeh Caves, Israel

JOHN J. SHEA

INTRODUCTION

In attempting to correlate changes in the hominid palaeontological and archaeological records, the Middle Palaeolithic of the Levant presents a seeming anomaly, having both archaic and anatomically-modern forms of *Homo sapiens* associated with the same archaeological industry. This paper explores the nature of behavioural variability among Upper Pleistocene hominids by describing patterns of tool use preserved in the Levalloiso-Mousterian assemblages from the Kebara and Qafzeh caves. The lack of behavioural correlates for physiological variability calls into question the behavioural significance of perceived phylogenetic relationships among the Upper Pleistocene hominids of the Levant.

While there is an expectation by palaeoanthropologists of concordance among the sources of information incorporated into models of hominid behavioural evolution, it is becoming increasingly clear that the hominid fossil and Palaeolithic archaeological records for the Upper Pleistocene responded to different sources of variability. Both archaic and anatomically-modern forms of *Homo sapiens* are associated with broadly similar lithic industries in Africa (Bräuer 1984; Rightmire 1984) and in the Levant (see Bar-Yosef, Vandermeersch this volume). This requires a reassessment of the relationship between physiological and technological adaptations of Upper Pleistocene hominids. Clarifying the phylogenetic and evolutionary relationships among Upper Pleistocene populations will require accurate information about the behavioural significance of Palaeolithic industrial variability.

While recent fossil and archaeological discoveries (Vandermeersch 1981; Bar-Yosef *et al.* 1986) suggest considerable physiological heterogeneity among the Middle Palaeolithic populations of the Levant, Trinkaus (1984, 1986) has argued for the regional continuity in the evolution of anatomically-modern *Homo sapiens* from their archaic predecessors. Early anatomically-modern humans, the fossils from Skhūl and Qafzeh, are described as very late (<40 000 BP) representatives of the populations associated with the Levalloiso-Mousterian industry. The archaic/Neanderthal fossils from Tabūn, Amūd and Kebara are described as the likely progenitors of these anatomically-modern humans. From functional studies of fossil anatomy,

Trinkaus has documented major changes in postcranial robusticity, reproductive strategies and the manipulative use of anterior dentition across the archaic-modern transition. Because differences in strength, mobility and metabolic demands should significantly affect the circumstances requiring tool use, it is surprising to find both archaic and modern human physiologies associated with the same Palaeolithic industry.

Technology extends the ability of an organism to persist in the face of adverse conditions with less cost and risk than for physiological responses originating on the genome. Tool use allows the physiology of an organism to remain constant until its exo-somatic adaptive strategies are overcome by environmental challenges. Within an evolutionary lineage, therefore, technological change should logically precede physiological reorganization. In order to understand why this does not seem to be the case in the Levant, the palaeoecological relations of Upper Pleistocene hominids must be reconstructed. This is only possible by examining their archaeological associations.

If the physiological variability of Upper Pleistocene hominids genuinely reflects differences in evolutionary ecological strategies, then one should expect correlation of the artifactual traces of those strategies with morphological variation. Because both archaic and modern forms of *Homo sapiens* are associated with the same Levalloiso-Mousterian technological industry, the archaeological record of the Levant challenges this hypothesis. If major behavioural discontinuities are not apparent among the industries associated with physiologically-distinct hominids, then it may be necessary to modify hypotheses of significant bio-behavioural distinctions among early Upper Pleistocene hominids.

BACKGROUND – THE LEVANTINE MOUSTERIAN

Historically, the Levantine archaeological record provided the first challenge to the equation of archaic/Neanderthal *Homo sapiens* with Middle Palaeolithic industries. The recovery of robust, anatomically-modern hominids from the excavation of the Wadi el-Mughara caves (Garrod and Bate 1937; McCown and Keith 1939) established the physiological variability of populations associated with the Levalloiso-Mousterian industry in Southwest Asia. Recent excavations at the caves of Kebara and Qafzeh have reaffirmed the diversity of Levalloiso-Mousterian populations.

Kebara Cave is located on the western face of Mount Carmel, 60 metres above sea level. The site has been the subject of excavations since its discovery by Stekelis in 1927 (Turville- Petre 1931; Shick and Stekelis 1977), most recently by the *Origins of Modern Humans in Southwest Asia Project* (Bar-Yosef *et al.* 1986). Typologically, the Levalloiso-Mousterian industries from Kebara Units VIII-XIII resemble Copeland's (1975) Phase 3 of Levantine Mousterian, characterized by the production of broad and thin lamellar flakes from both unipolar and discoidal cores. The assemblage from Unit XII is associated with the burial of an adult male Neanderthal. Unit XI, overlying the burial, has been dated by thermoluminescence to 56 000±3600 BP (Valladas *et al.* 1987).

Qafzeh Cave is located 2.5 km south of Nazareth in the Galilee Hills, 220 metres above sea level. Excavations by Neuville and Stekelis (Neuville 1951) first identified Levalloiso-Mousterian occupations at Qafzeh. More recent excavations of the Qafzeh terrace by Vandermeersch (1981) have exposed a sequence of Mousterian occupations stratigraphically associated with burials of anatomically-modern *Homo sapiens* in Units XV-XXII. The Qafzeh Mousterian collections have not been formally described, but Jelinek (1982) attributes to them a 'late' Mousterian age on the basis of comparisons with the Tabūn sequence. While microfaunal and lithostratigraphic evidence suggest an age of 80-100 000 BP for the Qafzeh Mousterian (Bar-Yosef and Vandermeersch 1981), a younger date of *c.* 40 000 BP has been proposed by Jelinek (1982) and Trinkaus (1984) to accommodate unilinear models of lithic industrial and hominid palaeontological successions. Recent dating of the Qafzeh Mousterian to 92 000±5000 BP by thermoluminescence analysis of burnt flints (Valladas *et al.* 1988) supports the earlier age estimates.

The discovery of the same Levalloiso-Mousterian industry in association with both the archaic/Neanderthal fossils of Tabūn, Amūd and Kebara, and with the anatomically-modern specimens from Skhūl and Qafzeh – all within a rather small region of Southwest Asia – requires explanation. This paper explores the behavioural significance of this archaeological association.

FUNCTIONAL ANALYSIS OF KEBARA AND QAFZEH MOUSTERIAN ASSEMBLAGES

In order to understand the relationship between prehistoric technological variability and hominid physiology,a clearer understanding is needed of the behavioural context of Levantine Mousterian industries. Humans are one of several species regularly employing a technological adaptive strategy (Beck 1980; Kortlandt 1986). While even the earliest lithic industries associated with the hominids are quite different from those of other vertebrates, the source of this distinctiveness lies not in the tools themselves but rather with the adaptive context in which the industries were formed, with the uses to which the tools were put (Reynolds 1981; Isaac 1983). All vertebrates confront the same problems in employing a technological strategy. Suitable materials must be collected and transported to the location of their use. If not so formed already, these materials must be shaped into useful implements. Finally, these tools must be used successfully to manipulate the environment. The range and relative frequencies of tasks to be performed, as well as the location of tool use, determine the formal properties of the implements manufactured. Therefore, the resolution of tool functions is the logical point of departure for understanding the structure of a prehistoric industry.

The most informative source on the functions of prehistoric stone tools is analysis of the wear traces resulting from their use (Semenov 1964). As most effectively practised, such use-wear analyses begin with an experimental program, replicating tasks likely to have occurred in prehistory (Keeley

Table 31.1. Size and functional composition of the lithic collections analysed from Kebara and Qafzeh caves.

	Kebara				Qafzeh
Variables	IX	X	XI	XII	XVII
Area excavated (m^2)	6	6	4	2	11
Total Artifacts	2127	1966	1315	306	1370
Used Artifacts	123	142	128	55	64
Employed Units	196	223	180	90	94
Ratio of Employed Units/Used Artifacts	1.6	1.6	1.4	1.6	1.5

1974, Odell 1975). The experimental reference collection employed for the present analysis numbers over 1000 specimens used for projectile points, and for butchery, skin-working, bone-working, processing soft-plant matter, woodworking, digging and stone-working. There also exists a series of taphonomic control specimens damaged from trampling, soil movement and fluviatile disturbance. The uses undertaken with prehistoric tools can be identified to the extent that their wear patterns are replicated in the experimental collection.

Wear, in the form of microscopic abrasion and bending fractures is observed under direct lighting with a stereoscopic microscope at magnifications between 5X and 160X, with most functional determinations being made between 40X and 80X (Odell and Odell-Vereecken 1980). Wear patterns are recorded on a numerical coding system for comparison with experimental reference specimens. Initial assessments of tool uses are retained only if the archaeological wear pattern is replicated in the experimental collection. As a further check on the accuracy of this analytical procedure, six independent blind tests were conducted incorporating 243 uses of both retouched and unretouched sections of 111 artifacts, together with 33 unused control specimens (Shea 1987). A full account of these tests is beyond the scope of the present paper, but the results showed correct identification of the *location* of use in 237/243 cases ($\bar{x}$ = 98%, sd = 1%), correct identification of the *action* employed (from a range of 16 possibilities) in 222/243 cases ($\bar{x}$ = 92%, sd = 5%), and correct identification of the *worked material* (from 12 possibilities) in 200/243 cases ($\bar{x}$ = 82%, sd = 5%). These results serve as indirect measures of the accuracy of this technique in archaeological applications.

The archaeological samples were drawn from the recent excavations of Kebara (Bar-Yosef *et al.* 1986) and Qafzeh (Vandermeersch 1981). The lithic collections from Kebara Units IX-XII and Qafzeh Unit XVII were scanned in their entirety. Smaller randomly-selected cluster samples were drawn from the microdebitage (<2 cm) of each excavation. Approximately 50 artifacts from Kebara and 20 artifacts from Qafzeh could not be analysed due to post-depostional surface modification.

Very few of the artifacts in the Kebara and Qafzeh collections exhibited traces of use (512/7084, $\bar{x}$ = 9%, sd = 5%) (see Table 31.1 and Figure

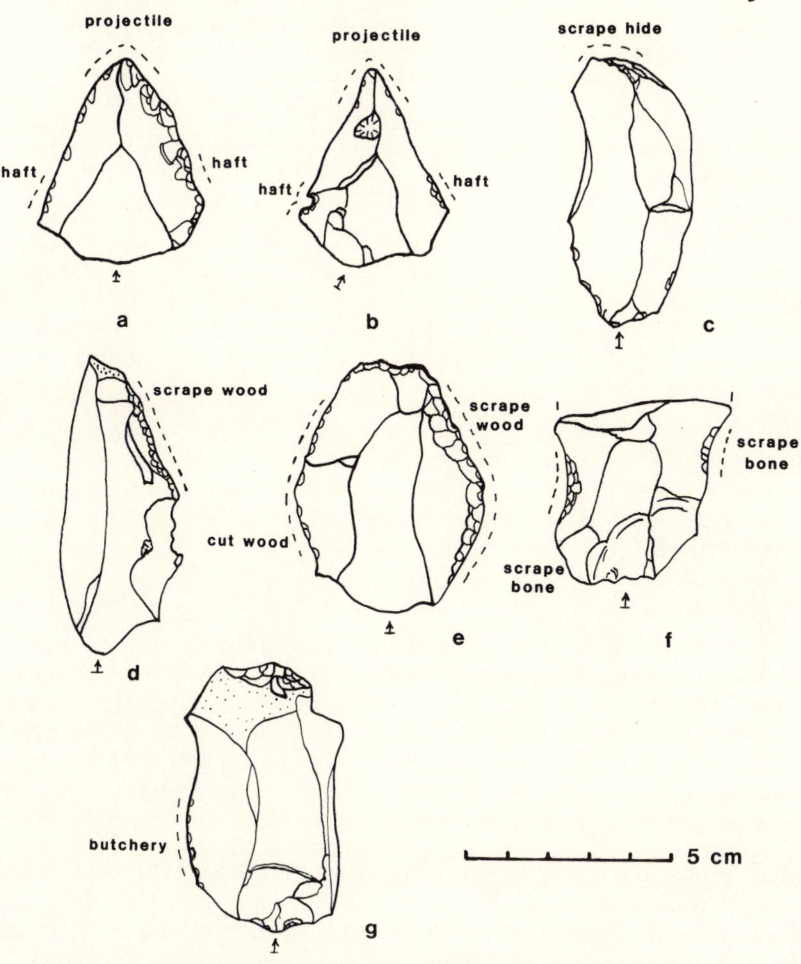

Figure 31.1. Utilized Levalloiso-Mousterian artifacts from Kebara Cave, indicating results of use-wear analysis. (a), (b) Levallois points; (c), (d) Levallois blades; (e) retouched Levallois flake; (f) flake fragment; (g) cortical flake with 'Nahr Ibrahim' truncation on distal and proximal ends.

31.1). While this fraction seems low, it is possible that many of the implements from the caves were used so briefly by prehistoric hominids that diagnostic wear patterns did not result. Experiments and blind test results suggest that for most activities continuous use of more than five minutes is required for characteristic wear patterns to form. On the other hand, if tool manufacture occurred in the caves, as attested by conjoining flakes and thousands of microdebitage fragments, then one might reasonably expect a large part of the assemblage to represent unused flint-knapping products.

In order to describe functional variability among these assemblages, tool uses were quantified by 'employed units', i.e. contiguous sections of tool

Table 31.2. Counts of employed units by activities for the collections from Kebara and Qafzeh caves.

				Kebara		Qafzeh	
Activities		IX	X	XI	XII	XVII	Total
Projectiles	n	15	15	7	13	3	53
	%	7.7	6.7	3.9	14.4	3.2	6.8
Butchery							
	n	30	19	43	12	11	115
	%	15.3	8.5	23.9	13.3	11.7	14.7
Skin-working							
	n	14	12	9		5	40
	%	7.1	5.4	5.0		5.3	5.1
Bone-working							
	n	25	33	18	7	16	99
	%	12.8	14.8	10.0	7.8	17.0	12.6
Soft Vegetal							
	n	4	10	10	8	8	40
	%	2.0	4.5	5.6	8.9	8.5	5.1
Wood-working							
	n	77	96	68	21	35	297
	%	39.3	43.1	37.8	23.3	37.2	37.9
Haft							
	n	26	24	11	26	8	95
	%	13.3	10.8	6.1	28.9	8.5	12.1
Other/Unknown							
	n	5	14	14	3	8	44
	%	2.6	6.3	7.8	3.3	8.5	5.6
Total:		196	223	180	90	94	783

edges, projections and surfaces preserving wear attributable to contact with a single worked material. Many used artifacts featured more than one employed unit. These were usually referable either to the same activity as the first employed unit or to hafting/prehensile damage (see Fig. 31.1).

The most common activity represented among the Qafzeh and Kebara collections was woodworking (38%), usually 'light-duty' tasks (cutting, shaving or scraping), rather than chopping or adzing. Butchery was the next-most-frequently-encountered activity (15%). While butchery may overlap with bone-working (13%) in the occasional contact of tools with bone during disarticulation, incised bone objects have been reported from the Mousterian levels of Kebara (Davis 1973). Damage attributable to the movement of the implement against a haft during use was often observed (12%). Impact damage from using stone tools as projectile points was also encountered (7%) (see Fischer *et al.* 1984; Odell and Cowan 1986 for descriptions of impact damage). Cutting soft vegetal matter (5%), skinworking (5%) and unidentifiable wear patterns (6%) accounted for the remainder of the uses detected. The counts of employed units for the excavation units are listed in Table 31.2. The Qafzeh assemblage resembles those from Kebara quite closely in the proportions of activities represented.

Table 31.3. The functional significance of variation in artifact shape in collections from Kebara and Qafzeh caves. [1]N.B. 'Other' includes broken and indeterminate specimens, as well as cores.

	Artifact Shapes			
Variables	Points	Blades	Flakes	Other[1]
Artifacts				
total	137	699	5930	368
used	69	70	270	104
% used	50.3	10.1	4.6	28.3
Action by Employed Units				
Cut/saw	34	51	151	52
Shave	6	4	34	6
Scrape	7	15	123	57
Plane		3	10	1
Shred			2	3
Adze			1	
Chop			3	1
Wedge		2	1	2
Projectile	33	9	10	1
Grave	2		8	6
Awl		5	15	5
Drill				
Hand	2	1	9	4
Haft	46	14	25	10
Unknown	1	2	1	2
Total Employed Units	131	104	393	152

The material traces of the tactics employed by a population in using tools constitute an 'archaeological signature' of that population (Gould 1980). Three tactics for tool use in the early Upper Pleistocene exert a strong influence upon techno-typological descriptions of Middle Palaeolithic industries. These include variation in the proportions of different debitage shapes in a collection, the use of the Levallois technique, and the patterned modification of implements by retouch (Bordes 1972: 48-54).

The Kebara and Qafzeh collections were classified during analysis as either points (convergent lateral edges), blades (parallel lateral edges), or flakes (excurvate lateral edges). The functional correlates of these distinctions lie mainly with the action employed, and are indicated in Table 31.3. Levallois points, pointed flakes and pointed blades were the most frequently used and were employed chiefly as hafted projectile points and as cutting tools for butchery. Blades were used mainly to cut or to saw, but were employed as frequently in woodworking as in butchery. Flakes were the most varied in their uses. The 'Other' category, comprised of cores and fragmentary debitage, mirrors the pattern for flakes. That the shapes of artifacts in some way conditions their use is not surprising, but the overlapping range of uses encountered for each of these morphological categories argues against any simple equations of tool shape with function.

Table 31.4. The functional significance of the Levallois technique in the collections from Kebara and Qafzeh caves.

	Kebara				Qafzeh	
	IX	X	XI	XII	XVII	Total
Levallois Artifacts						
total	223	412	169	89	139	1032
used	31	63	48	31	17	190
% used	13.9	15.3	28.4	34.8	12.2	18.4
modified	6	7	4	8	4	29
% modified	19.4	11.1	8.3	25.8	23.5	15.3
Non-Levallois Artifacts						
total	1904	1554	1146	217	1231	6052
used	92	79	80	24	46	321
% used	4.8	5.1	6.9	11.1	3.7	5.3
modified	26	17	22	2	34	101
% modified	28.3	21.5	27.5	8.3	73.9	31.5
Employed Units						
Levallois	72	113	72	51	31	339
Non-Levallois	129	110	108	39	63	444
Employed Units for Hafting						
Levallois	12	17	9	19	5	62
Non-Levallois	14	7	3	7	3	34
Employed Units per Used Artifact						
Levallois	2.3	1.8	1.5	1.6	1.8	1.8
Non-Levallois	1.4	1.4	1.4	1.6	1.4	1.4

At Kebara and Qafzeh, the Levallois technique seems mainly to have been a means of controlling variability in the shape of debitage products for more effective hafting. Implements struck with the Levallois technique (Bordes 1980; Boëda 1982; Copeland 1983) seem more frequently used, more frequently hafted, and less frequently modified than non-Levallois artifacts in the same collections (Table 31.4). Levallois artifacts also seem to have been more extensively used (more employed units per implement) than non-Levallois artifacts. With one important departure on the issue of retouch (discussed below), these parameters are precisely what one should expect for a set of implements made for use in a prefabricated handle (Keeley 1982). Consequently, the use of the Levallois index for assemblage description may in fact be monitoring changes in the frequency with which hafted implements were discarded at archaeological sites.

Modification by retouch is rare in all the collections examined, and seems mainly to have been employed for imposing particular shapes upon implements, rather than for resharpening use-dulled edges. Retouched edges were most often associated with woodworking, boneworking and hafting, activities in which buttressed edges are particularly useful (Table 31.5). As an additional check for the presence of curational retouching, 1500 pieces of microdebitage from each site were scanned for the presence of use-wear

Table 31.5. The functional significance of edge modification (retouch and burination) in the collections from Kebara and Qafzeh caves.

	Kebara				Qafzeh	
	IX	X	XI	XII	XVII	Total
Artifacts						
total	2127	1966	1315	306	1370	7084
retouched	108	109	66	11	72	366
% retouched	5.1	5.5	5.0	3.6	5.2	5.2
Used Artifacts						
total	123	142	128	55	64	512
retouched	32	24	26	10	41	133
% retouched	26.0	16.9	20.3	18.2	64.1	26.0
Employed Units						
total	196	223	180	90	94	783
retouched	32	27	29	10	48	146
% retouched	16.3	12.1	16.1	11.1	51.1	18.7
Activities by Employed Unit						
projectile			1		1	2
butchery	3	4	4		4	15
skin-working	3	2	2		7	14
bone-working	8	7	8	1	8	32
soft vegetal		1	1	1	3	6
wood-working	12	4	8	1	19	44
haft	6	4	2	7	2	21
other/unknown		5	3		4	12

on their striking platforms. Only three flakes from Kebara featured such wear. This may suggest that the conditions requiring tool use by the Levantine Mousterian populations were not sufficiently predictable for sufficient energetic payoff to accrue from curational retouch. Much larger samples of microdebitage will have to be sampled in order to test this hypothesis.

In functional terms, the Kebara and Qafzeh assemblages can hardly be separated. The range and relative frequencies of activities, the selection of shapes for use, and of techniques for implement manufacture seem quite similar. Only in the proportion of the utilized artifacts represented by retouched tools does the Qafzeh assemblage (64%) distinguish itself from those of Kebara ($\bar{x}$ = 20%, sd = 4%). Other than sample error or inadequacy, there are (at least) three possible sources for this difference:

1. Scarcity of local raw materials around Qafzeh may have encouraged more frequent use and resharpening of debitage. However, a seam of flint is exposed less than 1 km to the south of Qafzeh Cave in a modern quarry. While further sampling will be necessary to determine the proportion of the Qafzeh assemblage attributable to this source, a lack of use-damaged retouch flakes at Qafzeh also argues against this explanation.
2. The relatively lesser robusticity of the anatomically-modern populations may have required tool use in situations where Neanderthals employed unassisted bodily force. Arguing strongly against this hypothesis is the lack

of any significant difference between the Kebara and Qafzeh assemblages in either the range or relative frequencies of activities undertaken with stone tools.

3. The conditions for tool use in the vicinity of the Qafzeh site may have been more predictable than in the vicinity of Kebara, increasing the energetic payoff for specialized tool designs at Qafzeh. This could have taken the form of a less-diverse or seasonally-abundant range of prey (faunal or floral). This hypothesis may find support in the very high frequencies of red deer (*Cervus elaphus*) noted among the faunal collections from the Qafzeh Mousterian (Bouchud 1974) and the more diverse range of large mammals represented at Kebara (Davis 1977).

THE BEHAVIOURAL CONTEXT OF THE LEVALLOISO-MOUSTERIAN TECHNOLOGICAL ADAPTATION

Most functional analyses of Middle Palaeolithic artifacts have operated within a typological framework (Binford and Binford 1966; Beyries 1987; Anderson-Gerfaud, this Symposium). Identifying the functional correlates of morphological artifact-types can only inform on prehistoric hominid behaviour to the extent that those types were consistently used (Odell 1981). From the results described above, there is every reason to suspect that in using stone tools, Middle Palaeolithic populations had far less concern for the integrity of morphological artifact types than do Palaeolithic archaeologists. Given sufficiently large samples encompassing all artifacts recovered from excavation, a detailed resolution is possible of the technological strategies of prehistoric hominids and the behavioural context of Palaeolithic industries.

What does use-wear tell us about the behaviour of Upper Pleistocene hominids in the Levant? Judging from the numerical predominance of woodworking tools in the utilized assemblages from Kebara and Qafzeh, this activity – probably reflecting tool manufacture and repair – may have been a key reason for occupying these sites. The presence of impact-damaged projectile implements (mostly Levallois points and pointed flakes) firmly establishes the predatory capabilities of both archaic and anatomically-modern *Homo sapiens*, which have recently been challenged by Binford (1984: 248-266). The use of hafted implements is a sign of logistic planning, an ability to predict the likelihood of recurring tasks requiring a particular tool. That the frequency of hafting traces does not differ significantly between the Kebara and Qafzeh assemblages further suggests that such foresight was not a major behavioural discontinuity between archaic and anatomically-modern *Homo sapiens*. The presence of discarded projectile points and of artifacts bearing damage referable to butchery (or at least the division of edible animal tissues) supports a central-place provisioning model of meat procurement at both sites. While these inferences are possible from the analysis of wear traces alone, a complete picture of the behavioural significance of a prehistoric industry will require strategies for tool use to be related to patterns of materials procurement and implement manufacture.

The model of evolutionary continuity proposed for the Levant by Trinkaus (1984) involving an ancestor-descendant relationship between the archaic and anatomically-modern populations of *Homo sapiens* finds little support from this analysis. If the reduction of postcranial robusticity, increased locomotor efficiency, less-frequent manipulative use of the anterior dentition and decrease in birth spacing had significant ramifications for 'the technological and social aspects of food acquisition' (Trinkaus 1984: 269), then these differences – which Tattersall (1986) considers sufficient cause for inferring speciation – should be reflected in the lithic archaeological record. Quite to the contrary, the lithic industries associated with archaic and anatomically-modern populations are functionally almost identical.

Although for a brief period of geological time the activities of archaic and anatomically-modern *Homo sapiens* appear to be archaeologically indistinguishable, this does not mean that the behaviour of these populations was always indistinguishable. Gamble (1986: 367-383) has shown that considerable differences existed in the responses of European Neanderthal and anatomically-modern populations to the establishment of polar desert conditions of low terrestrial productivity. Whereas Neanderthals appear to have abandoned Central Europe during the glacial advances *c.* 70-50 000 BP, modern human populations seem to have persisted in the face of similar conditions between 30 000 and 12 000 BP. It may be that the behavioural differences between these populations are context-dependent, evident only in certain environments, particularly those of depressed primary productivity.

If the Qafzeh Mousterian occupations date to >80 000 BP (Bar-Yosef and Vandermeersch 1981) then the Southwest Asian Neanderthals may have arrived in the Levant by following the southward migration of vegetational belts and faunal communities at an early stage during the Last Glaciation, around 65-75 000 years ago (see Bar-Yosef, this volume). In such a scenario, the immigrant populations would adopt new technologies only to the extent required by the modification of their behavioural strategies to suit their new environment. If these Neanderthal populations were moving south along with an entire biological community, such modifications may have been minimal. Functional analysis of the Levallois-Mousterian industries from Kebara and Qafzeh Caves supports this scenario by demonstrating that no major differences existed between the technological adaptations of archaic and anatomically-modern humans in the Upper Pleistocene of the Levant. Those differences which have been observed can be explained by situational variability in the behavioural context for tool use at the sites in question.

If anatomically-modern human populations were present in the Levant >70-80 000 BP, before the appearance of Neanderthal populations, what does functional analysis of their lithic industries reveal about the co-evolutionary relationship between them? In view of the great similarity between the technological adaptations of the Kebara and Qafzeh populations, it would seem unlikely that the modern populations were displaced from the Levant entirely. Another possibility is that of niche partitioning. The relatively greater postcranial robusticity and lesser locomotor efficiency of archaic/Neanderthal populations may have required them to inhabit highly productive ecosystems, such as the coastal lowlands, where increased rainfall would have created highly-productive *maquis* forests. A wide variety of plant and animal foods would have been available on a year-round basis in this region. Anatomically-modern groups may have persisted in the less productive montane and desertic zones. During the wetter-than-present conditions of the early Last Glacial, roughly corresponding to oxygen-isotope stages 5d-4, these zones would have experienced an increase in vegetation cover and primary productivity (Horowitz 1979). The more effective locomotor adaptation and relatively lesser body mass of modern populations (Trinkaus 1984: 269) may have made possible more frequent movement between the seasonally-available resources as well as subsistence on the lower quality plant foods which would have been available in these regions. Radiocarbon dates for the earliest Mousterian occupations of the Negev fall within this early Last Glacial period *c*. 80 000 BP (Marks 1981). The earliest modern humans may have distinguished themselves by colonizing regions of relatively lower productivity than did their archaic predecessors and contemporaries.

CONCLUSION

General principles of evolutionary ecology suggest that tool use should be a more sensitive register of hominid adaptation than hominid anatomy. The use-wear analysis of archaeological industries associated with both archaic and anatomically-modern forms of *Homo sapiens* in the Upper Pleistocene of the Levant has revealed no corresponding difference between patterns of Levalloiso-Mousterian industrial variability which is not otherwise explicable as the effects of regional variation in the circumstances requiring tool use. These data challenge the assumption of major behavioural discontinuities beween these palaeontological taxa. More information must be sought on the palaeoecological context of Upper Pleistocene hominid behaviour and the ramifications of that behaviour for patterns of physiological variation. The behavioural differences among Upper Pleistocene hominids in the Levant may not lie so much with *how* they used their tools but rather with *where* they used them. The early Last Glacial hominid occupation of the Levant may have involved niche partitioning along lines of physiological variability, but whether such ecological specialization either constituted or reflected a speciation event remains an open question – but a question answerable only on behavioural terms. If archaic and anatomically-modern hominids of the Upper Pleistocene were truly different species, then the evidence from the Levantine archaeological record indicates that their behaviour with respect to tool use was less different than between many living human populations.

ACKNOWLEDGEMENTS

This paper is contribution No. 12 to the *Origins of Modern Humans in Southwest Asia Project.* I wish to thank D. Pilbeam, O. Bar-Yosef and P. Crawford for their constructive criticism of an earlier version of this paper.

REFERENCES

Bar-Yosef, O. and Vandermeersch, B. 1981. Notes concerning the possible age of Mousterian layers in Qafzeh Cave. In J. Cauvin and P. Sanlaville (eds) *Préhistoire du Levant.* Paris: Centre National de la Recherche Scientifique: 281-5.

Bar-Yosef, O., Vandermeersch, B., Arensburg, B., Goldberg, P., Laville, H., Meignen, L., Rak, Y., Tchernov, E. and Tillier, A.- M. 1986. New data concerning the origin of modern man in the Levant. *Current Anthropology* 27: 63-4.

Beck, B. 1980. *Animal Tool Behavior.* New York: Garland Press.

Beyries, S. 1987. *Variabilité de l'Industrie Lithique au Mousterien: Approche Fonctionelle sur Quelques Gisements Français.* Oxford: British Archaeological Reports International Series S238.

Binford, L. R. 1984. *Faunal Remains From Klasies River Mouth.* New York: Academic Press.

Binford, L. R. and Binford, S. R. 1966. A preliminary analysis of functional variability in the Mousterian of Levallois facies. *American Anthropologist* 68: 238-295.

Boëda, E. 1982. Etude expérimentale de la technologie des pointes Levallois. In D. Cahen (ed.) *Tailler! Pour Quoi Faire: Préhistoire et Technologie Lithique:* Vol. 2. Studia Praehistorica Belgica 2: 23-56.

Bordes, F. 1972. *A Tale of Two Caves.* New York: Harper and Row.

Bordes, F. 1980. Le débitage Levallois et ses variants. *Bulletin de la Société Préhistorique Français* 77: 45-49.

Bouchud, J. 1974. Etude préliminaire de la faune provenant de la Grotte de Qafzeh, près de Nazareth, Israël. *Paléorient* 2: 87-102.

Bräuer, G. 1984. A craniological approach to the origins of *Homo sapiens* in Africa and implications for the appearance of modern Europeans. In F. H. Smith and F. Spencer (eds) *The Origins of Modern Humans: a World Survey of the Fossil Evidence.* New York: Alan R. Liss: 327-410.

Copeland, L. 1975. The Middle and Upper Palaeolithic of Lebanon and Syria in the light of recent research. In J. Cauvin and P. Sanlaville (eds) *Préhistoire du Levant.* Paris: Centre National de la Recherche Scientifique: 317-350.

Copeland, L. 1983. Levallois/non-Levallois determinations in the early Levant Mousterian: problems and questions for 1983. *Paléorient* 9: 15-28.

Davis, S. 1973. Incised bones from the Mousterian of Kebara Cave (Mount Carmel) and the Aurignacian of Ha-Yonim Cave (western Galilee), Israel. *Paléorient* 2: 181-182.

Davis, S. 1977. The ungulate remains from Kebara Cave. *Eretz-Israel* 13: 150-163.

Fischer, A., Hansen, P. and Rasmussen, P. 1984. Macro and micro wear traces on lithic projectile points: experimental results and prehistoric examples. *Journal of Danish Archaeology* 3: 19-46.

Gamble, C. 1986. *The Prehistoric Settlement of Europe.* Cambridge: Cambridge University Press.

Garrod, D. A. and Bate, D. M. 1937. *The Stone Age of Mount Carmel,* Vol. 1: *Excavations in the Wady el-Mughara.* Oxford: Clarendon Press.

Gould, R. 1980. *Living Archaeology*. Cambridge: Cambridge University Press.

Horowitz, A. 1979. *The Quaternary of Israel*. New York: Academic Press.

Isaac, G. Ll. 1983. *Early Stages in the Evolution of Human Behaviour: the Adaptive Significance of Stone Tools*. Amsterdam: Stichting Museum voor Anthropologie en Prehistorie (Sixth Kroon Lecture).

Jelinek, A. 1982. The Middle Paleolithic in the southern Levant, with comments on the appearance of modern *Homo sapiens*. In A. Ronen (ed.) *The Transition from Lower to Middle Paleolithic and the Origin of Modern Man*. Oxford: British Archaeological Reports International Series S151: 57-104.

Keeley, L. 1974. Technique and methodology in microwear studies. *World Archaeology* 5: 323-336.

Keeley, L. 1982. Hafting and retooling: effects on the archaeological record. *American Antiquity* 47: 798-809.

Kortlandt, A. 1986. The use of stone tools by wild-living chimpanzees and earliest hominids. *Journal of Human Evolution* 15: 77-132.

Marks, A. 1981. The Middle Paleolithic of the Negev, Israel. In J. Cauvin and P. Sanlaville (eds) *Préhistoire du Levant*. Paris: Centre National de la Recherche Scientifique: 287-298.

McCown, T. and Keith, A. 1939. *The Stone Age of Mount Carmel*. Vol. 2: *The Fossil Human Remains from the Levalloiso-Mousterian*. Oxford: Clarendon Press.

Neuville, R. 1951. *Le Paléolithique et Mésolithique du Désert de Judée*. Paris: Archives de l'Institut de Paléontologie Humaine 24.

Odell, G. 1975. Microwear in perspective: a sympathetic response to Lawrence H. Keeley. *World Archaeology* 7: 226-240.

Odell, G. and Cowan, F. 1986. Experiments with spears and arrows on animal targets. *Journal of Field Archaeology* 13: 195-212.

Odell, G. and Odell-Vereecken, F. 1980. Verifying the reliability of lithic use-wear assessments by 'blind tests': the low power approach. *Journal of Field Archaeology* 7: 87-120.

Reynolds, P. 1981. *On the Evolution of Human Behavior: the Argument from Animals to Man*. Berkeley: University of California Press.

Rightmire, G. 1984. *Homo sapiens* in Sub-Saharan Africa. In F. H. Smith and F. Spencer (eds) *The Origins of Modern Humans: a World Survey of the Fossil Evidence*. New York: Alan R. Liss: 295-326.

Semenov, S. 1964. *Prehistoric Technology*. London: Corey Adams MacKay.

Shea, J. 1987. On accuracy and relevance in lithic use-wear analysis. *Lithic Technology* 16: 44-50.

Shick, T. and Stekelis, M. 1977. Mousterian assemblages in Kebara Cave, Mount Carmel. *Eretz-Israel* 13: 97-149.

Stringer, C. B., Hublin, J. J. and Vandermeersch, B. 1984. The origin of anatomically modern humans in Western Europe. In F. H. Smith and F. Spencer (eds.) *The Origins of Modern Humans: a World Survey of the Fossil Evidence*. New York: Alan R. Liss: 51-136.

Tattersall, I. 1986. Species recognition in human paleontology. *Journal of Human Evolution* 15: 165-175.

Trinkaus, E. 1984. Western Asia. In F. H. Smith and F. Spencer (eds) *The Origins of Modern Humans: a World Survey of the Fossil Evidence* New York: Alan R. Liss: 251-293.

Trinkaus, E. 1986. The Neandertals and modern human origins. *Annual Review of Anthropology* 15: 193-218.

Turville-Petre, F. 1932. Excavations in the Mugharet el-Kebarah. *Journal of the Royal Anthropological Institute* 62: 271-276.

Vandermeersch, B. 1981. *Les Hommes Fossiles de Qafzeh (Israël)*. Paris:

Centre National de la Recherche Scientifique.
Valladas, H., Joron, J. L., Valladas, G., Arensburg, B., Bar-Yosef, O., Belfer-Cohen, A., Goldberg, P., Laville, H., Meignen, L., Rak, Y., Tchernov, E., Tillier, A.-M., and Vandermeersch, B. 1987. Thermoluminescence dates for the Neanderthal burial site at Kebara (Mount Carmel), Israel. *Nature* 330: 159-160.
Valladas, H., Reyss, J. L., Joron, J. L., Valladas, G., Bar-Yosef, O. and Vandermeersch, B. 1988. Thermoluminescence dating of Mousterian 'Proto-Cro-Magnon' remains from Israel and the origins of modern man. *Nature* 331: 614-616.

32. The Case for Continuity: Observations on the Biocultural Transition in Europe and Western Asia

G. A. CLARK AND J. M. LINDLY

Unromantic C. Loring Brace
Thinks Cromagnons did only replace
Through genetic transmission
Not rapine and collision
To produce our now smooth browless face.
(M. Firestone, 1979)

INTRODUCTION

The last decade has seen major advances in both the data bases and the theoretical perspectives upon which understanding of the origins of modern humans depends. This spate of interest is manifest in symposia like this one, which bring together archaeologists and palaeoanthropologists interested in the transition, and which seek to untangle the cultural and biological aspects of change in those parts of the world where an Upper Pleistocene archaeological record is preserved. While it has been recognized for a long time that the biological transition to morphologically-modern humans (subsequently abbreviated to 'MMH') and the archaeological transition from the Middle to the Upper Palaeolithic were linked in subtle and complex ways, efforts to understand the nature of that relationship have so far not been entirely successful. There are many obvious reasons for this, among them the often-cited inadequacies in the temporal and spatial distribution of the data sets against which any proposed explanation must eventually be compared. However, the data do not speak for themselves (or, if they do, they say different things to different people!), and it might be the case that limitations imposed by our theoretical frameworks constitute more significant obstacles to understanding than do problems with the representativeness of our samples. So interwoven are the causal strands of human behavioural and biological evolution that archaeologists can only ignore the findings of palaeoanthropology at their peril. Conversely, given the enormous cultural component of human behaviour, palaeoanthropologists can often benefit from the results of archaeological research.

OUR PERCEPTIONS OF THE BIOLOGICAL TRANSITION

No consideration of the biological transition can be attempted without a 'position statement' on the Neanderthals themselves, since archaic *Homo sapiens* (subsequently abbreviated to 'AHS') forms a baseline against which morphologically-modern humans (*Homo sapiens sapiens*) must be compared. Until the very recent and controversial proposal of a molecular clock model based on the evolution of mitochondrial DNA (Cann *et al.* 1987), there was a near consensus that AHS evolved from *Homo erectus*. Since few anthropologists have had time to react to the evidence from molecular biology, continuity – overwhelmingly supported by both the archaeological and human fossil records – remains the majority view. In sharp contrast, however, the subsequent transformation of AHS into MMH is one of the most hotly debated topics in palaeoanthropology. We think this is due in large part to differences in paradigmatic biases, which put a 'loading' on basic concepts and terms like 'Neanderthal' itself (for an excellent discussion of the history of paradigm change in Neanderthal research, see Spencer 1984). An often-noted confounding of *grade* (shared features of a common organizational level) and *clade* (features deemed shared because of common descent) characteristics exacerbates paradigm differences in the case of Neanderthal (and much other palaeoanthropological) research since the samples are discontinuous in space and time and studies of them are often conducted within the confines of regional research traditions (which themselves tend to cleave along paradigm lines). Put bluntly, there is so little concern with epistemology in palaeoanthropological research designs that it is difficult to find any evidence of an awareness that epistemological issues lie at the heart of the ceaseless debates that characterize the field.

Biases

With Brace (1962, 1964, 1968), Wolpoff (1980a) and in the tradition of Schwalbe, Hrdlička, Weidenreich and others, we subscribe to a unilineal phylogenetic model for the Upper Pleistocene which implies the existence of an AHS grade in human evolution interposed between a pithecanthropine grade and that comprising 'modern' populations. Why we should favour a unilineal approach is bound up in our formal academic training. In matters taxonomic, we tend to be confirmed 'lumpers' – probably at least in part because of statistical backgrounds that impressed upon us the importance of coping with sample variability. As 'unscientific' as it might seem, paradigmatic biases usually reduce to matters of background and training (Kuhn 1974).

Adopting a unilineal perspective means rejecting 'presapiens' (Vallois 1954; Piveteau 1970; Vandermeersch 1981a) and 'preneanderthal' (Howell 1957; Kennedy 1975) phylogenetic models. While few modern workers would subscribe to these in their 'classic' forms, it is interesting to note that no consensus was reached regarding the phylogenetic aspects of the transition in the recent *Origins of Modern Humans volume* edited by Smith and Spencer (1984) – this in spite of much sophisticated functional and

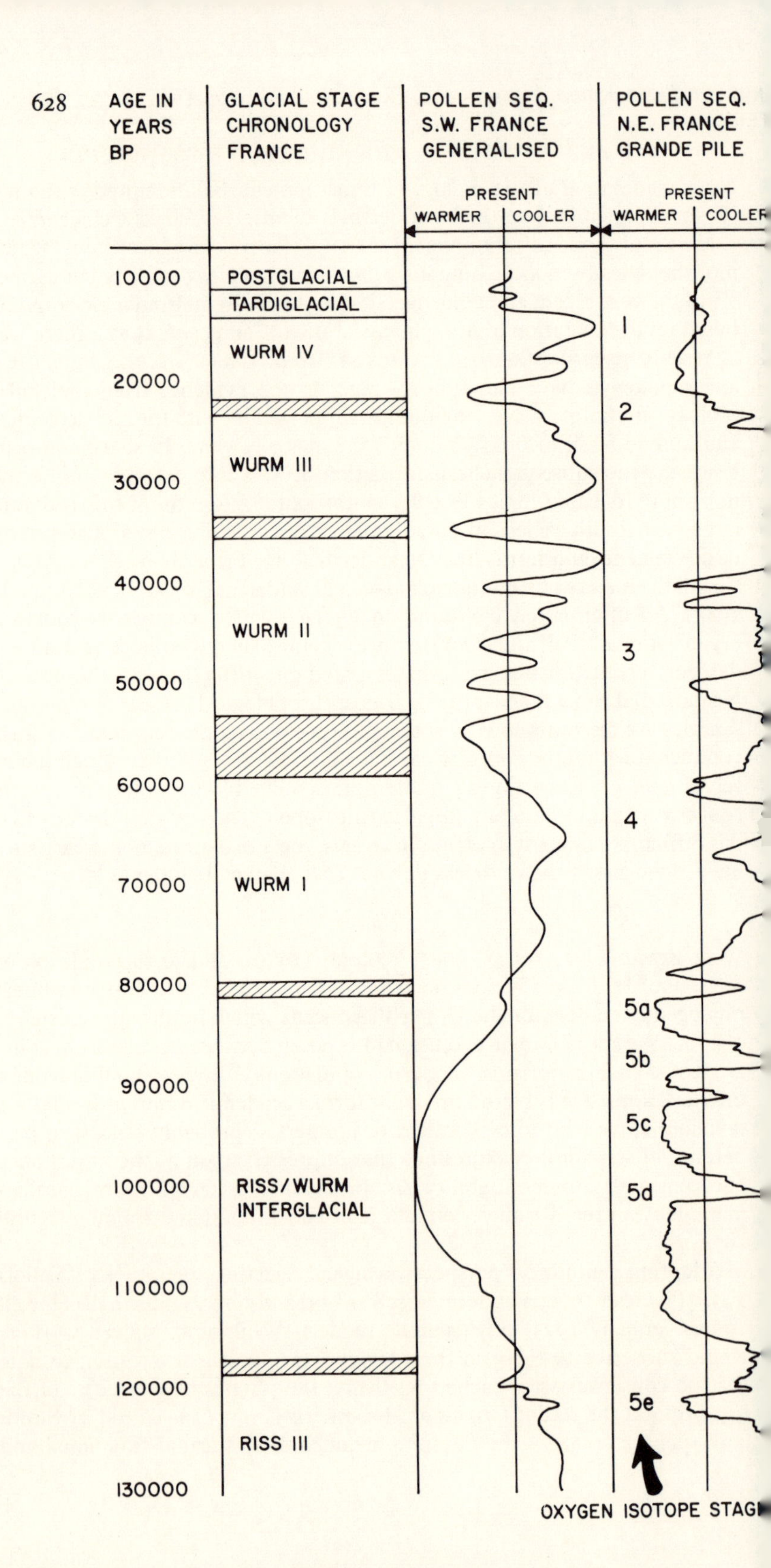
AGE IN YEARS BP
GLACIAL STAGE CHRONOLOGY FRANCE
POLLEN SEQ. S.W. FRANCE GENERALISED
POLLEN SEQ. N.E. FRANCE GRANDE PILE
PRESENT
WARMER
COOLER
PRESENT
WARMER
COOLER
10000
20000
30000
40000
50000
60000
70000
80000
90000
100000
110000
120000
130000
POSTGLACIAL
TARDIGLACIAL
WURM IV
WURM III
WURM II
WURM I
RISS/WURM INTERGLACIAL
RISS III
1
2
3
4
5a
5b
5c
5d
5e
OXYGEN ISOTOPE STAG

NID LOCALITIES IN TERN EUROPE		HOMINID LOCALITIES IN EASTERN & CENTRAL EUROPE		HOMINID LOCALITIES IN THE NEAR EAST & WESTERN ASIA	
UTE DATE	STAGE DATE	ABSOLUTE DATE	STAGE DATE	ABSOLUTE DATE	STAGE DATE
(12 kyr BP)					
de (15)					
g 2 (18)					
ole (18-20)					
19)					
21)	Dobritz				
ud (22-27)		Pavlov (25-26)			
23-25)		Predmosti (26)			
'18-27)	Engis ?	Dolni Vestoniče (26)	Brno		
ı (27)	Stetten		Svaty Prokop		Ein Gev
ch (31)	Combe Capelle ?		Podbaba		Antelias
vern (28-38)	Grimaldi	Istallosko (31)	Vindija		
(33)		Velika Pečina (>34)	Dzerava Skala		Kebara
Cure (33)	St. Cesaire		Šipka		
lole (34)			Mladeč		
srand (36 ?)			Willendorf		Starosel'e
			Miesslingstal		Darra-i-kur
			Cioclinova		Skhul B
37-45)		Kulna (38-46)	Zlaty Kun		
	La Chapelle				
er (40-56)	La Ferrassie ?	Bačo Kiro (>43)			
				Shanidar 1,3,5 (>45)	
e (>47)	La Quina				
	Devil's Tower				
	Cova Negra ?				
t (56)					Tabun C2
ore (60)				Kebara (60)	
	Monte Circeo				Zaskalnaya VI ?
	Wildscheuer		Silička Brezova		Kebara ?
	La Naulette ?		Ohaba Ponor		Shukbah ?
	Spy ?		Krapina		Bisitun ?
	Le Regordou		Ochoz		Teshik Tash ?
	Monsempron		Subalyuk		Tabun C1 ?
	Engis ?		Vindija		Kiik-Koba ?
					Amud ?
					Shanidar 2-4, 6-9
				Qafzeh (92)	
			Taubach		
	Fontêchevade		Ganovče		
	La Chaise (B-D)				
	Ehringsdorf ?				
			Krapina ?		
					Tabun E ?
	Bañolas ?				Azykh ?
					Gesher Banot Ya'acov
	Saccopastore ?				
	La Chaise (S)				
	Lazaret				
	Cova Negra ?				
	Ehringsdorf ?				
				Zuttiyeh (>148)	

morphometric analysis of AHS samples from different parts of the Old World (Howell 1984). At the same time, these is an increasingly widespread appreciation that (1) AHS samples are temporally and spatially variable (at least as variable as those of modern humans); (2) that the nature and rate of change varied geographically (and probably temporally as well); (3) that biological changes did not take place in a behavioural vacuum and must have reflected adaptation in ways that are, at present, only partly understood; and (4) that grade-clade distinctions are analytically essential if the elements of continuity that characterize the various regional samples are to be separated from those which reflect the general organizational level of Upper Pleistocene hominids. Hominid localities in Europe and West Asia are summarized in Figure 32.1.

A logical entailment of a unilineal perspective is that AHS grade characteristics would be globally intermediate between those of *Homo erectus* and those of *Homo sapiens sapiens*. Since any division between *Homo erectus* and archaic *Homo sapiens* must be an arbitrary one, AHS grade characteristics should exhibit a continuation of morphological trends seen first in *Homo erectus*. Those characteristics have been summarized by a number of authors (e.g. Brace 1967; Wolpoff 1980a, 1980b; Trinkaus 1984, 1986) and, so far as we can tell, are not much debated. They include (1) an expansion in brain-size (brain-case breadth approaches cranial breadth in some samples as the brain expands); (2) changes in cranial proportions (expansion of the upper part of the occiput, expansion of the frontal areas as the braincase becomes more 'inflated' relative to *Homo erectus*); (3) decreases in the surface areas of the posterior dentition; (4) increases in anterior tooth size and in the size of their support structures, indicating increasing use of the anterior

---

Figure 32.1. Upper Palaeolithic palaeoclimatic grid and hominid localities in western and east-central Europe, the Levant and Western Asia. All stage dates are crude approximations and most are contested. No ordering is implied within blocks of sites ordered by palaeoclimatic stages.

Claimed archaic *Homo sapiens*: St. Césaire (33 kyr BP), Shanidar (late sample, 45-60 kyr BP), St. Brelade (> 47), Devil's Tower (> 48), Tabūn C2 (50-60), Lebenstedt (> 55), Kebara 1 (60), Saccopastore (120), Zuttiyeh (< 148), La Chaise/Suard (151), Ehringsdorf (> 200); most French Neanderthals, St. Césaire (Würm II in southwest France), Spy (early Würm cold phase), La Naulette (early Würm), La Chaise/B-D (Würm I-II); Vindija G3, Kůlna, Šipka, Šala (?) (= Würm I-II); Tabūn C1, Amud, Bisitun, Kiik-Koba, Teshik Tash, Zaskalnaya VI (= early last glacial?); Cova Negra, Monte Circeo, Engis, Neanderthal, Carigüela (= Würm undifferentiated); Krapina, Ganovče, Ochoz, Subalyuk (= Riss/Würm Interglacial, Würm I); Bañolas, Ehringsdorf (Riss/Würm Interglacial); Shanidar (early sample), Tabūn E, Azykh, Gesher Banot Ya'acov (= last interglacial?); Lazaret, Fontechévade, Saccopastore (Riss III).

Claimed morphologically Modern Humans: Grimaldi (12 kyr BP), Chancelade (15), Neuessing 2 (18), Badger Hole (18-20), Paviland (18-27), Candide (19), Binshof (21), Abri Pataud (22-27), Paglicci (23-25), Paderborn (27), Kent's Cavern (28-38), Kelsterbach (31), Picken's Hole (34), Hahnöfersrand (36?); Qafzeh (92), Engis, Combe Capelle, Grimaldi (EUP), Stetten (middle Aurignacian), Dobritz (late Aurignacian/early Magdalenian); Zlatý Kün, Mladeč, Vindija, Velika Pečina, Brno, Předmostí', Dolní Vestonice, Pavlov (= mid/late Würm or local equivalent); Darra-i-kur, Starosel'e (early to mid last glacial); Kebara, Ein Gev, Antelias, Skhūl B.

dentition as an adjunct to technology; (5) decreases in the area of nuchal muscle attachment and in the volume of spongy bone in the basicranium; and (6) decreases in the skeletal indicators of cranial and post-cranial robusticity.

It is essential to keep in mind that these morphological trends are expressed to varying degrees in regional geographical samples of AHS, that they are only very broadly contemporaneous, that they are not irreversible and that they are only meaningful in relation to the grade characteristics of bracketing taxa. Superimposed on these grade features are clade features that reflect continuity in descent within relatively circumscribed geographical areas. One such area is, of course, Western Europe, which produced most of the Upper Pleistocene hominid fossils upon which the early systematics (*c.* 1900-1930) were based, and most of the early systematists as well (Spencer 1984). This historical precedence has so dominated concepts of AHS that it has been very difficult until recently to form a global perspective on variability that is not influenced to some extent by the skeletal hand of Marcellin Boule (1911-1913). As a practical matter, samples outside Western Europe (and the Near East) should at least be given equal consideration. This seems to be what has happened over the past decade.

THE BIOLOGICAL TRANSITION FROM AN ARCHAEOLOGICAL PERSPECTIVE

Whatever position is taken on phylogenetic issues, it seems to us that roughly the same general kinds of processes and morphoclinal trends can be documented over the AHS-MMH transition in both Europe and Western Asia. Where there are differences of opinion, they usually have to do with whether the *magnitude* of change is thought to be sufficient to warrant or invalidate the possibility of population replacement. We suggest that, given the present state of systematics in human palaeontology, these are unresolvable 'chicken and egg' arguments – dependent to some extent on matters of resolution and scale, but most significantly, on what variables different investigators choose to measure.

Our reading of the current literature on the archaic-modern human transition indicates widespread agreement on the following points (see especially papers in Smith and Spencer 1984): (1) the lower face reduces substantially, almost certainly because of a decrease in the habitual paramasticatory use of the anterior dentition; (2) incisors and canines themselves reduce, and for the same reasons; (3) the occlusal surface areas of the molar dentition decrease slightly; (4) brow-ridges shrink and change shape in response to reduction in anterior tooth loading and because higher, rounder foreheads were equally or better able to counteract such stress; (5) overall midfacial prognathism decreases; (6) the anteroposterior length of the jaws decreases as a consequence of decreases in the size of the anterior teeth, causing retromolar spaces to disappear and chins to develop; (7) nuchal muscle attachment areas decrease and 'migrate' to a lower position on the occiput; (8) the neurocranium becomes more spherical as a result of changes in the physics of muscle attachment and in the rate of development of different parts of the brain; (9) the proportions of the appendicular skeleton approach the modern condition; and, finally, (10) there is a decrease in the skeletal indicators of muscular robusticity.

Although the samples are poor because they are discontinuous in space and time, these changes are temporally vectored and seem to have a universal character. They correspond reasonably well to projections of morphoclinal change in the AHS grade characteristics summarized at the beginning of the essay. Many workers explain them by invoking behavioural changes and linked technological dependencies which, over the long run, required less reliance upon the maintenance of energetically-expensive physical strength and endurance. While this makes perfect biological sense, it is important to point out that there is *no correlation whatsoever* with archaeological industrial configurations (*contra* Foley 1987; see Clark 1989). Middle Palaeolithic industries are associated with MMHs at Skhūl and Qafzeh, and with a morphologically transitional hominid at Starosel'e. Early Upper Palaeolithic industries are found with a Neanderthal at Saint-Césaire and probably in South-Central Europe as well. While perceptions of differences and similarities in 'normative' characterizations of lithic assemblages tend to be heavily influenced by biases expressed through the typological 'filters' used to describe the material, the conclusion seems to be inescapable that the morphological transition was to a large extent (perhaps completely) independent of the technological transition between the 'Middle' and the 'Upper' Palaeolithic.

THE PHYLOGENETIC ISSUES

Some observations on the phylogenetic issues seem to be pertinent here, although in the light of our biases, most readers will probably guess where we stand on the question of genetic continuity across the transition. *We regard genetic continuity across the transition in both Europe and West Asia as a documented certainty.* So far as we can tell, there is a near-consensus that Neanderthals (and other archaic Upper Pleistocene hominids, for those who do not recognize a Neanderthal grade) pertain to *Homo sapiens*, being taxonomically distinct from modern humans only at the level of the subspecies. This implies acceptance of the idea that gene flow was possible across the transition, and simultaneously rules out the possibility that all Neanderthals (*sensu lato*) became extinct without issue.

The question then becomes one of demonstrating continuity in morphoclinal trends within clades in the face of (1) a lack of agreement about what constitutes clade characteristics; (2) a lack of agreement about which morphological complexes are 'significant' in the resolution of phylogenetic issues; (3) a persistent, variety-minimizing 'type-specimen' mentality on the part of some workers, reinforced by regional research traditions; (4) the inevitability of sampling error; (5) the likelihood of errors in chronological placement and in what to make of associated archaeological remains; and (6) an inability to distinguish derived from primitive features. We acknowledge that these are formidable obstacles, and that all of them figure in one way or another in assessments of morphological continuity both in general and within a particular region of interest.

If we dismiss the idea that AHS became extinct without issue, there remain three logical possibilities that still take a degree of gene flow across the transition into account. These have recently been summarized by Fred Smith (1982, 1983, 1984), and they can be expressed for maximum contrast as 'polar' positions, although in fact there is no reason to suppose them to be mutually exclusive. Either (1) there is an *in situ* transition between AHS and MMH within a given region, with neither significant gene flow nor population replacement; or (2) there is morphological change in the direction of MMHs accomplished *primarily* by gene flow, but with no population replacement; or (3) population replacement is the primary mechanism, with a minimal amount of gene flow (Smith 1982: 683).

Our reading of the 'transition' literature overwhelmingly supports the first alternative, not only in the Near East (where it seems to be the favoured explanation) but also in Europe (where the case is less clear) (Stringer 1982; Stringer *et al.* 1984; Trinkaus 1982a, 1982b, 1983a, 1983b). The second possibility would rest upon the identification of a regional sample which is both close enough geographically to the area of interest (so that gene flow would be a reasonable possibility) and where pene-contemporaneous or antecedent vectored change in the direction of MMHs could be demonstrated. In the present state of 'transition' research, it is probably impossible to meet these conditions. The palaeontological record is simply too 'coarse-grained' (low sample resolution, low integrity), an observation that applies to the archaeological record as well. Also, significant gene flow would seem to imply higher population densities than most workers would acknowledge for these remote time ranges. The third possibility – population replacement – seems least likely of all. While we acknowledge the near-certainty of extensive and overlapping hunter-gatherer annual territories, we can think of no behavioural mechanism that could account for population displacements on an order of magnitude large enough to support actual physical replacement given the *extremely low* population densities estimated for the Upper Pleistocene (see, for example, Clark and Yi 1983; Straus 1985; Clark and Straus 1986). Nor do significant technological differences coincide with the Middle-Upper Palaeolithic transition, as was formerly thought. Rather, archaeological evidence for the appearance of modern behavioural patterns post-dates both the Middle-Upper Palaeolithic transition and the appearance of MMHs by at least 15-20 000 (Gamble 1986).

THE CULTURAL TRANSITION

The interpretation of archaeological evidence for the Middle-Upper Palaeolithic transition in Europe and the Near East is also dependent upon one's academic background and paradigmatic biases. As noted above, the 'filters' created by these biases affect the nature of comparisons and the results obtained from them. For example, the different classificatory schemes used to describe Middle and Upper Palaeolithic stone tools in Europe emphasize typological differences between these periods at the expense of technological aspects that might clarify or explain the relationship between them. As a result, the Middle and Upper Palaeolithic tend to be artificially separated into two distinct conceptual units rather than viewed as points along a continuum of culture change. Many recent examinations of the transition have emphasized the differences between the Middle and Upper Palaeolithic by contrasting general 'traits' for each period (e.g. Klein 1973; Mellars 1973; White 1982) or by comparing a 'typical' Middle Palaeolithic configuration to that of a 'typical' Upper Palaeolithic one (e.g. Straus 1983). The problem with this approach is that the information compared is of such a general nature that it tends to conceal a great deal of variability within each period and at the same time downplays similarities between periods over the transition interval. The 'typical' sites or assemblages contrasted can be thousands of years removed from one another and from the approximate period of the transition. They merely serve to underscore major, expectable differences rather than illuminate the nature of the transition process itself.

Biases

As should be obvious, we bring our own biases to this topic as well, and they will become more apparent as this review progresses. For now, it should suffice to state that we think there is significant evidence for continuity of technology, subsistence and settlement patterns across the cultural transition in both Europe and the Near East – continuity to match that of the biological record discussed above. This is not to say that we believe the Upper Palaeolithic to be an *in situ* phenomenon wherever it occurs, with no evidence of outside 'influences'. However, arguments for intrusive 'cultures' borne by intrusive groups of modern humans during the early Upper Palaeolithic (EUP) must be founded on something more rigorous than a simple reference to the appearance of 'foreign' or 'alien' artifact types and industries. Given the exceptionally coarse 'grain' of the early archaeological record, it is probably true to state that, most of the time, we are not in a position to determine with certainty whether or not an artifact type or assemblage is of local or extralocal origin.

For the purposes of this review we will try to emphasize, whenever possible, data from the late Middle Palaeolithic (LMP) and the early Upper Palaeolithic (EUP). However, given the variable quality of the available evidence, it is almost impossible not to form what some might consider to be poorly-founded generalizations about these periods in their entirety. In our view, the deficiencies of areal data bases are more than offset by the long temporal perspective against which to examine change. A diachronic perspective is crucial for examining what was certainly a lengthy, complex process which began and ended outside the arbitrary boundaries of the transition itself. This review will emphasize data from Southwestern Europe and the Levant, since we are most familiar with these areas. However, we will try to incorporate data from Central and Eastern Europe, and from highland West Asia whenever possible, as the discussion should be applicable to areas with similar developmental sequences.

WESTERN EUROPE

Technological and Typological Considerations. The cultural transition as manifest in changes in the character of retouched stone tools has been interpreted in three major ways. Some workers see the transition as an *in situ* phenomenon with clear evidence of continuity between LMP and EUP assemblages – at least in Western Europe (e.g. Bahn 1984; Bricker 1976; Bordes 1972; Campbell 1986; Laville *et al.* 1980). Others argue that EUP industries, especially the Châtelperronian and the Szeletian, are 'adaptive responses' (whatever that means) by AHS populations to the arrival and influence of modern humans producing Aurignacian industries (e.g. Allsworth-Jones 1986; Butzer 1986; Harrold 1981, 1983). A third point of view suggests that no such transitional industries exist and, when EUP assemblages are present at the same site or in the same region as LMP assemblages, no relationship at all can be detected – the EUP (especially Aurignacian) must therefore be intrusive (e.g. Klein 1973; Kozlowski 1982). These different perspectives are related in large part to the classification and subsequent comparison of retouched stone tool inventories according to the entirely incompatible typological systems used for the Middle and Upper Palaeolithic (Binford 1973).

The principal Middle Palaeolithic typology in current use, *la méthode Bordes* (1954, 1961), is based on the relative frequencies of morphological tool types identified by the location and shape of retouched or modified edges irrespective of the overall form of the piece or of temporal considerations. In contrast, the most frequently used Upper Palaeolithic typology (Sonneville-Bordes and Perrot 1954-56) emphasizes 'chrono-stylistic' attributes within morphotypes – attributes that are thought to change over time. The differences in the attributes monitored, and in the resolution of these typological systems, create difficulties in assessing continuity or discontinuity across the transition in the areas where they are used. In short, they ensure that discontinuity will always be detected. A brief examination of interpretations based on these typologies might help clarify problems in analysing the cultural transition, at least as it exists in Southwestern Europe. Regardless of important differences in attributes selected to measure morphological variation, most workers assume that both typological systems monitor technology in some fairly direct way. This, in our opinion, is a dubious assumption.

The Mousterian Facies. As is well known, the Middle Palaeolithic of Southwestern Europe was partitioned, originally by Henri Breuil (1939), who thought that his subdivisions were 'evolutionary' in the sense that they were temporally ordered, and subsequently by François Bordes (1954), into four 'facies' based on the percentages of type groups with similar morphological characteristics. The present, generally recognized Mousterian facies are (1) Typical; (2) Denticulate; (3) Charentian (including Quina and Ferrassie variants); and (4) the Mousterian of Acheulian Tradition, or 'MTA' (including sub-types A and B, which are sometimes claimed to be temporally ordered). They occur together as a group only in southwestern

France, with various of them showing up in isolation elsewhere in Europe and, occasionally, as far afield as Eastern Asia (e.g. Sohn 1983).

Interestingly, the facies do not appear to change through time and according to Bordes demonstrated no clear temporal sequence, alternating throughout the course of the entire Middle Palaeolithic. In contrast, Mellars (1969, 1973, 1986) has argued that there is in fact vectored change through time at several French sites, with a Ferrassie/Quina/MTA progression from oldest to youngest. He has not, to our knowledge, offered an explanation for this phenomenon. Despite the different interpretations placed on the facies, the typology itself remains the *lingua franca* for the Mousterian and is used over large parts of the Old World, even in areas where the facies cannot be detected.

The question of the behavioural meaning of the facies has never been resolved. The legendary 'culture' (Bordes) *versus* 'function' (Binford) debate ended in stalemate in the mid-1970s (Binford and Binford 1966; Binford 1973; Bordes and Sonneville-Bordes 1970). Most workers opted for one or the other alternative in accordance with their biases. The meaning attributed to Bordes' types is not only critical to an understanding of Middle Palaeolithic technological organization but also to our ability to address the problem of the transition to the Upper Palaeolithic. Recent work constitutes a departure from normative typological characterizations and emphasizes the roles played by raw material procurement, reduction and use. It suggests that our understanding of Late Pleistocene technologies is far from complete, and provides a different perspective from which to view the transition.

Raw Material Variability. Variability in raw material procurement, reduction and use has become a recurrent theme in recent explanations of Mousterian lithic assemblage variability (Jelinek 1976; Fish 1979, 1981; Rolland 1981; Dibble 1984). Quality, size and availability of raw material will effect the choice of reduction strategies, blank dimensions and consequently assemblage composition. Many workers had noted variations in aggregate tool frequencies across the facies, and in the proportions of blanks which showed Levallois reduction, but the European tendency, at least, was to explain these in cultural terms. These findings have important implications because criteria such as scraper percentages, the incidence of bifaces, Levallois index etc. are used to identify modal variation in Middle Palaeolithic assemblages both in areas where the facies are recognized and in areas (e.g. the Levant) where they are not. They suggest that assemblage variability might simply be due to differences in raw material sources and in the intensity of site use (see below), and might have no 'cultural' implications whatsoever. Bordes' facies remain intact as analytical units only in southwestern France (and to some extent in northern Spain), but even in these areas there is a tendency to explain them in non-cultural terms. Some of these efforts go back more than 20 years (e.g. Freeman 1964).

Middle Palaeolithic Variability in Reduction and Intensity of Site Use. Connected with this shift in explanatory frameworks have been attempts to determine the extent to which Mousterian assemblages generally, and the facies in particular, have been 'reduced' as a result of the reworking and resharpening of tools in the course of use. The basic idea is that there are marked differences in reduction, and in the elaboration of tools, that correspond to the intensity with which a particular assemblage type was used. So far as we can determine, this idea originates in the early 1980s with Arthur Jelinek, who suggested that the most intensively reduced facies (Quina/Ferrassie and MTA) might be correlated with site contexts where habitation episodes were relatively intense and prolonged. The least reduced assemblages (Typical, Denticulate) – those dominated by expedient tools like notches, denticulates and continuously retouched pieces – might represent activities correlated with much more ephemeral site use.

Jelinek (pers. comm.) has suggested that in Western Europe (where late Pleistocene climatic fluctuations were marked), these modal differences in the intensity of site use might have had climatic correlates. Rigorous climatic intervals would have resulted in more intensive use of caves and rock-shelters, more reduction of assemblages, a higher incidence of formal retouched pieces, and a more predictable set of end products than those of milder climatic episodes. The casually-produced 'expedient' facies (Typical, but especially Denticulate) seem to represent a qualitatively different set of behaviours than do the more intensively reduced Quina/Ferrassie and MTA assemblages. The consistency in form, and the temporal persistence of the facies in southwestern France, is an interesting phenomenon. At least in that intensively studied region, it implies a consistency of behaviour and/or site formation processes over a very long period of time.

A direct result of this perspective has been to monitor the extent of reduction by examining the properties of Mousterian formal tool groups (especially side-scrapers). Dibble (1987, and this volume) has shown that many scraper types identified in the Bordes typology are probably stages in a reduction and re-use sequence, rather than formalized tool types in their own right. Considerable reduction of some of these tools suggests intensive re-use and possibly a longer 'use-life' than Binford's claims for expedient Mousterian lithic technologies would allow (Binford 1979, 1982, 1983). In addition, recent use-wear and edge analyses (Anderson 1980, 1981; Barton 1986; Beyries 1987; Panagopoulou 1985) suggest a variable and usually multipurpose function for most of these tools. Dibble (1987) thinks that Bordes' typology is still valid, but that it should be used to investigate patterns of manufacture and re-use within assemblages rather than directed toward interassemblage comparisons.

The Upper Palaeolithic Typology. The Upper Palaeolithic typology developed by Sonneville-Bordes and Perrot (1954-56) has not come under the same intense scrutiny as that of the Middle Palaeolithic (but see Straus 1978; Hemingway 1980; Bahn 1981). Reasons for this are probably related to the fact that the typology was initially very successful at partitioning the industries of the Upper Palaeolithic by using a *fossile directeur*-based system of time-sensitive 'stylistic markers'. Recently, archaeologists have come to realise that there are problems with the generality of this typology, and its original strength has become its greatest weakness, in that it 'weights' these stylistic index types more heavily than it does other assemblage characteristics. A concentration on *fossiles directeurs* masks a great deal of assemblage variability within and between the traditionally-defined Upper Palaeolithic culture-stratigraphic units. For example, Straus (1978) has argued that Solutrean assemblages are highly variable in Cantabrian Spain and are in consequence difficult to characterize in terms of the normative developmental schemes proposed by Jordá (1977) and his students. It is interesting to note that some prehistorians are coming to the realisation that there might be functional explanations for assemblage variability *within* parallel phyla such as the Périgordian and the Aurignacian, but refuse to acknowledge the possibility that these typologically-separate industries themselves could also signify functional aspects of the same EUP adaptation (Laville and Rigaud 1973; Laville *et al.* 1980; see also S. Binford 1972).

What one is left with, then, when attempting to examine the transition between LMP and EUP assemblages is more a comparison of typological systems than the technological changes they supposedly monitor. As will be seen below, it is exactly this typological comparison that lies at the heart of much disagreement concerning the Middle-Upper Palaeolithic transition.

The Lithic Transition: Lower Périgordian. In conventional interpretations of the transition in Southwestern Europe, the earliest typologically-identified Upper Palaeolithic industry, the Châtelperronian or Lower Périgordian, is thought to develop out of the Type B sub-phase of the Mousterian of Acheulian Tradition (Bordes 1972). This view is based on the presence of backed types in both assemblages, including the 'diagnostic' Châtelperron knife, and the consistent appearance of 'Middle Palaeolithic' elements in Châtelperronian contexts (Bordes 1972; Bricker 1976). A contrasting viewpoint is that the Châtelperronian is not in fact transitional because of strong 'Upper Palaeolithic tendencies' (i.e. higher lamellar indices, presence of bone/antler tools) in assemblages with type frequencies very different from those of the Middle Palaeolithic (Harrold 1981, 1983, and this volume). The disagreement turns on how the typologies are interpreted (i.e. are there more 'Middle Palaeolithic' types or more 'Upper Palaeolithic' ones?).

Stratigraphically, the Châtelperronian overlies Mousterian levels wherever the two occur together in a single site sequence and is apparently associated with the AHS skeleton at Saint-Césaire, thought to date to *c.* 33 000 BP (Lévêque and Vandermeersch 1980, 1981). However, most French sites contain an erosional episode thought to correspond with the Würm 2/3 interstadial which effectively separates Middle and Upper Palaeolithic deposits (e.g. Laville *et al.* 1980). This lack of stratigraphic continuity coupled with a primarily typological approach to comparison emphasizes the separation of LMP and EUP deposits (Brose and Wolpoff 1971). Studying the transition is made more difficult by a lack of technological comparisons of core types and debitage, analysis of platform morphology and metrical analysis of blank dimensions. Consequently the technological relationship between the Châtelperronian and the Mousterian is not at all clear (and certainly not as clear as the LMP/EUP transition in the Levant).

Despite the confounding effect of these problems, there seems to be

Table 32.1. Flake and blade counts as percent of debitage for selected early upper Palaeolithic assemblages in Europe.

Site	Blades		Flakes		F/B	Reference
	Number	Percent	Number	Percent	Dominant?	
B-Torle VII	67	21.8	99	32.2	F	Hahn 1977
B-Torle VI	71	20.9	71	20.9	Tie	Hahn 1977
H-Stadel IV	114	36.4	105	33.5	B	Hahn 1977
Vogelherd V	325	26.6	666	54.5	F	Hahn 1977
Vogelherd IV	648	39.1	727	43.8	F	Hahn 1977
Sirgenstein VI	37	13.9	162	60.9	F	Hahn 1977
Sirgenstein IV	95	22.4	260	61.3	F	Hahn 1977
Wildscheuer III	268	30.8	421	48.3	F	Hahn 1977
Lömmersum	41	3.8	206	19.3	F	Hahn 1977
Breitenbach	1417	31.9	1303	29.3	B	Hahn 1977
Willendorf 11/4	66	8.4	187	23.8	F	Hahn 1977
Senftenbeig	137	7.7	293	16.5	F	Hahn 1977
Langmannersdorf	437	32.4	242	17.9	B	Hahn 1977
Krepiče	50	16.9	77	26.0	F	Hahn 1977
M-Borky II	420	50.0	226	26.8	B	Hahn 1977
M-Občiny	81	21.7	143	38.2	F	Hahn 1977
Stranská Skálá	435	31.6	606	44.0	F	Hahn 1977
Zelesiče	136	24.3	161	28.7	F	Hahn 1977
Nova Dedina	253	23.9	249	23.5	B	Hahn 1977
Zlutava	55	10.3	80	15.0	F	Hahn 1977
Barča I/1-2	114	14.6	139	17.8	F	Hahn 1977
Barča I/3	280	21.5	119	9.1	B	Hahn 1977
Barča II	125	8.3	313	20.9	F	Hahn 1977
Tibava	165	36.6	161	30.8	B	Hahn 1977
Kostenki I/3	131	5.9	118	5.3	B	Hahn 1977
Bacho Kiro 11a	270	9.2	1874	63.6	F	Kozlowski 1982
Kent's Cavern	35	N/A	87	N/A	F	Campbell 1977
Paviland	109	N/A	1156	N/A	F	Campbell 1977

evidence for *in situ* development of EUP industries from the LMP in other areas of Europe. In Pyrenean France, for example, there are several sites thought to contain direct evidence of *in situ* development of Châtelperronian from Mousterian assemblages based on the presence of Middle Palaeolithic artifact types in the EUP (Bahn 1984). At La Ferrassie, in the Dordogne 'heartland' where the transition is supposedly obliterated by erosion, there is also a claim for an *in situ* development of Châtelperronian out of a Mousterian base (Tuffreau 1976). In Britain there is no apparent hiatus between late Mousterian industries and the 'Lincombian', an EUP leaf-point assemblage (Campbell 1986). In Eastern Europe, the Szeletian, another leaf-point industry, is argued to have developed out of the preceding Micoquian (Allsworth-Jones 1986). On balance it seems that the case for continuity is stronger than that for disjunction. To argue from negative evidence (i.e. the conveniently-placed Würm 2/3 erosional episode in south-western France) seems a weak form of argument, especially in the face of evidence for continuity elsewhere. However, it must be admitted that the nature of the technological transition from the late Middle to the earliest Upper Palaeolithic industries continues to be very poorly understood (Harrold 1983, and this volume).

The Lithic Transition: Aurignacian. The other EUP industry found in South-western Europe (and across Europe and into the Middle East) is the Aurignacian. What is probably the consensus view would see the origins of the Aurignacian outside Western Europe (usually in Eastern or Central Europe) based on the 'sudden' appearance of a fully-developed assemblage very different from both the preceding Mousterian and contemporaneous transitional industries (Bhattacharya 1977; Harrold 1981, 1983; Allsworth-Jones 1986; Butzer 1986; Campbell 1986; Davidson 1986; Gamble 1986). Typologically, the differences with the Middle Palaeolithic and the Châtelperronian cannot be denied. The occurrence of particular scraper types (especially carinated forms), a developed burin industry, Dufour bladelets, big blades with scalariform retouch ('Aurignacian' blades), and industries in bone, antler and ivory all underscore a consistent assemblage type very different from the Middle Palaeolithic or the Châtelperronian. However, it is not clear how early Aurignacian industries compare with the LMP and the Châtelperronian technologically, and there is some possibility of an *in situ* development of certain early facies of this industry such as the 'Castanet' variant in France (Laville *et al.* 1980).

Early Upper Palaeolithic Blank Frequencies. One of the major industrial characteristics used to differentiate the Aurignacian (and the EUP in general) from the Mousterian is the presence of a fully-developed blade technology. Are these so-called EUP blade industries really dominated by blanks with blade dimensions? In many cases the answer is no. Table 32.1 shows blade *versus* flake counts for 23 EUP sites in Europe. The debitage component from most of them is heavily dominated by pieces with flake dimensions, suggesting that too much emphasis might have been placed on this aspect of the early Upper Palaeolithic. It should also be pointed out that, for some sites with both LMP and EUP components, there is a discernible increase in laminar elements over the transition. However, this increase is usually slight, and is not a general characteristic of the debitage inventories of all sites where the transition is recorded.

On Raw Material and Blank Frequencies. It is widely believed that there was a change in lithic raw material procurement strategies over the transition, but the extent to which the change was a general phenomenon is debatable. This change in turn is argued to have affected the kinds of blanks that could have been produced. Middle Palaeolithic assemblages are made almost exclusively from local material, in many cases a substandard flint, chert or quartzite. Where relevant data are available, the Châtelperronian appears to continue this pattern and, in respect of raw material, these assemblages most closely resemble those of the Middle Palaeolithic. By contrast, European Aurignacian assemblages usually have substantial numbers of blanks produced on good quality flint which is apparently foreign to the site catchments in which it occurs, and there is usually a significant blade component. In support of these generalizations, we offer the following observations which, while not conclusive, at least suggest an empirical basis for questioning correlations between raw material procurement strategies and the normative culture-stratigraphic units of the LMP and EUP.

Demars (1982) reports that in the Brive Basin (Aquitaine, France), Mousterian raw material is derived from sources close to the sites (on the average *c.* 1 km distant) and is of poor quality. Aurignacian raw material comes from farther away (*c.* 10 km distant) and is also of poor quality. In Upper Périgordian contexts the stone comes from substantial distances (> 10 km), and is a high quality flint. At Bacho Kiro cave (Bulgaria), in a proposed archaic Aurignacian context (Level 11 – the 'Bachokirian'), most of the assemblage is produced on good quality flint foreign to the area. The underlying LMP levels are manufactured almost exclusively on poor quality local material (Kozlowski 1982). Could the differences between the LMP and the EUP correspond in these cases to organizational differences related to differences in mobility over the transition? It looks that way. It could be that procurement of good quality flint was required for the more efficient production of blades. Thus we would expect systematic correlations between assemblages in which blade blanks were important and high quality raw material, and no such correlations between assemblages which continue to exhibit flake technologies over the transition or that happen to be located in areas where high quality raw material was generally unavailable.

In the light of this proposal, it is interesting to note that in areas where raw material has remained constant due to geographic restrictions that precluded significant changes in mobility over time, overall assemblage characteristics themselves tend to be fairly stable across the transition. One such area is northern Spain, where there is no hiatus between Mousterian and Upper Palaeolithic settlement (Butzer 1986). The lithic assemblages of this region are very similar over time with 'Middle Palaeolithic' types occurring in high percentages throughout the Upper Palaeolithic and Mesolithic because raw material homogeneity (a fine-grained quartzite) seems to produce assemblage homogeneity almost regardless of 'cultural' input (Straus 1978). Analysis of raw material variability and origins at Palaeolithic sites is in its initial stages. However, it appears to be the case that much of what had been construed to be culturally-inspired changes in assemblage variability might simply be due to the ways in which stone was procured over time and what was done with it subsequently (Bricker 1975; Demars 1982; Larick 1986).

In sum, there is clear evidence, in both Eastern and Western Europe, for the *in situ* development of several EUP industries out of a LMP base. However, precise developmental relationships of these industries with older, contemporaneous and younger industries remain, for the most part, unclear due to a lack of detailed technological comparisons and to simple problems of resolution in the archaeological record.

Bone Technology. There is a clear and generally-recognized dichotomy in the manufacture of bone/antler/ivory implements and *objêts d'art* between the Middle and the Upper Palaeolithic. In our opinion there are very few unambiguous bone artifacts from Middle Palaeolithic contexts. Certainly they are not present either in the numbers or with the degree of standardization of form evident in late Upper Palaeolithic assemblages (for a different view, see Dewez 1982; Freeman 1983; Marshack 1982; Staesche 1983). In contrast, Upper Palaeolithic culture-stratigraphic units all contain at least some bone and antler artifacts. In fact, the presence of bone and antler artifacts is often used to distinguish the two periods and many of the index types (especially for various subdivisions of the LUP) comprise such objects (Klein 1973; Mellars 1973; White 1982). Yet there is a great deal of variability in the bone technologies of the Upper Palaeolithic, and EUP assemblages tend to have very poorly developed bone technologies (*contra* White, this volume). Châtelperronian levels in France and Spain invariably produce few bone artifacts, with very simple items like awls, pierced teeth, incised fragments and questionable miscellaneous tools comprising most of the inventory (Harrold 1981, and this volume). Early Aurignacian levels also contain relatively few worked bone objects, especially when compared to those found in the LUP (Bhattacharya 1977). Early Aurignacian levels at Bacho Kiro cave in Bulgaria have produced almost no bone implements (Kozlowski 1982). Szeletian assemblages in Eastern Europe contain relatively minor bone/antler industries, and these are usually 'explained away' by invoking 'contact' with contemporaneous Aurignacian 'peoples' of the region (Allsworth-Jones 1986). All in all, there is a recognizable increase in the size and diversity of bone and antler industries, and art objects, over time in the Upper Palaeolithic but the industries nearest the transition are not at all well developed. This clinal picture of change suggests gradual, *in situ* development of bone and antler technologies during the course of the Upper Palaeolithic and not something that was imported from elsewhere by modern *Homo sapiens*.

Subsistence: Taphonomic and Demographic Aspects. The faunal record across the transition is generally very homogeneous within regions except for the large carnivore component which seems to be much better represented in Middle Palaeolithic contexts (Straus 1982; Gamble 1983, 1984). The degree to which carnivores contributed to the formation of rock- shelter and cave faunas during the Middle and Upper Palaeolithic is still being assessed (Binford 1981; Gamble 1986; Straus 1982; Lindly, in press). In regions where diachronic studies are available (e.g. northern and southeastern Spain, Pyrenean France), carnivore involvement in the formation of cave faunas is always more marked during the Middle and early Upper Palaeolithic than it is during the LUP. Over the long run, this suggests a greater human involvement in the accumulation process as hunting strategies and technologies became more effective and, perhaps, as humans became more prevalent in the landscape. Increasing regional population densities might be indicated archaeologically by more frequent human use of rock-shelters and caves, and by a somewhat predictable diversification of the resource base (Clark and Yi 1983; Clark and Straus 1986; Clark 1987).

In the studies with which we are familiar, a kind of 'subsistence threshold' seems to have been crossed in the LUP, about 20 000 years ago. Prior to that time, changes are gradual, cumulative and largely explicable in terms of climatic perturbations that had an impact on the kinds and numbers of local fauna available in the landscape. After about 20 000 BP, change becomes much accelerated as a pattern of linked diversification and intensification increasingly seems to characterize regional diets. Although evidence for diversification and intensification is not everywhere apparent, in those areas where it does occur, it seems to be related to stress caused by regional population growth – growth that upset pre-existing balances between humans and resources and to which humans responded by redoubling their efforts to obtain food despite the increased costs necessitated by the addition of low-yield, labour intensive or dangerous species to the diet (Cohen 1977; Earle 1980; Christenson 1980). It is important to note that these changes occur as much as 20 000 years *after* the Middle-Upper Palaeolithic transition, and so are probably not related to the biological evolution of morphologically-modern humans.

Taphonomic and demographic considerations aside, the five main ungulate species found in Middle Palaeolithic contexts (red deer, reindeer, auroch, bison and horse) continue to occur in Châtelperronian, Aurignacian and Upper Périgordian contexts in Southwestern Europe (Bahn 1984; Chase 1986; Clark and Yi 1983; Freeman 1981; Straus 1977, 1985). Continuity is also apparent in the faunal remains across the transition in other areas of Europe (see Allsworth-Jones 1986 for Central Europe; Klein 1973 for the Ukraine; Gamble 1979 for Germany; Soffer 1985, and this volume, for the central Russian plain). This continuity underscores the fact that foragers are conservative in subsistence pursuits and will abandon a major subsistence strategy only under extreme duress. Thus it is likely that attempts to gain a better understanding of long term changes in the human food niche will depend more upon assessments of relative rather than absolute frequency changes. To the extent that these changes can be predicted by general models of human subsistence behaviour (e.g. Cohen 1977; Earle 1980), it is to that extent that more satisfactory explanations will be forthcoming in particular regional studies.

Subsistence: Hunting Strategies. The hunting strategies employed during these periods are more difficult to identify and evaluate. Chase (1986) suggests that Middle Palaeolithic hunters were practising a 'purposeful eclectic hunting strategy' whereby different selected species were exploited at different times (seasons, intervals) at different locations in the landscape. Straus (1977, 1985) and Clark (1987; Clark and Yi 1983) have argued that Middle Palaeolithic hunters were more opportunistic, procuring the more obvious game animals as they presented themselves in the environment. Given a mobile settlement/subsistence system, it is difficult to determine the difference between these two hypothesized strategies since we cannot generate the test implications required to subject them to a rigorous evaluation. Within the limits of available data, the EUP pattern is not distinguishable from that of the LMP.

The suggestion that Middle Palaeolithic hunters might have been less specialized than their Upper Palaeolithic counterparts is rendered less tenable in the light of data from sites like l'Hortus in France, with an ibex-dominated fauna (Pillard 1972), Starosel'e in the Crimea, with lots of horses (Klein 1969), Mauran in the Pyrenees, where the fauna is dominated by bovines (Bahn 1983) and Repolust Cave in Austria, with an ibex fauna (Chase 1986).

In sum, there does not appear to be a quantitative or a qualitative change in niche width or species diversity until the LUP when there is evidence for intensification in the procurement of food resources in Cantabrian Spain (Clark and Yi 1983; Clark and Straus 1986; Clark 1987), in the Périgord region of France (White 1985), on the central Russian plains (Soffer 1985), and perhaps in other areas as well. So far as we can tell, the pattern of utilization of animal resources is pretty much the same on either side of the Middle-Upper Palaeolithic transition.

Settlement Patterns. Southwestern European settlement pattern data from the Middle and Upper Palaeolithic are difficult to assess because of the historical bias toward the excavation of cave and rock-shelter sites in the archaeological research traditions of the area and a general absence of a regional perspective. In France, there is an opinion that Middle Palaeolithic open sites tend to be located on ridge crests and interfluves while Upper Palaeolithic open sites are more frequently found in valleys (White 1983, 1985). In southwest France, the number of sites with EUP deposits appears to *increase* over that of the Middle Palaeolithic (Mellars 1973; White 1982; Gamble 1986), but open-air locations have been all but ignored in these assessments (however, see Gaussen (1981) for France). In contrast, site numbers seem to *decrease* over the transition in southern Germany, possibly as a consequence of deteriorating climatic conditions in this part of Europe during Würm 3 (Gamble 1986).

Site areas also appear to increase on the average when the Upper is compared with the Middle Palaeolithic, but a poor understanding of formation processes and spatial arrangements within sites precludes an assessment of what this might mean in terms of regional population, local group size and duration of occupation. Many cave and rock-shelter sites that contain

Mousterian deposits also have superimposed EUP deposits suggesting continuity in the use of these locations over time, although not necessarily for the same purposes (for an example of shifting site functions over time at a single locale, see Straus and Clark 1986). There appears to be a significant break in patterns of site location during the LUP although there is also evidence of continuity in site use between the early and the late Upper Palaeolithic in some areas. Discontinuity is especially apparent beginning with the Solutrean (*c.* 21 000 BP) and continuing into the Magdalenian, with many of these LUP sites being in different locations than those of the preceding EUP (White 1985). It is significant that this change in settlement patterns appears to coincide in time with the evidence for subsistence intensification noted above.

THE NEAR EAST

Technology: Chronological Aspects of Assemblage Variability. In the Levant, Middle Palaeolithic assemblages have been separated into consecutive chrono-cultural industrial phases based on blank shape and tool frequencies found at the 'type site' of et-Tabūn, on Mount Carmel in Israel. Most workers recognize three phases, although the terminology used to identify them is not entirely consistent (cf. Copeland 1975; Jelinek 1981, 1982). The Bordes typology is used to classify and organize tool and debitage forms but, so far, it has not been possible to identify Near Eastern counterparts of the facies documented in southwestern France.

The early Levantine Mousterian (Phase 1) is characterized by elongated Levallois points, a preponderance of 'Upper Palaeolithic' tool types like burins and end-scrapers, and a blade technology. This phase has been dated to *c.* 90-80 000 BP at Tabūn, Layer D, and at open sites like Rosh ein Mor in the central Negev highlands (Crew 1976). Phase 2 Mousterian, found in Layer C at Tabūn, is an assemblage type dominated by broad, oval Levallois flakes, including fat little points, and classic 'Mousterian' tools such as flake side-scrapers. Phase 2 has been dated to *c.* 60-50 000 BP at the type site. The 'final' Phase 3 Mousterian, identified at Tabūn, Layer B, is not well known but has been characterized as a return to narrow point and blade forms. Phase 3 is frequently combined with Phase 2 (= Phase 2/3) by workers like Copeland (e.g. 1983) because of inadequate descriptions of the Layer B assemblage at the type site. The date for this phase at Tabūn would be of the order of 50 000 BP.

It has been recognized for about 10 years that not all Levantine Mousterian sites contain assemblage types that can be accommodated by the Tabūn sequence, thus calling into question its general currency as a descriptive device. Moreover, while the chronology of the phases is reasonably well established at the type site on the basis of sedimentological and geochronological work (Ferrand 1979) backed up to some extent by controversial amino-acid-racemisation dates (Masters 1982), whether the phases represent the same temporal order elsewhere as they do at Tabūn is highly questionable. Tabūn is the only site where all three phases occur together in a stratified sequence.

In the Northern Levant, the initial Mousterian corresponds typologically to Phase 2/Layer C at Tabūn. These assemblages come from open sites in Lebanon and Syria dated through study of marine transgression/regression cycles to the same interval (*c.* 90-80 000 BP) as Phase 1/Layer D at the type site (Copeland 1981, 1983). In the Southern Levant there is evidence from the Negev and from south Jordan for the persistence of Phase 1/Layer D type assemblages late in the Mousterian (Henry 1982; Marks 1983; Lindly and Clark 1987). These findings suggest both homogeneity and continuity in the Southern Levantine Mousterian in respect of assemblages with Phase 1/Layer D characteristics. Unfortunately, however, there is an erosional episode in the geological record between early assemblages (e.g. Rosh ein Mor) and late ones (e.g. Boker Tachtit, Level 1) so continuity cannot be demonstrated stratigraphically at any site. It is at least a possibility that there was an occupational hiatus between these periods as well (Marks 1981). Furthermore, Phase 2/3 assemblages have recently been described in northeastern Sinai (Gilead 1984) and in the northern Negev (Gilead and Grigson 1984), indicating that more variability probably exists in Southern Levantine Mousterian assemblages than was previously suspected.

Technology: Metric Studies of Blank Characteristics and their Chronological Implications. In addition to the phase division of Mousterian assemblages, the claim has been made that there is a metric basis for chronological separation of Phase 1/Layer D and Phase 2-3/Layer B-C type assemblages. At Tabūn, Jelinek (1977, 1981, 1982) believes that he has demonstrated vectored change in blank dimensions over time from the late Lower Palaeolithic through the Middle Palaeolithic. A comparison of width:thickness ratio variances from samples of whole unretouched flakes from these strata shows that Phase 1/Layer D flakes are significantly thicker than those of Phase 2-3/Layer B-C. In addition, metric comparisons of length:width ratios of Levallois points indicate that these distinctive artifacts are significantly more elongated in Phase 1/Layer D than in later assemblages (Jelinek 1982; cf. Munday 1979). Ideally, it should be possible to 'date' Middle Palaeolithic industries in the Levant on the basis of these metric criteria. Unfortunately, however, there are some glaring anomalies and also problems with the extent to which Jelinek's metric studies can be generalized beyond the type site.

First, as mentioned above, Phase 1/Layer D industries appear to be both early *and* late in the Southern Levant, and Phase 2/Layer C industries might also be present in the Northern and Central Levant both early and late in the Mousterian sequence. Second, the metric approach, utilizing width:thickness variance statistics, might retain some general chronological validity if comparisons are restricted to certain artifact classes like flake debitage and Levallois points (i.e. the standardized end products of standardized reduction sequences), rather than applied to the analysis of the entire debitage component (Jones 1985). The practice up until the present has been to analyse a sample of the entire debitage component. Third, the presence of substantial numbers of elongated Levallois points does not appear to be a useful chronological indicator, as was formerly thought, since such points are now known to occur (1) in *early* Levantine Mousterian assemblages (e.g. in Tabūn, Layer D); (2) in late ones (e.g. Tor Sabiha, Boker Tachtit, Ain Difla); and (3) in transitional Upper Palaeolithic industries (e.g. K'sar Akil, Phase A) (Copeland 1986). The capacity to produce elongated Levallois points was probably related to the size and quality of available raw material, and to pan-Levantine technological knowledge invoked differentially under the 'right' conditions (i.e. when there was a need and where suitable raw material was available). (In other words, everybody knew how to make them at least since the beginnings of the Middle Palaeolithic, and they made them when they needed them *and* when they had good flint in large enough nodules to allow for the implementation of a full-blown Levallois point technology!).

Finally, there are two sites (Far'ah II and Sefunim) which are thought to be late and which combine certain Phase 2/Layer C characteristics with very thick flake components and consequently smaller-than-expected width:thickness ratios (Gilead and Grigson 1984; Ronen 1984). This evidence suggests that the trend from thicker to thinner flakes seen at Tabūn and other sites might not occur in certain regions due to differences in raw material. A solution to problems with detecting vectored change in Middle Palaeolithic technologies depends on accurate dating of these assemblages by radiometric methods. Unfortunately, most of them appear to fall in the interval between the presently expanding capabilities of the radiocarbon tandem accelerator mass spectrometer (TAMS) methods (now up to *c.* 60 000 BP) and techniques developed for dating ancient geological deposits (e.g. potassium-argon and thorium-uranium - down to *c.* 200 000 BP).

Transition Scenarios. Problems with the Levantine Mousterian phase sequence carry over to models of the transition to the Upper Palaeolithic since the Mousterian levels that immediately precede the EUP are characterized in terms of the Tabūn sequence. Understanding the transition is exacerbated by an erosional episode between most Middle and Upper Palaeolithic deposits at cave and rock-shelter sites which effectively separates these assemblages stratigraphically and, in the opinions of some scholars, culturally as well (Bar-Yosef and Vandermeersch 1972). There is, however, good evidence of a developmental sequence from the Middle to the Upper Palaeolithic in the Levant, although the transition is only recorded at a few sites. Essentially three scenarios for the transition have been proposed, based on excavated data from three sites (Copeland 1986; Marks 1983, 1986; Ronen 1984).

1. *K'sar Akil:* The first is based on an analysis of the lithic assemblages from the K'sar Akil rock-shelter in Lebanon (Copeland 1986). Copeland believes that the Mousterian levels at this site are of a Phase 3/Layer B type – an assemblage of blade and point forms that she regards as 'late'. In addition, the Mousterian levels produced a number of chamfered pieces, regarded by some as an Upper Palaeolithic transition index type. However, a description and analysis of the Mousterian levels has never been fully published, and there is supposedly an occupational hiatus between them and the overlying Upper Palaeolithic sequence.

The initial Upper Palaeolithic deposits, the K'sar Akil Phase I (or A) transitional levels, contain standardized elongated point forms produced by a specialized Levallois technology, chamfered pieces like those in the Mousterian strata, an 'Upper Palaeolithic' retouched tool component (burins, end-scrapers), and a few flakes. Transitional Phase II (or B), which directly overlies Phase I levels, has two subunits. Phase IIA is characterized by backed blades along with a shift to the production of Upper Palaeolithic 'punch' blades, although the points continue to be manufactured by the Levallois method. Phase IIB is characterized by a retouched point made on blades struck from prismatic cores. Copeland argues that Phases I and II are typologically Upper Palaeolithic assemblages and are not transitional (despite the elongated points) in the sense of containing both Levallois Mousterian and Upper Palaeolithic tool forms. Technologically, blanks are produced by a specialized Middle Palaeolithic Levallois technique in Phase I, both Levallois and prismatic blade techniques are present in Phase IIA, and only a blade technique is found in Phase IIB.

There is another gap between the transitional levels and the overlying Aurignacian sequence. The early Aurignacian levels contain another specialized point form made on a retouched blade (the 'el Wad point'), twisted bladelets, and carinate and other burin types made on flakes. Blanks are manufactured by an 'Upper Palaeolithic' blade technology, including possible use of a punch technique. There thus appears to be some continuity in the transitional sequence at K'sar Akil in the production of point forms with a technological 'break' in the early Aurignacian levels (Bergman 1981). Continuity is also evident in the transition from a specialized Levallois point technique to an Upper Palaeolithic blade technology. A diagram of Copeland's (1986) conceptualization of the K'sar Akil sequence is given in Table 32.2. Figure 32.2 presents a generalized stratigraphy for the part of

---

Figure 32.2. The London Concordance of the Ksar Akil stratigraphies proposed by Bergman, Tixier and Copeland (March 1987). The Aurignacian classificatory framework of Copeland and Hours (1971), as developed from the old collections, is superimposed. The 20 shell dates from the site are incoherent and are rejected by these and most other workers. Dates given are Monaco dates on charcoal, except for the most recent, which is on bone.

According to Copeland (1986), Ewing's Levels XIII-VII are Aurignacian, divided into 'early' (XIII, XII) and 'later' (XI-VII) phases. The Middle/Upper Palaeolithic transition sequence comprises Ewing's Levels XXV-XIV, divided by Copeland (1986) into three phases (cf. Table 32.2).

To these workers, the Upper Palaeolitic at Ksar Akil appears to be comprised mainly of Levantine Aurignacian industries, typified by Tixier's Phases V-VII, and reckoned by tool frequencies and the presence/absence of *fossiles directeurs*. Inspection of blank forms and debitage from both collections indicates that small retouched blades and bladelets are common throughout (in Tixier's collections they average c. 40 percent of the tools, and can account for as much as 65 percent), raising the issues of whether or not Ahmarian levels might be present. Marks (pers. comm.), who has studied large parts of the old collections, thinks that there is a transitional Middle/Upper Palaeolithic sequence analogous to that of Boker Tachtit, followed by interstratified (and sometimes mixed) Levantine Aurignacian and Ahmarian deposits. The latter appear to be more common in the upper part of the sequence, where they grade into a kind of Kebaran (see Table 32.3).

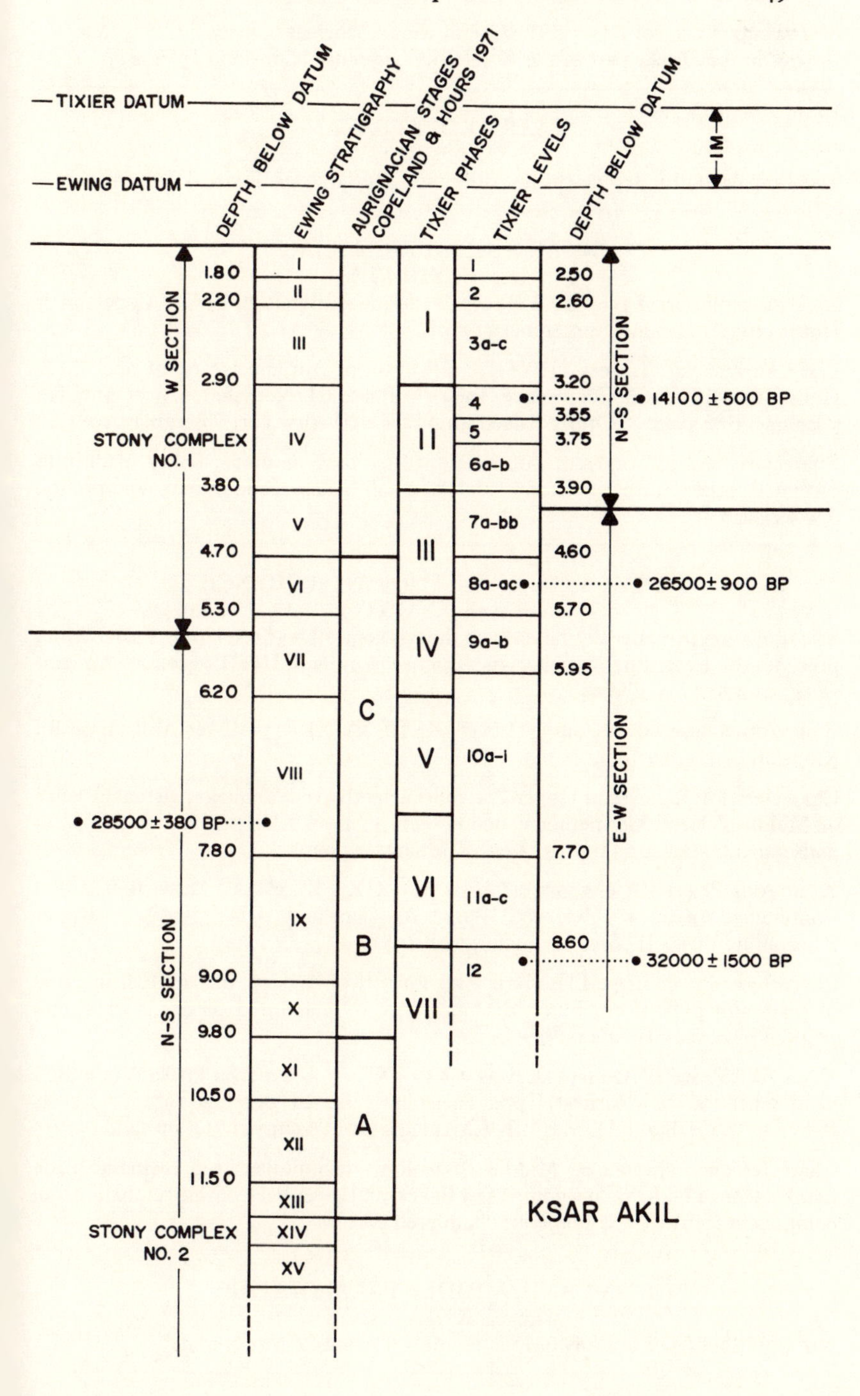

TIXIER DATUM
EWING DATUM
DEPTH BELOW DATUM
EWING STRATIGRAPHY
AURIGNACIAN STAGES COPELAND & HOURS 1971
TIXIER PHASES
TIXIER LEVELS
DEPTH BELOW DATUM
1M
W SECTION
STONY COMPLEX NO. 1
N-S SECTION
STONY COMPLEX NO. 2
E-W SECTION
14100 ± 500 BP
26500 ± 900 BP
28500 ± 380 BP
32000 ± 1500 BP
1.80
2.20
2.90
3.80
4.70
5.30
6.20
7.80
9.00
9.80
10.50
11.50
I
II
III
IV
V
VI
VII
VIII
IX
X
XI
XII
XIII
XIV
XV
C
B
A
1
2
3a-c
4
5
6a-b
7a-bb
8a-ac
9a-b
10a-i
11a-c
12
2.50
2.60
3.20
3.55
3.75
3.90
4.60
5.70
5.95
7.70
8.60
KSAR AKIL

Table 32.2 Concordance of the Classification Schemata for the Lithic Industries in the Upper part of the K'sar Akil Sequence (Copeland 1986: 4,5)

K'SAR AKIL KEBARAN
(Levels VI-I)

The Epipalaeolithic sequence: not fully published but see Tixier & Inizan (1981)

K'SAR AKIL AURIGNACIAN
(Levels XIII-VII)

Later Aurignacian (Levels XI-VII) : =Levantine Aurignacian B, C of Copeland & Hours (1971), London Conference (1969)

Early Aurignacian (Levels XIII-XII) : = Levantine Aurignacian A of Copeland & Hours (1971), London Conference (1969); = Phase III (Neuville), Lower Antelian (Garrod), first part; = Upper Palaeolithic Phase III (first part) (Bergman 1981)

Characteristics: co-occurrence of different knapping methods; flakes made into carinated, other burins; twisted bladelets, small blades made into el Wad points; few scrapers

K'SAR AKIL TRANSITIONAL SEQUENCE
(Levels XXV-XIV)

The transition in many subphases from the typologically earliest Upper Palaeolithic; precedes the Levantine Aurignacian sequence (Levels XIII-VII); grouped for convenience into three phases.

K'sar Akil Phase IIB (Azoury) (Levels XVIII-XV/XIV): = K'sar Akil Phase Bii (Copeland 1975, 1976)

Characterstics: increase in Upper Palaeolithic methods of debitage; virtual absence of Middle Palaeolithic methods; points (esp. in Lev. XVII); *pointes à face plan*; endscrapers; almost no burins; Lev. XIV sample poor.

K'sar Akil Phase IIA (Azoury) (Levels XX-XIX): K'sar Akil Phase B (London Conference 1969), = K'sar Akil Phase Bi (Copeland 1975), 1976), = Upper Palaeolithic Phase II (first part) (Bergman 1981)

Characteristics: debitage like K'sar Akil Phase I of Azoury, but gradual increase in plain and punctiform butts; disappearance of chamfered pieces; backed and pointed pieces; endscrapers; few burins

K'sar Akil Phase I (Azoury) (Levels XXV-XXI): = K'sar Akil Phase A (London Conference 1969), = (earliest) Upper Palaeolithic Phase I (Bergman 1981), = Levels If-IVe at Abu Halka and Levels VII-V at Antelias Cave (Azoury 1971, Copeland 1970)

Characteristics: specialised Middle Palaeolithic techniques used to make blade blanks, elongated Levallois points; few flakes; an Upper Palaeolithic retouched tool component (burins, endscrapers); chamfered pieces

K'SAR AKIL MIDDLE PALAEOLITHIC
(Levels XXXVI-XXVI)

Not published: = Levalloiso-Mousterian, = Levantine Mousterian the sequence bearing on the transition, and a reconciliation of the analytical units of previous workers. It is interesting to note that Copeland (1986) apparently does not recognize the existence of EUP 'Ahmarian' levels at K'sar Akil, while other workers (notably Marks) believe they are interstratified with Levantine Aurignacian levels after the Middle-Upper Palaeolithic transition sequence at some point below Ewing's Level XXV. The very recent (March, 1987) consensus definitions of these Levantine Upper Palaeolithic analytical units are given in Table 32.3, together with sites and levels where they are believed to occur.

2. *Sefunim:* The second transition scenario is based on data from Sefunim cave, on Mount Carmel in Israel (Ronen 1984). Here the Level 12 transitional industry develops out of a Phase 2-3/Layer B-C assemblage and is Mousterian in typology but Upper Palaeolithic in technology. The 'Upper Palaeolithic' affinity in Level 12 is particularly evident in the manufacture of blades, which compare well with those from other Upper Palaeolithic sites metrically, morphologically, and in the percentage of prepared platforms. This level also has very low technological and typological Levallois indices.

3. *Boker Tachtit:* The third transitional sequence is that recorded at the open site of Boker Tachtit, in the central Negev highlands of Israel. Boker Tachtit documents a sequential technological development from Middle to Upper Palaeolithic, but with little concurrent change in typology. The site is nearly unique in the Levant because of the presence of four intact 'living floors' that produced abundant lithic debris which, through extensive refitting of cores, allowed for the detailed examination of change in reduction sequences over time. This site has been well published by Marks and his colleagues (Marks 1983, 1985; Marks and Volkman 1983; Volkman 1983) so only a brief outline of the developmental sequence will be presented here.

Boker Tachtit Level 1 has been labelled 'terminal early Levantine Mousterian' due to the presence of both unidirectional and bidirectional Levallois point production (Volkman 1983; Marks 1985). The assemblage is also characterized by the presence of what are taken to be exclusively Upper Palaeolithic tools (burins, end-scrapers), along with retouched flakes. This level has a *terminus ante quem* date of >47 000 BP. Levels 2 and 3 overlie Level 1 and have been labelled 'transitional' industries because of technological heterogeneity in blank manufacturing techniques suggesting an interval of experimentation between the specialized Levallois reduction sequences of Level 1 and a single-platform prismatic blade technology found in Level 4. The blank shape remains consistent throughout these levels. Level 4 is considered Upper Palaeolithic because of the production of 'true blades' from uniform single platform blade cores.

Throughout the sequence, typologically-similar point forms are produced and it is worth noting that, in default of the core reconstructions, all these would have been classified as the same tool type (Levallois points). Yet they are manufactured by a specialized Levallois point technology in Level 1 and with an Upper Palaeolithic blade technology in Level 4. At Boker Tachtit, there appears to be a complete technological developmental

Table 32.3. Normative characterisations of the Levantine Upper Palaeolithic assemblage types (K'sar Akil Conference – London 1987 (ES = Endscraper; BU = Burin)

LEVANTINE AURIGNACIAN	AHMARIAN
dominated by flake debitage most large retouched pieces on flake blanks thick endscrapers relatively common ES + BU (45-65%) retouched bladelets variable, sometimes common twisted bladelets (some the by-products of retouching carinates) bone and antler tools	dominated by blade/bladelet debitage most large retouched pieces on blade blanks single and opposed platform blade/ bladelet cores with simple platforms direct percussion, soft hammer technique almost no carinates ES made on blades, thin flakes retouched (back, pointed) bladelets common (>35%) bladelets with abrupt to fine retouch ES + BU (<40%) few bone and antler tools
REPRESENTATIVE SITES AND LEVELS Sites are grouped subjectively into 'more' and 'less credible' according to observations at the Conference. No chronological order is implied.	
Ein Aqev (D31) (18-16 kyr BP) Hayonim D El Wad D Kebara D K'sar Akil VII, VIII (Ewing) K'sar Akil V-VII (Tixier) (32 kyr BP)	Ein Aqev East Boker BE II-IV (26 kyr BP) Lagama VII Abu Noshra II (33 kyr BP) Qadesh Barnea 9, 601A, 601B (34 kyr BP) Boker A (37 kyr BP) Qafzeh 7-9
El Wad C El Wad E Kebara E Arkov K'sar Akil XI-XIII (Ewing) D27A Sde Divshon (D27B) K9A G11 Boker C Boker BE I	K'sar Akil IX, X (Ewing) K'sar Akil XVI, XVII (Ewing) Lagama IIID, V, X, XV (34-31 kyr BP) Lagama VIII, XVI, XI

Differences of opinion turned on: (1) the relative importance of the retouched tool and debitage components as classificatory criteria; (2) on differences between old and modern excavation procedures and recovery techniques (and the impact these might have on assemblage composition); and (3) on differences in resolution among the several typological schemata currently used.

Some workers (esp. Marks) subscribe to the view that the Levantine Aurignacian and Ahmarian represent lithic reduction and manufacturing traditions characterised by temporal variability (as represented by phases which are, at present, poorly defined) and synchronic, site-functional variability (as represented by a Levantine equivalent to Bordes 'facies').

Levantine Aurignacian assemblages show a correlation with relatively mesic, more forested, Mediterranean phytogeographic associations. The Ahmarian seems to occur more frequently in xeric, open steppe/scrub, Irano-Turanian desert zones.

sequence from the late Mousterian to the early Upper Palaeolithic. The sequence cannot be related to earlier Levantine Mousterian assemblages, however, or to later Upper Palaeolithic ones, and since the site is so far unique, the question remains as to how general the vectored technological changes are.

On the surface, the three transitional scenarios suggest three somewhat distinct, although overlapping, technological transitions (or three aspects of the same transition) from the Middle to the Upper Palaeolithic. Since different variables are monitored in each case, it is difficult to untangle broader differences and similarities, nor is it clear to what degree raw material differences might have played a role. The transitional industries from K'sar Akil and Boker Tachtit appear to be very similar, with the Phase I-II sequence at K'sar Akil roughly matching the transition sequence at Boker Tachtit. Both sites record a transition from a specialized Levallois technology to one in which true blades were manufactured. The Sefunim sequence also records a transition from a Levallois to a non-Levallois technology, but instead of occurring in 'transitional' Level 12, the change had already taken place in the last Middle Palaeolithic level. Level 12 is also anomalous because it is typologically dominated by Middle Palaeolithic tool forms like side-scrapers and truncated flakes (Ronen 1984).

In sum, it seems clear that the technological transition in flaked stone industries varies in different parts of the Levant. In northern and central areas, there is a development of Tabūn Phase 2/Layer C or Phase 3/Layer B industries, which can be typologically dominated by either Upper or Middle Palaeolithic tool forms, into the Levantine Aurignacian. In the Southern Levant, there is a development of Tabūn Phase 1/Layer D Mousterian into both Ahmarian and Aurignacian industries. Modal variation within and between Upper Palaeolithic assemblage types is just beginning to be appreciated, and there is some disagreement in regard to normative characterizations of Levantine Aurignacian and 'Ahmarian' (after Erq el'Ahmar) analytical units (cf. Marks 1981; Gilead 1981; Henry 1982 and our Table 32.3). However, on the basis of the numerous core reconstructions from the Boker Tachtit sequence, there is clear technological continuity across the transition. Typological variation in these transitional assemblages might turn out to be irrelevant or misleading in efforts to understand exactly how the transition took place.

Bone Technology. There is little evidence for a well-developed bone industry in either the Middle or the Upper Palaeolithic of the Levant. The Middle Palaeolithic has few modified bone artifacts with examples like the incised shaft fragment from the Mousterian level at Kebara cave being about the extent of its development (Davis 1974). Bone industries are also rare in the Levantine EUP but, as in Europe, this technology becomes more important in the LUP and the Epipalaeolithic (Copeland and Hours 1977; Newcomer and Watson 1984). It is possible that bone artifacts were overlooked in earlier excavations due to rather primitive recovery procedures and to the simple, nondescript nature of the artifacts themselves, which might not have been noticed amongst the usually-meagre faunal remains. At present,

Table 32.4. Tentative Palaeoclimatic Sequence for the late Pleistocene of the Central Negev (Israel) (from Goldberg 1973, 1979, 1981; Horowitz 1976, 1979)

Archaeological Units	Sedimentological & Vegetational Characteristics	Avdat/Aqev Sites†	Macroclimatic Trends	Marks' (1981, 1983) Settlement Pattern Characterisitics
Epipalaeolithic-Early Natufian [ca. 17,000–13,000 BP]	periodic very slightly more humid oscillations 17-12,000 BP followed by contradictory evidence: drier conditions post-13,000 BP indicated by the sediments (Goldberg 1981), wetter condition post-14,000 BP by the pollen (Horowitz 1979); by the late Natufian (ca. 10,500 BP) a drier climate as indicated by pollen data from Rosh Zin (D-16)	D-5, Rosh Zin (D-16) D-101, Ein Aqev (D-31)	Drying Somewhat Wetter	climatic evidence equivocal; possible brief return to radiating configuration during the early Natufian, followed by circulating pattern in the late Natufian
Upper-Epipalaeolithic Transition [23-22,000–15,000 BP]	continuation of drying trend with arboreal fraction 7% (D-34), then 3% (D-31); NAP indicates slightly wetter conditions than present; erosion beginning ca. 23,000 BP becomes marked after ca. 15,000 BP; formation of colluvial silt lenses after ca. 18,000 BP; maximum aridity ca. 16-15,000 BP	Ein Aqev (D-31) D-34	Drying	
Upper Palaeolithic (ca. 45,000–20,000 BP]	complex sedimentary sequence with continued alluviation characterised by the accumulation of coarse, then fine terrace gravels, sands (until ca. 27,000 BP), then silts, clayey colluvium (until ca. 20,000 BP); decline in runoff energetics over time; climate somewhat more humid (and considerably more humid 32,000-27,000 BP) until ca. 27,000 BP, when a trend towards greater aridity begins; 16% AP at D-22, D-27; climatic belts 150-200 km S of present locations	D-22, D-27a,b D-100 D-34	Drying Wetter Drying	circulating pattern with no significant intersite variability (i.e., more difficult to distinguish between base camps, limited activity stations); repeated reoccupation of sites (but without spatial consistency in activity area placement); more mobile settlement/subsistence system tied to increased importance in scheduling in resource procurement in a more arid environment than during the Middle Palaeolithic
Middle-Upper Palaeolithic Transition [ca. 47,000–ca.45,000 BP]	new cycle of alluviation with formation of terraces up to 15 m thick; somewhat drier than previously with 17% AP at D-101; NAP much the same as early Mousterian	D-101	Dry Briefly Somewhat Wetter?	shift to circulating pattern with trend toward increased dessication; decline in site size, intersite variability and evidence of sedentism
Later Mousterian (ca. 65,000–ca. 45,000 BP]	drying trend; erosion (wadi downcutting with destruction of many early Mousterian sites), consequently few sedimentary traps for later Mousterian industries		Drying	radiating settlement/subsistence system with base camps and associated limited activity sites; base camps characterised by high artifact density, stratified deposits and the formation of middens, spatially-consistent tool kits; a relatively sedentary pattern or, alternatively, a pattern of reoccupation at regular intervals; logistical strategy possible due to optimal climatic conditions vis a vis the Upper Palaeolithic
Early Mousterian [90,000 +–ca. 65,000 BP]	wet; formation of gravel terraces and travertines in springs; 25% AP at D-35; channel aggradation by colluviation: climatic belts 200-250 km S of	D-35, D-15	Wet	

evidence for a technology in bone is so sparse that it cannot be used to differentiate the LMP from the EUP. There is, in consequence, no discernible change across the transition.

Subsistence. Unfortunately there are few large samples of faunal material that bear on the periods of interest here. The samples that do exist suggest that differences between the Middle and the Upper Palaeolithic are related more to environmental perturbations than to hypothetical changes in procurement strategies (Davis 1977; Garrard 1983; Lamdan 1984; Payne 1983; Tchernov 1984). As was the case in Europe, there is no real change in the species taken from the Middle to the Upper Palaeolithic that cannot be explained either in terms of micro-environmental differences between site locations or by macro-environmental change through time. The animals hunted during the Mousterian (*Capra, Dama, Gazella, Equus, Cervus*) continue to be exploited in more or less the same way during the EUP. The much-cited evidence of specialized hunting of wild goats in the Middle Palaeolithic levels at Shanidar Cave (e.g. Evins 1984) can probably be explained by the location of the site in the rocky terrain that these caprids favour.

Settlement Patterns. Unlike Europe, there is a tradition of regional survey research in the Near East that goes back at least to the early 1950s with Braidwood's pioneering efforts in Iraqi Kurdistan (e.g. Braidwood and Howe 1960; Braidwood *et al.* 1983). Settlement pattern studies in the Levant are of more recent vintage (essentially from the late 1960s) but have produced a wealth of data collected in intensive surveys of (1) the central Negev highlands (Marks 1976, 1977, 1983); (2) the northern Sinai Desert (Bar-Yosef and Belfer 1977; Gisis and Gilead 1977; Gilead 1977, 1981); (3) the south Jordan Plateau (Henry 1982, 1985; Henry *et al.* 1983); (4) the Black Desert of north-central Jordan (Betts 1983); (5) the Azraq Basin and its catchment (Garrard *et al.* 1977, 1985); and (6) the southern tributaries of the Wadi Hasa (west-central Jordan) (MacDonald *et al.* 1980, 1983; Coinman *et al.* 1986). Although of variable quality, this work allows for the examination of possible spatial relationships of Palaeolithic archaeological sites in these regions, and provides a data base that has no European counterparts.

Marks and Freidel's Circulating/Radiating Dichotomy. Probably the best-known settlement pattern model for the Middle and the Upper Palaeolithic is the circulating/radiating dichotomy developed for the Avdat/Aqev region of the Negev by Marks and his colleagues (Marks and Freidel 1977). A radiating pattern consists of one or two large sites or residential camps surrounded by a constellation of smaller special-purpose or 'limited-activity' sites. In a radiating pattern, residential mobility is relatively low, logistical mobility relatively high. A circulating pattern comprises many sites of approximately similar size and with approximately similar artifact inventories. Residential mobility in a circulating pattern is very high and there is little or no logistical mobility. In the Negev, the Middle Palaeolithic is argued to have been characterized by a radiating settlement pattern, and the Upper Palaeolithic a circulating one (Table 32.4). Mesic conditions that favoured the expansion of Mediterranean phytogeographic zones supposedly allowed for greater residential stability during the early Middle Palaeolithic. Environmental deterioration and the spread of xeric Irano-Turanian steppe/desert associations during the latter part of the Middle and throughout the Upper Palaeolithic resulted in smaller group sizes and more mobility. Readers will recognize the similarity of these configurations to Binford's (1980) forager/collector continuum. However, Marks and his colleagues are essentially arguing for modern, collector-like logistical behaviour during the Middle Palaeolithic while Binford (1982, 1983) is of the opinion that logistical strategies probably did not exist prior to the evolution of modern humans.

Henry's Transhumance Model. Recently, Donald Henry has developed a competing Levantine settlement/subsistence model based on contemporary and historical Bedouin land-use practices, which he believes to be of great antiquity. The 'transhumance model' has been partially tested with archaeological data from the Wadi Hisma, on the edge of the south Jordan Plateau (Henry 1984). It is applicable to areas of marked topographic relief and well-defined seasonal variation in the temperature and moisture regimes. Four distinct versions are presented (for the Middle and Upper Palaeolithic, Epipalaeolithic and Chalcolithic) which depict seasonal movement between the piedmont and the lowlands – movements that affected group size, composition, and activity patterns somewhat differently in each of the four chronological periods. Archaeological confirmation of the model is based on variability in site size and exposure, artifact density and permanency, number and diversity of features (Henry 1984: 14).

During the Middle, Upper and Epipalaeolithic, grossly similar patterns of transhumance are thought to have prevailed, although there are differences of degree amongst the three periods. The model predicts that large, semi-permanent Middle and Upper Palaeolithic winter sites would be located at intermediate elevations in the piedmont – the residential bases of aggregated groups. During the summer, these groups supposedly dispersed to more transitory encampments at lower elevations making temporary use of a series of small, low-density, limited-activity stations with relatively specialized toolkits.

Unfortunately, the Middle and Upper Palaeolithic archaeological data do not conform well to the predicted site distributions since large 'winter' sites are generally found in the lowlands, and small 'summer' camps are usually located in the piedmont. In any event, the pattern appears to remain the same across the transition. Henry has suggested that the lack of 'fit' is due to environmental changes since the late Pleistocene which affected the zones receiving maximal rainfall runoff. Today, the most mesic microenvironments are found in the Hisma lowlands. During the late Pleistocene, the piedmont might have received more runoff than it presently does (and relatively more than the Hisma lowlands).

The Composite Model of Coinman, Clark and Lindly. A third Levantine land-use model combines aspects of the circulating/radiating model, the forager/collector continuum and the transhumance model, and applies them to survey data from the southern tributaries of the Wadi Hasa (Coinman *et al.* 1986). Palaeolithic data from this survey are very variable in terms of reliability, but site size and site locations appear to be quite consistent on both sides of the transition, with primarily small sites found at medium-to-high elevations. The Middle/Upper Palaeolithic combined sample, comprising sites with artifacts 'diagnostic' of both periods, is characterized by somewhat smaller sites located at significantly lower elevations. The settlement pattern from the Middle and Upper Palaeolithic thus shows signs of continuity, discounting the combined sample which is anomalous *vis-à-vis* the samples where clear-cut Middle and Upper Palaeolithic components could be distinguished. The pattern more closely approximates a 'radiating' than a 'circulating' configuration, with fairly good site size and elevation dichotomies in both periods – dichotomies that become more clearly marked when the data sets from tributary wadis are considered individually. The Hasa data do not fit Henry's transhumance model. Although small sites at moderate-to-high elevations are recorded, the larger, more permanent residential sites at lower elevations are totally absent.

Settlement patterns in the Middle and Upper Palaeolithic are variable from region to region with major, although apparently clinal changes occurring in the Negev and no discernible change in Jordan. Mean site size appears to change across the transition in both areas, however, with significantly smaller sites found in the Upper Palaeolithic samples. Site numbers also decline fairly sharply in the Upper Palaeolithic, according to Ronen (1984) and MacDonald *et al.* (1982). When comparisons are restricted to cave and rock-shelter sites, locations remain fairly constant from the Middle to the Upper Palaeolithic, indicating continued, although sporadic use of these loci over very long periods of time. The size and locations of open sites vary greatly between the periods as criteria for site placement change with deteriorating environmental conditions, at least in the Negev (Marks and Freidel 1977).

Intrasite spatial analysis has been attempted for the Middle Palaeolithic open site of Rosh ein Mor in the Negev (Hietala and Stevens 1977). The study demonstrated the regular occurrence of certain artifact types at the same location in each level of the site, indicating either a degree of permanence in occupation (and almost compulsively patterned spatial behaviour!), or repeated episodes of site use separated by intervals short enough that customary use of space was retained in people's memories (and transmitted from generation to generation?) throughout the entire span of human use of the locus. A comparable study of Upper Palaeolithic intrasite patterns has not been undertaken, possibly because no sites analogous to Rosh ein Mor have been so extensively excavated.

THE CULTURAL TRANSITION IN PERSPECTIVE

The cultural transition as perceived in lithic industries across the Middle-Upper Palaeolithic boundary in the Levant is different from that of Europe. The difference is due to differences in biases and systematics which emphasize typological approaches and 'cultural' explanations in Europe and technological approaches and 'functional/behavioural' explanations in the Levant. At Boker Tachtit there is an unequivocal developmental sequence that conclusively demonstrates a technological transition. Unlike Europe, which has no Boker Tachtit, the case for a technological transition and continuity between the Middle and Upper Palaeolithic periods in the Levant is well established. Levantine Middle Palaeolithic systematics contribute to a view of change that is more comprehensive than most European approaches because assemblages are more often analysed in their entirety: retouched pieces, tool forms, debitage, metrical and raw material characteristics, and reduction sequences are all taken into account. This expands comparison beyond mere tool forms and consequently monitors more than typological change over time. Debitage analysis of this sort has only begun in Europe fairly recently (e.g. Fish 1979, 1981).

The structure of Levantine Upper Palaeolithic systematics also differs from that of Europe in that only two analytical units are recognized and these are largely contemporaneous throughout most of the period. These assemblage types, the Levantine Aurignacian and the Ahmarian, are not, so far as we can tell, perceived to be the remains of 'cultural' entities of any kind, although they are identified on the basis of blank and tool-type percentages, reflecting the influence of *la méthode Bordes* (although not its explanatory framework). Archaeological 'index types' like the Emireh point or the chamfered blade are only used to recognize the transition (Copeland 1986) and even for that purpose have been called into question (Marks 1983).

The bone/antler/ivory industries of Europe and the Near East are also very different. European bone industries, which practically do not exist in the Middle Palaeolithic, become very well developed in the Upper Palaeolithic (especially in the LUP). The LUP clearly has also produced most of the spectacular Upper Palaeolithic art. By contrast, Levantine bone industries never become very elaborate or important in either period. It is only during the Epipalaeolithic (i.e. after *c.* 20 000 BP) that there is good evidence for bone industries as elaborate and diversified as those of the European LUP.

Although the evidence is much better from Europe, the character and organizational features of subsistence systems appears to remain constant in both areas across the transition, with little change in the number or diversity of animals hunted until the LUP in Europe and the Epipalaeolithic in the Levant. Good regional samples of Levantine Upper Palaeolithic subsistence data are practically non-existent, however.

Settlement patterns evince no change in Southwestern Europe in respect of site locations, although site numbers might have increased when the Middle Palaeolithic as a whole is compared with the LUP. In the Levant, there is also little change in site locations, but site numbers appear to decline fairly dramatically during the Upper Palaeolithic. The scarcity of Upper Palaeolithic sites is usually explained by an absence of suitable depositional contexts related to predominantly xeric environments, and does not necessarily imply a reduction in population (e.g. Goldberg 1981).

CONCLUDING REMARKS

At the suggestion of Paul Mellars, we have attempted a broad comparison of the evidence for the biological *and* the cultural transition in Europe and the Near East – the two geographic areas where most of the transition research has been conducted. As is true of any synthetic effort, this was a formidable undertaking and one that, by its very nature, is likely to antagonize nearly everybody. However, we have tried throughout to make our own biases explicit, emphasizing the fact that data do not exist apart from the conceptual frameworks that define them (Clark 1982, 1987). Since variability in archaeological and palaeontological assemblages is potentially infinite, we can only apprehend pattern through the classificatory systems we choose to adopt. Although this may seem self-evident, paradigmatic biases underlie most of the differences of opinion registered at these meetings and in the literature. As Milford Wolpoff once put it: 'The data do not speak for themselves. I have been in rooms with data and listened very carefully. They never said a word' (1975: 15). Adopting this point of view meant that we had to try to come to grips with the biological evidence for the transition, an area in which both of us are admitted non-specialists. Since much is to be gained from trying to integrate the two fields, we hope we will be forgiven, or at least tolerated, for this invasion of intellectual turf (see also Clark and Lindly, in press).

Perceptions of Neanderthals and their relationships to early moderns influence perceptions of the cultural transition in complex and subtle ways. The transformation of AHS into MMH and the related mitochondrial DNA evidence are hotly debated topics in palaeoanthropology, and disagreements arise largely because of paradigm differences that put a 'loading' on basic concepts and terms like 'Neanderthal' itself. Perceptions of Neanderthals vary from those that invoke the essentially modern behavioural patterns implied by the consensus taxonomic designation (*Homo sapiens neanderthalensis*) to those that envision a pithecanthropine-like configuration, with only a rudimentary component of learned behaviour (e.g. Binford, among others). Whatever position is taken on taxonomic matters, AHS occupies a substantial chunk of time only bounded artificially, in our view, by workers interested in various aspects of what was clearly an evolutionary continuum.

The unilineal phylogenetic model for the Upper Pleistocene to which we subscribe implies the existence of a Neanderthal grade interposed between pithecanthropines and that of modern humans, and the rejection of various 'pre-sapiens' and 'pre-Neanderthal' phylogenetic models. Both positions are likely to be controversial in the context of these meetings. We submit, however, that the phylogenetic issue, while by no means a dead issue, is in the end an uninteresting one. No consensus was reached on it in the recent *Origins of Modern Humans* volume (Smith and Spencer 1984) despite much sophisticated functional and morphometrical analysis of AHS samples from different parts of the Old World. The irony in these disputes about phylogeny was that there was substantial agreement about the nature of the sample itself and how it should be apprehended, that the nature and rate of change varied geographically and through time, and that biological changes did not take place in a behavioural vacuum and must have reflected adaptation in ways that are, at present, only partly understood. There also seemed to be substantial agreement about AHS grade characters, and concordant interpretations of functional morphology, although the phylogenetic implications of these were much debated (Clark 1988).

Our reading of the palaeoanthropological literature indicated support for regional continuity across the transition both in Europe (where there is a diversity of opinion) and in the Near East (where there seems to be more consensus). Evidence for regional continuity is based on vectored change in brow-ridge morphology, the form of the cranial vault, midfacial reduction, the morphology of the mandibular symphysis, reduction in the anterior dentition and its support structures, and a decrease in the skeletal indicators of muscular robusticity. These changes are attributed to increasingly 'cultural' solutions to universal problems of survival – solutions that depended upon behavioural changes and linked technological dependencies that, *over the long run,* replaced sheer physical strength and endurance. We regard genetic continuity over the transition as a certainty, and see the transition as a largely *in situ* phenomenon in every area that has produced a late Pleistocene hominid sample. We reject population replacement arguments because we can conceive of no behavioural mechanism that could account for population displacements on an order of magnitude large enough to support actual physical movement of peoples, given the extremely low population densities estimated for the Upper Pleistocene. We also think that a persistent, variety-minimizing 'type-specimen' mentality on the part of some workers amplifies differences that would be considered trivial if we had better samples.

At this point, it would probably be appropriate to react to the discovery of the Saint-Césaire skeleton – an acknowledged 'Neanderthal' associated with an EUP industry (Vandermeersch 1984). To judge from the frequency with which it is cited, Saint-Césaire has caused something of a paradigm crisis, since it calls into question the venerable, although poorly supported, notion that Neanderthals were not responsible for the creation of Upper Palaeolithic industries (this in sharp contrast to the Near Eastern situation, where alleged 'moderns' are clearly associated with Middle Palaeolithic industries: Bar-Yosef and Vandermeersch 1981; Valladas *et al.* 1988).

Some reactions to Saint-Césaire call to mind what could be referred to as 'the Richard Leakey approach to palaeoanthropology' in which finds, claimed to be spectacular for one reason or another, are supposed to upset the entire theoretical and conceptual applecart and require the emergence of a comprehensive 'new synthesis' (Clark, 1988). With due account taken of Thomas Kuhn's (1977) model of relatively rapid paradigm change (a model we regard as essentially accurate), we do not think that science works the way that Leakey and some of his colleagues suggest that it does. If Saint-Césaire really necessitates a total restructuring of our conceptions of the transition, it would imply that the paradigms that govern this area of research are impoverished indeed, and conceptually bankrupt. Stringer

(1982, and this volume; Stringer *et al.* 1984) has argued for a break in continuity after the Neanderthals in Western Europe, using a suite of craniofacial indices, and has concluded that Neanderthals are more like modern Europeans than they are like the EUP sample. Smith (1982, 1984) and Wolpoff (1980a, 1980b; Wolpoff *et al.* 1981) see more clinal change – continuity in place – in respect of their analyses of Eastern European transition data. Trinkaus (1983, 1984, 1986) envisions an *in situ* pattern of gradual morphological change for the West Asian sample, but more evidence of discontinuity in Europe.

We contend that these different points of view are to a large extent paradigm dependent, that perceptions of differences and similarities are based exclusively on the suites of variables regarded as significant to measure, and on sample variability itself, which is a function of sample size. We suggest that we do not have anything like a representative sample of any conceivable biological population from either Europe or the Levant, that the time-space grid is so sparse that one can make a case for almost any postulated relationship, and that if we had a more representative sample, and a more complete time-space grid, that continuity would be demonstrated both in both areas.

Most research designs used in palaeoanthropology today are basically unconstrained 'pattern-search' approaches using variables that are selected more by convention or for convenience than for any demonstrated diagnostic utility (see Clark 1988) for an evaluation of systematics in human palaeontology). Explanations based on these approaches are what Binford has called *post-hoc accommodative arguments* – explanations developed after the fact to explain a pattern detected in a data set (1981: 31, 82, 83). They are only as convincing as the ingenuity of the investigator allows them to be. They can always be questioned by anyone inclined to reject the variables identified as 'significant to measure'.

Post-hoc accommodative arguments are a weak form of inference essentially because the research designs of which they are part lack a deductive component. Like archaeology, human palaeontology relies heavily upon methods borrowed from other fields that developed in the absence of general theory as a series of conventions for assigning meaning to the palaeontological record. These conventions exhibit a 'fad-like' quality in that they change within paradigms according to changes in commonly-recurring research situations. A typical research scenario involves a pattern search which, if at all competent, cannot fail to produce correlations among the variables examined. The question then becomes how to assign meaning to the patterns that have been isolated. One's imagination is typically engaged to identify the conditions that, if they actually occurred, would account for the observed pattern. Most people are sufficiently creative to be able to come up with a more-or-less plausible set of circumstances that could account for the observed facts. However, it is important to keep in mind that the degree of fit between the imagined conditions and the observed properties of the palaeontological record does not constitute a test of the accuracy of that reconstructed series of events. What usually happens is that warranting arguments are proposed to support the plausibility of the proposed explanation – to show that it is not unreasonable to suppose that it might have occurred the way the investigator suggests that it did (see Note 1).

Plausibility is frequently supported by an 'argument from elimination' that assumes that all potential causes of the pattern can be identified and enumerated (and ideally ranked, or assigned a probability), and that all but one can in fact be eliminated as the cause of the phenomenon in question. As noted earlier, however, the assertion that all possible causes were not in fact identified is sufficient to undermine the credibility of the argument (Binford 1981: 82-86). The case rests, so to speak, on the plausibility of the warranting arguments invoked in support of the explanation (or in some deplorable cases, by recourse to 'authority'). It must be acknowledged that there is no simple solution to this dilemma (Binford proposes an increased emphasis on 'middle range theory' – actualistic studies that will allow us to use arguments from elimination properly). However, little is to be gained by ignoring these epistemological issues. If we continue to do that, we will continue to fail to confront the fundamental ambiguity of pattern in both the archaeological and the palaeontological records. We will fail to develop a basis for making strong inferences about the past.

Acknowledging that the task of assigning meaning to observations is very difficult, the differences between the Eastern and Western European AHS samples seem to us to be differences in degree rather than kind with (as Smith has pointed out) a more even, regular pattern of change in the east and a 'change in the rate of change' – more accelerated gene flow – in the west (Smith *et al.*, this volume). We are not suggesting that there are no differences between the Eastern and Western European AHS samples, nor in the rate of gene flow between the two areas, but we see no evidence of population replacement and, on balance, much evidence for *in situ* evolutionary change.

So, what to make of Saint-Césaire? Saint-Césaire probably represents part of the range of morphological variation within a single evolving West European clade. Other representatives of that clade would include specimens characterized in normative terms as 'morphologically modern humans' (Brace 1979). It is scarcely coincidental that, at sites where we have relatively large skeletal series (e.g. Mladeč), the range of morphological variation includes specimens that, if they were found in isolation, would probably be considered both 'Neanderthal' *and* 'modern'.

Perceptions of the cultural transition are also dependent on paradigmatic biases since the 'filters' created by the biases affect the nature of comparisons and the results obtained from them. Our approach to the cultural transition was considerably more 'fine grained' than our attempt to cope with the biological transition. We looked at lithic typology and technology in the two areas, and the effects that the incompatible typological systems currently used on either side of the transition in Europe have had on perceptions of differences and similarities across it. We came to the conclusion that, for Europe, the classification and subsequent comparison of retouched tool inventories cannot fail to indicate disjunction, since the Middle and Upper

Palaeolithic typologies commonly used measure different things. The most common Middle Palaeolithic typology, developed by François Bordes, is based on relative frequencies of morphotypes defined by the location and shape of retouched edges irrespective of the overall form of the blank and of temporal considerations. The Upper Palaeolithic typology of Sonneville-Bordes and Perrot emphasizes chrono-stylistic attributes within morphotypes – attributes that are thought to change over time. Vectored technological change appeared to cross-cut assemblage boundaries as determined by typological variability, not only in the Levant, where (because of Boker Tachtit) the case is exceptionally strong, but in Europe as well.

We also looked at raw material variability and its effects on reduction strategies and blank forms, the intensity of site use, taphonomic and demographic aspects of subsistence, hunting strategies and settlement patterns on both sides of the transition.

Variability in raw material procurement, reduction and use is a recurrent theme in recent attempts to explain Mousterian lithic assemblage variability. Quality, size and availability of raw material will affect choice of reduction strategies, blank dimensions and consequently assemblage composition. Many workers had noted variations in aggregate tool frequencies across the Mousterian facies, and in the proportions of blanks which showed Levallois reduction, but the European tendency, at least, was to explain these in cultural terms. This shift in emphasis, and research conclusions resultant from it, have had important implications for tracing continuity because criteria such as scraper percentages, incidence of bifaces, Levallois indices and the like are used to identify modal characteristics of Middle Palaeolithic assemblages both in areas where the facies are recognized and in areas like the Levant where they are not. They suggest that assemblage variability might simply be due to differences in raw material sources and in the intensity of site use (see below), and might have no 'cultural' implications at all.

Connected to this shift in explanatory frameworks have been attempts to determine the extent to which Mousterian assemblages generally, and the facies in particular, have been reduced. The basic idea is that there are marked differences in reduction, and in the elaboration of tools, that correspond to the intensity with which a particular assemblage type was used. This idea originates in the early 1980s with Arthur Jelinek, who suggested that the most intensively reduced facies (Quina/Ferrassie and Mousterian of Acheulian Tradition) might be correlated with site contexts where habitation episodes were relatively intense and prolonged. The least reduced assemblages (Typical, Denticulate), with lots of expedient tools, might represent activities correlated with much more ephemeral site use.

Jelinek also suggested that in Western Europe, where Pleistocene climatic fluctuations were marked, these modal differences in the intensity of site use might have had climatic correlates. Rigorous climatic intervals would have resulted in more intensive use of caves and rock-shelters, more reduction of assemblages, a higher incidence of formal retouched pieces, and a more predictable set of end products than those of milder climatic episodes.

Another consequence of this change in perspective has been to monitor the extent of reduction by examining the properties of Mousterian formal tool groups, especially side-scrapers. One of Jelinek's students, Harold Dibble, has shown that many scraper types identified in the Bordes typology are probably stages in a reduction and re-use sequence, rather than formal tool types in their own right. Considerable reduction of some of these tools suggests intensive re-use and possibly a longer 'use-life' (i.e. curation) than Binford's claims for expedient Mousterian lithic technologies would allow.

The European Upper Palaeolithic typology has not come under the same intense scrutiny as that of the Middle Palaeolithic, probably because it was initially very successful at partitioning the industries of the Upper Palaeolithic by using a *fossile directeur*-based system of time-sensitive 'stylistic markers'. Recently, however, archaeologists have come to realise that there are problems with the generality of this typology and its original strength has become its greatest weakness. This is because it 'weights' these stylistic index types more heavily than it does other assemblage characteristics. Our work in Spain, and that of a number of other researchers, clearly shows that a concentration on *fossiles directeurs* conceals a great deal of assemblage variability within and between the traditionally- defined Upper Palaeolithic culture-stratigraphic units. It is interesting to note that some prehistorians are coming to the realisation that there might be functional explanations for assemblage variability *within* parallel phyla such as the Périgordian and the Aurignacian, but refuse to acknowledge the possibility that these typologically-separate industries themselves could also signify functional aspects of the same EUP adaptation (Clark 1983).

Because of these considerations, we came to the conclusion that perceptions of differences in the cultural transition as manifest in lithic industries in the two areas were largely owed to differences in biases and systematics. These have traditionally emphasized typological approaches and 'cultural' explanations in Europe, and technological approaches and 'functional-behavioural' explanations in the Levant. We think that Levantine systematics contribute to a view of change that is more comprehensive than most European approaches because assemblages are more frequently analysed in their entirety: retouched pieces, tool forms, debitage, metric and raw material characteristics, and reduction sequences are taken into account, expanding comparison beyond mere tool forms and monitoring more than typological change over time.

The structure of Levantine Upper Palaeolithic systematics also differs from that of Europe in that only two analytical units are recognized and these are largely contemporaneous, rather than sequent, throughout most of that period. These assemblage types, the 'Ahmarian' and the 'Levantine Aurignacian' are not perceived to be the remains of 'cultural' entities of any kind, although they are identified on the basis of blank and tool-type characteristics, reflecting the influence of the Bordesian method, although not its explanatory framework.

Although the evidence is much better from Europe, the character and organizational features of subsistence systems appear to remain constant in both areas over the transition, with little change in the number and the diversity of species hunted until the LUP in Europe and the Epipalaeolithic in the Levant. In the studies with which we are familiar, a kind of 'subsistence threshold' seems to have been crossed for the first time in the LUP, about 20 000 years ago. Prior to that time, changes are gradual, cumulative and largely explicable in terms of climatic perturbations that had an effect on the kinds and numbers of local fauna available in the landscape. After about 20 000 BP, change becomes much accelerated as a pattern of linked diversification and intensification increasingly seems to characterize regional diets. Although evidence for these trends is not apparent everywhere, in those areas where they do occur, they seem to be related to stress caused by regional population growth – growth that upset pre-existing balances between humans and resources and to which humans responded by redoubling their efforts to obtain food despite increased costs. It is important to note that these changes occur as much as 20 000 years *after* the Middle-Upper Palaeolithic transition.

The hunting strategies employed during these periods are more difficult to identify and evaluate. Chase has suggested that Middle Palaeolithic hunters were practising a 'purposeful eclectic hunting strategy' whereby different selected species were exploited at different seasons, or intervals at different locations in the landscape. By contrast, Straus and I have argued that Middle Palaeolithic hunters were more opportunistic, procuring the more obvious game animals as they presented themselves in the landscape. Given a mobile settlement-subsistence system, it is difficult to determine the difference between these two strategies since, so far, we have been unable to generate the test implications required to subject them to a rigorous evaluation. Within the limits of available data, the EUP pattern is indistinguishable from that of the LMP.

It might have been expected that settlement pattern data would either refute or confirm the evidence for a breakpoint at *c.* 20 000 BP, but we found it difficult to compare Europe with the Levant because European research traditions have not emphasized the collection of regional survey data. Survey information from the Near East is more common, probably because the arid environment allows for relatively easy discovery and recording of surface sites. To the extent that we could generalize, settlement patterns showed no apparent change over the transition in Southwestern Europe in respect of site locations. Many cave and rock-shelter sites that contain Mousterian deposits also have superimposed EUP levels suggesting continuity in the use of these locations over time, although not necessarily for the same purposes.

There appears to be a significant break in patterns of site location during the LUP, although there is also evidence of continuity in site use between the early and the late Upper Palaeolithic in some areas. Site numbers seem to have increased in Europe when the Middle Palaeolithic *as a whole* is contrasted with the LUP. In the Levant, the reverse pattern seems to hold. Again, there is little change in site location, but site numbers appear to decline dramatically during the Upper Palaeolithic. A climatic explanation has been proposed by Marks and his colleagues, which entails the replacement of the logistically-organized, radiating configuration of the early Middle Palaeolithic with a circulating foraging pattern later in the Mousterian – a pattern that continues throughout the Upper Palaeolithic.

In sum, we detected only vectored, clinal change over the transition in the biological and cultural variables that we examined, and recognized few differences between Europe and the Near East in regard to those variables. No correlations were found between archaeological industries and 'types' of early human remains, nor did important changes in technology, subsistence and settlement patterns coincide with the biological transition in either area. A 'change in the rate of change' and archaeological evidence for the appearance of fully modern behavioural patterns appears to post-date both the Middle-Upper Palaeolithic transition and the appearance of morphologically-modern humans by at least 15-20 000 years.

NOTE

1. We think that warranting arguments are more commonly developed in archaeological than in palaeontological research, probably reflecting the greater concern with epistemology found in the Anglo-american research tradition. They are not particularly common outside that tradition.

ACKNOWLEDGEMENTS

We wish to thank Paul Mellars for suggesting this broad and initially quite intimidating topic as our contribution to the Symposium volume. Like most anthropologists, we have tended to confine our investigations to more limited subjects, those which are strongly grounded empirically, the comfortable 'nuts and bolts' of our often rather circumscribed research interests. It is a healthy and intellectually stimulating thing to be joggled out of the confines of one's research to attempt a broad synthetic effort (in this case, exceptionally broad in terms of concepts and paradigms, temporal depth and geographic spread). While this sort of essay is practically guaranteed to engender disagreement, it will have served its purpose if it stimulates the kind of discussion that advances both prehistory and palaeoanthropology. We enjoyed doing it, and we hope we haven't made too many mistakes (although errors of commission and omission no doubt remain, as well as naïveté).

REFERENCES

Allsworth-Jones, P. 1986. *The Szeletian and the Transition from the Middle to the Upper Palaeolithic in Central Europe*. Oxford: Clarendon Press.

Anderson, P. 1980. A testimony of prehistoric tasks: diagnostic residues on stone tool working edges. *World Archaeology* 12: 181- 194.

Anderson-Gerfaud, P. 1981. *Contributions Méthodologiques à l'Analyse des Microtraces d'Utilisations sur les Outils Préhistoriques*. Unpublished Ph.D. Thesis, University of Bordeaux I.

Bahn, P. 1981. Review of M. Hemingway: *The Initial Magdalenian in France*. *Proceedings of the Prehistoric Society* 47: 325-326.

Bahn, P. 1983. Late Pleistocene Economies in the French Pyrenees. In G. N. Bailey (ed.) *Hunter-Gatherer Economy in Prehistory: a European Perspective*. Cambridge: Cambridge University Press: 167-185.

Bahn, P. G. 1983. *Pyrenean Prehistory: a Palaeoeconomic Survey of the French Sites*. Warminster: Aris and Phillips.

Barton, C. M. 1986. Patterns of variability in Middle Palaeolithic tools. Paper presented at the Annual Meeting of the Society for American Archaeology meeting, New Orleans, 1986.

Bar-Yosef, O. and Belfer, A. 1977. The Lagaman industry. In O. Bar-Yosef and J. Phillips (eds) *Prehistoric Investigations in Gebel Maghara, Northern Sinai*. Jerusalem: Institute of Archaeology of the Hebrew University: 42-84.

Bar-Yosef, O. and Vandermeersch, B. 1972. The stratigraphic and cultural problems of the passage from Middle to Upper Palaeolithic Paleolithic in Palestinian caves. In F. Bordes (ed.) *The Origin of Homo Sapiens*. Paris: UNESCO: 221-225.

Bar-Yosef, O. and Vandermeersch, B. 1981. Notes concerning the possible age of the Mousterian layers in Qafzeh Cave. In J. Cauvin and P. Sanlaville (eds) *Préhistoire du Levant*. Paris: Centre National de la Recherche Scientifique: 281-285.

Bergman, C. 1981. Point types in the Upper Palaeolithic sequence at K'sar Akil, Lebanon. In J. Cauvin and P. Sanlaville (eds) *Préhistoire du Levant*. Paris: Centre National de la Recherche Scientifique: 319-330.

Betts, A. 1983. Black Desert survey: second preliminary report. *Levant* 15: 1-17.

Beyries, S. 1987. *Variabilité de l'Industrie Lithique au Mousterien: Approche Fonctionelle sur Quelques Gisements Français*. Oxford: British Archaeological Reports International Series S238.

Bhattacharya, D. K. 1977. *Paleolithic Europe*. Oosterhout (The Netherlands): Anthropological Publications.

Binford, L. R. 1973. Interassemblage variability: the Mousterian and the functional argument. In C. Renfrew (ed.) *The Explanation of Cultural Change*. London: Duckworth: 227-254.

Binford, L. R. 1979. Organization and formation processes: looking at curated technologies. *Journal of Anthropological Research* 35: 255-273.

Binford, L. R. 1980. Willow smoke and dogs' tails: hunter-gatherer settlement systems and archaeological site formation. *American Antiquity* 45: 4-20.

Binford, L. R. 1981. *Bones: Ancient Men and Modern Myths*. New York: Academic Press.

Binford, L. R. 1982. Comment on R. White: 'Rethinking the Middle/Upper Paleolithic transition'. *Current Anthropology* 23: 177-181.

Binford, L. R. 1983. *In Pursuit of the Past*. London: Thames and Hudson.

Binford, L. R. and Binford, S. R. A preliminary analysis of functional variability in the Mousterian of Levallois facies. *American Anthropologist* 68: 238-295.

Binford, S. R. 1972. The significance of variability: a minority report. In F. Bordes (ed.) *The Origin of Homo sapiens*. Paris: UNESCO: 199-210.

Bordes, F. 1954. Notules de typologie paléolithique III: pointes moustériennes, racloirs convergents et déjétés, limaces. *Bulletin de la Société Préhistorique Française* 51: 336-339.

Bordes, F. 1961. *Typologie du Paléolithique, Ancien et Moyen*. Institut de Préhistoire de l'Université de Bordeaux, Mémoire 1: Vol. 1-2. Paris: Delmas.

Bordes, F. 1972. Du paléolithique moyen au paléolithique supérieur: continuité ou discontinuité? In F. Bordes (ed.) *The Origin of Homo*

Sapiens. Paris: UNESCO: 211-218.

Bordes, F. and Sonneville-Bordes, D. de. 1970. The significance of variability in paleolithic assemblages. *World Archaeology* 2: 61-73.

Boule, M. 1911-1913. L'homme fossile de La Chapelle-aux-Saints. *Annales de Paléontologie* 6: 111-172; 7: 21-56, 85-192; 8: 1-70.

Brace, C. 1962. Refocusing on the Neanderthal problem. *American Anthropologist* 64: 729-741.

Brace, C. 1964. The fate of the 'classic' neanderthals: a consideration of hominid catastrophism. *Current Anthropology* 5: 3-43.

Brace, C. 1967. *The Stages of Human Evolution*. Englewood Cliffs (NJ): Prentice-Hall.

Brace, C. 1968. Neanderthal. *Natural History* 77 (5): 38-45.

Brace, C. 1979. Krapina, 'classic' Neanderthals, and the evolution of the European face. *Journal of Human Evolution* 6: 527-550.

Braidwood, R. and Howe, B. 1960. *Prehistoric Investigations in Iraqi Kurdistan*. Studies in Ancient Oriental Civilizations 31. Chicago: University of Chicago Press.

Braidwood, L., Braidwood, R., Howe, B., Reed, C. and Watson, P. J. 1983. *Prehistoric Archaeology along the Zagros Flanks*. Oriental Institute Publications 105. Chicago: University of Chicago Press.

Breuil, H. 1912. Les subdivisions du paléolithique supérieur et leur signification. *Comptes-Rendus du 14eme Congrès International d'Anthropologie et d'Archéologie Préhistorique* (Genève, 1912): 165-238.

Breuil, H. 1939. The Pleistocene succession in the Somme valley. *Proceedings of the Prehistoric Society* 5: 33-38.

Bricker, H. 1975. The provenience of flint used for the manufacture of tools. In H. Movius Jr. (ed.) *Excavations of the Abri Pataud, Les Eyzies (Dordogne)*. American Schools of Prehistoric Research Bulletin 30. Harvard: Peabody Museum: 194- 197.

Bricker, H. 1976. Upper Paleolithic archaeology. *Annual Review of Anthropology* 5: 133-148.

Bricker, H. 1978. Lower to Middle Périgordian continuity. In M. Giardino *et al.* (eds) *Codex Wauchope (Human Mosaic):* Vol. 11: 165-182.

Brose, D. and Wolpoff, M. 1971. Early Upper Paleolithic man and late Middle Paleolithic tools. *American Anthropologist* 73: 1156-1194.

Butzer, K. 1986. Paleolithic adaptations and settlement in Cantabrian Spain. In F. Wendorf and A. Close (eds) *Advances in World Archaeology* Vol. 5. New York: Academic Press: 201-252.

Campbell, J. 1986. Hiatus and continuity in the British Upper Paleolithic: a view from the antipodes. In D. Roe (ed.) *Studies in the Upper Paleolithic of Britain and Northwest Europe*. Oxford: British Archaeological Reports International Series S296: 7-42.

Cann, R., Stoneking, M. and Wilson, A. 1987. Mitochondrial DNA and human evolution. *Nature* 325: 31-36.

Chase, P. 1986. *The Hunters of Combe Grenal*. Oxford: British Archaeological Reports International Series S286.

Christenson, A. 1980. Change in the human niche in response to population growth. In T. Earle and Christenson, A. (eds) *Modeling Change in Prehistoric Subsistence Economies*. New York: Academic Press: 31-72.

Clark, G. 1982. Quantifying archaeological research. In M. B. Schiffer (ed.) *Advances in Archaeological Method and Theory* Vol. 5. New York: Academic Press: 217-273.

Clark, G. 1983. Una perspectiva funcionalista en la prehistoria de la región Cantábrica. In A. Balíl *et al.* (eds) *Homenaje al Profesor Martín Almagro Basch*: Vol. 1. Madrid: Ministerio de Cultura: 155-170.

Clark, G. 1984. The Negev model for paleoclimatic change and human adaptation in the Levant. *Annual of the Department of Antiquities of Jordan* 28: 225-248.

Clark, G. 1987. From the Mousterian to the Metal Ages: long-term change in the human diet of northern Spain. In O. Soffer (ed.) *The Pleistocene Old World: Regional Perspectives*. New York: Plenum: 293-316.

Clark, G. 1988. Observations on the Black Skull (WT-17000, *A. boisei*): an archaeological assessment of systematics in human paleontology. *American Anthropologist* 90: 357-371.

Clark, G. 1989. Alternative models of Pleistocene biocultural evolution: a response to Foley. *Antiquity* 63: 153-161.

Clark, G. and Lindly, J. (in press). The biocultural transition and the origin of modern humans in the Levant and Western Asia. In Aurenche, O., Cauvin, M.-C. and Sanlaville, P. (eds) *The Prehistory of the Levant: Culture Change from the Beginning until the Sixth Millennium* BC. Paris: Centre National de la Recherche Scientifique. In Press.

Clark, G. and Straus, L. 1983. Late pleistocene hunter-gatherer adaptations in Cantabrian Spain. In G. N. Bailey (ed.) *Hunter-Gatherer Economy in Prehistory: a European Perspective*. Cambridge: Cambridge University Press: 131-148.

Clark, G. and Straus, L. 1986. Synthesis and conclusions I: Upper Paleolithic and Mesolithic hunter-gatherer subsistence in northern Spain. In L. Straus and G. Clark (eds) *La Riera Cave: Stone Age Hunter-Gatherer Adaptations in Northern Spain*. Arizona State University Anthropological Research Papers 36. Tempe: Arizona State University Press: 351-366.

Clark, G. and Yi, S. 1983. Niche-width variation in Cantabrian archaeofaunas: a diachronic study. In J. Clutton-Brock and C. Grigson (eds) *Animals and Archaeology*. Vol. 1: *Hunters and their Prey*. Oxford: British Archaeological Reports International Series S163: 183-208.

Cohen, M. 1977. *The Food Crisis in Prehistory*. New Haven: Yale University Press.

Coinman, N., Clark, G. A. and Lindly, J. 1986. Prehistoric hunter-gatherer settlement in the Wadi el'Hasa, west-central Jordan. In L. Straus (ed.) *The End of the Paleolithic in the Old World*. Oxford: British Archaeological Reports International Series S284: 129-269.

Copeland, L. 1975. The Middle and Upper Palaeolithic of Lebanon and Syria in the light of recent research. In F. Wendorf and A. Marks (eds) *Problems in Prehistory: North Africa and the Levant*. Dallas: Southern Methodist University Press: 317-360.

Copeland, L. 1981. Chronology and distribution of the Middle Paleolithic as known in 1980 in Lebanon and Syria. In J. Cauvin and P. Sanlaville (eds) *Préhistoire du Levant*. Paris: Centre National de la Recherche Scientifique: 239-263.

Copeland, L. 1986. Introduction to Volume 1. In I. Azoury *K'sar Akil, Lebanon*. Vol. 1: *Levels XXV-XII*. Oxford: British Archaeological Reports International Series S289: 1-24.

Copeland, L. and Hours, F. 1977. Engraved and plain bone tools from Jiita (Lebanon) and their early Kebaran context. *Proceedings of the Prehistoric Society* 43: 295-301.

Crew, H. 1976. The Mousterian site of Rosh Ein Mor. In A. Marks (ed.) *Prehistory and Paleoenvironments in the Central Negev*, Israel. Vol. 1: *The Avdat/Aqev Area (Part 1)*. Dallas: Southern Methodist University Press: 75-112.

Davidson, I. 1986. The geographical study of late Paleolithic stages in eastern Spain. In G. N. Bailey and P. Callow (eds) *Stone Age Prehistory:*

Studies in Memory of Charles McBurney. Cambridge: Cambridge University Press: 95-118.

Davis, S. 1974. Incised bones from the Mousterian of Kebara Cave (Mt. Carmel) and the Aurignacian of Ha-Yonim cave (western Galilee), Israel. *Paléorient* 2: 181-182.

Davis, S. 1977. The ungulate remains from Kebara Cave. *Eretz-Israel* 13: 150-163.

Demars, P. 1982. *L'Utilisation du Silex au Paléolithique supérieur: Choix, Approvisionnement, Circulation – l'Exemple du Bassin de Brive*. Paris: Centre National de la Recherche Scientifique.

Dewez, M. 1982. Comment on R. White: 'Rethinking the Middle/Upper Paleolithic transition'. *Current Anthropology* 23: 182.

Dibble, H. 1984. Interpreting typological variation of Middle Paleolithic scrapers: function, style, or a sequence of reduction? *Journal of Field Archaeology* 11: 431-436.

Dibble, H. 1987. The interpretation of Middle Paleolithic scraper morphology. *American Antiquity* 52: 109-117.

Earle, T. 1980. A model of subsistence change. In T. Earle and A. Christenson (eds) *Modeling Change in Prehistoric Subsistence Economies*. New York: Academic Press: 1-30.

Evins, M. 1982. The fauna from Shanidar cave: Mousterian wild goat exploitation in northeastern Iraq. *Paléorient* 8: 37-58.

Farrand, W. 1979. Chronology and paleoenvironment of Levantine prehistoric sites as seen from sediment studies. *Journal of Archaeological Science* 6: 369-392.

Fish, P. 1979. *The Interpretive Potential of Mousterian Debitage*. Arizona State University Anthropological Research Papers 16. Tempe: Arizona State University Press: 1-167.

Fish, P. 1981. Beyond tools: Middle Paleolithic debitage analysis and cultural inference. *Journal of Anthropological Research* 38: 347-386.

Foley, R. 1987. Hominid species and stone-tool assemblages: how are they related? *Antiquity* 61: 380-392.

Freeman, L. 1964. *Mousterian Developments in Cantabrian Spain*. Unpublished Ph.D. Dissertation, Department of Anthropology, University of Chicago.

Freeman, L. 1973. The significance of mammalian faunas from Paleolithic occupations in Cantabrian Spain. *American Antiquity* 38: 3-44.

Freeman, L. 1981. The fat of the land: notes on Paleolithic diet in Iberia. In R. S. O. Harding and G. Teleki (eds) *Omnivorous Primates: Gathering and Hunting in Human Evolution*. New York: Columbia University Press: 104-165.

Freeman, L. 1983. More on the Mousterian: flaked bone from Cueva Morín. *Current Anthropology* 24: 366-372.

Gamble, C. 1979. Hunting strategies in the Central European Paleolithic. *Proceedings of the Prehistoric Society* 45: 35-52.

Gamble, C. 1983. Caves and faunas from last glacial Europe. In J. Clutton-Brock and C. Grigson (eds) *Animals and Archaeology*. Vol. 1: *Hunters and their Prey*. Oxford: British Archaeological Reports International Series S163: 163-172.

Gamble, C. 1984. Regional variation in hunter-gatherer strategy in the Upper Pleistocene of Europe. In R. Foley (ed.) *Hominid Evolution and Community Ecology: Prehistoric Human Adaptation in Biological Perspective*. London: Academic Press: 237-260.

Gamble, C. 1986. *The Palaeolithic Settlement of Europe*. Cambridge: Cambridge University Press.

Garrard, A. 1983. The Paleolithic faunal remains from Adlun and their ecological context. In D. Roe (ed.) *Adlun and the Stone Age:* Vol. 2.

Oxford: British Archaeological Reports International Series S159: 397-409.

Garrard, A., Stanley-Price, N. and Copeland, L. 1977. A survey of prehistoric sites in the Azraq Basin of eastern Jordan. *Paléorient* 3: 109-126.

Garrard, A., Harvey, P., Hivernel, F. and Byrd, B. 1985. The environmental history of the Azraq Basin. In A. Hadidi (ed.) *Studies in the History and Archaeology of Jordan:* Vol. 2. London: Routledge and Kegan Paul: 109-115.

Gaussen, J. 1981. Le paléolithique supérieur de plein air en Périgord. 14th Supplement to *Gallia Préhistoire*. Paris: Centre National de la Recherche Scientifique.

Gilead, I. 1977. Lagama X. In O. Bar-Yosef and J. Phillips (eds) *Prehistoric Investigations in Gebel Maghara, Northern Sinai*. Jerusalem: Institute of Archaeology of the Hebrew University: 102-114.

Gilead, I. 1981. Upper Paleolithic tool assemblages from the Negev and Sinai. In J. Cauvin and P. Sanlaville (eds) *Préhistoire du Levant*. Paris: Centre National de la Recherche Scientifique: 331-343.

Gilead, I. 1984. Paleolithic sites in northeastern Sinai. *Paléorient* 10: 135-142.

Gilead, I. and Grigson, C. 1984. Far'ah II: a Middle Paleolithic open air site in the northern Negev. *Proceedings of the Prehistoric Society* 50: 71-98.

Gisis, I. and Gilead, I. 1977. Lagama III. In O. Bar-Yosef and J. Phillips (eds) *Prehistoric Investigations in Gebel Maghara, Northern Sinai*. Jerusalem: Institute of Archaeology of the Hebrew University: 85-102.

Goldberg, P. 1981. Late Quaternary stratigraphy of Israel: an eclectic view. In J. Cauvin and P. Sanlaville (eds) *Préhistoire du Levant*. Paris: Centre National de la Recherche Scientifique: 55-66.

Harrold, F. 1981. New perspectives on the Châtelperronian. *Ampurias* 43: 35-85.

Harrold, F. 1983. The Châtelperronian and the Middle-Upper Paleolithic transition. In E. Trinkaus (ed.) *The Mousterian Legacy: Human Biocultural Change in the Upper Pleistocene*. Oxford: British Archaeological Reports International Series S164: 123-140.

Hemingway, M. 1980. *The Initial Magdalenian in France*. Oxford: British Archaeological Reports International Series S90.

Henry, D. 1982. The prehistory of southern Jordan and relationships with the Levant. *Journal of Field Archaeology* 9: 417-444.

Henry, D. 1982. Paleolithic adaptive strategies in southern Jordan: results of the 1979 field season. In A. Hadidi (ed) *Studies in the History and Archaeology of Jordan:* Vol. 1. London: Routledge and Kegan Paul: 41-48.

Henry, D. 1984. Patterns of transhumance in southern Jordan: ethnographic and archaeological analogs. Paper presented at the Annual Meeting of the American Schools of Oriental Research, Chicago.

Henry, D. 1985. Late Pleistocene environment and Paleolithic adaptations in the Southern Levant. In A. Hadidi (ed.) *Studies in the History and Archaeology of Jordan:* Vol. 2. London: Routledge and Kegan Paul: 67-78.

Henry, D., Hassan, F., Henry, K. and Jones, M. 1983. An investigation of the prehistory of southern Jordan. *Palestine Exploration Quarterly* 83: 1-24.

Hietala, H. and Stevens, D. 1977. Spatial analysis: multiple procedures in pattern recognition studies. *American Antiquity* 42: 539-559.

Horowitz, A. 1979. *The Quaternary of Israel*. New York: Academic Press.

Howell, F. 1957. The evolutionary significance of variation and varieties of 'Neanderthal' man. *Quarterly Review of Biology* 32: 330-347.
Howell, F. 1984. Introduction. In F. H. Smith and F. Spencer (eds) *The Origins of Modern Humans: a World Survey of the Fossil Evidence*. New York: Alan R. Liss: xiii-xxii.
Jelinek, A. 1976. Form, function and style in lithic analysis. In C. Cleland (ed.) *Culture Change and Continuity*. New York: Academic Press: 19-33.
Jelinek, A. 1977. A preliminary study of flakes from the Tabūn Cave, Mount Carmel. *Eretz-Israel* 13: 87-96.
Jelinek, A. 1981. The Middle Paleolithic in the Southern Levant from the perspective of the Tabūn Cave. In J. Cauvin and P. Sanlaville (eds) *Préhistoire du Levant*. Paris: Centre National de la Recherche Scientifique: 265-280.
Jelinek, A. 1982. The Tabūn Cave and Paleolithic man in the Levant. *Science* 216: 1369-1375.
Jelinek, A. 1982. The Middle Paleolithic of the Southern Levant, with comments on the appearance of modern *Homo Sapiens*. In A. Ronen (ed.) *The Transition from the Lower to Middle Paleolithic and the Origins of Modern Man*. Oxford: British Archaeological Reports International Series S151: 57-104.
Jones, M. 1985. The use of technological indices: a case study for the Levantine Mousterian. In C. Carr (ed.) *For Concordance in Archaeological Analysis*. Kansas City: Westport: 540-565.
Jordá, F. 1977. *Historia de Asturias: Prehistoria*. Oviedo: Ayalga.
Kennedy, K. 1975. *Neanderthal Man*. Minneapolis: Burgess.
Klein, R. 1969. *Man and Culture in the Late Pleistocene*. San Francisco: Chandler.
Klein, R. 1973. *Ice Age Hunters of the Ukraine*. Chicago: University of Chicago Press.
Kozlowski, J. (ed.) 1982. *Excavation at the Bacho Kiro Cave, Bulgaria (Final Report)*. Warsaw: Paristwowe Wydarunictwo, Naukowe.
Kuhn, T. 1974. *The Structure of Scientific Revolutions*. Chicago: University of Chicago Press.
Kuhn, T. 1977. *The Essential Tension*. Chicago: University of Chicago Press.
Lamdan, M. 1984. Faunal remains in Sefunim Shelter. In A. Ronen (ed.) *Sefunim Prehistoric Sites, Mount Carmel, Israel*. Oxford: British Archaeological Reports International Series S230: 475-486.
Larick, R. 1986. Périgord cherts: an analytical frame for investigating the movement of Paleolithic hunter-gatherers and their resources. In G. Sieveking and M. Hart (eds) *The Scientific Study of Flint and Chert*. Cambridge: Cambridge University Press: 111-120.
Laville, H. and Rigaud, J. P. 1973. The Périgordian V industries in Périgord: typological variations, stratigraphy, and relative chronology. *World Archaeology* 4: 330-338.
Laville, H., Rigaud, J. P. and Sackett, J. 1980. *Rock Shelters of the Périgord*. New York: Academic Press.
Lévêque, F. and Vandermeersch, B. 1980. Découverte de restes humains dans un niveau castelperronien à Saint Césaire (Charente-Maritime). *Comptes-Rendus de l'Académie des Sciences de Paris (Série D)* 291: 187-189.
Lévêque, F. and Vandermeersch, B. 1981. Le neandertalien de Saint Césaire. *La Recherche* 12: 242-244.
Lindly, J. (in press). Hominid and carnivore activity at Middle and Upper Paleolithic cave sites in eastern Spain. *Munibe* 39. In Press.
Lindly, J. and Clark, G. 1987. A preliminary lithic analysis of the

Mousterian site of Ain Defla (WHS site 634) in the Wadi Ali, west-central Jordan. *Proceedings of the Prehistoric Society* 53: 279-292.

MacDonald, B., Banning, E. and Pavlish, L. 1980. The Wadi el'Hasa Survey 1979: a preliminary report. *Annual Report of the Department of Antiquities of Jordan* 24: 169-184.

MacDonald, B., Rollefson, G., Banning, E., B. Byrd and d'Annibale, C. 1983. The Wadi el'Hasa archaeological survey, 1982: a preliminary report. *Annual of the Department of Antiquities of Jordan* 27: 311-324.

Marks, A. (ed.). 1976. *Prehistory and Paleoenvironments in the Central Negev, Israel*: Vol. 1. Dallas: Southern Methodist University Press.

Marks, A. (ed.). 1977. *Prehistory and Paleoenvironments in the Central Negev, Israel*: Vol. 2. Dallas: Southern Methodist University Press.

Marks, A. 1981. The Upper Paleolithic of the Negev. In P. Sanlaville and J. Cauvin (eds) *Préhistoire du Levant*. Paris: Centre National de la Recherche Scientifique: 343-352.

Marks, A. 1983. The Middle to Upper Paleolithic transition in the Levant. In F. Wendorf and A. Close (eds) *Advances in World Archaeology* Vol. 2. New York: Academic Press: 51-98.

Marks, A. The Levantine Middle to Upper Paleolithic transition: the past and present. In M. Liverani and R. Peroni (eds) *Studi di Paletnologia in Onore di Salvatore M. Puglisi*. Rome: Universitá di Roma: 123-136.

Marks, A. and Freidel, D. 1977. Prehistoric settlement patterns in the Advat/Aqev area. In A. Marks (ed.) *Prehistory and Paleoenvironments in the Central Negev, Israel*: Vol. 2. Dallas: Southern Methodist University Press: 131-159.

Marks, A. and Volkman, P. 1983. Changing core reduction strategies: a technological shift from the Middle to the Upper Paleolithic in the Southern Levant. In E. Trinkaus (ed.) *The Mousterian Legacy: Human Biocultural Change in the Upper Pleistocene*. Oxford: British Archaeological Reports International Series S164: 13-33.

Marshack, A. 1982. Non-utilitarian fragment of bone from the Middle Paleolithic layer. In J. Kozlowski (ed.) 1982. *Excavation at the Bacho Kiro Cave, Bulgaria (Final Report)*. Warsaw: Paristwowe Wydarunictwo: 117.

Masters, P. 1982. An amino acid racemization chronology for Tabūn. In A. Ronen (ed.) *The Transition from the Lower to the Middle Paleolithic and the Origin of Modern Man*. Oxford: British Archaeological Reports International Series S151: 43-56.

Mellars, P. A. 1969. The chronology of Mousterian industries in the Périgord region of south-west France. *Proceedings of the Prehistoric Society* 30: 199-244.

Mellars, P. A. 1973. The character of the Middle-Upper Paleolithic transition in south-west France. In C. Renfrew (ed.) *The Explanation of Culture Change: Models in Prehistory*. London: Duckworth: 255-276.

Mellars, P. A. 1986. A new chronology for the French Mousterian period. *Nature* 322: 410-411.

Munday, F. 1979. Levantine Mousterian technological variability: a perspective from the Negev. *Paléorient* 5: 87-104.

Newcomer, M. and Watson, J. 1984. Bone artifacts from K'sar Aqil (Lebanon). *Paléorient* 10: 143-147.

Papagopoulou, H. 1985. *Form vs. Function: Lithic Use-wear Analysis and its Application to a Class of Levantine Mousterian Tools*. Ph.D. Dissertation, Columbia University. Ann Arbor: University Microfilms.

Payne, S. 1983. The animal bones from the 1984 excavations at Douara Cave. In K. Hanihara and T. Akazawa (eds) *Paleolithic Site of the Douara Cave and Paleogeography of Palymra Basin in Syria*. Part 3:

Animal Bones and Further Analysis of Archaeological Materials. Tokyo: University of Press: 1-108.

Pillard, B. 1972. La faune des grands mammifères du Würmien II. In H. de Lumley (ed.) *La Grotte de l'Hortus*. Marseilles: Etudes Quaternaires 1: 163-205.

Piveteau, J. 1970. Les grottes de La Chaise (Charente). Paléontologie Humaine 1. l'Homme de l'Abri Suard. *Annales de Paléontologie Vertèbres* 56: 175-225.

Rolland, N. 1981. The interpretation of Middle Paleolithic variability. *Man* 16: 15-42.

Ronen, A. (ed.). 1984. *Sefunim Prehistoric Sites, Mount Carmel, Israel*. 2 Vols. Oxford: British Archaeological Reports International Series S230.

Smith, F. H. 1982. Upper Pleistocene hominid evolution in South-Central Europe: a review of the evidence and analysis of trends. *Current Anthropology* 23: 667-704.

Smith, F. H. 1983. A behavioral interpretation of changes in craniofacial morphology across the archaic/modern *Homo sapiens* transition. In E. Trinkaus (ed.) *The Mousterian Legacy: Human Biocultural Change in the Upper Pleistocene*. Oxford: British Archaeological Reports International Series S164: 141-164.

Smith, F. H. 1984. Fossil hominids from the Upper Pleistocene of Central Europe and the origin of modern Europeans. In F. H. Smith and F. Spencer (eds) *The Origins of Modern Humans: a World Survey of the Fossil Evidence*. New York: Alan R. Liss: 137-210.

Smith, F. H. and F. Spencer (eds). 1984. *The Origins of Modern Humans: a World Survey of the Fossil Evidence*. New York: Alan R. Liss.

Soffer, O. 1985. *The Upper Paleolithic of the Central Russian Plain*. Orlando: Academic Press.

Sohn, P. 1983. Comment on Yi and Clark: 'The Lower Paleolithic of Northeast Asia'. *Current Anthropology* 24: 196.

Sonneville-Bordes, D. de and Perrot, J. 1954-1956. Lexique typologique du paléolithique supérieur. *Bulletin de la Société Préhistorique Française* 51: 327-335; 52: 76-79; 53: 408-412, 547- 559.

Spencer, F. 1984. The Neanderthals and their evolutionary significance: a brief historical survey. In F. H. Smith and F. Spencer (eds) *The Origins of Modern Humans: a World Survey of the Fossil Evidence*. New York: Alan R. Liss: 1-50.

Staesche, U. 1983. Aspects of life of Middle Paleolithic hunters in the northwest German lowlands, based on the site of Salzgitter-Lebenstedt. In J. Clutton-Brock and C. Grigson (eds.) *Animals and Archaeology*. Vol. 1: *Hunters and their Prey*. Oxford: British Archaeological Reports International Series S163: 173- 181.

Straus, L. 1977. Of deerslayers and mountain men: Paleolithic faunal exploitation in Cantabrian Spain. In L. R. Binford (ed.) *For Theory Building in Archaeology*. New York: Academic Press: 41-76.

Straus, L. 1978. Of Neanderthal hillbillies, origin myths and stone tools: notes on Upper Paleolithic assemblage variability. *Lithic Technology* 7: 36-39.

Straus, L. 1982. Carnivores and cave sites in Cantabrian Spain. *Journal of Anthropological Research* 38: 75-96.

Straus, L. 1983. From Mousterian to Magdalenian: cultural evolution viewed from Vasco-Cantabrian Spain and Pyrenean France. In E. Trinkaus (ed.) *The Mousterian Legacy: Human Biocultural Change in the Upper Pleistocene*. Oxford: British Archaeological Reports International Series S164: 73-111.

Straus, L. 1985. Stone Age prehistory in northern Spain. *Science* 230:

501-507.

Stringer, C. 1982. Toward a solution to the Neanderthal problem. *Journal of Human Evolution* 11: 431-438.

Stringer, C., Hublin, J. and Vandermeersch, B. 1984. The origin of anatomically modern humans in Western Europe. In F. H. Smith and F. Spencer (eds) *The Origins of Modern Humans: a World Survey of the Fossil Evidence*. New York: Alan R. Liss: 51-136.

Tchernov, E. 1984. The fauna of Sefunim Cave. In A. Ronen (ed.) *Sefunim Prehistoric Sites, Mount Carmel, Israel*. Oxford: British Archaeological Reports International Series S230: 401-419.

Trinkaus, E. 1982a. Evolutionary continuity among archaic *Homo sapiens*. In A. Ronen (ed.) *The Transition from Lower to Middle Paleolithic and the Origin of Modern Man*. Oxford: British Archaeological Reports International Series S151: 301-314.

Trinkaus, E. 1982b. Comment on F. Smith: 'Upper Pleistocene hominid evolution in South-Central Europe'. *Current Anthropology* 23: 691-692.

Trinkaus, E. 1983a. Neanderthal postcrania and the adaptive shift to modern humans. In E. Trinkaus (ed.) *The Mousterian Legacy: Human Biocultural Change in the Upper Pleistocene*. Oxford: British Archaeological Reports International Series S164: 165-200.

Trinkaus, E. 1983b. *The Shanidar Neanderthals*. New York: Academic Press.

Trinkaus, E. 1984. Western Asia. In F. H. Smith and F. Spencer (eds) *The Origins of Modern Humans: a World Survey of the Fossil Evidence*. New York: Alan R. Liss: 251-294.

Trinkaus, E. 1986. The Neanderthals and modern human origins. *Annual Review of Anthropology* 15: 193-218.

Tuffreau, A. 1976. Les civilisations du paléolithique moyen dans le bassin de la Somme et en Picardie. In H. de Lumley (ed.) *La Préhistoire Française*. Paris: Centre National de la Recherche Scientifique: 1105-1109.

Valladas, H., Reyss, J. L., Joron, J. L., Valladas, G., Bar-Yosef, O. and Vandermeersch, B. 1988. Thermoluminescence dating of Mousterian 'Proto-Cro-Magnon' remains from Israel and the origin of modern man. *Nature* 331: 614-616.

Vallois, H. 1954. Neanderthals and praesapiens. *Journal of the Royal Anthropological Institute* 84: 1-20.

Vandermeersch, B. 1981a. *Les Hommes Fossiles de Qafzeh (Israël)*. Paris: Centre National de la Recherche Scientifique.

Vandermeersch, B. 1981b. A Neanderthal skeleton from a Châtelperronian level at Saint-Césaire (France). Paper presented at the Annual Meeting of the American Association of Physical Anthropologists, Detroit.

Vandermeersch, B. 1984. A propos de la découverte du squelette Néandertalien de Saint-Césaire. *Bulletins et Mémoires de la Société Anthropologique de Paris* 14: 191-196.

Volkman, P. 1983. Boker Tachtit: core reconstructions. In A. Marks (ed.) *Prehistory and Paleoenvironments in the Central Negev, Israel*. Vol. 3: *The Avdat/Agev Area (Part 3)*. Dallas: Southern Methodist University Press: 127-190.

White, R. 1982. Rethinking the Middle/Upper Paleolithic transition. *Current Anthropology* 23: 169-76.

White, R. 1983. Changing land-use patterns across the Middle/Upper Paleolithic transition: the complex case of the Périgord. In E. Trinkaus (ed.) *The Mousterian Legacy: Human Biocultural Change in the Upper Pleistocene*. Oxford: British Archaeological Reports International Series S164: 113-121.

White, R. 1985. *Upper Paleolithic Land Use in the Périgord: a Topographic Approach to Subsistence and Settlement*. Oxford: British Archaeological Reports International Series S253.

Wolpoff, M. 1975. Discussion. In R. Tuttle (ed.) *Paleoanthropology, Morphology, and Paleoecology*. The Hague: Mouton: 15.

Wolpoff, M. 1980a. *Paleoanthropology*. New York: Knopf.

Wolpoff, M. 1980b. Cranial remains of Middle Pleistocene European hominids. *Journal of Human Evolution* 9: 339-358.

Wolpoff, M., Smith, F., Malez, M. Radovčic, J. and Rukavina, D. 1981. Upper Pleistocene hominid remains from Vindija Cave, Croatia, Yugoslavia. *American Journal of Physical Anthropology* 54: 499-545.

33: Mousterian, Châtelperronian and Early Aurignacian in Western Europe: Continuity or Discontinuity?

FRANCIS B. HARROLD

INTRODUCTION

It is a commonplace in the study of Pleistocene prehistory that the contrasts between the Middle and Upper Palaeolithic periods are both numerous and striking (e.g. Mellars 1973; White 1982). In brief, it is widely agreed that the Upper Palaeolithic, when compared to the Mousterian, is characterized by: (1) the adoption of blade technology for many or most stone tools; (2) lithic assemblages with more (and more complex) recognized tool types, whose formal variation in space and time is greater, and more clearly patterned; (3) the widespread use of bone, antler, and ivory artifacts shaped by complex new methods; (4) a proliferation of artifacts of apparent symbolic (often non-utilitarian) significance, such as items of personal ornament, and both mobiliary and parietal art; (5) evidence for more sophisticated subsistence practices; and (6) indications of changed human settlement patterns and increased populations.

These changes in the archaeological record are generally interpreted as reflecting dramatic changes in human technology, subsistence, and social organisation. Because this change in cultural remains roughly parallels the human fossil transition in Western Europe from Neanderthals (*Homo sapiens neanderthalensis*) to anatomically-modern humans (*Homo sapiens sapiens*), it is possible that the archaeological record is monitoring not only changes in cultural systems, but changes in human cultural capacities. This possibility adds interest to our task of understanding this period, but also makes it even more complicated.

When the Mousterian is compared globally to the Upper Palaeolithic, or especially, to a late manifestation like the Magdalenian (e.g. Straus 1983), these contrasts are among the most impressive to be seen in Pleistocene prehistory. When we focus attention on the times just before, during, and after the Middle-Upper Palaeolithic transition, however, the differences are somewhat less salient. Furthermore, it is at this point that we face the reality that our understanding of this transition in terms of Last Glacial human behaviour and adaptations, and their relationship to the evolution of modern humanity, is still seriously incomplete.

This paper will be concerned with only certain aspects of the biological and behavioural changes which are the subject of this Symposium, and only in the corner of Europe with which I am most familiar. More specifically, it will examine continuities and discontinuities across the Middle-Upper Palaeolithic transition and in the early Upper Palaeolithic, in certain parts of France and Spain. There, the transition saw the Mousterian give way to the Châtelperronian industrial tradition and, essentially contemporaneously, the early Aurignacian.

Geographically, then, this study is restricted to a territory stretching from Cantabrian Spain through the western and central Pyrenees to include southwestern and much of central and north-central France (see Figs 33.1-3). Topically, it is restricted to the continuity and change over time seen in the artifacts associated with these three culture-stratigraphic units. It will not deal with evidence relating to past subsistence-settlement systems or intrasite spatial analysis, which is treated elsewhere in this Symposium (see papers by Straus and Chase). It will be more concerned with delineating patterning in the archaeological record than with detailed interpretations of it in terms of particular forms of social organisation and subsistence.

This deliberately narrow focus precludes a comprehensive overview here of our understanding of the transitional period in the Châtelperronian geographical sphere. However, this understanding is in any event so limited that there is probably more to be gained by a relatively intensive examination of the patterning now apparent in this segment of the archaeological record. Such structure can then serve as part of the base for the focused future research which is so much needed before we can begin to explain the Middle-Upper Palaeolithic transition in Western Europe.

CHATELPERRONIAN, MOUSTERIAN, AND EARLY AURIGNACIAN

Preliminary Considerations

Before treating the continuities and discontinuities among these traditions, some preliminary points should be made concerning all three.

The Mousterian: While the Mousterian is sometimes spoken of as an undifferentiated block, it is important to recognize its significant internal variability – not only in the form of the well-known Mousterian facies (whatever their nature may be), but also significant temporal and geographical variability. Temporally, for instance, there is the replacement of the Mousterian of Acheulian Tradition type A (MAT-A) in the Périgord by the MAT-B. Furthermore, there is the possibility on the basis of recent thermoluminescence dates from Le Moustier, of a greater temporal element to inter-facies variability than has previously been demonstrable (Mellars 1986). Spatially, one can point at the regional level to the absence of the MAT from Provence, or the 'Vasconian' Mousterian with cleaver flakes, restricted to Spain and the Pyrenees. At a more local scale, there is almost exclusive occurrence of the MAT in open-air contexts in the Périgord.

Nevertheless, the Mousterian in general shows far less artifactual variability over space and time than the Upper Palaeolithic.

The Châtelperronian: Despite impressions to the contrary, due primarily to disturbed and mixed assemblages from La Ferrassie and Le Moustier, the Châtelperronian is a true early Upper Palaeolithic industry (Harrold 1981,

1983, 1986; Farizy, this Symposium). Also known as the 'Castelperronian' and 'Lower Périgordian', it is characterized by (in decreasing order of importance) Châtelperron knives, end-scrapers, burins, truncated and retouched pieces, and such 'Mousterian' types as side-scrapers, notches, and denticulates. High proportions of these tools are made on blades struck from prismatic cores.

The Early Aurignacian: There is some disorder in the time-space systematics of this second industrial component of the early Upper Palaeolithic in France and Spain. A brief historical review might be in order:

Peyrony (1933) established the well-known sequence of five Aurignacian stages (I-V) primarily on the basis of the succession of bone-point 'fossil directors' found in his excavations at La Ferrassie, with split-base bone points identifying the Aurignacian I, and so on. But as the Aurignacian data base has expanded, the neat succession of stages has broken down – rather analogously to the putatively successive substages (a, b, c) of Peyrony's 'Périgordian V' (Rigaud 1980), when fossil directors began to be found in the 'wrong' stratigraphic sequence. The typological integrity of the Aurignacian industry is not in question, nor is the fact that it has some temporal patterning (classic 'Aurignacian I' assemblages, for instance, seem reliably older than 'evolved' ones).

That an earlier phase of the Aurignacian preceded the classic Aurignacian I with split-base bone points was initially suggested by Sonneville-Bordes (1960). She established that Peyrony's 'Périgordian II' with Dufour bladelets at sites like La Ferrassie layer E, Dufour, and Chanlat, was actually Aurignacian in assemblage composition. Since it stratigraphically underlay the classic Aurignacian I at La Ferrassie, it became an Aurignacian '0' (also variously called 'Archaic Aurignacian', 'Corrèzian' or 'Proto-Aurignacian'). Since then, various assemblages have been attributed to a phase of the Aurignacian which (a) preceded the Aurignacian I, and (b) differed typologically from it, lacking the characteristic bone points and perhaps other fossil directors (such as Aurignacian blades), and (in most formulations) containing small, semi-abruptly retouched 'Dufour' bladelets. However, this typological unity, accepted previously by myself among others (e.g. Harrold 1983), is not apparent in all cases.

It no longer appears that a typologically coherent 'Archaic Aurignacian' precedes the Aurignacian I throughout our geographical sphere of concern (e.g. Leroyer and Leroi-Gourhan 1983; Rigaud 1982: 384-89, 440-43; Sonneville-Bordes 1980b; Delporte and Mazière 1977; Bernaldo de Quiros 1980).

Some 'Archaic' Aurignacian assemblages are fairly close to Sonneville-Bordes' original notion, and indeed antedate the Aurignacian I – as for example at sites in Languedoc (Bazile 1976, 1984) and, in our area, at El Pendo, Gatzarria, and still essentially unpublished assemblages at Saint-Césaire and Abri Pataud. Others, however, are in overall typology quite close to Aurignacian I assemblages, except that they lack split-base bone points and/or contain Dufour bladelets (e.g. Chanlat and Dufour (Sonneville-Bordes 1960) or Le Piage (Champagne and Espitalié 1980). Furthermore, assemblages of both sorts do not always antedate the Aurignacian I; they may be contemporaneous with it (e.g. Le Piage), or of unknown age (Dufour).

Thus an 'Archaic Aurignacian' unified in space, time, and typology cannot clearly be distinguished. I will instead use the term 'Early Aurignacian' in a strictly chronological sense. Early Aurignacian assemblages are those which are roughly contemporary in a given region with the Châtelperronian – that is, those assigned with reasonable certainty to the same chronostratigraphic periods as the Châtelperronian (see Table 33.1 and discussion below).

Two sets of relations will be examined – those between the Mousterian and Châtelperronian, and those between the Châtelperronian and Aurignacian:

MOUSTERIAN-CHATELPERRONIAN

There has long been disagreement over the nature of the transition between these two industries, which occurred some 37 000 to 35 000 years ago (see below for discussion of temporal and chronostratigraphic context). Three main views have been put forward:

1. *The Châtelperronian represents a tradition or culture intrusive into France and Spain, not derived from the local Mousterian.* This view was more popular formerly than now (e.g. Breuil 1913; Peyrony 1933), though there are still some authors uneasy with assertions of Mousterian-Châtelperronian filiation (Ashton 1983). In support of this point of view, Mousterian- Châtelperronian *discontinuities* can be stressed – for example, the lack of well-documented 'transitional' assemblages intermediate between the two, and the fact that the Châtelperronian is distinctively Upper Palaeolithic in overall technology and typology.

2. *The Châtelperronian developed independently from the local Mousterian.* This is the majority position among researchers (e.g. Laville, Rigaud, and Sackett 1980: 267; Mellars 1973; Bahn 1983; Bricker 1976). François Bordes strongly supported this hypothesis; indeed, he saw the Châtelperronian as deriving specifically from one particular Mousterian facies, the Mousterian of Acheulian Tradition, or MAT (Bordes 1972a, 1972b). In this formulation, the relatively high occurrence in the MAT of blades and 'Upper Palaeolithic' types such as burins, backed knives, and end-scrapers is seen as a harbinger of their elaboration during the Châtelperronian. In short, the theme of Mousterian-Châtelperronian continuity is stressed. Indeed, it has been suspected that such MAT-Châtelperronian industrial evolution may have been accompanied by Neanderthal-*Homo sapiens sapiens* biological evolution. Bordes (1968) has pointed out the lack of skeletal material associated with the MAT – with the implication that, for all we know, advanced or transitional Neanderthals may have made MAT assemblages. Elsewhere, a Denticulate-Châtelperronian link has been postulated where all late Mousterian is Denticulate – notably at Arcy-sur-Cure (Girard 1980) and in Cantabria (Butzer 1986).

The question of the derivation of the Châtelperronian specifically from the MAT depends upon the much-debated issue of the nature of Mousterian facies variation. If facies represent cultural traditions associated with distinct human populations (e.g. Bordes 1973), then Bordes' hypothesis is at least possible; if they instead reflect different combinations of tool-making, tool-use, and discard behaviour unrelated to ethnic identity (e.g. Binford 1973), and/or if they are in part artificial creations which partition continuous interassemblage variability (Freeman 1980), then Bordes' derivation is impossible. And if, as Mellars (1986) has argued, MAT is simply late Mousterian, then the Châtelperronian-MAT link means something quite different from what is usually envisioned. I shall not seek to settle that issue here, but will deal below with the question of Mousterian-Châtelperronian evolution.

A third possibility is intermediate between the first two:

3. *The Châtelperronian represents a heavily acculturated derivation of the Mousterian.* In this formulation there are important roles for both continuity and discontinuity. It was first cogently suggested by Klein (1973) that the Châtelperronian resulted from a sort of 'bow wave' effect, as cultural diffusion, and probably also population movement of anatomically-modern humans, brought the Upper Palaeolithic to Western Europe in the form of the Aurignacian (see Note 1). This position has gained popularity in the wake of the the discovery (Lévêque and Vandermeersch 1980, 1981; Vandermeersch 1984) of Neanderthal remains in a Châtelperronian level at Saint-Césaire (e.g. Stringer, Hublin, and Vandermeersch 1984; Dibble 1983; Butzer 1986; Harrold 1983, 1986).

Given the present quantity and quality of data, and state of theory, concerning this issue, none of the above positions has won universal acceptance, or is susceptible of conclusive testing. However, we could suggest certain predictable consequences of each hypothesis, and compare them to the patterning currently visible in the archaeological record. This might suggest which (if any) scenario provides a 'best fit' to such patterning, and suggest avenues for future research (see Note 2):

1. For instance, if the first hypothesis presented above were true, we should expect to see an abrupt replacement of the Mousterian by the Châtelperronian, and strong technological and typological discontinuities in stone and other artifacts. 'Transitional assemblages' intermediate in technology and typology between the two should not be found. We could not expect discontinuities to be absolute, since artifactual change in the Palaeolithic was additive as well as substitutive (after all, chopping tools are found in the Upper Palaeolithic); but the more dramatic such change is, the stronger our confidence in this hypothesis would be.

This hypothesis is not *necessarily* connected with human biological evolution. It is true nonetheless that strong morphological discontinuity between human remains associated with the Mousterian and those associated with the Châtelperronian would support it – for example, if Neanderthals making Mousterian artifacts were abruptly replaced by modern humans making Châtelperronian ones.

2. If the Châtelperronian were autochthonously derived from the Mousterian, we should expect to see in the archaeological record a gradual enrich-

ment of Mousterian assemblages in Upper Palaeolithic technological and typological elements, culminating in the Châtelperronian. Transitional assemblages should be present; if the transformation lasted as long as several thousand years, a graded series of such assemblages should be available. The case for this hypothesis would be especially strengthened if such change occurred well before there was evidence of diffusion or incursion of the Upper Palaeolithic from elsewhere.

In this case, the cultural and biological transformations of the mid-Last Glacial in the geographical area of our concern would be essentially localized phenomena, and presumably linked. This proposal would thus be supported by evidence for the gradual in situ evolution of local Neanderthals into *Homo sapiens sapiens,* rather than their replacement. Transitional or modern human remains should be associated with both the Mousterian and the Châtelperronian.

3. If the third hypothesis were true, we should expect considerable Mousterian-Châtelperronian continuity in assemblage composition, but the relatively abrupt introduction of Upper Palaeolithic elements at about the same time as the appearance of the other local manifestation of the Upper Palaeolithic, the Aurignacian.

Furthermore, if the Châtelperronian represented a 'bow wave' phenomenon, then any associated human remains should be those of Neanderthals or of 'hybrid' populations showing the effects of gene flow.

Unfortunately, these hypotheses are not currently definitively testable. We are unable satisfactorily to quantify and specify them, and their outcomes in the archaeological record may not always be mutually-exclusive. Just how much Mousterian-Châtelperronian continuity, how measured, is sufficient to support hypothesis (2) over hypothesis (3)? At what rate must artifactual change proceed to be judged as 'abrupt'?

Especially vexing in this context is the crude temporal resolution of the dating methods available to us in this time range – notably radiocarbon dating and litho- and pollen-stratigraphic frameworks. Assemblages which could *appear* to us to be essentially synchronous might in fact be separated in true age by several centuries – and the crucial processes we are trying to elucidate could conceivably operate over just such a timespan. Nonetheless, I would argue that the process of pattern-seeking and hypothesis-clarification is a useful one, especially in identifying those areas most in need of clarification and future research.

CHATELPERRONIAN-EARLY AURIGNACIAN

These two culture-stratigraphic units of the early Upper Palaeolithic have long been recognized as distinctive and sequential. Even after Peyrony's (1933) schematization of the Périgordian (including the Châtelperronian) and Aurignacian as two parallel cultural traditions associated with two different races, they were perceived as mutually exclusive in any one region.

Since the 1960s, however, interstratification and other evidence (discussed below) has shown that the Châtelperronian and Aurignacian are at least partly contemporaneous. How do we interpret the phenomenon of two distinct but contemporary industries occupying the same geographical region?

1. Here again, there is a clear majority opinion among prehistorians familiar with the region: that the Châtelperronian and Aurignacian indeed represent two distinct cultural traditions associated with different ethnic (and perhaps racial) groupings (e.g., Sonneville-Bordes 1960, 1980a; Laville, Rigaud and Sackett 1980: 285; Howell 1984). According to this view, differences between the two sets of assemblages, especially in the mutually-exclusive occurrence of such fossil directors as Châtelperron knives, or Aurignacian blades, are too great to explain otherwise than as the result of different culturally-transmitted traditions of artifact manufacture.

2. In an echo of the Bordes/Binford debate over the Mousterian, a minority opinion interprets Châtelperronian-Aurignacian differences as unrelated to sociocultural boundaries, but instead reflecting functionally differentiated sets of behaviour at different sites, perhaps due to seasonal variations in behaviour (S. Binford 1972; Ashton 1983).

Here again, we can propose that each hypothesis would entail different patterning in archaeological residues. In the case of the first, we would expect on close examination to see considerable discontinuity in artifact technology and typology, especially in terms of stylistically-distinct 'fossil directors' characteristic of each industry (Note 3). We would not expect to find significant 'contamination' of Châtelperronian assemblages with Aurignacian fossil directors, and *vice-versa*. Associated human remains might or might not differ markedly between the two industries.

In the case of the second hypothesis, we would expect just such contamination, as well as stylistic continuity between the two industries. Differences in assemblage content would be quantitative rather than qualitative, dependent on differing combinations of activities involving the making, use, modification, and discard of tools at different sites. Such interassemblage differences might even form a continuum, ranging from 'pure Châtelperronian' assemblages with little Aurignacian content through a spectrum to 'pure Aurignacian' ones. Skeletal remains associated with the two industries should necessarily represent the same human population.

Two caveats are appropriate here. First, it is vital to emphasize that the classes of data considered here are only some of those relevant to these hypotheses. Information on faunas, settlement patterns, and the like would be absolutely necessary as well – particularly since advocates of hypothesis (2) have proposed that it is seasonal human movements and shifts in faunal exploitation which underlie Châtelperronian-Aurignacian interassemblage differences.

Second, as noted above for the Châtelperronian and Mousterian, we cannot yet neatly quantify the degrees of interassemblage variability which would confirm one explanation or the other. Again, however, the patterning found can be compared to our hypotheses for goodness-of-fit.

The rest of this paper will review the evidence for continuity and discontinuity – first temporal, then artifactual – among the three industries with which we are concerned.

TEMPORAL CONTINUITY

The evidence for temporal relationships among these industries falls into three categories: stratigraphic superposition, radiocarbon chronology and chronostratigraphic frameworks.

Stratigraphic Superposition

Stratigraphic relationships of Mousterian, Châtelperronian, and Aurignacian assemblages found at the same sites give, in aggregate, good indications of their relative ages.

At 24 sites, Châtelperronian assemblages have been found in at least minimally documented stratigraphic contexts (see Note 4). The following information is based on this sample.

Mousterian levels are known at 17 of these 24 sites. They are directly overlain by Châtelperronian levels at 16 of them (by 'directly' I mean here that there is no other archaeological level between Mousterian and Châtelperronian levels; there may or may not be an interposed sterile stratigraphic unit). In one case (El Pendo), the Mousterian is directly overlain by an Aurignacian level. There are no documented cases of Mousterian overlying either Châtelperronian or Aurignacian levels, or interstratified with them.

Aurignacian levels are documented at 19 of these sites. In 16 of them, the Aurignacian simply directly overlies one or more Châtelperronian levels. These overlying Aurignacian levels are variously characterized as Aurignacian 0, I, II, or 'evolved' Aurignacian; some are well-dated, others are not.

In three cases, Châtelperronian-Aurignacian interstratification has been documented:

1. At Roc de Combe (Lot) (Bordes and Labrot 1967), an Aurignacian level (layer 9), lies sandwiched between two Châtelperronian ones (layers 10 and 8), with an Aurignacian I level (layer 7) above them all;
2. At nearby Le Piage (Lot) (Champagne and Espitalié 1981), a Châtelperronian level (F1) is underlain by three Aurignacian levels (K, J, G-I), and overlain by another (F);
3. At El Pendo in Cantabria (González Echegaray 1980), two Aurignacian levels (VIIIb and VIIIa), one of them quite poor, lie under a scanty Châtelperronian level (VIII), and then an Aurignacian I level (VII).

Thus, Châtelperronian overlies Aurignacian in three cases, and underlies it in 19 (the 16 mentioned above, plus the three with interstratification). Interestingly, in no case does a classic Aurignacian I assemblage with split-base bone points underlie the Châtelperronian; the underlying Aurignacian at Roc de Combe is too scanty to characterize; at El Pendo, it is described as 'archaic' ('Aurignacian 0'), and at Le Piage, it is like Aurignacian I, but lacks the split-base points, and in one case (layer K), has many Dufour bladelets.

The stratigraphic information strongly suggests that both the Aurignacian and the Châtelperronian postdate the Mousterian, and further that Aurignacian assemblages are *generally* more recent than Châtelperronian ones. It may be that fully-fledged classic Aurignacian I levels always post-

Table 33.1. Radiocarbon dates for Châtelperronian and other relevant contexts. Dates are excluded which are merely minima (e.g. >35,000) or dubious due to contamination or other causes, or without available documentation. Designations following site names indicate layer numbers. The dates are derived from the following sources: (1) Vogel and Waterbolk 1967; (2) Delibrias and Evin 1980; (3) Moure Romanillo and Garcia Soto 1983; (4) Vogel and Waterbolk 1963; (5) Stuckenrath 1978; (6) Delibrias 1984; (7) Mellars *et al.* 1987.

Provenience	Years BP	Sample Reference	Source
Mousterian (Youngest dates)			
La Rochette 7	36,000±500	GrN-4362	(1)
La Quina, Final Mousterian	35,250±530	GrN-2526	(1)
	34,100±700	GrN-4494	(1)
Les Cottés I	37,600±700	GrN-4421	(1)
Renne (Arcy) XII	34,600±850	GrN-4217	(1)
Camiac	$35{,}100^{+2000}_{-1500}$	Ly-1104	(2)
Cueva Millan (Burgos, Spain) 1a	37,600±700	GrN-11021	(3)
Châtelperronian			
Renne (Arcy) VIII	33,500±400	GrN-1736	(4)
	33,860±250	GrN-1742	(4)
Les Cottés G	33,300±500	GrN-4333	(1)
Cueva Morín 10	36,950±6777	SI-951	(5)
Basal Aurignacian			
Abri Pataud 14	34,250±675	GrN-4507	(1)
	33,330±410	GrN-4720	(1)
	33,300±760	GrN-4610	(1)
Abri Pataud 12	33,000±500	GrN-4327	(1)
Early Aurignacian			
Cueva Morín 8a	28,435±556	SI-952	(6)
	28,155±757	SI-952A	(6)
	28,515±1324	SI-956	(6)
Aurignacian I			
Abri Pataud 11	32,600±550	GrN-4309	(1)
	32,000±800	GrN-4326	(1)
La Ferrassie K6	33,200±800	GrN-5751	(6)
La Quina, 1	31,400±350	GrN-1493	(4)
Les Cottés E	30,800±500	GrN-4258	(1)
	31,000±320	GrN-4296	(1)
	31,200±410	GrN-4509	(1)
Cueva Morín 7	29,515±865	SI-955	(6)
	28,055±1535	SI-955A	(6)
Cueva Morín 7/6	32,415±901	SI-954	(6)
Other Aurignacian			
Le Flageolet XI	33,800±1800	OxA-598	(7)
La Ferrassie K4	31,300±300	Gif-4277	(6)
	28,000±1050	OxA-409	(7)

date Châtelperronian ones.

Radiocarbon Chronology

The radiocarbon dates relevant to the Middle-Upper Palaeolithic transition in our region, listed in Table 33.1, are fewer than might be expected four decades after the development of the method. But the use of this method in European Palaeolithic contexts developed slowly, and the period under scrutiny is so near to its extreme range that problems of contamination loom large (see Mellars *et al.* 1987).

In any event, the overall pattern visible in Table 33.1 is similar to that just noted. There is some overlap in the error ranges of dates among the three industries, but that between the Châtelperronian and Aurignacian industries is far greater. The dates are consistent with a Mousterian-Châtelperronian- Aurignacian sequence, with the latter two clearly overlapping. Whether the Mousterian in this geographic sphere actually temporally overlaps with the early Upper Palaeolithic is not yet clear on the basis of radiocarbon dates.

Chronostratigraphic Frameworks

As outlined in Table 33.2, pollen and litho-stratigraphic frameworks, especially those of Laville, Butzer, and Leroi- Gourhan, can be integrated with radiocarbon chronologies to produce a series of dated climatic periods to which many archaeological levels can be assigned, allowing for seriation of many sites which lack radiocarbon dates. The suggested correlations in Table 33.2 are tentative, but what is particularly noteworthy here is the broad convergence of results from different areas by workers in both palynology and geomorphology.

The patterning apparent in Table 33.2 is consonant with that derived from the data sources already mentioned; the Mousterian in any one area is succeeded by the Upper Palaeolithic, with little or no temporal overlap, while the Aurignacian succeeds the Châtelperronian with a distinct overlap.

It is important to point out that the contemporaneity of assemblages assigned to the same units in Table 33.2 is very rough (within blocks of about 1500 years), but is the best that can be achieved at present. A number of observations are worth noting:

1. The Mousterian apparently persisted in Cantabria later than elsewhere, into the 'Würm II/III' (or Cottés/Hengelo) Interstadial. However, as elsewhere, it is not known to have overlapped with the Châtelperronian there.
2. Several Châtelperronian levels are now dated to the Würm II/III Interstadial. With one possible exception (Note 5), no Châtelperronian occurrences are yet reported from this interstadial from the Périgord, the richest Franco-Cantabrian prehistoric province; according to Laville, heavy erosion at the close of the interstadial typically destroyed whatever deposits may have accumulated during it.
3. In contrast, no Aurignacian occupations in our region of interest are known to date to the Würm II/III, although the Aurignacian is found at this time to the southeast in the Languedoc region (Leroyer and Leroi-

			Laville and Tuffreau 1984	Mazière & Raynal 1983; Laville 1976		
30,000	Würm III, Phase III, Temperataure	Arcy Interstadial	*Aurignacian I, II, "evolué"*	*Aurignacial I, II, "evolué"*		Unit 33: *Aurignacian I* Morín VI, VII El Pendo VII La Flecha
31,500	Cold	Würm III, Phase II	*Aurignacian I & II:* La Ferrassie K4, 5, 6 Roc de Combe 7, 6 (base) Le Piage F Caminade-Est G, F Pataud 12, 11? Font-de-Gaume 3	*Châtelperronian:* Arcy (Renne) VIII *Aurignacian:* St.-Césaire 5 Cottés E inf.? (Aur. I) Gatzarria Cf (Aur. I) Grotte du Loup 2	← Grande-Roche Em, Ej; La Côte III →	Unit 32: *Aurignacian:* Morín VIII (Aur. "O") La Flecha (Aur. I)
33,000	Fluctuating Unstable	Würm III, Phase I	*Châtelperronian:* La Ferrassie L3a? Roc de Combe 10, 8 Le Piage F1 La Chèvre 18-15 Font-de-Gaume 4-5 *Aurignacian:* Roc de Combe 7 (Aur. I), 9 Le Piage G-I, J, K Pataud 14, 13? La Chèvre 14 (Aur. I) Flageolet XI (?)	Arcy (Renne) IX Grotte du Loup 5, 4 Les Cottés G Châtelperron B La Quina 4 Tambourets? Gatzarria Cj? Basté 3bm? *Aurignacian:* St.-Césaire 6? Gatzarria Cj?		Unit 31: *Châtelperronian:* El Pendo VIII *Aurignacian "O"* El Pendo VIIIa Morín IX Unit 30: *Châtelperronian:* Morín X *Aurignacian "O"* El Pendo VIIIb
34,500	Temperate	*Interstadial:* Würm II/III Cottés, Hengelo	? (Deposits typically removed by erosion) La Ferrassie L3b?	*Châtelperronian:* Arcy (Renne) X Grande-Roche Eg, En St.-Césaire 8, 9		Unit 29: Mousterian: Morín XI (Denticulate)
37,000 to 40,000	Cold	End of Würm II Stadial (Laville); Unit 28 (Butzer)	Mousterian: MAT and Typical at Combe-Grenal MAT (?) at Pech de l'Azé I	Mousterian, including "Post-Mousterian" at Arcy-sur-Cure, Quina at Les Cottés?		Unit 28: Mousterian: Morín XII (Denticulate) El Pendo VIIId (Denticulate)

Note: Grande-Roche Em, Ej, and La Côte III are assigned to Phase I and/or Phase 2 of Würm III.

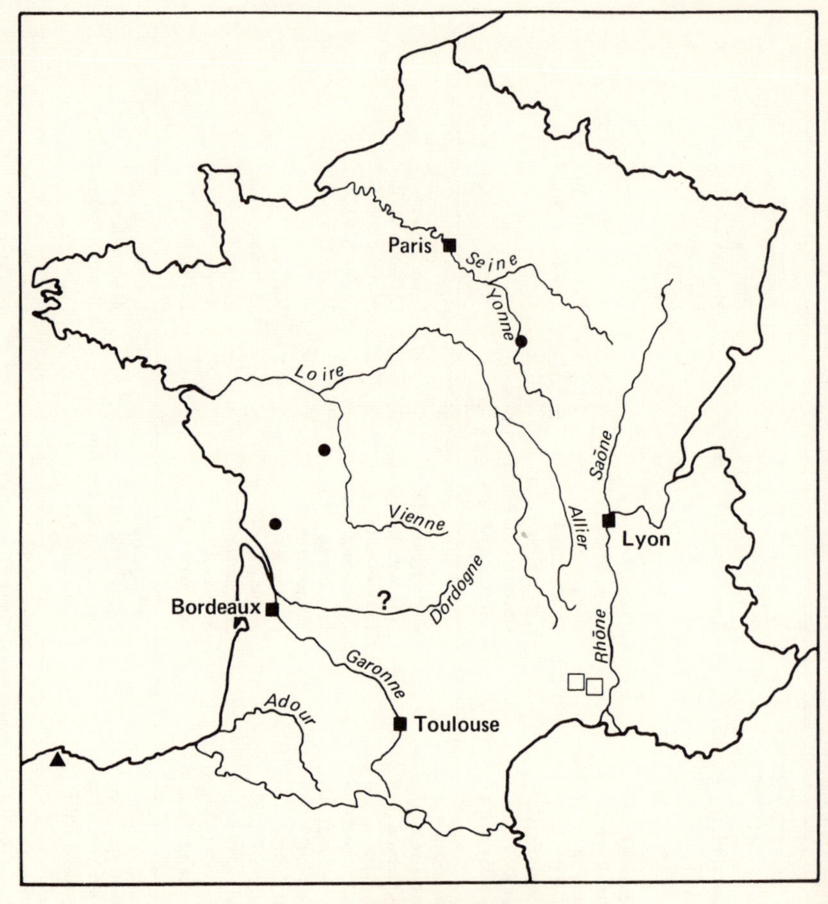

Figure 33.1. Distribution of dated Mousterian, Châtelperronian and Aurignacian sites during Würm II/III (= Hengelo/Cottés) Interstadial.

Gourhan 1983).

4. Most Châtelperronian occupations date to the cool, fluctuating period identified by Laville as 'Phase I' of the Würm III stadial. However, this industry persists in its northern periphery (at Arcy-sur-Cure) into the period equivalent to Laville's Phase II.

When these successive industrial replacements are considered from a geographical perspective, clear spatial-temporal patterning is apparent, as Leroyer and Leroi-Gourhan (1983) have suggested – despite the crudeness and shortcomings of our dating methods, and the small numbers of sites involved (see Figs 33.1-3).

At the end of the cold Würm II stadial, only Mousterian occupations are

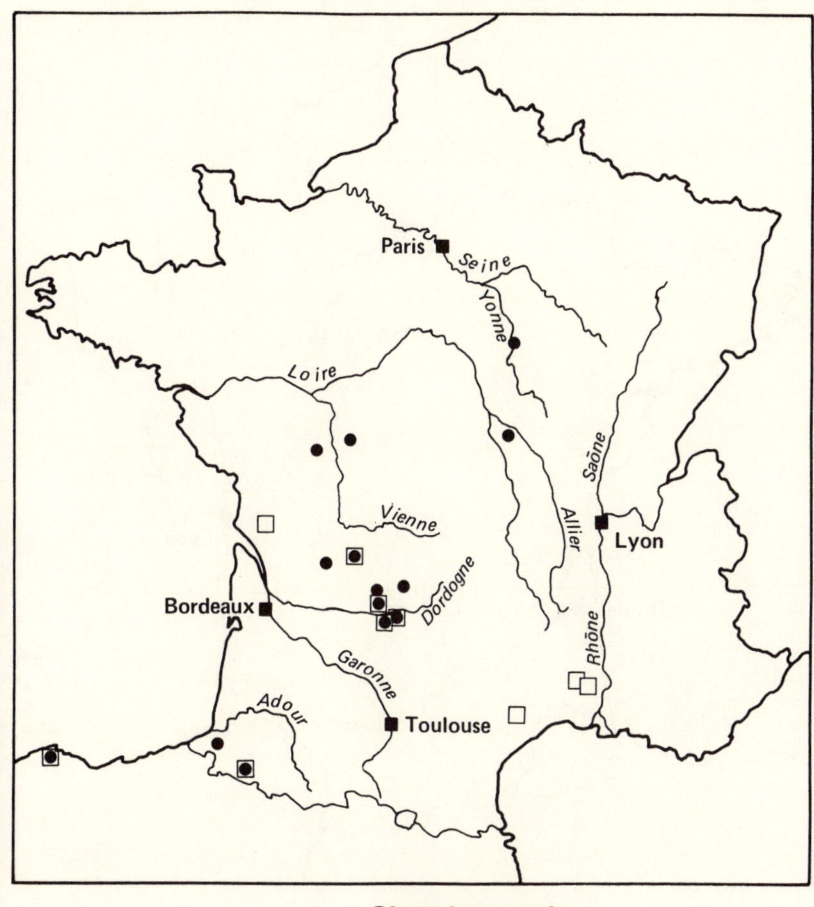

Figure 33.2. Distribution of dated Châtelperronian and Aurignacian sites during Laville's Würm III, Phase I.

known in our area of interest. Then, during the succeeding interstadial (see Fig. 33.1), we find Mousterian on the periphery (Cantabria), Châtelperronian in the centre and north (with uncertain developments in the Périgord) and, just outside our sphere, archaic Aurignacian reported to the southeast in Languedoc. During the ensuing phase of fluctuating climate (see Fig. 33.2), numerous Châtelperronian levels are found throughout the sphere, while early Aurignacian occupations are found in its southern and central parts; overlapping distribution and interstratification of the two industries occur. Then, in the following cold phase (see Fig. 33.3), only Aurignacian levels are found, except for Châtelperronian in the northern periphery. Thereafter the Châtelperronian disappears, while the Aurignacian persists for several millennia. In sum, various lines of evidence suggest that the

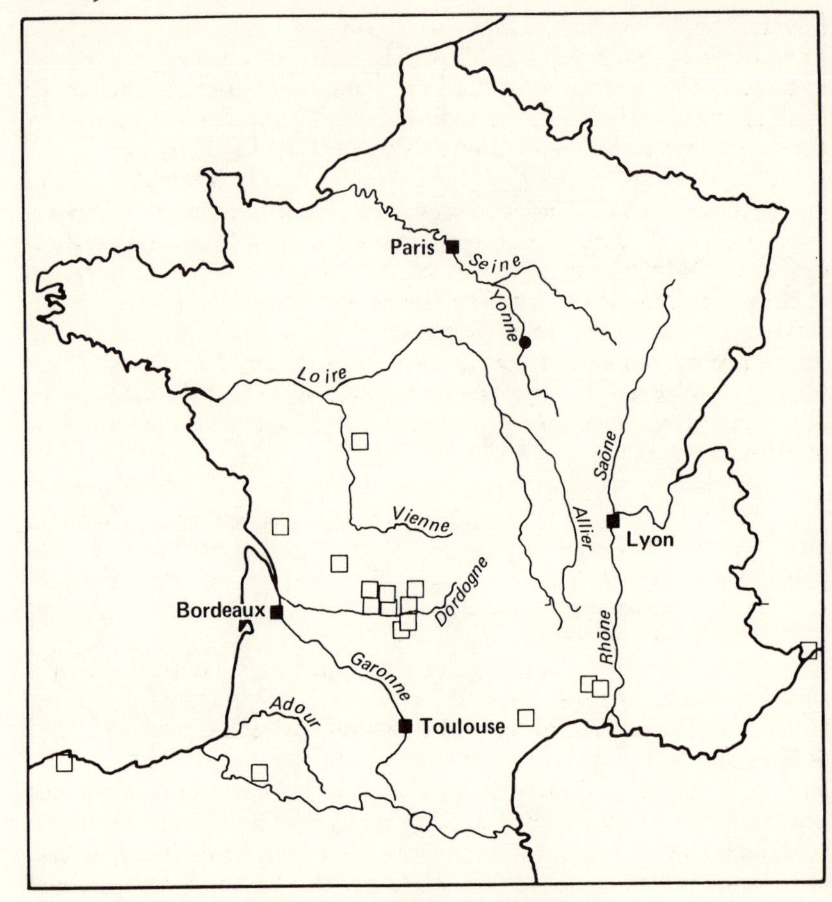

Figure 33.3. Distribution of dated Châtelperronian and Aurignacian sites during Laville's Würm III, Phase II.

Mousterian in our area of interest was replaced, to all appearances rather abruptly, by the Châtelperronian, which co-existed with the Aurignacian before giving way to it. Furthermore, spatial distributions of adequately dated sites suggest that the gradual replacement of the Châtelperronian by the Aurignacian was time-transgressive, from the south and east towards the north and west.

ARTIFACTUAL CONTINUITY AND DISCONTINUITY

Mousterian and Châtelperronian

We will here consider continuities and discontinuities between the artifacts associated with these industries, as documented in my own examination of Châtelperronian and some Mousterian assemblages, and in the available literature. First, we will consider stone tools.

Lithic artifacts. One striking Middle/Upper Palaeolithic contrast is in respect to blank-production methods – i.e., the common occurrence of blades. Of course, ordinary flakes, and the discoid or irregular cores used to produce them, continue to be found right through the Upper Palaeolithic.

Strictly speaking, a blade is a flake twice as long as it is wide, and one can produce a blade without necessarily trying – certainly without a specialized blade core. But Upper Palaeolithic assemblages are generally characterized by numerous blades, struck (though not necessarily punch-struck) from prismatic cores. The Châtelperronian is no exception in this respect; the 14 sizeable assemblages which I examined (see Table 33.3) were characterized by a laminar or blade index (I Lam) for retouched tools ranging from 39.3 (i.e. 39.3% of tools were on blades) to 87.2, with a mean of 57.7; most of these blades had the parallel flake-scar ridges characteristic of prismatic cores.

Unfortunately, systematic studies of Mousterian and Châtelperronian assemblages in terms of lithic reduction practices are quite rare. Exactly how many of these blades were produced from prismatic cores is not known, although it is my impression that such 'true' blades are the rule in the Châtelperronian, and very rare in the Mousterian.

Turning to the Mousterian, one finds that the laminar index is often not reported in assemblage descriptions. What data were available, though, show that blades, broadly defined, are indeed far less common than in the Upper Palaeolithic. Data for 35 Mousterian assemblages from 11 sites (Note 6) indicate a mean blade index of 13.3, and a range from 0 to 35.5. This latter peak figure (at Goderville: *série mate*) is still lower than that of any Châtelperronian assemblage which I examined. And the Mousterian sample used here may be somewhat biased toward a high blade index, in that collections are well-represented from two contexts unusually rich in blades – the loess deposits of the Paris basin (four assemblages), and the Grotte de l'Hyène and Grotte du Renne at Arcy-sur-Cure (11 assemblages) (Note 7). Furthermore, in all these Mousterian cases, published illustrations of blades struck from Upper Palaeolithic-style prismatic cores are quite rare.

It is also worth noting that these data provide no support for the notion that the late Mousterian tended toward higher blade production. Most of these Mousterian assemblages are not datable more precisely than as 'Last Glacial'; but those which are better dated (at Pech de l'Azé, Combe-Grenal, Le Moustier, Grotte du Renne, and El Pendo) show no temporal trend towards a higher blade index. In the Pech de l'Azé, Arcy-sur-Cure, and El Pendo cases, superimposed series of assemblages likewise fail to exhibit such a trend.

Technologically, then, there is continuity in basic flake production methods between Mousterian and Châtelperronian, but also a notable discontinuity in the importance and regularity of blade production. Further work, however, is needed to clarify the nature of this technical contrast.

Châtelperronian/Mousterian lithic artifact comparisons can also be made in terms of the morphology of retouched tools – i.e. typologically. Two

Table 33.3. Assemblage diversity: number of Sonneville-Bordes/Perrot types represented

	Mean No. of Types Represented	Range	Mean Assemblage Size
Mousterian[1] (4 assemblages)	14.0	10-21	121.5 tools
Châtelperronian[2] (14 assemblages)	33.6	23-42	281.8 tools[3]
Early Aurignacian[4] (13 assemblages)	42.1	32-54	617.6 tools

Notes:
1. Pech de l'Azé B and 7, Goderville *série mate* and *série lustrée*.
2. Roc de Combe 8, Le Piage F1, La Chèvre 1, 1a, 2, 2a, La Côte III, Grotte du Loup 4, 5, Basté 3bm, Tambourets (1873 excavation only), Cottés G, Châtelperron B, Morín 10. For bibliographic and other details see Harrold 1978, 1981, 1986.
3. When two large assemblages from the Grotte du Renne at Arcy-sur-Cure (Xb and Xc) are included (Farizy and Schmider 1985), this figure rises to 415.6. However, the number of types represented in these assemblages is unreported.
4. These are assemblages climatostratigraphically contemporary with the Châtelperronian, with typological information available: Le Piage K, J, G-I, F (Champagne and Espitalié 1981), La Chèvre 3 (Jude and Arambourou 1964), Caminade-Est G, F (Sonneville-Bordes 1970), Cottés E (Pradel 1961), La Ferrassie K6, K5, K4 (Delporte 1984), Morín 9 (González Echegaray and Freeman 1978), Pendo VIIIa (González Echegaray 1980).

questions arise here: First, is the Châtelperronian in fact characterized by greater formal variability than the Mousterian? And second, is there a discernible trend in the late Mousterian towards more Châtelperronian-like typology?

In dealing with the first question, an immediately apparent problem is that direct interassemblage comparisons are difficult; different typologies are used to describe Middle Palaeolithic (Bordes 1961) and Upper Palaeolithic (Sonneville-Bordes and Perrot 1954-55) assemblages (Note 8).

A minor exception to this disjunction came about during my study of Châtelperronian assemblages some years ago (Harrold 1978), when I classified four Mousterian collections in terms of the Upper Palaeolithic typology. The four collections – Goderville *série mate* and *série lustrée*, and Pech de l'Azé levels B and 7 – were chosen as assemblages which Bordes had described as evolved MAT (or in the case of Goderville *série mate*, incipient Châtelperronian). It seemed that if any Mousterian collections should show typological continuity with the Châtelperronian, these should.

However, I found very little resemblance at all. The Mousterian assemblages did not fit well into the Sonneville-Bordes-Perrot system, exhibiting high proportions of 'divers' (i.e. unclassifiable) tools. Furthermore, overall assemblage typological diversity was notably lower for these

Mousterian assemblages than in the Châtelperronian. On the average, their tools fell into only 14 Upper Palaeolithic types, compared to a mean of 33.6 types for 14 Châtelperronian collections (see Table 33.3) (Note 9). The four Mousterian assemblages are of course too few to constitute a definitive sample, but this result is consistent with other sources of typological information.

A similar conclusion can be drawn within the confines of the Middle Palaeolithic typology itself. Bordes' typological 'Group III', includes the 'Upper Palaeolithic' types: end-scrapers, burins, perforators, backed knives, and truncated pieces (type nos. 30-37 and 40). This group rarely exceeds 15% in Mousterian assemblages, even in the MAT, which is putatively ancestral to the Châtelperronian. For instance, the eight MAT assemblages from Pech de l'Azé I range in Group III percentage from 4.9 to 15.8%. The great majority of these tools are made on typically Mousterian flakes, and while most would be recognized in the Upper Palaeolithic typology, they would with very few exceptions be classified as 'atypical' examples.

I also typed 11 Châtelperronian assemblages according to the Middle Palaeolithic typology (those listed in Table 33.3, less La Chèvre 2 and 2a). Again, results were anomalous; these were unlike any Mousterian collections. Type lists were dominated by Bordes' Group III, which averaged 59.8% of all tools, and by unclassifiable pieces ('Divers').

Thus considerable overall typological discontinuity can be found between the Châtelperronian and Mousterian. But to go back to the second question raised above, is there evidence of change over time in the late Mousterian in a Châtelperronian direction?

One obstacle to a satisfactory answer to this question is the poor dating of many Mousterian assemblages. However, those in stratigraphically superimposed sequences, or which have been dated chronostratigraphically by Laville (1975), do not show such vectorial change. The latest Mousterian assemblages at Combe-Grenal, Pech de l'Azé I, and Le Moustier, for instance, are not particularly 'evolved', while the four Mousterian assemblages mentioned above (examined because of their putatively evolved status) are all now known to date from the middle part of the Würm II stadial, not its end. Leroi-Gourhan and Leroi-Gourhan (1964) mention an evolved 'Post-Mousterian' at the Grotte du Renne, underlying the Châtelperronian there. But this has been described in detail by Girard (1980) as 'Denticulate Mousterian'.

What about claimed 'transitional' assemblages, i.e. those with characteristics intermediate between Mousterian and Châtelperronian which would demonstrate typological continuity? The classic, supposedly transitional assemblages (La Ferrassie E and Le Moustier K) are now known to be artificial mixtures.

Another assemblage needs to be mentioned, deriving from Delporte's re-excavation of La Ferrassie (Delporte 1984). The assemblage, from level L3b, has 200 retouched tools, and seems indeed to involve a mixture of Middle and Upper Palaeolithic types and technology (Tuffreau 1984). However, it derives from a site where cryoturbation is known to have disturbed deposits dug by Peyrony. Several of the side-scrapers which I saw in a brief examination of this assemblage carried Quina or demi-Quina retouch, like those from the underlying Charentian (Ferrassie) Mousterian; this could indicate some degree of mixture of deposits. Nonetheless, this assemblage is clearly the best candidate yet for an industry of 'transitional' status. Finally, excavations by Rigaud and Simek currently underway at Grotte XVI (Dordogne) may have located a transitional assemblage; here again, fuller details must be awaited.

The claim for vectorial change is sometimes extended into the Châtelperronian from the Mousterian. Châtelperronian assemblages with more Middle Palaeolithic types such as side-scrapers and denticulates, or cruder handiwork, are described as 'early', (only recently emerged from the Mousterian), while those with long, well-made blades and/or few side-scrapers are assumed to be 'evolved'. As I have discussed elsewhere, this assumption is not consistently upheld by available dating evidence (Harrold 1981: 32-35). At two multi-component Châtelperronian sites (Grotte du Renne and Grande-Roche), for instance, the youngest Châtelperronian levels are the most 'primitive' (or 'regressive') in typological/technological terms.

In sum, a picture of abrupt, rather than gradual and additive lithic change seems warranted by the available data. Of course, there was significant lithic continuity between the two industries. The Middle Palaeolithic provided a technological and typological base to which the Upper Palaeolithic added; and this base included the irregular appearance of forms which were to become standardized and elaborated later on. But insofar as we can tell, the late Mousterian cannot be described as 'evolving' in a Châtelperronian or Upper Palaeolithic direction in our area of interest.

More geographically-circumscribed elements of continuity across the Middle-Upper Palaeolithic transition have been pointed to; the continuing emphasis on notches and denticulates in Cantabria, for instance, and the persistence of small 'pediform' side-scrapers at Arcy-sur-Cure (Leroi-Gourhan 1968). However, other explanations besides persistent, traditionally-transmitted craft norms are plausible here, including the demands of local raw material (which in both these cases is not very good). Detailed studies of these assemblages would be necessary to test competing explanations.

Bone artifacts. We will consider here utilitarian items of antler and ivory as well as bone.

Bone was utilized and modified in the Mousterian, but usually without extensive shaping (Mellars 1973). Bones were broken, and sometimes flaked in a manner analogous to stone tools. Because the resultant items may be very difficult to distinguish from those gnawed by animals, there is strong disagreement over just how common this practice was (e.g. Binford 1982, 1983; Freeman 1983). Bone was also very occasionally shaped into recognizable forms such as punches and, rarely, 'wands' or '*baguettes*' which resemble some of those encountered in Upper Palaeolithic sites. Several sites, generally in the Charente and Vienne, have yielded a number of such pieces (e.g. Debénath and Duport 1971; Pradel and Pradel 1954). However, Mousterian bone tools can broadly be characterized as unstandardized, and sporadic in occurrence.

In the Châtelperronian, by contrast, bone tools are far more numerous (see the discussion below of Châtelperronian and Aurignacian bonework, and Table 33.7). They make common use of Upper Palaeolithic bonework-ing techniques, such as groove-and-splinter extraction, whittling, and polishing, and appear in the usual forms, including *poinçons,* points, *lissoirs,* and *baguettes*.

Overall, the Mousterian-Châtelperronian contrast is greater with regard to bone tools than for lithic ones. Perhaps because the bone artifact record is much less rich than the lithic one in both industries, no hint of a 'trans-itional' bone tool assemblage is found.

Artifacts of symbolic significance. This class includes such categories as incised bone and stone, decorative items (e.g. pierced teeth), colouring materials, and figurative art. All seem to involve the imposition of arbitrary form on objects for non-utilitarian purposes.

First let us consider incised bones, i.e. those carrying incisions made by stone tools which are apparently not due to butchering, skinning, or tool manufacture. The qualification 'apparently' should be emphasized here, for to my knowledge, we lack controlled experimental studies, of the sort profitably applied to Plio-Pleistocene African assemblages (e.g. Potts 1984), to determine reliably the agents and processes which produced marks on Middle and Upper Palaeolithic bone.

However, careful observers have noted on both Middle and Upper Palaeolithic bone examples of incisions without apparent utilitarian causes. Such marks are usually linear and grouped (*'marques de chasse'*), and most often short. They may be roughly parallel, or arranged in intersecting pat-terns, even in crude chevrons. Sometimes they are quite wide and deep. These marks occur occasionally in Mousterian sites, generally those which also furnish bone tools. Debénath and Duport (1971) report six such pieces from three sites in the Charente; several others are also known from L'Er-mitage (Pradel and Pradel 1954), La Quina (Camps-Fabrer 1976), and La Ferrassie (Peyrony 1934).

Quite comparable items are found in the Châtelperronian, sometimes additionally carrying punctate and curvilinear marks. Incised stones, unre-ported from the Mousterian, are also known from the Grotte du Renne (Leroi-Gourhan 1976) and Grotte du Loup (White 1986: 108) (see Table 33.8 and further discussion below regarding the Aurignacian).

Items of adornment, typically pierced or grooved for suspension, consti-tute a category whose existence in the Mousterian has not yet been estab-lished beyond a reasonable doubt (Mellars 1973; White 1982). The only reported examples are a pierced tooth and a pierced reindeer phalange from the early excavations at La Quina (Henri-Martin 1907-1910: 130-139), and perhaps the fragment of pierced bone from Pech de l'Azé II (Bordes 1969) – which may also be a utilitarian object.

Such artifacts are notably more common in the Châtelperronian, where four sites have yielded pierced or grooved teeth, two sites bone pendants, and another a piece of stalactite grooved for suspension (see Table 33.8).

Fragments of colouring materials – red and yellow ochre (iron oxide in various forms) and manganese – are not uncommon in the Mousterian (Wreschner 1980). They sometimes show scratches and polish from application to various surfaces (Bordes 1972b: 92-95). These materials are very frequently found in the Châtelperronian, in at least ten sites; and another site (Roche-au-Loup, Yonne; Poplin 1986) has produced a fragment of galena, which can be ground to produce a colouring material. From several sites, ochre is reported in prodigious quantities which deeply stain the sediments. At the Grotte du Renne, fire had been used to achieve colour variations in the ochre, ranging from red to violet (Leroi-Gourhan 1976).

We might also mention that in both Mousterian and Châtelperronian sites, occasional 'curios' are also found – fossils, pyrites, and concretions, apparently collected because of their unusual appearance.

No examples of figurative art are known from either Mousterian or Châtelperronian contexts.

Human remains. In view of the impressive overall contrasts in complexity between Middle and Upper Palaeolithic burials (Harrold 1980), a comparison of mortuary features between the Mousterian and Châtelperronian would be of great interest. Unfortunately, no burials are known from the Châtelperronian. The intentionally- buried anatomically-modern man from Combe-Capelle may or may not derive from the Châtelperronian level there (Harrold 1978: 252-58); and there is no published indication that the Neanderthal individual from Saint-Césaire represents a burial. The only other human remains known from Châtelperronian contexts are several teeth, large but undiagnostic, from the Grotte du Renne (Leroi-Gourhan 1959), and a sole undescribed tooth from Font-de-Gaume (Prat and Sonneville-Bordes 1969). However, available remains do indicate one strong element of continuity between the Mousterian and Châtelperronian: so far as we know, both are associated in our region of interest only with *Homo sapiens neanderthalensis*.

Conclusions. We can see important elements of continuity between the Mousterian and Châtelperronian. In stone tools, there is a common fund of flake production methods, and types such as side-scrapers and denticulates (and perhaps highly localized types) persist across the Middle-Upper Palaeolithic transition. There is also the occurrence in both industries of simple bone tools, incised bones, colouring materials, and curios. Neither tradition is known to be associated with figurative art, and both are associated with Neanderthal man.

On the other hand, important discontinuities are apparent. Blade technology is important in the Châtelperronian, and numerous standardized Upper Palaeolithic types such as Châtelperron points dominate its assemblages. New techniques and new tool forms also revolutionized bone working, which was clearly more frequent in the Châtelperronian. Items of adornment, which had been extremely rare at best in the Mousterian, became fairly common – and occasional traces of other 'new' non-utilitarian behaviour are found, such as incised stone plaques and burned ochre.

These discontinuities are notable, and they show no indication of having developed gradually in the late Mousterian. They appear, as far as we can now tell, abruptly in the archaeological record. Transitional assemblages remain elusive; the most economical explanation is that the transitional period was brief. To return to the three hypotheses concerning the Mousterian-Châtelperronian relationship put forward above, I would suggest that the third hypothesis best accords with the current evidence. Considering the important roles for both continuity and abrupt change which we have seen, as well as the early appearance of the Aurignacian in Languedoc and its gradual spread to the north and west, and the association of both Mousterian and Châtelperronian with Neanderthals, it seems most likely that the Châtelperronian represents an indigenous development of the local Mousterian under the impact of diffusion and probably migration. For reasons given above this inference is tentative, but it represents a 'best fit' to the available evidence.

Châtelperronian and Aurignacian

In this section artifactual continuities and contrasts will be reviewed for these two early Upper Palaeolithic industries in the light of the hypotheses noted above concerning the nature of their relationship.

Lithic artifacts. Let us first consider technological comparisons. In terms of blank production technology, both industries are broadly characterized by blades, but more detailed information on, for example, laminar indices of early Aurignacian assemblages, is difficult to obtain. As noted above for the Mousterian, such data are often not included in the literature. Laminar indices for four early Aurignacian collections were found: 39.7 for level Cbf (Aurignacian I?) at Gatzarria, and respectively, 35.3 and 48.1 for levels Cjn1 and Cjn2 ('Proto-Aurignacian') at the same site (Lévêque 1966); and 23.8 at El Pendo VIIIa (González Echegaray 1980). This sample is too small to be more than suggestive, but its mean index of 36.7 is actually lower than the Châtelperronian mean of 57.7.

Other indications of the importance of early Aurignacian blade technology are available from information from Sonneville-Bordes' (1960) synthesis of the Upper Palaeolithic of the Périgord region. In it are found percentages of blades, not among retouched tools, but among unretouched artifacts, for nine Aurignacian 'o' or I assemblages (Note 10). It must be stressed that most of these assemblages are from old excavations and are not necessarily representative samples; furthermore, few are well-dated, either relative to the Châtelperronian, or absolutely. Thus they can be at most suggestive. Reported percentages of blades and bladelets among unretouched pieces range from 28.8 to 78.8, with a mean of 63.2 – not far from the Châtelperronian mean of 57.7 for retouched tools.

Finally, data on cores are available for the same nine collections. Prismatic blade cores totalled between 0% and 56% of all cores in each collection (the mean is 29.8%). The comparable figures for nine Châtelperronian assemblages – from generally more recent excavations – is a range from

35.8% to 75% and a mean of 55.3%.

Not too much can be inferred from these various indicators, but there is so far no indication that early Aurignacian assemblages were more heavily laminar than Châtelperronian ones.

Another technological variable that should be considered here is an oft-noted characteristic of the Aurignacian, the distinctive scalar 'Aurignacian' retouch. Since the occurrence of this retouch was incorporated into several type definitions in the Upper Palaeolithic typology (e.g. no. 6, end-scrapers on blades with Aurignacian retouch), details on differential occurrence of this mode of retouch will be found below in the discussion of typological differences. Suffice it to note here that such retouch is of variable frequency in Aurignacian assemblages, but almost totally lacking in the Châtelperronian.

It is in regard to Aurignacian retouch that we will briefly discuss the question of direct Mousterian-Aurignacian comparison. This paper has not been organized in terms of such comparisons, because only rarely have the two industries been compared to each other. Each has been seen universally as having more points of comparison with the Châtelperronian than with the other. It has occasionally been suggested, though, that the similarity between Aurignacian retouch and 'Quina' retouch could indicate an evolution of the Aurignacian from the Quina Mousterian, analogous to the putative evolution of the Châtelperronian from the MAT (e.g. Laville, Rigaud and Sackett 1980: 267).

However, this notion has never been seriously pursued. Aurignacian retouch is described by Sonneville-Bordes (1960: 808) as *'écailleuse'*, or scalar, with broad and overlapping flake scars. Bordes (1961: 8) characterized Quina retouch as *'en écaille scalariforme'*, or scalar and scalariform (stair-step), noting that Quina scrapers are typically made on thick flakes with the retouch giving a stair-step profile. However, scalariform retouch on thick pieces is not characteristic of Aurignacian retouch. Comparative study establishing a close degree of similarity between these two modes of retouch would be needed before Aurignacian-Quina connections can be entertained.

Turning to typological comparisons between the two industries, the comparative data base is richer here; Aurignacian assemblages described in the Sonneville-Bordes/Perrot system abound, and 14 Châtelperronian assemblages of adequate sample size (Harrold 1983) can be compared to them. The exact choice of Aurignacian assemblages, however, was a problem. Many, of course, are from old excavations, with serious questions concerning sample bias and stratigraphic integrity. Many others lack these problems, but are either not yet well dated, or are known to postdate the Châtelperronian. I eventually chose 13 Early Aurignacian assemblages (listed in Table 33.3) with adequate credentials of provenience and sample size, which have been fitted into the chronostratigraphic framework outlined in Table 33.2, and found to be broadly contemporary with the local Châtelperronian. Those definitely postdating the Châtelperronian, such as those from Laville's Phase III or IV of Würm III, were excluded. This

Table 33.4. Summary statistics for Châtelperronian and Early Aurignacian assemblages. For details of the assemblages, see Table 33.3

GA = Aurignacian Group (Sonneville-Bordes-Perrot Types 4, 6, 11-13, 32, 67, 68)
GP = Perigoridan Group (Types 45-64, 85-87)
GM = "Mousterian Group' (Types 74, 75, 77) (after Harrold 1978)

		GA (%)	GP (%)	GM (%)
For 14 Châtelperronian Assemblages				
	Mean	2.6	31.8	20.6
	Range	(0–7.2)	(7.5–55.8)	(5.1–39.5)
For 13 Early Aurignacian Assemblages				
	Mean	19.8	3.8	12.0
	Range	(11.8–41.7)	(0.1–8.5)	(1.5–39.1)
For 13 *French* Châtelperronian Assemblages only	Mean	2.3	33.7	19.3
	Range	(0–7.2)	(14.5–55.8)	(5.1–39.5)
For 11 *French* Early Aurignacian Assemblages only	Mean	20.9	3.4	8.2
	Range	(13.4–41.7)	(0.1–8.0)	(1.5–19.0)

process reduced sample size, but had the advantage of eliminating Châtelperronian-Aurignacian contrasts which might be due to developments arising *after* the disappearance of the Châtelperronian.

Various measures of assemblage variability can be used to compare these two traditions. We may first look at assemblage diversity, as measured by numbers of tool types in each assemblage of retouched implements (Table 33.3). The Aurignacian assemblages have on the average 8.5 more tool types than the Châtelperronian ones (42.1 to 33.6); however, their mean assemblage size is also far greater. Given the positive relationship between assemblage size and diversity, is this difference a real one? If we consider only those eight Aurignacian assemblages with sizes in the same range as the Châtelperronian ones (89-583 tools), the mean number of types is reduced to 37.6, closer to the Châtelperronian mean, but still greater.

A summary measure of typological contrast between the two industries involves the typological groups devised by Sonneville-Bordes to combine the totals of implements considered characteristic of each. The 'GA' *(Groupe Aurignacien)* is an index totalling percentages of Aurignacian end-scrapers and blades, keeled and nosed scrapers, and busked burins. The 'GP' *(Groupe Périgordien)* includes primarily backed and truncated pieces, as well as some Upper Périgordian fossil directors which we may ignore here, since they are not found in the Châtelperronian. To this can be added a Mousterian group ('GM'; Harrold 1978), totalling the percentages of side-scrapers, notches and denticulates.

Table 33.4 shows that there is considerable variability in these indices within each industry, but that in the GA there are strong mean differences between Châtelperronian and Aurignacian, and no overlap at all; in the

Table 33.5. Aurignacian fossil directors in Early Upper Palaeolithic assemblages. (See Table 33.3 for list of assemblages.)

	13 Early Aurignacian Assemblages			14 Châtelperronian Assemblages		
	Mean Frequency	Range	Occurring in Assemblages	Mean Frequency	Range	Occurring in Assemblages
Enscrapers on Aurignacian blades (Sonneville-Bordes-Perrot Type 6)	3.0%	0.5–8.1%	13	0.2%	0-2.9%	1
Keeled scrapers (Type 11-12)	5.0%	1.4-11.6%	13	0.8%	0-2.2%	8
Nosed scrapers (Type 13-14)	7.4%	1.8-27.3%	13	1.6%	0-6.2%	6
Busked burins (Type 32)	0.6%	0-1.2%	11	0.1%	0-0.4%	2
Font-Yves points (Type 52)	0.2%	0-2.5%	1	0	0	0
Aurignacian blades (Type 67)	4.4%	0.4-17.9%	13	0	0	0
Strangulated blades (Type 68)	0.5%	0-2.4%	8	0	0	0
Dufour bladelets (Type 90)	2.6%	0-21.3%	8	0.3%	0-1.4%	4

Table 33.6. Occurrence of Châtelperronian fossil directors in Early Upper Palaeolithic assemblages. (See Table 33.3 for list of assemblages.)

	13 Early Aurignacian Assemblages			14 Châtelperronian Assemblages		
	Mean Frequency	Range	Occurring in Assemblages	Mean Frequency	Range	Occurring in Assemblages
Châtelperron Knives (Types 46–47)	0.2%	0–2.2%	3	18.4%	1.9–46.1%	14
Backed blades (Types 58-59)	0.7%	0–5.3%	4	5.3%	0–11.2%	13
Truncated blades (Types 60-64)	2.3%	0.5–5.1%	13	6.7%	2.1–15.4%	14

GP, very strong mean differences, and a bare overlap; and in the GM, moderate differences and a great overlap. Leaving aside the GM, then, the two groups differ notably in frequency in the two industries. Interestingly, if we temporarily remove the Cantabrian assemblages from consideration, Châlperronian-Aurignacian contrasts become sharper (Table 33.4). In particular, there is no longer an overlap in the GP between them. The two industries are less clearly differentiated from each other in Cantabria; this seems to be a general feature of the Upper Palaeolithic in that region (Clark and Straus 1983; Butzer 1986).

Let us look in more detail at the individual fossil director types associated with each industry, and their occurrence in the Châtelperronian and Aurignacian. Table 33.5 displays the mean frequencies and ranges of Aurignacian fossil directors in both Aurignacian and Châtelperronian assemblages. Interestingly, most of these occur in rather modest frequencies in the Aurignacian, and some only occasionally (notably Font-Yves points). Just as interestingly, several of them, such as keeled scrapers, occur in some Châtelperronian collections, though never in great numbers. Aurignacian scrapers, however, are found in only one Châtelperronian assemblage, and Aurignacian blades, strangulated blades, and Font-Yves bladelets are never found. It is noteworthy that among the three types characterized by Aurignacian retouch (nos. 6, 67, 68), exactly four examples of only one type are known from 14 Châtelperronian assemblages – and those are from a level (Le Piage layer F1) sandwiched between two Aurignacian levels. Given what is known about mixing of materials from different levels which can occur in sites (e.g. Villa 1982), one can agree with Sonneville-Bordes (1980a: 117) that certain characteristic Aurignacian tools with a particular retouch technique do *not* occur in the Châtelperronian.

Table 33.6 examines occurrences of types claimed as characteristic of the Châtelperronian. Truncated blades are fairly common in the Aurignacian, but Châtelperron points and backed blades are so rare that their occurrence is most probably again due to mixing. Furthermore, these are types characterized by a particular retouch technique, that of steep backing.

In opposition to Ashton's (1983) claim that Aurignacian-Châtelperronian typological differences are simply those of frequency variation, I agree with Rigaud (1982: 381) that the very rare occurrence in one industry of characteristic types and retouch forms in the other is insufficient to infer typological continuity.

In sum, Aurignacian lithic assemblages seem to be larger, more typologically diverse, and poorer in 'Mousterian' type than Châtelperronian ones. The two industries differ both in relative tool type frequencies, and in the absolute occurrence of a few key fossil directors.

Bone artifacts. Utilitarian bone industries from Châtelperronian and Early Aurignacian assemblages are compared in Table 33.7. Information is less than complete, but the overall range and complexity of the two sets of bone assemblages seem quite comparable. It may be noted, however, that while 12 of our 13 Aurignacian assemblages (plus two others) report bone artifacts, only 6 of 14 Châtelperronian ones do – and that furthermore, bone points at least are clearly less numerous in the Châtelperronian (while split-base bone points are lacking entirely).

Despite this difference, none of the Aurignacian bone assemblages listed in Table 33.7 is really numerous, while the unquantified information available on the Châtelperronian bone industry from the Grotte du Renne describes it as rich. Furthermore, the bone industries of the Aurignacian I and II are described generally as rather poor (e.g. Sonneville-Bordes 1980b: 258-259; Rigaud 1982: 384).

We do know of some Aurignacian I bone industries in the Périgord which were very abundant, notably at Abri Blanchard and Abri Cellier (Leroy-Prost 1975, 1979; Sonneville-Bordes 1960: 98-100, 83-88), and La Ferrassie (Peyrony 1934). Blanchard and La Ferrassie, furthermore, contained examples of a tool type unknown in the Châtelperronian, the pierced baton. Unfortunately, though, dating of these assemblages is still imprecise, and the stratigraphic record of the first two is crude by later standards. Some Aurignacian I levels, at least, have yielded far richer bone assemblages than any Châtelperronian ones; but whether they were contemporary with the Châtelperronian is not yet known. In any event, the Aurignacian bone industry on current evidence seems to have been somewhat more abundant.

Artifacts of symbolic significance. Table 33.8 compares the distributions of these objects in our two samples. There are problems with imprecise information from sites not yet fully published, but the overall impression is of comparable – and not terribly plentiful – amounts and types in the two industries.

Several points deserve mention in this respect. Pierced teeth are rather numerous in one Aurignacian level (Le Piage G-I); they are also said to be numerous in the Châtelperronian levels at the Grotte du Renne, however. Also at the Grotte du Renne, Leroi-Gourhan (1967: 77) has noted the continuity of a specific type of ornament, the ring-shaped bone pendant, reported from both Châtelperronian and Aurignacian levels there. The frequent and often heavy Châtelperronian use of ochre is reflected in the Table. Also noted is the absence from the Table of any items from Cantabrian Spain, although the sample of levels there is small.

As with Table 33.7, though, Table 33.8 may reflect an incomplete picture for the Early Aurignacian. Once again, there are some relatively early Aurignacian levels which have yielded some very impressive artifacts in this class; but their exact temporal relationship to the Châtelperronian is unknown.

Several sites (Blanchard, Cellier, Lartet) contained numerous bones with patterns of incisions and punctuations more complex than anything represented in Table 33.8 (Sonneville-Bordes 1960, 1972; Leroi-Gourhan 1967: 300; White 1986: 92-96, and this volume). Some of these pieces have been interpreted as representing complex notational systems (Marshack 1972). A flute of bird bone from Blanchard may come from the Aurignacian I level, and beads of stone, bone, and ivory are known from the Vallon des Roches in the Vézère valley, where Blanchard and Castanet are located (Sonneville-Bordes 1972, pl. 56; White 1986: 92-93).

Several examples of figurative art are also reported from Aurignacian I

Table 33.7. Early Upper Palaeolithic bone industries. The assemblages correspond with those listed in Table 33.3, with the addition of Roc de Combe 6 and 7. The sources are as in Table 33.3, plus Leroy-Prost 1979: 213-214, 365. 'X' = present.

	Points	Poinçons	Lissoirs-Spatulas	Baguettes	Miscel-laneous
Early Aurignacian					
Le Piage, K	6	7	1		1
Le Piage, J	6	1			
Le Piage, G-I	10	3		1	9
Le Piage, F	1				
Caminade-Est, G	1(?)				1
Caminade-Est, F	2				2
Cottés, E	3	1		1	X
Trou de la Chèvre, 3	X				
La Ferrassie, K6	X				
La Ferrassie, K5	1				X
La Ferrassie, K4	X	X			X
Roc de Combe, 7				1	
Roc de Combe, 6	7	2		2	
Morín, 9	1				
Chatelperronian					
La Chèvre, 1-2a	1				2
Roc de Combe, 8		4			2
Châtelperron, B		X	X		X
Cottès, G		X			
Grotte du Renne, 9, 10	X	X		X	X
Laussel	X		X		

Table 33.8. Symbolic/Decorative items in Early Upper Palaeolithic contexts. Sources as in Table 33.7, plus Poplin 1986; White 1986: 108; Dance 1975; Bouchud 1975. X = Present.

	Incised Bone	Incised Stone	Pendants	Pierced/ Grooved Teeth	Shells	Fossils	Colouring Materials
Early Aurignacian[1]							
Le Piage K				2			
Le Piage J				1			
Le Piage G-I	2			17		2	
Caminade-Est G							X
Caminade-Est F					1		X
Cottés E	3			≥3			X
La Ferrassie K6	1						
La Ferrassie K4	X			1			
Pataud, 14					X[3]		
Pataud, 12						1	
Pataud, 11				2			
Chatelperronian:[1]							
La Chèvre, 1-2a						X	
Roc de Combe 8	2			1			X
Châtelperron B			2	2			X
Cottés G							X
Grotte du Renne, 9, 10	X	1	X	X		X	X
Grotte du Loup, 4,5		1					X
Grande-Roche				X			
Gargas[2]					1		
Roche-au-Loup (Yonne)							X
Basté 3 bm							X

[1] Among assemblages listed in Table 33.7. [2] Association not certain. [3] One is pierced for suspension.

contexts: two sculpted bone pendants from Blanchard and Cellier (White 1986: 94, 96), and engravings of apparent vulvas, on stone blocks from Cellier, Blanchard, and Abri du Poisson. There is even indirect evidence for parietal art, in the form of vault fragments bearing traces of paint and/or engraving incorporated into deposits at Castanet and La Ferrassie (Delluc and Delluc 1978).

These items of art and adornment are not on the lavish scale or in the sophisticated style of the Magdalenian, but are more numerous and elaborate than anything the Châtelperronian has to offer, and in the case of figurative art, involve a whole new universe of symbolic expression. Were they known to be contemporary with the Châtelperronian, they would demonstrate a more complex symbolic repertoire – perhaps even enhanced cultural capabilities – on the part of the makers of Aurignacian assemblages relative to their Châtelperronian contemporaries. Without more precise dating, however, it is also possible that these developments (though still impressively early) all postdate the Châtelperronian.

Human remains As noted above, the Châtelperronian is associated with Neanderthal remains. None of the Aurignacian assemblages examined here is associated with diagnostic human fossils. However, there is good reason for inferring the association of the Early Aurignacian with anatomically-modern humans. First, the modern skeletons from the famous site of Cro-Magnon probably derived from a fairly early Aurignacian (?Aurignacian II) level (Movius 1969). Secondly, five other French sites have yielded modern skeletal remains which were probably associated with one stage or another of the Aurignacian (see Gambier, this volume). Finally, the 'pseudomorph' burials from Cueva Morín (González Echegaray and Freeman 1978) were almost certainly of modern humans. They derived from levels postdating our Early Aurignacian level 9 at Morín, but which were typologically, like it, Aurignacian '0'.

Conclusions. There is clearly considerable continuity or similarity between the material remains of the Aurignacian and Châtelperronian considered here – greater than that between the Mousterian and Châtelperronian. The two industries draw on the same basic repertoire of Upper Palaeolithic artifact production techniques and morphologies – blades, end-scrapers and burins, and Middle Palaeolithic types like side-scrapers. Certain lithic types occur in both industries, but in different proportions. Whether there are subtle differences between the two industries (generally or locally) in the techniques for producing these common types remains to be investigated. Regarding both utilitarian bone artifacts and those of decorative or artistic nature, the situation is ambiguous. Among contemporary assemblages from the two industries, the available inventories are comparable in both departments, though such artifacts are somewhat more frequent in the Aurignacian. However, the existence of much more numerous and elaborate examples of both these categories in poorly-dated Aurignacian contexts leaves open the possibility of a far more complex Aurignacian development of these aspects of material culture, with important implications for past human behaviour and behavioural capacities.

In certain respects, clear artifactual discontinuity can be documented between the two industries: in the differential occurrence of steep backing and Aurignacian retouch, and accordingly of several lithic fossil directors, as well as split-base bone points, the two industries are effectively mutually exclusive.

To return to the two hypotheses advanced concerning Châtelperronian-Aurignacian relations, there can be little doubt that the first one – postulating separate traditions of material culture – better fits the evidence currently available. This is partly because the second, 'functional-variant' hypothesis has never been developed in detail, and especially because evidence absolutely crucial to it – evidence relating to faunal exploitation, settlement patterns, and seasonality – is beyond the scope of this paper. However, even with these caveats in mind, the first hypothesis is stronger. Its provisional acceptance certainly entrains unanswered questions – for example, why would most Aurignacian technical practices have been borrowed in the Châtelperronian, but not Aurignacian retouch? But it better fits the mutually exclusive occurrence of certain fossil directors; the time-transgressive spread of the Aurignacian at the expense of the Châtelperronian (Figs 33.1-3); the probable association of the two industries with two different human populations; and the persistence of the Aurignacian thousands of years after the disappearance of the Châtelperronian.

CONCLUDING REMARKS

The two hypotheses favoured here concerning the relations among Mousterian, Châtelperronian, and Aurignacian in the region of our concern combine to suggest a coherent sequence of events which, if not established beyond doubt, fits the available data. It should be stressed that this account is meant to apply to a restricted area of Western Europe; as various papers in this volume make clear, very different developments seem to have taken place in other parts of the world (although to the east, in Central Europe, the Szeletian may well provide a close analogy to the Châtelperronian).

I interpret the Châtelperronian as the handiwork of Neanderthals, developing rather rapidly from the local Mousterian. Its rise is probably due to diffusion and ultimately migration, of the Aurignacian and *Homo sapiens sapiens,* from points south and east of our area. Neanderthals thus seem to have been capable of producing much (if perhaps not all) of the range of Upper Palaeolithic material culture. However, they seem also to have been at an adaptive disadvantage relative to modern humans, though maybe only a slight one (see Zubrow, this volume). Thus, over a period of 2000-3000 years, the Châtelperronian gradually lost ground to the Aurignacian, culminating in its disappearance (and, presumably, that of the Neanderthals) before 30 000 BP. The biological and cultural details of this slow and doubtless complex process – such as the extent of gene flow involved, or the nature of competing subsistence-settlement systems – still elude us entirely. If we seem to have a general picture of *what* happened and *when,* we still do not know *how.*

This paper has been as much an exercise in pointing out the limitations of the available data base as in interpreting it. It thus seems fitting that it should end in emphasizing, not the inferences to which current data seem to point, but rather, ways to improve that data base.

One way to improve the present data base is to enlarge it in very traditional terms. The relatively few sites and assemblages analysed in this paper, and the fewer still which are fully published, are an obvious handicap. For instance, the three arguably most important Châtelperronian sites (the Grotte du Renne at Arcy-sur-Cure, the rich multi-component Grande-Roche de la Plématrie, and Saint-Césaire) are only partly published.

Secondly, dating methods need to be more widely applied if we are to identify trends and processes with precision. More relevant stratigraphic units need to be integrated into chronostratigraphic frameworks, and radiocarbon-dated.

Third, systematic attention, of the sort exemplified by the papers by Chase and Straus in this Symposium, needs to be paid to faunal, settlement, and subsistence data, which in turn must be integrated with the data sets discussed here.

Finally, relatively new methods which have been successfully applied elsewhere to relate artifactual variation to the behaviour producing it, should be brought to bear on the problems of the Middle-Upper Palaeolithic transition in Western Europe. These include microwear polish and other edge-damage assessment techniques; scanning-electron-microscope studies of modified bone; and detailed studies of lithic reduction sequences and patterns of artifact use, recycling and discard such as have been used by Dibble and others (e.g. Dibble 1984, 1987; McCartney 1985), primarily in Southwest Asia.

Much of this suggested research could be performed using existing collections, without having to await a new generation of excavations, which unfortunately become slower and more expensive as they become more methodical and informative. It is thus within the realm of possibility for us, within the next decade, to improve notably the quality and quantity of our data base concerning the Middle-Upper Palaeolithic transition in the Châtelperronian sphere, and to begin formulating and testing hypotheses which will lead to a much-improved understanding of it.

ACKNOWLEDGEMENTS

I have profited in the preparation of this paper's final version from discussions with other participants at the Cambridge Symposium, especially Paul Mellars, Lawrence Straus, Catherine Farizy, Harold Dibble, Thomas Volman, Harvey Bricker, and Jan Simek. Any shortcomings in the paper are, of course, my responsibility. I thank the American Council of Learned Societies and the L. S. B. Leakey Foundation for travel grants which allowed my participation at the Cambridge Symposium. I also thank Fay Self and Jane Nicol for their help with the tables, and Yafit Avizemal for drafting the figures.

NOTES

1. It has been suggested (Howell 1964; Meiklejohn 1982) that the Châtelperronian was essentially a 'Mousterian' industry with a few Upper Palaeolithic blade tools added, perhaps due to acculturation. But it is now abundantly clear (Harrold 1981, 1983) that it is indeed an early Upper Palaeolithic industry, typologically and technologically.
2. These hypotheses are framed narrowly in terms of only those classes of evidence considered in this paper. Obviously, data regarding subsistence, settlement, and the like would be crucial in an overall consideration.
3. Whether stylistic differences are conceived here in Sackett's (1982, 1986) terms, or Binford's (1972), I refer in this context to functionally more or less equivalent artifacts with distinctively different morphologies due to differences in traditionally-transmitted craft norms.
4. The sites are: Roc de Combe, Le Piage, Trou de la Chèvre, La-Côte, Le Moustier, La Ferrassie, Laussel, Combe-Capelle, Font-de-Gaume, Grotte du Loup, La Quina, Pair-non-Pair, Basté, Belleroche, Gatzarria, Les Tambourets, Gargas, Les Cottés, La Grande-Roche de la Plématrie, Saint-Césaire, Châtelperron, Grotte du Renne (Arcy-sur-Cure), Cueva Morín, and El Pendo (for bibliographic information and details, see Harrold 1978, 1981, 1986).
5. This is the possibly transitional assemblage from level L3b from the re-excavation of La Ferrassie (Laville and Tuffreau 1984; Tuffreau 1984).
6. Goderville *série lustrée, série mate,* Houpeville *série claire* and Oissel (Bordes 1952); Pech de l'Azé I, layers 3, 4, 5, 6, 7, A, B, and C (Bordes 1954-55); Fontmaure, lower level; Le Moustier B and J; Combe-Grenal G, N, and X; La Ferrassie C (Bordes 1974); El Pendo XVI, XIV, XIII, XII-XI, and VIIId (Freeman 1980); Grotte de l'Hyène IVa, IVb1, IVb2, IVb3, IVb4, IVb5, and IVb6 (Girard 1978); Grotte du Renne XI, XII, XIII, XIV (Girard 1980).
7. The impression of the Arcy assemblages' laminarity derived from the '*Indice Laminaire*' (I Lam) is probably exaggerated (Girard 1978, 1980). Girard calculated an '*Indice Laminaire restreint*' (I Lam r.), including only those pieces with length/width ratios classed as 3:1 and 4:1, but excluding those classed as 2:1 (pieces included in the I Lam). This new index ranges from 0 to 7.7 for the Arcy Mousterian assemblages, averaging 3.3, against the mean I Lam of 22.2.
8. White (1982) has made this point, arguing further that Middle-Upper Palaeolithic continuity may have been thus masked, because the two typologies are measuring different things (roughly, function *versus* style). However, the point can be made (and is supported by my attempts to type Mousterian assemblages in the Upper Palaeolithic typology) that this dichotomy in typologies is due to the recognition, based on the extensive experience of Bordes and Sonneville-Bordes, that different sorts of variability obtain in Middle Palaeolithic assemblages than in Upper Palaeolithic ones.
9. It is true that assemblage size is a factor here; the four Mousterian

assemblages are smaller on average than the Châtelperronian ones (see Table 33.3), and there is a positive relationship between assemblage size and number of types represented. However, we can to some extent control for size by selecting only assemblages with between 100 and 200 tools. In that case, the two qualifying Mousterian collections contain 10 and 15 types, with an average of 12.5; the five Châtelperronian ones average 30 types. Thus the same relationship holds.

10. Caminade-Ouest *inférieur*, Castanet I, Chanlat I and II, Les Cottés E, Dufour, La Ferrassie E, Abri Lartet, Abri du Poisson.

REFERENCES

Ashton, N. M. 1983. Spatial patterning in the Middle-Upper Palaeolithic transition. *World Archaeology* 15: 224-235.

Bahn, P. G. 1983. *Pyrenean Prehistory: a Palaeoeconomic Survey of the French Sites*. Warminster: Aris and Phillips.

Bazile, F. 1976. Nouvelles données sur le Paléolithique supérieur ancien en Languedoc oriental. *Congrès Préhistorique de France* 20, Provence, 1974: 24-28.

Bazile, F. 1984. Les industries du Paléolithique supérieur en Languedoc oriental. *L'Anthropologie* 88: 77-88.

Bernaldo de Quiros, F. 1980. The early Upper Paleolithic in Cantabrian Spain. In L. Banesz and J. K. Kozlowski (eds) *L'Aurignacien et le Périgordien dans leur Cadre Ecologique*. Nitra: Institut Archéologique de l'Academie Slovaque des Sciences: 53-64.

Binford, L. R. 1972. Contemporary model building: paradigms and the current state of Palaeolithic research. In D. L. Clarke (ed.) *Models in Archaeology*. London: Methuen: 109-166.

Binford, L. R. 1973. Interassemblage variability – the Mousterian and the 'functional' argument. In C. Renfrew (ed.) *The Explanation of Culture Change*. London: Duckworth: 227-254.

Binford, L. R. 1982. Comment on R. White: 'Rethinking the Middle/Upper Paleolithic transition'. *Current Anthropology* 23: 177-181.

Binford, L. R. 1983. Reply to L. G. Freeman: 'More on the Mousterian: flaked bone from Cueva Morín'. *Current Anthropology* 24: 372-77.

Binford, S. R. 1972. The significance of variability: a minority report. In F. Bordes (ed.) *The Origin of Homo Sapiens*. Paris: UNESCO: 207-210.

Bordes, F. 1952. Stratigraphie du Loess et évolution des industries paléolithiques dans le ouest du Bassin de Paris. *L'Anthropologie* 56: 1-39, 405-452.

Bordes, F. 1954-55. Les gisements du Pech de l'Azé. *L'Anthropologie* 58: 401-432; 59: 1-38.

Bordes, F. 1961. *Typologie du Paléolithique ancien et moyen*. Bordeaux: Delmas.

Bordes, F. 1968. La question périgordienne. In F. Bordes and D. de Sonneville-Bordes (eds) *La Préhistoire: Problèmes et Tendances*. Paris: Centre National de la Recherche Scientifique: 59-70.

Bordes, F. 1969. Os percé moustérien et os gravé acheuléen du Pech de l'Azé II. *Quaternaria* 11: 1-6.

Bordes, F. 1972a. Du Paléolithique moyen au Paléolithique supérieur: continuité ou discontinuité? In F. Bordes (ed.) *The Origin of Homo Sapiens*. Paris: UNESCO: 211-218.

Bordes, F. 1972b. *A Tale of Two Caves*. New York: Harper and Row.

Bordes, F. 1973. On the chronology and contemporaneity of different

Palaeolithic cultures in France. In A. C. Renfrew (ed.) *The Explanation of Culture Change*. London: Duckworth: 217-226.

Bordes, F. 1974. *Le Paléolithique en Europe*. Mimeographed course text, University of Bordeaux I.

Bordes, F. and Labrot, J. 1967. La stratigraphie du gisement de Roc de Combe et ses implications. *Bulletin de la Société Préhistorique Française* 64: 15-28.

Breuil, H. 1913. Les subdivisions du Paléolithique supérieur et leur signification. *Congrès International d'Anthropologie et d'Archéologie Préhistorique*, Geneva, 1912: 165-238.

Bricker, H. M. 1976. Upper Paleolithic archaeology. *Annual Review of Anthropology* 5: 133-148.

Butzer, K. W. 1981. Cave sediments, Upper Pleistocene stratigraphy and Mousterian facies in Cantabrian Spain. *Journal of Archaeological Science* 8: 133-184.

Butzer, K. W. 1986. Paleolithic adaptations and settlement in Cantabrian Spain. In F. Wendorf and A. Close (eds) *Advances in World Archaeology* Vol. 5: 201-252. New York: Academic Press.

Camps-Fabrer, H. 1976. Le travail de l'os. In H. de Lumley (ed.) *La Préhistoire Française:* Vol. 1. Paris: Centre National de la Recherche Scientifique: 717-722.

Champagne, F. and Espitalié, R. 1981. *Le Piage: Site Préhistorique du Lot*. Paris: Mémoires de la Société Préhistorique Française 15.

Clark, G. A. and Straus, L.G. 1983. Late Pleistocene hunter- gatherer adaptations in Cantabrian Spain. In G. Bailey (ed.) *Hunter-Gatherer Economy in Prehistory: a European Perspective*. Cambridge: Cambridge University Press: 131-148.

Debénath, A. and L. Duport. 1971. Os travaillés et os utilisés de quelques gisements préhistoriques charentais. *Bulletins et Mémoires de la Société Historique et Archéologique de la Charente* 1971: 189-202.

Delibrias, G. 1984. La datation par le carbone 14 des ossements de la Ferrassie. In H. Delporte (ed.) *Le Grand Abri de La Ferrassie: Fouilles 1968-1973*. Paris: Etudes Quaternaires, Université de Provence 7: 105-107.

Delibrias, G. and Evin, J. 1980. Sommaire des datations 14C concernant la préhistoire en France, II. *Bulletin de la Société Préhistorique Française* 77: 215-244.

Delluc, B. and Delluc, G. 1978. Les manifestations graphiques aurignaciennes sur support rocheux des environs des Eyzies (Dordogne). *Gallia Préhistoire* 21: 213-438.

Delporte, H. 1976. Le gisement de La Ferrassie, commune de Savignac-de-Miremont. In J.-P. Rigaud and B. Vandermeersch (eds) *Livret-Guide à l'Excursion A4 (Aquitaine-Charente)*. Nice: UISPP, 9th Congress: 88-91.

Delporte, H. (ed.) 1984. *Le Grand Abri de La Ferrassie: Fouilles 1968-1973*. Paris: Etudes Quaternaires, Université de Provence 7.

Delporte, H. and G. Mazière. 1977. L'Aurignacien de la Ferrassie: observations préliminaires à la suite de fouilles récentes. *Bulletin de la Société Préhistorique Française* 74: 343-361.

Dibble, H. L. 1983. Variability and change in the Middle Paleolithic of Western Europe and the Near East. In E. Trinkaus (ed.) *The Mousterian Legacy: Human Biocultural Change in the Upper Pleistocene*. Oxford: British Archaeological Reports International Series S164: 53-72.

Dibble, H. L. 1984. Interpreting typological variation of Middle Paleolithic scrapers: function, style, and sequence of reduction. *Journal of Field Archaeology* 11: 431-36.

Dibble, H. L. 1987. The interpretation of Middle Paleolithic scraper variability. *American Antiquity* 52: 109-117.

Farizy, C. and Schmider, B. 1985. Contribution à l'identification culturelle du Châtelperronien: les données de la couche X de la Grotte du Renne à Arcy-sur-Cure. In M. Otte (ed.) *La Signification Culturelle des Industries Lithiques*. Oxford: British Archaeological Reports International Series S239: 149-169.

Freeman, L. G. 1980. Ocupaciones musterienses. In J. González Echegaray (ed.) *El Yacimiento de la Cueva de 'El Pendo'*. Madrid: Bibliotheca Praehistorica Hispana 17: 29-74.

Freeman, L. G. 1983. More on the Mousterian: flaked bone from Cueva Morín. *Current Anthropology* 24: 366-372.

Girard, C. 1978. *Les Industries Moustériennes de la Grotte de l'Hyène à Arcy-sur-Cure*. 11th Supplement to *Gallia Préhistoire*. Paris: Centre National de la Recherche Scientifique.

Girard, C. 1980. Les industries moustériennes de la Grotte du Renne à Arcy-sur-Cure (Yonne). *Gallia Préhistoire* 23: 1-36.

González Echegaray, J. 1980. El Paleolítico superior. In J. González Echegaray (ed.) *El Yacimiento de la Cueva de 'El Pendo'*. Madrid: Bibliotheca Praehistorica Hispana 17: 75-148.

González Echegaray, J. and Freeman, L. 1978. *Vida y Muerte en Cueva Morín*. Santander: Institución Cultural de Cantabria.

Harrold, F. 1978. *A Study of the Châtelperronian*. Unpublished Ph.D. Dissertation, University of Chicago.

Harrold, F. 1980. A comparative analysis of Eurasian Paleolithic burials. *World Archaeology* 12: 195-211.

Harrold, F. 1981. New perspectives on the Châtelperronian. *Ampurias* 43: 35-85.

Harrold, F. 1983. The Châtelperronian and the Middle-Upper Paleolithic transition. In E. Trinkaus (ed.) *The Mousterian Legacy: Human Biocultural Change in the Upper Pleistocene*. Oxford: British Archaeological Reports International Series S164: 123-140.

Harrold, F. 1986. Une réévaluation du Châtelperronien. *Bulletin de la Société Préhistorique Ariège-Pyrenées* 41: 151-169.

Henri-Martin, Dr. 1907-1910. *Recherches sur l'Evolution du Moustérien dans le Gisement de la Quina:* Vol. 1: *Industrie Osseuse*. Paris: Schleicher.

Howell, F. C. 1964. Comment. *Current Anthropology* 5: 25-26.

Howell, F. C. 1984. Introduction. In F.H. Smith and F. Spencer (eds) *The Origins of Modern Humans: a World Survey of the Fossil Evidence*. New York: Alan R. Liss: xiii-xxii.

Jude, P. and Arambourou, R. 1964. *Le Gisement de La Chèvre*. Périgueux: Magne.

Klein, R. G. 1973. *Ice-Age Hunters of the Ukraine*. Chicago: University of Chicago Press.

Laville, H. 1975. *Climatologie et Chronologie du Paléolithique en Périgord*. Paris: Etudes Quaternaires, Université de Provence 4.

Laville, H. 1976. Le remplissage de grottes et abris sous roche dans le Sud-Ouest. In H. de Lumley (ed.) *La Préhistoire Française:* Vol. 1. Paris: Centre National de la Recherche Scientifique: 250-270.

Laville, H., Rigaud, J.-P. and Sackett, J. 1980. *Rock Shelters of the Périgord*. New York: Academic Press.

Laville, H. and Tuffreau, A. 1984. Les dépots du grand abri de La Ferrassie: stratigraphie, signification climatique et chronologie. In H. Delporte (ed.) *Le Grand Abri de La Ferrassie: Fouilles 1968-1973*. Paris: Etudes Quaternaires, Université de Provence 7: 25-50.

Leroi-Gourhan, A. 1959. Etudes des restes humains fossiles provenant des grottes d'Arcy-sur-Cure. *Annales de Paléontologie* 44: 87-148.

Leroi-Gourhan, A. 1967. *Treasures of Prehistoric Art*. New York: Abrams.
Leroi-Gourhan, A. 1968. Le petit racloir châtelperronienne. In F. Bordes and D. de Sonneville-Bordes (eds) *La Préhistoire: Problèmes et Tendances*. Paris: Centre National de la Recherche Scientifique: 274-282.
Leroi-Gourhan, A. 1976. Les réligions de la préhistoire. In H. de Lumley (ed.) *La Préhistoire Française:* Vol. 1. Paris: Centre National de la Recherche Scientifique: 755-759.
Leroi-Gourhan, A. and Leroi-Gourhan, Arl. 1964. Chronologie des grottes d'Arcy-sur-Cure (Yonne). *Gallia Préhistoire* 7: 1-64.
Leroi-Gourhan, Arl. 1984. La place du Néandertalien de St.-Césaire dans la chronologie würmienne. *Bulletin de la Société Préhistorique Française* 81: 196-198.
Leroi-Gourhan, Arl. and Renault-Miskovsky, J. 1977. La palynologie appliquée à l'archéologie: méthodes, limites et résultats. *Bulletin de l'Association Française pour l'Etude du Quaternaire* 14 (Supplément): 35-49.
Leroy-Prost, C. 1975. L'industrie osseuse aurignacienne: essai régional de classification: Poitou, Charentes, Périgord. *Gallia Préhistoire* 18: 65-156.
Leroy-Prost, C. 1979. L'industrie osseuse aurignacienne: essai régional de classification: Poitou, Charentes, Périgord. *Gallia Préhistoire* 22: 205-370.
Leroyer, C. and Leroi-Gourhan, Arl. 1983. Problèmes de chronologie: le castelperronien et l'aurignacien. *Bulletin de la Société Préhistorique Française* 80: 41-44.
Lévêque, F. 1966. *La Grotte Gatzarria de Suhare, Basses Pyrenées*. Mémoire de Diplôme d'Etudes Supérieurs de Sciences Naturelles, Faculté de Sciences de Poitiers.
Lévêque, F. and J.-C. Miskovsky. 1983. Le Castelperronien dans son environnement géologique. *L'Anthropologie 87:* 369-391.
Lévêque, F. and Vandermeersch, B. 1980. Découverte de restes humains dans un niveau castelperronien à Saint-Césaire (Charente-Maritime). *Comptes-Rendus de l'Académie des Sciences de Paris (Série II)* 291: 187-189.
Lévêque, F, and B. Vandermeersch. 1981. Le Néandertalien de Saint-Césaire. *La Recherche* 12: 242-44.
Marshack, A. 1972. *The Roots of Civilization*. New York: McGraw-Hill.
Mazière, G. and Raynal, J.-P. 1983. La grotte du Loup (Cosnac, Corrèze), nouveau gisement à Castelperronien et Aurignacien. *Comptes Rendus de l'Académie des Sciences de Paris (Série II)* 296: 1611-1614.
McCartney, P. 1985. Changes in behavioral organization during the Late Pleistocene: preliminary evidence from chipped tools. In M. Thompson, M. T. Garcia and F. J. Kense (eds) *Status, Structure, and Stratification*. Calgary: University of Calgary Archaeological Association: 269-275.
Meiklejohn, C. 1982. Comment on R. White: 'Rethinking the Middle/Upper Paleolithic transition'. *Current Anthropology* 23: 183-184.
Mellars, P. A. 1973. The character of the Middle-Upper Palaeolithic transition in south-west France. In A. C. Renfrew (ed.) *The Explanation of Culture Change*. London: Duckworth: 255- 276.
Mellars, P. A. 1986. A new chronology for the French Mousterian period. *Nature* 322: 410-411.
Mellars, P. A., Bricker, H., Gowlett, J. and Hedges, R. 1987. Radiocarbon accelerator dating of French Upper Paleolithic sites. *Current Anthropology* 28: 128-133.

Moure Romanillo, J. A. and Garcia Soto, E. 1983. Radiocarbon dating of the Mousterian at Cueva Millan (Hortigüela, Burgos, Spain). *Current Anthropology* 24: 232-33.

Movius, H. L. 1969. The Abri de Cro-Magnon, Les Eyzies (Dordogne) and the probable age of the contained burials on the basis of the nearby Abri Pataud. *Anuario de Estudios Atlánticos 15:* 323-344.

Peyrony, D. 1933. Les industries 'aurignaciennes' dans le bassin de la Vézère. *Bulletin de la Société Préhistorique Française* 30: 543-559.

Peyrony, D. 1934. La Ferrassie. *Préhistoire* 3: 1-92.

Poplin, F. 1986. Découverte et utilisation probable de galène dans le châtelperronien de Merry-sur-Yonne (Yonne). *Bulletin de la Société Préhistorique Française* 83: 132.

Potts, R. 1984. Home bases and early hominids. *American Scientist* 72: 338-347.

Pradel, L. 1961. La grotte des Cottés, commune de Saint-Pierre-de-Maillé (Vienne). *L'Anthropologie* 65: 229-258.

Pradel, L. and Pradel, J. 1954. Le Moustérien évolué de l'Ermitage. *L'Anthropologie* 58: 433-443.

Prat, F. and Sonneville-Bordes, D. de. 1969. Découvertes récentes de Paléolithique supérieur à la grotte de Font-de-Gaume (Dordogne). *Quaternaria* 11: 115-131.

Rigaud, J.-P. 1980. Données nouvelles sur l'Aurignacien et le Périgordien en Périgord. In L. Banesz and J. Kozlowski (eds) *L'Aurignacien et le Gravettien (Périgordien) dans leur Cadre Ecologique*. Nitra: Institut Archéologique de l'Academie Slovaque des Sciences: 213-241.

Rigaud, J.-P. 1982. *Le Paléolithique en Périgord: les Données du Sud-Ouest Sarladais et leur Implications*. Unpublished Doctoral Thesis, University of Bordeaux I.

Sackett, J. R. 1982. Approaches to style in lithic archaeology. *Journal of Anthropological Archaeology* 1: 59-112.

Sackett, J. R. 1986. Isochrestism and style: a clarification. *Journal of Anthropological Archaeology* 5: 266-277.

Sonneville-Bordes, D. de. 1960. *Le Paléolithique Supérieur en Périgord*. Bordeaux: Delmas.

Sonneville-Bordes, D. de. 1970. Les industries aurignaciennes de l'abri Caminade-Est, commune de la Canéda (Dordogne). *Quaternaria* 13: 77-131.

Sonneville-Bordes, D. de. 1972. *La Préhistoire Moderne*. Périgueux: Pierre Fanlac.

Sonneville-Bordes, D. de. 1980a. Cultures et milieux d'*Homo sapiens sapiens* en Europe. In *Les Processus d'Hominisation*. Paris: Centre National de la Recherche Scientifique 599: 115-129.

Sonneville-Bordes, D. de. 1980b. L'évolution des industries aurignaciennes. In L. Banesz and J.K. Kozlowski (eds) *L'Aurignacien et le Gravettien (Périgordien) dans leur Cadre Ecologique*. Nitra: Institut Archéologique de l'Academie Slovaque des Sciences: 255-273.

Sonneville-Bordes, D. de and Perrot, J. 1954-56. Lexique typologique du Paléolithique supérieur. *Bulletin de la Société Préhistorique Française* 51: 327-335; 52: 76-79; 53: 408-412, 547- 559.

Straus, L. G. 1983. From Mousterian to Magdalenian: cultural evolution viewed from Vasco-Cantabrian Spain and Pyrenean France. In E. Trinkaus (ed.) *The Mousterian Legacy: Human Biocultural Evolution in the Upper Pleistocene*. Oxford: British Archaeological Reports International Series S164: 73-112.

Stringer, C. B., Hublin, J. J. and Vandermeersch, B. 1984. The origin of anatomically modern humans in Western Europe. In F. H. Smith and F. Spencer (eds) *The Origins of Modern Humans: a World Survey*

of the Fossil Evidence. New York: Alan R. Liss: 51-136.
Stuckenrath, R. 1978. Dataciones de carbono-14. In J. González Echegaray and L. G. Freeman (eds) *Vida y Muerte en Cueva Morín*. Santander: Institucion Cultural de Cantabria: 215.
Tuffreau, A. 1984. Les industries moustériennes et castelperroniennes de la Ferrassie. In H. Delporte (ed.) *Le Grand Abri de la Ferrassie: Fouilles 1968-1973*. Paris: Etudes Quaternaires, Université de Provence 7: 111-144.
Vandermeersch, B. 1984. A propos de la découverte du squelette néandertalien de Saint-Césaire. *Bulletins et Mémoires de la Société d'Anthropologie de Paris (série 14)* 1: 191-196.
Villa, P. 1982. Conjoinable pieces and site formation processes. *American Antiquity* 47: 276-290.
Vogel, J. C. and Waterbolk, H. T. 1963. Groningen radiocarbon dates IV. *Radiocarbon* 5: 163-202.
Vogel, J. C. and Waterbolk, H. T. 1967. Groningen radiocarbon dates VII. *Radiocarbon* 9: 107-155.
White, R. 1982. Rethinking the Middle/Upper Paleolithic transition. *Current Anthropology* 23: 169-192.
White, R. 1986. *Dark Caves, Bright Visions*. New York: American Museum of Natural History.
Wreschner, E. 1980. Red ocher and human evolution: a case for discussion. *Current Anthropology* 21: 631-644.

34. The Middle to Upper Palaeolithic Transition on the Russian Plain

OLGA SOFFER

INTRODUCTION

In the scant one hundred years that we have accepted the antiquity of our species, our reconstructions have undergone many shifts in the way we view ancestral lifeways (Binford 1985; Cartmill 1983; Grayson 1983; Landau 1984; Soffer and Gray 1985). These shifts in paradigms are especially evident when dealing with recent hominids – those who lived during the Late Pleistocene and left behind archaeological remains classified as Middle and Upper Palaeolithic. Our reconstructions of their cultural practices have changed from seeing Middle Palaeolithic hominids as qualitatively different from anatomically-modern people, to views, especially prevalent in the 1960s, which claimed that the differences were quantitative at best (Binford 1985; Landau 1984; Soffer and Gray 1985, with references). The comparability of cultural practices at the Middle-Upper Palaeolithic transition is currently receiving a good deal of attention in the literature (Ashton 1983; Binford 1981, 1982a, 1984, 1985; Chase 1986; Orquera 1984; Otte, 1988; Trinkaus 1983; White 1982, with comments) and, once again, we appear to be on the threshold of a change in our perceptions of the past.

While palaeoanthropological data used in these discussions came from a variety of regions, archaeological data, by and large, came from a remarkably small segment of the occupied Old World and concentrated heavily on information recovered from Western Europe and the Near East. This paper offers information on Late Pleistocene adaptations extant on the Russian Plain – an area which, unfortunately, continues to receive little beyond anecdotal attention in discussions about the Middle-Upper Palaeolithic transition. Since one cannot cogently discuss some 70 000 years of prehistory of 42 per cent of the European landmass within the spatial confines of a brief paper, I will not deal exhaustively with this record but rather focus on a few selected issues – specifically the occupation history of the region and the evidence for hominid behaviour at the Middle to Upper Palaeolithic transition. Finally, because no hominid remains have been found on the Russian Plain in a clear stratigraphic context which date to the period under discussion, I leave open the difficult issue of which hominid types were responsible for generating which archaeological records found on the Russian Plain.

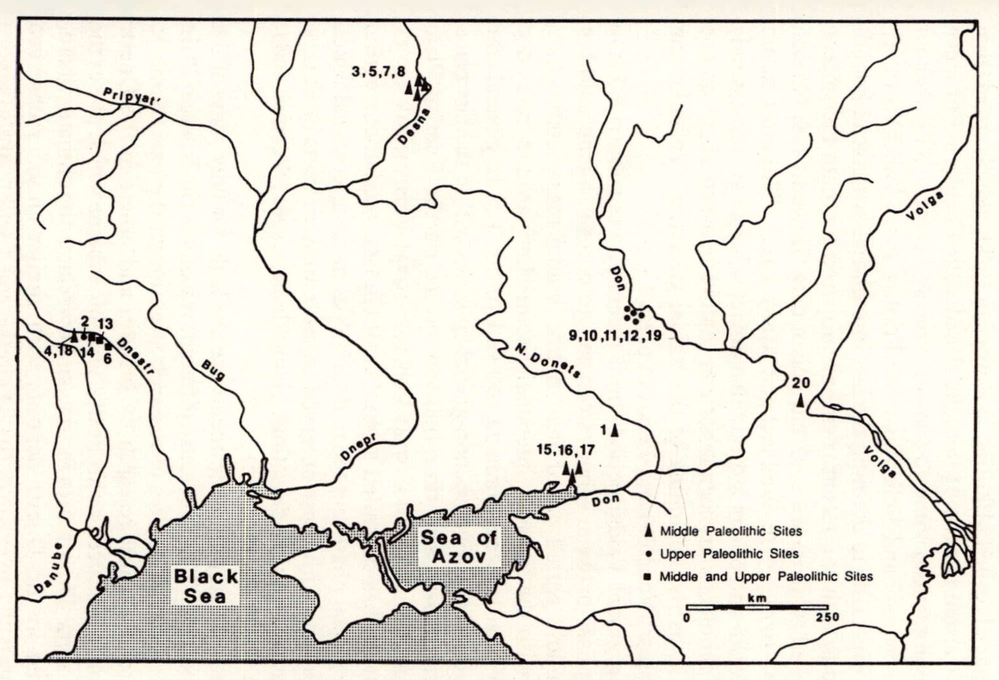

Figure 34.1. Late Pleistocene sites on the Russian Plain: (1) Antonovka; (2) Babin; (3) Betovo; (4) Ketrosy; (5) Khotylevo I; (6) Korman' IV; (7) Korshevo I; (8) Korshevo II; (9) Kostenki XI; (10) Kostenki 11; (11) Kostenki 12; (12) Kostenki 17; (13) Molodova I; (14) Molodova V; (15) Nosovo; (16) Rozhok I; (17) Rozhok II; (18) Stinka; (19) Streletskaya 2; (20) Sukhaya Mechetka (Stalingradskaya; Volgogradskaya).

CHRONOLOGY

The chronology of Middle Palaeolithic sites on the Russian Plain, as elsewhere, is problematic. Both they and the early Upper Palaeolithic sites date to the Late Pleistocene (Fig. 34.1) (Boriskovskij 1984; Chernysh 1977a; Goretskij and Ivanova 1982; Goretskij and Tseitlin 1977; Klein 1973; Praslov 1981; Praslov and Rogachev 1982; Zarrina *et al.* 1980). Until quite recently the majority of Soviet scholars segmented this last climatic cycle into the interglacial, called Mikulino in European USSR, which lasted from some 125 000 to around 72 000 BP, and the glacial, called Valdai in European USSR, which commenced some 70 000 years ago and came to an end about 10 000 BP (Gerasimov and Velichko 1982; Praslov 1984; Velichko 1973). This scheme, especially in defining the last interglacial-last glacial boundary, was at odds with the recent revisions proposed for Late Pleistocene chronology based on deep-sea core and pollen data (Bowen 1978; Nilsson 1983; Woillard 1978, 1979; Woillard and Mook 1982). The most recent works by Velichko and others indicate that Soviet scholars are increasingly accepting the Smolensk cryogenic horizon, dating between about 90 000 and 80 000 BP, as the beginning of the Valdai glaciation (Velichko and Nechaev, in press; Velichko and Pécsi, in press).

The ensuing glacial Valdai period, until recently, was subdivided into an early cold phase between some 72 000 and 50 000 BP, a long mild but fluctuating middle period called the 'middle Valdai mega-interstadial', which ended with the Bryansk interstadial somewhere around 29-25 000 BP, and late Valdai between some 24 000 and 10 000 BP. In general, the first two periods were viewed as non-glacial episodes within the last glaciation with long expanses of time that saw no ice on the Russian Plain (Chebotareva and Makaricheva 1974; Velichko 1973; Voznyachuk 1973). More recent work has delimited a number of climatic oscillations within both the early glacial phase and the middle Valdai mega-interstadial which are generally in line with those observed in more western parts of Europe but which also show local deviations (Dolukhanov 1982; Praslov 1984; Veklich 1974; Zarrina *et al.* 1980).

Firm chronostratigraphic assignments are available for only some of the sites considered here. This is because different regions of the Russian Plain have received different degrees of research attention in the past. Some 30 years of geoarchaeological research by Ivanova and others in the Dnestr basin has recently produced a chronostratigraphic scheme which places the occupation of some Dnestr sites in time and correlates the climatic fluctuations observed in the Dnestr deposits with similar episodes observed elsewhere (Goretskij and Ivanova 1982; Goretkij and Tseitlin 1977; Ivanova 1965, 1969, 1977a, 1977b, 1981, 1982). A long history of geologic research along the Don permits a tentative correlation of this sequence with that observed at the Dnestr sites as well (Fig. 34.2) (Ivanova 1977b; Kozlowski 1986; Praslov 1984, 1985; Velichko 1981). Unfortunately no correlations exist for points in between the Dnestr and the Don, and Middle and early Upper Palaeolithic sites found between these river basins, as well as those

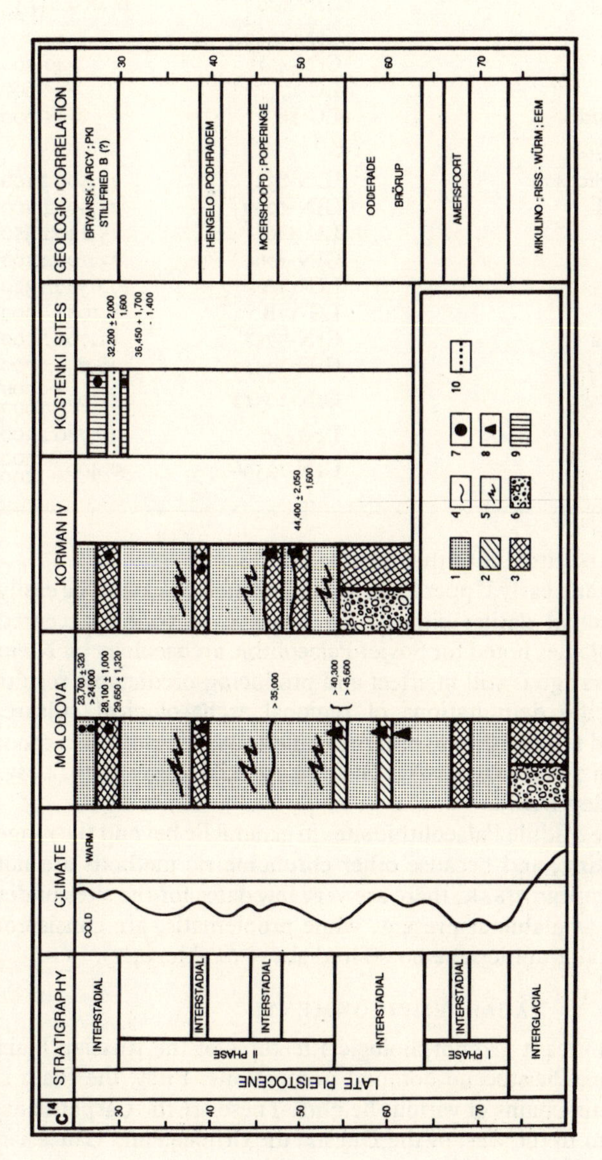

Figure 34.2. Chronostratigraphy of some Late Pleistocene sites on the Russian Plain (modified after Praslov 1984, Figure 7V). Symbols are as follows: (1) loess-like loam; (2) traces of soil formation; (3) relic soils; (4) ashy bands; (5) periglacial deformations; (6) gravels; (7) Upper Palaeolithic sites; (8) Middle Palaeolithic sites; (9) humic layers; (10) bands of volcanic ash.

Table 34.1 Radiocarbon dates for Palaeolithic sites on the Russian Plain.

	Sample Reference	Date BP
Middle Palaeolithic Sites		
Korman IV-layer XI	GrN-6807	$44{,}000^{+2050}_{-1630}$
Molodova I-layer IV	GrN-3659	>44,000
Molodova V-layer XI	GrN-4017	>40,300
	LG-17	>45,600
Molodova V-ashy band	LG-16	>35,600
Upper Palaeolithic Sites		
Korman IV-upper relict soil	GIN-832	27,500±100
Korman IV-layer VII	GIN-1099	24,500±500
	LU-586	25,140±350
Molodova V-layer X	GIN-106	23,100±400
Molodova-layer IX	LG-17A	29,650±1320
	LG-17B	28,100±1000
Kostenki XII-layer 1a	GrN-7758	32,700±700
Kostenki XVII-layer 1	GrN-10511	26,750±700
Kostenki XVII-layer 2	GrN-10512	$32{,}200^{+2000}_{-1600}$
	Le-1436	32,780±300
	GrN-12596	$36{,}400^{+1700}_{-1400}$

in Crimea, cannot be fitted into this scheme at present.

Issues of Middle and early Upper Palaeolithic chronology cannot be easily resolved by artifactual dating either. The absence of widely accepted uniform lithic typologies noted for Soviet Palaeolithic archaeology by Klein (1973) over 15 years ago is still in effect and producing predictable results of researcher-specific delimitations of regional archaeological cultures whose integrity and relationship to others in time and space remains a moot point (Boriskovskij 1984; Davis 1983; Gladilin 1976; Kolosov 1979, 1983; Velichko and Praslov 1978).

Finally, since the Middle Palaeolithic sites in general lie beyond the range of radiocarbon dating, and because other chronometric methods are not widely used in European USSR, there are very few dates for the sites under discussion. Those available at present, while problematic, are consistent with the chronostratigraphic scheme offered above (Table 34.1).

PALAEOENVIRONMENTS

There are two significant geomorphological features of the Russian Plain which affected Late Pleistocene hominid occupations. First, the Plain is rimmed by mountain chains of various heights. These are the Carpathians, found some 100 km to the west of the Dnestr; the Crimean and Caucasian foothills and mountains that lie roughly in the south; and, on the eastern part of the Plain, the Urals. The second significant feature which affected the Russian Plain during the Late Pleistocene was the extreme fluctuation in sea levels. During warmer interglacial times rises in sea levels isolated

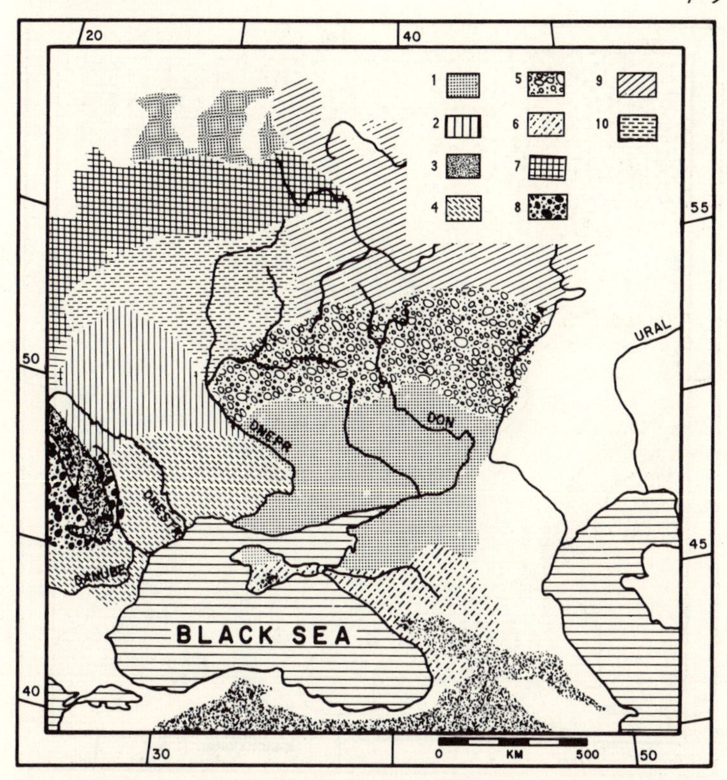

Figure 34.3. Vegetation cover during the Mikulino Interglacial (after Gerasimov and Velichko 1982, Map 9): (1) forest-steppe/meadow steppes combined with oak forest; (2) hornbeam and pine/broadleaf forest; (3) mountain regions; (4) meadow steppes combined with mixed broadleaf forest; (5) broadleaf forest with hornbeam, lime and oak; (6) broadleaf and coniferous/broadleaf mountain forest; (7) hornbeam forest with lime and oak; (8) mixed broadleaf and hornbeam forest; (9) mixed broadleaf (hornbeam, lime, oak) forest with spruce; (10) hornbeam forest with oak and pine.

the area of the present Crimea into an island, while falls in sea levels during colder glacial periods made Crimea a part of the Russian Plain rather than the peninsula it is today (Boriskovskij 1984; Soffer 1985).

Palaeoenvironmental reconstructions are available for two points of Late Pleistocene time – the Mikulino Interglacial and the Valdai glacial maximum around 20 000 BP (see Figs 34.3 and 34.4). While, admittedly, neither of these is ideal for fine-tuned analysis of human adaptations during the various stadials and interstadials between 90 000 and 30 000 BP, they can be used as heuristic devices for information about inter-regional differences in the environmental context at various points of the Late Pleistocene.

Recent research suggests that climatic conditions during the Last Interglacial climatic optimum were differentially distributed across Europe and that regions in the east, including the Russian Plain, witnessed significantly more pronounced warming while western and southern continental regions

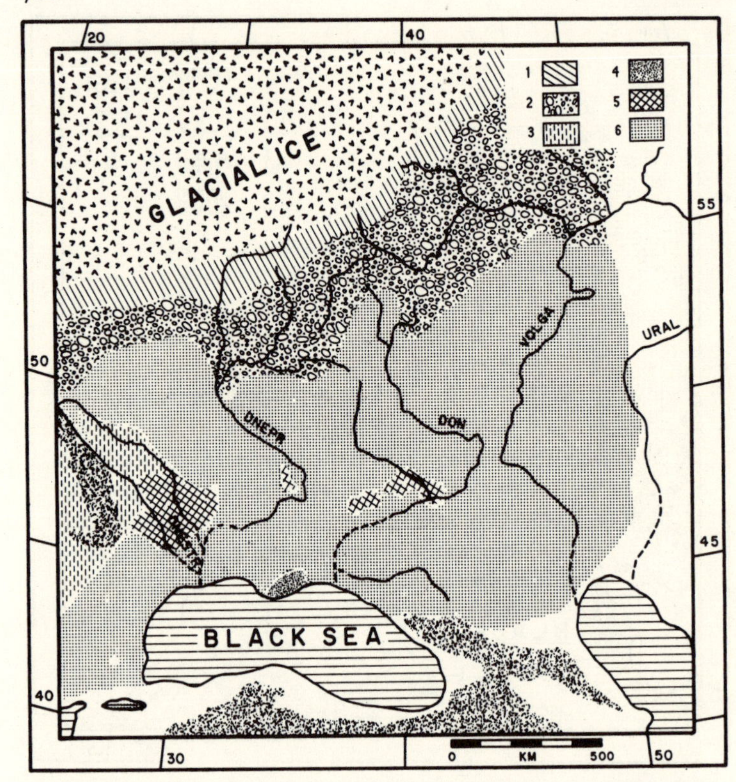

Figure 34.4. Vegetation at the Valdai Glacial Maximum (20 000-18 000 BP) (after Praslov 1984, Figure 6): (1) southern variant of periglacial vegetation – combination of tundra and steppe communities with birch and pine open woodlands; (2) periglacial forest-steppe/meadow-steppes with pine, larch and birch gallery forest; (3) mixed pine, larch, birch and coniferous forests; (4) mountain regions; (5) southern periglacial forest-steppe/meadow-steppe with birch, pine and broadleaf gallery forests; (6) periglacial steppe.

saw more significant increases in precipitation (Gerasimov and Velichko 1982; Velichko 1982a; Velichko and Morozova 1982; Velichko and Pécsi, in press). The Russian Plain during this climatic optimum witnessed a modest increase in annual precipitation as well as a more significant increase in mean annual temperatures. The latter were 4°-6°C warmer than at the present day. This warming resulted from higher temperatures during the shorter cold season than observed today (Velichko and Morozova 1982; Velichko and Pécsi, in press).

During the Mikulino interstadial the northern half of the Plain was under a continuous broadleaf nemoral forest cover (Fig. 34.3). At the same time the southern half featured a somewhat more open mix of deciduous forest and meadow steppe formations. These forests were characterized by a particular composition of dominant species which has no present-day analogues (Grichuk 1982; Grichuk, in press a.). Sub-xerophilous and xerophilous herb-grass and sagebrush grass steppes were present only in small patches in the extreme south near the Black Sea and Sea of Azov.

At the glacial maximum, and presumably in a similar although less dramatic manner during stadial episodes before 20 000 BP, the Plain changed to an open landscape covered by a periglacial steppe where arboreal growth was restricted to gallery forest formations (Fig. 34.4). During this time, when the Scandinavian ice sheet extended down to about the 52°N, the mean annual temperatures were from 7°C to 10°C lower than today at the periphery of the periglacial zone (Velichko 1982a, 1982b; Velichko and Pécsi, in press). This reduction in temperatures was primarily achieved by significant drops in winter temperatures. Reconstructed mean July values deviated only by 2° to 4°C from those of today. These decreases in temperatures were accompanied by equally significant falls in annual rates of precipitation as well.

Stadials saw extremely cold and dry continental climates with a dramatic reduction in latitudinal biotic differentiation and diversification. These periods were characterized by biotic hyperzonality (Grichuk 1982, in press b.; Soffer 1985; Velichko 1982a, 1982b; Velichko and Pécsi, in press). There were, however, exceptions to this hyperzonality and, as Figure 34.4 shows, these exceptions occurred in regions with vertical differentiation in elevations such as found in the Dnestr-Prut region and in Crimea. The close proximity of the plain, foothills, and mountain ranges here created more complex and diverse biotic communities and resulted in the existence of a number of more productive ecotones.

MIDDLE AND UPPER PALAEOLITHIC SETTLEMENT OF THE REGION

The oldest archaeological remains found in primary context on the Russian Plain and in Crimea are Middle Palaeolithic ones and come from deposits dating probably no earlier than the Last Interglacial (Boriskovskij 1984; Hoffecker 1986; Klein 1966, 1973). Middle Palaeolithic artifact inventories come from both rock-shelters (Dnestr-Prut region and Crimea) and open-air sites (the Russian Plain proper). These inventories, in contrast to a number of Upper Palaeolithic ones (especially on the central and eastern parts of the Plain), usually come from stratified deposits with a number of Middle Palaeolithic layers (e.g. Korman' IV, Molodova I and IV, Rozhok I, and various Crimean sites.

In the Dnestr-Prut area these Middle Palaeolithic layers are often overlaid by Upper Palaeolithic occupations (e.g. Korman' IV, Molodova I, IV). A similar regional continuity of occupation, albeit not in the same cave or rock-shelter, is observed across the Middle-Upper Palaeolithic boundary in Crimea as well (Boriskovskij 1984; Klein 1966; Kolosov 1979, 1983). Stratification of occupations is likewise in evidence at the Kostenki-Borschevo sites along the Don. Here, however, only Upper Palaeolithic occupations are represented (Boriskovskij 1984; Praslov 1985; Praslov and Rogachev 1982).

Sufficient biostratigraphic data are available to tentatively assign the published Middle Palaeolithic sites to two periods of occupation: (1) an early

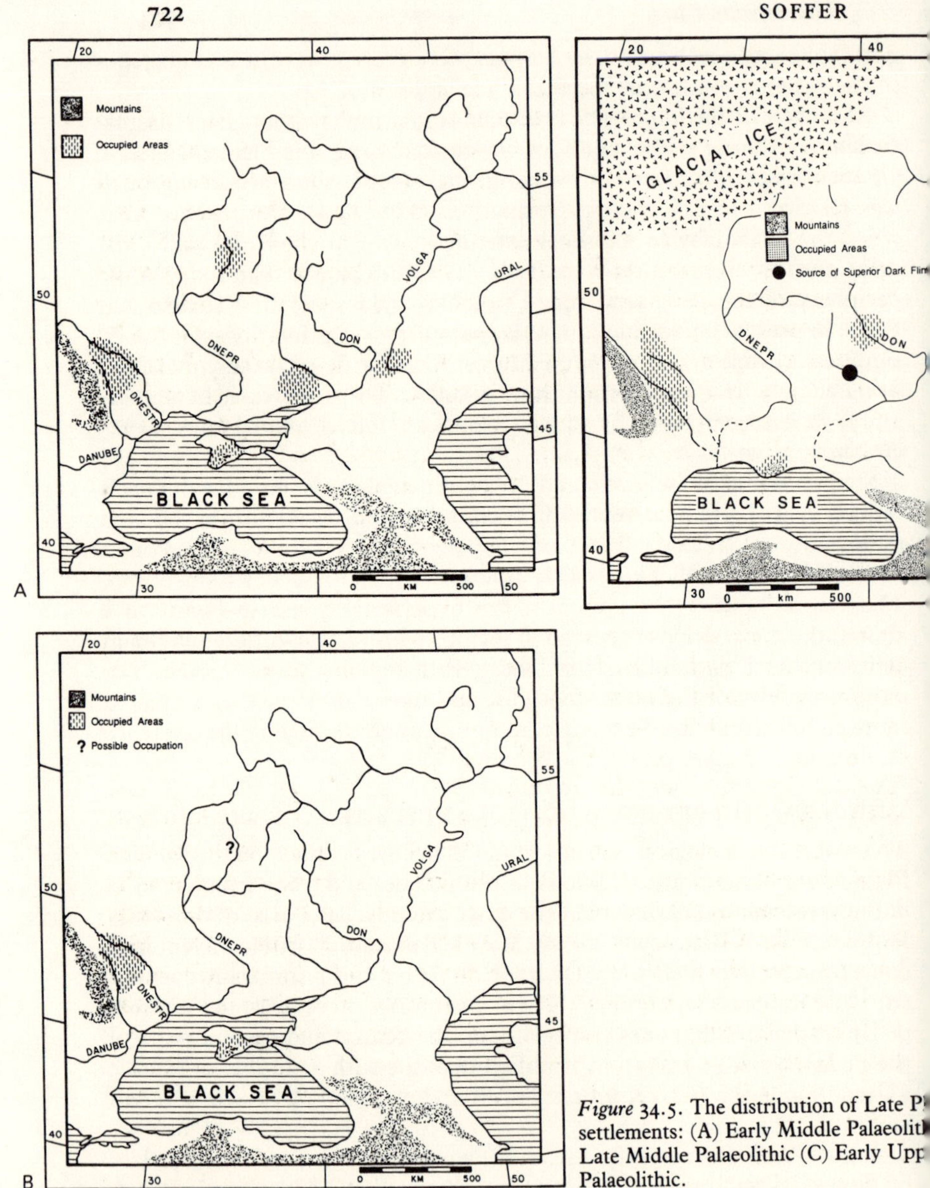

Figure 34.5. The distribution of Late P settlements: (A) Early Middle Palaeolit Late Middle Palaeolithic (C) Early Upp Palaeolithic.

one with warmer and milder conditions, corresponding, perhaps, to the end of the last interglacial/beginning of early glacial (somewhere prior to 80 000-90 000 BP); and (2) a considerably later period of deteriorating climate which occurred after 65 000 BP and which lasted perhaps as late as 38 000 BP.

Early Middle Palaeolithic

The early Middle Palaeolithic sites (Fig. 34.5A) include such sites as Sukhaya Mechetka (also known as Stalingradskaya and Volgogradskaya) (Boriskovskij 1984; Lazukov *et al.* 1981; Klein 1966, 1973), Antonovka (Boriskovkij 1984; Gladilin 1966, 1976), Nosovo (Praslov 1968, 1972), Rozhok I and II (Boriskovskij 1984; Boriskovskij and Praslov 1964; Praslov 1968) and Khotylevo I (Velichko *et al.* 1981, Zavernyaev 1978). Biological data suggest that these sites were occupied during mild and wet times which saw the presence of birch, fir, hornbeam, oak, and pine even in what today is the most arid southern steppe zone (e.g. pollen spectra from Rozhok I and Sukhaya Mechetka). Warm interglacial molluscan fauna has been identified at Khotylevo I. The sparseness of published information on the stratigraphic context of materials from Korshevo I and II precludes an assignment of these neighbouring sites to this early period of occupation (Boriskovskij 1984; Tarasov 1976, 1977b, 1986).

All of the early Middle Palaeolithic sites are found south of approximately the 52°N, and occur both in Crimea and in discontinuous patches across the Plain. The relatively small number of Middle Palaeolithic sites depicted in Figure 34.1 in the Dnestr region is deceptive and a product of present-day administrative boundaries. Many more not considered here are known just south-south-west in Moldavia and Rumania.

Late Middle Palaeolithic

The distribution of late Middle Palaeolithic sites (Fig. 34.5B), in general, shows a decrease in the occupied regions through time. While dating admittedly is problematic, and various post-depositional factors clearly at play, the Dnestr-Prut area appears to show not only a continuation of occupation but an increase in the number of sites throughout the Valdai stage (Boriskovskij 1984; Chernysh 1965, 1973, 1977a; Grigor'ev 1968, 1970). Sites become especially numerous along the narrow canyon-like middle section of the Dnestr and include such well known stratified localities as Korman' IV (Chernysh 1977b; Goretskij and Tseitlin 1977), Molodova I (Chernysh 1982; Goretskij and Ivanova 1982; Ivanova 1977b, 1982), Molodova V (Chernysh 1961, 1977a), Ketrosy (Anisyutkin 1980, 1981; Boriskovskij 1984; Praslov 1981), and Stinka (Anisyutkin 1969, Boriskovskij 1984, Ivanova 1969).

Biological and pedological data suggest that these sites were occupied during various different climatic episodes of the middle Valdai. Stratigraphic and pollen data at Ketrosy indicate a predominance of boreal forest growth and suggest that this site was occupied either at the end of the preceding colder episode or at the onset of the Brørup interstadial (Ivanova 1981; Levkovskaya 1981; Praslov 1981). Palaeoclimatic reconstructions at Molodova I suggest that Middle Palaeolithic groups settled here during a warm and relatively moist interstadial period, possibly also the Brørup (Bolikovskaya 1981; Bolikhovskaya and Pashkevich 1982). At the time, mean July temperatures were comparable to those in the area today while

January ones were depressed by some 10°-15°C. A complex mix of vegetation is indicated in the Molodova I pollen spectra which included a mix of forest-steppe and steppe zones together with arboreal growth that consisted of mixed boreal and broadleaf forms. Data from Korman' IV indicate initial occupation during a somewhat later mild interstadial and reflect a deterioration of climate throughout the occupation sequence (Ivanova 1977a, 1977b). Arboreal pollen, in general, predominate in all the middle Dnestr spectra and boreal elements increase through time at the expense of warmth-loving deciduous growth.

It is precisely this part of the Russian Plain, one with the greatest vertical differentiation in elevations and the most complex and diversified mix of biotic resources, which contains evidence for continuous human occupation throughout the Valdai glacial. Although chronostratigraphic controls are still to be established, a similar pattern seems to be present in the second comparable region with altitudinal and resource differentiation and diversification – Crimea (Boriskovskij 1984; Klein 1966; Kolosov 1979, 1983; Velichko and Praslov 1978).

At the same time, evidence of continued occupation disappears from more eastern parts of the Plain. While the Khotylevo I lithic workshop was apparently utilized during the Mikulino interglacial, a recently-obtained radiocarbon date of 36,500±100 BP (Tarasov, pers. comm.), if correct, places the occupation of Betovo at the other end of the Middle Palaeolithic sequence (Tarasov 1976, 1977a, 1977b, 1980). This date makes the site contemporaneous with Upper Palaeolithic occupations present in the more south-south-easterly region of the Plain. Although the rather sparse Middle Palaeolithic materials at Korshevo I and II lithic workshops (Tarasov 1976, 1977b, 1986), found in similar geologic context near Betovo, still remain to be dated, I suspect that they will similarly date to the close of the middle Valdai.

I interpret this regionally disparate archaeological data for Middle Palaeolithic occupations to indicate a more continuous human presence in some parts of European USSR and a more sporadic and discontinuous one in others. I suggest that since the Russian Plain repeatedly underwent increases in hyperzonality and concomitant decreases in resource diversity during various stadial periods of the early and middle Valdai, this repeated environmental degradation made resources on the Plain proper far more unpredictable than in the Dnestr-Prut region and in Crimea. Thus, data from European USSR indicate that only some areas, specifically those with the greatest regional diversification and differentiation of biotic resources, saw continuous occupation during the Middle Palaeolithic, and that other parts witnessed a pattern of sporadic and discontinuous colonization and abandonment.

Early Upper Palaeolithic

The transition between the Middle and Upper Palaeolithic is poorly documented and poorly dated in European USSR, and, in general, associated with erosional rather than depositional episodes (Anikovich 1977,

1983; Boriskovskij 1984; Goretskij and Ivanova 1982; Goretskij and Tseitlin 1977; Hoffecker 1986; Klein 1973; Praslov 1985). Those Upper Palaeolithic sites which can be dated to an early period of occupation show an occupation pattern different from that of the Middle Palaeolithic sites (Fig. 34.5C). Regions occupied are localized in two parts of the Plain: in the middle Dnestr and middle Don. While Crimean Upper Palaeolithic sites still remain undated, it is likely that this area was occupied at the onset of the Upper Palaeolithic as well (Boriskovskij 1984; Kolosov 1979, 1983; Velichko and Praslov 1978).

The continued presence of sites along the middle Dnestr has been used to argue for uninterrupted occupation throughout the Middle-Upper Palaeolithic transition, and the purported genetic relationship of archaeological cultures through time offered as proof for an *in situ* evolution of the Upper Palaeolithic here (Boriskovskij 1984; Chernysh 1965, 1973; Grigor'ev 1968, 1970; Rogachev 1970). A depositional hiatus, however, is in evidence at Korman' IV (Chernysh 1977b) and at both of the Molodova sites (Chernysh 1961, 1982; Ivanova 1977b, 1982). An average of 3 metres of sterile deposits separate Middle from Upper Palaeolithic layers at these sites. This contrasts with the patterning within both the Middle and Upper Palaeolithic stratigraphic columns at the sites, where mean depths of sterile layers average between 35 and 40 cm. While these colluvial, loess-like loams clearly reflect geomorphological realities, I would also suggest that they may signal a break in human occupation before the appearance of Upper Palaeolithic tool-makers in this area.

While some continuity of human occupation, albeit perhaps an interrupted one, is in evidence in the Dnestr region, the appearance of Upper Palaeolithic sites in the Kostenki-Borschevo area marks a new recolonization of the Russian Plain. The earliest evidence for this recolonization comes from the Don region where at least seven sites are dated between 36 000 and 32 000 BP (Figs 34.1 and 34.5C; Table 34.1) (Boriskovskij 1963, 1984; Praslov 1985; Praslov and Rogachev 1982; Rogachev 1957; Vereschagin and Kuzmina 1977; Zarrina *et al.* 1980). Once begun, this colonization, especially after *c.* 26 000 BP, expands through time and, in contrast to the Middle Palaeolithic occupation, is a continuous one replete with stratified sites in the middle Dnestr and middle Don regions and with increasing numbers of sites on the eastern, central, and southern parts of the Plain (Boriskovskij 1984; Praslov and Rogachev 1982; Rogachev 1957; Soffer 1985; 1986).

The exact dating of the appearance of the earliest Upper Palaeolithic industries is problematic. Dates on *in situ* remains from the mid-Dnestr show some inversions (Boriskovskij 1984). The oldest shows a maximal age of about 31 000 BP (Molodova V, layer XI) and, like others, is probably too young – but too young by a magnitude of not much more than some 5000 years (Grigor'ev, pers. comm.). Somewhat older and more consistent dates are available for the Kostenki sites. These dates, like those from the middle Dnestr are, however, not nearly as old as dates reported from South-Central Europe (e.g. Bacho Kiro).

Furthermore, the great spatial discontinuity between the two regions with early Upper Palaeolithic sites on the Russian Plain (i.e. the Don and the Dnestr), as well as the absence of stylistic affinities between their inventories, possibly indicate that the two areas were re-peopled from different regions during the early Upper Palaeolithic. The earliest Upper Palaeolithic sites on the mid-Dnestr, like their late Middle Palaeolithic predecessors, contain inventories which show highly regionalized stylistic affinities confined to the Dnestr-Prut and adjacent regions of Moldavia and Rumania (Chernysh 1977a, 1977b, 1982; Borkiskovskij 1984). Some Soviet scholars have argued that early Upper Palaeolithic inventories with pronounced Middle Palaeolithic elements (assigned to the 'Streletskaya culture') in the Kostenki region show affinities to Middle Palaeolithic inventories found in the south-south-eastern steppe zone and in Crimea (Boriskovskij 1984, with references). Contemporaneous Kostenki assemblages which lack these Middle Palaeolithic elements (assigned to the 'Spitsynskaya culture') likewise show evidence for contact with the southern steppe zone. Stone tools at these Spitsyn sites are almost exclusively made of superior dark flint which came from considerable distances to the south-south-west of Kostenki (Fig. 34.5C) (Boriskovskij 1963, 1984).

In brief, the available data suggest that the Russian Plain should not be considered a part of the core area which saw the earliest appearance of Upper Palaeolithic industries, nor a locus where we can trace an uninterrupted *in situ* evolution from the Middle to the Upper Palaeolithic.

LATE MIDDLE AND EARLY UPPER PALAEOLITHIC FOOD MANAGEMENT

The nature of Pleistocene hominid occupation of Europe has received some recent attention in the literature and arguments have been offered that the whole continent did not see continuous occupation much before the advent of anatomically-modern people some 35 000 to 40 000 years ago (Binford 1982a, 1985; Butzer 1982; Gamble 1986). Various reasons have been offered to account for these disparate occupation records (Binford 1980; Butzer 1982; Dennell 1983; Gamble 1986). Data from the Russian Plain similarly indicate that a patchy and discontinuous occupation existed during the Middle Palaeolithic and was replaced by a more continuous one in the Upper Palaeolithic. A comparison of faunal remains of species used as food at the Middle and early Upper Palaeolithic sites suggests that a significant change in food management strategies occurred in the Upper Palaeolithic, and may have been one of the factors responsible for this difference in occupation. Specifically, I suggest that the change involved a switch from opportunistic exploitation of what was encountered (whether through active hunting, various forms of collecting, or both) to strategies which took into account seasonal and longer-term fluctuations in the availability and abundance of the different resources.

Data on density-controlled food remains found at the relevant Middle and Upper Palaeolithic sites are presented in Table 34.2. Figures 34.6 and 34.7 summarize these remains for different periods of occupation and dif-

Table 34.2 Densities of remains of different food taxa recorded at Middle and Upper Palaeolithic sites on the Russian Plain. In each column the figure on the left indicates the Numbers of Identified Specimens ('NISP') and that on the right the Minimum Number of Individuals ('MNI'), per metre square of the excavated areas. 'n.d.' = no data available.

MIDDLE PALAEOLITHIC SITES

	Mammuthus	*Coelodonta*	*Equus*	*Bison priscus*	*Bos primig.*	*Caprinae*	*Saiga*	*Cervus elaphus*	*Cervus sp.*	*Megaloceros*	*Alces alces*	*Rangifer*	*Lepus sp.*	*Marmota*
Betovo	n.d./0.01	n.d./0.01	n.d./–	n.d/–	n.d./–	n.d./–	n.d./–	n.d./–	n.d./–	n.d./–	n.d./–	n.d./–	n.d./–	n.d./0.22
Ketrosy I	0.82/0.04	0.02/0.01	0.10/0.01	0.57/0.02	–	–	–	–	–	0.01/0.01	–	–	–	0.02/0.01
Ketrosy II	1.92/0.04	0.06/0.01	0.51/0.02	0.10/0.02	–	0.01/0.01	–	0.06/0.01	–	0.02/0.01	–	–	–	0.02/0.01
Korman IV–XII	0.03/0.02	0.13/0.03	1.07/0.03	0.95/0.03	–	–	–	–	–	0.08/0.02	–	–	–	–
Korman XI	0.27/0.03	0.01/0.01	0.48/0.03	–	–	–	–	0.58/0.03	–	–	–	0.32/0.03	–	–
Korman X	–	–	–	0.14/0.02	–	–	–	0.20/0.02	–	–	–	–	–	–
Korman IX	–	–	1.04/0.04	–	–	–	–	0.94/0.04	–	–	0.04/0.02	–	–	–
Korman VII	–	0.06/0.01	–	–	–	–	–	–	–	–	–	–	–	–
Molodova I–IV	n.d./0.02	n.d./0.01	n.d./0.01	n.d./0.01	n.d./–	n.d./–	n.d./–	n.d/–	n.d./0.01	n.d./–	n.d./0.01	n.d./0.01	n.d./0.01	n.d./–
Sukhaya Mechetka	0.08/n.d.	–/n.d.	0.06/n.d.	0.56/n.d.	–/n.d.	–/n.d.	0.05/n.d.	0.02/n.d.	–/n.d.	–/n.d.	–/n.d.	0.01/n.d.	–/n.d.	–/n.d.

UPPER PALAEOLITHIC SITES

	Mammuthus	*Coelodonta*	*Equus*	*Assinus*	*Bison priscus*	*Bos primig.*	*Bison/ Bos*	*Saiga*	*Cervus elaphus*	*Megaloceros*	*Rangifer*	*Lepus sp.*	*Aves sp.*
Babin I	0.01/0.01	–	0.45/0.05	–	–	–	–	–	–	–	6.15/0.13	–	–
Korman IV–VI	0.03/0.01	–	0.08/0.01	–	–	0.01/0.01	–	–	–	–	0.01/0.01	–	–
Kostenki I–V	–	–	0.04/0.02	–	–	–	0.02/0.02	–	–	–	–	–	–
Kostenki 11–V	1.20/0.10	–	0.10/0.10	–	–	–	–	–	–	–	–	–	–
Kostenki 12–III	0.94/0.12	0.06/0.06	7.13/0.19	–	–	–	–	–	0.25/0.06	–	0.87/0.12	–	–
Kostenki 12–II	0.02/0.06	0.44/0.02	0.03/1.00	0.44/0.06	–	–	0.02/0.06	0.02/0.06	0.75/0.06	–	0.02/0.06	0.02/0.12	0.06/0.06
Kostenki 12 Ia	–	–	0.02/0.02	–	–	–	–	–	–	0.02/0.02	–	–	–
Kostenki 17–II	0.14/0.02	–	0.32/0.03	–	–	–	0.18/0.02	0.02/0.02	–	–	0.17/0.02	0.02/0.02	–
Molodova I–III	0.04/n.d.	0.08/n.d.	0.21/n.d.	–/n.d.	–/n.d.	–/n.d.	–/n.d.	–/n.d.	–/n.d.	–/n.d.	0.43/n.d.	–/n.d.	–/n.d.
Molodova V–X	0.01/0.01	0.01/0.01	0.07/0.01	–	0.01/0.01	–	–	–	–	–	0.01/0.01	–	–
Molodova V–IX	0.01/0.01	0.01/0.01	0.17/0.01	–	0.01/0.01	–	–	–	–	–	0.08/0.01	–	–
Streletskaya 2	0.15/0.01	0.01/0.01	1.11/0.03	–	–	–	0.08/0.01	–	0.01/0.01	–	0.15/0.07	0.03/0.01	0.02/0.01

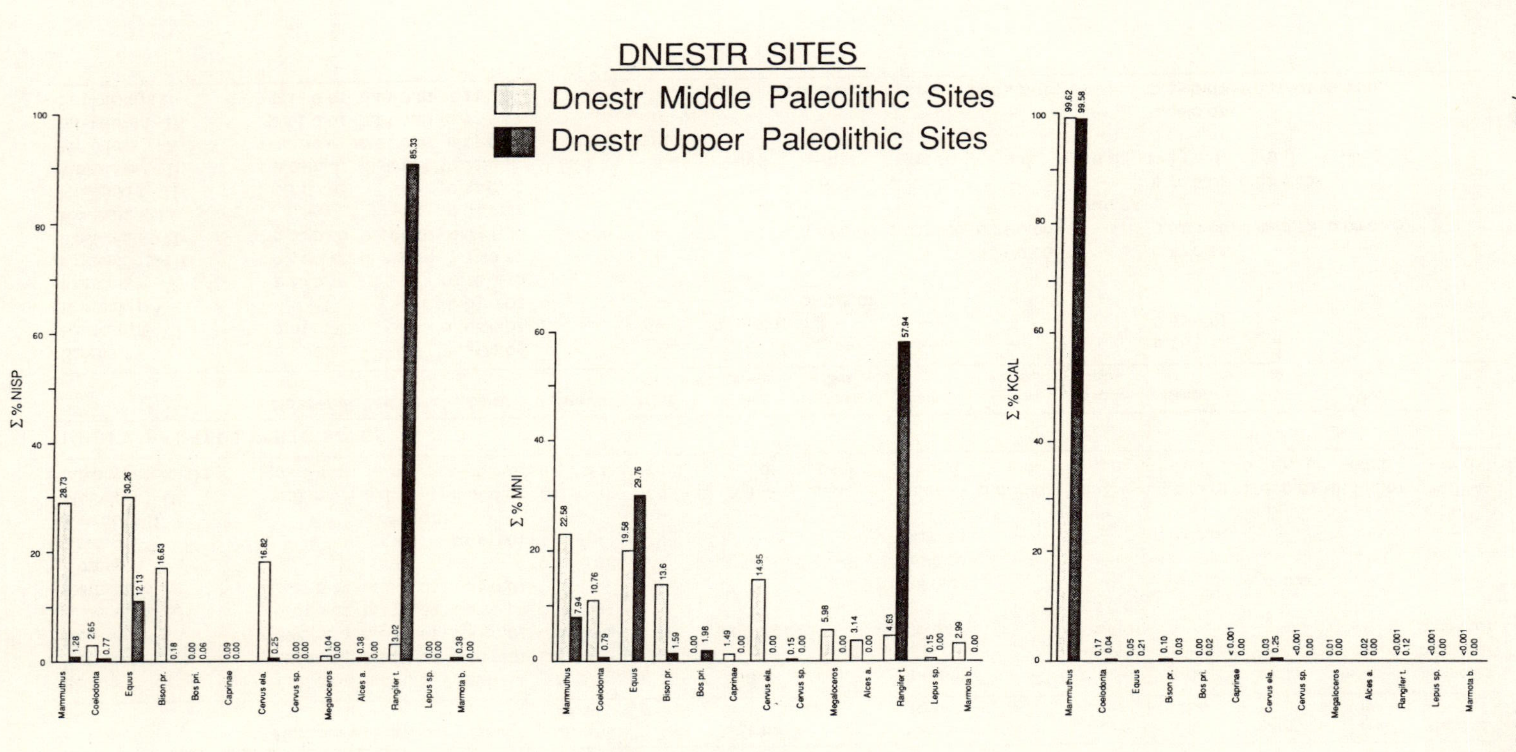

Figure 34.6. Summed density-controlled values for taxa used as food at the Dnestr sites (from data in Table 34.2). The graphs represent percentages based respectively on: Numbers of Identified Specimens (NISP) (left); Minimum Numbers of Individuals (MNI) (centre); and Kilocalories (KCAL) (right).

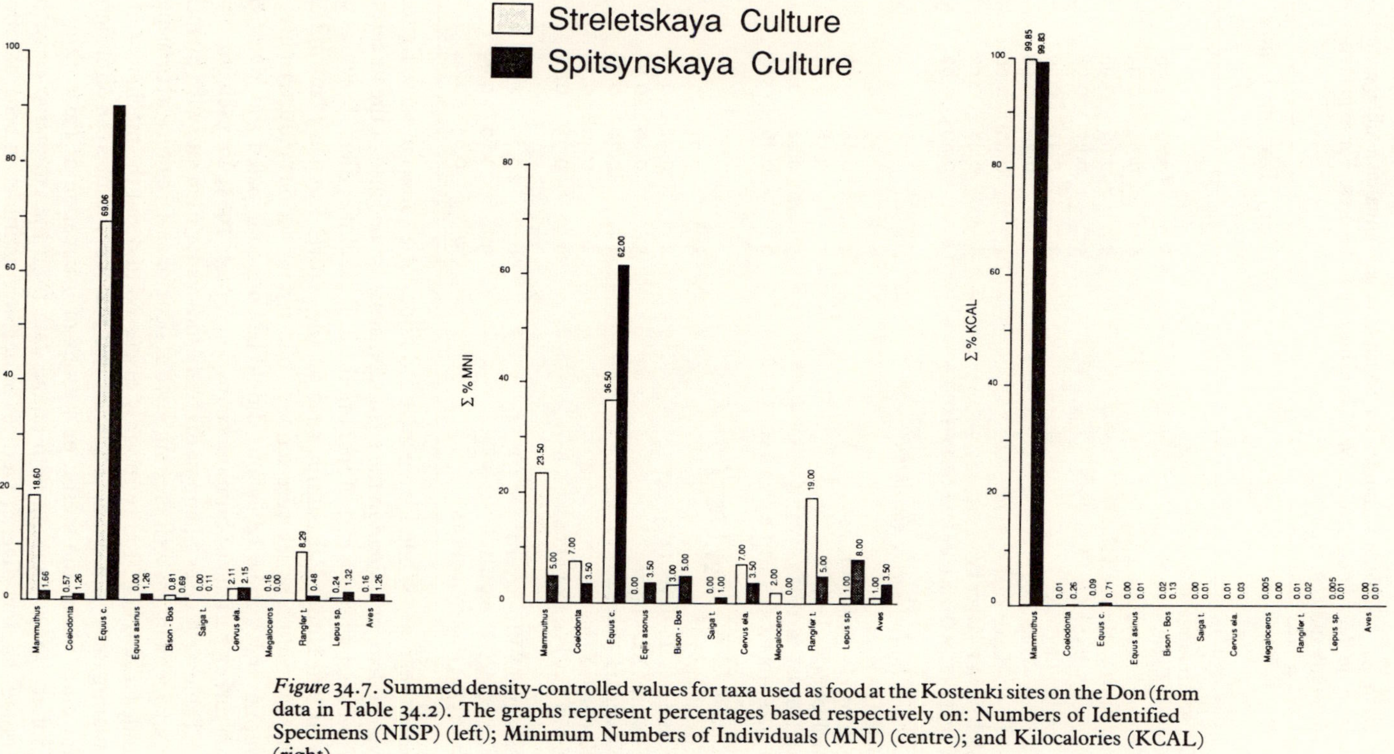

Figure 34.7. Summed density-controlled values for taxa used as food at the Kostenki sites on the Don (from data in Table 34.2). The graphs represent percentages based respectively on: Numbers of Identified Specimens (NISP) (left); Minimum Numbers of Individuals (MNI) (centre); and Kilocalories (KCAL) (right).

Table 34.3. Diversity Indices for South Russian faunal assemblages, calculated on density-controlled M.N.I. values of herbivores. For explanation of diversity indices, see text.

	Number of species	Diversity Indices 'H'	'J'
Dnestr-Middle Palaeolithic			
Ketrosy Excavation 1	6	0.70	0.90
Ketrosy Excavation 2	8	0.84	0.93
Korman IV – Layer XII	5	0.69	0.98
– layer XI	5	0.67	0.96
– layer X	2	0.31	0.99
– layer IX	3	0.46	0.96
– layer VIII	1	0.00	0.00
Molodova 1 – layer IV	8	0.48	0.53
Σ DNESTR MIDDLE PALAEOLITHIC	12	0.89	0.82
Dnestr – Early Upper Palaeolithic			
Babin 1 – layer 1	3	0.33	0.70
Korman IV – layer VI	4	0.58	0.96
Molodova V – layer X	5	0.56	0.81
Molodova – layer IX	5	0.62	0.89
Σ DNESTR EARLY UPPER PALAEOLITHIC	6	0.46	0.59
Russian Plain – Early Upper Palaeolithic			
Streletskaya II	8	0.71	0.79
Kostenki XII – layer 1a	2	0.30	0.38
Kostenki XII – layer 3	5	0.66	0.94
Kostenki I – layer 5	2	0.30	0.38
Kostenki XI layer 5	2	0.30	0.38
Σ STRELETSKAYA SITES	9	0.73	0.77
Kostenki XVII – layer 2	6	0.79	1.00
Kostenki XII – layer 2	9	0.60	0.63
Σ SPITSYNSKAYA SITES	10	0.64	0.64

ferent regions. While I have included kilocalorie values in these figures as well, I disregard them in the ensuing discussion because of the uncertainty associated with the role of mammoths in Palaeolithic diets. Previous work with late Upper Palaeolithic sites on the central Russian Plain indicated that mammoth bones were actively collected for various industrial uses and implies that similar behaviour should be both anticipated and investigated for earlier periods of human occupation (Soffer 1985; Vereschagin 1979).

Table 34.3 gives Shannon diversity indices for each site with quantified faunal remains, as well as cumulative indices for each region and time-period under discussion. I have recently discussed in detail the mechanics of calculating these indices and demonstrated that the 'J' values can offer significant clues about the season or seasons of occupation at Palaeolithic sites (Soffer 1985). Working with faunal data from the ethnographic record I argued that low-diversity indices in northern latitudes in general reflected cold-season occupations, while high ones were symptomatic of occupations during the warm seasons.

Information on Numbers of Identified Specimens ('NISP') and Minimum Numbers of Individuals ('MNI') values, together with the diversity indices tentatively indicate that a change in food procurement and management strategies occurred between the Middle and the Upper Palaeolithic. These conclusions, as well as all subsequent ones I draw from the faunal data, are based on visual comparisons of the different values rather than on statistical evaluations. The sparsity and nature of the available data, which can be characterized as incomplete 'kitchen lists' at best, at present only allow tenuous descriptive rather than rigorous quantitative evaluations.

Data from the Dnestr indicate broad-spectrum feeding behaviour during the late Middle Palaeolithic, when 12 taxa contributed to the diet, and a narrowing of this spectrum to six species during the Upper Palaeolithic period. Both the bar graphs and the diversity indices indicate not only a reduction in the number of food taxa, but also a significant change in the rate of exploitation of the different species. The greater evenness in the distribution of the NISP and MNI values between the different food species suggests a pattern of opportunistic exploitation of taxa as encountered. The different patterning of faunal data from early Upper Palaeolithic layers, where greater unevenness and selectivity of specific taxa are observed, suggests a food-management strategy which took into account the temporal and spatial variability in the prey taxa.

Recently I have argued that that analyses of diversity indices are most productive when the nature of occupation at the sites in question is well understood and site typologies established (Soffer 1985). To date, this unfortunately is not the case with the Dnestr sites and precludes an unambiguous assessment of the significance of differences between the diversity indices at different sites. The indices in Table 34.2 appear to indicate a greater range in diversity at the Middle Palaeolithic sites and a reduction in this range at the Upper Palaeolithic ones. The presence of both low and high values at the Middle Palaeolithic sites tentatively suggests that groups were present in the middle Dnestr region all year around, while the greater uniformity of Upper Palaeolithic values may indicate only seasonal occupation of the area.

Since the Dnestr data presented here telescope a huge segment of time and reflect a number of directional changes in climate towards increasing cold and dry conditions which were accompanied by an overall reduction in resource diversity, it could be argued that the changes in faunal assemblages discussed above may just reflect a directional reduction in the diversity of the resources themselves. This proposition can be tested with data from the early Upper Palaeolithic sites on the middle Don. Kostenki sites dating before 30 000 BP consist of two groups of roughly contemporaneous sites which were occupied during a period when the area was covered by cold taiga formations (Boriskovskij 1963, 1984; Fedorova 1963; Klein 1969; Levkovskaya 1977; Praslov 1985; Praslov and Rogachev 1982). Soviet scholars assign these sites to two distinct disparate early Upper Palaeolithic archaeological cultures. The Streletskaya culture contains many Middle Palaeolithic elements, is characterized by bifacially-retouched hollow-base points, and features a complete absence of bone tools (Anikovich 1977; 1983; Praslov and Rogachev 1982). Sites assigned to the Spitsynskaya (Spitsyn) culture have prismatic cores, tools made on blades, and some worked bone (Boriskovskij 1963, 1984; Praslov and Rogachev 1982). Streletskaya inventories are exclusively made of locally-available, inferior lithic materials. The vast majority of tools in the Spitsyn assemblages, as noted above, are of exotic, superior dark flint which came from 150-300 km to the south-south-west of Kostenki (Fig. 34.5C).

Data on species used as food at the Kostenki sites are presented in Table 34.2. The differences between the NISP and MNI histograms at these two groups of sites (Fig. 34.7) show similar, although admittedly more muted, patterning to that observed between the late Middle and early Upper Palaeolithic sites on the Dnestr. Specifically, differences in the values (especially MNI figures) indicate a greater diversity in harvested resources at the Streletskaya sites and a reduction in this diversity at the Spitsyn sites. These data suggest that even though we are dealing with two early Upper Palaeolithic cultures here, the one with the stronger presence of Middle Palaeolithic elements also shows closer affinities in the structure of its organic remains to patterns observed at Middle Palaeolithic sites elsewhere. I tentatively interpret this patterning to indicate that food management strategies at Streletskaya sites were closer to the opportunistic feeding patterns suggested for Dnestr late Middle Palaeolithic sites, and that selective exploitation of specific seasonally and locally abundant resources is far more in evidence at the Spitsyn sites.

Diversity indices calculated for the Kostenki sites are similarly difficult to evaluate in the absence of site typologies. Those offered in Table 34.3 show a wider range of values at the Streletskaya sites and a narrower range at the Spitsyn sites. This difference may reflect differences in subsistence strategies and settlement systems which were more regionally circumscribed in the former case. The narrower range of values at the Spitsyn sites (if not solely a product of sample size) may reflect foraging adaptations which involved 'mapping-on' to resources over wider areas, and entailed seasonal group mobility.

Finally, since the two groups of sites are (1) roughly contemporaneous; (2) both classified as early Upper Palaeolithic; and (3) since no hominid remains have been found there – the preliminary suggestions of disparate food management strategies offered above may indicate that the change from 'opportunistic' exploitation to 'mapping-on' to resources neither occurred overnight at the Middle-Upper Palaeolithic rubicon, nor was necessarily associated with different types of hominids.

DISCUSSION

Similar changes previously noted between the Middle and Upper Palaeolithic faunal assemblages recovered elsewhere in the Old World have been used to argue for the rise of specialized exploitation of one or two species during the Upper Palaeolithic (Binford 1968; Mellars 1973). Other scholars have argued that data from the Near East in particular indicate an expansion of the resource base and an advent of broad-spectrum feeding in the Upper Palaeolithic (Hole and Flannery 1967). Some researchers have combined both of these hypotheses into a model which sees Upper Palaeolithic food procurement as both specialized in the exploitation of one or two key resources as well as dependent on the harvesting of a broader range of species (Freeman 1973; Straus 1983). Arguments have also been offered that these and other changes reflect a switch from foraging to logistical organization of food management (Binford 1982a).

In general, most of these models are regionally specific, mutually exclusive, and are, explicitly or implicitly, based on differences observed in the economic behaviour of ethnographically known groups of anatomically-modern people. I would suggest that Middle and early Upper Palaeolithic data from the Russian Plain in particular, and those from elsewhere in general, do not mirror a switch from one extreme to the other of ethnographically known behaviour (i.e. from forager to logistically organized strategies), but reflect the evolution of ethnographically observed food procurement. Specifically, I would argue that the faunal data may reflect a switch from 'opportunistic encounter' within a limited occupied area to purposeful 'mapping-on' to resources over a larger region. The former strategy, hypothesized for the Middle Palaeolithic, has no analogues among ethnographically known groups. The latter is characteristic of ethnographically known people and is based on (1) planning on a group-wide basis, and (2) on residential mobility to position consumers near to the available foods (Binford 1980, 1982a, 1982b, 1985; Whallon, this volume). A change of this kind would necessarily have different archaeozoological signatures in different environments. It would appear as a narrowing of the resource base and concentration on one or two species in areas with great seasonal fluctuation in the availability of key resources. In regions with less seasonal variability but greater spatial differentiation in both the amount and kinds of resources, on the other hand, it would appear as an expansion of the resource base and broad spectrum feeding.

Admittedly my hypotheses are based on very problematic data sets – a collection of published 'kitchen lists' rather than on first-hand examination of inventories. They are speculative first approximations rather than definitive conclusions. Overall, however, differences in the observed between the Middle and early Upper Palaeolithic faunal data sets are repeated in other categories of archaeological remains such as lithic inventories and structural features found at the occupation sites.

Middle Palaeolithic tool inventories, whether from early or late sites, and whether from the Dnestr, Crimea, or the Volga, all show the redundant use of locally-available resources. Claimed exotic materials – a few limestone blocks from Nosovo and some materials from Rozhok – come from no further than 10 to 12 km away (Praslov 1968, 1972). A similar use of local materials is in evidence at the Streletskaya early Upper Palaeolithic sites as well. This use of locally-available materials is, in fact, a feature of Middle Palaeolithic stone-tool-making behaviour in most parts of the Old World (Bahn 1984; Klein 1976; Kozlowski 1972-1973; Marks 1987;

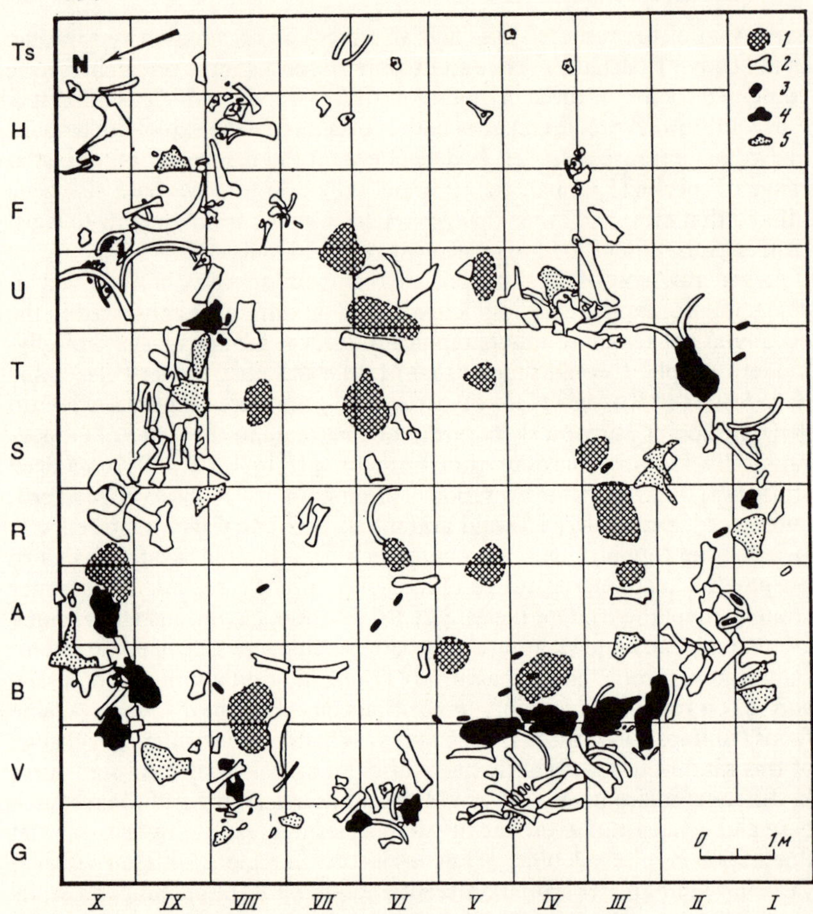

Figure 34.8. Supposed remains of a mammoth-bone dwelling at Molodova I, Layer IV (after Chernysh 1982, Figure 9). Symbols are as follows: (1) hearths; (2) large bones; (3) mammoth teeth; (4) mammoth crania; (5) mammoth crania and scapulae.

Mellars 1973; White 1982, with references).

Early Upper Palaeolithic inventories on the Russian Plain, with the noted exception of Streletskaya ones, all contain exotic lithic materials. The percentages of non-local materials found vary considerably in different regions. In the middle Don, an area lacking good flint sources, nearly all lithic materials recovered at the Spitsyn sites are non-local in origin. Sites on the middle Dnestr, an area with locally-available good-quality flint outcrops, contain non-local lithics in small quantities (e.g. Carpathian obsidian at Voronovitsa, Central European radiolarite at Babin I, and mountain crystal from the northern Volyn' area at Molodova V) (Chernysh 1956, 1973, 1977a). The presence of these exotic materials at the early Upper Palaeolithic sites suggests both familiarity with and systematic exploitation of larger territories than during the preceding Middle Palaeolithic period. This

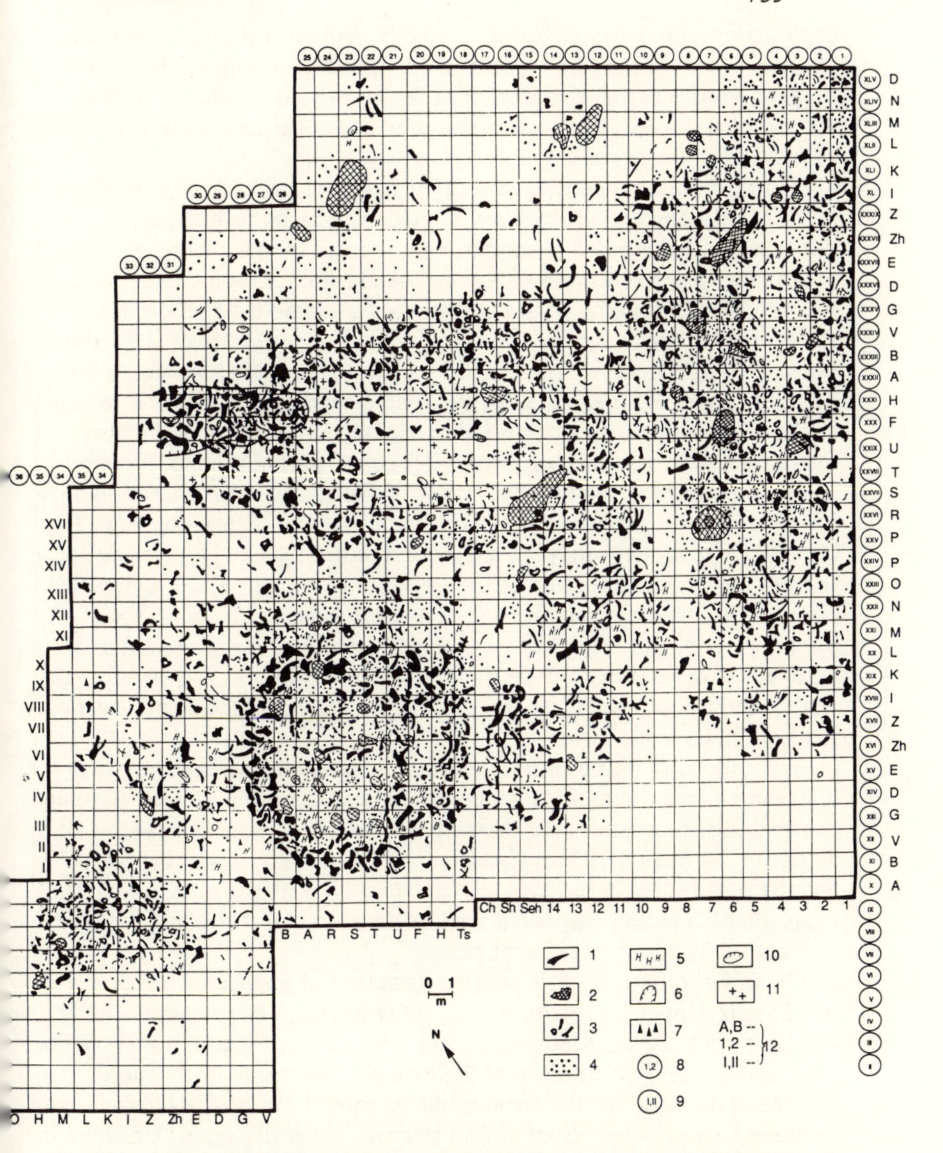

Figure 34.9. Patterning of cultural remains at Molodova I, Layer IV (after Chernysh 1982, Figure 8). Symbols are as follows: (1) mammoth tusks; (2) remains of hearths; (3) bones; (4) worked stone; (5) nuclei; (6) accumulation of bones in a natural pit; (7) stone tools; (8, 9) combined grid numbering systems; (10) mammoth teeth; (11) engraved marks on large bones; (12) field grid numbers.

exploitation may have occurred indirectly through various exchange networks, as was the case during the late Upper Palaeolithic (Soffer 1985). I suspect, however, that at the onset of the Upper Palaeolithic it more probably resulted directly from patterned mobility of the tool makers, possibly as part of seasonal residential moves.

Significant differences also exist in the nature of structural features found at the Middle and Upper Palaeolithic sites. Although the literature abounds with claims for supposed unambiguous structure and patterning of features at Middle Palaeolithic sites, these claims are quite problematic. Some of the strongest data come from the supposed mammoth-bone 'dwellings' at the Molodova sites – specifically at Molodova I, layer IV (Fig. 34.8) (Boriskovskij 1984; Chernysh 1961, 1965, 1973, 1977a, 1982; Rogachev 1970; Shimkin 1978; Sklenař 1975, 1976). The interpretation of these remains as a mammoth-bone dwelling is based on a telescoped view of the distribution of just some remains on one small (110 m^2) part of the site. A recently-published complete plan for the whole of the excavated area at Molodova I-IV (1200 m^2) considerably reduces the clarity of such patterning (Fig. 34.9). The complete site plan suggests that claims for the existence of a roofed-over mammoth-bone structure with 15 hearths and a living floor strewn with over 29 000 pieces of flint have been vastly exaggerated. As Păunescu (1978) has argued, it is more likely that the structures at Molodova I-IV were more akin to a series of diachronically-utilized windbreaks analogous to those he reconstructed for the Middle Palaeolithic layers at Ripiceni Izvor in Rumania.

Oval concentrations of lithic and bone material measuring from 30 to 50 m^2 in area, sometimes with shallow hearths, and often at least partially overlapping one another, are characteristic of late Middle Palaeolithic sites along the Dnestr (Boriskovskij 1984; Chernysh 1961, 1965, 1973, 1977a, 1977b, 1982; Sklenař 1975, 1976). However, the spatial integrity of these ovals is far from clear cut, and it is more than likely that they resulted from palimpsests of repeated occupations.

The organization of space and the structure of features is far clearer at the Dnestr Upper Palaeolithic sites (Boriskovskij 1984; Chernysh 1973, 1977a, 1977b, 1982; Klein 1973; Rogachev 1970; Sklenař 1975, 1976). Post-moulds suggest outlines of lightweight structures with one or two hearths. This clear-cut patterning, like its muted Middle Palaeolithic predecessor, is redundant from site to site and suggests, again, a change in time in the organization and use and reuse of living space at sites.

In conclusion, then, I suggest that the transition from Middle to Upper Palaeolithic adaptations, at least in European USSR, involved: (1) a qualitative change in both the perception and in the utilization of nature; (2) a change in food management strategies and in the accompanying settlement patterns; and finally (3) that this change clearly did not happen overnight nor at a particular chronological or physiological rubicon. Delving into the realm of still further speculation, I propose that it is precisely the change in food management strategies from opportunistic encounter to 'mapping-on' foraging which permitted the successful colonization and continuous occupation of the Russian Plain proper, and that this occurred irreversibly only in the Upper Palaeolithic.

ACKNOWLEDGEMENTS

I would like to thank Paul Mellars for inviting me to cast this first glance at the possible differences between the Middle and Upper Palaeolithic adaptations, and my colleagues Stanley Ambrose and Randall White for inciteful comments on my various preliminary conclusions. Financial support of the University of Illinois Scholar's Travel Fund and Russian and East European Center is likewise gratefully acknowledged.

REFERENCES

Anikovich, M. V. 1977. *Pamyatniki Streletskoj Kul'tury v Kostenkakh.* Avtoreferat Dissertatsii na Soiskaniye Uchenoj Stepeni Kandidata Istorichskikh Nauk. Moscow: Institut Arkheologii AN SSSR.

Anikovich, M. V. 1983. K probleme sinkhronizatsii nekotorikh Pozdnepaleoliticheskikh pamyatnikov Kostenkovsko-Borschevskogo rajona. *Kratkiye Soobscheniya Instituta Arkheologii* 173: 16-23.

Anisyutkin, N. K. 1969. Must'erskaya stoyanka Stinka na Srednem Dnestre. *Arkheologicheskij Sbornik* 11: 5-17.

Anisyutkin, N. K. 1980. Zhiloj kompleks must'erskogo poseleniya Ketrosy v Podnestrov'ye. In I. I. Artemenko (ed.) *Pervobytnaya Arkehologiya – poiski i nakhodki.* Kiev: Naukova Dumka: 38-46.

Anisyutkin, N. K. 1981. Arkheologicheskoye uzuchenie must'erskoj stoyanki Ketrosy. In N. D. Praslov (ed.) *Ketrosy: Must'erskaya stoyanka na Srednem Dnestre.* Moscow: Nauka: 7-52.

Ashton, N. M. 1983. Spatial patterning in the Middle-Upper Palaeolithic transition. *World Archaeology* 15: 224-235.

Bahn, P. 1984. *Pyrenean Prehistory.* Warminster: Aris and Philips.

Binford, L. R. 1980. Willow smoke and dogs tails: hunter-gatherer settlement systems and archaeological site formation. *American Antiquity* 45: 4-20.

Binford, L. R. 1981. *Bones: Ancient Men and Modern Myths.* New York: Academic Press.

Binford, L. R. 1982a. Comment on R. White: 'Rethinking the Middle/Upper Paleolithic transition'. *Current Anthropology* 23: 372-377.

Binford, L. R. 1982b. The archaeology of place. *Journal of Anthropological Archaeology* 1: 5-31.

Binford, L. R. 1984. *Faunal Remains From Klasies River Mouth.* New York: Academic Press.

Binford, L. R. 1985. Human ancestors: changing views of their behavior. *Journal of Anthropological Archaeology* 4: 292-327.

Binford, S. R. 1968. Early Upper Pleistocene adaptations in the Levant. *American Anthropologist* 70: 707-717.

Bolikhovskaya, N. S. 1981. Rastitel'nost' i klimat Srednego Pridnestro'ya v pozdnem Plejstotsene. Rezulj'taty palinologicheskogo analiza otlozhenij Kashlyanskogo Yara. In N. D. Praslov (ed.) *Ketrosy: Must'erskaya Stoyanka na Srednem Dnestre.* Moscow: Nauka: 103-124.

Bolikhovskaya, N. S. and Pashkevich, G. A. 1982. Dinamika rastitel'nosti v okrestnostyakh stoyanki Molodova I v pozdnem Plejstotsene (po materialam palinologicheskogo issledovaniya). In G. I. Goretsky and I. K. Ivanova (eds) *Molodova I: Unikal'noye Must'erskoe Polelenie na Srednem Dnestre.* Moscow: Nauka: 120-144.

Boriskovskij, P. I. 1963. Ocherki po Paleolitu basseina Dona. *Materialy i Issledovaniya po Arkheologii* SSR 121.

Boriskovskij, P. I. and Praslov, N. D. 1964. *Paleolit basseina Dnepra i Priazov'ya*. Svod Arkheologicheskikh Istochnikov A 1-5. Moscow and Leningrad: Nauka.

Boriskovskij, P. I. (ed.) 1984. *Paleolit* SSSR. Moscow: Nauka.

Bowen, Q. D. 1978. *Quaternary Geology: a Stratigraphic Framework for Multidisciplinary Work*. Oxford: Pergamon Press.

Butzer, K. W. 1982. *Archaeology as Human Ecology*. Cambridge: Cambridge University Press.

Cartmill, M. 1983. Four legs good, two legs bad. *Natural History* 11: 65-79.

Chase, P. G. 1986. *The Hunters of Combe Grenal: Approaches to Middle Palaeolithic Subsistence in Europe*. Oxford: British Archaeological Reports International Series S286.

Chebotareva, N. S. and Makaricheva, I. A. 1974. *Poslednee Oledenenie Evropy i ego Geokhronologiya*. Moscow: Nauka.

Chebotareva, N. S. and Makaricheva, I. A. 1982. Geokhronologiya prirodnykh izmenenij lednikovoj oblasti Vostochnoj Evropy v Valdajskuyu epokhu. In. I. P. Gerasimov and A. A. Velichko (eds) *Paleogeografiya Evropy za Poslednie Sto Tysyach Let*. Moscow: Nauka: 16-27.

Chernysh, A. P. 1953. Paleoliticheskaya stoyanka Babin I po materialam raskopok 1949-50 goda. *Kratkiye Soobscheniya Instituta Istorii Material'noj Kul'tury* 49: 56-65.

Chernysh, A. P. 1956. Paleoliticheskaya stoyanka Voronovitsa l. *Kratkiye Soobscheniya Instituta Istorii Material'noj Kul'tury* 63: 40-48.

Chernysh, A. P. 1961. *Paleolitichna Stoyanka Molodove* V. Kiev: Vidavnitstvo AN URSR.

Chernysh, A. P. 1965. Rannij i srednij paleolit Pridnestrov'ya. *Trudy Komissii po Izucheniyu Chetvertichnogo Perioda* 25.

Chernysh, A. P. 1973. *Paleolit i Mezolit Pridnestrov'ya*. Moscow: Nauka.

Chernysh, A. P. 1977a. Nekotoriye itogi mnogoletnikh issledovanij paleolita i mezolita Prikarpat'ya. In N. D. Praslov (ed.) *Problemy Paleolita Vostochnoj i Tsentral'noj Evropy*. Leningrad: Nauka: 197-208.

Chernysh, A. P. 1977b. Mnogoslojnaya paleoliticheskaya stoyanka Korman' IV i ee mesto v Paleolite. In G. I. Goretskij and S. M. Tseitlin (eds) *Mnogoslojnaya Stoyanka Korman' IV na Srednem Dnestre*. Moscow: Nauka: 7-77.

Chernysh, A. P. 1982. Mnogoslojnaya paleoliticheskaya stoyanka Molodova I. In G. I. Goretskij and I. K. Ivanova (eds) *Molodova I: Unikal'noye Must'erskoye Poseleniye na Srednem Dnestre*. Moscow: Nauka: 6-102.

Davis, R. S. 1983. Theoretical issues in contemporary Soviet Paleolithic archaeology. *Annual Reviews of Anthropology* 12: 403-428.

Dennell, R. 1983. *European Economic Prehistory*. London: Academic Press.

Dolukhanov, P. M. 1982. Upper Pleistocene and Holocene cultures of the Russian Plain and Caucasus: ecology, economy, and settlement patterns. In F. Wendorf and A. E. Close (eds) *Advances in World Archaeology* Vol. 1. New York: Academic Press: 323-358.

Fedorova, R. V. 1963. Prirodnye usloviya v period obitaniya verkhnepaleoliticheskogo cheloveka v rajone s. Kostenok Voronezhskoj oblasti. In P. I. Boriskovskij, Ocherki po Paleolitu Basseijna Dona. *Materialy i Issledovaniya po Arkheologii* SSSR 121: 220-230.

Freeman, L. G. 1973. The significance of mammalian faunas from Paleolithic occupations in Cantabrian Spain. *American Antiquity* 38: 3-44.

Gamble, C. 1986. *The Palaeolithic Settlement of Europe*. Cambridge: Cambridge University Press.
Gerasimov, I. P. and Velichko, A. A. (eds) 1982. *Paleogeografiya Evropy za Posledniye Sto Tysyach Let*. Moscow: Nauka.
Gladilin, V. N. 1976. *Problemy Rannego Paleolita Vostochnoj Evropy*. Kiev : Naukova Dumka.
Gladilin, V. N. 1966. Vidkrittya Must'erskoj stoyanki na Donechchini. *Arkheologia* 20: 135-142.
Goretskij, G. I. and Ivanova, I. K. (eds) 1982. *Molodova I: Unikal'noe Must'erskoe Polelenie na Srednem Dnestre*. Moscow: Nauka.
Goretskij, G. I. and Tseitlin, S. M. (eds) 1977. *Mnogoslojnaya Paleoliticheskaya Stoyanka Korman' IV*. Moscow: Nauka.
Grayson, D. 1983. *The Establishment of Human Antiquity*. New York: Academic Press.
Grichuk, V. P. 1982. Rastitel'nost' Evropy v pozdnem Pleistotsene. In I. P. Gerasimov and A. A. Velichko (eds) *Paleogeografiya Evropy za Posledniye Sto Tysyach Let*. Moscow: Nauka: 92-109, Map 10.
Grichuk, V. P., in press a. Vegetation of the Late Pleistocene interglacial (Eemian-Kazantsevo-Mikulino-Sangamon-stage 5e), about 125 000 BP. In A. A. Velichko and M. Pécsi (eds) *Paleogeographic Atlas of the Quaternary: Northern Latitudes*. Budapest: Kiado. In Press.
Grichuk, V. P., in press b. Main types of vegetation (ecosystems) of Eurasia during the maximum cooling of the last glaciation (explanatory text to Map 10). In A. A. Velichko and M. Pécsi (eds) *Late Pleistocene Paleogeographic Atlas: Northern Latitudes*. Budapest: Kiado. In Press.
Grigor'ev, G. P. 1968. *Nachalo Verkhnego Paleolita i Proiskhozhdenie Homo Sapiens*. Leningrad: Nauka.
Grigor'ev, G. P. 1970. Verkhnij Paleolit. In A. A. Formozov (ed.) *Kamennij Vek na Territorii SSSR*. Moscow: Nauka: 43-64.
Hole, F. and Flannery, K. V. 1967. The prehistory of south-western Iran: a preliminary report. *Proceedings of the Prehistoric Society* 33: 147-206.
Hoffecker, J. F. 1986. *Upper Paleolithic Settlement on the Russian Plain*. Unpublished Ph.D. Dissertation, Department of Anthropology, University of Chicago.
Ivanova, I. K. 1965. Stratigraficheskoe polozhenie Molodovskikh paleoliticheskikh stoyanok na Srednem Dnestre v svete obshchikh voprosov statigrafiii absolytnoj geokhronologii verkhnego Plejstotsena Evropy. In O. N. Bader, I. K. Ivanova and A. A. Velichko (eds) *Stratigrafiya i Periodizatsiya Paleolita Vostochnoj i Tsentral'noj Evropy*. Moscow: Nauka: 123-140.
Ivanova, I. K. 1969. Geologicheskoe stroyenie doliny r. Dnestra v rajone Must'erskogo mestonakhozhdeniya Stinka. *Byulleten' Komissii po Izucheniyu Chetvertichnogo Perioda* 36: 129-136.
Ivanova, I. K. 1977a. Geologiya i paleogeografiya stoyanki Korman' IV na obschem fone geologicheskoj istorii kamennogo veka Srednego Pridnestov'ya. In G. I. Goreskij and S. M. Tseitlin (eds) *Mnogoslojnaya Paleoliticheskaya Stoyanka Korman' IV*. Moscow: Nauka: 126-181.
Ivanova, I. K. 1977b. Prirodniye uslov'ya obitaniya l'yudej kamennogo veka v basseine reki Dnestr. In I. K. Ivanova and N. D. Praslov (eds) *Paleoekologiya Drevnego Cheloveka*. Moscow: Nauka: 7-18.
Ivanova, I. K. 1981. Geologiya i geomorfologiya okrestnostej stoyanki Ketrosy. In N. D. Praslov (ed.) *Ketrosy, Must'erskaya stoyanka na Srednem Dnestre*. Moscow: Nauka: 59-80.
Ivanova, I. K. 1982. Geologiya i paleogeografiya Must'erskogo poseleniya Molodova I. In G. I. Goretskij and I. K. Ivanova (eds) *Molodova I: Unikal'noe Must'erskoye Poselenie na Srednem Dnestre*. Moscow: Nauka: 188-235.

Klein, R. G. 1966. *The Mousterian of European Russia*. Unpublished Ph.D. Dissertation, Department of Anthropology, University of Chicago.
Klein, R. G. 1969. *Man and Culture in the Late Pleistocene: a Case Study*. San Francisco: Chandler.
Klein, R. G. 1973. *Ice-Age Hunters of the Ukraine*. Chicago: University of Chicago Press.
Klein, R. G. 1976. The mammalian fauna of the Klasies River Mouth sites, southern Cape Province, South Africa. *South African Archaeological Bulletin* 31: 75-98.
Kolosov Yu. G. 1983. *Must'erskie Stoyanki Rajona Belogorska*. Kiev: Naukova Dumka.
Kolosov, Yu. G. (ed.) 1979. *Issledovanie Paleolita v Krymu*. Kiev: Naukova Dumka.
Kozlowski, J. K. 1972-1973. The origin of lithic raw materials used in the Palaeolithic of the Carpathian countries. *Acta Arcaheologica Carpathica* 13: 5-19.
Kozlowski, J. K. 1986. The Gravettian in Central and Eastern Europe. In F. Wendorf and A. E. Close (eds) *Advances in World Archaeology* Vol. 5. New York: Academic Press: 131-200.
Landau, M. 1984. Human evolution as narrative. *American Scientist* 73: 262-268.
Lazukov, G. I., Gvozdover, M. D., Roginskii, Y. Y., Urynson, M. I., Kharitonov, V. M. and Yakimov, V. P. 1981. *Priroda i Drevnij Chelovek*. Moscow: Mysl'.
Levkovskaya, G. M. 1977. Palinologicheskaya kharakteristika razrezov Kostenkovsko-Borschevskogo rajona. In I. K. Ivanova and N. D. Praslov (eds) *Paleoekologiya Drevnego Cheloveka*. Moscow: Nauka: 74-83.
Levkovskaya, G. M. 1981. Palinologicheskaya kharakteristika Must'erskogo kul'turnogo sloya stoyanki Ketrosy. In N. D. Praslov (ed.) *Ketrosy: Must'erskaya Stoyanka na Srednem Dnestre*. Moscow: Nauka: 125-135.
Marks, A. E. 1988. The Middle to Upper Paleolithic transition in the Southern Levant: technological change as an adaptation to increasing mobility. In M. Otte (ed) *L'Homme de Néandertal*, Vol. 8: *La Mutation*. Liège: Etudes et Recherches Archéologique de l'Université de Liège 35: 108-123.
Mellars, P. A. 1973. The character of the Middle-Upper Palaeolithic transition in south-west France. In C. Renfrew (ed.) *The Explanation of Culture Change*. London: Duckworth: 255-276.
Nilsson, T. 1983. *The Pleistocene: Geology and Life in the Quaternary Ice Age*. Dordrecht: Reidel.
Orquera, L. A. 1984. Specialization and the Middle/Upper Paleolithic transition. *Current Anthropology* 25: 73-93.
Otte, M. (ed.) 1988. *L'Homme de Néandertal*. Etudes et Recherches Archéologique de l'Université de Liège.
Păunescu, A. 1978. Complexe de lucuire Musteriene descoperite in asezarea de la Ripiceni-Izvor (Jud. Botosani) si unele consideratii privind evolutia tipului de locuinta Paleolitica. *Sciva* 29 (3): 317-333.
Praslov, N. D. 1968. Rannij Paleolit Severo-vostochnogo Priazov'ya i Nizhnego Dona. *Materialy i Issledovaniya po Arkehologii SSSR* 157.
Praslov, N. D. 1972. Must'erskoye poselenie Nosovo I v Prazov'e. *Materialy i Issledovaniya po Arkheologii SSSR* 185: 75-82.
Praslov, N. D. (ed.) 1981. *Ketrosy: Must'erskaya stoyanka na Srednem Dnestre*. Moscow: Nauka.
Praslov, N. D. 1984. Geologicheskiye i paleogeograficheskiye ramki

Paleolita i razvitie prirodnoj sredy na territorii SSR i problemy khronologii i periodizatsii Paleolita. In P. I. Boriskovskij (ed.) *Paleolit SSSR*. Moscow: Nauka: 17-40.

Praslov, N. D. 1985. Kostenkovskaya gruppa Paleoliticheskikh stoyanok. In S. M. Shik (ed.) *Kraevye Obrazovaniya Materikovykh Oledenenij: Putevoditel' Ekskursii VII Vsesoyuznogo Soveshchaniya*. Moscow: Nauka: 24-28.

Praslov, N. D. and Rogachev, A. N. (eds) 1982. *Paleolit Kostenkovsko-Borschevskogo Rajona na Donu* 1879-1979. Nauka: Leningrad.

Rogachev, A. N. 1957. Mnogoslojnye stoyanki Kostenkovsko-Borschevskogo rajona na Donu i problema razvitiya kul'tury v epokhu verkhnego Paleolita na Russkoj ravnine. In *Materialy i Issledovaniya po Arkheologii SSSR* 59: 9-134.

Rogachev, A. N. 1970. Paleoliticheskie zhilishcha i poseleniya. In A. A. Formozov (ed.) *Kamennyj Vek na Territorii SSSR*. Moscow: Nauka: 64-78.

Shimkin, E. M. 1978. The Upper Paleolithic in North-Central Eurasia: evidence and problems. In L. G. Freeman (ed.) *Views of the Past: Essays in Old World Prehistory and Paleoanthropology*. The Hague: Mouton: 193-315.

Sklenař, K. 1975. Palaeolithic and Mesolithic dwellings: problems of interpretation. *Pamatky Archeologicke* 66: 266-304.

Sklenař, K. 1976. Palaeolithic and Mesolithic dwellings: an essay in classification. *Pamatky Archeologicke* 67: 249-340.

Soffer, O. 1985. *The Upper Paleolithic of the Central Russian Plain*. Orlando: Academic Press.

Soffer, O. 1986. Radiocarbon accelerator dates for Upper Paleolithic sites in European USSR. In J. A. J. Gowlett and R. E. M. Hedges (eds) *Archaeological Results from Accelerator Dating*. Oxford: Oxford University Committee for Archaeology Monograph 11: 109-115.

Soffer, O. and Gray, P. 1985. Constructing, deconstructing, and reconstructing the ancestors. Paper presented at the Annual Meeting of the American Anthropological Association. Washington (DC), December 1985.

Straus, L. G. 1983. From Mousterian to Magdalenian: cultural evolution viewed from Vasco-Cantabrian Spain and Pyrenean France. In E. Trinkaus (ed.) *The Mousterian Legacy: Human Biocultural Evolution in the Upper Pleistocene*. Oxford: British Archaeological Reports International Series S164: 73-112.

Tarasov, L. M. 1976. Issledovaniye Paleolita na Desne v rajone Betovo. *Arkheologicheskiye Otkritiya 1975 goda*. Moscow: Nauka: 91-92.

Tarasov, L. M. 1977a. Must'erskaya stoyanka Betovo i ee prirodnoe okruzhenie. In I. K. Ivanova and N. D. Praslov (eds) *Paleoekologya Drevnego Cheloveka*. Moscow: Nauka: 18-31.

Tarasov, L. M. 1977b. Raskopki Paleoliticheskikh sotyanok na verkhnej Desne. *Arkheologicheskiye Otkritiya 1976 goda*. Moscow: Nauka: 74-75.

Tarasov, L. M. 1980. Issledovaniye Must'erskoj stoyanki Betovo. *Arkheologicheskiye Otkritiya 1979 goda*. Moscow: Nauka: 81-82.

Tarasov, L. M. 1986. Mnogoslojnaya stoyanka Korshevo I. In V. P. Lyubin (ed.) *Paleolit i Neolit*. Leningrad: Nauka: 46-53.

Trinkaus E. (ed.) 1983. *The Mousterian Legacy: Human Biocultural Evolution in the Upper Pleistocene*. Oxford: British Archaeological Reports International Series S164.

Veklich, M. F. 1974. *Stratigrafiya Lyossovikh Otlozhenij Ukrainii*. Kiev: AN USSR

Velichko, A. A. 1973. *Prirodnij Process v Pleistotsene*. Moscow: Nauka.

Velichko, A. A. (ed.) 1981. *Arkheologiya i Paleogeografiya Pozdnego Paleolita Russkoj Ravnini*. Moscow: Nauka.

Velichko, A. A. 1982a. Osnovniye osobennosti poslednego klimaticheskogo makrotsykla i sovremennoye sostoyaniye prirodnoj sredy. In I. P. Gerasimov and A. A. Velichko (eds) *Paleogeografiya Evropy za Posledniye Sto tysyach Let*. Moscow: Nauka: 131-143.

Velichko, A. A. 1982b. Paleogeografiya Pozdnepleistotsenovoj periglyatsial'noj oblasti. In I. P. Gerasimov and A. A. Velichko (eds) *Paleogeografiya Evropy za Posledniye Sto tysyach Let*. Moscow: Nauka: 67-70.

Velichko, A. A. and Morozova, T. D. 1982. Izmeneniye prirodnoj sredy v pozdnem Pleistotsene po dannim uzucheniya lyossov, kriogennikh yavlenij, iskopaemykh pochv, i fauni. In I. P. Gerasimov and A. A. Velichko (eds) *Paleogeografiya Evropy za Posledniye Sto Tysyach Let*. Moscow: Nauka: 115-120.

Velichko, A. A. and Nechaev, V. P., in press. Ancient cryogenic regions (explanatory text to Map 6B). In A. A. Velichko and M. Pecsi (eds) *Paleogeographic Atlas of the Quaternary: Northern Latitudes*. Budapest: Kiado. In Press.

Velichko, A. A. and Pecsi, M. (eds), in press. *Paleogeographic Atlas of the Quaternary: Northern Latitudes*. Budapest: Kiado. In Press.

Velichko, A. A. and Praslov, N.D. (eds) 1978. *Arkheologiya i Paleogeografiya Rannego Paleolita Kryma i Kavkaza. Putevoditel'*. Moscow: Nauka.

Velichko, A. A., Zavernyaev, F. M., Gribchenko, Yu. N., Gubonina, Z. P., Zelikson, E. M., Markova, A. K. and Udartsev, V. P. 1981. Khotylevskiye Stoyanki. In A. Velichko (ed.) *Arkheologiya i Paleogeografiya Pozdnego Paleolita Russkoj Ravnini*. Moscow: Nauka: 57-69.

Vereschagin, N. K. 1979. *Pochemu Vymerli Mamonty*. Leningrad: Nauka.

Vereschagin, N. K. and Kuzmina, I. E. 1977. Ostatki mlyekopitayuschikh iz Paleoliticheskikh stoyanok na Donu i verkhnej Desne. In O. Skarlato (ed.) *Mamontovaya Fauna Russkoj Ravnini*. Trudy Zoologicheskogo Instituta 72: 77-110.

Voznyachuk, L. M. 1973. K stratigrafii i paleogeografii Neopleistotsena Byelorussii i smezhnikh territorii. In E. A. Levkoe (ed.) *Problemy Paleogeografii Antrapagena Belarussii*. Minsk: Nauka i Tekhnika: 45-75.

White, R. 1982. Rethinking the Middle/Upper Paleolithic transition. *Current Anthropology* 23: 169-192.

Woillard, G. M. 1978. Grand Pile Peat Bog: a continuous pollen record for the last 140 000 years. *Quaternary Research* 9: 1-21.

Woillard, G. M. 1979. Abrupt end of the last interglacial in north-east France. *Nature* 281: 558-562.

Woillard, G. M. and Mook, W. G. 1982. Carbon-14 dates at Grande Pile: correlation of land and sea chronologies. *Science* 215: 159-161.

Zarrina, E. P., Krasnov, I. I. and Spiridonova, I. A. 1980. Klimatostratigraficheskaya korrelyatsiya i khronologiya pozdnego Pleistotsena severozapada i tsentra Russkoj ravnini. In I. I. Krasnov (ed.) *Chetvertichnaya Geologiya i Geomorfologiya*. Moscow: Nauka: 46-50.

Zavernyaev F. M. 1978. *Khotylevskoe Paleoliticheskoe Mestonakhozhdenie*. Leningrad: Nauka.

35. East of Wallace's Line: Issues and Problems in the Colonisation of the Australian Continent

RHYS JONES

INTRODUCTION

> The number of Mammalia known to inhabit the Indo-Malay region is very considerable, exceeding 170 species. With the exception of the bats, none of these have any regular means of passing arms of the sea many miles in extent, and a consideration of their distribution must therefore greatly assist us in determining whether these islands have ever been connected with each other or with the continent since the epoch of existing species (Wallace 1890: 108).

During the final break-up of Gondwanaland some 120 million years ago, the Indian Tectonic Plate separated from Eastern Antarctica and drifted northwards, eventually to collide with Asia and form the Himalayas. Some 50 million years later, Australia followed suite, colliding with the Pacific Plate and the subduction zone of the Indonesian arc during the Pliocene. Much of New Guinea was uplifted, with its central mountain chain rising to 5 000 m (the highest mountains between the Himalayas and the Andes) and there was a concomitant sinking of the Timor Trough (Veevers and Evans 1975: 593, 601; Collerson and Sheraton 1986: 49-51). Although New Guinea is an island, its southern portion forms part of the Australian Plate and the sea straits between it and Australia are shallow. During most of the Pleistocene, except in full interglacial phases, as at present, there has been a land bridge connecting it to northern Australia. The Torres ridge was only broken some 6.5-8.0 kyr ago (Jennings 1971: 10). During full Last Glacial times, a plain 1000 km wide extended northwards from present-day Australia to the foot of the New Guinea cordillera. A closed-basin lake, 160 000 km^2 in area, one of the largest in the world and draining an area of 1 million km^2, was situated in the floor of the present-day Gulf of Carpentaria (Nix and Kalma 1972: 88; Löffler 1977: 5; Torgerson *et al.* 1983). On the southeastern extremity of the continent, Tasmania forms an extension of the Great Dividing Range, separated from the mainland by an island-studded Bass Strait, but during Last Glacial times connected to it by the exposed floor of the Strait.

We may thus consider a continent of 'Greater Australia' with an area of 10 million km^2, extending latitudinally from the Equator to 43° south. It supported a land mammalian fauna consisting almost entirely of marsupials and monotremes, the few placental mammals being restricted to mice, rats and bats which presumably had migrated there from Asia. However the bulk of the Asiatic fauna never reached Australia. Separating the Sahul (Australian) and the Sunda (Asian) shelves are deep ocean troughs and a series of island arcs which the biogeographer G. G. Simpson (1977) has pointed out were true 'oceanic islands', whose faunal colonisation had to take place across water and were subject to the rules of classic island biogeography (MacArthur and Wilson 1967). These posit for any species that there is an inverse relationship between colonising success and cross-water distance; and also another inverse one between island area and the chances of local extinction. Some successful colonisation into this island region had occurred from both adjacent continents, most being from Sundaland eastwards. However, even the first narrow water barrier between Bali and Lombok (25 km) in the lesser Sunda Chain filtered out so many of the oriental fauna, both land animals and birds, that A. R. Wallace (1890) postulated a biogeographic 'line' at this location. Later, Huxley (1868) extended it to coincide with what today is recognized as the edge of the Sundaland-Asian continental shelf. Lydekker (1896) proposed a similar boundary along the edge of the Sahul- Australian shelf. Between these are a series of islands separated by oceanic troughs, most oriented north-south, each acting as a separate filter for any colonisation southeastwards from Asia to Australia and *vice versa*.

On the Sunda shelf, islands such as Sumatra, Borneo, Java and Bali were periodically joined as a single landmass extension of Southeast Asia. The Javanese fossil record shows a rich oriental fauna with several phases of incursion since Pliocene times. The earliest had Indian (Siwalik) affinities related distantly to the Villafranchian fauna (Koenigswald 1949), and this was added to and partially replaced in early Pleistocene times by other faunal elements with affinities with southern China. The oldest deposits containing this 'Sino-Malayan' fauna (Koenigswald 1949; Bellwood 1985: 27) are the Djetis beds. At Sangiran a recent report listed the following among species recovered: hare, panther, tiger, cuon (cf. dog), *Homotherium,* the elephant-like Stegodon, elephant, tapir, rhinoceros of at least three species, pigs of three species, hippopotamus, deer of five species, antelope, water buffalo and bos (Aimi and Aziz 1985: 166-67). In addition, amongst the Primates were macaques and colobine monkeys. It is likely that there were phases of local extinction of some elements of this fauna during interglacial high sea-level phases, due to the isolation of a once-continental fauna on a smaller island land-mass. Many elements survived to historic or modern times.

In contrast to this faunal abundance, we can consider islands separated only by a single deep-water barrier from Sundaland. The Pleistocene fauna of Luzon in the northern Philippines contained two species of the extinct family of Stegodontidae and one extinct species of elephant; one extinct rhinoceros; one cervid and one bovid (Koenigswald 1949). On the Celebes, with only one strait between 40 and 100 km wide (depending on sea level) between it and the Sunda shelf, there were two stegodons and one elephant species; amongst the suids, an extinct species, an extinct genus and the still living *Babyrousa*, and from the primates, macaques of perhaps seven species – all derived ultimately from Asia and some perhaps in late Pliocene times (Simpson 1977: 111; Groves 1976: 207-8). From the Sahul shelf came one colonist, the marsupial phalangerid, cuscus, probably during the late Pleistocene or later. T. Flannery (pers. comm.) has suggested to me that this may even have occurred due to human actions, paralleling the carriage of small wallabies and a phalanger (*P. orientalis*) to the large oceanic islands of the Bismarck Archipelago (e.g. New Ireland) in early Holocene times (cf. White *et al.* 1988). Smaller or more isolated islands had a corresponding pauperization of species – Mindanao in the southern Philippines supported murids, two species of *Tarsius* (*T. spectrum* and *T. Pumilus*) and two species of extinct *Stegodon*. On Flores and Timor along the Lesser Sunda Chain there were fossils of the same two stegodons (*S. trigonocephalus* and *S. sompoensis*) and both underwent a process of dwarfism due to the selective pressures of being parts of small island ecosystems. These fossils came from undated but presumed Middle (or Early) Pleistocene deposits, and none have been found in stratified archaeological excavations, which in Timor extend back to some 14 000 years (Glover 1969).

Between Timor and Australia, there lay but one further water barrier, which during glacial low-sea level episodes would have been some 90 km wide to the exposed Sahul shelf (Birdsell 1977: 125). However, no land animal larger than a rat ever made this crossing since the early Pliocene, when the land masses were effectively in their present relative locations. Experiments have shown that rats have negligible swimming abilities in open seas (Spennemann and Rapp 1987). In the context of the Pacific islands, such colonisation must have taken place with the aid of man, i.e. in boats. Concerning the migration of the Australian/New Guinea Muridae, presumably their colonisation took place via rafting with natural vegetation of some kind. Almost all of these Muridae (rats and mice) are of endemic species and some even of endemic subfamilies (Simpson 1977: 115), so their ancestors had arrived on the Australian continent a long time ago and there is no compelling evidence that many or any crossed during the time frame of our purview (i.e. the Middle and Upper Pleistocene).

These biogeographic facts, backed by the fossil record since Pliocene times, indicate what a stupendous barrier to the migration of land fauna these seemingly insignificant sea straits have been. Diamond (1987) has shown that cross-water colonisation by natural rafting in warm or temperate seas has strongly favoured small animals. On the large Sepik River in northern New Guinea I have seen many rafts of floating vegetation, scores of metres long and often supporting sago palm and coconut trees, being carried downstream in late wet season floods; and some of these are swept several kilometres out to sea. This would also have been the situation on the large tropical rivers of Asia. Despite this, even primates such as the colobine monkeys, or macaques, endemic to the forests of Sundaland, never managed to cross all the straits of Wallacea.

A key indicator species in this regard might be the proboscis monkey *Nasalis larvatus* whose habitat at present consists of the mangrove and deltaic forests of Borneo. During Pleistocene low-sea-level episodes, its range is believed to have been more extensive along the edges of the palaeo-South China Sea (Weitzel 1987). Despite this adaptation to the coastal ecotone, this monkey did not manage any significant cross-sea traverses. The canopy tops of the New Guinea rainforest are silent, except for the whooshing wings of hornbill birds; no chattering of monkey immigrants, no radiation of arboreal primates following some ancient episode of colonisation. Man alone of all the larger animals living along the river valleys and sea shore of Sundaland, gained the capacity to cross these straits and colonise the continent of marsupials. It was an act of immense biological significance.

JAVANESE HOMINIDS

Dubois' (1894) discovery at Trinil on the banks of the Solo River in 1891 in central Java, of a skull cap assigned by him to *Pithecanthropus erectus*, and later recognized as the holotype of *H. erectus*, established the southeastern edge of Sundaland as one of the cradles of mankind (Koenigswald 1956: 25-6). More recent research now postulates the ultimate human ancestors in tropical Africa, with a spread of *H. erectus* in Early and Middle Pleistocene times across the tropical zones of Asia to be included within the Djetis fauna of Java at Sangiran and also at Choukoutien in China. Concerning the early Pleistocene Javanese hominids, some were recovered from the upper part of the Pucangan beds in the Sangiran dome dated most recently by the fission track method to 1.16 myr (Suzuki *et al.* 1985: 330). Others, including the important Sangiran 17 *H. erectus* (Pithecanthropus VIII) cranium, come from the lower part of the Kabuh beds dated by fission track to 0.78-0.71 myr, a value which is consistent with some earlier estimates (Sartono 1971; Ninkovich and Burckle 1978; Suzuki *et al.* 1985: 330; Itihara *et al.* 1985: 376).

In 1931, the geologist ter Haar discovered a bone-bearing upper terrace of gravel and tufaceous sand, some 20 m above the present Solo River near the village of Ngandong, where the river cuts through the limestone Kendeng Hills. Excavations into it resulted in the discovery of pieces of the crania of eleven individual hominids, known collectively as the 'Solo' or 'Ngandong' series. These represent a late and evolved *Erectus* population often referred to as *H. erectus soloensis* (e.g. Bartstra *et al.* 1988: 326). On the basis of the associated vertebrate fossils, this Solo High Terrace was assigned an Upper Pleistocene age (Koenigswald 1939), a dating which is supported by geomorphological evidence. Several recent observers have considered a general age of some 200 000 years or more for it (e.g. Santa Luca 1980). The original excavators left what von Koenigswald called 'a small pillar standing to enable future geologists to check our statement' (1956: 75).

This was recently relocated (with some difficulty) by Bartstra and coworkers, who in 1986 were able to excavate a small, stratigraphically controlled pit, 1.5 m^2 and 2.5 m deep into it, the base of the pit reaching the marly bedrock (Bartstra *et al.* 1988: 327). Bone fragments were collected and carefully dated by the $^{230}Th/^{234}Ur$ method in the Groningen laboratory.

Six samples recovered from the surface and from between depths of 1.10 m and 2.20 m gave age estimates of between 56 ± 10 and 31 ± 3 kyr, with no correlation with depth. Five of these samples were statistically indistinguishable at the 5% level, and may be combined to give an average value of 48 ± 2.8 kyr.

Between depths of 2.30 and 2.32 m were two statistically indistinguishable values averaging at 76 ± 5.3 kyr, and from the base at 2.50 m below the surface was a value of 101 ± 12 kyr. In their excavation, Bartstra *et al.* found bone fragments from a depth of just over one metre down to the base of the excavation, which corresponds to von Koenigswald's observation (1956: 68) that 'the bones lay in the bottom two feet'. The precise stratigraphic location of the hominid remains are not known, but it is likely that they occurred with the bulk of the other bones. The implications of this new study is that Solo people might have lived as recently as some 75-100 000 years ago, or even more recently. It is clear that Bartstra and his colleagues are aware of the revolutionary nature of their data, and they clearly recognize the potential problems with their dating method. On the other hand, the dates are coherently correlated with depth and are also independent of absolute uranium concentrations within the stratigraphic profile (Bartstra *et al.* 1988: 329).

Santa Luca (1980) in his study of the Solo hominids, perhaps influenced by the archaic morphological characteristics of the fossils, wonders whether or not they may have been derived from older original deposits. The study by Bartstra *et al.* also dated bones from some stratigraphic contexts which they believed were roughly equivalent to the 'High Terrace' elsewhere along the Solo River valley – for example at a place called Matar on the inner bend of the river opposite the classic Ngandong location, from which a Th/U date of 165 ± 30 kyr was obtained (Bartstra *et al.* 1988: 328). Clearly in such a dynamic landscape environment, with tectonic uplift, volcanism and consequent fluviatile action, there is every possibility of exposure and subsequent re-burial of fossils within younger deposits. Nevertheless, the original excavators of the Ngandong human material had no doubts that it was integrally situated with the rest of the fauna from the excavations (e.g. Koenigswald 1956: 74), and Bartstra and his co-authors point to the fact that the large fossil human calvaria and calottes would have been unlikely to have been transported over a long distance in a coarse channel without severe degradation or other obvious physical effects. Their view is that the Solo human material must still be considered autochthonous to the High Terrace (Bartstra *et al.* 1988: 332). We may have to face the fact that *Homo erectus* people survived within the forest edges of Southeast Asia until a hundred thousand years ago or less. Von Koenigswald, somewhat laconically, saw Solo man as a 'Javanese Neanderthaler' (1956: 70). At least in chronological terms, he may have been more accurate than he thought.

Within Java it is likely that there has existed a single lineage of archaic *erectus* hominids over the best part of a million years (Weidenreich 1951; Wood 1978: 56; Thorne and Wolpoff 1981: 337). This branch of mankind was situated on the very southeastern corner of Sundaland, with only the water barriers of Wallacea separating it from the southern continent.

THE PROBLEM OF THE ANTIQUITY OF THE JAVANESE STONE TOOL ASSEMBLAGES

Despite earlier influential claims, stone tool assemblages of any proven high antiquity have been notoriously difficult to obtain in Java or indeed elsewhere in the Sundaland area. This may be partly because of the nature of sediment-forming conditions of the hominid-bearing beds. These are estuarine tending to lacustrine, with some reworked volcanic ashes, and therefore not conducive to the primary deposition of stone tools nor their subsequent transport. Another possibility is that stone tools were rarely (if ever) made and used by the Middle Pleistocene Javanese hominids, alternative cutting materials being possibly supplied by split bamboo. The issue runs deeper than this however, since if we follow the view of Bartstra (1982, 1983, 1984) discussed below, there are no known stone tool assemblages from the region older than the Upper Pleistocene. In other words, were it not for the human skeletal material itself, there would not be any archaeological record for perhaps the first three quarters of a million years of hominid occupation. This places Sundaland in a different category in this regard compared to the situation in Africa, temperate Europe and probably China.

During the 1930s, archaeologists believed that they had found the stone artifacts made by *erectus* people (e.g. Koenigswald 1956: 117-122). These were referred to as the 'Patjitanian' assemblages, first found by von Koenigswald and Tweedie in 1935 in the bed of the Baksoka River near the south coast of central Java (Koenigswald 1936a, 1936b). These tools included large flaked cobbles referred to in the literature as 'choppers', large flat flakes with lateral retouch, and high-domed core tools, some of which were elongated in the shape of a keel. There were a few pieces with bifacial retouch looking somewhat like a hand-axe, and it was on typological grounds that this assemblage was assigned to an early or Lower Palaeolithic status and antiquity (Koenigswald 1936b; see Bartstra 1983: 426-9, 1984: 253-6). The Patjitanian was included by Movius (1944, 1948) as an industry within his famous 'Chopper/chopping tool complex', seen as Middle Pleistocene in antiquity and forming the Eastern Asian equivalent of the Acheulian of Africa and Europe. Subsequent field research led to a softening of the claim for such a high antiquity, with von Koenigswald (1978) referring to the Patjitanian as being 'perhaps Upper Pleistocene'), and van Heekeren (1975) assigning an antiquity within the 'early Upper Pleistocene'. Typologically, the published description of this industry has overstressed the 'hand-axe' and cobble tool components – even the original accounts stating that such tools were in a great minority. The bulk of the Patjitanian tools consist of large flat marginally retouched flakes (e.g. Mulvaney 1970). Systematic research by Bartstra has indicated that the tools come from young terrace and clay mantle deposits which at the very oldest are terminal Pleistocene, and probably extend into the Holocene. He sees the Patjitanian in fact as the Javan equivalent of the Hoabinhian. Thus if one were to try and attribute the makers of the Patjitanian to any fossil human group 'there is only one

Javanese fossil hominid that now merits consideration, and that is Wajak man. . . a representative *Homo sapiens*. . . part of a new population group on Java'. If these speculations are confirmed by further systematic work, they have profound implications for the general questions under review in this symposium.

At the classic hominid location of the Sangiran dome, stone tools were found by von Koenigswald in 1934 within fluviatile gravels that capped a prominent hill near the village of Ngebung (Koenigswald 1936a: 52). In his initial reports, he did not show much interest in these small and typologically undifferentiated tools, since at that time he was convinced that it was the Patjitanian artifacts which had been made by *H. erectus* people. Later, as it became obvious that the stratigraphic situation of the Patjitanian could not support such a high antiquity, von Koenigswald renewed his interest in the Ngebung site, seeing it as being of Middle Pleistocene age and adducing support for this claim from the presence of some bones of the Trinil fauna within the gravels (Koenigswald and Ghosh 1973; Koenigswald 1978). Other key workers have never supported a great age for these gravels, seeing them as part of the youngest sediment units in the Sangiran sequence and pointing out that the Trinil fossils, being few in number and heavily rolled, are almost certainly derived from older deposits (Movius 1949: 90; Heekeren 1972: 48). Bartstra (1985: 103-108) undertook a detailed field study of the site. He was able to define a 'Young River Gravel' which he assigned to the terminal Pleistocene or Holocene, and an underlying 'Old River Gravel'. The latter contained numerous stone implements, consisting for the most part of small flakes and casually struck cores, made mostly from silicified limestone and some from a softer rhyolite. The artifacts within this gravel are mostly heavily rounded. This 'Old River Gravel' lay unconformably on the Notoporo and Kabuh units, and Bartstra considers, on general stratigraphical evidence, that this gravel dates from the middle part of the Upper Pleistocene (Bartstra 1985: 106, Plate 1).

At Ngandong, within the Solo High Terrace, Bartstra (*et al.* 1988: 332-333) believes that there are also stone implements, usually showing heavy rolling by river action. The artifacts from this terrace are described as being small, mostly less than 5 cm in maximum dimension, and typologically nondescript, being somewhat similar to those from Ngebung. They are made from river-transported chalcedony, with some chert and jasper. As discussed previously, this is the terrace from which the Ngandong Solo hominid fossil remains were derived and which, on present evidence, may date from *c.* 100 kyr to 45 kyr BP. Bartstra considers that the Solo High Terrace and the Ngebung gravels may both belong to the same general epoch (1985: 110). No one working in this field has denied the possibility that older tool assemblages may be found in future, and indeed they expect them (cf. Bartstra 1985: 111). The point is that convincing evidence for such older stone artifact assemblages in Java and other areas of Sundaland does not as yet exist.

A claim has recently been made from northern Thailand for stone tools *in situ* within river gravels which are said to be overlain with the Lampang basalt dated by K/Ar and palaeomagnetism to at least 0.73 million years (K/Ar dates of 0.8 ± 0.3 and 0.6 ± 0.2 myr) (Pope *et al.* 1986, 1987). Originally, artifacts were found on the surface of the gravel, but two small excavations were made at a place called Ban Don Mun at what appears to be an outlier of the main basalt, several hundred metres from the main outcrop. A flaked cobble and two cortical flakes were found within a gravel layer, 0.65m below the surface (Pope *et al.* 1987: 750-751). The stratigraphic sequence showed clay overlain by gravel, which itself was overlain by pieces of basalt. Whether the latter were in primary position, or had somehow been reworked onto this part of the site is not clear. For such an important claim and given the absence of comparable-aged evidence elsewhere in Southeast Asia – we need unambiguous proof of the primary stratigraphic position of implementiferous gravels directly beneath the full basalt unit. My caution in this regard is shared by Anderson (1987: 184).

LIMESTONE CAVES AND THE OLDEST PRIMARY OCCUPATION EVIDENCE

The oldest well documented evidence in this region for stratified archaeological deposits, indicating clear occupation layers with stone tools and remains of food refuse, comes from three limestone caves: Tabon on Palawan Island (the Philippines), Niah on Borneo Island (Sarawak) and Lang Rongrien in peninsular southern Thailand. Both Palawan and Borneo were joined to the rest of Southeast Asia during glacial low-sea level times. At Tabon, occupation deposits extending back to some 26 kyr were obtained. Stone artifacts consisted of a variety of scrapers, including steep edged ones and small domed-shaped cores or core-scrapers (Fox 1970).

Niah

At the West-Mouth Cave of Niah, between 1954 and 1967, T. Harrisson carried out extensive excavations which removed almost all of the archaeologically significant deposit. Only outlines of this work were published, and no stratigraphic sections produced. A series of ^{14}C dates were obtained from Pleistocene levels containing stone tools, charcoal and charred bones reaching back to 39 600 ± 1000 (Grol-339) and 41 500 ± 1000 (Grol-338) at a depth of between 2.44 m and 2.54 m below the surface (Harrisson 1959, 1970; Majid 1982: 39). A human skull was excavated from this level and was described by Brothwell (1960) as being fully sapient, rather gracile, and best compared to a Tasmanian Aboriginal form. However in the absence of detailed stratigraphic information, there is always the possibility that the skull had been inserted from a higher level within a burial pit (cf. Bellwood 1985: 89). To resolve this potentially important question, a direct date on the bone by means of AMS dating is urgently needed (cf. Stringer 1986: 48). A careful re-excavation in the little that remained of the deposit was made by Majid in 1977, who obtained a further series of ^{14}C dates extending back to 21 410 ± 760 BP (GX-4834) from a depth of about 2.20 m (Majid 1982: 48-9). Whereas some of her dates were somewhat younger than those from equivalent depths in Harrisson's excavations, there may have been lateral variation in the thicknesses of stratigraphic units, and the very deepest of Harrisson's units (between 2.20 and 2.54 m) were not represented in Majid's excavations (1982: 49). From an analysis of her own excavated material, together with a re-examination of Harrisson's museum collections and field notes, Majid was able to reconstruct a cultural sequence of five units defined broadly as to stratigraphic position and, as far as possible, fixed to a chronometric time scale.

The oldest unit 1, dating from probably 40 kyr down to 20 kyr ago, contained a stone assemblage consisting only of unretouched flakes selected from shattered pebbles (Majid 1982: 90, 97). A few bone tools consisting of shafts with ends ground into spatulate shapes were also found. Bones of a few extinct species were found at these levels including the giant pangolin (*Manis palaeojavanica*) and the Sumatran tapir (*Tapirus indicus*), but most of the bones were of the bearded pig, monkey, deer and porcupine (1982: 134-5; Medway 1979). Many of these were burnt and it is considered that at least some were the debris of human middens. Shells of fresh water molluscs consisting mostly of an edible gastropod (*Clea*) and a few *Unionid* mussels indicated exploitation of the river environment (1982: 108-109). Also found were numerous seed cases, some charred, of the tree fruit *Pangium edule* (1982: 110-113, 64-5). This is edible, but in a raw state contains a poison (hydrocyanic acid) which causes severe nausea and sometimes even death (Burkill 1966: 1681). This toxin can be removed by dicing large nuts and leaching them for several days in an open weave bag in running water. Clearly, this preparation technology was known to the earliest inhabitants at this cave. A sense of how their subsistence might have articulated on a regional scale is indicated by the presence of some *Cyrena,* shells in the lowest levels (1982: 108). This is a mangrove bivalve, and must have been carried to the cave over a distance of about 20 km (under present sea levels) and probably several times as far during the time period being considered. The stone flake raw materials consisted of metamorphic sandstone, jasper and chert, all exotic to the immediate area, with the nearest potential source being the cobble beds of the Tinjar River some 45 km to the southeast (1982: 185). Thus these first people of whom we have evidence in the cave were exploiting a wide range of environments, from local forest and river to the estuarine coast downstream, and valley resources over a water-shed towards the hill country behind.

During the succeeding Phases 2 and 3, there occurred the first introduction of flaked pebble tools (or 'choppers') and of axe- adzes. At least one of the latter appears to be an authentic 'waisted blade' of the type found in New Guinea and discussed further below. This has two waisted indentations on either side of a roughly flaked flat tabular stem which was broken in antiquity (Majid 1982: Plate V 3: 150). Groube (1984) independently agrees with this identification, adding that it is the only one that he knows from Southeast Asia that closely matches the Sahul ones in dimension and shape. Flake tools include well retouched notched scrapers, and the economic evidence indicates a wide range of hunting activities. A date for Phase 3 is indicated by the measurement of 14 930 ± 460 BP (GX-4839) for a sample obtained from a depth of *c*. 1.35 m below the surface. Phases 2 and 3 thus span the general period corresponding to the height and end of the Last Glacial maximum, and are succeeded by Phase 4 dated to early Holocene times. The Niah sequence, although revealing only a fraction of what might have been obtained by modern field techniques, nevertheless is still one of the key sites documenting a fundamental change in the archaeological record which occurred in Southeast Asia at about 40 kyr ago, and which may reflect changes in human behaviour on a global scale.

Lang Rongrien

Confirmation of some of the key elements in the Niah sequence comes from the recently-excavated Lang Rongrien rock-shelter which (having been excavated to exacting modern standards) has a precisely defined stratigraphic and chronometric sequence (Anderson 1987). The site is situated on the western end of the Thai-Malay Peninsula in a small valley of tower karst clothed in thick primary rain forest, about 26 km from the present coast. There are two main phases in the sequence; an upper series of occupation deposits (Units 5 and 6) dated by five radio carbon dates to between 7500 and 9600 BP, and a lower one, again of superimposed charcoal-rich deposits with hearths, bones and stone tools (Units 8-10). These two sets were isolated stratigraphically by a layer of weathered limestone up to a metre thick (the result of a roof fall) which totally sealed the lower units. Unit 8 was dated by three samples: 27 350 ± 570 (Sl-6217); 27 110 ± 615 (Sl-6816) from a depth of 2.35-2.45 m, and 32 180 ± 1330 (Sl-6818) from a depth of 3.08 m. Unit 9 was dated to 37 000 ± 1780 BP (Sl-6819). The underlying Unit 10 also contained stone artifacts, bone fragments and charcoal, but has not yet been dated. The site thus contains what is so far the oldest directly dated primary occupation evidence in the whole of mainland Southeast Asia.

The stone artifacts from the basal Units 8-10 consisted of mostly small scrapers made from chert. These were of fairly amorphous shape but with neatly retouched edges. Two illustrated examples (Anderson 1987: 195) referred to as 'gouges' look like side-struck rejuvenation flakes removed from cores or steep-edged scrapers. Only three worked pebbles were found, and of these only one example corresponded to a formal flaked pebble tool (Anderson 1987: 193). Three antler or bone tools were also found in these lowest levels.

Above the limestone collapse within the early Holocene levels was a distinctive Hoabinhian stone assemblage. The tools were almost all made from a locally available carbonaceous shale and consisted of large bifacially-flaked discoids and long side-retouched knives. Typologically, these were similar to other contemporary early Holocene assemblages from western Malaysia such as at the large limestone rock-shelter of Gua Cha (Sieveking 1954; Adi 1981; Bellwood 1985: 162-6). At the classic Hoabinhian sites in Vietnam, the unifacially flaked core tools were thicker, having been manufactured from river cobbles (Pham 1980). The key point is that the Hoabinhian, a distinctive archaeological complex found geographically throughout

Southeast Asia, is now securely dated as a terminal Pleistocene and early Holocene phenomenon (see also Bellwood 1985: 162). The typological similarity of the flaked cobbles and discoids from Hoabinhian contexts, with the 'chopper/chopping tools' which Movius (1944, 1948) and others saw as the original tool kits of Middle Pleistocene age in East and Southeast Asia, led to two erroneous theories. First that the Hoabinhian itself had a substantial Pleistocene antiquity, and second that there had been a vast continuity of technological tradition within the Eastern Asian region involving a highly conservative mode of manufacturing stone tools. Indeed it was even postulated that this continuity of stone artifact production had spanned the transition from *Homo erectus* to modern *H. sapiens*. Explanations for such conservatism sometimes appealed to the presumed unchanging nature of the tropical forests of Southeast Asia during the bulk of the Pleistocene, and that a presumably successful adaptation had had little selective pressure to change under such circumstances.

WALLACEA: THE EARLIEST EVIDENCE

Within Wallacea itself, there are several cave excavations which document roughly the same time period in the late Upper Pleistocene. In Sulawesi, Leang Burung 2 (excavated in 1975) was occupied from about 31 kyr to 10 kyr BP. Tools consisted of flat pointed flakes and steep-edge scrapers, many showing silica polish, there being only minor typological changes from the earliest levels into the post-Pleistocene (Glover and Presland 1985). In eastern Timor, the cave of Uai Bobo 2 yielded a basal date of 13.5 kyr BP, and within this and other sites is documented a continuous technological tradition of stone tool manufacture down to as recently as some 2 kyr BP (Glover 1969: 110). The characteristic tool type, often made from high quality non-crystalline flint, is a narrow elongated steep-sided scraper, with a deep concave profile on one and sometimes both margins. Surface stone tool collections have been made on Timor, Flores and in the Cagayan Valley in northern Luzon, and claims of great antiquity have sometimes been made for them. In Timor and Flores, these tools manufactured from rough chert seem similar to the Javanese Patjitanian (Glover and Glover 1970) and included large flat flakes with some lateral retouch, and flaked cobbles including van Heekeren's (1957: 30) 'flat-iron chopper' type. Although there were original claims that these tools were associated with the Stegodon-bearing gravels (Verhoeven 1964), none have been found *in situ*, so their stratigraphic correlation is unclear. Their age is unknown except that it is likely to be older than the typologically different industries found in Uai Bobo 2 and other caves. Similarly, there is no proven association of the surface stone tools at Cagayan with any of the stratified extinct fauna.

While further fieldwork might change the picture, the situation at present is that there is no evidence of any archaeological site on the islands of Wallacea older than the oldest Australian evidence to be presented below. Indeed the oldest firm evidence that we have of man crossing Wallace's line comes from the greater Australian continent itself. If this picture holds, then there may not have been a slow but steady trickle of early man into the various islands of Wallacea in Middle and early Upper Pleistocene times, with a final crossing to Australia. Rather the reverse appears to be true – that Wallace's line was crossed reasonably late in the Pleistocene, but that once having happened, movement was rapid through the entire archipelago, onto the Sahul shelf itself, and even beyond to some of the large islands of the Southwest Pacific.

WATERCRAFT

The technological development that made this colonisation possible must have been adequate watercraft. Birdsell (1977) has painstakingly documented the various possible routes across Wallacea, with inter-island distances, target visibility, and angles of direction of travel to hit the targets. He did this for a series of different Pleistocene sea-level scenarios. Even the best case required at least one open sea crossing of *c.* 90 km, and many of the order of 30 km, in some cases with no target mountains to be seen over the horizon. He postulated that the watercraft used in the first voyages were better than any used ethnographically by the Australian Aborigines. Thorne has followed this view (1980: 36), but contradicts himself slightly in that he also posited, quite reasonably, that ever since initial colonisation, there has also been a steady trickle of later immigrants up to presumably modern times. It would be perverse to assign to all of these a superior maritime technology than that used by their descendants once they had arrived at Australia/New Guinea. Both Birdsell and Thorne consider rafts made of bamboo as the most likely vehicles, and Thorne has carried out replication experiments based on ethnographic examples in South China. Bamboo is endemic and common in Southeast Asia. It is also common in New Guinea, but only a few species grow as small isolated pockets on the coastal plains of northern Australia (e.g. *Bambusa arnhemica*). In northwestern Australia, Aborigines usually made small rafts out of light mangrove woods and used these to ride tides – often over distances of several kilometres. Other suitable materials were *Pandanus* and palm trunks, and bundles of paper bark as floats.

On the Sepik River in northern New Guinea I have seen people making large rafts 8-10 m long and 4 m wide, made from lashed criss-crossed frameworks of poles four layers thick. These would be capped by a deck made from strips of black-palm bark; on the deck might be a bough shelter and a clay hearth. The lashings come from split kanda cane, which grows as a vine in the forest. The entire craft could be made with the minimum cutting technology – tree trunks of soft woods chopped at the base and pushed over, and the cane lengths cut with a sharp stone flake. The craft are used to carry people and goods downstream over distances of more than a hundred kilometres. The lashings could be strong enough to survive the torsion of moderate seas over a considerable period of time, and such rafts might be capable of surviving a strait-crossing of perhaps scores of kilometres given suitable combinations of weather and current.

The ethnographic literature of Tasmania shows that canoe-shaped rafts made from rolls of Melaleuca paper bark or reeds were used to visit islands up to 5-8 km offshore, but that beyond that, islands further away were never visited, even over periods of several thousand years (Jones 1977). However the tides are strong and the waters and weather of Bass Strait are cold, and hypothermia might have been the greatest danger. In the tropical north, distances of *c.* 10-15 km separating such islands as Bathurst/Melville Islands, Bentinct Island and Keppel Island from the adjacent mainland, had a formidable effect in the partial or total isolation of the islanders from the mainland people – isolation to be seen in differences of material culture and in genetic markers.

Nevertheless, I still hold to the general position that I advocated in 1968, that taking a parsimonious view, the colonisation of Australia could be accounted for by some version of G. G. Simpson's (1977) 'sweep stake' process. Viewed from Timor through to the islands in the northeast corner of the Wallacean archipelago, the Sahul shelf presents a massive target, some 3000 km long aligned northeast to southwest. During the southern hemisphere wet season, the northwest monsoon winds (the famous 'Barra' of Arnhem Land) blow directly onto this coast. Computer simulation models have shown that under certain combinations of sea and weather conditions, there can be a reasonable chance of a raft travelling from off the Timor coast to that of northern Australia in about seven days (Thorne and Raymond 1989). The wet season, although dangerous from the point of view of tropical cyclones, does however have the advantage of offering the chance of a shower to provide drinking water to any castaway. My own scenario is that in the period just prior to the colonisation of Australia – say 40 kyr ago – there were people living on the shores of Sundaland, in the mangrove swamps and using the river mouth resources. They had an adequate technology of inshore watercraft, perhaps rafts made of bamboo palm or other suitable materials. Random events such as storms and currents sometimes swept people off into the ocean, where under suitable conditions of wind and current they made new land falls. The odds against any one such episode being successful might have been high, yet given enough time the entire archipelago could be colonised.

These voyages, voluntary or random, mark the earliest evidence we have of man's capacity to cross stretches of open seas beyond the sight of land. It is perhaps not coincidental that they occurred at the junction of the world's greatest archipelago with two of the world's oceans. In the Southeast Asian region, perhaps the ability to make and use watercraft of a certain capacity was one of the distinctive skills of sapient man.

ON COLONISING

For a successful colonisation, one needs not merely a landfall but also the genetic capacity to continue the population. While John Calaby's (1976: 24) playful scenario of a single young pregnant woman provides a minimalist and incest-free possibility (Noah and his daughters faced the same problem after the Flood) this is hardly likely to have been the origin of the peopling of the Australian continent. A successful colonising group would probably have to consist of several individuals of both genders. McArthur *et al.*

(1976) carried out simulation modelling of the chances of long-term reproductive success of extremely small founding groups, taking into account such factors as the proportions of male or female offspring, death rates in 'primitive' conditions, random disasters, rules of marriage and incest etc. As the founding group becomes extremely small, the odds of success lessen markedly. It is likely that there were many landfalls on the Australian/New Guinea shores, and even some groups surviving several generations before eventual extinction. Eventually, however, the correct combinations can come together and a viable population established. Once this happened, Birdsell (1957) has modelled that a founding group, expanding only at ethnographically observed rates, could have filled the Australian continent at a certain density in only a few thousand years. This density would have depended on the technological capacity of the society – technology in terms of both material culture and knowledge. In these scenarios of founding populations from extremely small original sources, no account has been taken of genetic issues such as recessive genes, the lack of genetic variability, 'in-breeding' ratios etc. Any new arrival on the Australian coast breeding into the existing population would have provided great genetic benefits.

Preliminary results from mitochondrial DNA analyses of Papua New Guinean and Australian populations suggest that for both regions there were some 15-20 separate lineages, indicating a minimum of those numbers of separate colonising females (Stoneking *et al.* 1986; Stoneking, pers. comm.). The study also shows little shared lineage between both areas (Stoneking *et al.* 1986: 434), suggesting that the history of colonisation may have been much more complex than that discussed in our simple model. It is indeed likely that other people over thousands of years followed the first arrivals, but the situation of landing on a shore already occupied by another population is quite different from that of being the primary colonists. Social behaviour might have led to new arrivals being killed, or at least the men being killed and the women incorporated as wives. The newcomers might have had little cultural or technological influence, though their genetic inheritance might have been incorporated into the host population. The arrival in Australia of the dingo, a semi-domesticated dog, almost certainly a companion of people, at about 4 kyr ago, proves that at least one other successful incursion took place.

AUSTRALIA: A FIELD OF DISCOVERY

Twenty-five years ago Grahame Clark wrote of Australia that 'so far there is no convincing evidence for human occupation beyond neothermal times'. In 1962, John Mulvaney obtained the first convincing stratified date of *c.* 10 000 years from a deep level at Keniff Cave in the southeastern highlands of Queensland (Mulvaney 1964) and then later in 1964 a near basal date of 15 000 years was obtained (since extended back to *c.* 20 kyr). From 1962 onwards there has been a flowering of field research in Australia and New Guinea, where excellent stratigraphic field methods backed by first-class radiocarbon and other absolute dating techniques have transformed our picture of the prehistory of the continent. The rate of progress can be

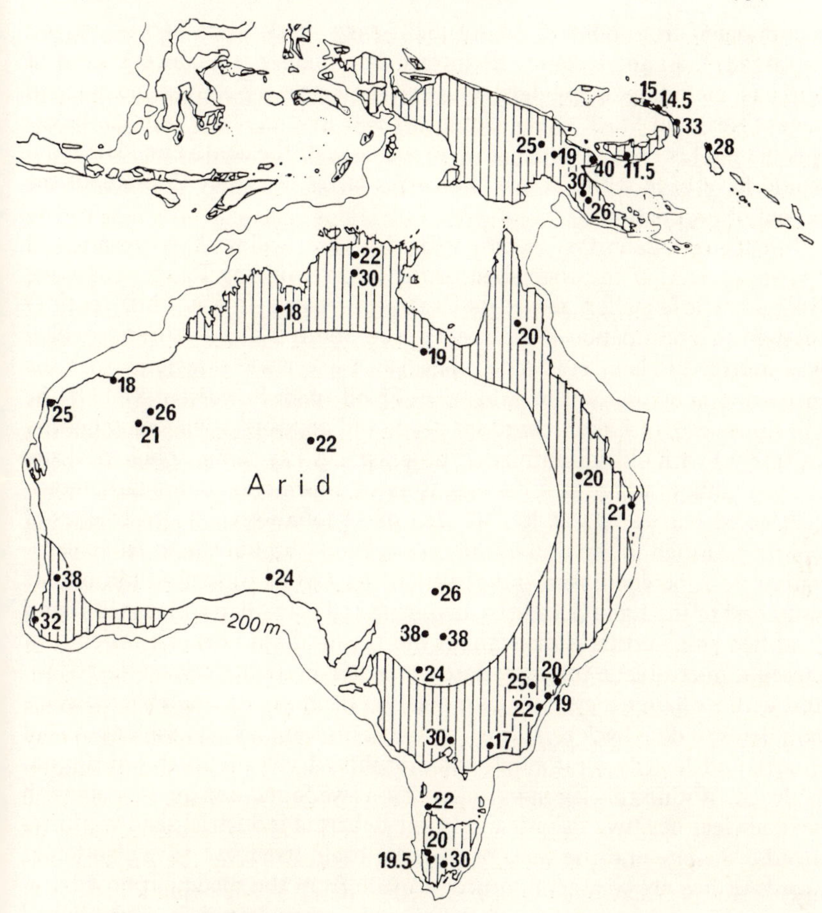

Figure 35.1. Map showing the oldest dates for human occupation in various areas of Australia and New Guinea (in thousands of years).

followed in successive reviews (e.g. Mulvaney 1964, 1969, 1975; Jones 1968, 1973, 1979; Flood 1983; White and O'Connell 1982), and I will not repeat the data included there.

Figure 35.1 shows the oldest ^{14}C dates so far obtained in securely stratified contexts from various regions of Australia/New Guinea. It can be seen that there was human presence in all major ecological zones by the time period 20-25 kyr ago. I must stress that the pace of research continues, and to exemplify his, I will refer briefly to research results from several regions, all completed within the past 3-4 years and either published recently or in preparation.

On the east coastline of southern Queensland on what is now North Stradbroke Island, a stratified sequence of stone tools has been excavated from a 1.5 m column of fossil dune sands underlying 0.8 m of shell midden. A series of six radiocarbon dates on charcoal taken from positions ranging from 1.00 to 2.10 m below the surface establishes that these sands have accumulated from 6950 ± 80 BP (SUA-2343) back to 20 560 ± 250 BP (SUA-2341). At the level of the lowest date there was a concentration of artifacts, including steep-edged scrapers, showing steep convex and notched edges (Neal and Stock 1986). Some artifacts may exist up to 0.4 m below this date. This site is located close to the edge of the continental shelf and would have been within a few kilometres of the sea shore throughout the period of occupation.

Inland near Canberra, in an enclosed plateau valley at an elevation of 750 m, located at the watershed of the Great Dividing Range, lies Lake George, whose pollen record is discussed below. It was until recently thought that occupation of the eastern New South Wales Tablelands region was confined to later Holocene times (say 3-4 kyr BP) associated with the introduction of microblade industries (Flood 1980; Bowdler 1981). On the hill slopes east of Lake George are deep sand deposits, some being heavily weathered with old soil formation processes. A key geomorphological site is Fern Gulley, where R. Coventry isolated a formation dated radiometrically to be between 22-26 kyr BP. In 1980 I found several small flakes of quartz stratified in the wall of the type section (unpublished). At an excavation on a perched sand dune site on the top of Butmaroo hill on the southeast of the Lake George basin during 1983, J. Allen and myself found stratified stone artifacts throughout the 1.5 m of sand deposit which contained a micro-blade industry dated to 4 kyr in its uppermost 12-20 cm. and with its base believed on extrapolation from the six available radiocarbon dates to date back to at least 10 kyr. This sand lay on a lag deposit, of quartz and heavily weathered metamorphosed volcanic rocks resting on bedrock. Within this lag in our excavation, we found one quartz core with several clear negative flake scars. In the debris of industrial sand workings around the site and the floor of the lake basin itself, we have also found scores of heavily weathered artifacts made from the metamorphosed volcanics. These consist of large flakes with rough lateral retouch, flaked cobbles and dome-shaped 'horse-hoof' cores (or core tools). There is strong presumptive evidence that these date from the terminal Pleistocene if not earlier, and it is possible that they may be derived from the clay-rich weathered sands older than *c*. 20-25 kyr. Nearby in a small montane rock-shelter (called Birrigai) Flood *et al*. (1987) have recently found stratified stone tools at the base dated securely to 21 kyr, showing colonisation of the mountain valleys just prior to the Last Glacial maximum.

Continuing our east-to-west transect of the continent, we may briefly consider the arid and semi-arid region, which has been the subject of considerable debate – Bowdler (1977) stating that human occupation of interior Australia had to wait until terminal Pleistocene times. Most of the inland and western Australia is classified as desert, the reasonably watered part of the continent forming a crescent-shaped region along the northern, eastern and southeastern coasts, with an isolated enclave in southwestern Australia. While Mungo and other lower Darling and lower Murray sites discussed below are within the eastern part of the semi-arid zone, the situation is complicated by rivers and lakes fed originally from the eastern, well-watered highlands. Recent research has documented early occupation of several rock-shelters in the Hamersley Range of the Pilbara in Western Australia – as for example at Mount Newman (22 000 BP) and Ethel Gorge (26 000 BP), with the possibility of older assemblages stratified below the lowest ^{14}C dates (Brown 1987). Finally in the very centre of the arid heart of Australia near the Cleland Hills, a huge rock-shelter (Puritjarra) has been found and excavated by Smith (1987). A stratified sequence containing numerous stone tools in all levels has been dated with a top date of *c.* 6 kyr, a middle one of *c.* 12 kyr and basal occupation at least *c.* 22 kyr BP – the sequence spanning the full dry period of the Last Glacial maximum. This site clearly demonstrates that people had penetrated into the centre of the arid zone of Australia in Pleistocene times, and had made the necessary ecological adjustments for this (Jones 1987a). Smith (1988) has argued that the distinctive seed grinding technology used by desert Aborigines in modern times only extends back in the archaeological record to the mid-Holocene. For plant foods, the first colonists may have had to rely on tubers, especially *Ipomoea* spp. (D. Yen, pers. comm).

Finally, to complete our transect to the shore of the Indian Ocean, there is the Mandu Mandu rock-shelter in coastal limestone at North-West Cape, close to the edge of the continental shelf. Excavations revealed a lower unit, its top (spit 9) dated by a corrected shell date of 19 590 ± 440 BP (SUA-2614) and its base (spit 17) by a carbonate date of 25 200 ± 250 BP (SUA-2354) (Morse 1988: 83-84). Stone tools in the Pleistocene unit were made from silcretes and hard limestone and consisted of large flakes and a steep-edged horse-hoof core tool. A striking aspect of this lower unit is the fauna, which included the remains of marine and shoreline food, such as edible molluscs, crab bones, fish teeth of several species and a baler shell fragment. This is the first direct evidence that we have of the Pleistocene exploitation of marine resources in the whole of Australia, though such a scenario has long been predicted. At the time of occupation of the lower unit at Mandu Mandu, between 20 and 25 kyr ago, the coastline would then have been about 6 km away (Morse 1988: 87). The site became abandoned at the height of the Last Glacial Maximum, when the shoreline retreated a further 10 km and desert conditions intensified in the hinterland behind.

Parallel patterns of recent search and discovery have been repeated in other areas of the continent, such as in the savanna regions of northern Australia (Schrire 1982; Jones 1985), the Gulf of Carpentaria and Cape York; Tasmania; the Nullarbor Plain and the coast of South Australia/Victoria; the Highlands and the Sepik lowlands of New Guinea; and in the large islands of the Bismarck archipelago and the northern Solomons. They serve to illustrate not only the revolutionary extent of the new information about the Pleistocene colonisation of the continent, but also how the expansion of the data-base is still outstripping our capacity to incorporate it into our synthetic historical frameworks.

THE OLDEST EVIDENCE

Upper Swan terrace and Devil's Lair, southwestern Australia

Commercial clay-pit digging into the second oldest (Guilford) terrace of the Upper Swan River northeast of Perth revealed stone artifacts stratified some 0.7-0.9 m below the original surface in a red sandy clay. An extensive and carefully-controlled campaign of excavations yielded more than 200 stone artifacts *in situ*, including small flakelets and pebble fragments which conjoined (Pearce and Barbetti 1981) The first radiocarbon date on charcoal gave value of 39 500 $^{+2300}_{-1800}$ BP (SUA-1500), and because of this antiquity another sample was run with a small amount of charcoal from the same stratified patch, which gave a result of >31 500 BP (SUA-1500x) A following season's excavation was carried out in the same deposit, and a sample 13 m away from original one was dated to 37 100 $^{+1600}_{-1300}$ (SUA-1665). A sample 5 cm higher than this gave a value of 35 100 $^{+1500}_{-1300}$ (SUA-1704). Pearce and Barbetti consider that these data clearly indicate the manufacture of stone artifacts at this site at *c.* 38 kyr BP, and they also consider the question that 'the observed count rates are very close to the lower limit of our best counter' (Pearce and Barbetti 1981: 177), which will be returned to later. The artifacts were mostly made from quartz or quartzite, with some from chert and a heavily weathered dolerite, and 5 per cent from a bryozoan chert, a source exposed during Pleistocene low sea level episodes. Typologically, they seemed to be miscellaneously retouched flakes.

To the south of Perth, in the limestone cave of Devil's Lair, a meticulous excavation led by C. Dortch revealed almost five metres of stratified deposit dated by a series of 27 radiocarbon dates (Dortch 1979a, 1984; Balme 1980). Evidence for reasonably regular human use of the site with hearths, stone and bone artifacts and charred and smashed bones of hunted macropods (kangaroos/wallabies) were found down to layer 30 dated to between 30 590 $^{+2220}_{-1420}$ (GX-7255) and 32 480 ± 1250 (SUA-585) BP. However some artifacts, such as a small struck flake of opal, came from layer 33/34. Layer 31 has a date of 35 160 ± 1800 BP (SUA-586), and the lowest dated level (layer 39) a date of 37 750 ± 2500 BP (SUA-698). It is hard to imagine how charcoal could get into the deposit without some form of human agency, and Dortch claims some sparse stone and bone objects modified by man down to layer 38. Below this level (layer 39 to 51) there is no evidence for any human presence.

The Willandra and lower Darling River, western New South Wales:

In 1969, I was involved in the original discovery of the Lake Mungo hominid and archaeological site (Bowler *et al.* 1970). This was within late Pleistocene sand deposits blown up on the lee, or eastern shore of large lakes at the lower end of the drainage system of the Willandra River, which once fed into the lower Murrumbidgee in southwestern New South Wales. The region is now semi-arid and the lake beds are dry, but between 40 and 25-20 kyr they were full, due to wetter and cooler conditions than now, (Jones and Bowler 1980) and westerly winds blew sand off the beaches to form crescent-shaped dunes or lunettes on their lee shores (Bowler 1971). The original Mungo I site was an eroded core of the lunette of Lake Mungo, containing stratified hearths with freshwater mussels, fish and small mammal bones, a variety of stone tools, and also the cremated remains of a young woman (Mungo I hominid), many of the finds being covered in calcium carbonate. Radiocarbon dates on hearth charcoal, burnt bone and the carbonate showed this site to date from *c.* 26 kyr BP – at the time of the research the oldest evidence for people on the Australian continent.

Later research between 1970 and 1973 on the various lunettes of the lower Willandra River system led to the discovery of freshwater mussel middens containing fish bones and stone artifacts dated back to 32 kyr (Barbetti and Allen 1972) and further hominid fossils (e.g. Mungo 3) dated to the general 30 kyr time period (Bowler *et al.* 1972, Bowler and Thorne 1976). A major excavation at the Mungo I site has since been carried out; the oldest firm date comes from a thin stratified shell midden near Lake Mungo dated to between 34 and 37 kyr (Bowler 1976: 59).

On the western edge of the lower Darling River valley, some 100 km west of the Willandra, are other fossil or attenuated stream systems which had a similar fluviatile history during the Pleistocene. These fed into a series of lakes, the most important of which in archaeological terms is Lake Tandou. On the Tandou lunette, the first phase of research dated eleven sites believed on stratigraphic grounds to belong to the late Pleistocene, and the ^{14}C evidence confirmed this. Four of the sites were dated to between 14 850 ± 190 BP (ANU-2309) and 16 100 ± 180 BP (ANU-2755); and the other seven between 22 050 ± 440 BP and 26 900 ± 590 BP (ANU-3000) (Hope *et al.* 1983: 48). Further research has revealed middens of freshwater *Unionids* dated back to 36 000 years BP (J. Hope pers. comm., 13/3/87). Hope is carrying out an analysis of all of the radiocarbon dates so far obtained in the country covering the confluences of the lower Willandra, Murrumbidgee and Darling Rivers with the Murray – roughly a triangle with each side 100 km long. She estimates that about 320 dates have been assayed (cf. Hope 1978), about 150 being dates of archaeological sites. Of the latter about 30 per cent date from the period 15-27 kyr, with a few extending back to the time period discussed here. Perhaps only a fifth of the sites confidently predicted on geomorphic grounds to date from this period have actually been dated by radiocarbon. Thus there may already be known of the order of 200 separate sites dating from this late Pleistocene period.

Prominent amongst the faunal remains at these sites are fish otoliths and gastroliths of freshwater crayfish. The main fish species was the Golden Perch (*Macquaria ambigua*) and a detailed analysis by Balme (1983) has shown not only large numbers of individual fish at the sites (each believed to result from a single event) but also that the fish came from tightly restricted size ranges. This strongly suggests the use of gill nets at some sites and traps at others (Balme 1983: 30). Such sophisticated fishing methods within the time range 22-26 kyr BP also imply the knowledge to make plant fibre cordage and a familiarity with river conditions and the seasonal pattern of fish behaviour.

AT THE LIMITS OF THE RADIOCARBON METHOD

It can be seen that these oldest directly dated sites in Australia give values close to the practical limits of the radiocarbon method. The question naturally arises whether or not we are limited in our chronological perspective by our methodology (Jones 1982: 30-2). There are in fact several sites in Australia where artifacts have been found securely *in situ* below the lowest perceptible charcoal. This is the case at the Mungo lunette itself where rolled flakes were found *in situ* in a gravel unit at the base of the Mungo sand unit, at a depth of 1.8 m below the 32 000 BP dated archaeological finds. Small specks of charcoal in the section gave a ^{14}C value indistinguishable from background (Shawcross 1975: 30; Shawcross and Kaye 1980). One might speculate that these lowest artifacts have a general antiquity of *c.* 40-45 kyr at least.

In terrace deposits of the Nepean River at the foot of the Blue Mountains, 30 km west of Sydney, a few stone artifacts consisting of flaked cobbles and steep-edged scrapers, have been claimed to be in genuine stratigraphic association with late Pleistocene gravels (Stockton and Holland 1974). These were originally believed to be dated by a ^{14}C measurement of 26 700 ± 1700 (Gak-3014), but recent detailed geomorphological work on these gravels with radiocarbon and thermoluminescence programmes have shown the gravels to have an antiquity of 40-47 kyr (Nanson *et al.* 1987). Given the potential importance of the claim for human occupation in southeast Australia at this time, the stratigraphic association of these artifacts with the deposit being dated has to be unimpeachable, and supported by direct excavation *in situ*.

On the west Victorian coastline near Warrnambool, a calcarenite cemented fossil sand dune associated with a previous high sea level had cemented within it scores of *Subninella* gastropod shells. This is an edible species, and the appearance of the shells, including their operculae, looked similar to a humanly deposited midden. All radiometric studies done so far seem to confirm an antiquity for this feature (as predicted from the sea level curves) to be of at least a previous interstadial age (J. Sherwood, pers. comm.; Bowler 1987: 51). Again, given the potential implications of such a discovery, the onus of proof rests on establishing beyond all doubt a human origin for the shells.

The most promising new research approach will be the application of the TL dating method to naturally deposited sands. Technical success of the method in the field has been achieved in the Kakadu area of the Northern Territory, where sandsheets in excess of 35 kyr year have been successfully dated (East *et al.* 1987). Currently, a pilot project is being carried out by B. Roberts and myself on the dating of sand columns in occupied rock-shelters where artifacts have been found stratified beneath the oldest charcoal dates – in many cases themselves with values of 20-24 kyr. A key site in this regard is Nauwalabila 1, the Lindner Site, where a 2.5 m thick sandsheet resting on 0.4 m of heavily weathered rubble, had numerous stone artifacts in every one of 80 excavation units, including within the rubble itself (Jones

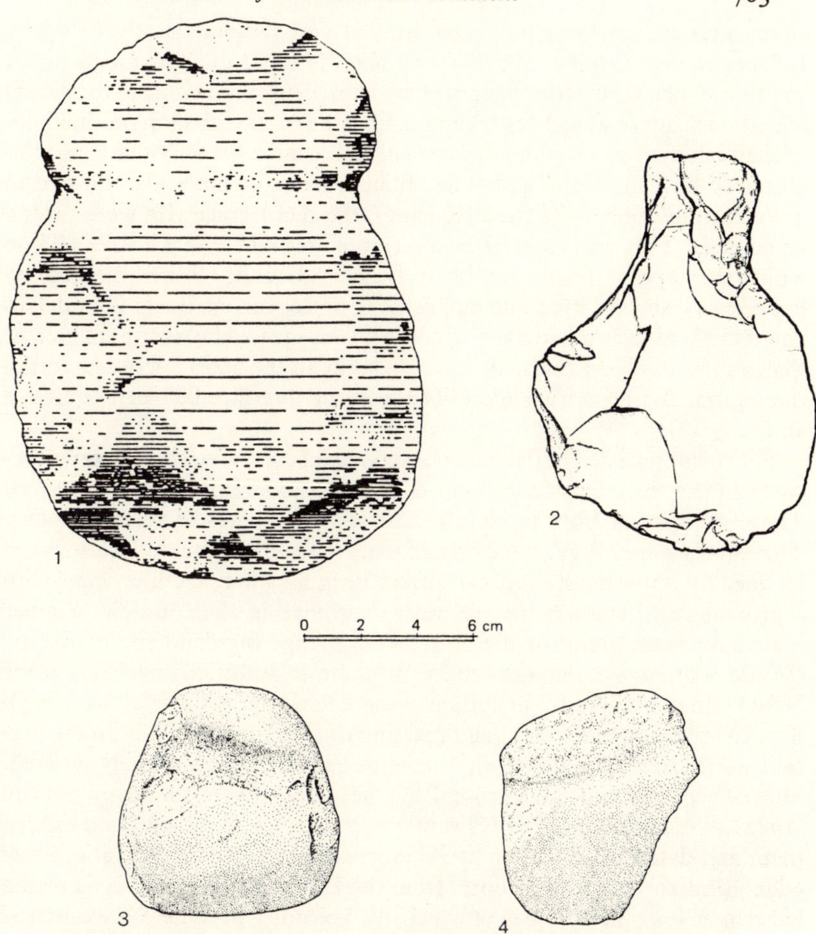

Figure 35.2. Pleistocene waisted blades and edge-ground axes from Australia: (1) The Huon Terrace IIIa (Groube et al. 1986); (2) Nombe rock shelter (Mountain 1983); (3) Malangangerr (Schrire 1982); (4) Nawamoyn (Schrire 1982).

and Johnson 1985). Extrapolation of the age-depth correlations to the base of the sand suggested for it an antiquity of at least 30 kyr. There is every possibility that the artifacts stratified within the rubble, consisting of flakes and a flaked cobble, are considerably older. In the lowest levels of the overlying sand were implements such as a horse-hoof core-scraper and steep-edged scrapers with well-worked small notches and noses. These tools are typically representative of what has been loosely termed the Australian 'core tool and scraper' tradition.

THE HUON TERRACES, PAPUA NEW GUINEA

The northeastern coast of Papua New Guinea at the Huon Peninsula is rapidly uplifting, due to over-riding of the Australian Plate onto the Pacific Plate. A succession of raised coral terraces up to a height of 400 meters above present sea level has been studied and radiometrically dated by J.Chappell (1974). The uplift was sufficiently rapid to allow independent analysis of glacio-eustatic changes of sea level. During the time of formation of each reef, there was a back lagoon against the coast with high productivity of shellfish, fish and lagoon/estuarine plants, conducive to potential exploitation by human hunter/gatherers. Found *in situ* in the walls of a stream gully cutting the small palaeo-lagoon on the reef terrace IIIa were several stone tools. This was covered by a series of volcanic tephras. Reef IIIa on which the tephras rested has been dated by Uranium-series methods to between 45 and 53 kyr, but tephra does not occur on Reefs II and IIIb dated to 28-30 kyr and 40 kyr respectively. An age for the tephras of greater than 40 kyr has been confirmed by TL dating of the quartz particles within the tephra. Artifacts were found *in situ* at the interface between tephras 2 and 3.

The tools consisted of flakes and steep-edged cores, but the most significant artifacts were large core-tools called 'waisted blades' (see Fig. 35.2). These were made from large flakes struck off cobbles of volcanic rocks. These tools were heavy, some being up to 2.9 kg in weight, and were flaked on one face opposite the cortical surface forming a flat, axe-like shape with a pronounced hammer-dressed waist or groove in their middle. Similar tools have been found in the open, swamp-edge highland site of Kosipe (White *et al.* 1970), dated to 26 kyr BP, and in stratified cave sites in the New Guinea highlands, including some dated to *c.* 25 kyr at Nombe in deposits containing extinct giant marsupials (Mountain 1983). On the present northern Australian plain, analogous smaller tools, namely waisted, bifacially-flaked and edge-ground hatchet heads, were found by Schrire (1982) at Malangangerr and Nawamoyn, close to the Arnhem Land escarpment and dated to 20-24 kyr BP. Confirmation of Pleistocene antiquity of edge-grinding technology comes from the Lindner Site in the same region back to at least 14 kyr, below which the volcanics become too weathered to be diagnostic (Jones 1985). Waisted blades have also been found as surface tools on Kangaroo Island south of the continent (Lampert 1981).

Groube *et al.* (1986) have discussed the probable functions of the waisted tools of the Huon sites and suggest that they were used for ring-barking and otherwise modifying the vegetation – thus breaking the tree canopy and allowing the penetration of sunlight down to the ground plants. Such an ecological manipulation would lead to the enhanced growth of wild food plants such as yams, banana, sugar-cane, perhaps taro, tree fruits, pandanus etc. It must be remembered that by 9000 years ago in the highlands of New Guinea, there were major water-control ditch systems, almost certainly associated with the cultivation of taro. In northern Australia today, many of the plants which are the staples of Melanesian and Southeast Asian agriculture are collected wild (Meehan and Jones, in press). The development of horticulture in the New Guinea constant-wet tropical zone in early post-Pleistocene times, with its late Pleistocene antecedents, would have led to a large relative growth in population compared with that of the hunter/gatherers on the rest of the continent. The postglacial flooding of the Arafura Plain and Torres Ridge would have increased the cultural and perhaps genetic differences between the hunters of the Australian plain and the emerging horticulturalists of the New Guinea cordillera, whereas during the previous 30 000 years of known human existence on the continent, their ancestors had practised elements of a contiguous cline of subsistence systems.

FIRE AND LANDSCAPE EVOLUTION

Ethnographic observations from the time of first contact with Europeans through to modern studies indicate that over most of the continent, Aboriginal groups used fire in a systematic way to modify the landscape vegetation (e.g. Jones 1969; Hallam 1975; Haynes 1978, 1985). This firing was done for a variety of reasons, from burning-off dead grass, or otherwise 'cleaning the country' to allow easier movement, to more specialized uses in hunting and signalling. Within Aboriginal territorial estates, this tended to produce a mosaic of regeneration states, the full ecological implications of which are only recently being realised in terms of plant distribution and animal populations (e.g. Johnson 1988; Flannery, in prep.).

Ever since the Australian continent drifted northwards to its mid latitudinal position, lightning-induced fire has been an element of the natural ecosystem. However, the arrival of people and their firesticks would have radically altered the fire regime over many or most parts of the continent. The probability of a fire occurring in any particular area would have been increased by several orders of magnitude. Those of us working from an ethnographic perspective have long predicted that such a human impact would be capable of being perceived within pollen and geomorphic sequences (Jones 1968, 1973). Recent important pollen sequences have features which may be interpreted as confirming this.

A key sequence is that obtained by Singh and co-workers from the bed of Lake George, near Canberra (Singh and Geissler 1985). The full sequence extends back beyond the Brunhes-Matuyama Reversal at 700 000 years ago, and it documents a series of glacial and interglacial events. At this altitude, glacials were characterized by relatively treeless conditions, whereas the interglacials had high tree pollen. Within the upper part of the sequence, in Zone F, two probably interrelated and fundamental changes occurred. First, there was the appearance of large amounts of charcoal fragments, whose presence persisted in the sequence from then on, even in the relatively treeless Last Glacial maximum. Secondly, the relative percentage of tree species changed. Before this event, the trees were dominated by a now extinct *Casuarina* species, podocarps (a southern family distantly related to conifers) and *Cyathea* tree-ferns. These are all fire-sensitive species, and from F zone times onward, they declined and became extinct, even though present climatic conditions are similar to previous interglacial cycles that had favoured them. They were replaced by fire-adapted eucalypts, acacias and grasses, forming the open eucalypt savanna which characterized the region at the time of European contact. Singh believes that these floristic changes were causally related to increased fire frequency, and can find no plausible climatic explanations. He favours the view that these phenomena were due to human impact, through the use of fire (Singh and Geissler 1985: 437-439).

There still remains the problem of the age of zone F which Singh and Geissler, on the basis of correlations with deep-sea palaeotemperature curves, have assigned to the end of the Last Interglacial – i.e. some 120 000 years ago. As has been discussed above, this is significantly older than any believable archaeological claim for human antiquity in Australia so far. Wright (1986) has carried out a re-analysis of the possible age of this zone. He considered the series of direct carbon dates obtained from the upper part of the sequence (back to the oldest of 20 000 BP) and showed that an age-depth curve gave an almost perfect correlation ($R = 0.986$, $N = 6$). Extending this to Zone F would give a value of the order of 60 000 years, i.e. an event, which might have correlated to an interstadial within the main glacial period. Such an antiquity is not beyond the bounds of possibility as a date for the arrival of humans, based on other archaeological evidence.

A deep pollen sequence showing a similar story has been obtained from the western edge of the north Queensland rainforest on the Atherton Tableland (Kershaw 1985). Here there was also a major retreat of rainforest and other fire-sensitive species, with a local spread of eucalypt, which occurred quickly at a period dated to about 40 000 BP. In the highlands of New Guinea, near the Kosipe Swamp, Hope (1982) has dated a major phase of tree retreat, instability of the local ground soil, and the sudden appearance of many pieces of charcoal in deposits dated to 30 000 BP and attributed by him directly to humans, their firesticks and plant manipulation.

The human impact on the indigenous fauna is beyond the scope of this paper. However, it can be noted that in late Upper Pleistocene times, there was a major and rapid phase of extinctions within the Australian fauna, particularly the larger animals, when up to one third of species became extinct. These included families, such as the rhinoceros-sized Diprotodontids. Many of these extinct animals did not seem to be particularly restricted in ecological range, Diprotodontid fossils being found in such diverse ecological zones as salt lakes in central Australia, swamps in northwestern Tasmania, and the inter-montane valleys of New Guinea. My own view has been that it was the arrival of humans which was the ultimate cause of this phase of extinction, either through direct hunting or by modification of the landscape through fire (Jones 1968, 1973). Direct archaeological associations of giant marsupial bones within occupation sites have been hard to find, but good cases for such associations are slowly being developed in (for example) some New South Wales lake-edge sites (Hope 1978) and at the Nombe limestone cave in the New Guinea highlands (Mountain 1983).

THE BISMARCKS AND THE SOLOMONS: OCEANIC ISLANDS

A few degrees south of the equator are the large islands of the Bismarck Archipelago, off the northeast New Guinea coast. These, consisting of New Ireland and New Britain, are truly oceanic islands, separated from the continental shelf by deep troughs. A date of 12 kyr was obtained from Misisil Cave in southern New Britain in 1981 and shows that at least one water strait had been crossed in late Pleistocene times (Specht *et al.* 1981). In 1985, as part of the Australian National University sponsored Lapita Project, a series of uplifted coral limestone caves were excavated in central and northern New Ireland. The basal occupation at Baloff 2 has been dated to 14.5 kyr BP; a near basal date at Panakiwuk, situated inland in primary tropical forest, has an AMS date of 15.5 kyr; and the coastal cave of Matenkupkum contains a shell midden of reef molluscs with four basal dates of between 31 and 33 kyr BP (Allen *et al.* 1988b). Thus people had managed at that time to cross two further water barriers (going via New Britain) beyond those separating Asia and Australia.

The straits between New Guinea (near the Huon terraces) and New Britain are some 90 km wide, with a large island in between: those between New Ireland and New Britain being 30 km wide, in both cases with land in sight. New Ireland, being a true oceanic island, is extremely depauperate in land fauna – the first occupants would have had access only to murids, reptiles and endemic birds. The tree possum and wallabies, now on the island, came there later, carried by people (White *et al.* 1988). Matenkupkum showed the exploitation of reef molluscs (mostly the gastropod *Turbo argyrostoma*), those from the basal layers being much bigger than living examples there today. There were also numerous fish bones, still under analysis. The site is nowadays situated some 50 m from the present shore. Because of the extremely steep offshore bathymetry of this coast, the reef and marine resources would never have been far from the cave, even during Last Glacial maximum low sea levels. Stone tools at Matenkupkum consisted of smashed cobbles and large flakes made from local dolerites. Those from the Pleistocene layers in Panakiwuk were flat flakes which had been struck from large steep-edged cores, the stone material being silicified volcanics brought into the cave from specialized sources. In contrast to the paucity of land animals, the forests of New Ireland may have had indigenous food sources such as yams, sugar cane, wild bananas and tree fruits.

The sea-going capacity of these Pleistocene colonists has been amply demonstrated by the recent fascinating discovery of human occupation back to 28 kyr at Kilu Cave, a limestone rock-shelter on Buka Island situated at the northern end of the Solomon Islands chain (Wickler and Spriggs 1988). The sequence had a depth of 2.4 metres of stratified deposit which included food bone, flaked stone tools and marine shell midden. Between 1.8 and 2.4 metres depth, there was a shell midden dated between about 20 kyr and 28 kyr. The shells were edible coastal gastropods and there were numerous fish bones. The land fauna of the island is depauperate, so only the bones of rodents and lizards were found. Flaked stone tools consisted of simple retouched flakes. Residue analysis showed that some of these Pleistocene tools retained well-preserved starch grains, showing they had been used for scraping or cutting some kind of tubers. Given the limited terrestrial animal resources of this small island, a substantial component of the diet from vegetable foods would have been necessary to sustain life, in addition to the marine resources.

The sea distance from southern New Ireland southeast to Buka is some 180 km wide and the latter is out of sight from the former. The angular projection of the entire Buka-Bougainville islands group back onto the nearest point on the New Ireland coast is 30°. Two small island groups are situated in between. The Feni group, situated to the northeast on the oceanic side of the direct route, would still require one crossing of 145 km, with accurate capacity to direct one's craft onto a small target. The other islands are the Green Islands of Nissan and Pinipir, two coral atolls of probable Pleistocene age. It is not known whether or not these existed during the period under review. In any case using these islands as stop-off points would also require open sea voyaging across minimum distances of 70 km onto angular targets which subtended as little as 8° for one of the legs, to an observer starting the voyage in south New Ireland. Once on Buka, the rest of the Northern Solomons chain would have been easily accessible, either with narrow straits, or with some of the main islands such as Bougainville, perhaps even having been joined during the glacial low sea level episode. The limit of this region was probably San Cristobal Island. Waisted blades similar to the Huon examples have been found as surface finds on several of these islands – the entire group with probable Pleistocene occupation measuring some 1100 km along the northwest-southeast axis. The sea passages southeast into the Pacific towards Santa Cruz were of a different order of magnitude, and this probably formed the southeastern boundary of Pleistocene people during their first great expansion into the Pacific region (Jones 1989).

TASMANIA: A CONTINENTAL ISLAND

Tasmania, some 64 000 km² in area, is part of the Australian continental shelf, but the Bassian Bridge connecting northeastern Tasmania to Wilson's Promontory via the Furneaux Group has a sill with a depth between 55 and 65 meters below present sea level. As global sea levels began to drop with the onset of the main phase of the last glaciation, the critical depth was reached about 24 kyr ago, and a land bridge some 100 km wide remained open until deglaciation *c.* 12 kyr ago (Jones 1977b: 360). That people had taken advantage of this new space is shown by Bowdler's (1984) discovery of occupation levels at Cave Bay Cave (an old sea cave in metamorphosed siltstone) on what is now Hunter Island off northwest Tasmania, dated back to 22.7 kyr BP. Inland in the Florentine Valley of southern Tasmania, a limestone cave ('Beginner's Luck' Cave) had stone flakes embedded in a breccia dated to 20 kyr (Goede and Murray 1977).

In 1981, limestone exposures were explored in the extremely rugged rain forested wilderness valleys of the Franklin and Gordon Rivers of southwest Tasmania. At Kutikina Cave (formerly Fraser Cave), a series of rich occupation levels were excavated, dated between 14 kyr and 20 kyr BP (Kiernan *et al.* 1983). These contained stone flakes and tools at a density of about 70 000 pieces per cubic metre. In the lower half of the sequence, tools made mostly from quartzite obtained from local river cobbles, consisted of core-scrapers and steep-edged scrapers some with well worked notches. Between excavation units 12 and 13, dated at about 16-17 kyr BP, there was a marked typological change, with almost all stone artifacts above it consisting of tiny 'thumb- nail' shaped scrapers made from quartz (Jones, Fullagar and Ranson, in prep.; Jones 1987b). In this upper unit were also found flakes and a few retouched pieces made from Darwin Glass, a natural glass formed by the impact of a meteorite at a site called Darwin Crater, some 25 km northwest of the Kutikina Cave, and only recent discovered by science (Fudali and Ford 1979). Given the deeply broken terrain, this distance would involve a walk of several days, even if generally open glacial tundra conditions prevailed. There were thousands of burnt and smashed animal bones, mostly of the red-necked wallaby (*M. rufogriseus*) and also a few stout bone points made from the fibulae of wallabies. Between 1981 and 1984, a series of further explorations in the Franklin-Gordon river system have led to the discovery of about twenty more occupied caves along these rivers, all dated within the same age range, with the large Deena Reena cave having a basal occupation date of 19.5 kyr, the lowest hearths being underlain by two metres of sterile clays (Blain *et al.* 1983; Jones 1984, 1987c).

Fieldwork during 1987 and 1988 has expanded our picture of Late Glacial occupation in the area slightly to the east, within what are called the Southern Forests, a rugged and uninhabited wet sclerophyll and rainforest region at the watershed between the western, eastern and southern rivers of Tasmania (Jones *et al.* 1988). On the Weld River, a small limestone cave, called Bone Cave, contained rich occupation deposits, spanning the period from 13 kyr to 17 kyr, with the base of the deposit not yet having been excavated (Allen *et al.* 1988a: 81). A newly-discovered limestone site, Bluff Cave on the Florentine not far from the original find at Beginner's Luck, yielded dates for the uppermost layers of 11.5 and 13 kyr and occupation deposits dated to beyond 24 kyr ago (Cosgrove, in press). At both sites, there were also thumb-nail scrapers and a few flakes of Darwin Glass, the latter involving travel over straight-line distances of 100 km to both sites, with actual distances across feasible terrain routes possibly doubling that figure. The typological similarities with the sites on the Franklin, and also the shared quarry source for the Darwin Glass, indicate that the occupants of all of these Pleistocene southern and southwestern Tasmanian sites were, at some level, integrated into a single social system. Hunted fauna at Bone and Bluff caves indicated a greater diversity than at Kutikina, with eastern grey kangaroo and wombat common in Bone Cave, and with emu egg, kangaroo, dasyurid (native cat) and birds in addition to red-necked wallaby, in Bluff Cave. The emu egg would indicate late winter or early spring occupation (Cosgrove, in press), an additional facet to the model of seasonal summer occupation of these intermontane valleys which I had previously proposed (1984).

The basal date within the artifact-bearing deposits at Bluff Cave was 30.4 kyr at a depth of 0.6 m, with other dates of 23.6 kyr at 0.53 m and 27.8 kyr at 0.42 m (Cosgrove, in press). Directly underneath the basal date was a sterile clay which may date from a previous interstadial. Cosgrove obtained another date around 30 kyr for basal occupation at a sandstone rock-shelter (ORS 7), situated some 60 km northeast of Bluff Cave in upland dry sclerophyll country on the Shannon River, a tributary of the same Derwent River catchment as the Florentine. This was an AMS date of 30.8 kyr from a small sample obtained from what he described as a hearth *in situ,* 0.45 m below the surface. However another small sample of scattered charcoal from a depth 5 cm below it, gave a date of only 16.2 kyr. Cosgrove believes that the latter may have been contaminated by younger charcoal during excavation. A sample at a depth of 0.25 m gave a value of 17.7 kyr. I have a reservation about one aspect of the Bluff Cave sequence. Within it, there is a parallel to the Kutikina cultural sequence, in that thumb-nail scrapers and Darwin Glass both appear in the upper levels, but they do so at Bluff Cave some 7 kyr and 10 kyr respectively earlier than at Kutikina (Cosgrove, in press). The introduction of thumb-nail scrapers, the associated core-reduction technology and raw materials used, was stratigraphically precisely isolated at Kutikina to within two centimetres – what I called an archaeologically 'instantaneous event' (Jones 1987c: 36). I find it hard to believe that such a dramatic technological change could have occurred in two sites only 60 km away from each other, but separated in time by thousands of years. In this regard, it is likely that either the Bluff Cave sequence is incorrectly dated or that of Kutikina is. These will be questions to be resolved by further active fieldwork.

If, however, the dates of 30 kyr can be confirmed for the basal occupations at Bluff and 'O.R.S. 7' sites, they would have important implications for our scenario concerning the initial colonisation of Tasmania, since this would then have occurred during a low sea episode older than that associated with the Last Glacial maximum. Cosgrove suggests by extrapolation from sea level curves that the Bassian Bridge was exposed between *c.* 37 kyr and 29 kyr ago. Of course, if it turns out that the human colonisation of the Australian continent had occurred some 50-60 kyr ago as discussed previously, then the possibility arises of the potential use of the clear land bridge to Tasmania between 50 and 58 kyr ago, as pointed out by Orchiston (1979). Research continues.

During the period 14-19 kyr, the southern Tasmanian valleys contained glaciers at their heads with a tundra vegetation and alpine herb fields. Their occupants were then the most southerly humans on earth. This pattern of dense occupation ceased abruptly at about 13-12.5 kyr, possibly associated with the re-encroachment of the rain forest vegetation in warmer and moister times. Since the occupation was probably seasonal, with winter ranges either on the floor of the Bassian plain or even on the present Australian mainland, the swift flooding of the final remnants of the Bassian bridge might have had a catastrophic effect on the social and economic systems of these people.

The Tasmanian data have a special importance, since Tasmania became cut off from the mainland at this time, and the people isolated on the island maintained a cultural trajectory of late Pleistocene origin, separated from the cultural changes that occurred on the adjacent continent. In terms of stone-tool technology, the tool kit of hand-held steep-edge scrapers and core-scrapers used ethnographically by the Tasmanians can be linked in a single, fairly stable technological tradition to that used in terminal Pleistocene times, when Tasmania and southern Australia were parts of the same land mass (Jones 1977a). On the mainland in Mid-Recent times, stone technology was transformed by the introduction or development of backed, micro-blade industries, or assemblages of hafted chisel/adzes and unifacial and bifacial points.

PLEISTOCENE ART

Between 1986 and 1988, three fascinating art sites have been found in the wilderness of southwest Tasmania, which add a new dimension to the Pleistocene cultural record of the region. The sites are all limestone caves and within them, 20 metres or more from the entrance, in darkness, have been found panels of hand stencils and areas with continuous red pigment, and also 'blazes' of red at the cave entrance or on prominent cave features. At Judd's Cavern (Jones *et al.* 1988: 13-20), much of the cave walls are covered by thick stalagmitic layers, the calcite extending over the art. This is also the case at the two others, Ballawinne Cave on the Maxwell River, only about a dozen kilometres to the east of Kutikina, and the as-yet unnamed site (TASI No. 3614) on the Weld, near Bone Cave, which also shows calcite layers covering the art (Harris *et al.* 1988; Allen *et al.* 1988a: 85). For this reason, and due to the fact that no evidence of prehistoric occupation in any of the caves of these valleys dates to more recently than about 12 kyr ago, and are sealed with surface soft stalagmitic layers, we believe that the art dates to the same late Pleistocene phase as the occupation layers of the caves. Tests on the pigment of panels within Judd's Cavern have shown two of them contained blood, which when tested against a standard mammal Immunoglobin G test, showed this blood to be of mammalian origin (Jones *et al.* 1988). T. Loy, who did the original identifications, is carrying out further tests to see whether or not this blood is human.

Koonalda Cave on the limestone Nullabor Plain is a large sinkhole some 80 metres deep where flint modules were mined between 15 and 22 kyr BP. Deeper inside the cave in total darkness are spectacular panels consisting of wavy lines made by dragging the outstretched fingers of the hand across the soft chalk walls. Where the limestone is harder, patterns of incisions have been made, and there were panels of rectilinear lines. At a distance of 280 metres from the cave mouth is a small tight squeeze, only 30 cm high, leading onto a ledge overlooking an underground lake. At the entrance to this squeeze are numerous deeply incised lines, and on the floor just under it, were flint flakes and fragments of old fire torches, their shape still recognizable. One of these was dated to *c.* 20 kyr BP (Wright 1971).

Within the arid zone of the central part of the continent, there are probably thousands of what are termed 'rock engraving' sites. These consist of designs, usually pecked or abraded onto rock surfaces. Motifs, often referred to as the 'Panaramitee' style after a major site in South Australia, characteristically consist of circles and other geometric forms (such as stars and crosses) and the tracks of birds, macropods and other animals. Some sites have anthropomorphic-like creatures, often with complex head-dresses and other features. The sets of motifs have broad similarity to sand paintings and other designs made by desert Aborigines of the present day, and are ritually embedded into their totemic belief system (Munn 1973; Sutton 1988: 182).

At many of these sites, the decorated rocks are covered with silicate and other skins referred to as 'desert varnish'. Dragovich (1987: 31) obtained a date of 7090 ± 310 BP (SUA-2011) on calcium carbonate which covered such varnish at a site in western New South Wales, and she considers this to be an underestimate for its age. Direct dating of varnish using the cation ratio method has been developed for the western United States by Dorn. Recently Dorn and Nobbs obtained silicate samples overlying engravings on dolomite outcrops at the South Australian site of Mannahil. Dates ranging from 16 kyr back to 31 kyr, have been reported (Sutton 1988: 184), and the published scientific account is keenly awaited (Dorn *et al.* 1988).

The back wall of Early Man rock-shelter in Cape York Peninsula was decorated with extensive panels of these designs, and stratified occupation deposit against them gave a minimum date of 13 kyr BP for this art (Rosenfeld *et al.* 1981).

Finally, we may consider the Kakadu region, where, on many grounds, Chaloupka (1985) has argued that some of the rock art there has considerable antiquity, probably extending back into late Pleistocene times. These reasons include the fact that what appear to be old, faded depictions in weathered red ochre show land animals with none of the fish and crocodiles of the most recent styles, which may reflect the present estuarine conditions. There are depictions of extinct animals, such as thylacines and giant anteaters. The human figures are shown using spears, which appear not to be tipped by stone points. Perhaps, ultimately, the most convincing argument is that many of these panels are covered in a silicate skin, often of considerable thickness. Watchman (1985) has analysed some of these, and shown them to have complex mineral compositions, including in some cases calcium oxylate salts. The carbon within the oxyalic acid was derived by an organic pathway, probably lichens, from atmospheric carbon. In a virtuoso display of conventional radiocarbon dating by J. Head of the Australian National University laboratory, dates of up to 8880 ± 590 BP (ANU-4271) were obtained (Watchman 1987: 38-39). These were from multi-layered coatings, and thus give a mean age for the crust. The age of the first layer to be deposited would, of course, be older than these values, and they point the way towards further dating by AMS techniques.

In most levels of the Lindner Site, Nauwalabila I, pieces of ochre of various sorts were found, including, right at the base of the sands, high quality haematite pieces with carefully shaped ground facets, showing they had been intensively used to obtain iron ore powder. Such haematite is not found locally, the nearest source being several scores of kilometres away. The stratigraphic position of the oldest ones would indicate an age of some

30 kyr at least. Whether this pigment was used for making rock art or for other uses, such as body decoration, is of course not known (Jones and Johnson 1985: 223).

It can be seen that high quality data are being rapidly accumulated in Australia, which point to a Pleistocene antiquity for art sites in many parts of the continent. With the oldest evidence so far at about 30 kyr, this is of the same order of magnitude as the first appearance of art in the Western European record, and within Australia too, art may be one of the hallmarks of modern man. Another may be ritual processes in the disposal of the dead. The young Mungo female hominid, dated at about 26 kyr, had been cremated, the charred bones smashed and placed in a small pit. The extended Mungo III hominid, of about the same antiquity, was buried in a pit that contained powdered red ochre. The somewhat younger Nitchie hominid, of terminal Pleistocene/early Holocene age, was buried with a necklace of over 100 bored canine teeth of Tasmanian devil (*Sarcophilus harrissi*), an animal now extinct on the Australian mainland (Macintosh 1971).

Bone ornaments, such as beads made from worked segments of macropod fibuli, have also been found in late Pleistocene layers at Devil's Lair, south-western Australia (Dortch 1984). Clearly, these data are evidence of other facets of human behaviour associated with an aesthetic sense, perhaps social differentiations and beliefs in the religious sphere. Within Australia, it is possible that we may be able to demonstrate archaeologically broads strands of continuity of artistic and religious belief extending back from the ethnographically described Aboriginal culture to terminal Pleistocene times (ses also Sutton 1988; Chaloupka 1985).

CONCLUSION

The data thus presented and foreshadowed, clearly establishes the arrival of people on the Australian continent at *c.* 40 kyr BP. By 30 kyr, they had occupied every single major ecological zone of the continent and the adjacent major islands. They survived through the climatic changes of the main phase of the last Ice Age. In their capacity for technological skills, ecological success, and in such essential human traits as art, ornament and religion, they manifested the same degree of sophistication as their contemporaries in, for example, Europe and Africa. If the data remain roughly as they might seem today, then one might postulate that the colonisation of the Australian continent was one of the fundamental events associated with the emergence of modern humans.

What happens to our arguments, however, if it were to be demonstrated that the initial occupation of the continent had occurred say 60 to 80 kyr ago – a time when it is possible that Solo, or Solo-derived people were still living in adjacent Java? Was Australia, like Europe, one of those places where the modern sapientization process had a long and complex history? Finally the distant descendants of the Pleistocene colonists to Australia are the contemporary Aboriginal Australian citizens. In Australia, unlike Europe, these debates have a great political significance.

REFERENCES

Adi, H. T. 1981. *The Re-Excavation of the Rock Shelter of Gua Cha, Ulu Kelantan, West Malaysia*. Unpublished MA Thesis, Department of Prehistory and Anthropology, Australian National University, Canberra.

Aimi, M. and Aziz, F. 1985. Vertebrate fossils from the Sangiran, Mojokerto, Trinil and Sambungmacan areas. In N. Watanabe and D. Kadar (eds) *Quaternary Geology of the Hominid Fossil Bearing Formations in Java*. Geological Research and Development Centre, Special Publication 4. Ministry of Mines and Energy, Bandung, Republic of Indonesia: 155-197.

Allen, J., Cosgrove, R. and Brown, S. (1988a). New archaeological data from the Southern Forests region, Tasmania: a preliminary statement. *Australian Archaeology* 27: 75-88.

Allen, J., Gosden, C., Jones R., and White, J. P. 1988b. Pleistocene dates for the human occupation of New Ireland, northern Melanesia. *Nature* 331: 707-9.

Anderson, D. D. 1987. A Pleistocene-early Holocene rock shelter in Peninsular Thailand. *National Geographic Research* 3: 184-198.

Balme, J. 1980. An analysis of charred bone from Devil's Lair, Western Australia. *Archaeology and Physical Anthropology in Oceania* 15: 81-85.

Balme, J. 1983. Prehistoric fishing in the lower Darling, western New South Wales. In C. Grigson and J. Clutton-Brock (eds) *Animals and Archaeology*. Vol. 2: *Shell Middens, Fishes and Birds*. Oxford: British Archaeological Reports International Series S183: 19-32.

Barbetti, M. and Allen, H. R. 1972. Prehistoric man at Lake Mungo, Australia, by 32,000 years BP. *Nature* 240: 46-48.

Bartstra, G.-J. 1982. *Homo erectus:* the search for his artefacts. *Current Anthropology* 23: 318-320.

Bartstra, G.-J. 1983. Some remarks upon fossil man from Java, his age and his tools. *Bijdragen Tot de Taal-, Land- en Volkenkunde* 139: 421-34.

Bartsrta, G.-J. 1984. Dating the Pacitanian: some thoughts. *Courier Forschungsinstitut Senkenberg* 69: 253-258.

Bartsrta, G.-J. 1985. Sangiran, the stone implements of Ngebung, and the Paleolithic of Java. *Modern Quaternary Reseach in Southeast Asia* 9: 99-113.

Bartstra, G.-J. 1986. A note on the Paleolithic of Java. In R. P. Soejono (ed.) *Pertemuan Ilmiah Arkeologi IV*. Vol. 1: *Evolusi Manusia, Lingkungan Hidup dan Teknologi*. Pusat Penelitian Arkeologi Nasional, Jakarta, Republic of Indonesia: 77-83.

Bartstra, G.-J., Soegondho, S. and van der Wijk, A. 1988. Ngandong man: age and artifacts. *Journal of Human Evolution* 17: 325-337.

Bellwood, P. 1985. *Prehistory of the Indo-Malaysian Archipelago*. Sydney: Academic Press.

Birdsell, J. B. 1957. Some population problems involving Pleistocene man. *Cold Spring Harbor Symposium on Quantitative Biology* 122: 47-69.

Birdsell, J. B. 1977. The recalibration of a paradigm for the first peopling of greater Australia. In J. Allen, J. Golson and R. Jones (eds) *Sunda and Sahul: Prehistoric studies in South-East Asia, Melanesia and Australia*. London: Academic Press: 113-167.

Blain, B., Fullagar, R., Allen, J., Harris, S., Jones, R., Stadler, E., Cosgrove, R., and Middleton, G. 1983. The Australian National University-Tasmanian National Parks and Wildlife Service Expedition to the Franklin and Gordon rivers, 1983: a summary of results. *Australian Archaeology* 16: 71-83.

Bowdler, S. 1977. The coastal colonisation of Australia. In J. Allen, J. Golson and R. Jones (eds) *Sunda and Sahul: Prehistoric studies in*

South-East Asia, Melanesia and Australia. London: Academic Press: 205-246.

Bowdler, S. 1981. Hunters in the highlands: Aboriginal adaptations in the eastern Australian uplands. *Archaeology in Oceania* 16: 99-111.

Bowdler, S. 1982. Prehistoric archaeology in Tasmania. In F. Wendorf and A. E. Close (eds) *Advances in World Archaeology* Vol 1. New York: Academic Press: 1-49.

Bowdler, S. 1984. *Hunter Hill, Hunter Island*. Terra Australis 8. Canberra: Department of Prehistory, Research School of Pacific Studies, Australian National University.

Bowler, J. M. 1971. Pleistocene salinities and climatic change: evidence from lakes and lunettes in southeastern Australia. In D. J. Mulvaney and J. Golson (eds) *Aboriginal Man and Environment in Australia*. Canberra: Australian National University Press: 47- 65.

Bowler, J. M. 1976. Recent developments in reconstructing late Quaternary environments in Australia. In R. L. Kirk and A. G. Thorne, (eds) *The Origin of the Australians*. Canberra: Australian Institute of Aboriginal Studies: 55-77.

Bowler, J. 1987. Edmund Gill 1908-1986. *Australian Archaeology* 24: 48-52.

Bowler, J. M., Jones, R., Allen, H. R. and Thorne, A. G. 1970. Pleistocene human remains from Australia: a living site and human cremation from Lake Mungo, western New South Wales. *World Archaeology* 12: 39-60.

Bowler, J. M., Thorne, A. G. and Polach, H. 1972. Pleistocene man in Australia: age and significance of the Mungo skeleton. *Nature* 240: 48-50.

Bowler, J. M. and Thorne, A. G. 1976. Human remains from Lake Mungo: discovery and excavation of Lake Mungo III. In R. L. Kirk and A. G. Thorne (eds) *The Origin of the Australians*. Canberra: Australian Institute of Aboriginal Studies: 127-138

Brothwell, D.R. 1960. Upper Pleistocene human skull from Niah Caves, Sarawak. *Sarawak Museum Journal* 9: 323-349.

Brown, S. 1987. *Towards a Prehistory of the Hamersley Plateau, Northwest Australia*. Canberra: Department of Prehistory, Research School of Pacific Studies, Australian National University. Occasional Papers in Prehistory 5.

Burkill, I. H. 1966. *A Dictionary of the Economic Products of the Malay Peninsula*. Kuala Lumpur: Ministry of Agriculture and Co- operatives on behalf of the Governments of Malaysia and Singapore.

Calaby, J. H. 1976. Some biogeographical factors relevant to the Pleistocene movement of man to Australasia. In R. L. Kirk and A. G. Thorne (eds) *The Origin of the Australians*. Canberra: Australian Institute of Aboriginal Studies: 23-28.

Chaloupka, G. 1985. Chronological sequence of Arnhem Land rock art. In R. Jones (ed.) *Archaeological Research in Kakadu National Park*. Canberra: Australian National Parks and Wildlife Service Special Publication 13: 269-280.

Chappell, J. 1974. Geology of coral terraces, Huon Peninsula, New Guinea: a study of Quaternary tectonic movements and sea-level changes. *Bulletin of the Geological Society of America* 85: 553- 570.

Clark, G. 1961. *World Prehistory: an Outline*. Cambridge: Cambridge University Press.

Collerson, K. D. and Sheraton, J. W. 1986. Bedrock geology and coastal evolution of the Vestfold Hills. In J. Pickard (ed.) *Antarctic oasis: Terrestrial Environments and History of the Vestfold Hills*. Sydney: Academic Press: 21-62.

Cosgrove, R. (in press). Thirty thousand years of human colonization in Tasmania – new Pleistocene d[illegible]. *Science*. In Press.

Diamond, J. M. 1977. Distributional strategies. In J. Allen, J. Golson and R. Jones (eds) *Sunda and Sahul: Prehistoric studies in Southeast Asia, Melanesia and Australia*. London: Academic Press: 295-316.

Diamond, J. M. 1987. How do flightless mammals colonise oceanic islands? *Nature* 327: 374.

Dorn, R. I., Nobbs, M. and Cahill, T. A. 1988 Cation-ratio dating of rock engravings from the Olary Province of arid South Australia. *Antiquity* 62: 681-689.

Dortch, C. E. 1979a. 33,000 year old stone and bone artefacts from Devil's Lair, Western Australia. *Records of the Western Australian Museum* 7: 329-367.

Dortch, C. 1979b. Devil's Lair: an example of prolonged cave use in south western Australia. *World Archaeology* 10: 258-279.

Dortch, C. 1984. *Devil's Lair: a Study in Prehistory*. Perth: Western Australian Museum.

Dragovich, D. 1987. Desert varnish and problems of dating rock engravings in western New South Wales. In W. R. Ambrose and J. M. J. Mummery (eds) *Archaeometry: Further Australasian Studies*. Canberra: Department of Prehistory, Research School of Pacific Studies, Australian National University: 28-35.

Dubois, E. 1894. *Pithecanthropus erectus: eine Menschenaehnliche Uebergangsform aus Java*. Batavia: Landesdruckerei.

East, T. J., Murray, A. S., Nanson, G. C. and Clark, R. L. 1987. Late Quaternary evolution of Magela Creek sand-bed channels. In *Alligator Rivers Research Institute Annual Research Summary 1985-86*. Canberra: Australian Government Publishing Service.

Flannery, T. F. (in prep.). Historic extinctions in Australia: aftershock of megafaunal loss?

Flood, J. M. 1980. *The Moth Hunters*. Canberra: Australian Institute of Aboriginal Studies.

Flood, J. M. 1983. *Archaeology of the Dreamtime*. Sydney: Collins.

Flood, J. M., David, B., Magee, J. and English, B. 1987. Birrigai: a Pleistocene site the the south-eastern highlands. *Archaeology in Oceania* 22: 9-26.

Fox, R. B. 1970. *The Tabon Caves*. Manila: National Museum of the Phillipines Monograph 1.

Fudali, R. F. and Ford, R. J. 1979. Darwin glass and Darwin crater: a progress report. *Meteoritics* 14: 283.

Glover, I. C. 1969. Radiocarbon dates from Portuguese Timor. *Archaeology and Physical Anthropology in Oceania* 4: 107-112.

Glover, I. C. and Glover, E. A. 1970. Pleistocene flaked stone tools from Timor and Flores. *Mankind* 7: 188-190.

Glover, I. C. and Presland, G. 1985. Microliths in Indonesian flaked stone industries. In V. N. Misra and P. Bellwood (eds) *Recent Advances in Indo-Pacific Prehistory*. Leiden: Brill: 185-195.

Goede, A. and Murray, P. 1977. Pleistocene man in south central Tasmania: evidence from a cave site in the Florentine Valley. *Mankind* 11: 2-10.

Groube, L., Chappell, J., Muke, J. and Price, D. 1986. A 40,000 year old occupation site at Huon Peninsula, Papua New Guinea. *Nature* 324: 453-455.

Groube, L. 1984. Waisted axes of Asia, Melanesia and Australia. In G. K. Ward (ed.) *Archaeology at ANZAAS, Canberra*. Canberra: Canberra Archaeology Society: 168-177.

Groves, C. P. 1976. The origin of the mammalian fauna of Sulawesi

(Celebes). *Zeitschrift für Saugetierkunde* 41: 201-216.
Haar, C. ter 1934. Homo-soloënsis. *De Ingenieur in Nederlands-Indie, Mijnbouw en Geologie de Mijningenieur* 1: 51-7.
Hallam, S. 1975. *Fire and Hearth.* Canberra : Australian Institute of Aboriginal Studies.
Harris, S., Ranson, D., and Brown, S., 1988. Maxwell River archaeological survey 1986. *Australian Archaeology* 27: 89-97.
Harrisson, T. 1959. Radio-carbon C-14 datings from Niah: a note. *Sarawak Museum Journal* 13: 136-138.
Harrisson, T. 1970. The prehistory of Borneo. *Asian Perspectives* 13: 17-46.
Haynes, C. D. 1978. Gunret, gundulk dja bining. *Commonwealth Forestry Review* 57: 99-106.
Haynes, C. D. 1985. The pattern and ecology of *munwag:* traditional Aboriginal fire regimes in north-central Arnhem Land. *Proceedings of the Ecological Society of Australia* 13: 203-214.
Heekeren, H. R. van 1957. *The Stone Age of Indonesia.* The Hague: Martinus Nijhoff.
Hooijer, D. A. 1975. Quaternary mammals east and west of Wallace's Line. *Modern Quaternary Research in Southeast Asia* 1: 37-46.
Hope, G. 1982. Pollen from archaeological sites: a comparison of swamp and open archaeological site pollen spectra at Kosipe Mission, Papua New Guinea. In W. Ambrose (ed.) *Archaeometry: an Australasian Perspective.* Canberra: Department of Prehistory, Research School of Pacific Studies, Australian National University: 211-219.
Hope, J. H. 1978. Pleistocene mammal extinctions: the problem of Mungo and Menindee, New South Wales. *Alcheringa* 12: 65-82.
Hope, J. H., Dare-Edwards, A. and Mclntyre, M. L. 1983. Middens and megafauna: stratigraphy and dating of Lake Tandau lunette, western New South Wales. *Archaeology in Oceania* 18: 38-45.
Huxley, T. H. 1868. On the classification and distribution of the Alectoromorphae and Heteromorphae. *Proceedings of the Zoological Society London* 1868: 294-319.
Itihara, M., Kadar, D. and Watanabe, N. 1985. Concluding remarks. In N. Watanabe and D. Kadar (eds) *Quaternary Geology of the Hominid Fossil Bearing Formations in Java.* Special Publication 4. Geology Research and Development Centre, Ministry of Mines and Energy, Bandung, Republic of Indonesia: 367-78.
Jennings, J. N. 1971. Sea level changes and land links. In D. J. Mulvaney and J. Golson (eds) *Aboriginal Man and Environment in Australia.* Canberra: Australian National University Press: 1-13.
Johnson, K. 1988. Rare and endangered rufous hare-wallaby. *Australian Natural History* 22: 406-407.
Jones, R. 1968. The geographical background to the arrival of man in Australia and Tasmania. *Archaeology and Physical Anthropology in Oceania* 3: 186-215.
Jones, R. 1969. Firestick farming. *Australian Natural History* 16: 224-228.
Jones, R. 1971. *Rocky Cape and the Problem of the Tasmanians.* Unpublished Ph.D. thesis, University of Sydney.
Jones, R. 1973. Emerging picture of Pleistocene Australians. *Nature* 246: 278-281.
Jones, R. 1973. The neolithic palaeolithic and the hunting gardeners: man and land in the Antipodes. In R. P. Suggate and M. M. Cresswell (eds) *Quaternary Studies: Royal Society of New Zealand Bulletin* 13: 21-34.
Jones, R. 1977a. The Tasmanian paradox. In R. V. S. Wright (ed.) *Stone Tools as Cultural Markers: Change, Evolution and Complexity.* Canberra: Australian Institute of Aboriginal Studies: 189-204.

Jones, R. 1977b. Man as an element of a continental fauna: the case of the sundering of the Bassian bridge. In J. Allen, J. Golson and R. Jones (eds) *Sunda and Sahul: Prehistoric Studies in Southeast Asia, Melanesia and Australia*. London: Academic Press: 317-386.

Jones, R. 1979. The fifth continent: problems concerning the human colonization of Australia. *Annual Review of Anthropology* 8: 445-466.

Jones, R. 1982. Ions and eons: some thoughts on archaeological science and scientific archaeology. In W. Ambrose (ed.) *Archaeometry: an Australasian Perspective*. Canberra: Department of Prehistory, Research School of Pacific Studies, Australian National University: 22-35.

Jones, R. 1984. Hunters and history: a case study from western Tasmania. In C. Schrire (ed.) *Past and Present in Hunter Gatherer Studies*. Orlando: Academic Press: 27-65.

Jones, R. (ed.) 1985. *Archaeological Research in Kakadu National Park*. Canberra: Australian National Parks and Wildlife Service Special Publication 13.

Jones, R. 1987a. Pleistocene life in the dead heart of Australia. *Nature* 328: 666.

Jones, R. 1987b. Ice-age hunters of the Tasmanian wilderness. *Australian Geographic* 8: 26-45.

Jones, R. 1987c. Hunting forbears. In M. Roe (ed.) *The Flow of Culture: Tasmanian Studies*. Canberra: Australian Academy of the Humanities: 14-49.

Jones, R., 1989. Island occupation. *Nature* 337: 605-6.

Jones, R. and Bowler, J. 1980. Struggle for the savanna: Northern Australia in ecological and prehistoric perspective. In R. Jones (ed.) *Northern Australia: Options and Implications*. Canberra: Research School of Pacific Studies, Australian National University: 3-31.

Jones, R., Cosgrove, R., Allen, J., Cane, S., Kiernan, K., Webb, S., Loy, T., West, D. and Stadler, E. 1988. An archaeological reconnaissance of karst caves within the Southern Forests region of Tasmania, September 1987. *Australian Archaeology* 26: 1-23.

Jones, R. and Johnson, I. 1985. Deaf Adder Gorge: Lindner Site, Nauwalabila I. In R. Jones (ed.) *Archaeological Research in Kakadu National Park*. Canberra: Australian National Parks and Wildlife Service Special Publication 13: 165-227.

Jones, R. and Meehan, B. (in press). Plant foods of the Gidjingali: ethnographic and archaeological perspectives from northern Australia on tuber and seed exploitation. In D. R. Harris and G. C. Hillman (eds) *Foraging and Farming: the Evolution of Plant Exploitation*. London: Allen and Unwin. In Press.

Kershaw, A. P. 1985. An extended late Quaternary vegetation record from north-eastern Queensland and its implications for the seasonal tropics of Australia. *Proceedings of the Ecological Society of Australia* 13: 179-189.

Kiernan, K., Jones, R. and Ranson, D. 1983. New evidence from Fraser Cave for glacial age man in south west Tasmania. *Nature* 301: 28-32.

Koenigswald, G. H. R. von 1936a. Early Palaeolithic stone implements from Java. *Bulletin of the Raffles Museum, Singapore,* B 1: 51-60.

Koenigswald, G. H. R. von 1936b. Über altpaläolithische artifakte von Java. *Tijdschrift Koninklijk Nederlands Aardrijkskundig Genootschap* 2nd Series 53: 41-44.

Koenigswald, G. H. R. von 1939. Das Pleistocän Javas. *Quartär* 2: 28-53.

Koenigswald, G. H. R. von 1949. Vertebrate stratigraphy. In R. W. van Bemmelen (ed.) *The Geology of Indonesia*. The Hague: Government Printing Office: 91-93.

Koenigswald, G. H. R. von 1956. *Meeting Prehistoric Man*. London: Scientific Book Club.
Koenigswald, G. H. R. von 1978. Lithic industries of *Pithecanthropus erectus* of Java. In F. Ikawa-Smith (ed.) *Early Palaeolithic in South and East Asia*. The Hague: Mouton: 23-27.
Koenigswald, G. H. R. von and Ghosh, A. K. 1973. Stone implements from the Trinil beds of Sangiran, central Java. *Proceedings of the Koninklijke Nederlandse Akademie van Wetenschappen B78*, 5: 403-407.
Lampert, R. 1981. *The Great Kartan Mystery*. Terra Australia 5. Canberra: Department of Prehistory, Research School of Pacific Studies, Australian National University.
Löffler, E. 1977. *Geomorphology of Papua New Guinea*. Canberra: Australian National University Press.
Lydekker, R. 1896. *A Geographical History of Mammals*. Cambridge: Cambridge University Press.
MacArthur, R. H. and Wilson, E. O. 1967. *The Theory of Island Biogeography*. Princeton: Princeton University Press.
Macintosh, N. W. G. 1971. Analysis of an Aboriginal skeleton and a pierced tooth necklace from Lake Nitchie, Australia. *Anthropologie Brno* 9: 49-62.
McArthur, N., Saunders, I. W. and Tweedie, R. L. 1976. Small population isolates: a micro-simulation study. *Journal of the Polynesian Society* 85: 307-326.
Majid, Z. 1982. The West Mouth, Niah, in the Prehistory of Southeast Asia. *Sarawak Museum Journal* 31: 1-200 (Special Monograph 3, The Museum, Kuching).
Medway, G. 1979. The Niah excavations and an assessment of the impact of early man on mammals in Borneo. *Asian Perspectives* 20: 51-69.
Mountain, M. J. 1983. Preliminary report of excavations at Nombe rockshelter. *Bulletin of the Indo-Pacific Prehistory Association* 4: 84-99.
Movius, H. L. 1944. *Early Man and Pleistocene Stratigraphy in South and East Asia*. Cambridge (Mass): Papers of the Peabody Museum, Harvard University 19 (3).
Movius, H. L. 1948. The lower Palaeolithic culture of Southern and Eastern Asia. *Transactions of the American Philosophical Society* 38: 329-420.
Mulvaney, D. J. 1964. The Pleistocene colonization of Australia. *Antiquity* 38: 263-267.
Mulvaney, D. J. 1969. *The Prehistory of Australia*. London: Thames and Hudson.
Mulvaney, D. J. 1975. *The Prehistory of Australia* (2nd edition). Harmondsworth: Penguin.
Munn, N. D. 1973. *Walbiri Iconography: Graphic Representation and Cultural Symbolism in a Central Australian Society*. Ithaca: Cornell University Press.
Nanson, G. C., Young, R. W. and Stockton, E. D. 1987. Chronology and palaeoenvironment of the Cranebrook Terrace (near Sydney) containing artefacts more than 40,000 years old. *Archaeology in Oceania* 22: 72-78.
Neal, R. and Stock, E. 1986. Pleistocene occupation in the southeast Queensland coastal region. *Nature* 323: 618-621.
Ninkovich, D. and Burckle, L. H. 1978. Absolute age of the base of the hominid bearing beds in eastern Java. *Nature* 275: 306-308.
Nix, H. A. and Kalma, J. D. 1972. Climate as a dominant control in the biogeography of northern Australia. In D. Walker (ed.) *Bridge and Barrier*. Canberra: Department of Biogeography and Geomorphology, Research School of Pacific Studies, Australian National University: 69-92.

Orchiston, D. W. 1979. Pleistocene sea level changes, and the initial Aboriginal occupation of the Tasmanian region. *Modern Quaternary Research in Southeast Asia* 5: 91-103.

Pearce, R. H. and Barbetti, M. 1981. A 38,000 year old archaeological site at Upper Swan, Western Australia. *Archaeology in Oceania* 16: 173-178.

Pham, T. H. 1980. Con Moong Cave: a noteworthy archaeological discovery in Vietnam. *Asian Perspectives* 23: 17-21.

Pope, G. G., Barr, S., Macdonald, A. and Nakabanlang, S. 1986. Earliest radiometrically dated artifacts from Southeast Asia. *Current Anthropology* 27: 275-279.

Pope, G. G., Nakabanlong, S. and Pitragool, S. 1987. Le Paléolithique du nord de la Thaïlande: découvertes et perspectives nouvelles. *L'Anthropologie* 91: 749-754.

Rosenfeld, A., Horton, D. R. and Winter, J. W. 1981. *Art and Archaeology in the Laura Area, North Australia.* Terra Australis 6. Canberra: Department of Prehistory, Research School of Pacific Studies, Australian National University.

Santa Luca, A. P. 1980. *The Ngandong Fossil Hominids.* Yale University Publications in Anthropology 78. New Haven: Yale University Press.

Sartono, S. 1971. Observations on a new skull of *Pithecanthropus erectus* (Pithecanthropus VIII) from Sangiran, central Java. *Proceedings Koninklijk Nederlands Akademie van Wetenschappen, Amsterdam (Series B)* 74: 185-194.

Schrire, C. 1982. *The Alligator Rivers: Prehistory and Ecology in Western Arnhem Land.* Terra Australis 7. Canberra: Prehistory Department, Research School of Pacific Studies, Australian National University.

Shawcross, F. W. 1975. Thirty thousand years and more. *Hemisphere* 19: 26-31.

Shawcross, F. W. and Kaye, M. 1980. Australian archaeology: implications of current interdisciplinary research. *Interdisciplinary Science Reviews* 5: 112-128.

Sieveking, G. de G. 1954. Excavations at Gua Cha, Kelantan. 1954, Part 1. *Federation Museums Journal* 1 and 2: 75-143.

Simpson, G. G. 1977. Too many lines: the limits of the Oriental and Australian zoogeographic regions. *Proceedings of the American Philosophical Society* 121: 107-120.

Singh, G. and Geissler, E. A. 1985. Late Cainozoic history of vegetation, fire, lake levels and climate, at Lake George, New South Wales, Australia. *Philosophical Transactions of the Royal Society of London (Series B)* 311: 379-447.

Smith, M. A. 1987. Pleistocene occupation in the arid heart of Australia. *Nature* 328: 710-711.

Smith, M. A. 1988. Central Australian seed grinding implements and Pleistocene grindstones. In B. Meehan and R. Jones (eds) *Archaeology with Ethnography: an Australian Perspective.* Canberra: Department of Prehistory, Research School of Pacific Studies, Australian National University: 94-108.

Specht, J., Lilley, I. and Norman, J. 1981. Radiocarbon dates from west New Britain, Papua New Guinea. *Australian Archaeology* 12: 13-15.

Spennermann, D. H. R. and Rapp, G. 1987. Swimming capabilities of the Black Rat (*Rattus rattus*) in tropical lagoonal waters in Tonga. *Alafia Agricultural Bulletin* (Western Samoa) 12: 17-19.

Stockton, E. and Holland, W. 1974. Cultural sites and their environment in the Blue Mountains. *Archaeology and Physical Anthropology in Oceania* 9: 36-65.

Stoneking, M., Bhatia, K. and Wilson, A. C. 1986. Rate of sequence divergence estimated from restriction maps of Mitochondrial DNAs from Papua New Guinea. *Cold Spring Harbor Symposia on Quantitative Biology* 51: 433-439.

Stringer 1986. Direct dates for the fossil hominid record. In J. A. J. Gowlett and R. E. M. Hedges (eds) *Archaeological Results from Accelerator Dating*. Oxford: Oxford University Committee for Archaeology Monograph 11: 45-50.

Sutton, P. (ed.) 1988. *Dreamings: the Art of Aboriginal Australia*. New York: Viking Press in association with the Asia Society Galleries.

Suzuki, M., Budisantoso, W., Saefudin, I. and Itihara, M. 1985. Fission track ages of pumice tuff, tuff layers and javites of hominid fossil bearing formations in Sangiran area, central Java. In N. Watanabe and D. Kadar (eds) *Quaternary Geology of the Hominid Fossil Bearing Formations in Java*. Special Publication 4. Geological Research and Development Centre, Ministry of Mines and Energy, Bandung, Indonesia: 309-357.

Thorne, A. G. 1980. The longest link: human evolution in Southeast Asia and the settlement of Australia. In J. J. Fox, R. G. Garnaut, P. T. McCawley and J. A. C. Mackie (eds) *Indonesia: Australian Perspectives*. Canberra: Research School of Pacific Studies, Australian National University: 35-43.

Thorne, A. G. and Raymond, R. 1989. *Man on the Rim: The Peopling of the Pacific*. Sydney: Angus and Robertson.

Thorne, A. G. and Wolpoff, M. H. 1981. Regional continuity in Australasian Pleistocene hominid evolution. *American Journal of Physical Anthropology* 55: 337-349.

Torgersen, T., Hutchinson, M. F., Searle, P. E. and Nix, H. A. 1983. General bathymetry of the Gulf of Carpentaria and the Quaternary physiography of Lake Carpentaria. *Palaeogeography, Palaeoclimatology, Palaeòecology* 41: 207-225.

Veevers, J. J. and Evans, P. R. 1975. Late Palaeozoic and Mesozoic history of Australia. In K. S. W. Campbell (ed.) *Gondwana Geology*. Canberra: Australian National University Press: 579-607.

Verhoeven, T. 1964. Stegodon-fossilien auf der Insel Timor. *Anthropos* 59: 634.

Wallace, A. R. 1890. *The Malay Archipelago*. London: Macmillan.

Watchman, A. 1985. Mineralogical analysis of silica skins covering rock art. In R. Jones (ed.) *Archaeological Research in Kakadu National Park*. Canberra: Australian National Parks and Wildlife Service Special Publication 13: 281-289.

Watchman, A. 1987. Preliminary determinations of the age and composition of mineral salts on rock art surfaces in the Kakadu National Park. In W. R. Ambrose and J. M. J. Mummery (eds) *Archaeometry: Further Australasian Studies*. Canberra: Department of Prehistory, Research School of Pacific Studies, Australian National University: 36-41.

Weidenreich, F. 1951. *Morphology of Solo Man*. New York: Anthropological Papers of the American Museum of Natural History 43: 205-290.

Weitzel, V. 1987. The mangrove men. *Australian Primatology* 2: 1.

White, J. P., Allen, J. and Specht, J. 1988. The Lapita homeland project. *Australian Natural History* 22: 410-414.

White, J. P., Crook, K. A. W. and Ruxton, B. P. 1970. Kosipe: a late Pleistocene site in the Papuan Highlands. *Proceedings of the Prehistoric Society* 36: 152-170.

White, J. P., and O'Connell, J. F. 1982. *A Prehistory of Australia, New Guinea and Sahul*. Sydney: Academic Press.

Wickler, S. and Spriggs, M., 1988. Pleistocene human occupation of the Solomon Islands, Melanesia. *Antiquity* 62: 703-706.

Wood, B. A. 1978. *Human Evolution*. London: Chapman and Hall.

Wright, R. V. S. 1971. *The Archaeology of the Gallus Site, Koonalda Cave*. Canberra: Australian Institute of Aboriginal Studies.

Wright, R. V. S. 1986. How old is Zone F at Lake George? *Archaeology in Oceania* 21: 138-139.

Author Index

Site and Topic Index